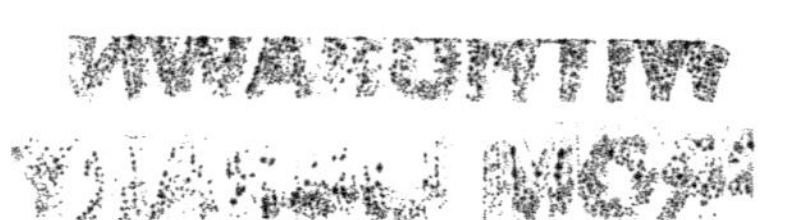

Olszewski and Baxter's
Cytoarchitecture of the Human Brainstem

Olszewski and Baxter's
Cytoarchitecture of the Human Brainstem

Editors

J.A. Büttner-Ennever, Munich
A.K.E. Horn, Munich

3rd, revised and extended edition

604 figures and 1 table, 2014

KARGER Basel · Freiburg · Paris · London · New York · New Delhi · Bangkok · Beijing · Tokyo · Kuala Lumpur · Singapore · Sydney

Dr. Jean A. Büttner-Ennever
Dr. Anja K.E. Horn
Institute of Anatomy and Cell Biology I
Ludwig Maximilian University Munich
DE–80336 Munich (Germany)

Library of Congress Cataloging-in-Publication Data

Olszewski and Baxter's cytoarchitecture of the human brainstem / editors,
J.A. Büttner-Ennever, A.K.E. Horn. -- 3rd, revised and extended edition.
p. ; cm.
Cytoarchitecture of the human brainstem
Preceded by: Cytoarchitecture of the human brain stem / by Jerzy Olszewski
and Donald Baxter. 1982.
Includes bibliographical references and index.
ISBN 978-3-318-02367-1 (hardcover : alk. paper) -- ISBN 978-3-318-02368-8
(e-ISBN)
I. Büttner-Ennever, Jean A., editor. II. Horn, A. K. E. (Anja K. E.),
editor. III. Olszewski, Jerzy, 1913-1964. Cytoarchitecture of the human
brain stem. Preceded by (work): IV. Title: Cytoarchitecture of the human
brainstem.
[DNLM: 1. Brain Stem--cytology--Atlases. WL 17]
QP376.8
612.8'26--dc23
2013023998

www.karger.com
Printed in Switzerland on acid-free and non-aging paper (ISO 9706) by Reinhardt Druck, Basel
ISBN 978–3–318–02367–1

To Ulrich Büttner

Contents

Reticular Formation

Neuromodulatory Systems

Limbic Nuclei

Precerebellar Nuclei

Nuclei of Unknown Function

Foreword to the 1st Edition

Recent physiological studies have shown the importance of a system of cells, distributed throughout the brainstem, and somewhat loosely called 'the reticular formation'. But up till now no atlas of the human brainstem has described, in at all adequate fashion, either the distribution or the histological appearance of the nerve cells which constitute this formation. The need to fill this gap was the primary stimulus to the study of the cytology of the brainstem which has resulted in this atlas.

The authors have not only brought together what was already known about the formatio reticularis, but have added several fresh observations on its structure. While this may be the only original part of the atlas, in many other ways it helps to fill the constantly growing need for more detailed knowledge of the structure of the central nervous system. It is a truism that the acquisition of this knowledge is the most difficult part of the training of a neuropathologist; indeed there are few, if any, whose knowledge is so complete that they can dispense with frequent references to atlases of the normal. Most of those which deal with the brainstem are based on sections stained for myelin, and although the position of the various nerve cell nuclei is indicated, neither the peculiar grouping of the cells nor the special characters of the individual cells in the different nuclei are described. The present atlas is arranged to supply both these needs. It includes photomicrographs in three or more magnifications; the lowest indicates the position and extent of the nuclei in sections at right angle to the long axis of the brainstem; a medium magnification shows the details of grouping of the cells in the various nuclei, and higher magnifications give detailed views of the structure of the individual cells with Nissl's stain. The importance of being able to recognise the cells belonging to a certain system, not only by their position but also by their peculiarities of structure, has always been stressed by Cecile and Oskar Vogt, in whose laboratory the senior author was trained. It has long been recognised for such special cells as the pigmented nerve cells in the floor of the 4th ventricle and surrounding the nucleus ruber, and those of the mesencephalic trigeminal nucleus, but it is here applied also to nuclei in which the nerve cells have less obvious characteristics. This aspect of the work is of special value as it helps both in distinguishing nuclei in sections cut in unusual planes, and in assessing departures from the normal in pathological material.

The production of a first class atlas of the central nervous system requires from its authors both profound anatomical knowledge, brought up to date as far as possible, and a high degree of artistic skill. This atlas meets both requirements, satisfying the eye as well as the mind, and it may be hoped that it will soon take its place in most libraries alongside the classical atlases of the nervous system.

J.G. Greenfield

Foreword to the 2nd Edition

The closing sentence of the foreword to the first edition of this atlas in 1954 by Greenfield has been fully confirmed in the past years. Greenfield wrote: 'This atlas meets both requirements, satisfying the eye as well as the mind, and it may be hoped that it will soon take its place in most libraries alongside the classical atlases of the nervous system'. Indeed, the Olszewski atlas means a hallmark and became an indispensable auxiliary of the neuroscientist and neuropathologist in their daily work. For description of the site, extension and intensity of focal lesions in the brainstem of any kind it is necessary to identify exactly the many individual cell groups which only can be done appropriately on sections stained for cells. The recognition of defined nuclei in not serial sections and on atypical plans also for the experienced scientist requires certain help by neurohistological documentation. Furthermore it has become relevant to perceive and to distinguish the kind of changes of the fine cellular structure for defining the quality or peculiarity of a pathological process. That, too, is only possible on the basis of comparison with the normal appearance of the affected nerve cell. These pathohistological functions of the atlas recently became still more important because of the increased application of electron-microscopical studies also in human neuropathology. For evaluation and interpretation of so-called semithin sections regarding the selection of relevant sites for high power preparations the cytological details are fundamental. This means a new field of use of the Olszewski atlas in human neuropathology.

The arrangement of the atlas represents the localization and the size of a nucleus in the plan perpendicular to the brainstem axis as well as by higher magnifications the kind of cell grouping and the particular structure of individual cells: this pattern has equally proved very practicable.

The undertaking of a reprint to the atlas will be highly appreciated by all workers in the field of neuroanatomy and neuropathology because of the great need of that topographic guide through the complicated three-dimensional cytological structures of the human brainstem.

F. Seitelberger
August 1981

Foreword to the 3rd Edition

At the outbreak of World War II, Jerzy (George) Olszewski had just returned from the Salpêtrière Hospital, Paris, to his home town Vilnius, then part of Poland, to work in the University Neurology Clinic. As the tides of war changed, Olszewski was forced to work 'underground' in a civilian army hospital, where he was relatively safe from enemy search parties. Firm ties between the Polish neurological community at Vilnius and the Vogt Brain Research Institute of Buch-Berlin had been established before the war, through Prof. Maximilian Rose and Prof. Wlodzimierz Godlowski, former chairmen of the Stefan Batory University, Department for Neurology and Psychiatry. Drawn by these connections, Olszewski wrote to Vogt in 1943, and with incredible luck received an invitation from Vogt, and furthermore the official permission, to go and work as an assistant at Vogt's Brain Research Institute, which by then had been forced out of Berlin and moved to Neustadt in the Black Forest [Olszewski, 1950; Baxter et al., 1987]. When Jerzy Olszewski arrived in Neustadt from Poland, Cécile and Oskar Vogt welcomed him warmly into their close scientific community [Klatzo, 2002]. As a research project they suggested to him the need for an investigation of the reticular formation of the brainstem, and the first plans for the present atlas *Cytoachitecture of the Human Brainstem* were discussed. Due to the general economic difficulties after the war Olszewski emigrated to Canada, and with the help of Dr. Wilder Penfield took up a fellowship that included teaching in the Department of Neuroanatomy at the Montreal Neurological Institute [Klatzo, 2002]. In 1952 he resumed work on the cytoarchitectural subdivisions of the human brainstem, reinforced with the feeling that cytoarchitecture was an important tool for neuropathology. With the help of Donald Baxter and several others, the atlas was published in 1954.

Olszewski and Baxter's Atlas *Cytoachitecture of the Human Brain Stem* has proved to be a unique and useful atlas for over 50 years. One reason for this is that very few atlases consider cytoarchitecture of the brain in such detail. Furthermore, the subdivision of the brainstem reticular formation, and its associated cell groups, was based on very precise and astute observations after many years of experience in the field. These same cytoarchitectural features were found to be recognizable across the whole spectrum of mammalian species. But perhaps the most important reason for the lasting value of the atlas lies in the fact that 'cytoarchitectural differences reflect functional differences' – a premise not well established at the time when the atlas was compiled, but it was something that Olszewski firmly believed in, as a student of the 'Vogt School'. He argues this point carefully in the 'Introduction', to convince the reader that the anatomical subdivisions will also provide a key to as yet unrecognized functional areas. That 'cytoarchitectural differences reflect functional differences' is a principle accepted now by every neuroanatomist: a cytoarchitectural difference in any nucleus is taken to reflect a functional difference, whether in the red nucleus, vestibular nucleus, or oculomotor nucleus. With a little practise at the microscope, and good histology, it is easy to recognize from which part of the brain a section is taken, just by the cytoarchitectural character of the individual neurons: abducens motoneurons, the mesencephalic trigeminal nucleus, the inferior olive, Purkinje cells, pyramidal cells of the cortex or the omnipause neurons in the pontine reticular formation, all have a very individual and characteristic appearance, and of course different functions.

Since the first edition of this atlas was compiled, neuroscience has progressed a long way, and revealed countless functional areas of the brainstem. In this edition we have combined the cytoarchitectural details of the brainstem nuclei described by Olszewski and Baxter with the current concepts of their function and connectivity. The atlas falls into 2 main parts: 42 plates of low-power views of the brainstem, followed by over a hundred descriptions or chapters of individual nuclei.

In this 3rd edition we have added new drawings and photographs to illustrate the rostral pole of the midbrain, and we have included several 'new' chapters to cover more recently recognized nuclei. Each chapter deals with a single nucleus, and starts with the original description of the 'Location and cytoarchitecture' from the previous edition. Any editing is indicated by square brackets, [...] indicates the omission of some phrases in the original edition, and [text] indicates the insertion of comments into the first edition manuscript. This is followed by a new section 'Functional neuroanatomy', which lays out current views on the *function* of the nucleus, and a short review of its *connections*. In addition, a selection of recent references is included to assist the reader in searching the literature for more details.

The order of the chapters in the last part of the book is based on the following global concept of the brainstem. The brainstem and its nuclei have a similar organization throughout vertebrate species. In lower species they form almost the whole brain and can be considered as a complete nervous system, which accomplishes the main aim in life – survival. Higher centers, like the cerebral cortex, hippocampus and amygdala, are, from an evolutional point of view, later additions that extend and modify the brainstem functions. In order to operate, the brainstem needs afferent inputs or *sensory systems,* to which the brain responds and then interacts with the environment through its *motor systems;* here, the somatic domain can be differentiated from the visceral domain. The input and output systems are interconnected in the brain by numerous control systems such as the *reticular formation* – a center for coordination –, or the cerebellum and *precerebellar nuclei* for motor coordination, or the *neuromodulatory nuclei* for modulating the intensity of signals, or level setting. Nuclei associated with emotionally relevant information are grouped under *limbic relay nuclei,* leaving a few brainstem nuclei with an *unknown function.* In this edition of the atlas, the brainstem nuclei have been grouped into these functional systems, sometimes with difficulty, due to lack of knowledge or because of multifunctional subdivisions (e.g. parvocellular and magnocellular parts of the red nucleus; or substantia nigra, pars compacta and reticulata).

Our special thanks are due to Thomas Karger who created the project, and whose friendly encouragement and patience enabled us to complete this edition. Second, we are extremely grateful to Rita Büttner, and are convinced that without her hard work and loyal persistence this book would not have been finished. Finally the support of our families, in particular Maximillian, Ludwig and Leopold Bochtler, has been greatly appreciated.

Jean A. Büttner-Ennever
Anja K.E. Horn
Munich 2013

Introduction

Part 1 – Cytoarchitecture and the Reticular Formation (Olszewski and Baxter, 1954)

This work had its inception in 1945 when Jerzy Olszewski arrived from Poland to work with Cécile and Oskar Vogt at the Brain Research Institute, Neustadt, Germany. They suggested to him the need for investigation of the reticular formation of the brainstem. The Vogts felt that their studies of the extrapyramidal diseases, limited to an investigation of the basal ganglia, had illuminated only one aspect of the problem. In their opinion, thorough knowledge of the morphological and functional organization of the reticular formation was indispensable for the understanding of the regulation and coordination of motor activity by the extrapyramidal systems. Accordingly, a study of the normal cytoarchitecture of the reticular formation was commenced with the intention of applying this knowledge in the investigation of pathological material at a later date. For various reasons, this work was interrupted for a period of several years, and it was not until 1952 that it was resumed at the Montreal Neurological Institute.

It soon became apparent that it was advisable to extend the scope of the project to include the cytoarchitecture of all the gray masses of the lower brainstem. This conclusion was based largely on the lack of a precise definition of the boundaries of the reticular formation and the insufficient data available concerning the cytoarchitecture of many nonreticular nuclei.

At that time the most comprehensive descriptions of the cytoarchitecture of the human brainstem were to be found in Jacobsohn's *Über die Kerne des menschlichen Hirnstamms* [1909] and in Ziehen's *Anatomie des Zentralnervensystems* [1934]. Jacobsohn's widely utilized drawings of cross-sections of the brainstem constituted the most accurate guide available for the delineation of the various nuclear masses. These drawings are necessarily overschematized due to their small size, and neither they nor the descriptions of nuclei are supplemented by photomicrographs. Ziehen's exhaustive description of the cyto- and myeloarchitecture and fiber connections of the brainstem, supplemented by an extensive bibliography, suffers from the lack of representative serial cross-sections stained for nerve cells. In addition the photomicrographs presented are limited to a few myelin-stained preparations.

Marburg devoted a considerable part of his atlas to the description of the cytoarchitecture of the human brainstem but presents only 5 schematic representations of Nissl-stained cross-sections and includes no photomicrographs of this region.

More cytoarchitectural detail of various regions of the human brainstem may be found in the monographs of Gagel and Bodechtel [1930], Stern [1936] and Crosby and Woodburne [1943], all of which are illustrated by excellent photomicrographs. Riley's *Atlas of the Basal Ganglia, Brain Stem and Spinal Cord* [1943], although not directly concerned with cytoarchitecture, should be mentioned as an invaluable guide to any student of these regions. Apart from these comprehensive reviews, innumerable other investigators have confined their studies to the anatomy of individual brainstem nuclei.[1]

The present work is an attempt to portray adequately and objectively the cytoarchitecture of all the nuclear structures of the medulla oblongata, pons and midbrain. In addition to the presentation of formerly recognized nuclei, several previously undecided cell groups have been delineated on the basis of cytoarchitectonic criteria. The majority of these lie within the reticular formation.

The contents of the monograph fall naturally into two parts. The first consists of a series of 19 semi-schematic representative cross-sections of the brainstem, accompanied by low-power photomicrographs. Descriptions of the individual nuclei, supplemented by photomicrographs of higher magnifications, compose the second part. It is hoped that this atlas will prove of value to the neuroanatomist who is interested in the position and morphology of the individual nuclei, and to the neuropathologist in his attempts to localize pathological processes and to distinguish abnormal cell forms from the confusing array of morphologically different normal cells found within various regions of the brainstem. Further, the neurophysiologist may find it useful to have available detailed human morphological data which can be correlated with that of experimental animals, and with which his functional concepts may be integrated.

Cytoarchitecture and Function

The term 'cytoarchitectonics' is applied to a method of anatomical investigation which is primarily concerned with patterns of arrangement and morphological details of nerve cells as revealed by magnifications within the range of the ordinary light microscope. The staining method almost exclusively used is the Nissl technique or one of its innumerable variants.[2]

[1] Some recent descriptions of the brainstem have concentrated on the organization of nuclei [Koutcherov et al., 2004], others on the chemoarchitecture [Paxinos and Huang, 1995]. The atlas of Nieuwenhuys et al. [2008] combines semicytoarchitectural drawings with excellent diagrams and reviews of connectivity. In addition, nowadays there are several interactive atlases on the web that allow an excellent correlation of Nissl-, or fiber-stained human material with MRI scans. These offer exciting new features like manipulation of the plane of viewing or sweeping from low to amazingly high magnification at high resolution (http://www.brainmaps.org or The Human Brain Atlas put together by Sudheimer et al. at https://www.msu.edu or from Johnson et al. at https://www.msu.edu/~brains/brains/human/brainstem/ from Michigan State University. However, none of these discuss the detailed cytoarchitecture of each nucleus.

[2] The Nissl technique stains ribosomal ribonucleic acid of the rough endoplasmic reticulum, called Nissl substance, giving the cytoplasm a mottled blue appearance. Other staining methods have been used in some atlases; for example, acetylcholinesterase staining was used by Paxinos and Huang [1995]. It reveals different patterns in the brainstem, which may or may not differ from the cytoarchitectural picture seen with the Nissl stain.

The primary objective of the cytoarchitectonic method is the subdivision of the cellular masses of the nervous system into regions with distinctive morphological characteristics. Such regions are referred to as 'areas' in the cortical gray matter, and as 'nuclei' in the subcortical gray matter. The value of the method rests on the hypothesis that the criteria used for cytoarchitectonic subdivisions are of biological significance. In other words, groups of cells delineated by this method presumably possess certain properties of biological importance which distinguish them from their neighbors. Such properties include a differential reaction to disease processes, a characteristic developmental and involutionary cycle, and a distinctive functional import.

It follows from the above that, by definition, a nucleus is characterized not only by distinctive morphological features, but also by a distinctive function. Accordingly, a major value of the cytoarchitectonic approach is that it allows the anatomist to postulate that nuclei possessing different morphological characteristics must differ in their function and other biological properties. This point is clearly illustrated in the discussion below, and brings us to the important problem of the correlation of the morphology of a neuron and its function.

The concept of the function of a neuron differs, depending upon whether it is considered as a single isolated unit (basic function) or as a constituent of an integrated cell population (specialized function). In the light of present knowledge, it seems probable that the basic function of all neurons is essentially the same. This function is the generation of nerve impulses when the cell is adequately stimulated. However, several recent concepts, if proved correct, may make modification of this statement necessary. These include: (1) the possibility that chemical transmitters released by different cell types of the central nervous system may differ; (2) the suggestion that certain neurons may exert inhibitory influences as contrasted to excitatory influences of other neurons, and (3) the possibility that the physicochemical processes essential for the generation and propagation of a nerve impulse may be different in different neurons. In regard to this third point, it must be borne in mind that the identity of electrical side effects of nerve impulses need not, of necessity, indicate the identity of the basic underlying physicochemical phenomena.

The concept of a specialized neuronal function is applied when reference is made to the activity of cells as constituents of a particular nucleus. For example, the cells of the lateral geniculate body are concerned with the reception, integration and transmission of visual impulses, those of the hypoglossal nucleus with the motor innervation of the tongue, and those of the main sensory nucleus of the trigeminal nerve with the reception, integration and transmission of tactile impulses from the face. It is apparent that the function of neurons in this sense is determined predominantly by their efferent and afferent connections.

Returning now to the discussion of the relationship between structure and function, it is probable that both the basic and the specialized function of a cell may be correlated with its structure. At the moment, however, most emphasis is placed on the specialized functions, and many instances of such correlations may be found within the brainstem. Thus, the nuclei which give rise to fibers supplying the motor innervation to somatic and branchiomeric musculature are composed of neurons which are morphologically similar; the cells composing the nuclei of the visceral efferent column, i.e. the dorsal motor nucleus of the vagus and the nucleus of Edinger-Westphal, belong to the same morphological type, and the cells of the nucleus of the mesencephalic trigeminal root resemble closely those found in the dorsal root ganglia. The question now arises whether this correlation is sufficiently valid to allow us to postulate that cell groups of similar morphology must possess similar functional significance. For example, is it permissible to assume that the dorsal raphe nucleus (formerly called nucleus supratrochlearis), the cells of which possess morphological features similar to those of the visceral efferent nuclei, must give rise to preganglionic autonomic fibers? In our opinion such suggestions, although valuable, must be advanced very cautiously. The final answer to such questions may be obtained only by the application of other neuroanatomical as well as neurophysiological methods.

Nevertheless, the postulate that morphological differences between two nuclei invariably indicate a difference in the function of the constituent cells remains valid. The apparent contradictions which one occasionally encounters should not be accepted with indifference but rather should serve as incentive to further study. The observation that the cells of the ventral and dorsal cochlear nuclei are different in structure, in spite of the fact that their efferent and afferent connections appear to be similar, is one such example. It is conceivable that different integrator mechanisms within each of these nuclei, or different and as yet unknown connections, may be responsible for this morphological dissimilarity.[3]

Cytoarchitectural Features of Aging

The cytoarchitectural features characteristic of the individual nuclei which compose the human brainstem remain remarkably constant from one brainstem to another. This is true not only of the overall appearance of the nucleus, but also of those morphological features which characterize the various cell types such as the size and shape of the cell, the position of the nucleus, the length and stainability of the processes and the pattern of distribution of Nissl granules. One does occasionally, however, note quite marked differences in the size of the

[3] It is now over 50 years ago that Olszewski and Baxter wrote these paragraphs in support of the interrelationship between cytoarchitecture and function – differences in one reflect differences in the other. We now have some answers to the questions they posed: the dorsal and ventral cochlear nuclei, which were seen to differ in structure but not in connectivity (function), are now known to have different functions and connectivity (see nuclei No. 21 and 22). The mesencephalic trigeminal nucleus neurons are indeed displaced dorsal root ganglia (nucleus No. 11). Present knowledge shows that the dorsal raphe nucleus (nucleus No. 68) is part of a diffuse modulatory system utilizing serotonin (see 'Neuromodulatory systems: overview') and does indeed look similar to the Edinger-Westphal nucleus in the human. This is not because the dorsal raphe nucleus is a visceral efferent nucleus, as Olszewski and Baxter proposed above, but because specifically in man the cell group called the Edinger-Westphal nucleus has recently been shown to be, like the dorsal raphe nucleus, part of a diffuse modulatory system (nucleus No. 49). Thus, the authors were remarkably astute in their observations, to be able to predict that the dorsal raphe nucleus and the nucleus Edinger-Westphal (specifically in man) have functional similarities, alone from their similar cytoarchitecture. Their observations prove that even in the light of modern techniques cytoarchitectural observations are still of great value.

cells of homologous nuclei in the brainstems of adult humans of comparable ages. It seems possible that slight variations in the technique of preserving and embedding the blocks may partially account for this observation.

If one examines the brainstems of individuals at the extremes of life, certain cytoarchitectural features characteristic of both the infant and the elderly adult may be observed. The most notable difference, as viewed through a microscope, between a baby's brainstem and that of an adult is the impression of compactness that one gains from the former. Here the cells composing the nuclei are relatively small and closely arranged and the glial background is much denser than it is in the adult. This compact arrangement of both cells and glia, which is probably largely due to the unmyelinated state of the majority of fibers in the infant's brainstem, serves to emphasize the borders of structures characterized by the accumulation either of cells or of glial nuclei (fig. 1–7).

In general, cells composing the nuclei of an infant's brainstem are similar to those which compose the homologous nuclei in the adult. One notable exception to this statement is the lack of visible melanin pigment in the cytoplasm of any nerve cells of the infant or young child – a feature noted particularly in the cells of the locus coeruleus and the nucleus substantia nigra (fig. 8–11). This pigment, the significance of which is not known, does not begin to accumulate until the fourth or fifth year of life.[4]

The most striking characteristic of the aging nerve cell is the accumulation of the pigment lipofuscin. This pigment, which first appears in some cells in early adult life, becomes more and more abundant in an ever increasing number of nerve cells as the individual ages. It accumulates in the form of small yellowish-brown, intracytoplasmic granules which ... displace ... the Nissl substance (fig. 12). Certain nerve cells appear particularly susceptible to the accumulation of lipofuscin. This is true of the cells of the inferior olive where this pigment appears at a relatively early age and, with increasing age, accumulates in some cells to such a degree that the cytoplasm is completely replaced, and the nerve cell is no longer visible in Nissl preparations. This accounts for the relative acellularity of the inferior olive of an elderly human when compared with that of a young adult. Different degrees of accumulation of lipofuscin in the cells of the inferior olive are represented in figures 16–18. On the other hand, cells such as the Purkinje cells of the cerebellum rarely, if ever, accumulate lipofuscin.[5]

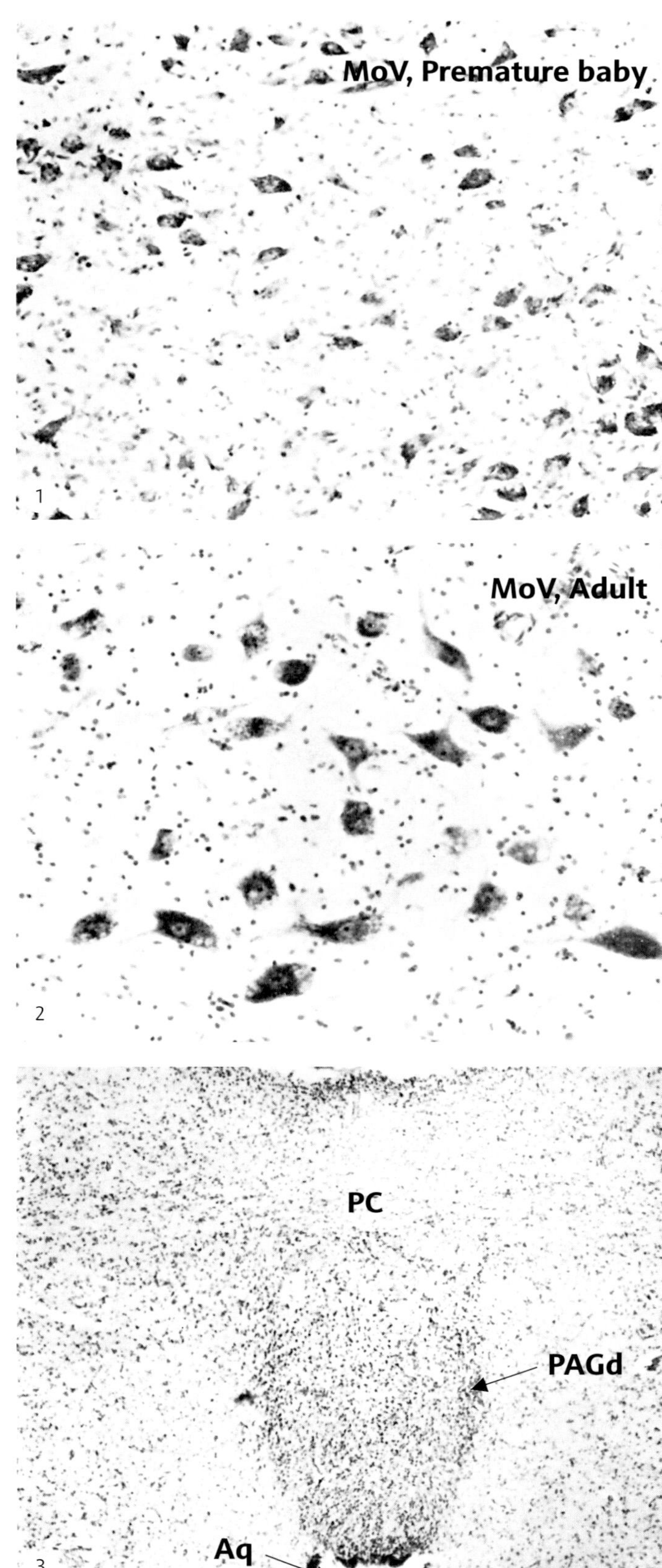

Fig. 1. Motor trigeminal nucleus (MoV) of a premature baby (7th month). Magnification ×150.

Fig. 2. Motor trigeminal nucleus (MoV) of a 47-year-old male. Magnification ×150.

Fig. 3. Dorsal portion of the periaqueductal gray of a premature baby (7th month). Note the clear delineation of the dorsal subnucleus (PAGd) compared with an adult one in plate 39. Aq = Aqueduct; PC = posterior commissure.

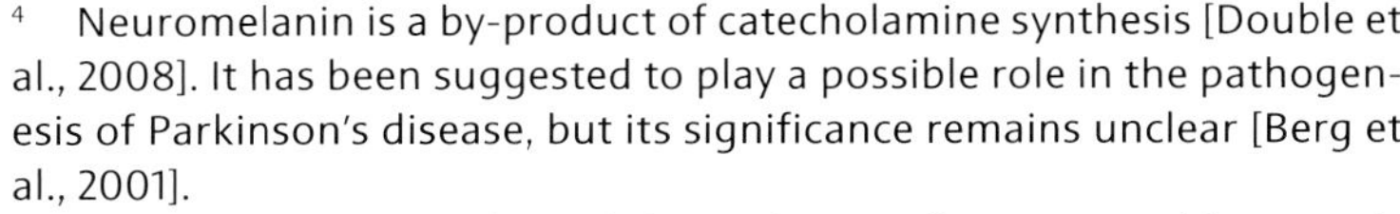

[4] Neuromelanin is a by-product of catecholamine synthesis [Double et al., 2008]. It has been suggested to play a possible role in the pathogenesis of Parkinson's disease, but its significance remains unclear [Berg et al., 2001].

[5] Lipofuscin is the product of the oxidation of unsaturated fatty acids and may be symptomatic of membrane damage, or damage to mitochondria and lysosomes. Recently it was suggested that amassed lipofuscin may be hazardous to cellular functions [Brunk and Terman, 2002], and may accumulate in degenerating neurons and senile plaques [Mirra and Hyman, 2002].

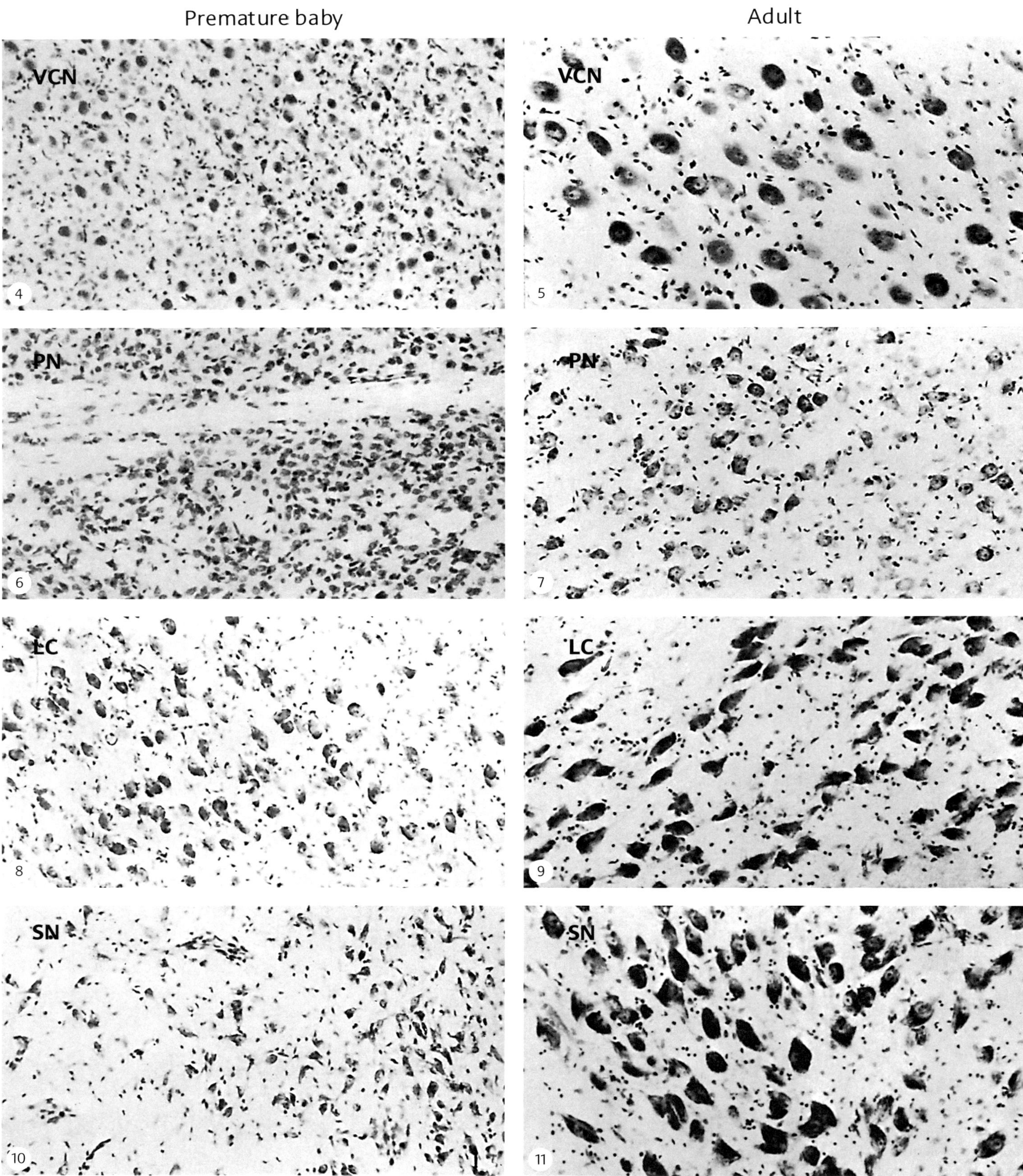

Fig. 4–11. In the left column, the following nuclei of a premature baby (7th month) are represented: ventral cochlear nucleus (VCN, **4**), pontine nuclei (PN, **6**), nucleus of the locus coeruleus (LC, **8**) and substantia nigra (SN, **10**). In the right column, the corresponding nuclei of an adult are pictured. Magnification of all photomicrographs ×150.

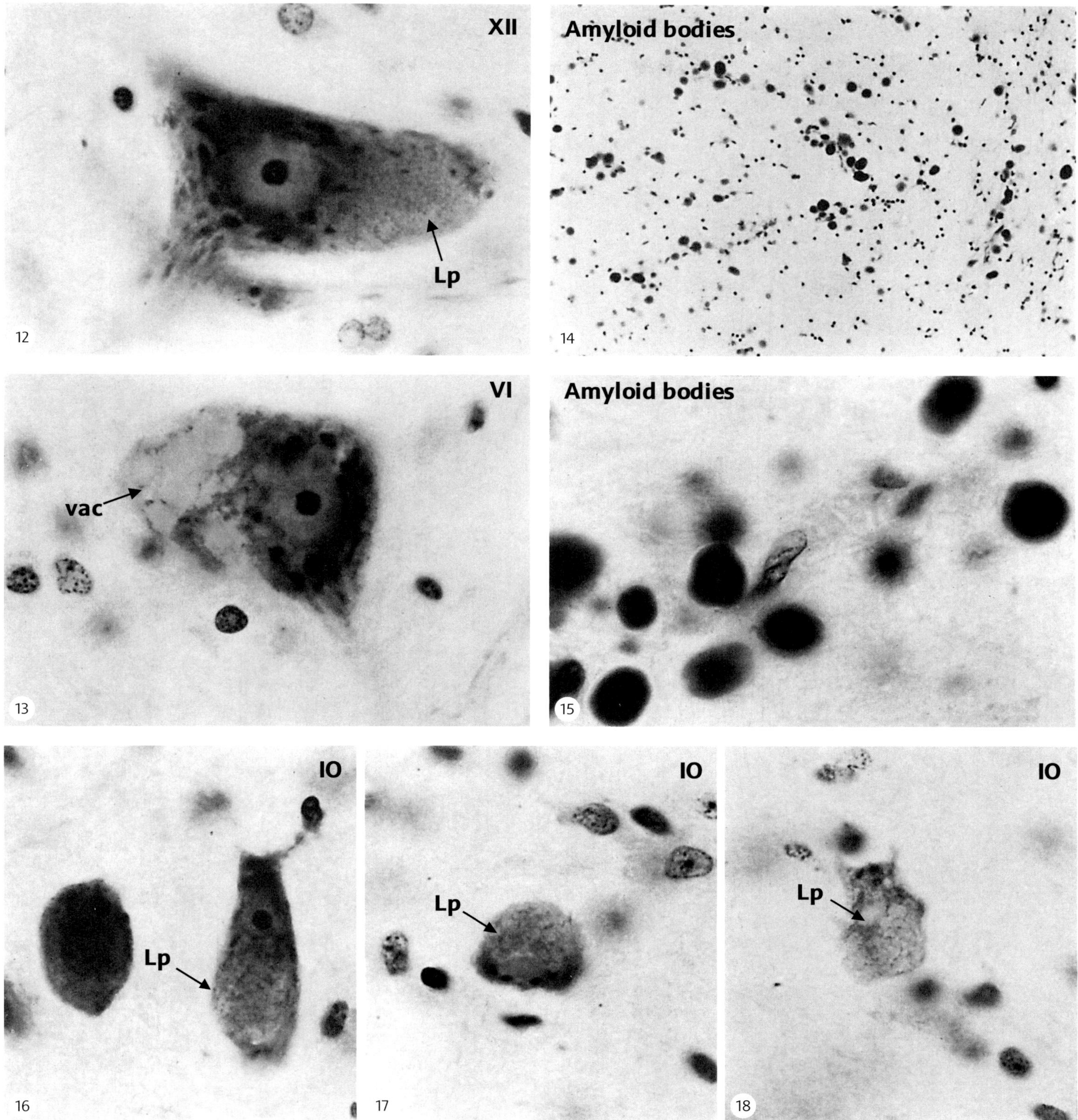

Fig. 12. Nerve cell from the hypoglossal nucleus (XII) of a 77-year-old male. Note the accumulation of lipofuscin (Lp) in the right third of the perikaryon. Magnification ×1,000.

Fig. 13. Nerve cell from the abducens nucleus (VI) of a 77-year-old male. Note the vacuolization (vac) of the perikaryon. Magnification ×1,000.

Fig. 14. Amyloid bodies in the dorsal medulla of a 75-year-old male. Magnification ×150.

Fig. 15. Amyloid bodies. Magnification ×1,000.

Fig. 16–18. Nerve cells from the inferior olive (IO) of a 75-year-old male. Note the accumulation of lipofuscin (Lp), the progressive replacement of Nissl substance and the apparent disappearance of the nucleus and nucleolus.

Lipofuscin is usually considered insoluble in fat solvents such as toluol and benzol. However, it is our impression that, if blocks of nervous tissue are subjected to procedures involving the prolonged use of fat solvents, the lipofuscin may be dissolved out of an occasional cell. This would serve to explain the appearance of vacuolated nerve cells in persons in whom there is no reason to suspect a pathological process (fig. 13). The pattern of distribution of the remaining Nissl substance in such cells remains normal and thus allows their differentiation from cells exhibiting pathological vacuolization.

A further peculiarity of aging nervous tissue is the presence of amyloid bodies (fig. 14, 15). These are small round structures of unknown constitution and significance. They stain purplish blue in Nissl preparations and usually exhibit a darkly stained central portion surrounded by a more lightly stained halo. Amyloid bodies tend to accumulate particularly in the nervous tissue immediately underlying pial surfaces.[6]

Reticular Formation

The vital importance of the reticular formation of the brainstem has long been recognized on the basis of innumerable observations that interference with the function of this region by stimulation or coagulation produces respiratory, cardiovascular, postural and autonomic responses. Within the past several years even more attention has been focussed on the reticular formation by the observations of Magoun and his associates that by stimulating various areas of the reticular formation they were able either to facilitate or to inhibit cortically or reflexly induced movements. Further, it has been demonstrated that collaterals from all the known sensory systems enter the central core of the brainstem reticular formation and that here a multisynaptic pathway is formed over which impulses are conducted to wide areas of the cortex. On entering this multisynaptic system, the sensory impulses are believed to lose their specific modalities. The ascending system from reticular formation to cortex 'does not function specifically ..., but its ascending influences in initiating and maintaining the conscious state provide the necessary background of activity without which no integrated sensory, motor or adaptive function is possible. Moreover, the medial system may be involved in the management of gradations of attention superimposed upon inattentive wakefulness' [French et al., 1953].

Despite this obvious fundamental physiological significance of the reticular formation, surprisingly little detail is available concerning its morphology. It is regrettable that the intensive physiological investigations of this region have not stimulated a corresponding degree of anatomical interest. The inadequate information which the anatomist is prepared to offer concerning the reticular formation would serve as poor ground upon which to base the simplest hypothesis, let alone the complicated concepts of modern neurophysiologists.

One need not look far to determine several factors which have contributed to the present lack of knowledge in regard to the anatomical organization of the reticular formation.

[6] They are thought to be important in the trafficking of amyloid-β precursor protein [Verbeek et al., 2002]. These, and many other different types of protein aggregates, are abundant in Alzheimer's, Parkinson's, Huntington's and prion disease, and there is evidence that they may be involved in causing the disorders [Haass and Selkoe, 2007].

Perhaps the most significant of these is the simple fact that the term 'reticular formation' represents no precise anatomical concept. To the anatomist the term refers to those areas of a myelin-stained section which are characterized by an interlacing network of myelinated fiber bundles, and the 'reticular neurons' are those cells which lie between such bundles. Such a definition, however, is actually of very little value for it is most difficult, even on a myelin-stained section, to delineate accurately the borders of the reticular formation. And if, by means of a little imagination, one does succeed in accomplishing such a feat, examination of the corresponding Nissl-stained section will show, in many instances, that the boundary line has subdivided areas of apparently uniform cytoarchitecture. The study of the myeloarchitecture of such regions as the reticular formation is probably of very little value for purposes of delineation and subdivision of the gray matter. This is due to the fact that innumerable 'fibres de passage' obscure the intrinsic fiber pattern which forms the basis for myeloarchitectonic subdivisions.

When the physiologist, on the other hand, speaks of the reticular formation, he is almost invariably referring only to those areas which, on the basis of physiological observations, he knows to be concerned with the fundamental functions referred to above. This is a much more precise and limited concept than that of the anatomist, and the two by no means correspond. The lateral reticular nucleus of the medulla, for example, is from the anatomical point of view one of the most reticulated of all the reticular nuclei, and yet it has been demonstrated that this nucleus functions predominantly as a relay station in pathways between spinal cord, brainstem and cerebellum. Thus, the lateral reticular nucleus falls within the anatomist's but not within the physiologist's concept of a reticular nucleus.

A third factor which has contributed to this confused picture has been the tendency on the part of anatomists to describe and name certain morphologically distinct groups within the reticular formation without prefixing the proper names of such nuclei with the term 'reticularis'. The nucleus of Roller and the nucleus interfascicularis hypoglossi, among others, certainly lie within the boundaries of the reticular formation yet rarely if ever does one hear them referred to as reticular nuclei.

A fourth and very important factor, which has retarded the anatomical investigation of the reticular formation, has been the pessimistic attitude which has prevailed regarding the practical value of any further subdivision of this region. This attitude is the product of two observations. (a) It is apparent even to the casual observer that cells of many different sizes, shapes, staining qualities and Nissl patterns are to be found within the reticular formation and that there appears to be little tendency for cells of a single type to congregate into a compact group. Thus, the delineation of the most reticular nuclei is much more difficult than the delineation of cranial nerve nuclei, the majority of which are composed predominantly of morphologically similar cells. (b) It has long been observed that similar physiological responses can be obtained by interfering with the function of widely divergent regions of the reticular formation and that such regions may differ markedly in their cytoarchitecture. Thus, the notion has arisen that the cells of many different functional types are intermingled haphazardly throughout the entire reticular formation and that this simple fact renders any attempt to delineate and classify reticular nuclei on the basis of morphological criteria a waste of time.

Having reviewed the major difficulties in the way of a clear conception of the morphology of the reticular formation, it is well to consider the steps necessary to rectify the situation.

To begin with, an attempt should be made to subdivide the reticular formation into regions of specific cytoarchitecture and to classify all such regions as nuclei. Many nuclei of the reticular formation are easily delineated since they are composed predominantly of compactly arranged cells of a single type. The major part of the reticular formation, however, is populated by loosely distributed cells of more than one type. The subdivision of such areas into nuclei is possible by the application of strict cytoarchitectonic criteria with particular attention paid to the morphological features of the individual cell types and to the general pattern of cell arrangement. For example, region A may be composed of cell types y and z, whereas the neighboring region B may differ only in that cells of type w replace those of type y. We believe that the difference in the morphological patterns of the regions A and B, due to the substitution of cells of type w for those of type y, is indicative of a difference in functional organization and thus justifies the delineation and classification of such regions as nuclei.

Due to the extreme complexity of organization of the reticular formation, physiological procedures probably elucidate only a part of the total functional significance of any area investigated. It is quite conceivable that regions of different structure may possess common functional properties. However, more detailed physiological investigations of such regions would certainly disclose functional differences as well. It should also be borne in mind that the actual functional organization of the reticular formation may be different from that of the current concept. If such proves to be the case; present physiological data will require reinterpretation.

Once a subdivision of the reticular formation is accomplished, the problem of nomenclature must be considered. As mentioned above, the prefix 'reticularis' is included in the proper name of many but not of all the nuclei of the brainstem reticular formation. In order to avoid such inconsistencies one should either utilize the prefix in the proper name of each of the reticular nuclei or else avoid it altogether. We believe, for several reasons, that the latter choice is the most practical. First, it is often difficult to decide from a strictly anatomical point of view whether or not a given nucleus lies within the boundaries of the reticular formation. Second, the prefix 'reticularis' at the present time carries a physiological implication which is not justified in regard to some nuclei. Third, and most importantly, since the anatomical and physiological concepts of the reticular formation do not coincide, it would be extremely difficult to consistently utilize the prefix 'reticularis' in a manner compatible with both points of view.

Further, it is hoped that once an adequate classification of all nuclei of the brainstem based on their connections and functions is established, the term 'reticular formation' will become obsolete. In the meantime, the use of specific names for nuclei, the functions and connections of which remain unknown, should prove helpful in attaining this goal.

Finally, it must be stressed that the subdivision of uncharted regions and the precise delineation of nuclei are only introductory steps in the anatomical investigation of any region. The elucidation of fiber connections is the second step and one which must precede an adequate understanding of the functional significance of any region.

Jerzy Olszewski and *Donald Baxter* (1954)

Part 2 – The Reticular Formation 60 Years Later (Büttner-Ennever and Horn, 2013)

Sixty years have passed since the above description of the reticular formation, and several of the problems that plagued Olszewski and Baxter, and even now, such as the diffuse physiological characteristics thought to be associated with the reticular formation, have been clarified. The reticular formation and its adjacent nuclei have been the subjects of many studies, and it is now clear that they can have very different properties and functions. So to complete this discussion of the reticular formation, we will consider its current nomenclature and definition, also the principle of 'diffuse modulatory systems', all of which are essential to understand before the functional neuroanatomy of the brainstem is clear.

A useful definition of the reticular formation based on cytoarchitecture was formulated by August Forel [1877]. He had the good fortune to be one of the first scientists to use Bernhard von Gudden's newly designed sliding microtome to make histological sections of the brainstem [Akert, 1993]. With the help of these high-quality, thin and undamaged sections, he could examine the cytoarchitecture of large areas of the human brainstem. Forel identified a loose network of multipolar neurons which formed a central core to the brainstem, filling the spaces that were not occupied by clearly defined structures such as the red nucleus, the cranial nerve nuclei or the raphe nuclei. These core areas all had a similar, but not identical, cytoarchitectural appearance, which was diffuse and without conspicuous boundaries. He called the core 'the reticular formation' and could trace it from the rostral tip of the midbrain down into the intermediate zone of the spinal cord.

In the 1950s, Moruzzi, Magoun and associates found that by stimulating various areas of the reticular formation in experimental animals they were able to simulate an aroused state, whereas lesions of the midline brainstem structures caused a state similar to non-REM sleep [Moruzzi and Magoun, 1949]. They interpreted this as evidence for an 'ascending reticular activating system' associated with the reticular formation – unfortunately the diffuse nature of the reticular formation cytoarchitecture and the work of Scheibel and Scheibel [1958] appeared to fit well with the diffuse concept of wakefulness and led to the misconception that the reticular formation generated the 'ascending reticular activation' – it is now clear that this is not the case [Saper et al., 2001].

While Moruzzi – a Nobel laureate –, Magoun and the Scheibels were publishing their results, Olszewski was attempting at the same time to divide the reticular formation up into discrete cytological areas. His strict subdivisions of the reticular formation did not fit well with the popular concepts like attention or the state of arousal. The word 'reticular' at that time had quickly become a synonym for 'diffuse'. For this, and other, reasons, Olszewski decided to avoid the term 'reticular' in his atlas. He suggested that the term 'reticular formation' should be dropped by the anatomists altogether, and consequently he himself only used names such as 'nucleus pontis centralis caudalis' and 'nucleus pontis centralis oralis'.

Nowadays, since the description of the 'ascending reticular activating system', research has been able to define far more exactly which specific nuclei participate in the changes of alertness found by Moruzzi and Magoun [1949]. The nuclei are compact cell groups with truly extensive, and massively widespread, axonal projections, such as the locus coeruleus (nucleus No. 69, as described in the chapters on the individual nuclei), the raphe nuclei (nucleus No. 64) and the pedunculopontine tegmental nucleus (nucleus No. 72) [Pahapill and Lozano, 2000]. These cell groups are now recognized as belonging to a 'diffuse modulatory system'. They tend to lie around the borders of the reticular formation, and they look very different in terms of cytoarchitecture to the reticular formation defined by Forel. The extraordinarily extensive projections of all these cell groups are exemplified by individual locus coeruleus neurons, whose axon branches cover all the cerebral cortex as well as the gray matter of the spinal cord and the cerebellum. They were first demonstrated by Kjell Fuxe and colleagues in the 1960s, to the initial disbelief of many neuroanatomists [Dahlstroem and Fuxe, 1964]. However, the identification of similar neurons in other nuclei has led to the concept of 'diffuse modulatory systems'. These are the neuroanatomical correlates (caricatures) of the 'ascending reticular activating system' [Yu and Dayan, 2005]. They are collected into discrete nuclei, which serve to *modulate* specific neuronal signals, much as the volume knob of a radio modulates the sound intensity of the words spoken, but it does not change the message itself. There are several diffuse modulatory systems each arising from small individual nuclei, each using a different neural transmitter: the raphe nuclei use serotonin, the locus coeruleus uses catecholamines, the pedunculopontine tegmental nucleus and the basal nucleus of Meynert use acetylcholine, the ventral tegmental nucleus uses dopamine and the tuberomammillary nucleus uses histamine [Saper et al., 2001]. The activity of neurons in some 'diffuse modulatory systems' has been recorded and has been found to correlate well with different states of arousal, attention or sleep. For example, the locus coeruleus is highly active during attention, active during wakefulness, weakly active during non-REM sleep and silent in REM sleep (nucleus No. 69).

The reticular formation (as defined cytoarchitecturally by Forel) is *not* part of a 'diffuse modulatory system'. The vital importance of different parts of reticular formation of the brainstem outlined here by Olszewski and Baxter has long been recognized on the basis of numerous stimulation or coagulation studies. The lesions produce functional deficits in vital processes such as respiratory, cardiovascular, postural, oculomotor and autonomic responses. Although the old ideas associating the 'ascending reticular activating system' with the reticular formation persist in many textbooks, it is now well established that the reticular formation is a highly specific coordination network. Its multipolar neurons receive specific premotor afferents and generate a highly coordinated output signal to motor or sensory nuclei usually lying close by. For example, a relatively compact group of medium-sized reticular neurons in the paramedian pontine reticular formation, located in the nucleus reticularis pontis caudalis, receive specific premotor afferents from the superior colliculus and cerebellum, and these reticular neurons in turn accurately drive the extraocular motoneurons (monosynaptically) for a precise saccadic eye movement, and a lesion of this small part of the reticular formation will lead to permanent ipsilateral horizontal gaze palsy (nucleus No. 56). Similarly inconspicuous regions of the medullary reticular formation coordinate the activity of the muscle groups essential for swallowing, sneezing or the blink reflex. These areas are very hard to identify on a cytoarchitectural basis in Nissl sections, but in some cases they can be visualized using other histochemical stains. For example, the premotor region for vertical and torsional saccades in the rostral mesencephalic reticular formation can be highlighted using either parvalbumin, or even better chondroitin sulfate which stains perineuronal nets (nucleus No. 61). The function of such specific regions of the reticular formation has been verified experimentally by using neuroanatomical tract-tracing methods [Holstege, 1991; Horn and Büttner-Ennever, 1998; Nieuwenhuys et al., 2008].

Olszewski and Baxter end their treatise on the reticular formation with the remark: 'For these reasons we have decided not to use the prefix "reticularis" in our nomenclature.' They wrote these words 60 years ago, and since then the term 'ascending reticular activating system' has become recognized as being made up of several nuclei collectively called 'diffuse modulatory systems' which lie outside the reticular formation. The reticular formation, as defined cytoarchitecturally by Forel and mapped by Olszewski and Baxter, is now recognized by neuroanatomists as a coordination network for the brainstem nuclei.

Modern brainstem neuroanatomists are not plagued, like Olszewski and Baxter, by the same 'reticular' doubts. Scientists now happily adopt the authors' delineation of the nuclei of the reticular formation and have also slipped the word 'reticular' back into the nomenclature [Taber, 1961]. The terms 'nucleus reticularis pontis caudalis' and 'nucleus reticularis pontis oralis' have become universally accepted nowadays. For these reasons, we hope Olszewski and Baxter will forgive us for reintroducing this 'reticular' terminology into the third edition of their atlas.

Jean A. Büttner-Ennever and *Anja K.E. Horn* (2013)

References

Akert, K: August Forel – cofounder of the neuron theory (1848–1931). Brain Pathol 1993;3:425–430.

Berg D, Gerlach M, Youdim MBH, Double KL, Zecca L, Riederer P, Becker G: Brain iron pathways and their relevance to Parkinson's disease. J Neurochem 2001; 79:225–236.

Brunk UT, Terman A: Lipofuscin: mechanisms of age-related accumulation and influence on cell function. Free Radic Biol Med 2002;33:611–619.

Crosby EC, Woodburne RT: The nuclear pattern of the non-tectal portions of the midbrain and isthmus in primates. J Comp Neurol 1943;78:441–482.

Dahlstroem A, Fuxe K: Evidence for the existence of monoamine neurons in the central nervous system. I. Demonstration of monoamines in the cell bodies of brain stem neurons. Acta Physiol Scand 1964;62:1–55.

Double KL, Dedov VN, Fedorow H, Kettle E, Halliday GM, Garner B, Brunk UT: The comparative biology of neuromelanin and lipofuscin in the human brain. Cell Mol Life Sci 2008;65:1669–1682.

Forel A: Untersuchungen über die Haubenregion und ihre oberen Verknüpfungen im Gehirn des Menschen und einiger Säugetiere, mit Beiträgen zu den Methoden der Gehirnuntersuchungen. Arch Psychiatrie 1877;7:1–495.

French JD, Verzeano M, Magoun HW: An extralemniscal sensory system in the brain. Arch Neurol Psychiatry 1953;69:505–518.

Gagel O, Bodechtel G: Die Topik und feinere Histologie der Ganglienzellgruppen in der Medulla oblongata und im Ponsgebiet mit einem kurzen Hinweis auf die Gliaverhältnisse und die Histopathologie. Z Gesamte Anat 1930;91:130–250.

Haass C, Selkoe DJ: Soluble protein oligomers in neurodegeneration: lessons from the Alzheimer's amyloid β-peptide. Nat Rev 2007;8:101–112.

Holstege G: Descending motor pathways and the spinal motor system: limbic and non-limbic components. Prog Brain Res 1991;87:307–421.

Horn AKE, Büttner-Ennever JA: Premotor neurons for vertical eye-movements in the rostral mesencephalon of monkey and man: the histological identification by parvalbumin immunostaining. J Comp Neurol 1998;392:413–427.

Jacobsohn L: Über die Kerne des menschlichen Hirnstamms. Abh Preuss Akad Wiss 1909, pp 1–70.

Koutcherov Y, Huang X-F, Halliday G, Paxinos G: Organization of human brain stem nuclei; in Paxinos G, Mai JK (eds): The Human Nervous System, ed 2. Amsterdam, Elsevier Academic Press, 2004.

Mirra SS, Hyman BT: Ageing and dementia; in Graham DI, Lantos PL (eds): Greenfield's Neuropathology. London, Arnold, 2002, vol 2, pp 195–271.

Moruzzi G, Magoun HW: Brainstem reticular formation and activation of the EEG. Electroencephalogr Clin Neurophysiol 1949;1:455–473.

Nieuwenhuys R, Voogd J, Van Huijzen C: The Human Central Nervous System. Berlin, Springer, 2008.

Pahapill PA, Lozano AM: The pedunculopontine nucleus and Parkinson's disease. Brain 2000;123:1767–1783.

Paxinos G, Huang X-F: Atlas of the Human Brainstem. San Diego, Academic Press, 1995.

Riley HA: An Atlas of the Basal Ganglia, Brain Stem and Spinal Cord. Baltimore, Williams & Wilkins, 1943, p 708.

Saper CB, Chou TC, Scammell TE: The sleep switch: hypothalamic control of sleep and wakefulness. Trends Neurosci 2001;24:726–731.

Scheibel ME, Scheibel AB: Structural substrates for integrative patterns in the brain stem reticular core; in Jasper HH, Proctor LD, Knighton RS, Noshay WC, Costello RT (eds): Reticular Formation of the Brain. Boston, Little Brown, 1958, pp 31–55.

Stern K: Der Zellaufbau des menschlichen Mittelhirns. Z Gesamte Neurol Psychiatrie 1936;154:521–598.

Taber E: The cytoarchitecture of the brain stem of the cat. I. Brain stem nuclei of the cat. J Comp Neurol 1961;116:27–70.

Verbeek MM, Otte-Höller I, Fransen JAM, De Waal RMW: Accumulation of the amyloid-β precursor protein in multivesicular body-like organelles. J Histochem Cytochem 2002;50:681–690.

Yu AJ, Dayan P: Uncertainty, neuromodulation, and attention. Neuron 2005;46:681–692.

Ziehen T: Zentralnervensystem. Jena, Fischer, 1934.

Materials and Methods

Materials

The material used in this investigation consisted of [16 brainstems; 15 of these were] obtained through the courtesy of Dr. Kenneth Earle[1]. The pertinent data related to this material are summarized in table 1.

As may be noted from the varying pathological diagnoses listed in the above table, no attempt was made to select so-called normal material. However, no material was used in which evidence of significant pathological changes was found on microscopic examination of the brainstem and in only one instance (31-51) were signs or symptoms referable to the central nervous system recorded prior to the death of the patient. It is believed that the brainstems studied represent an average cross-section of the material which the general pathologist routinely encounters.

Methods

Any attempt to represent the general arrangement of the nuclei of the human brainstem by means of direct photomicrography is thwarted by the fact that photomicrographs of even low magnification are of such a size as to be for practical purposes irreproducible. The photomicrographic representation of Nissl-stained sections at magnifications lower than ×30 is very unsatisfactory, since at such magnifications the smaller neurons tend to become indistinguishable from the background. If one half of a cross-section through the human pons is photographed at a magnification of ×30, the resulting picture measures approximately 60 × 40 cm. Pictures of such a size can hardly be used for illustrations in an atlas. In order to avoid this difficulty, semischematic representations of cross-sections were prepared in the following manner.

Nineteen Nissl-stained cross-sections through representative regions of the brainstem were selected from set 33-51. These sections were then photographed at a magnification of ×20 on the Zeiss two-unit horizontal photomicrographic camera. The resulting pictures were enlarged 3 times, to attain a final magnification of ×60. At such a magnification, the majority of neurons may be visualized with the naked eye, and can readily be distinguished from the glial nuclei. All such neurons were silhouetted with India ink, constant reference being made to the original histological sections during this procedure. Then, by comparing the photomicrographs with the actual cross-sections, boundaries of the various nuclei were drawn with thin India ink lines. A solid line indicates a definite nuclear border; a dotted line is used in those cases where the nuclear borders are indistinct; and a broken line indicates a subdivision within a nucleus. The subpial and subependymal borders of the sections are outlined with thick and medium-width India ink lines, respectively. The background of the photomicrographs prepared in this manner was then bleached in photographic reducer. Since the India ink is neither dissolved nor chemically attacked by this solution, only those structures outlined or silhouetted by the India ink remained visible. The resulting pictures were then rephotographed to reduce the magnification from ×60 to ×15.

No pretence is made that every neuron within the cross-sections selected is represented by a dot on the semischematic drawings. Nevertheless, we believe that this method does allow accurate representation of the relative size, orientation and density of neurons in different regions. Further, the elimination of the glial background and the enhancement of contrast by the use of India ink allow clear representation of even the smallest neurons at relatively low magnification.

In order to demonstrate objectively the appearance of the cross-sections as viewed under the low power of the microscope, various regions of the 19 representative cross-sections were photographed at a magnification of ×40. These appear on pages adjacent to the corresponding semischematic drawings.

[An attempt was made in the 3rd edition to reproduce the 'semischematic sections' described above, with modern methods. The result is shown in figure 1 of the chapter on nucleus No. 61. The section was taken from brain FM-03 (Munich) and cut at a slightly different angle to that of plates 1–3. The section was digitalized with a slide scanner (NanoZoomer 2.0, Hamamatsu) using a ×20 objective lens and stored on a computer. The slide was viewed at a magnification of ×2.5 (NPD view), and the appropriate detail was stored as TIF file. With Photoshop software, the TIF file of the digitalized overview was converted into black and white, the contrast and brightness were enhanced, blood vessels were removed and lines were added using CorelDraw to resemble the original semischematic drawing. Conversion into black and white, adjustments of brightness and contrast were conducted using Adobe Photoshop 7.0. The nuclear boundaries were added to the digitalized pictures using drawing software (CorelDraw).]

Descriptions of the anatomical features of individual nuclei, accompanied by photomicrographs of higher magnification compose the remainder of the monograph. In the description of each nucleus the original text from the 2nd edition describes its position within the brainstem and its cytoarchitecture. [...] [In the 3rd edition, this is followed by a new section outlining the current functional concepts connected with the nucleus, its neural connections, and a brief review

[1] 1952, an Associate Professor of Pathology and Neuropathology, University of Texas, Medical Branch, Galveston, Tex., USA.

Table 1. Brain specimens

Autopsy No.	Age	Sex	Pathological diagnosis	Interval between death and autopsy	Fixation	Em-bedding	Thick-ness of sections	Interval between mounted sections[1]	Plane of sections
FM-03	19 years	F	self-administered overdose	15 h	phosphate-buffered 4% paraformaldehyde	frozen	20 μm	400 μm	transverse
18-51	8 weeks	M	bronchopneumonia; partial atelectasis	4.5 h	10% formalin immersion	paraffin	20 μm	1 mm	transverse
25-51	47 years	M	sudden death – no demonstrable pathology	11 h	10% formalin perfusion	paraffin	20 μm	2 mm	transverse
26-51	6 months	F	fibrocystic disease of pancreas; bilateral bronchiectasis; bilateral bronchopneumonia	4 h	10% formalin immersion	paraffin	20 μm	1 mm	transverse
27-51	68 years	M	adenocarcinoma of stomach	12 h	10% formalin immersion	paraffin	20 μm	2 mm	transverse
28-51	75 years	M	postgastrectomy gastrointestinal hemorrhage	11 h	10% formalin immersion	paraffin	20 μm	2 mm	transverse
29-51	50 years	F	carcinoma of rectum; metastases in liver	21 h	10% formalin immersion	paraffin	20 μm	1 mm	transverse
31-51	48 years	M	subarachnoid hemorrhage	4 h	10% formalin immersion	paraffin	20 μm	2 mm	transverse
32-51	42 years	M	tuberculous empyema; septic splenitis	3.5 h	10% formalin immersion	paraffin	20 μm	0.5 mm	sagittal
33-51	20 years	F	hypertensive heart disease; malignant nephrosclerosis	11 h	10% formalin immersion	paraffin	20 μm	1 mm	transverse
34-51	premature baby, 7th month	M	fetal atelectasis	3 days	10% formalin immersion	paraffin	20 μm	1 mm	transverse
1-52	3.5 months	F	multiple lung abscesses	4 h	10% formalin immersion	paraffin	20 μm	1 mm	transverse
4-52	31 years	F	acute disseminated lupus erythematosus	1 h	10% formalin immersion	paraffin	20 μm	1 mm	transverse
6-52	77 years	M	old and recent myocardial infarctions	13 h	10% formalin immersion	paraffin	20 μm	1 mm	sagittal
9-52	32 years	M	rheumatic heart disease; congestive failure	8 h	10% formalin immersion	paraffin	20 μm	0.5 mm	transverse
X-52[2]					10% formalin immersion	frozen	50 μm	all sections mounted	transverse

[1] At the indicated intervals 5 sections were mounted. From these, 2 adjacent sections were stained, one with cresyl violet for nerve cells and the other with the Heidenhain technique for myelin.
[2] Brainstem X-52 was sectioned serially and all sections were mounted and stained with cresyl violet. No clinical data regarding this specimen could be obtained.

of the literature, followed by a short bibliography. Photomicrographs of individual nuclei accompanying the anatomical descriptions were taken with the Leitz Panphot camera at magnifications of ×150 and ×1,000, or taken with a digital camera (Pixera Pro 600 ES; Klughammer, Markt Indersdorf, Germany) mounted on a Leica DMRB. For this, images were captured on a computer with Pixera Viewfinder software converted in black and white and processed for contrast and brightness with Photoshop 7.0. Magnification of ×150 allows demonstration of the various sizes and shapes of cells constituting a nucleus, the orientation of the cells with respect to one another, and the density of cell distribution within the nucleus.] Photomicrographs of representative individual cells within a nucleus at a magnification of ×1,000 permit the demonstration of the size and shape of the cell body, the position of the nucleus and nucleolus, the distribution and intensity of staining of the Nissl substance, and the presence or absence of pigment within the cytoplasm.

Alphabetical List of Nuclei, Abbreviations and Original Names

Current name	Abbre-viation	Location (chapter N, plate P or page Pg)	Original name
Abducens nucleus	VI	N44	Nucleus nervi abducentis
Accessory facial nucleus	VIIac	N41	Nucleus nervi fascialis accessorius
Ambiguus nucleus	AMB	N38	Nucleus ambiguus
Anterior pretectal nucleus	APN	N33	Not shown
Arcuate nucleus	ARC	N89	Nucleus arcuatus
Area postrema	AP	N35	Area postrema
Brachium of the inferior colliculus	BIC	P32–37	Brachium colliculi inferioris
Brachium of the superior colliculus	BSC	P38	Brachium colliculi superioris
Burster-driving neurons	BDN	N55	
Caudal linear nucleus	CLi	N75	In substantia perforata posterior
Central caudal nucleus	CCN	N47	Nucleus oculomotorius caudalis centralis
Central gelatinous substance	CGS	P4–9	Substantia gliosa centralis
Central gray matter	CGr	N61, 85	Substantia grisea subependymale
Central mesencephalic reticular formation	cMRF	N59	Nucleus subcuneiformis
Central nucleus of the medulla oblongata	CN	N51	Nucleus medullae oblongatae centralis
Central nucleus of the medulla oblongata, ventral subnucleus	CNv	N51	Nucleus medullae oblongatae centralis, subnucleus ventralis
Central nucleus of the medulla oblongata, dorsal subnucleus	CNd	N51	Nucleus medullae oblongatae centralis, subnucleus dorsalis
Central tegmental tract	CTT	P32–39	Tractus tegmentalis centralis
Cerebral peduncle	CP	P30–42	Pes pedunculi
Compact interfascicular nucleus	CIF	N107	Nucleus compactus interfascicularis
Conterminal nucleus	CT	N103	Nucleus conterminalis
Cuneiform nucleus	CNF	N58	Nucleus cuneiformis, caudal part
Decussation of the superior cerebellar peduncle	dSCP	P30–33	Decussatio pedunculorum cerebellorum superiorum
Descending vestibular nucleus	DVN	N16	Nucleus vestibularis spinalis
Dorsal accessory olive	DAO	N92	Nucleus olivaris inferior accessorius dorsalis
Dorsal cap of Kooy	dc	N92	Nucleus olivaris inferior groups
Dorsal cochlear nucleus	DCN	N22	Nucleus cochlearis dorsalis
Dorsal motor nucleus of the vagal nerve	DMX	N50	Nucleus dorsalis motorius nervi vagi
Dorsal nucleus of the lateral lemniscus	DLL	N26	Nucleus lemnisci lateralis dorsalis
Dorsal paragigantocellular nucleus	PGiD	N55	Nucleus paragigantocellularis dorsalis
Dorsal raphe nucleus	DR	N68	Nucleus supratrochlearis
Dorsal raphe nucleus, caudal part	DRc	N68	Nucleus centralis superior, subnucleus dorsalis
Dorsal raphe nucleus, dorsal part	DRd	N68	Nucleus supratrochlearis
Dorsal raphe nucleus, interfascicular part	DRif	N68	Nucleus supratrochlearis
Dorsal raphe nucleus, ventral part	DRv	N68	Nucleus supratrochlearis
Dorsal raphe nucleus, ventrolateral part	DRvl	N68	Nucleus supratrochlearis
Dorsal tegmental nucleus (of Gudden)	DTG	N82	Nucleus compactus suprafascicularis
Dorsolateral pontine nuclei	DLPN	N100	Processus griseum pontis tegmentosus lateralis
Dorsomedial cell column	dmcc	N92	Nucleus olivaris inferior groups
Dorsomedial pontine nuclei	DMPN	P24	
Edinger-Westphal nucleus	EW	N49	Nucleus Edinger-Westphal
Edinger-Westphal central projecting nucleus	EWcp	N49	Nucleus Edinger-Westphal
Edinger-Westphal preganglionic neurons	EWpg	N49	Not described
Facial nerve	NVII	P20–24	Nervus facialis
Facial nucleus	VII	N40	Nucleus nervi facialis
Gigantocellular nucleus	Gi	N52	Nucleus gigantocellularis
Gigantocellular nucleus proper	Gi proper	N52	Nucleus gigantocellularis
Gigantocellular nucleus, pars α	Giα	N52	Nucleus gigantocellularis, pars ventralis rostralis
Gigantocellular nucleus, ventral part	Giv	N52	Nucleus gigantocellularis, pars caudalis
Gracile nucleus	GR	N1	Nucleus gracilis
Group β of the medial accessory olive	β	N92	Nucleus olivaris inferior groups
Group e	E	N19	Not described
Group f	F	N19	Not described
Group l	L	N19	Not described
Group m, of vestibular complex	M	N19	Not described
Group x	X	N19	Not described
Group y (see Y group)			
Group z	Z	N19	Not described
H_1 fields of Forel	H_1	N59, 61	Not described
H_2 fields of Forel	H_2	N59, 61	Not described
Habenular commissure	HC	P42	Not shown
Hypoglossal nerve	NXII	P15	Nervus hypoglossi
Hypoglossal nucleus	XII	N36	Nucleus nervi hypoglossi

Current name	Abbre-viation	Location (chapter N, plate P or page Pg)	Original name
Hypothalamus	Hypoth.	P42, N61	Hypothalamus
Inferior cerebellar peduncle	ICP	P12–20	Pedunculus cerebelli inferior
Inferior colliculus	IC	N27	Nucleus colliculi inferioris
Inferior colliculus, central nucleus	ICc	N27	Nucleus colliculi inferioris
Inferior colliculus, dorsal cortex	ICd	N27	Nucleus colliculi inferioris
Inferior colliculus, external nucleus	ICx	N27	Nucleus colliculi inferioris
Inferior olive	IO	N92	Nucleus olivaris inferior
Inferior olive, dorsal lamina	d	N92	Nucleus olivaris inferior
Inhibitory burst neurons	IBN	N55	
Intercalated nucleus (of Staderini)	INSt	N90	Nucleus intercalatus
Intercollicular nucleus	ICOL	N5	Nucleus intercollicularis
Interfascicular hypoglossal nucleus	IFH	N93	Nucleus interfascicularis hypoglossi
Interfascicular nucleus	IFN	N74	Not described
Intermediate reticular nucleus	IRt	Pg 164, N53	
Interpeduncular nucleus	IPN	N86	Nucleus interpeduncularis
Interstitial nucleus of Cajal	INC	N20	Nucleus interstitialis (Cajal)
Interstitial nucleus of the superior fascicle, posterior fibers	ITN	N32	
Interstitial nucleus of the vestibular nerve	INV	N17	β-Cell groups of the vestibular nerve
Intracuneiform nucleus	ICUN	N60	Nucleus intracuneiformis
Islands of lateral cuneate nucleus	ISL	N4, P10	Insulae nuclei cuneati lateralis
Kölliker-Fuse nucleus	KF	N80, P26	
Lateral cuneate nucleus	LCU	N4	Nucleus cuneatus lateralis
Lateral lemniscus	LL	P24–31	Lemniscus lateralis
Lateral nucleus of the inferior olive	LSO	N92	Not described
Lateral parabrachial nucleus	PBL	N81	Nucleus parabrachialis lateralis
Lateral paragigantocellular nucleus	PGiL	N53	Nucleus paragigantocellularis lateralis
Lateral reticular nucleus	LRN	N88	Nucleus medullae oblongatae lateralis
Lateral reticular nucleus, principal part	LRNp	N88	Nucleus medullae oblongatae lateralis, subnucleus ventralis
Lateral reticular nucleus, subtrigeminal part	LRNst	N88	Nucleus medullae oblongatae lateralis, subnucleus dorsalis, nucleus medullae oblongatae subtrigeminalis
Lateral superior olivary nucleus	LSO	N24	Nucleus olivaris superior lateralis
Lateral terminal nucleus	LTN	N32	Not described
Lateral vestibular nucleus	LVN	N15	Nucleus vestibularis lateralis
Laterodorsal tegmental nucleus	LDT	N71	Not described
Locus coeruleus	LC	N69	Nucleus locus coeruleus
M group of rostral mesencephalon	M	N61	Not described
Mammillary body	MB	P42, N101	Corpora mammillaria
Mammillary nucleus, lateral	MBl	N61	
Mammillary nucleus, medial	MBm	N61	
Marginal zone of the medial vestibular nucleus	MZ	N13	Nucleus interpositus
Medial accessory nucleus of Bechterew	NB	P41, N61, 101	Not described
Medial accessory olive	MAO	N92	Nucleus olivaris inferior accessorius medialis
Medial cuneate nucleus	MCU	N3	Nucleus cuneatus medialis
Medial geniculate body	MGB	P38–42	Corpus geniculatum mediale
Medial lamina of the principal olive	ml	N92	Nucleus olivaris inferior groups
Medial lemniscus	ML	P8–38	Lemniscus medialis
Medial longitudinal fasciculus	MLF	P14–34	Fasciculus longitudinalis medialis
Medial nucleus of the trapezoid body	MNTB	N23	
Medial parabrachial nucleus	PBM	N80	Nucleus parabrachialis medialis
Medial pretectal nucleus	MPN	N33	Not shown
Medial superior olivary nucleus	MSO	N24	Nucleus olivaris superior medialis
Medial tegmental tract	MTT	N101	Not described
Medial vestibular nucleus	MVN	N13	Nucleus vestibularis medialis
Medial vestibular nucleus, magnocellular part	MVNm	N13	
Medial vestibular nucleus, parvocellular part	MVNp	N13	
Median raphe nucleus	MnR	N67	Nucleus centralis superior, subnucleus medialis
Median raphe nucleus, ventral part	MnRv	N67	Nucleus centralis superior, subnucleus ventralis
Mesencephalic reticular formation	MRF	N59	Nucleus subcuneiformis + nucleus cuneiformis, rostral part
Mesencephalic reticular formation, ventral part	MRFv	N59	Nucleus subcuneiformis
Mesencephalic tract of the trigeminal nerve	TMesV	N12	Tractus nervi trigemini mesencephalicus
Mesencephalic trigeminal nucleus	MesV	N11	Nucleus nervi trigemini mesencephalicus
Middle cerebellar peduncle	MCP	P22–26	Pedunculus cerebelli medialis
Motor trigeminal nucleus	MoV	N42	Nucleus nervi trigemini motorius
Nuclei pararaphales	PRA	N95	Nucleus pararaphales
Nucleus limitans	NL	P42, N33	Not shown
Nucleus of Bechterew (see medial accessory nucleus of Bechterew)			
Nucleus of Darkschewitsch	ND	N102	Nucleus of Darkschewitsch
Nucleus of Perlia	NP	N48	Nucleus Perlia
Nucleus of Roller	Ro	N91	Nucleus Roller
Nucleus of the brachium of the inferior colliculus	nBIC	N29	Nucleus paralemniscalis
Nucleus of the optic tract	NOT	N33	Not shown
Nucleus of the pontobulbar body	PBu	N97	Nucleus corporis pontobulbaris
Nucleus of the posterior commissure	NPC	N62	Not described
Nucleus of the trapezoid body	NTB	N23	Not described
Nucleus paramedianus dorsalis	PMD	N106	Nucleus paramedianus dorsalis
Nucleus paramedianus dorsalis, caudal subnucleus	PMDc	N106	Nucleus paramedianus dorsalis caudalis
Nucleus paramedianus dorsalis, oral subnucleus	PMDo	N106	Nucleus paramedianus dorsalis oralis

Current name	Abbreviation	Location (chapter N, plate P or page Pg)	Original name
Nucleus pararaphales	PRA	N95	Nucleus pararaphales
Nucleus parvocellularis compactus	PVC	N2	Nucleus parvocellularis compactus
Nucleus raphe interpositus	RIP	N56	Nucleus pontis centralis caudalis, medial part
Nucleus raphe magnus	RMg	N66	Mistakenly called nucleus raphe pallidus
Nucleus raphe obscurus	ROb	N65	Nucleus raphe obscurus, subnucleus intraraphalis, nucleus raphae obscurus
Nucleus raphe obscurus, subnucleus extraraphalis	RObe	N65	Nucleus raphe obscurus, subnucleus extraraphalis
Nucleus raphe obscurus, subnucleus intraraphalis	RObi	N65	
Nucleus raphe pallidus	RPa	N64	Not described
Nucleus reticularis pontis caudalis	NRPC	N56	Nucleus pontis centralis caudalis
Nucleus reticularis pontis oralis	NRPO	N57	Nucleus pontis centralis oralis
Nucleus reticularis tegmenti pontis	NRTP	N99	Nucleus papillioformis
Nucleus subcoeruleus	SubC	N70	Nucleus subcoeruleus, subnucleus dorsalis
Nucleus subcoeruleus, ventral part	SubCv	N70	Nucleus subcoeruleus, subnucleus ventralis
Oculomotor nerve	NIII	P36–38, N101	Nervus oculomotorius
Oculomotor nucleus	III	N46	Nucleus oculomotorius principalis
Olivocochlear bundle	OCB	N23	Not described
Parabigeminal nucleus	PBG	N31	Not described
Parabrachial pigmented nucleus	PBP	N78	Nucleus parabrachialis pigmentosus
Paramedian pontine reticular formation	PPRF	N56	
Paramedian raphe nucleus	PMR	N67	Nucleus centralis superior, subnucleus lateralis
Paramedian tract	PMT	N96	Not described
Paramedian tract (cell groups)	PMT	N96	Not described
Paranigral nucleus	PNg	N77	Nucleus paranigralis
Parasolitary nucleus	PSol	N19	Not described
Parvocellular reticular nucleus	PCR	N54	Nucleus parvocellularis
Pedunculopontine nucleus	PPT	N72	Nucleus tegmenti pedunculopontinus
Pedunculopontine nucleus, subnucleus compactus	PPTc	N72	Nucleus tegmenti pedunculopontinus, subnucleus compactus
Pedunculopontine nucleus, subnucleus dissipatus	PPTd	N72	Nucleus tegmenti pedunculopontinus, subnucleus dissipatus
Periaqueductal gray	PAG	N85	Griseum centrale mesencephali
Periaqueductal gray, dorsal subnucleus	PAGd	N85	Griseum centrale mesencephali, subnucleus dorsalis
Periaqueductal gray, lateral part	PAGl	N85	Griseum centrale mesencephali, subnucleus lateralis
Periaqueductal gray, medial part	PAGm	N85	Griseum centrale mesencephali, subnucleus medialis
Periolivary complex	POC	N23	Nucleus trapezoidalis, nucleus of the trapezoid body
Peripeduncular nucleus	PPD	N87	Nucleus peripeduncularis
Perirubral capsule	cRN	P39–42, N102	Capsule of the nucleus ruber
Pineal body	PIN	N61	Corpus pinealis
Pontine central gray	PCG	N67, 84	Griseum centrale pontis
Pontine nuclei	PN	N100	Griseum pontis
Pontine nuclei, morphologically similar islands	α	N100	Islands of cells which morphologically resemble the cells of the griseum pontis
Pontine nuclei, possible subgroups	γ	N100	Griseum pontis possible subgroups
Posterior commissure	PC	N63	Commissura posterior
Posterior pretectal nucleus	PPN	N33	Not shown
Posterior trigeminal nucleus	PoV	N43	Nucleus retrotrigeminalis
Prepositus nucleus	PrP	N94	Nucleus prepositus hypoglossi
Pretectal olivary nucleus	PON	N33	Pretectal olivary nucleus
Pretectum	Pretectum	N33	Regio pretectalis
Principal olive	PO	N92	Nucleus olivaris inferior principalis
Principal sensory trigeminal nucleus	PrV	N10	Nucleus nervi trigemini sensibilis principalis
Pulvinar	PUL	P42	Not described
Pyramidal decussation	PD	P4, 5	Decussatio pyramidum
Pyramidal tract	PT	P6–28	Tractus pyramidalis
Red nucleus	RN	N101	Nucleus ruber
Red nucleus, magnocellular subnucleus	RNm	N101	Nucleus ruber
Red nucleus, parvocellular subnucleus	RNp	N101	Nucleus ruber
Reticulospinal neurons	RSN	N55	
Retroambiguus nucleus	RAm	N79	Nucleus retroambigualis
Retrofacial nucleus	RFN	N39	Nucleus retrofacialis
Retrorubral fields	RRF	P36, 37	Not used
Rostral interstitial nucleus of the medial longitudinal fasciculus	RIMLF	N61	Not described
Rostral linear nucleus	RLi	N76	Nucleus intracapsularis
Sagulum nucleus	SAG	N28	Nucleus sagulum
Solitary nucleus	SOL	N34	Nucleus tractus solitarii
Solitary nucleus, rostral tip	SOLo	N34	Nucleus ovalis
Solitary nucleus, gelatinous subnucleus	SOLg	N34	Nucleus tractus solitarii, subnucleus gelatinosus
Solitary tract	TSOL	N34, P10–15	Tractus solitarius
Spinal trigeminal nucleus, caudal part	SpVc	N6	Nucleus tractus spinalis trigemini caudalis
Spinal trigeminal nucleus, caudal part, marginal zone	mz	N6	Nucleus tractus spinalis trigemini caudalis, subnucleus zonalis
Spinal trigeminal nucleus, caudal part, magnocellular zone	mg	N6	Nucleus tractus spinalis trigemini caudalis, subnucleus magnocellularis
Spinal trigeminal nucleus, caudal part, substantia gelatinosa	sg	N6	Nucleus tractus spinalis trigemini caudalis, subnucleus gelatinosus
Spinal trigeminal nucleus, interpolar part	SpVi	N7	Nucleus tractus spinalis trigemini interpolaris
Spinal trigeminal nucleus, oral part	SpVo	N8	Nucleus tractus spinalis trigemini oralis
Spinal trigeminal tract	TSpV	N9	Tractus nervi trigemini spinalis
Subependymal layer	SEL	P10–42	Stratum gliosum subependymale
Substantia nigra	SN	N73	Nucleus substantiae nigrae

Current name	Abbreviation	Location (chapter N, plate P or page Pg)	Original name
Substantia nigra, pars compacta	SNc	N73	Nucleus substantiae nigrae, subnucleus compactus
Substantia nigra, pars compacta, dorsal tier	SNd	N73	Pars β of the subnucleus compactus
Substantia nigra, pars compacta, ventral tier	SNv	N73	Pars α of the subnucleus compactus
Substantia nigra, pars medialis	SNm	N73	
Substantia nigra, pars reticulata	SNr	N73	Nucleus substantiae nigrae, subnucleus reticularis
Substantia perforata posterior	SPP	P36–38	Substantia perforata posterior
Subthalamic nucleus	STH	N61	Not described
Subventricular nuclei	SBV	N105	Nucleus subventricularis (not on plates)
Superior cerebellar peduncle	SCP	N61, 101, P24–36	Pedunculus cerebelli superior
Superior colliculus	SC	N30	Colliculus superior
Superior colliculus, deep gray layer	SGP	N30	Stratum griseum profundum
Superior colliculus, deep white layer	SAP	N30	Stratum album profundum
Superior colliculus, intermediate gray layer	SGI	N30	Stratum griseum intermediale
Superior colliculus, intermediate white layer	SAI	N30	Stratum album intermediale
Superior colliculus, optic layer	SO	N30	Stratum opticum
Superior colliculus, superficial gray layer	SGS	N30	Stratum griseum superficiale
Superior colliculus, zonal layer	SZ	N30	Stratum zonale
Superior olivary complex	SOC	N24	Nucleus olivaris superior
Superior vestibular nucleus	SVN	N14	Nucleus vestibularis superior
Supragenual nucleus	SG	N98	Nucleus suprageniculatus
Supralemniscal process of the pontine nuclei	SLPN	N99, 100	Processus griseum pontis supralemniscalis
Supraspinal and spinal accessory nucleus	XI	Pg 18, N37	
Supraspinal nucleus	SSp	N37	Nucleus supraspinalis
Supravestibular nucleus	SPV	N104	Nucleus supravestibularis
Tractus retroflexus	TR	N61, P42	Tractus habenulo-interpeduncularis
Trapezoid body	TZ	N23	
Trigeminal nerve	NV	P24, 25	Nervus trigeminus
Trochlear nerve	NIV	P28–30	Nervus trochlearis
Trochlear nucleus	IV	N45	Nucleus nervi trochlearis
Velum medullare anterior	VELa	P28, 29	Velum medullare anterior
Ventral cochlear nucleus	VCN	N21	Nucleus cochlearis ventralis
Ventral lamina of the principal olive	v	N92	Nucleus olivaris inferior
Ventral nucleus of the lateral lemniscus	VLL	N25	Nucleus lemnisci lateralis ventralis
Ventral tegmental area (of Tsai)	VTA	N74	Not used
Ventral tegmental nucleus (of Gudden)	VTG	N83	Not described
Ventrolateral outgrowth	vlo	N92	Nucleus olivaris inferior groups
Vestibular nerve	NVIII	P20	Nervus vestibularis
Vestibular nucleus	VIII	Pg 18, Fig. 1	
Y group	Y	N18	Not described (plate 27)
Zona incerta	ZI	N61, fig. 1	Not described

Plates of Serial Sections through the Human Brainstem

Organization of the Brainstem Nuclei

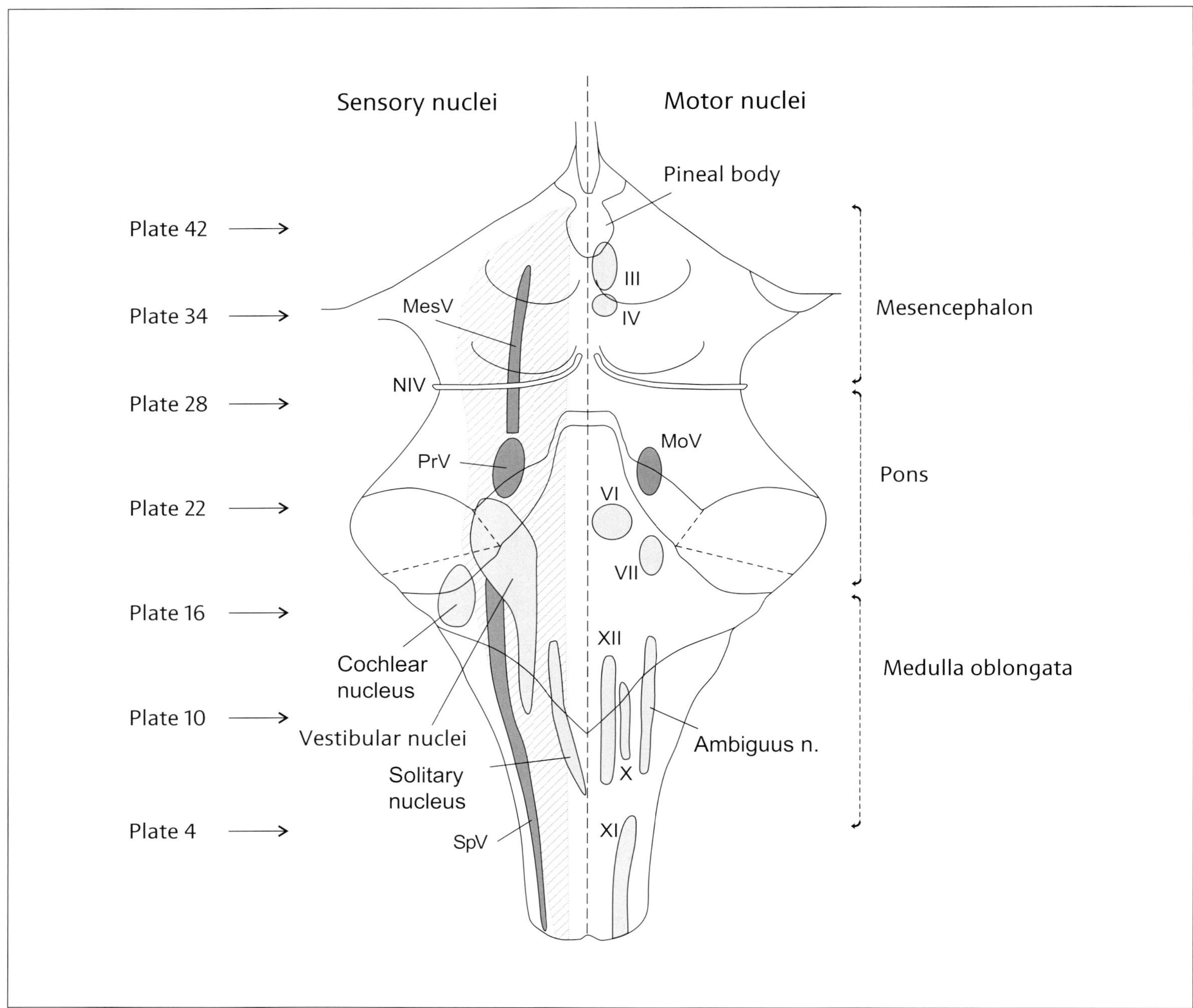

Fig. 1. Organization of the brainstem. A schematic drawing of the dorsal aspect (cerebellum removed) to show the general organization of the brainstem cranial nuclei and the reticular formation (cross-hatching). It indicates the main subdivisions of the brainstem – the medulla oblongata, the pons and the mesencephalon – and shows their relationship to the cranial nerve nuclei in humans and the conventional nomenclature. The sensory cranial nerve nuclei are indicated on the left side and the motor nuclei on the right side; the reticular formation is shown on the left side only. The subsequent plates illustrating these nuclei are indicated by arrows. Abbreviations: III = oculomotor nucleus; IV = trochlear nucleus; MesV = mesencephalic trigeminal nucleus; MoV = motor trigeminal nucleus; NIV = trochlear nerve; PrV = principal sensory trigeminal nucleus; SpV = spinal trigeminal nucleus; VI = abducens nucleus; VII = facial nucleus; X = dorsal motor nucleus of the vagus nerve; XI = supraspinal nucleus and spinal accessory nucleus; XII = hypoglossal nucleus.

Plates 1–42 Serial Semischematic Drawings and Photomicropgraphs

Plate 1

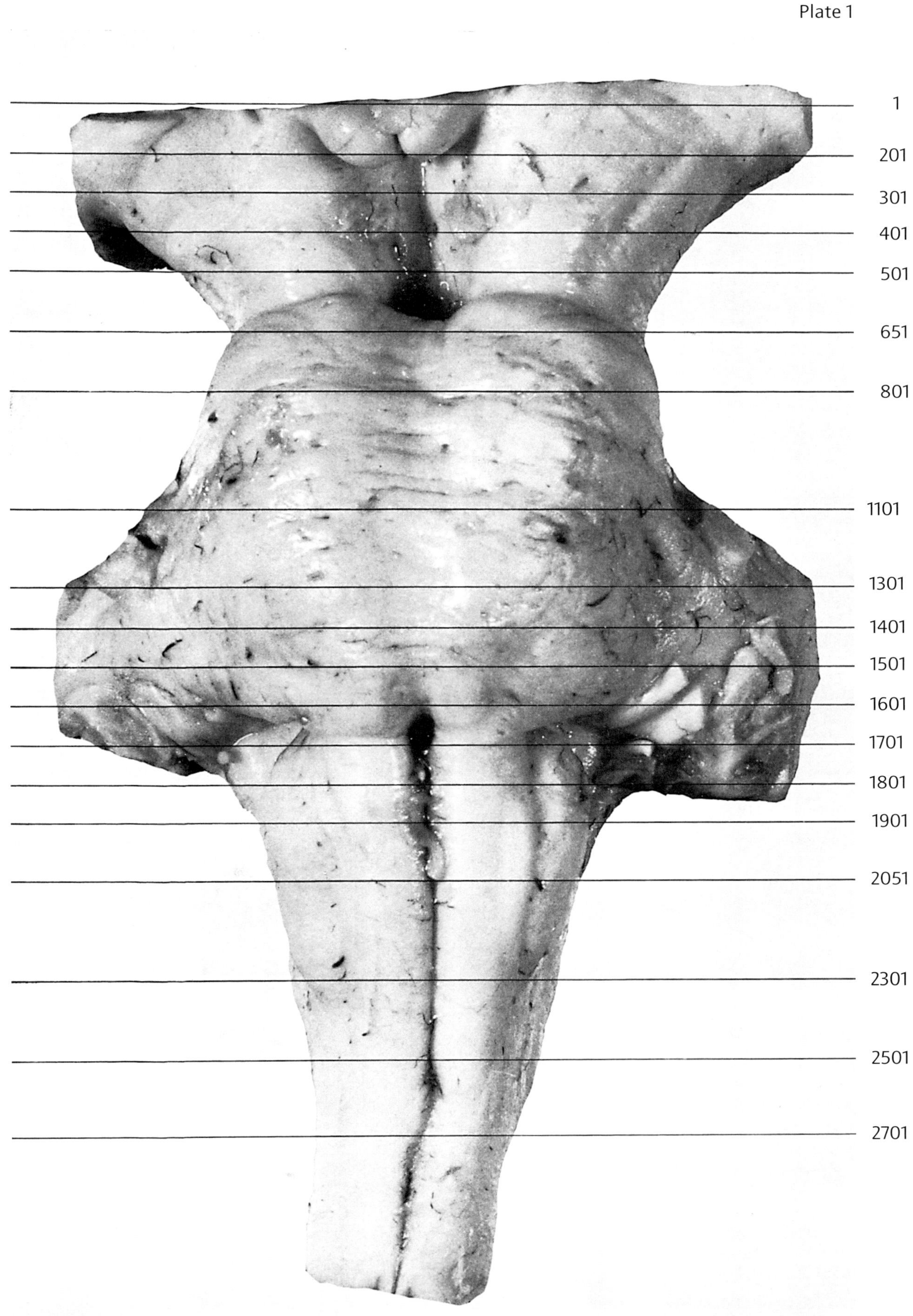

Plates 1–3. Dorsal, lateral and ventral views of the brainstem specimen 33-51. Transverse sections of this brain are used in the succeeding semischematic drawings and photomicrographs. The block was sectioned in an orocaudal direction, and the horizontal lines indicate the levels at which the sections were chosen for representation. All sections (20 µm thick) were counted and numbered. The numerals above each horizontal line correspond to the number of the section.

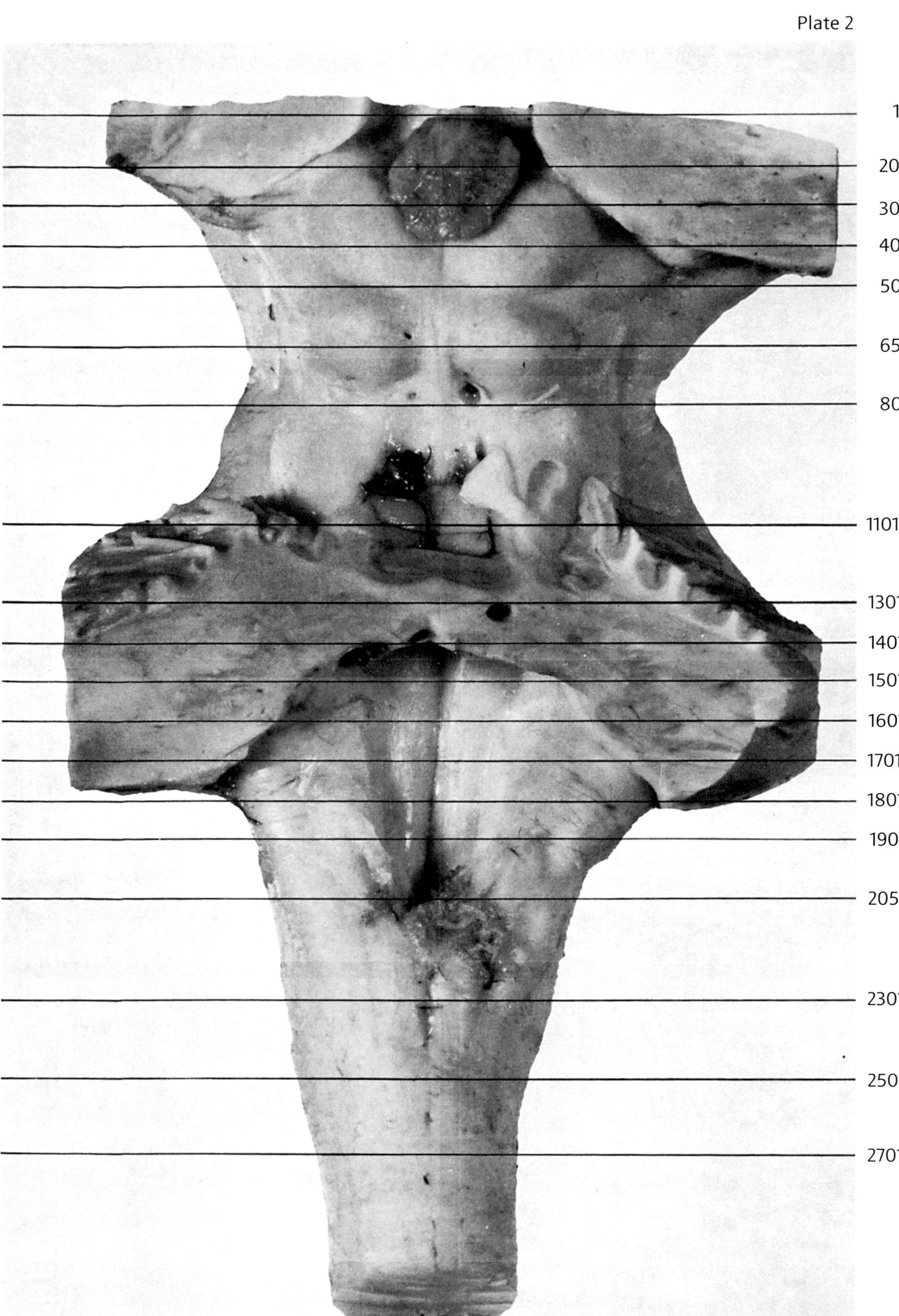
1
201
301
401
501
651
801
1101
1301
1401
1501
1601
1701
1801
1901
2051
2301
2501
2701

Plate 3

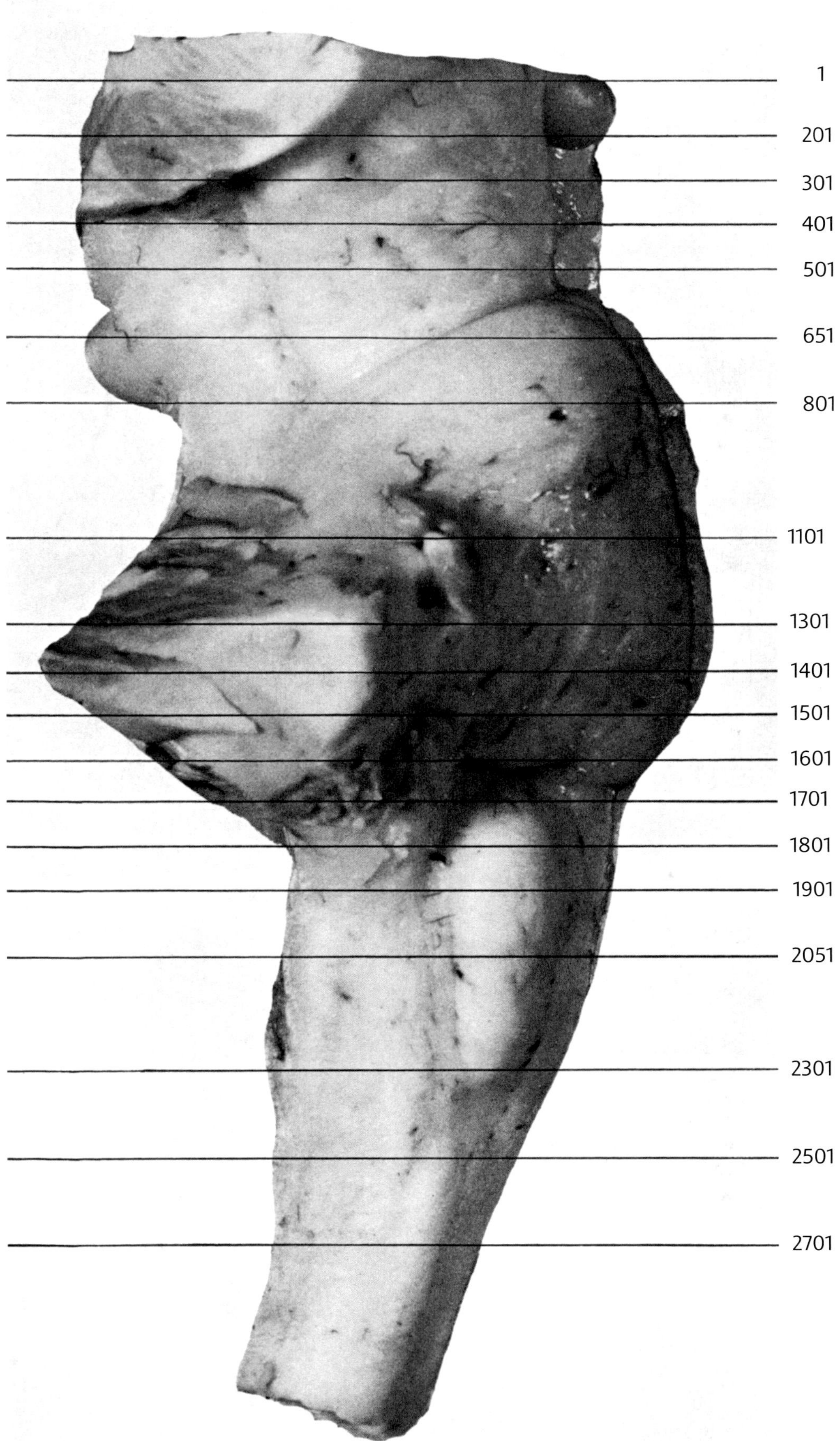
1
201
301
401
501
651
801
1101
1301
1401
1501
1601
1701
1801
1901
2051
2301
2501
2701

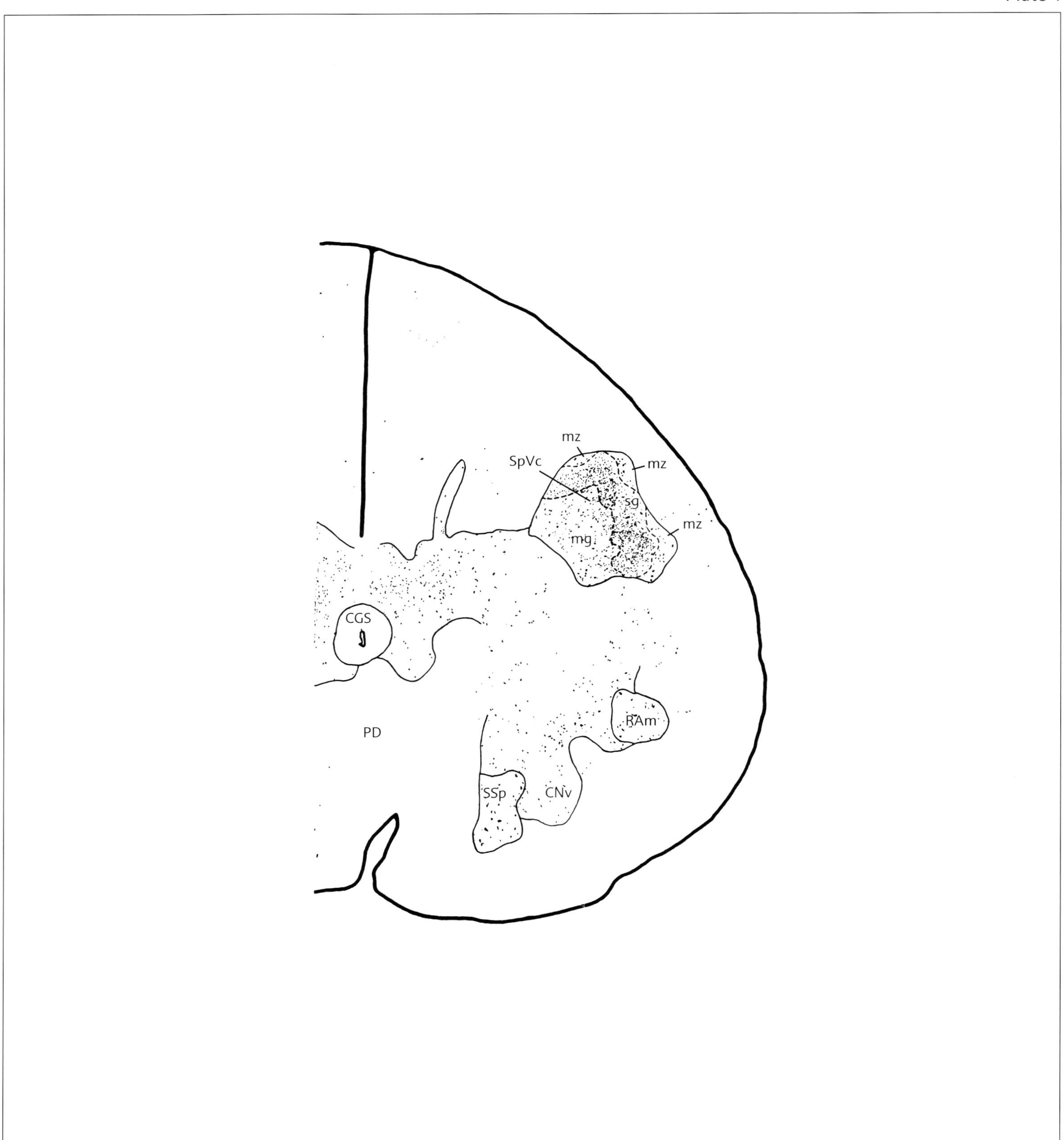
mz
SpVc
mz
sg
mz
mg
CGS
RAm
PD
SSp
CNv

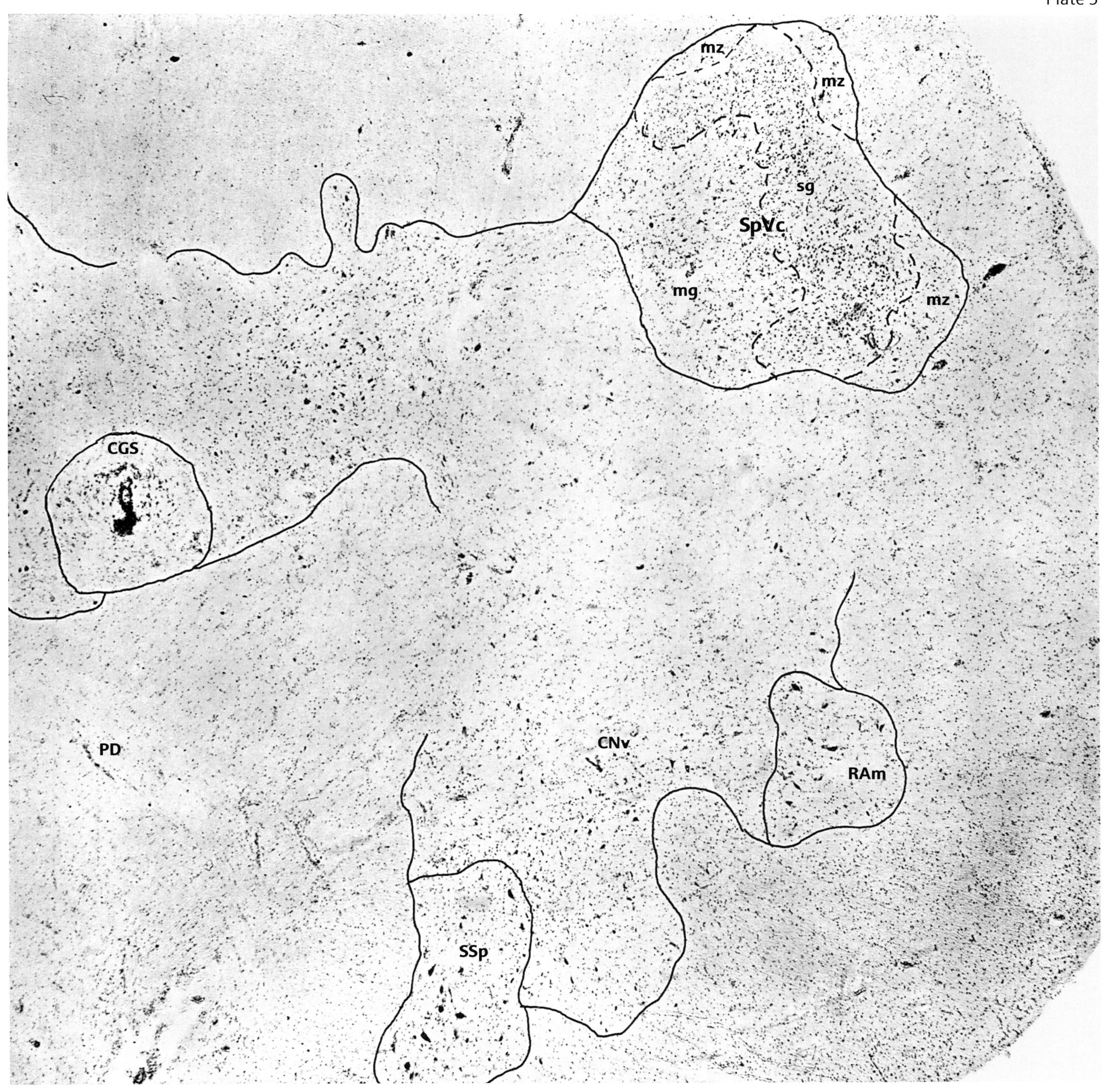

Plate 4. Semischematic representation of cross-section 2701.
Plate 5. Photomicrograph of cross-section 2701. Magnification ×40.

Abbreviations:

CGS	Central gelatinous substance	SpVc	Spinal trigeminal nucleus, caudal part	mz	Spinal trigeminal nucleus, caudal part marginal zone
CNv	Central nucleus of the medulla oblongata, ventral subnucleus	sg	Spinal trigeminal nucleus, caudal part substantia gelatinosa	SSp	Supraspinal nucleus
PD	Pyramidal decussation	mg	Spinal trigeminal nucleus, caudal part magnocellular zone		
RAm	Retroambiguus nucleus				

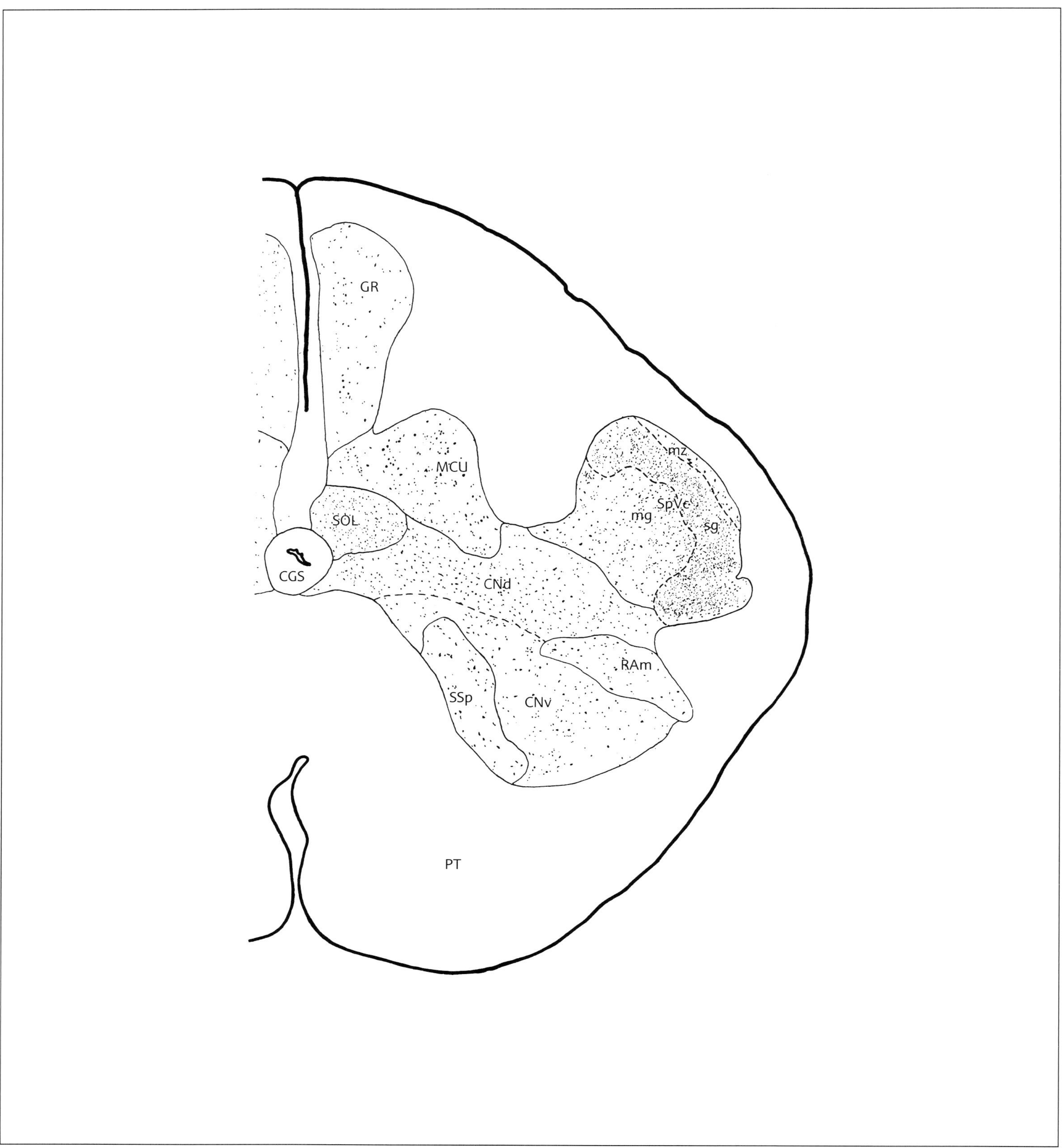
GR
MCU
SOL
CGS
CNd
mz
SpVc
mg
sg
RAm
SSp
CNv
PT

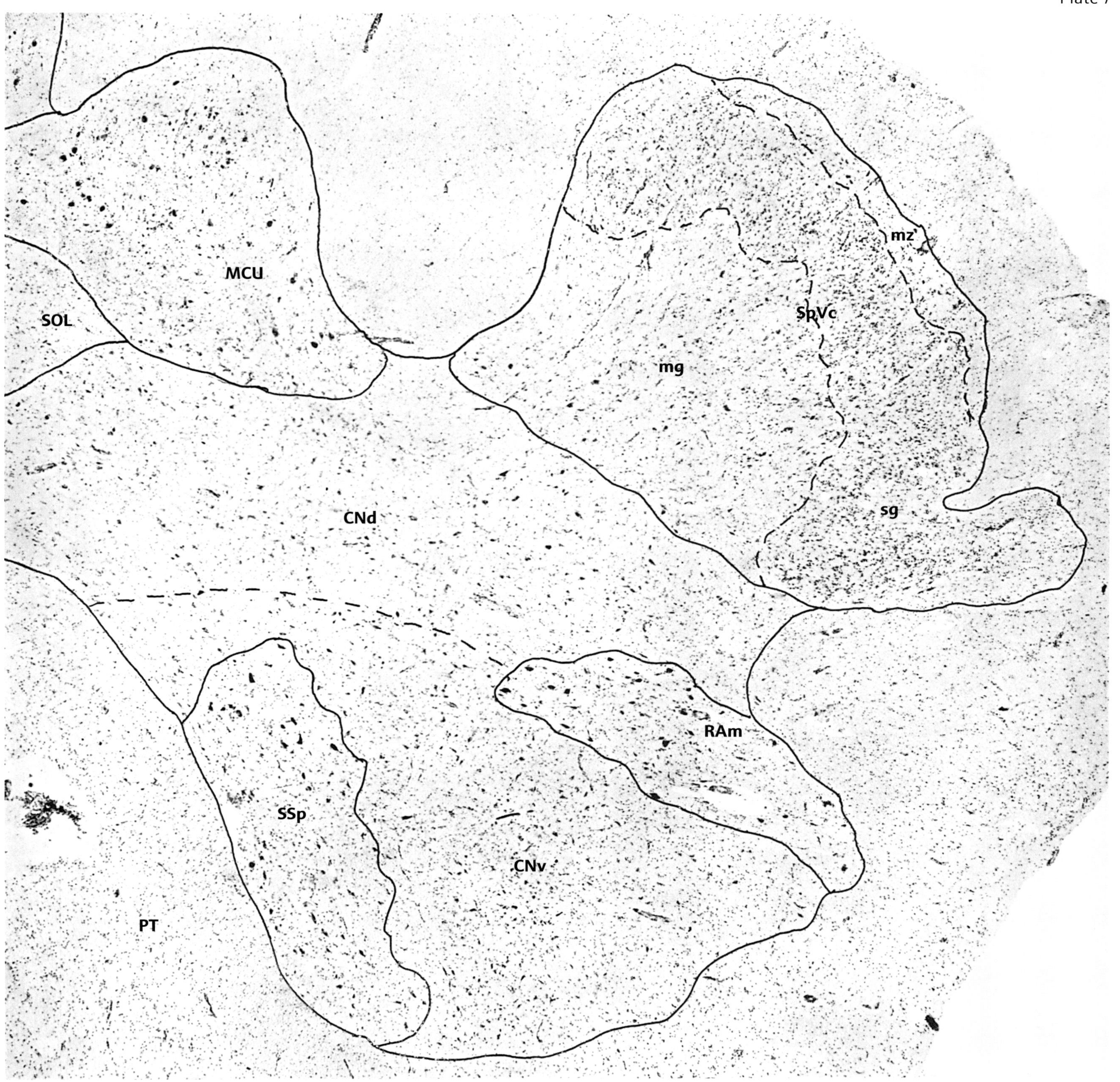

Plate 6. Semischematic representation of cross-section 2501.
Plate 7. Photomicrograph of cross-section 2501. Magnification ×40.

Abbreviations:

CGS	Central gelatinous substance
CNd	Central nucleus of the medulla oblongata, dorsal subnucleus
CNv	Central nucleus of the medulla oblongata, ventral subnucleus
GR	Gracile nucleus
MCU	Medial cuneate nucleus
PT	Pyramidal tract
RAm	Retroambiguus nucleus
SOL	Solitary nucleus
SpVc	Spinal trigeminal nucleus, caudal part
mg	Spinal trigeminal nucleus, caudal part magnocellular zone
mz	Spinal trigeminal nucleus, caudal part marginal zone
sg	Spinal trigeminal nucleus, caudal part substantia gelatinosa
SSp	Supraspinal nucleus

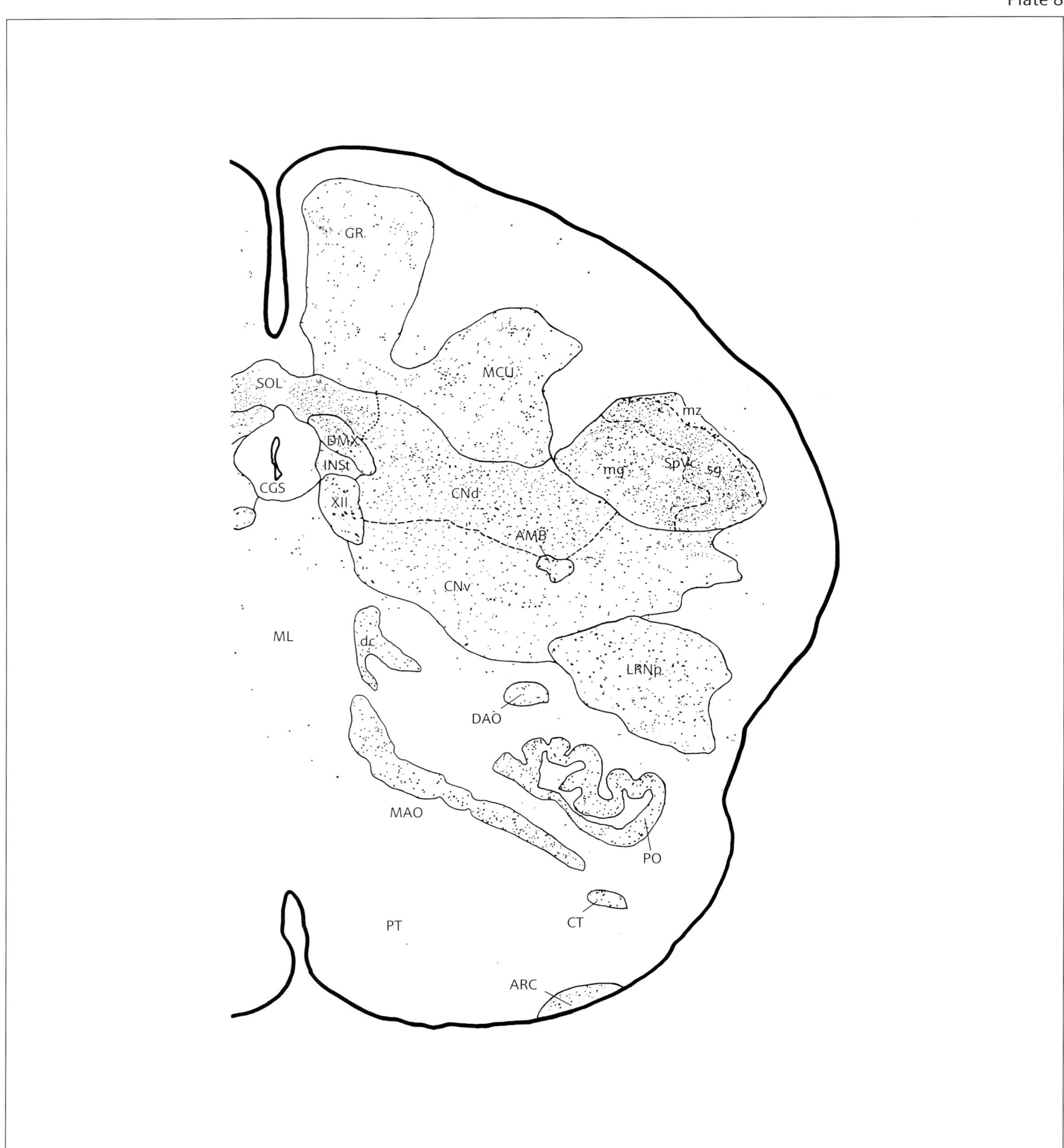
GR
MCU
SOL
mz
DMX
INSt
mg
SpVc
sg
CGS
CNd
XII
AMB
CNv
ML
dc
LRNp
DAO
MAO
PO
PT
CT
ARC

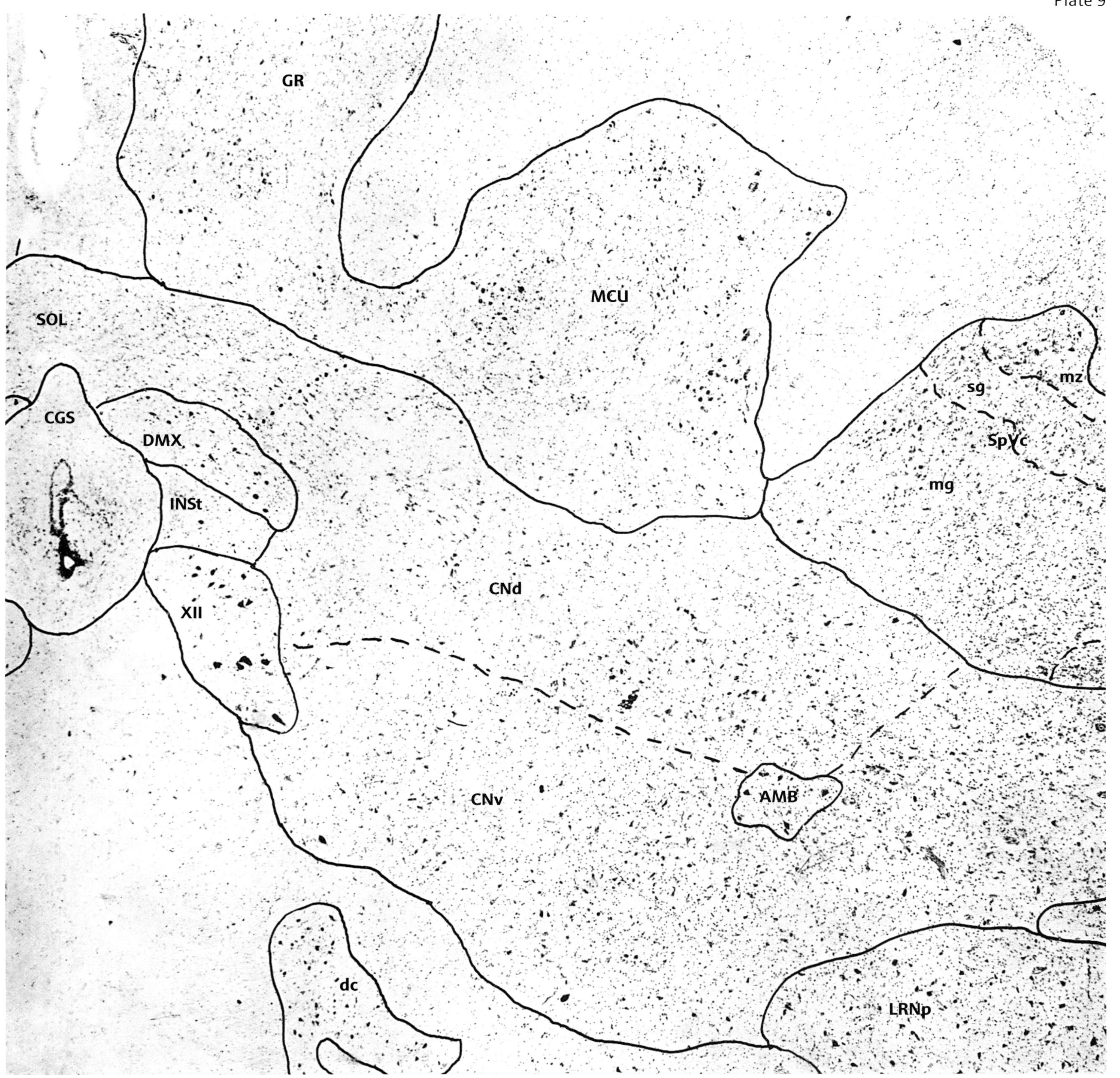

Plate 8. Semischematic representation of cross-section 2301.
Plate 9. Photomicrograph of cross-section 2301. Magnification ×40.

Abbreviations:

AMB	Ambiguus nucleus	DMX	Dorsal motor nucleus of the vagus nerve	SpVc	Spinal trigeminal nucleus, caudal part
ARC	Arcuate nucleus	GR	Gracile nucleus	sg	Spinal trigeminal nucleus, caudal part substantia gelatinosa
CGS	Central gelatinous substance	INSt	Intercalated nucleus of Staderini	mg	Spinal trigeminal nucleus, caudal part magnocellular zone
CNd	Central nucleus of the medulla oblongata, dorsal subnucleus	LRNp	Lateral reticular nucleus, principal part	mz	Spinal trigeminal nucleus, caudal part marginal zone
CNv	Central nucleus of the medulla oblongata, ventral subnucleus	MAO	Medial accessory olive	XII	Hypoglossal nucleus
CT	Conterminal nucleus	MCU	Medial cuneate nucleus		
DAO	Dorsal accessory olive	ML	Medial lemniscus		
dc	Dorsal cap	PO	Principal olive		
		PT	Pyramidal tract		
		SOL	Solitary nucleus		

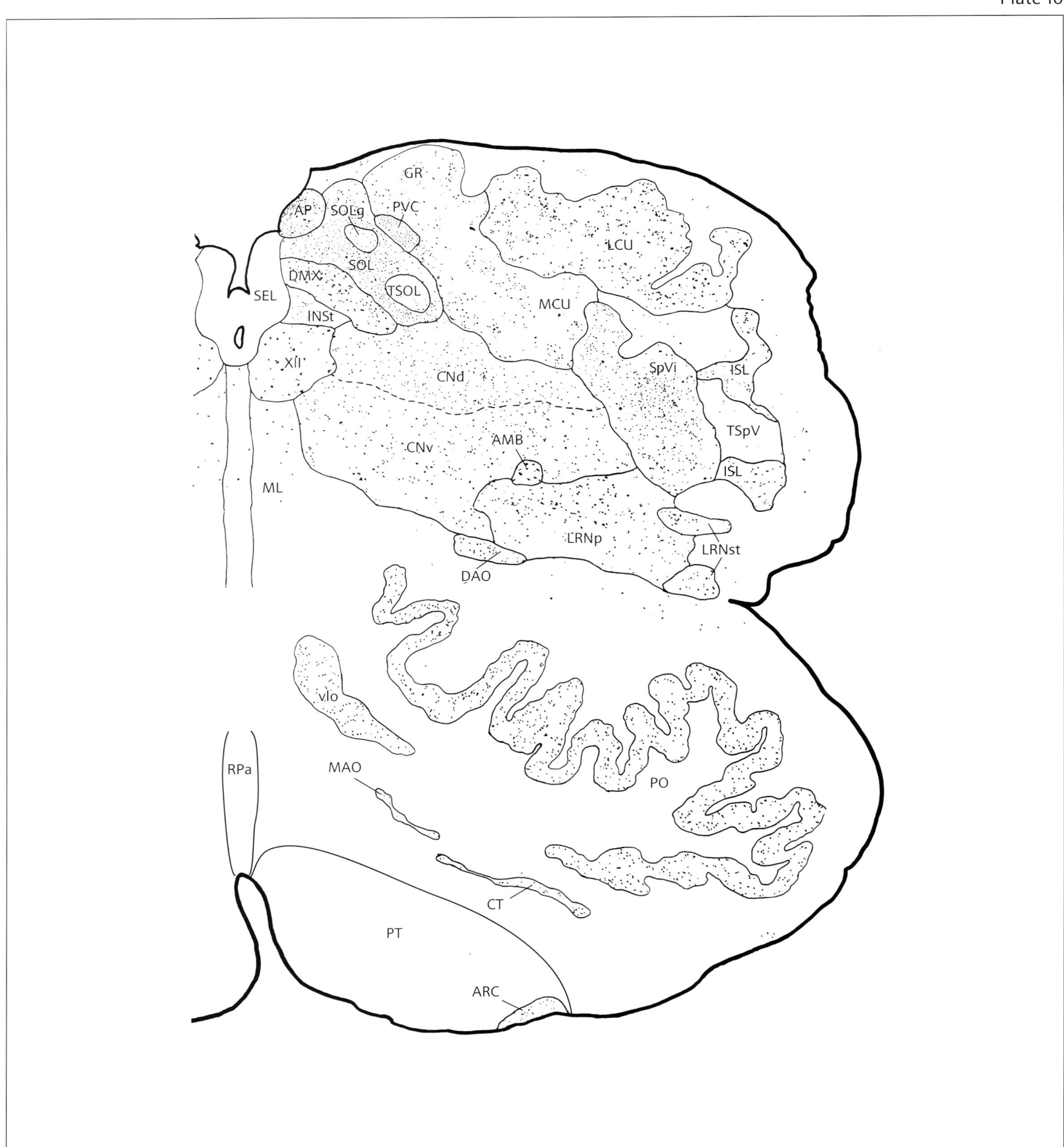
GR
AP
SOLg
PVC
LCU
SOL
DMX
TSOL
SEL
MCU
INSt
XII
SpVi
ISL
CNd
TSpV
AMB
CNv
ISL
ML
LRNp
LRNst
DAO
vlo
RPa
MAO
PO
CT
PT
ARC

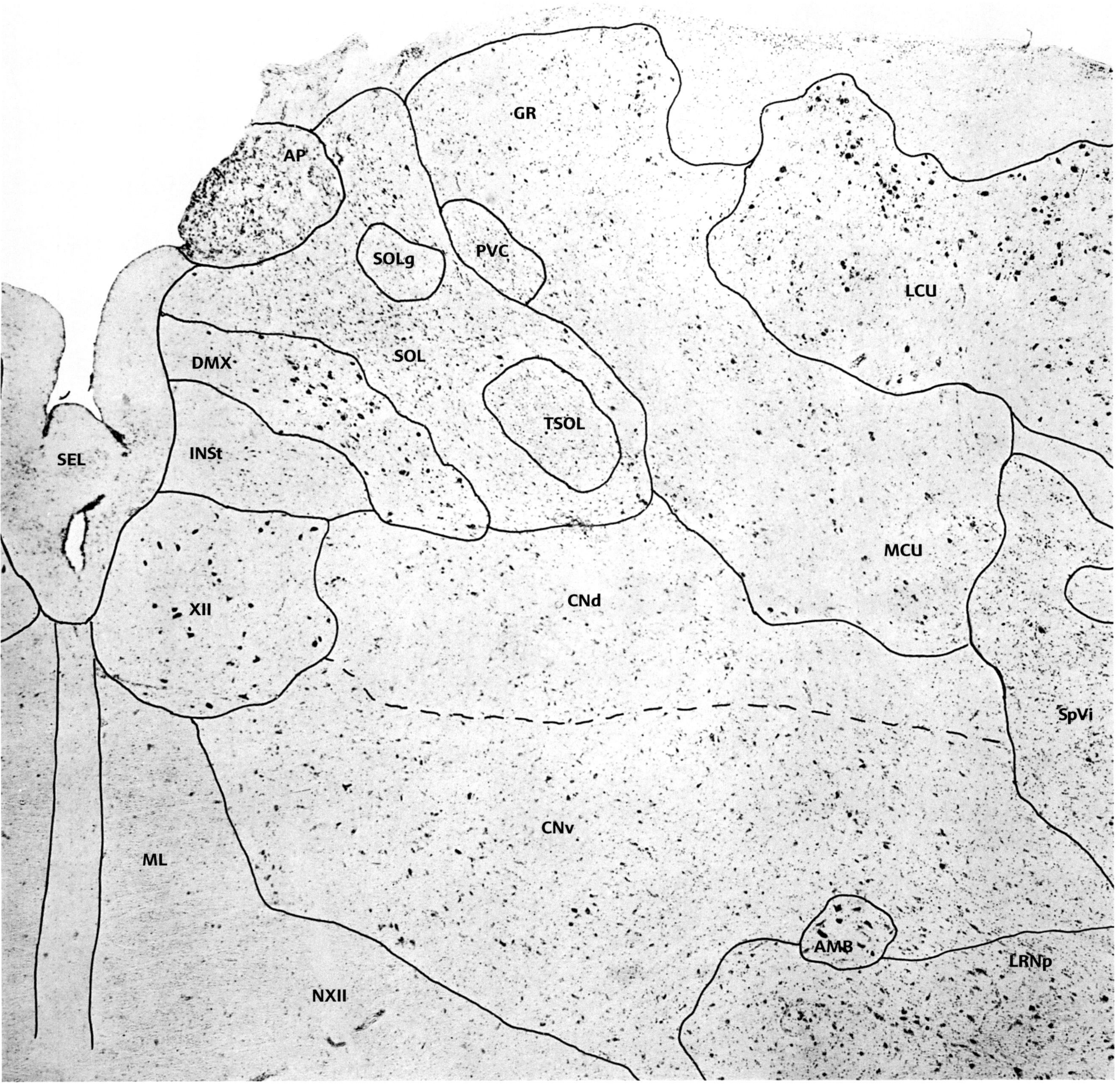

Plate 10. Semischematic representation of cross-section 2051.
Plate 11. Photomicrograph of cross-section 2051. Magnification ×40.

Abbreviations:

AMB	Ambiguus nucleus
AP	Area postrema
ARC	Arcuate nucleus
CNd	Central nucleus of the medulla oblongata, dorsal subnucleus
CNv	Central nucleus of the medulla oblongata, ventral subnucleus
CT	Conterminal nucleus
DAO	Dorsal accessory olive
DMX	Dorsal motor nucleus of the vagus nerve
GR	Gracile nucleus
INSt	Intercalated nucleus of Staderini
ISL	Islands of the lateral cuneate nuclei
LCU	Lateral cuneate nucleus
LRNp	Lateral reticular nucleus, principal part
LRNst	Lateral reticular nucleus, subtrigeminal part
MAO	Medial accessory olive
MCU	Medial cuneate nucleus
ML	Medial lemniscus
NXII	Hypoglossal nerve
PO	Principal olive
PT	Pyramidal tract
PVC	Nucleus parvocellularis compactus
RPa	Nucleus raphe pallidus
SEL	Subependymal layer
SOL	Solitary nucleus
SOLg	Solitary nucleus, gelatinous subnucleus
SpVi	Spinal trigeminal nucleus, interpolar part
TSOL	Solitary tract
TSpV	Spinal trigeminal tract
vlo	Ventrolateral outgrowth
XII	Hypoglossal nucleus

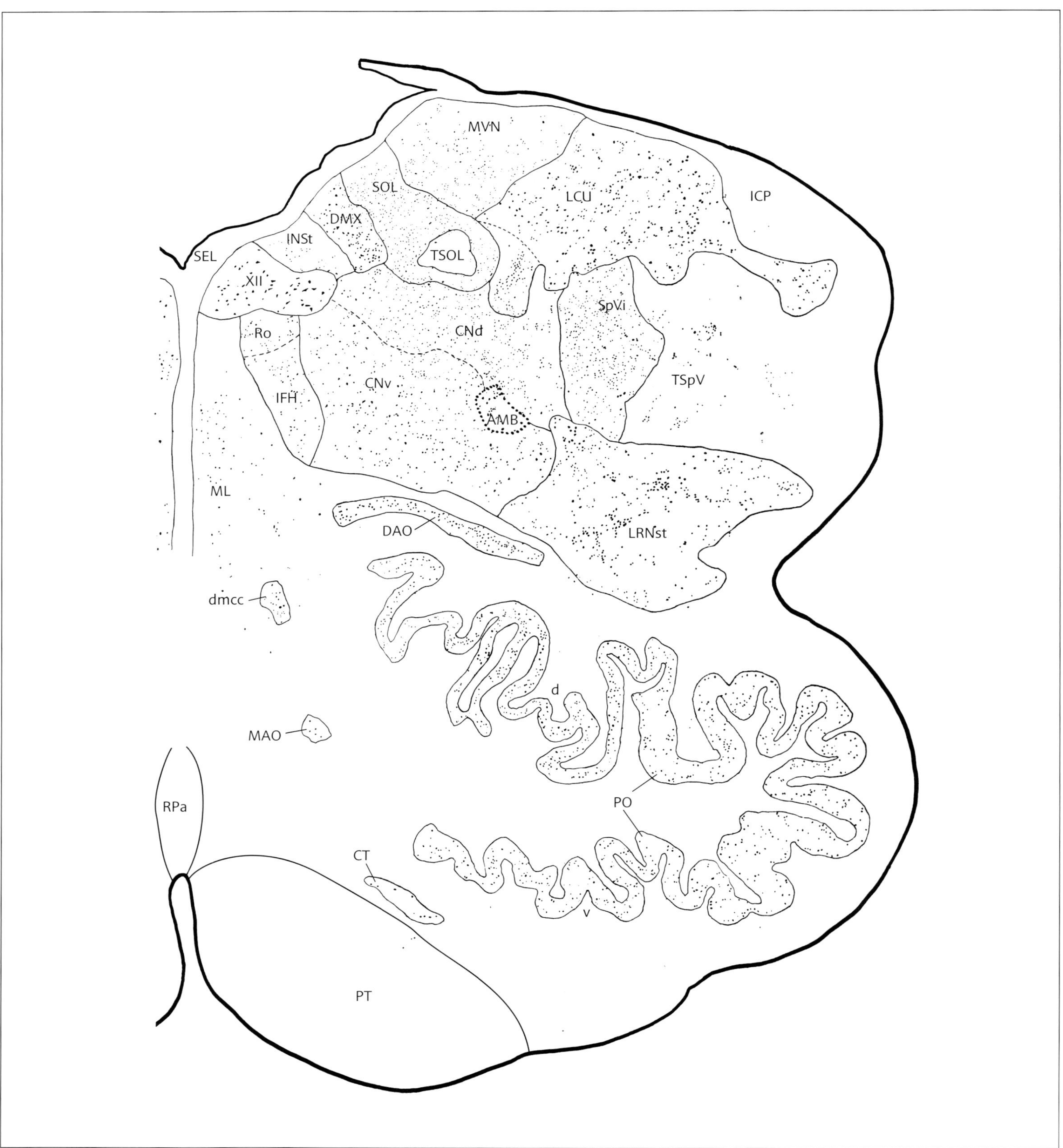
MVN
SOL
LCU
ICP
DMX
INSt
TSOL
SEL
XII
SpVi
CNd
Ro
TSpV
CNv
IFH
AMB
ML
DAO
LRNst
dmcc
d
MAO
PO
RPa
CT
v
PT

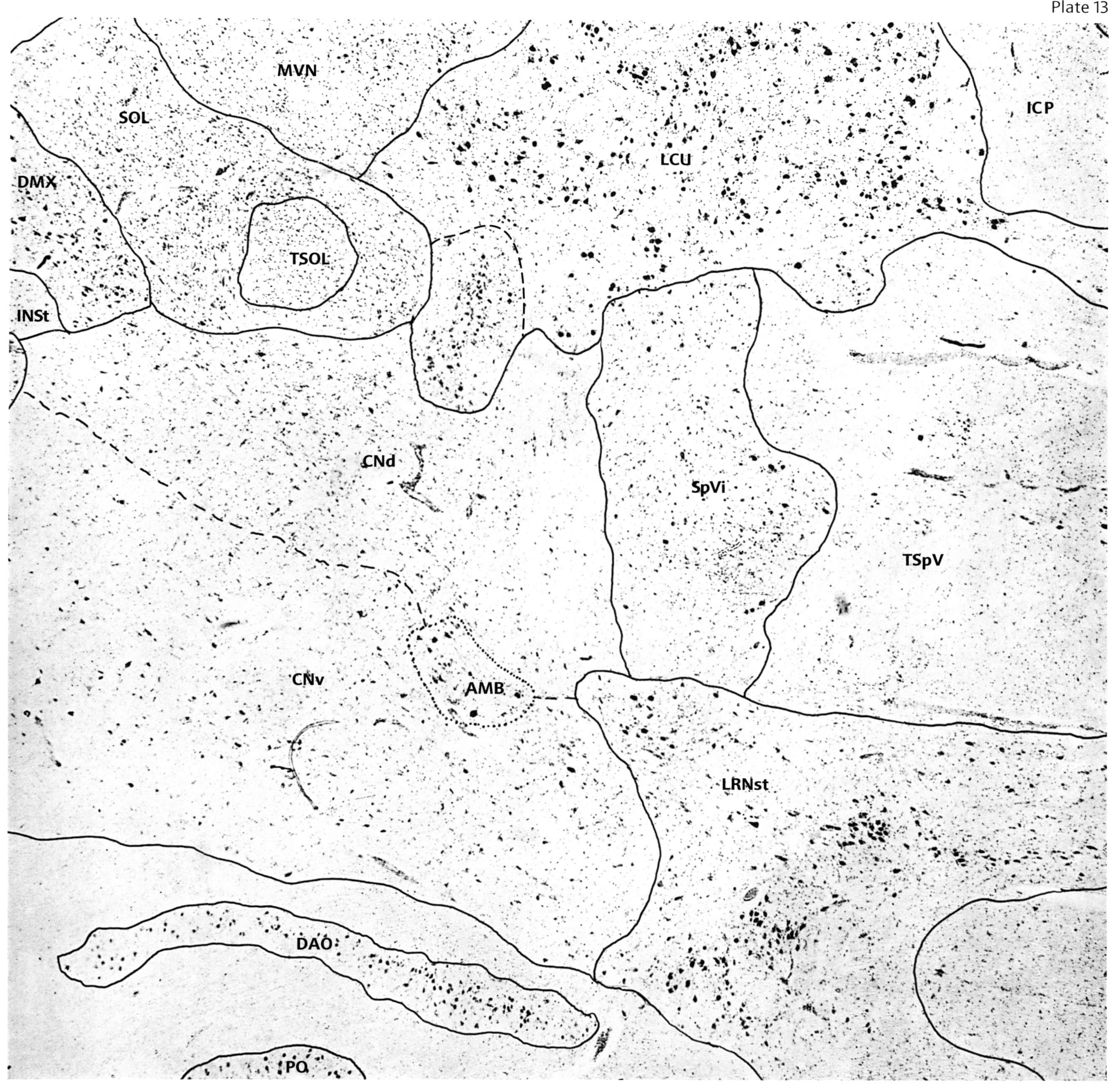

Plate 12. Semischematic representation of cross-section 1901.
Plate 13. Photomicrograph of cross-section 1901. Magnification ×40.

Abbreviations:

AMB	Ambiguus nucleus	ICP	Inferior cerebellar peduncle	v	ventral lamina
CNd	Central nucleus of the medulla oblongata, dorsal subnucleus	IFH	Interfascicular hypoglossal nucleus	PT	Pyramidal tract
CNv	Central nucleus of the medulla oblongata, ventral subnucleus	INSt	Intercalated nucleus of Staderini	RPa	Nucleus raphe pallidus
CT	Conterminal nucleus	LCU	Lateral cuneate nucleus	Ro	Nucleus of Roller
DAO	Dorsal accessory olive	LRNst	Lateral reticular nucleus, subtrigeminal part	SEL	Subependymal layer
dmcc	Dorsomedial cell columns of the inferior olive	MAO	Medial accessory olive	SOL	Solitary nucleus
DMX	Dorsal motor nucleus of the vagus nerve	ML	Medial lemniscus	SpVi	Spinal trigeminal nucleus, interpolar part
		MVN	Medial vestibular nucleus	TSOL	Solitary tract
		PO	Principal olive	TSpV	Spinal trigeminal tract
		d	dorsal lamina	XII	Hypoglossal nucleus

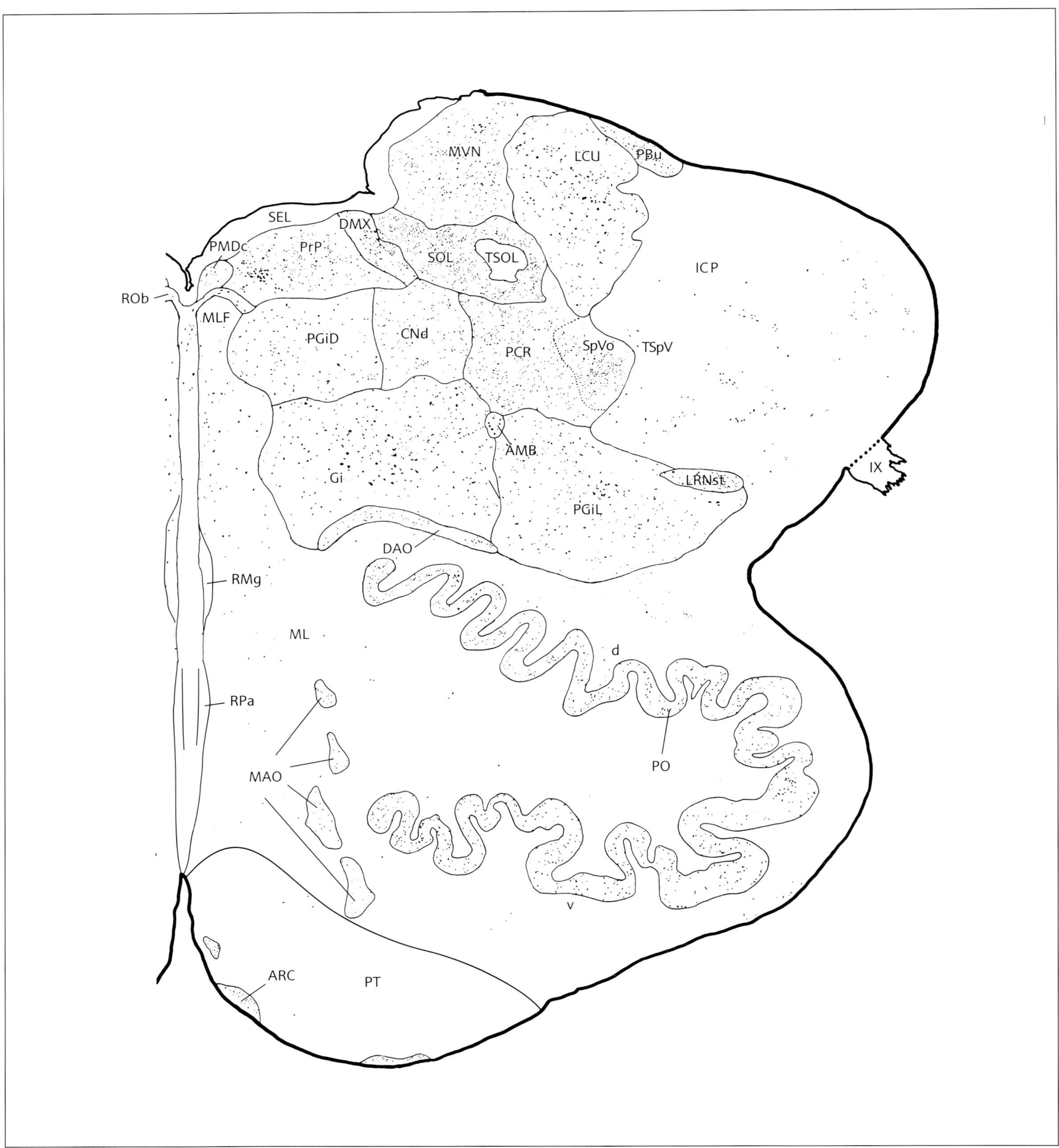
MVN
LCU
PBu
SEL
DMX
PMDc
PrP
SOL
TSOL
ICP
ROb
MLF
PGiD
CNd
PCR
SpVo
TSpV
AMB
IX
Gi
LRNst
PGiL
DAO
RMg
ML
d
RPa
PO
MAO
v
ARC
PT

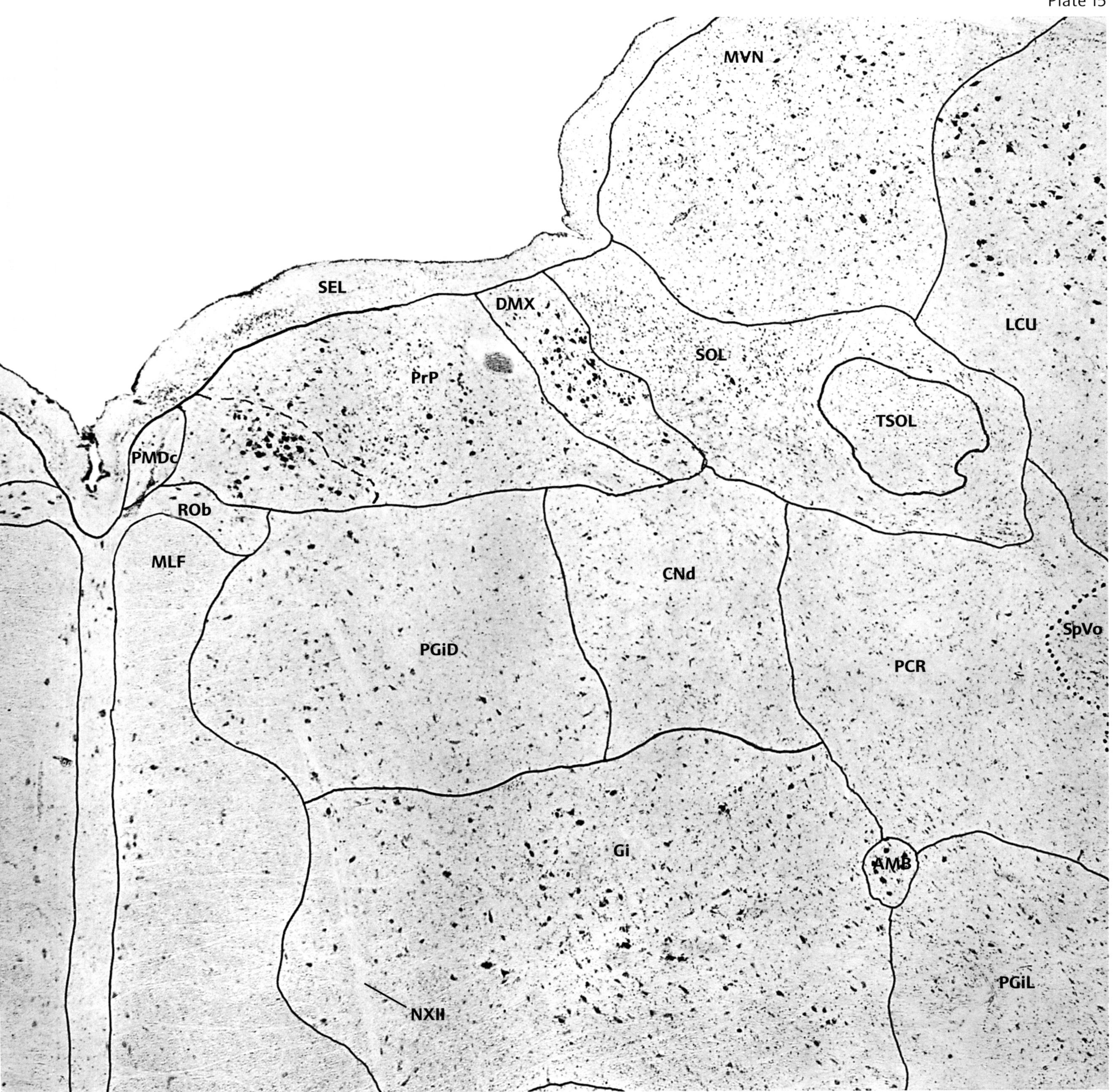

Plate 14. Semischematic representation of cross-section 1801.
Plate 15. Photomicrograph of cross-section 1801. See also chapter No. 65 on nucleus raphe obscurus for a photomicrograph taken slightly further rostrally. Magnification ×40.

Abbreviations:

AMB	Ambiguus nucleus
ARC	Arcuate nucleus
CNd	Central nucleus of the medulla oblongata, dorsal subnucleus
DAO	Dorsal accessory olive
DMX	Dorsal motor nucleus of the vagus nerve
Gi	Gigantocellular nucleus
ICP	Inferior cerebellar peduncle
IX	Glossopharyngeal nerve
LCU	Lateral cuneate nucleus
LRNst	Lateral reticular nucleus, subtrigeminal part
MAO	Medial accessory olive
ML	Medial lemniscus
MLF	Medial longitudinal fasciculus
MVN	Medial vestibular nucleus
NXII	Hypoglossal nerve
PBu	Nucleus of the pontobulbar body
PGiD	Dorsal paragigantocellular nucleus
PGiL	Lateral paragigantocellular nucleus
PMDc	Nucleus paramedianus dorsalis, caudal subnucleus
PO	Principal olive
d	dorsal lamina
v	ventral lamina
PrP	Prepositus nucleus
PT	Pyramidal tract
PCR	Parvocellular reticular nucleus
RMg	Nucleus raphe magnus
ROb	Nucleus raphe obscurus
RPa	Nucleus raphe pallidus
SEL	Subependymal layer
SOL	Solitary nucleus
SpVo	Spinal trigeminal nucleus, oral part
TSOL	Solitary tract
TSpV	Spinal trigeminal tract

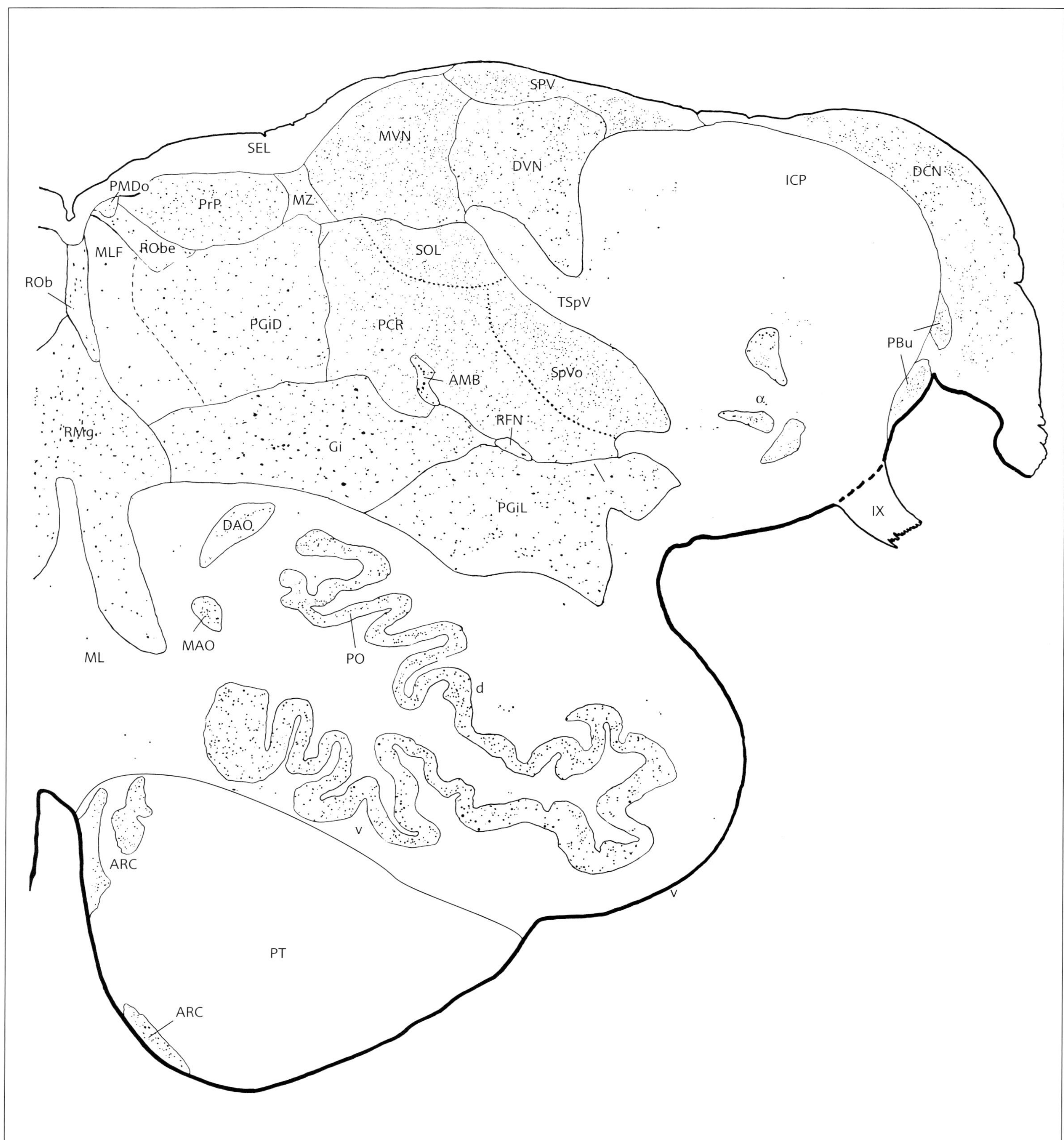
SPV
MVN
SEL
DVN
ICP
DCN
PMDo
PrP
MZ
MLF
RObe
SOL
ROb
TSpV
PGiD
PCR
PBu
AMB
SpVo
α
RFN
RMg
Gi
PGiL
IX
DAO
MAO
ML
PO
d
v
ARC
v
PT
ARC

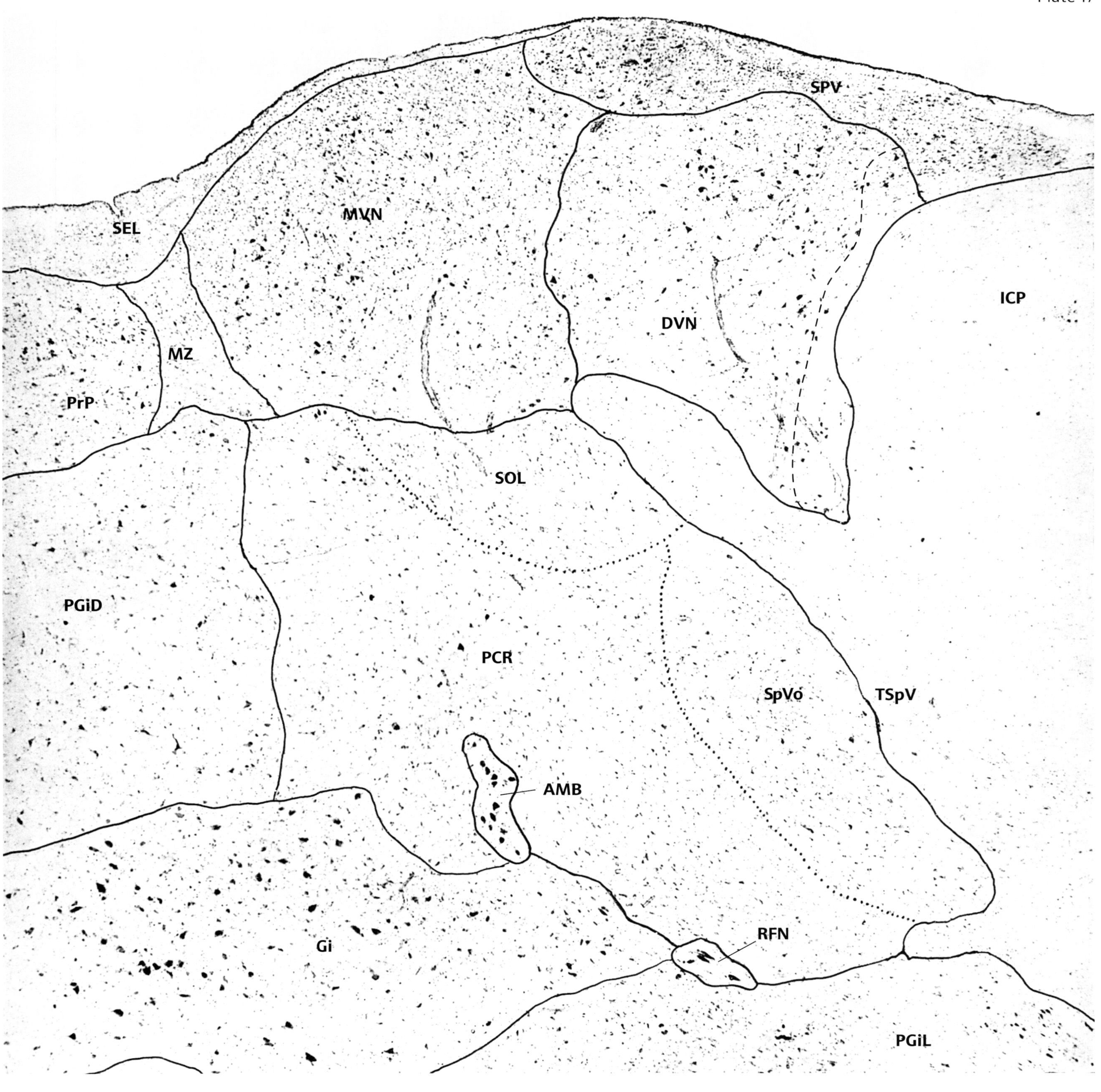

Plate 16. Semischematic representation of cross-section 1701.
Plate 17. Photomicrograph of cross-section 1701. Magnification ×40.

Abbreviations:

α	Islands morphologically similar to pontine nuclei
AMB	Ambiguus nucleus
ARC	Arcuate nucleus
DAO	Dorsal accessory olive
DCN	Dorsal cochlear nucleus
DVN	Descending vestibular nucleus
Gi	Gigantocellular nucleus
ICP	Inferior cerebellar peduncle
IX	Glossopharyngeal nerve
MAO	Medial accessory olive
ML	Medial lemniscus
MLF	Medial longitudinal fascicle
MVN	Medial vestibular nucleus
MZ	Marginal zone
PBu	Nucleus of the pontobulbar body
PGiD	Dorsal paragigantocellular nucleus
PGiL	Lateral paragigantocellular nucleus
PMDo	Nucleus paramedianus dorsalis, oral subnucleus
PO	Principal olive
d	dorsal lamina
v	ventral lamina
PrP	Prepositus nucleus
PT	Pyramidal tract
PCR	Parvocellular reticular nucleus
RFN	Retrofacial nucleus
RMg	Nucleus raphe magnus
ROb	Nucleus raphe obscurus
RObe	Nucleus raphe obscurus, subnucleus extraraphalis
SEL	Subependymal layer
SOL	Solitary nucleus
SpVo	Spinal trigeminal nucleus, oral part
SPV	Supravestibular nucleus
TSpV	Spinal trigeminal tract

Y
ICP
SEL
MVN
DVN
PMDo
PrP
MZ
ROb
MLF
PGiD
SpVo
SOLo
AMB
TSpV
PCR
Gi
α
VII
α
VCN
α
RMg
PGiL
PBu
ML
PO
ARC
PT
ARC

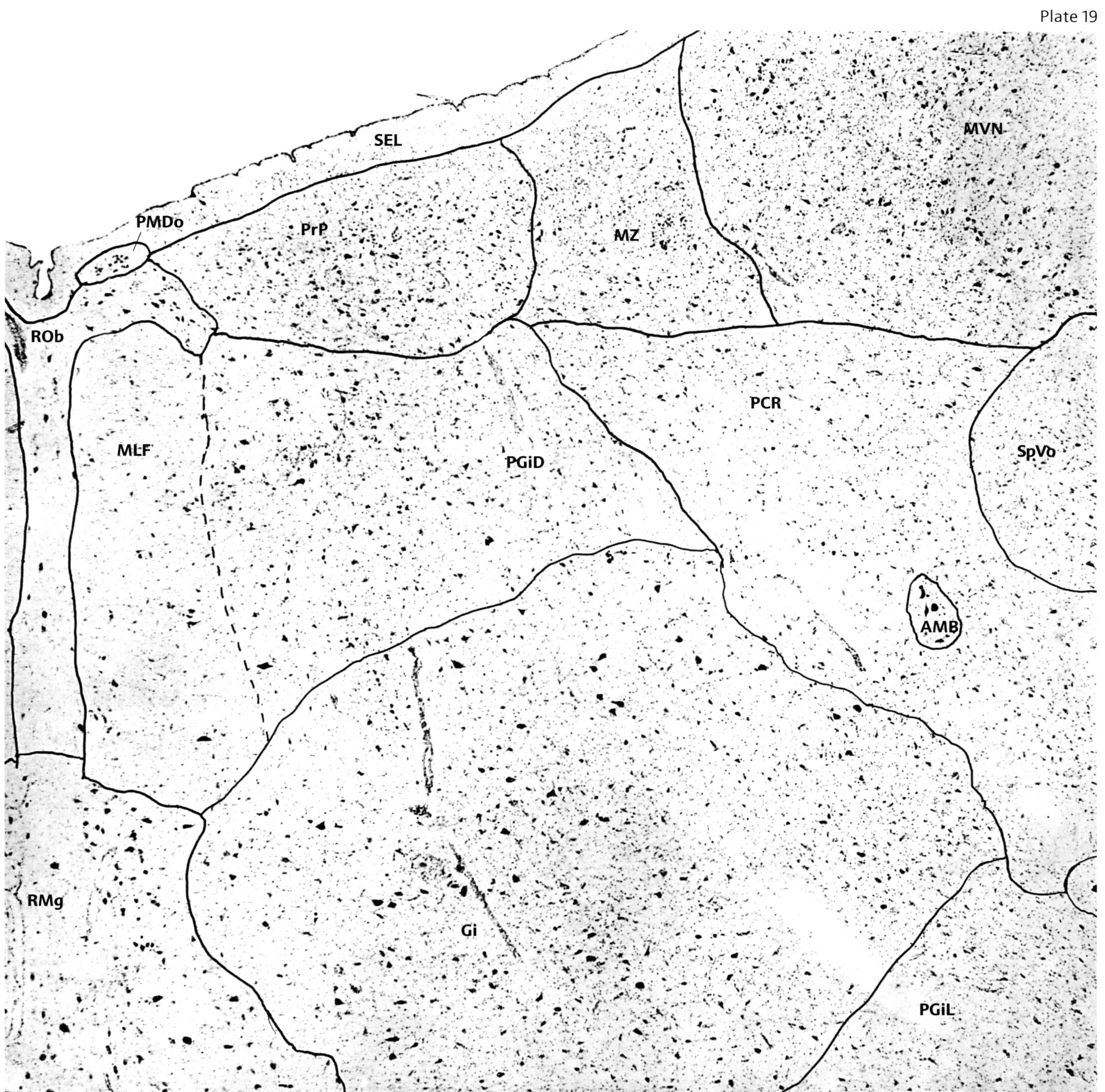

Plate 18. Semischematic representation of cross-section 1601.
Plate 19. Photomicrograph of cross-section 1601. Magnification ×40.

Abbreviations:

α	Islands morphologically similar to pontine nuclei
AMB	Ambiguus nucleus
ARC	Arcuate nucleus
DVN	Descending vestibular nucleus
Gi	Gigantocellular nucleus
ICP	Inferior cerebellar peduncle
ML	Medial lemniscus
MLF	Medial longitudinal fasciculus
MVN	Medial vestibular nucleus
MZ	Marginal zone
PBu	Nucleus of the pontobulbar body
PGiD	Dorsal paragigantocellular nucleus
PGiL	Lateral paragigantocellular nucleus
PMDo	Nucleus paramedianus dorsalis, oral subnucleus
PO	Principal olive
PrP	Prepositus nucleus
PT	Pyramidal tract
PCR	Parvocellular reticular nucleus
RMg	Nucleus raphe magnus
ROb	Nucleus raphe obscurus
SEL	Subependymal layer
SOLo	Solitary nucleus, rostral tip
SpVo	Spinal trigeminal nucleus, oral part
TSpV	Spinal trigeminal tract
VCN	Ventral cochlear nucleus
VII	Facial nucleus
Y	Y group

ICP
LVN
SEL
PrP
MVN
SOLo
NVIII
MLF
INV
SpVo
PGiD
PCR
TSpV
VII
NVII
LSO
Gi
NVII
MSO
POC
TZ
ML
NVI
PN
PN
PT
PT

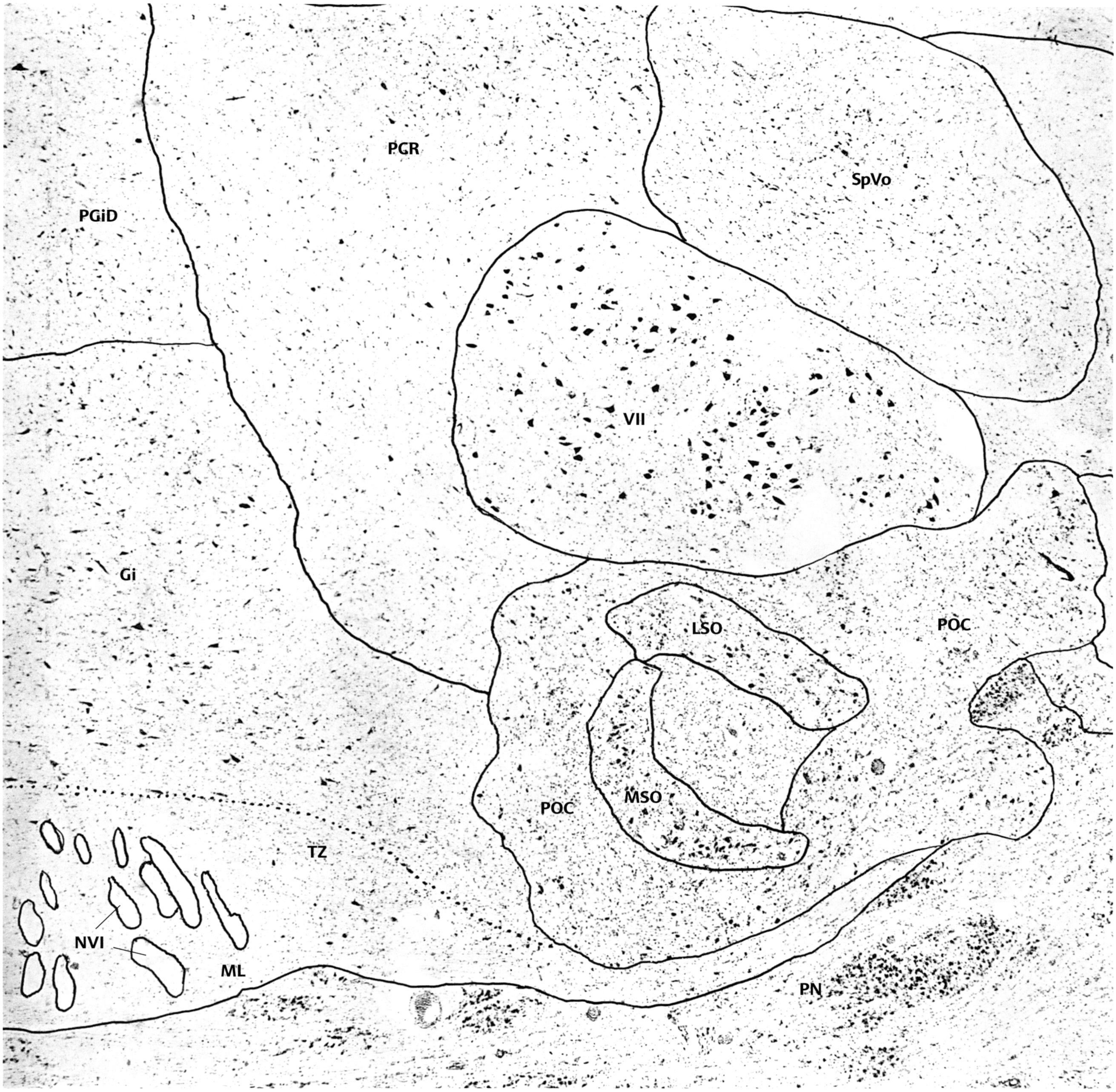

Plate 20. Semischematic representation of cross-section 1501.
Plate 21. Photomicrograph of cross-section 1501. Magnification ×40.

Abbreviations:

Gi	Gigantocellular nucleus	MVN	Medial vestibular nucleus	PCR	Parvocellular reticular nucleus
INV	Interstitial nucleus of the vestibular nerve	NVI	Abducens nerve	PGiD	Dorsal paragigantocellular nucleus
LSO	Lateral superior olivary nucleus	NVII	Facial nerve	SEL	Subependymal layer
LVN	Lateral vestibular nucleus	NVIII	Vestibular nerve	SOLo	Solitary nucleus, rostral tip
ICP	Inferior cerebellar peduncle	PN	Pontine nuclei	SpVo	Spinal trigeminal nucleus, oral part
ML	Medial lemniscus	POC	Periolivary complex	TSpV	Spinal trigeminal tract
MLF	Medial longitudinal fascicle	PrP	Prepositus nucleus	TZ	Trapezoid body
MSO	Medial superior olivary nucleus	PT	Pyramidal tract	VII	Facial nucleus

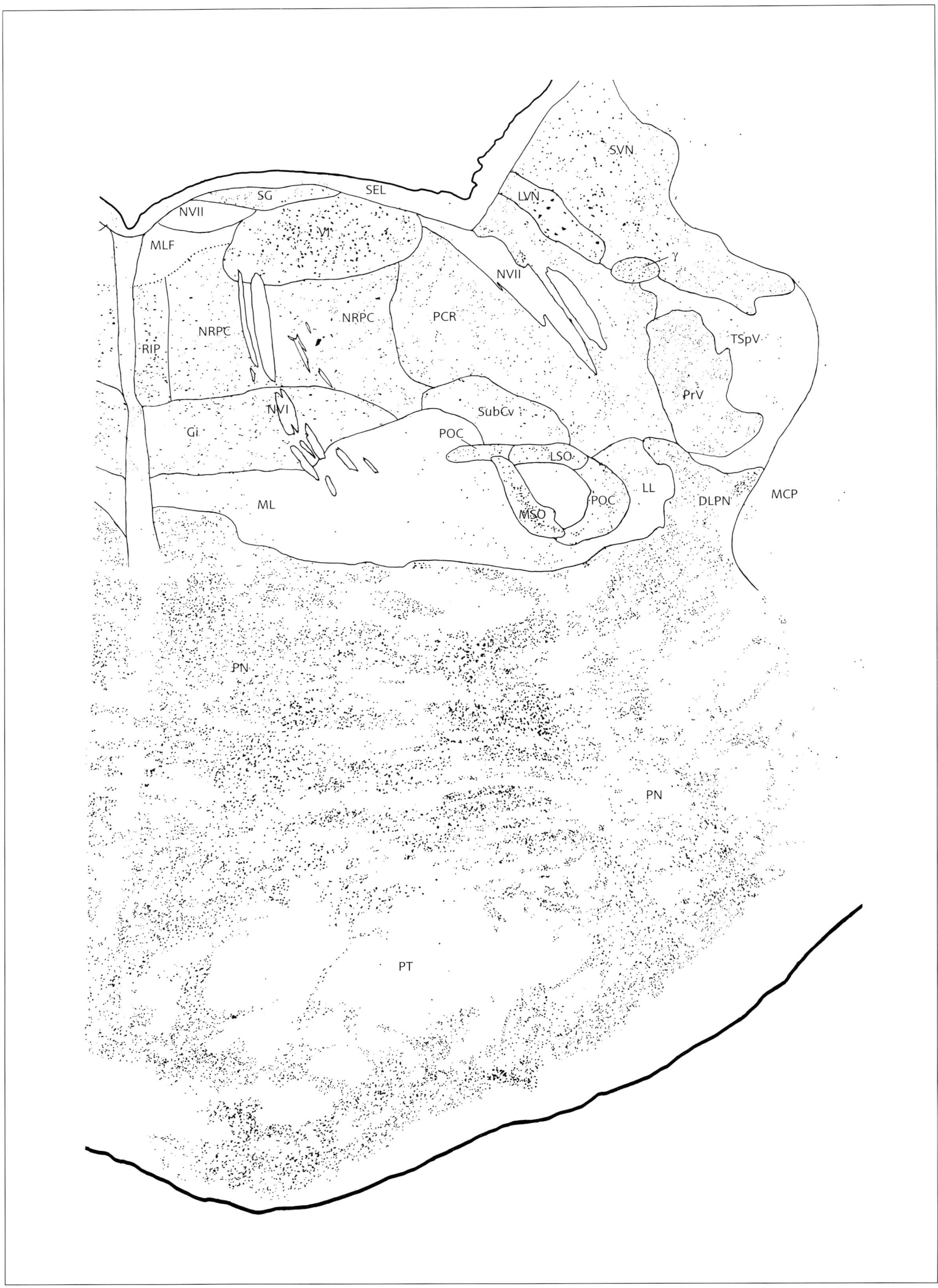
SVN
SG
SEL
LVN
NVII
VI
MLF
γ
NVII
NRPC
NRPC
PCR
TSpV
RIP
PrV
NVI
SubCv
Gi
POC
LSO
LL
POC
DLPN
MCP
ML
MSO
PN
PN
PT

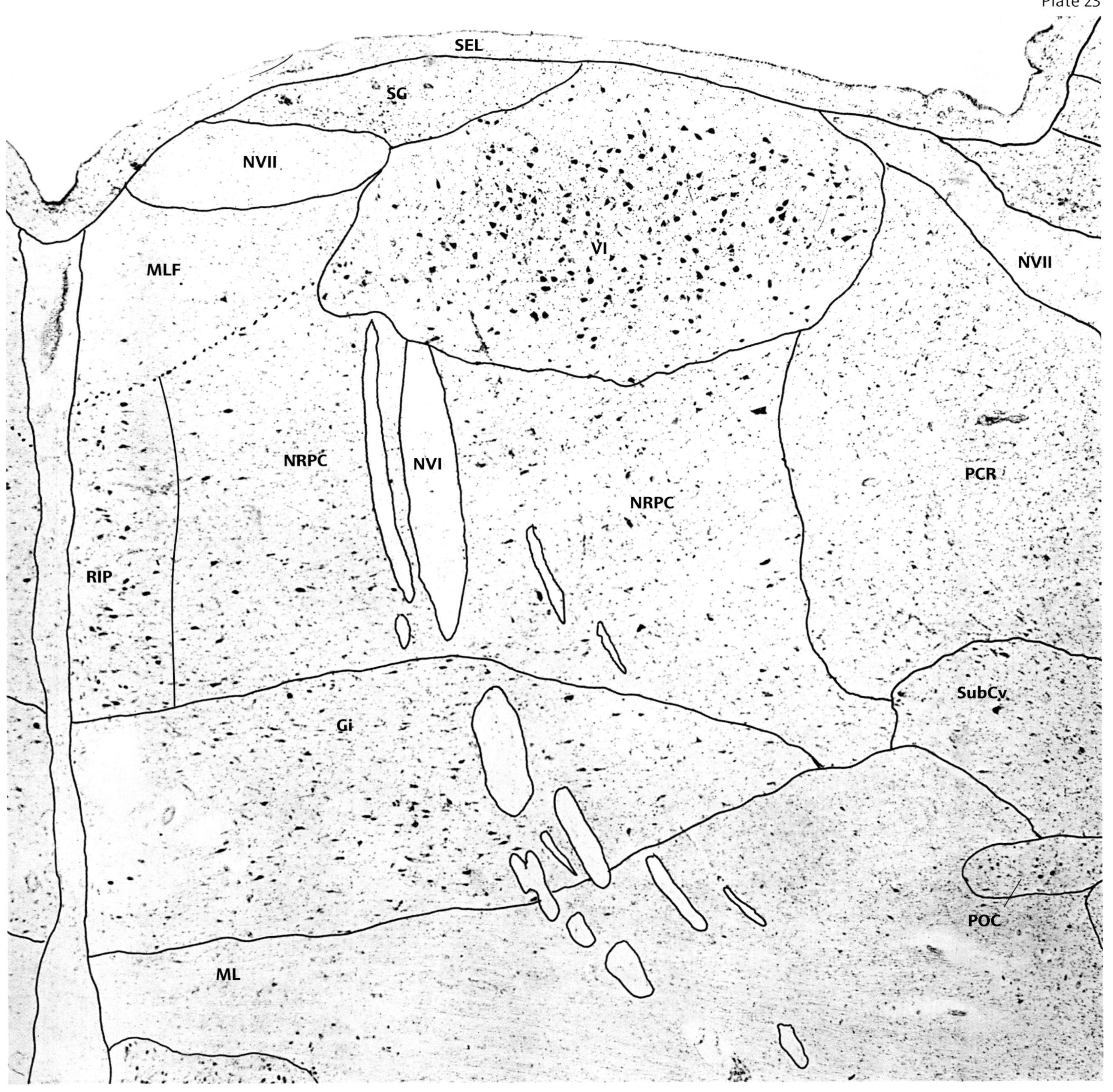

Plate 22. Semischematic representation of cross-section 1401.
Plate 23. Photomicrograph of cross-section 1401. Magnification ×40.

Abbreviations:

DLPN	Dorsolateral pontine nuclei	MSO	Medial superior olivary nucleus	RIP	Nucleus raphe interpositus
γ	Pontine nuclei, possible subgroups	NRPC	Nucleus reticularis pontis caudalis	SEL	Subependymal layer
Gi	Gigantocellular nucleus	NVI	Abducens nerve	SG	Supragenual nucleus
LL	Lateral lemniscus	NVII	Facial nerve	SubCv	Nucleus subcoeruleus, ventral part
LSO	Lateral superior olivary nucleus	PCR	Parvocellular reticular nucleus	SVN	Superior vestibular nucleus
LVN	Lateral vestibular nucleus	PN	Pontine nuclei	TSpV	Spinal trigeminal tract
MCP	Medial cerebellar peduncle	POC	Periolivary complex	VI	Abducens nucleus
ML	Medial lemniscus	PrV	Principal trigeminal nucleus		
MLF	Medial longitudinal fascicle	PT	Pyramidal tract		

MesV
SCP
SVN
PCG
SG
SEL
γ
TMesV
NVII
MLF
NV
MoV
NRPC
PrV
SubCv
NRTP
ML
VLL
NV
DMPN
DLPN
LL
PN
MCP
PN
PT

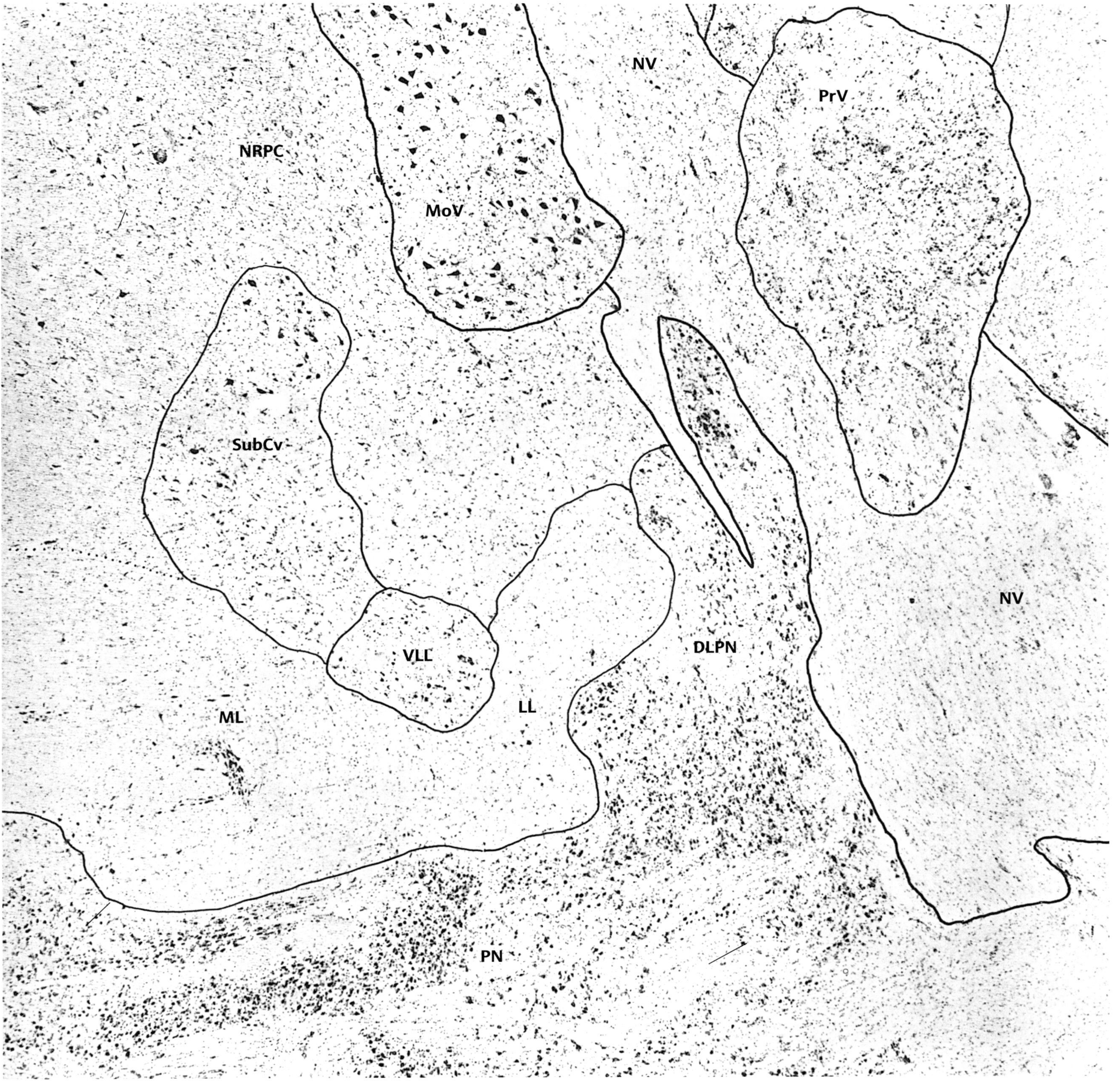

Plate 24. Semischematic representation of cross-section 1301.
Plate 25. Photomicrograph of cross-section 1301. Magnification ×40.

Abbreviations:

DLPN	Dorsolateral pontine nuclei	MoV	Motor trigeminal nucleus	PT	Pyramidal tract
DMPN	Dorsomedial pontine nuclei	NV	Trigeminal nerve	SCP	Superior cerebellar peduncle
γ	Pontine nuclei, possible subgroups	NVII	Facial nerve	SEL	Subependymal layer
LL	Lateral lemniscus	NRPC	Nucleus reticularis pontis caudalis	SG	Supragenual nucleus
MCP	Medial cerebellar peduncle	NRTP	Nucleus reticularis tegmenti pontis	SubCv	Nucleus subcoeruleus, ventral part
MesV	Mesencephalic trigeminal nucleus	PCG	Pontine central gray	SVN	Superior vestibular nucleus
ML	Medial lemniscus	PN	Pontine nuclei	TMesV	Mesencephalic tract of the trigeminal nerve
MLF	Medial longitudinal fascicle	PrV	Principal trigeminal nucleus	VLL	Ventral nucleus of the lateral lemniscus

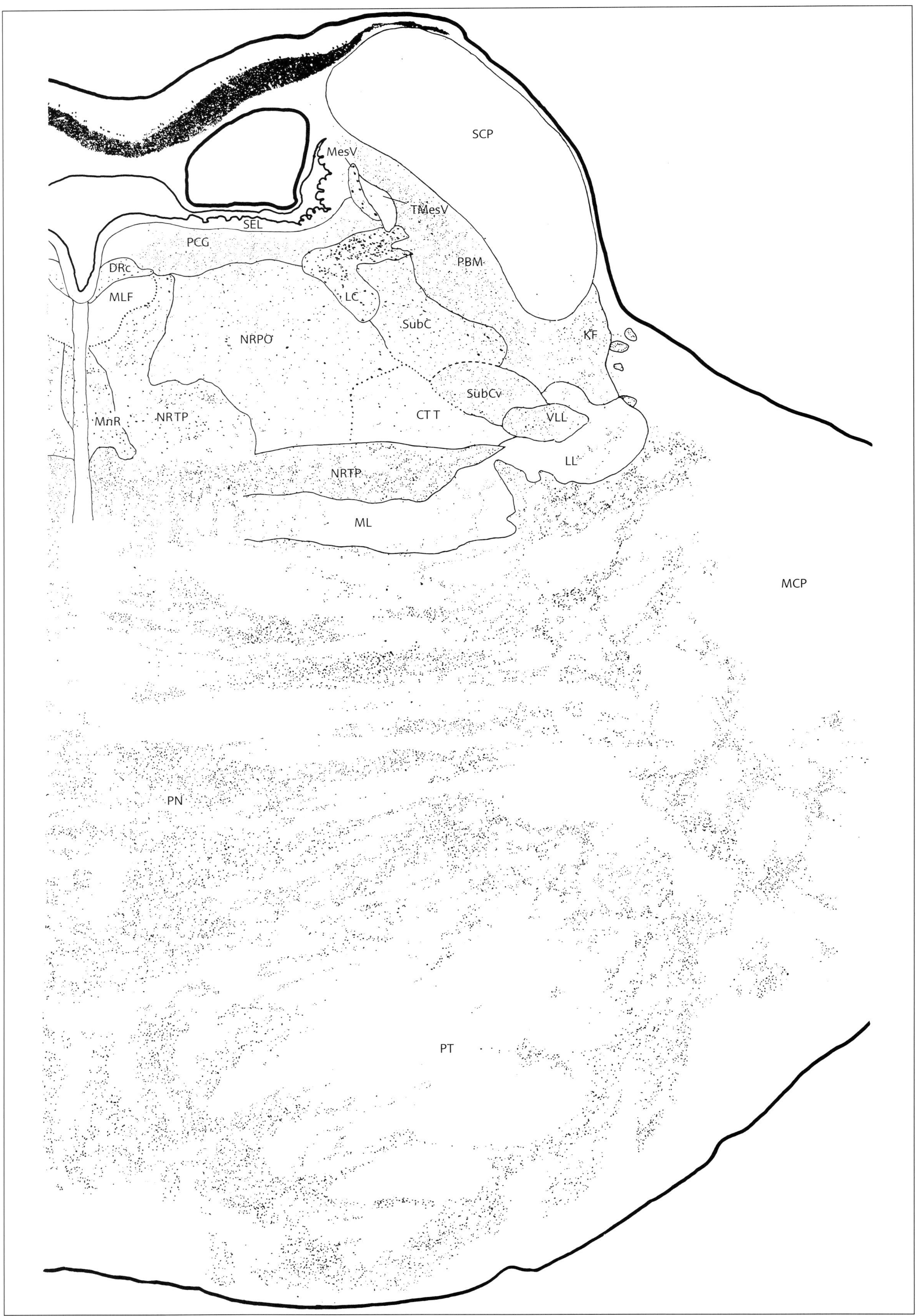
SCP
MesV
TMesV
SEL
PCG
PBM
DRc
MLF
LC
SubC
KF
NRPO
SubCv
MnR
NRTP
CT T
VLL
LL
NRTP
ML
MCP
PN
PT

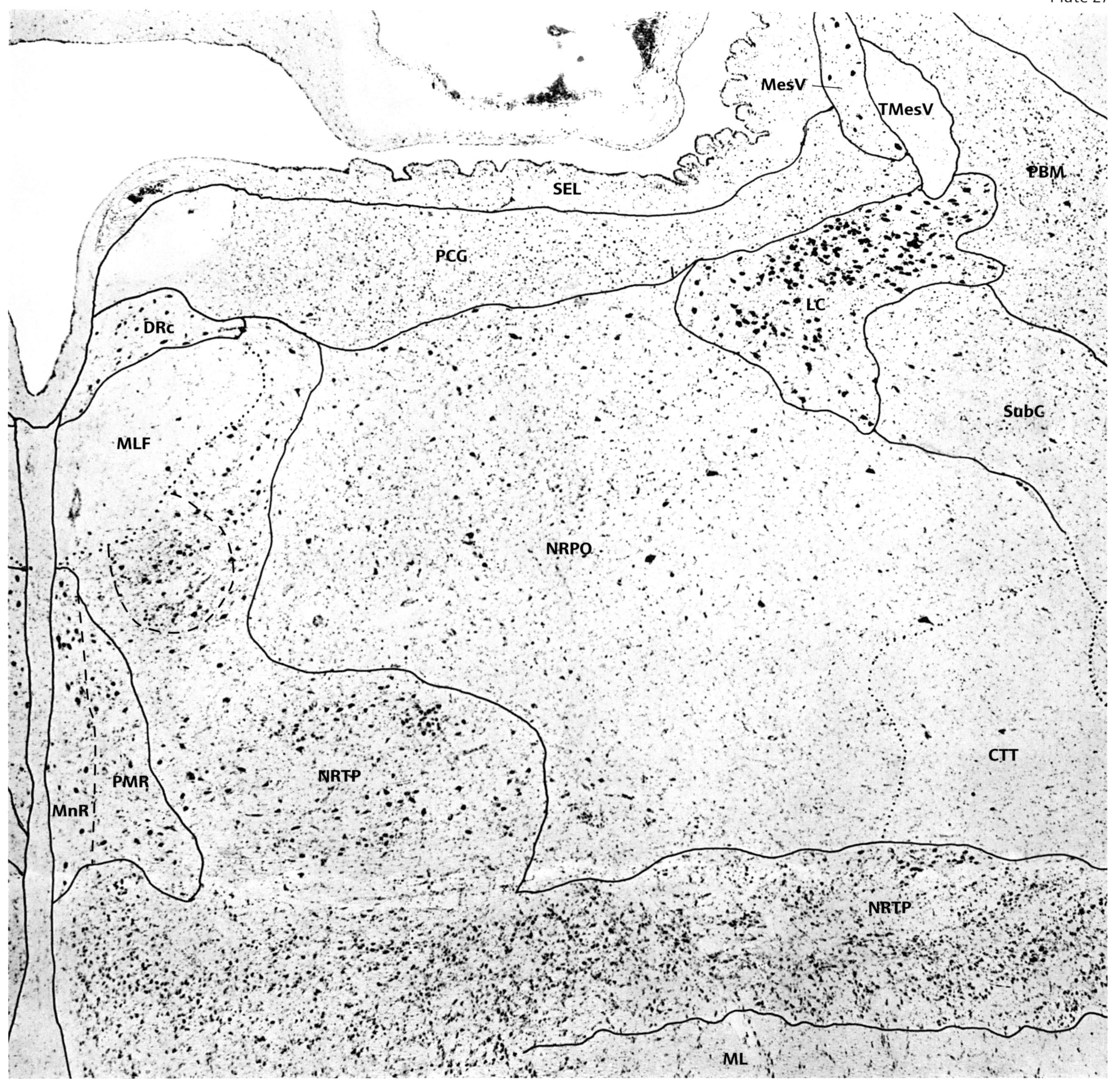

Plate 26. Semischematic representation of cross-section 1101.
Plate 27. Photomicrograph of cross-section 1101. Magnification ×40.

Abbreviations:

CTT	Central tegmental tract	MLF	Medial longitudinal fascicle	PT	Pyramidal tract
DRc	Dorsal raphe nucleus, caudal subnucleus	MnR	Median raphe nucleus	SCP	Superior cerebellar peduncle
KF	Kölliker-Fuse nucleus	NRPO	Nucleus reticularis pontis oralis	SEL	Subependymal layer
LC	Locus coeruleus	NRTP	Nucleus reticularis tegmenti pontis	SubC	Nucleus subcoeruleus
LL	Lateral lemniscus	PBM	Medial parabrachial nucleus	SubCv	Nucleus subcoeruleus, ventral part
MCP	Medial cerebellar peduncle	PCG	Pontine central gray	TMesV	Mesencephalic tract of the trigeminal nerve
MesV	Mesencephalic trigeminal nucleus	PMR	Paramedian raphe nucleus		
ML	Medial lemniscus	PN	Pontine nuclei	VLL	Ventral nucleus of the lateral lemniscus

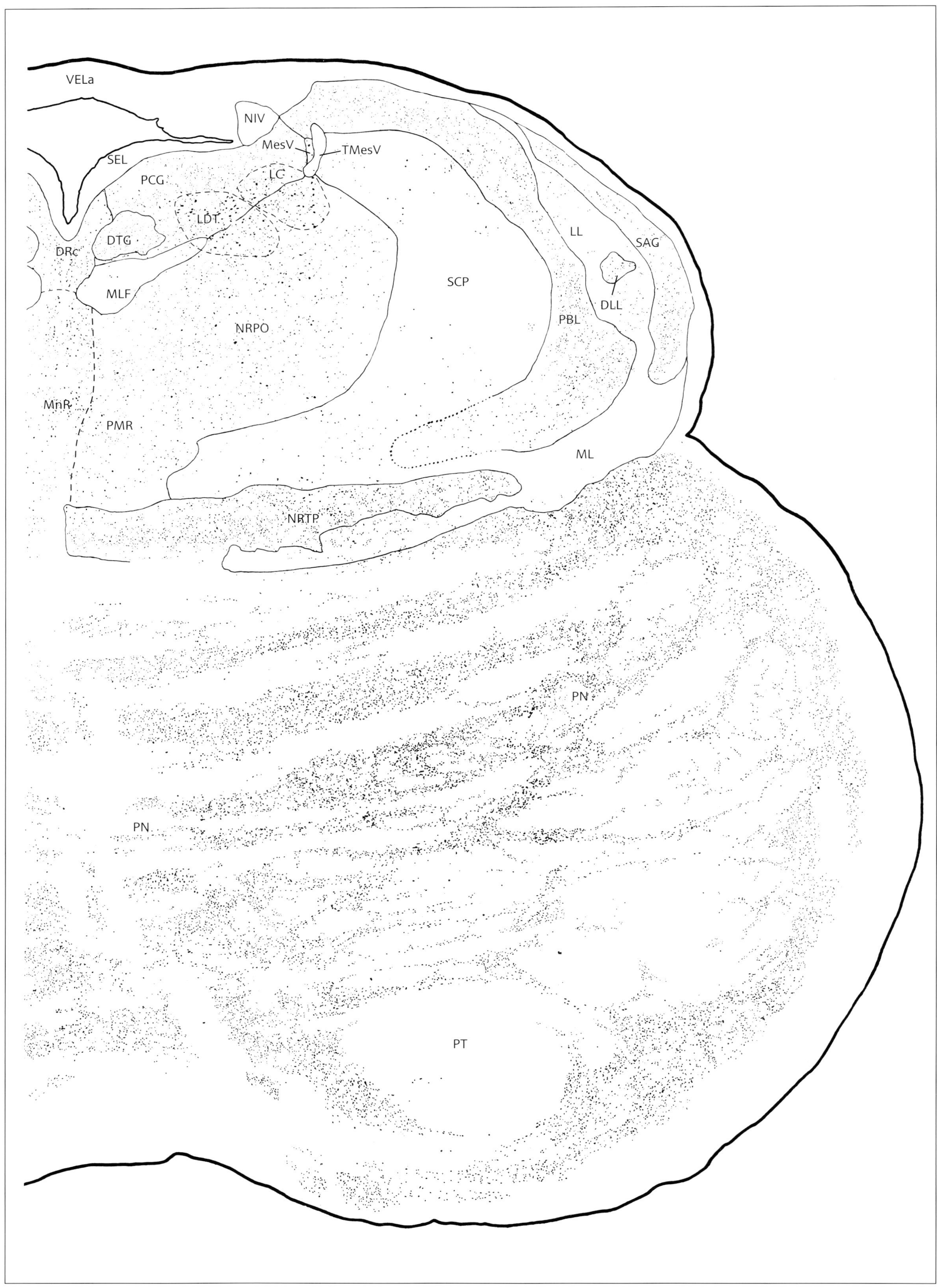
VELa
NIV
MesV
TMesV
SEL
PCG
LC
LDT
DTG
DRc
MLF
NRPO
SCP
LL
SAG
DLL
PBL
MnR
PMR
ML
NRTP
PN
PN
PT

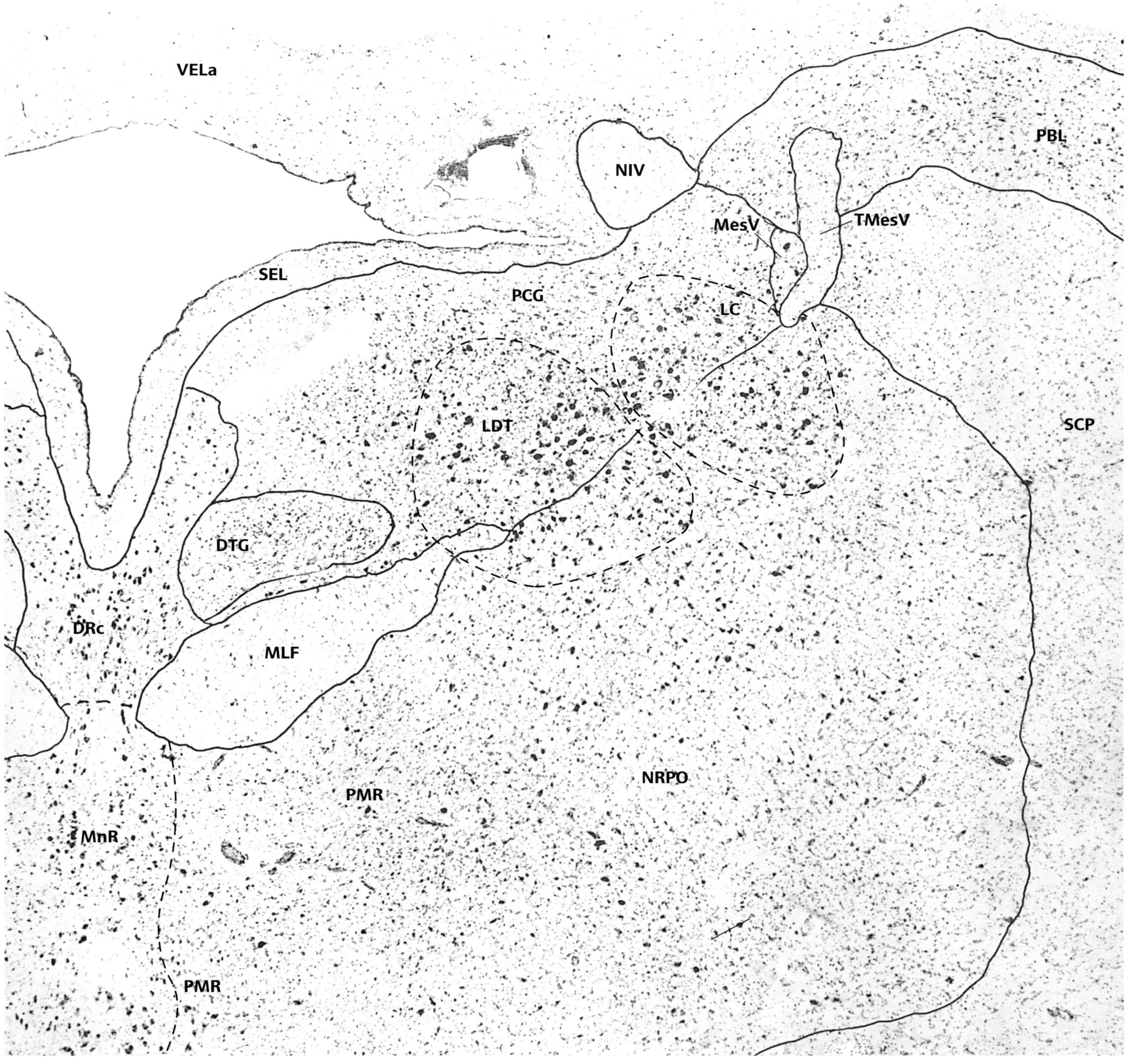

Plate 28. Semischematic representation of cross-section 801.
Plate 29. Photomicrograph of cross-section 801. See also the chapter on nucleus No. 67, the median raphe nucleus, for a photomicrograph taken slightly further caudally. Magnification ×40.

Abbreviations:

DLL	Dorsal nucleus of the lateral lemniscus
DRc	Dorsal raphe nucleus, caudal subnucleus
DTG	Dorsal tegmental nucleus of Gudden
LC	Locus coeruleus
LDT	Laterodorsal tegmental nucleus
LL	Lateral lemniscus
MesV	Mesencephalic trigeminal nucleus
ML	Medial lemniscus
MLF	Medial longitudinal fascicle
MnR	Median raphe nucleus
NIV	Trochlear nerve
NRPO	Nucleus reticularis pontis oralis
NRTP	Nucleus reticularis tegmenti pontis
PBL	Lateral parabrachial nucleus
PCG	Pontine central gray
PMR	Paramedian raphe nucleus
PN	Pontine nuclei
PT	Pyramidal tract
SAG	Sagulum nucleus
SCP	Superior cerebellar peduncle
SEL	Subependymal layer
TMesV	Mesencephalic tract of the trigeminal nerve
VELa	Velum medullare anterior

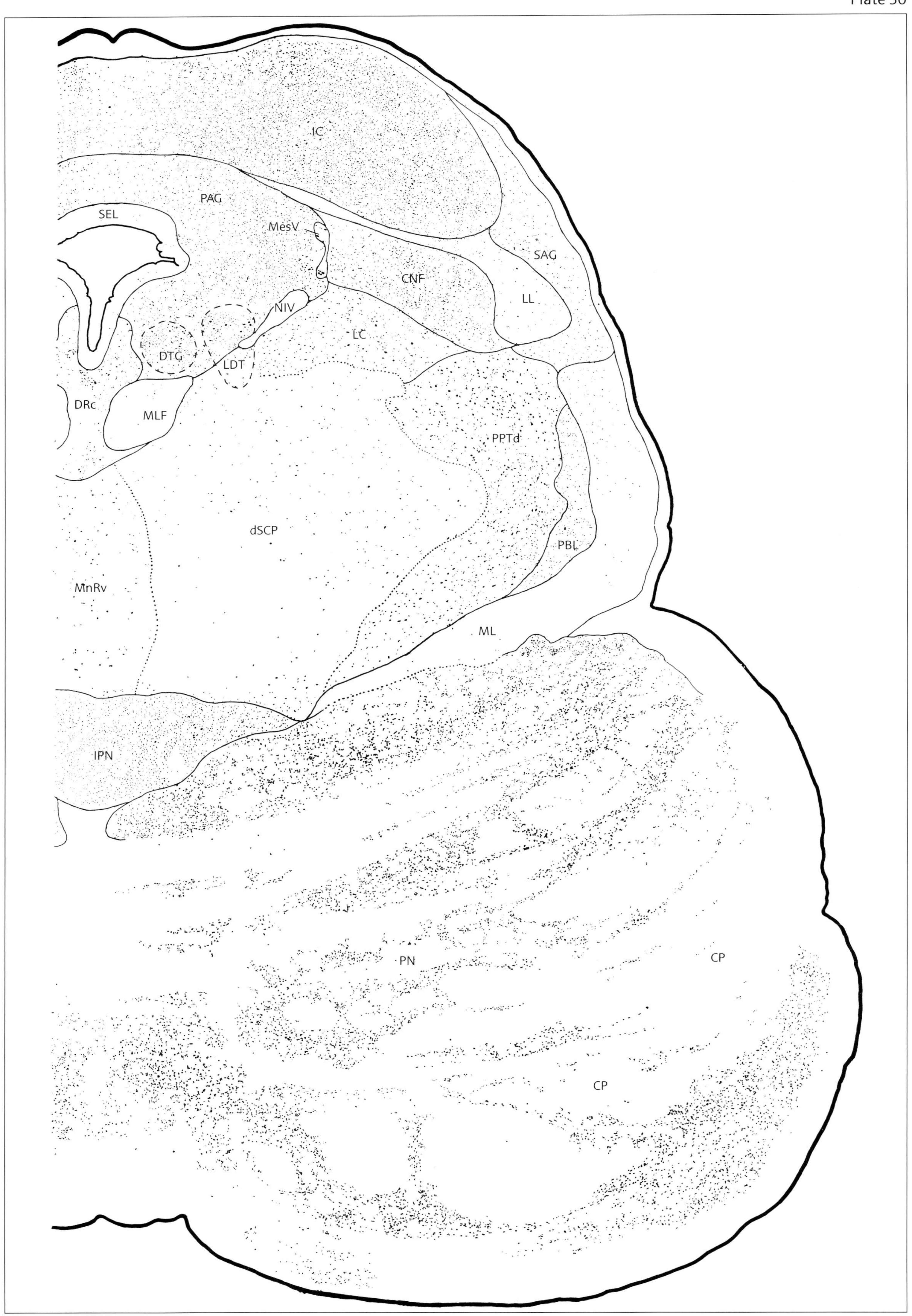
IC
PAG
SEL
MesV
SAG
CNF
LL
NIV
LC
DTG
LDT
DRc
MLF
PPTd
dSCP
PBL
MnRv
ML
IPN
PN
CP
CP

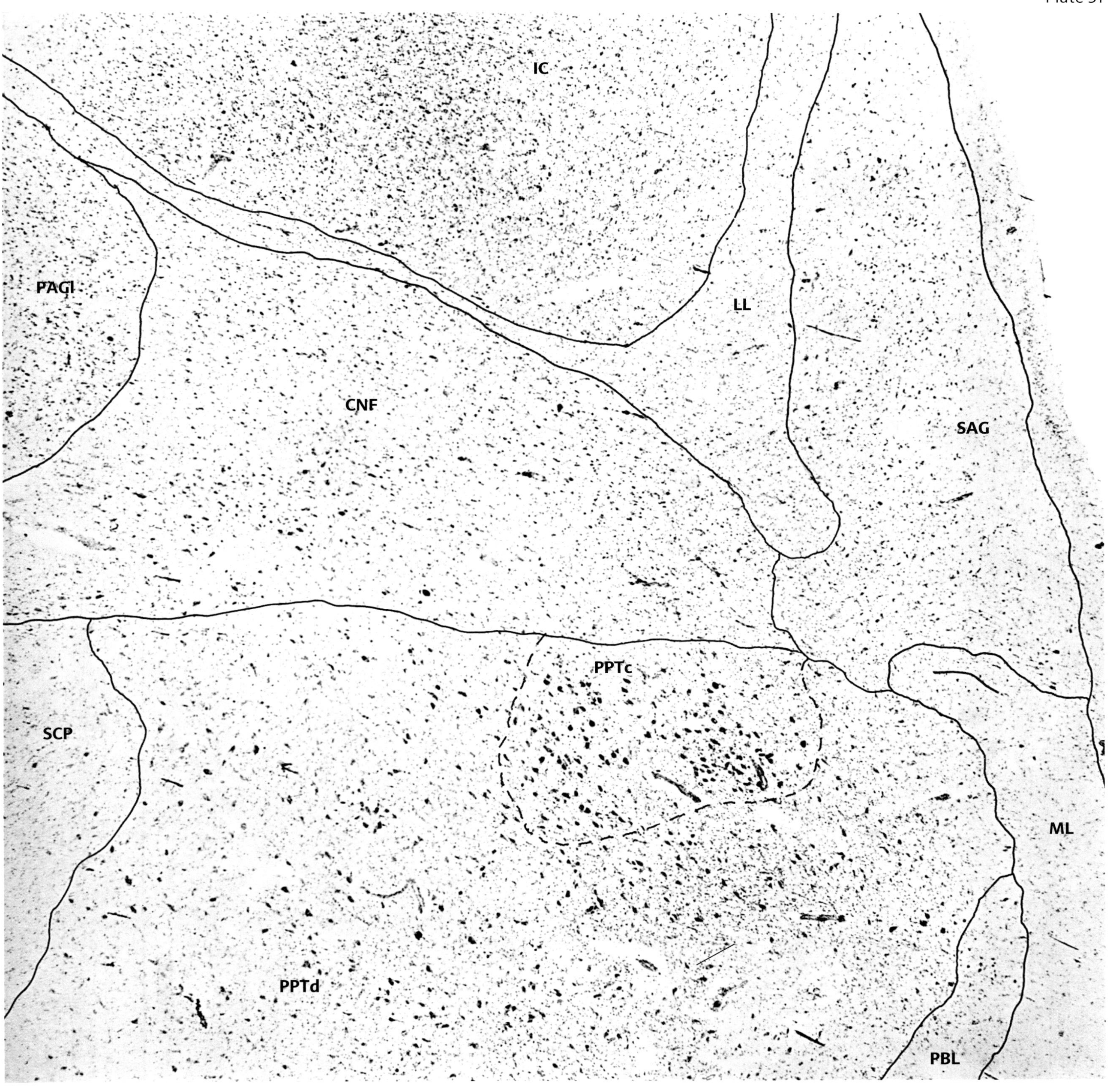

Plate 30. Semischematic representation of cross-section 651.

Plate 31. Photomicrograph of cross-section 601. Magnification ×40. Note that this photomicrograph is of a section 1 mm oral to that represented in plate 30. Note the accumulation of large cells in the dorsolateral part of the pedunculopontine nucleus to form the subnucleus compactus (PPTc). At this level the locus coeruleus has disappeared and the cuneiform nucleus is larger than in the preceding section. At this level the compact interfascicular nucleus would be present on the midline (see nucleus No. 107).

Abbreviations:

CNF	Cuneiform nucleus
CP	Cerebral peduncle
DRc	Dorsal raphe nucleus, caudal subnucleus
dSCP	Decussation of the superior cerebellar peduncle
DTG	Dorsal tegmental nucleus of Gudden
IC	Inferior colliculus
IPN	Interpeduncular nucleus
LC	Locus coeruleus
LDT	Laterodorsal tegmental nucleus
LL	Lateral lemniscus
MesV	Mesencephalic trigeminal nucleus
ML	Medial lemniscus
MLF	Medial longitudinal fascicle
MnRv	Median raphe nucleus, ventral part
NIV	Trochlear nerve
PAG	Periaqueductal gray
PAGl	Lateral periaqueductal gray
PBL	Lateral parabrachial nucleus
PN	Pontine nuclei
PPTc	Pedunculopontine nucleus, pars compacta
PPTd	Pedunculopontine nucleus, pars dissipata
SAG	Sagulum nucleus
SCP	Superior cerebellar peduncle
SEL	Subependymal layer

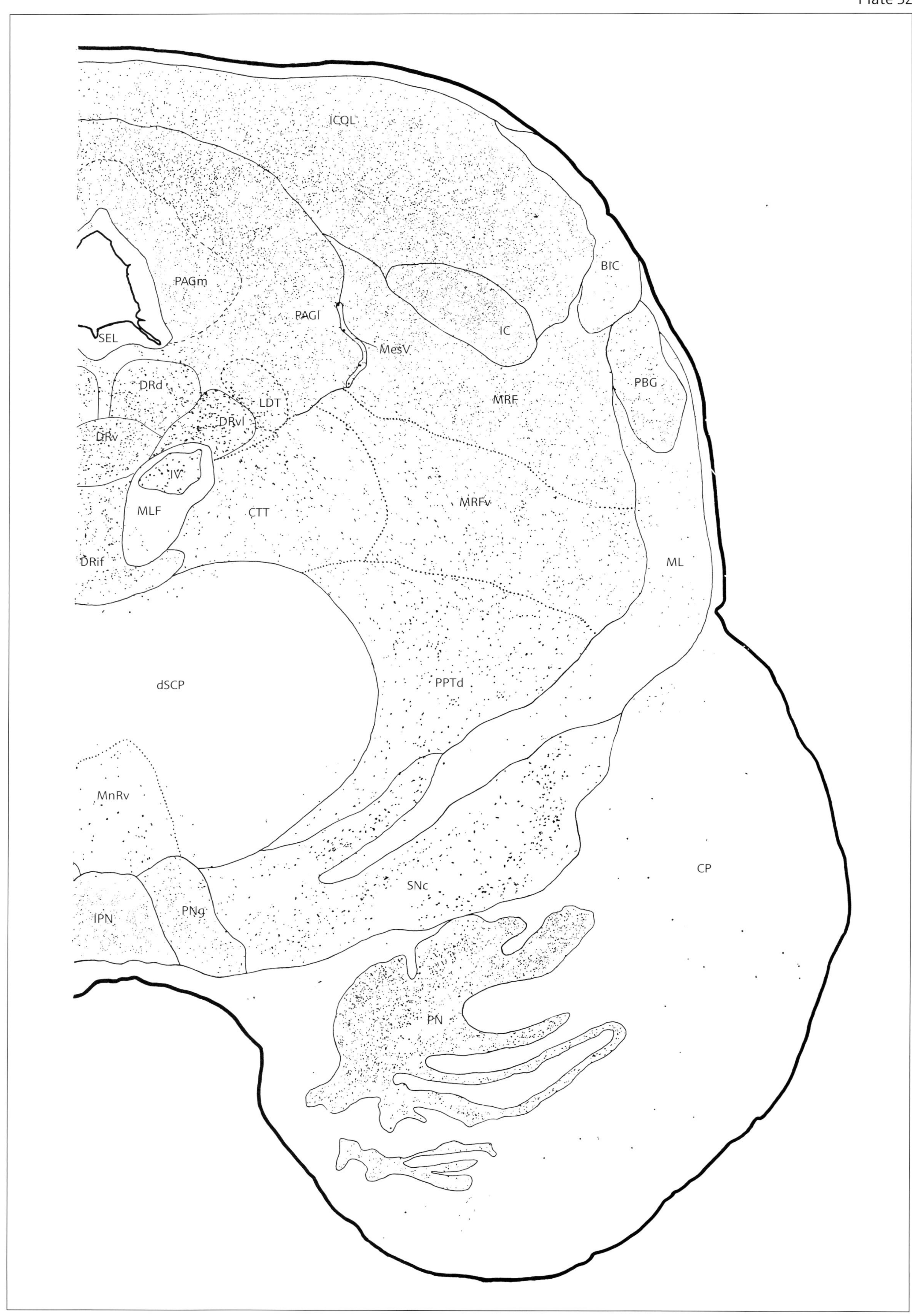
ICQL
BIC
PAGm
PAGl
IC
SEL
MesV
PBG
DRd
LDT
MRF
DRvl
DRv
IV
MLF
CTT
MRFv
DRif
ML
dSCP
PPTd
MnRv
CP
IPN
PNg
SNc
PN

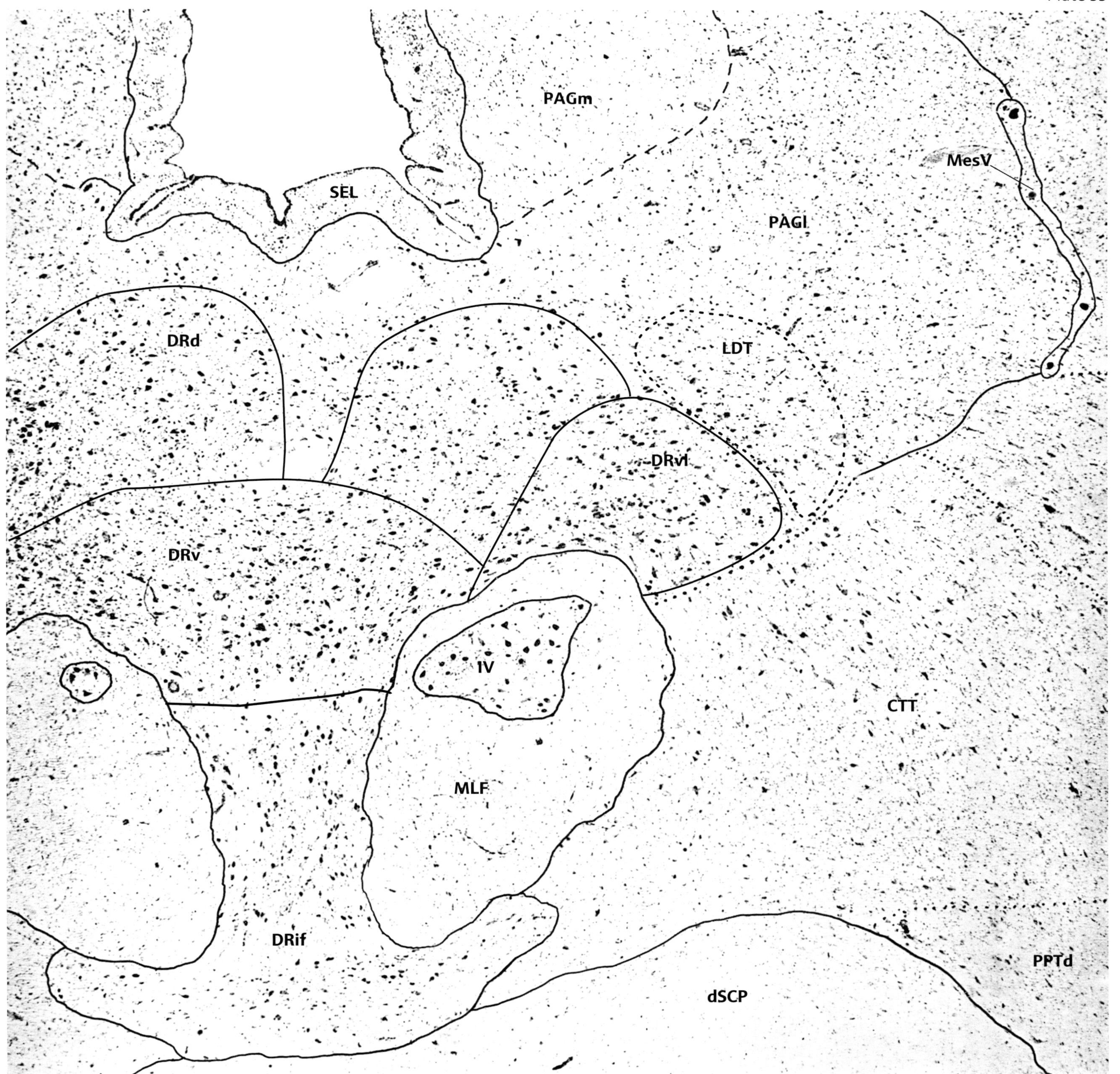

Plate 32. Semischematic representation of cross-section 501.
Plate 33. Photomicrograph of cross-section 501. Magnification ×40.

Abbreviations:

BIC	Brachium of the inferior colliculus
CP	Cerebral peduncle
CTT	Central tegmental tract
DRd	Dorsal raphe nucleus, dorsal part
DRif	Dorsal raphe nucleus, interfascicular part
DRv	Dorsal raphe nucleus, ventral part
DRvl	Dorsal raphe nucleus, ventrolateral part
dSCP	Decussation of the superior cerebellar peduncle
IC	Inferior colliculus
ICOL	Intercollicular nucleus
IPN	Interpeduncular nucleus
IV	Trochlear nucleus
LDT	Laterodorsal tegmental nucleus
MesV	Mesencephalic trigeminal nucleus
ML	Medial lemniscus
MLF	Medial longitudinal fascicle
MnRv	Median raphe nucleus, ventral part
MRF	Mesencephalic reticular formation
MRFv	Mesencephalic reticular formation, ventral part
PAGl	Periaqueductal gray, lateral part
PAGm	Periaqueductal gray, medial part
PBG	Parabigeminal nucleus
PN	Pontine nuclei
PNg	Paranigral nucleus
PPTd	Pedunculopontine nucleus, pars dissipata
SEL	Subependymal layer
SNc	Substantia nigra, pars compacta

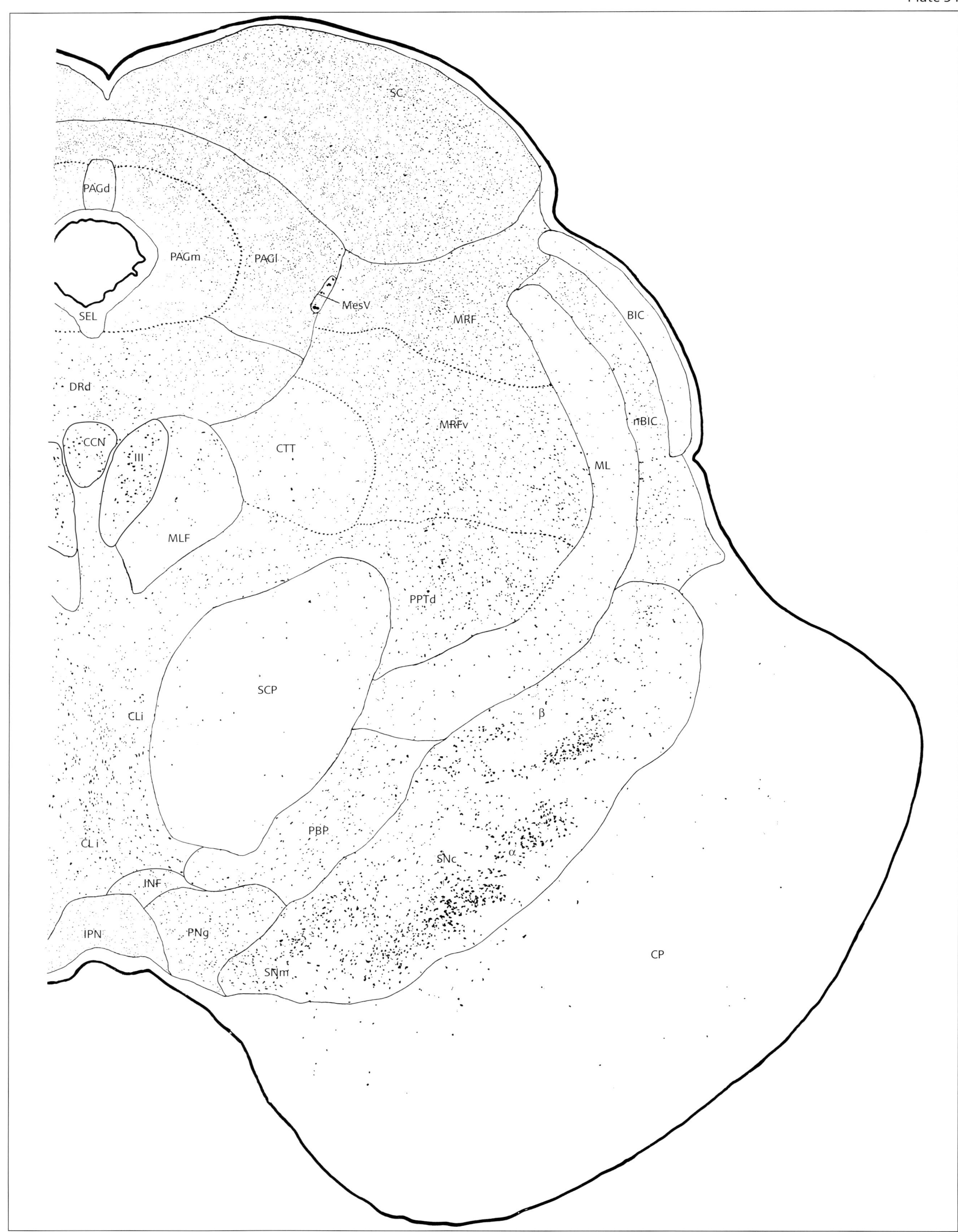
SC
PAGd
PAGm
PAGl
SEL
MesV
MRF
BIC
DRd
MRFv
nBIC
CCN
III
CTT
ML
MLF
PPTd
SCP
CLi
β
PBP
CLi
SNc
α
INF
IPN
PNg
SNm
CP

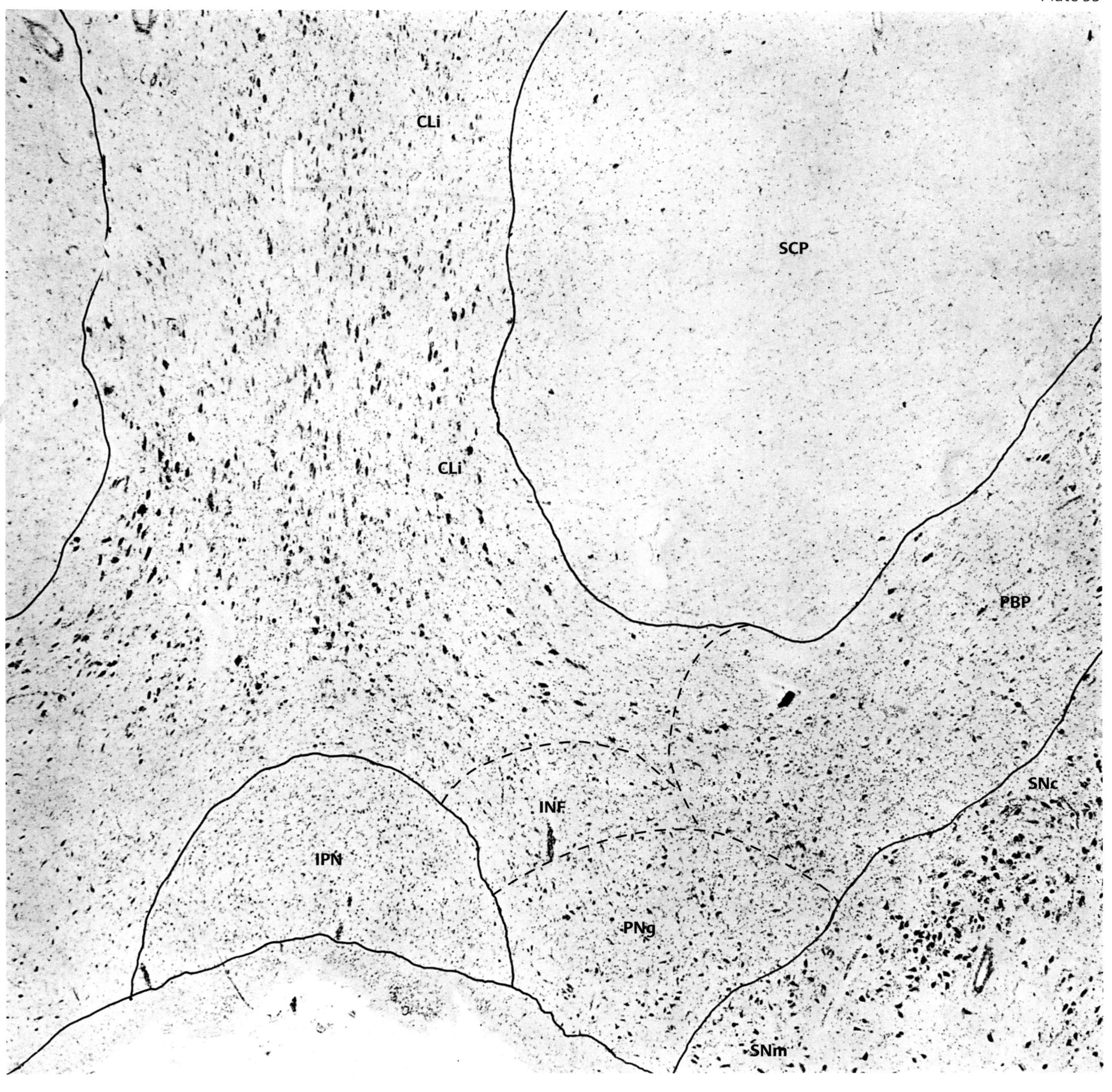

Plate 34. Semischematic representation of cross-section 401.
Plate 35. Photomicrograph of cross-section 401. Magnification ×40.

Abbreviations:

BIC	Brachium of the inferior colliculus
CCN	Central caudal nucleus
CLi	Caudal linear nucleus
CP	Cerebral peduncle
CTT	Central tegmental tract
DRd	Dorsal raphe nucleus, dorsal part
III	Oculomotor nucleus
INF	Interfascicular nucleus
IPN	Interpeduncular nucleus
MesV	Mesencephalic trigeminal nucleus
ML	Medial lemniscus
MLF	Medial longitudinal fascicle
MRF	Mesencephalic reticular formation
MRFv	Mesencephalic reticular formation, ventral part
nBIC	Nucleus of the brachium of the inferior colliculus
PAGd	Periaqueductal gray, dorsal subnucleus
PAGl	Periaqueductal gray, lateral part
PAGm	Periaqueductal gray, medial part
PBP	Parabrachial pigmented nucleus
PNg	Paranigral nucleus
PPTd	Pedunculopontine nucleus, pars dissipata
SC	Superior colliculus
SCP	Superior cerebellar peduncle
SEL	Subependymal layer
SNc	Substantia nigra, pars compacta (ventral tier pars α and dorsal tier pars β)
SNm	Substantia nigra, pars medialis

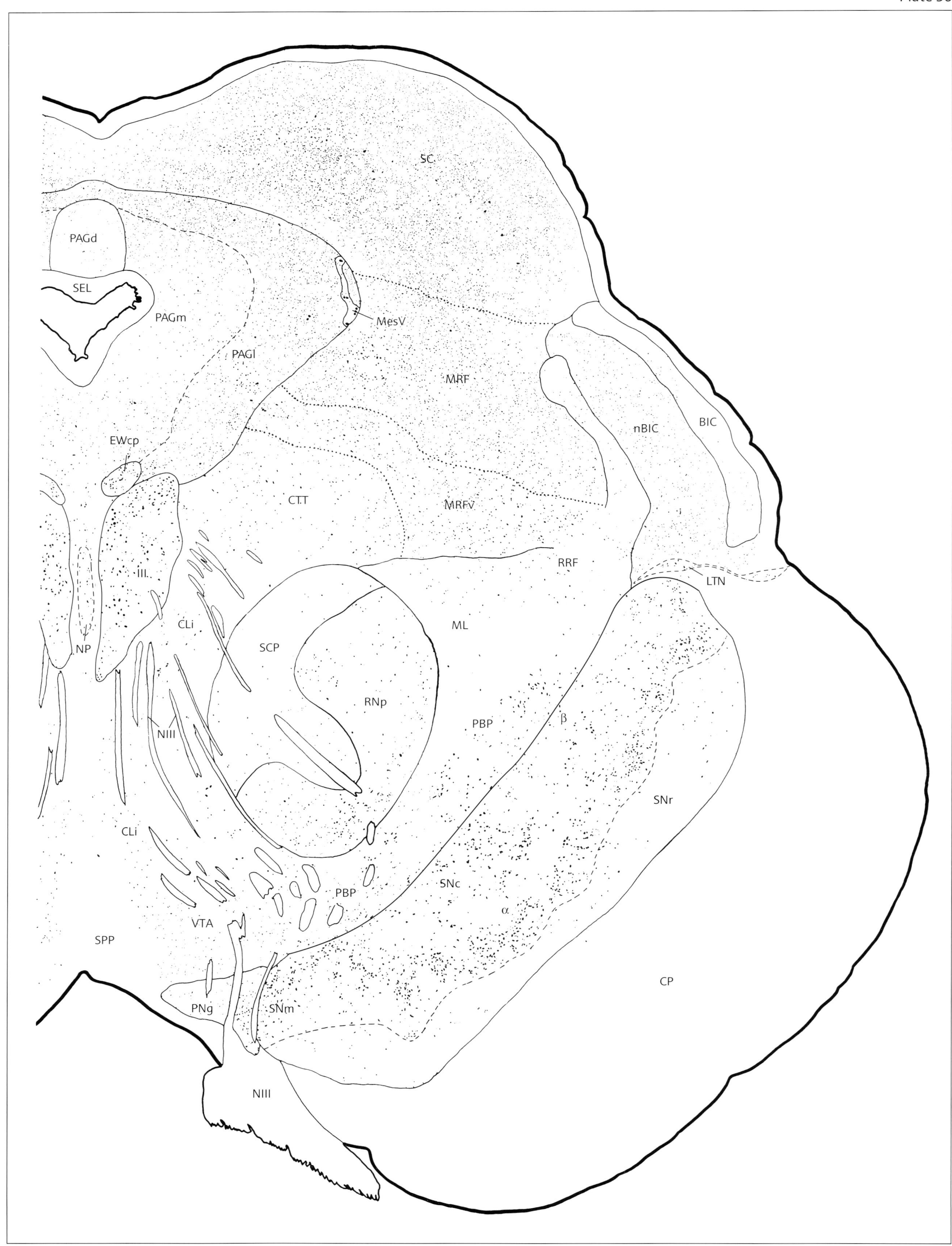
SC
PAGd
SEL
PAGm
MesV
PAGl
MRF
EWcp
nBIC
BIC
CTT
MRFv
III
RRF
LTN
CLi
ML
NP
SCP
RNp
PBP
β
NIII
SNr
CLi
SNc
PBP
α
VTA
SPP
CP
PNg
SNm
NIII

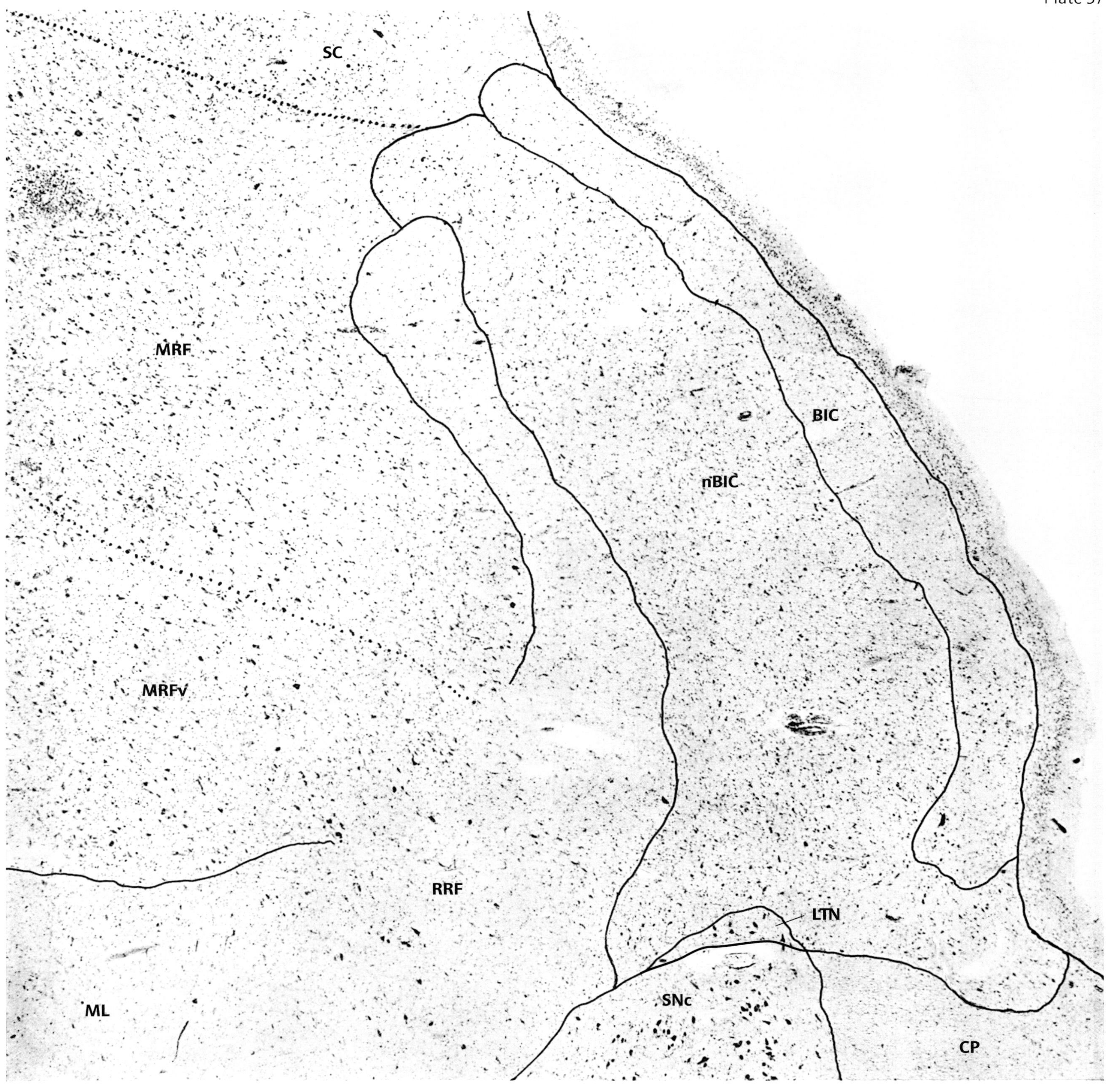

Plate 36. Semischematic representation of cross-section 301.
Plate 37. Photomicrograph of cross-section 301. Magnification ×40.

Abbreviations:

BIC	Brachium of the inferior colliculus
CLi	Caudal linear nucleus
CP	Cerebral peduncle
CTT	Central tegmental tract
EWcp	Edinger-Westphal nucleus, centrally projecting subnucleus
III	Oculomotor nucleus
LTN	Lateral terminal nucleus
MesV	Mesencephalic trigeminal nucleus
ML	Medial lemniscus
MRF	Mesencephalic reticular formation
MRFv	Mesencephalic reticular formation, ventral part
nBIC	Nucleus of the brachium of the inferior colliculus
NIII	Oculomotor nerve
NP	Nucleus of Perlia
PAGd	Periaqueductal gray, dorsal subnucleus
PAGl	Periaqueductal gray, lateral part
PAGm	Periaqueductal gray, medial part
PBP	Parabrachial pigmented nucleus
PNg	Paranigral nucleus
RNp	Red nucleus, parvocellular subnucleus
RRF	Retrorubral fields
SC	Superior colliculus
SCP	Superior cerebellar peduncle
SEL	Subependymal layer
SNc	Substantia nigra, pars compacta (ventral tier pars α and dorsal tier pars β)
SNm	Substantia nigra, pars medialis
SNr	Substantia nigra, pars reticulata
SPP	Substantia perforata posterior
VTA	Ventral tegmental area of Tsai

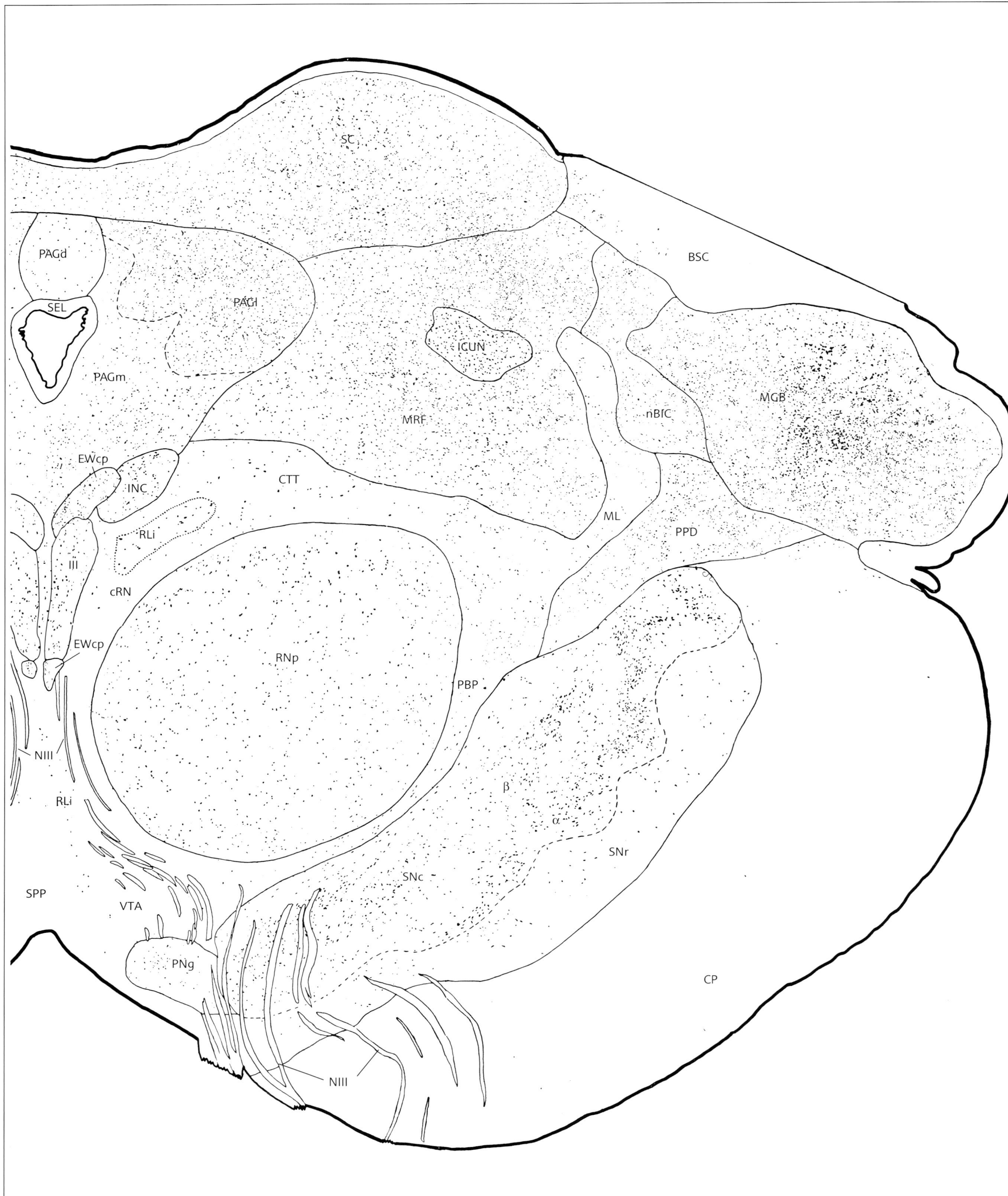

SC
BSC
PAGd
PAGl
SEL
ICUN
PAGm
MGB
nBIC
MRF
EWcp
CTT
INC
ML
RLi
PPD
III
cRN
EWcp
RNp
PBP
NIII
β
RLi
α
SNr
SNc
SPP
VTA
PNg
CP
NIII

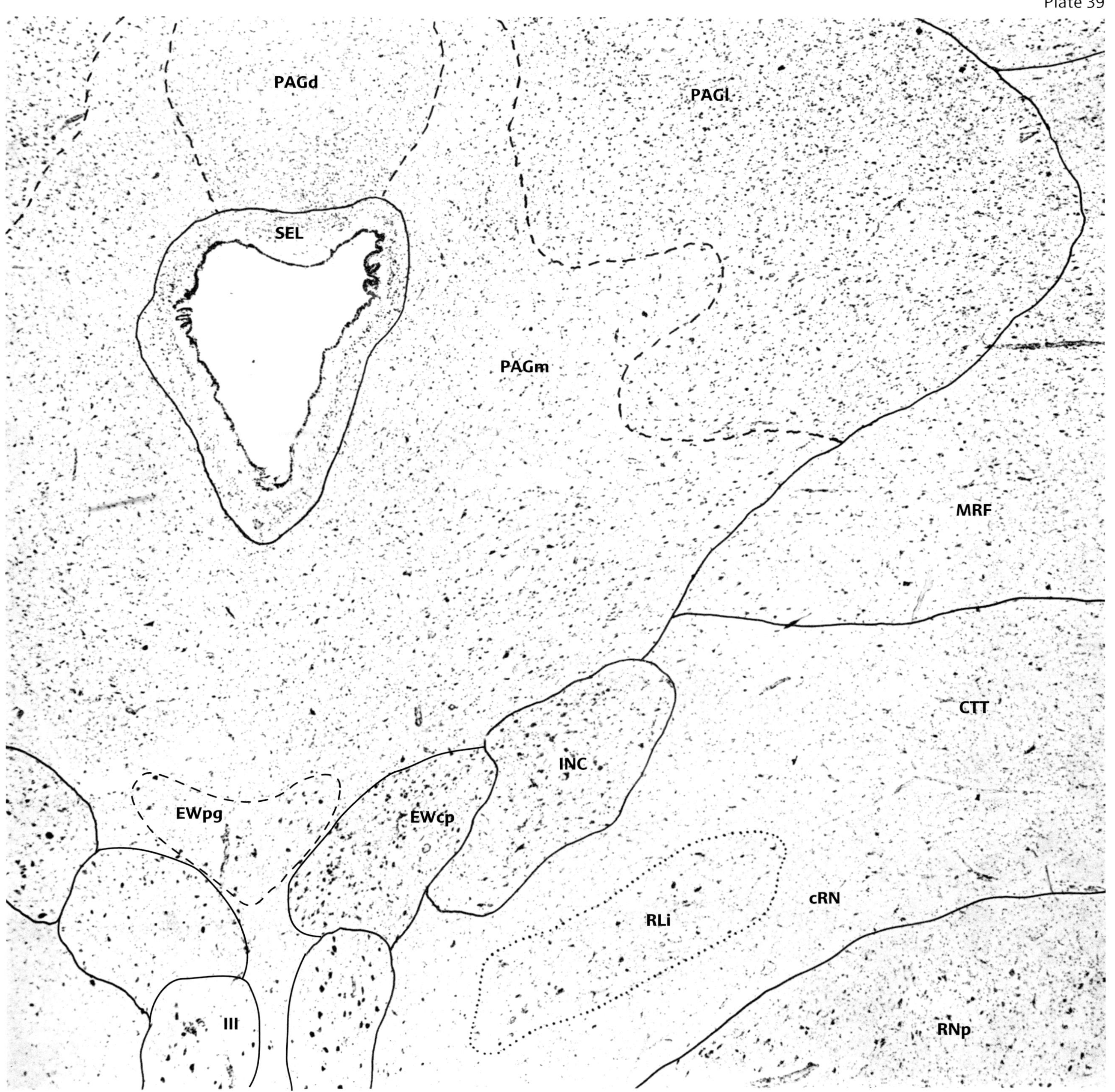

Plate 38. Semischematic representation of cross-section 201.
Plate 39. Photomicrograph of cross-section 201. Magnification ×40.

Abbreviations:

BSC	Brachium of the superior colliculus
CP	Cerebral peduncle
cRN	Capsule of the red nucleus
CTT	Central tegmental tract
EWcp	Edinger-Westphal nucleus, centrally projecting subnucleus
EWpg	Edinger-Westphal nucleus, preganglionic subgroup
ICUN	Intracuneiform nucleus
III	Oculomotor nucleus
INC	Interstitial nucleus of Cajal
MGB	Medial geniculate body
ML	Medial lemniscus
MRF	Mesencephalic reticular formation
nBIC	Nucleus of the brachium of the inferior colliculus
NIII	Oculomotor nerve
PAGd	Periaqueductal gray, dorsal subnucleus
PAGl	Periaqueductal gray, lateral part
PAGm	Periaqueductal gray, medial part
PBP	Parabrachial pigmented nucleus
PNg	Paranigral nucleus
PPD	Peripeduncular nucleus
RLi	Rostral linear nucleus
RNp	Red nucleus, parvocellular subnucleus
SC	Superior colliculus
SEL	Subependymal layer
SNc	Substantia nigra, pars compacta (ventral tier pars α and dorsal tier pars β)
SNr	Substantia nigra, pars reticulata
SPP	Substantia perforata posterior
VTA	Ventral tegmental area of Tsai

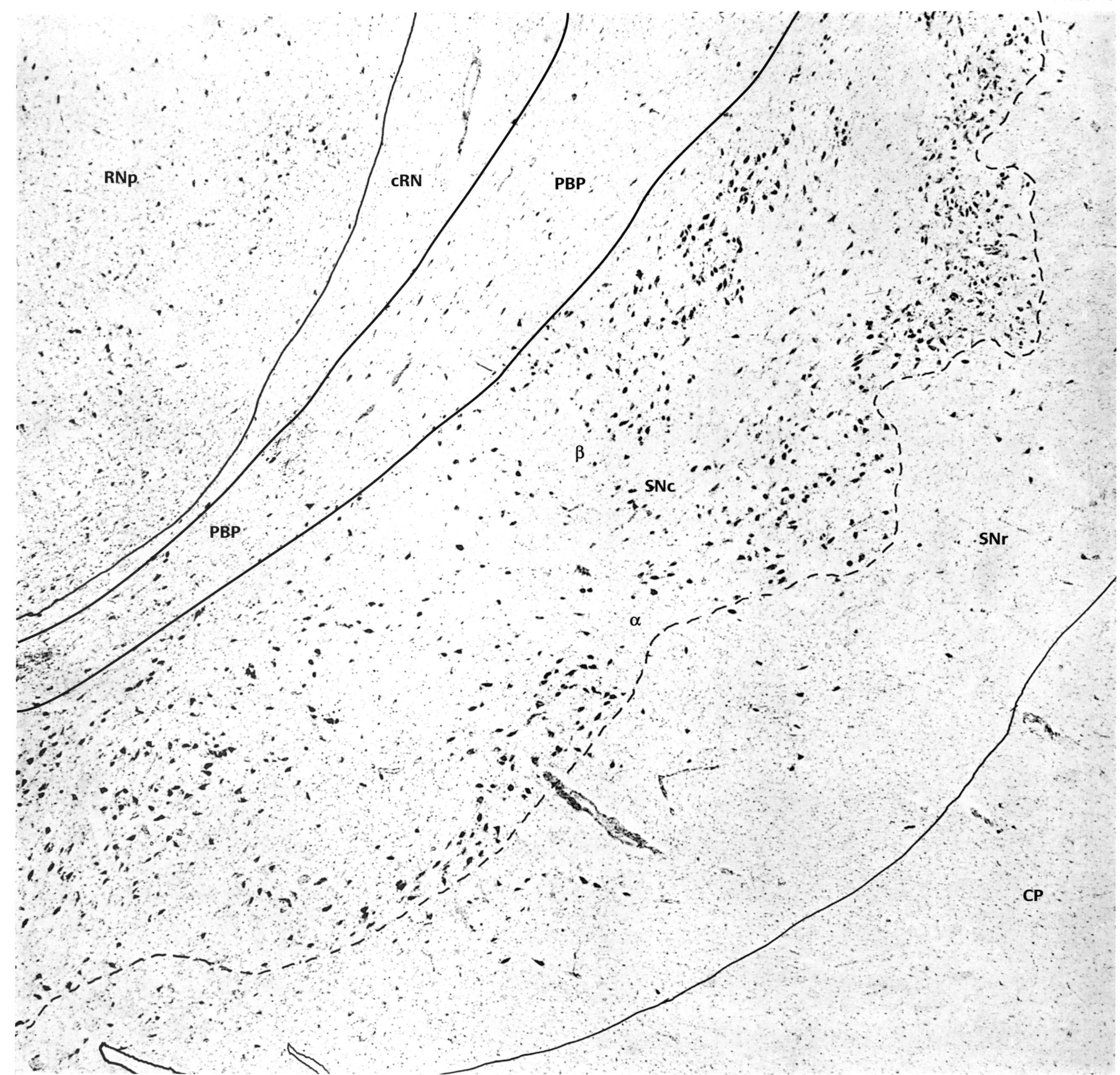

Plate 40. Photomicrograph of cross-section 201. Magnification ×40.

Abbreviations:

CP	Cerebral peduncle	RNp	Red nucleus, parvocellular subnucleus	SNr	Substantia nigra, pars reticulata
cRN	Capsule of the red nucleus	SNc	Substantia nigra, pars compacta (ventral tier pars α and dorsal tier pars β)		
PBP	Parabrachial pigmented nucleus				

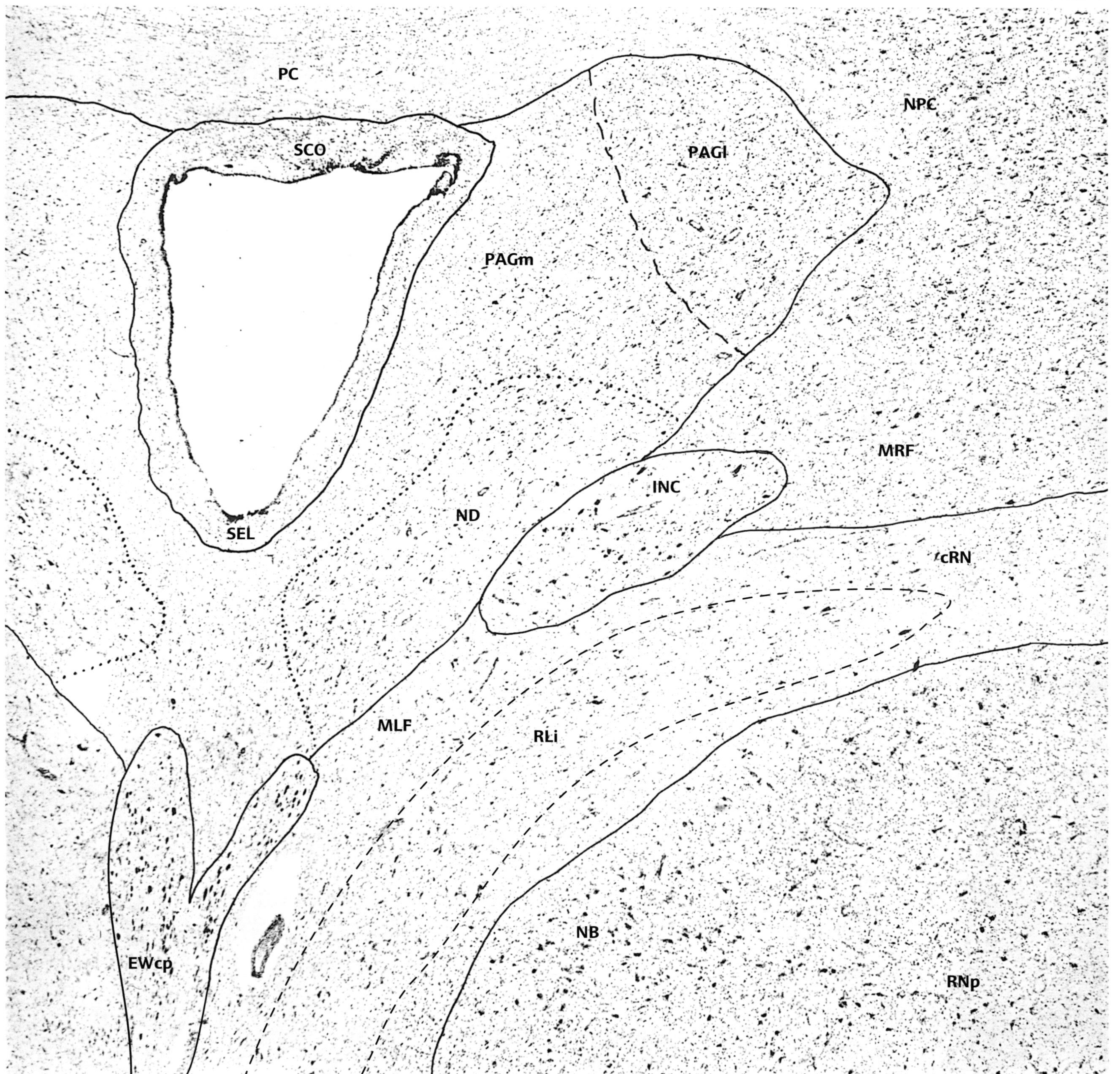

Plate 41. Photomicrograph of cross-section 101. This is a photograph of a section 2 mm oral to section 201 pictured on the preceding pages. Note that the oculomotor nucleus has disappeared and the nucleus of Darkschewitsch is now present. A photomicrograph of a more typical nucleus of Darkschewitsch appears in chapter No. 102. Magnification ×40.

Abbreviations:

cRN	Capsule of the red nucleus
EWcp	Edinger-Westphal nucleus, central projecting subnucleus
INC	Interstitial nucleus of Cajal
MLF	Medial longitudinal fascicle
MRF	Mesencephalic reticular formation
NB	Medial accessory nucleus of Bechterew
ND	Nucleus of Darkschewitsch
NPC	Nucleus of the posterior commissure
PAGl	Periaqueductal gray, lateral part
PAGm	Periaqueductal gray, medial part
PC	Posterior commissure
RLi	Rostral linear nucleus
RNp	Red nucleus, parvocellular subnucleus
SCO	Subcommissural organ
SEL	Subependymal layer

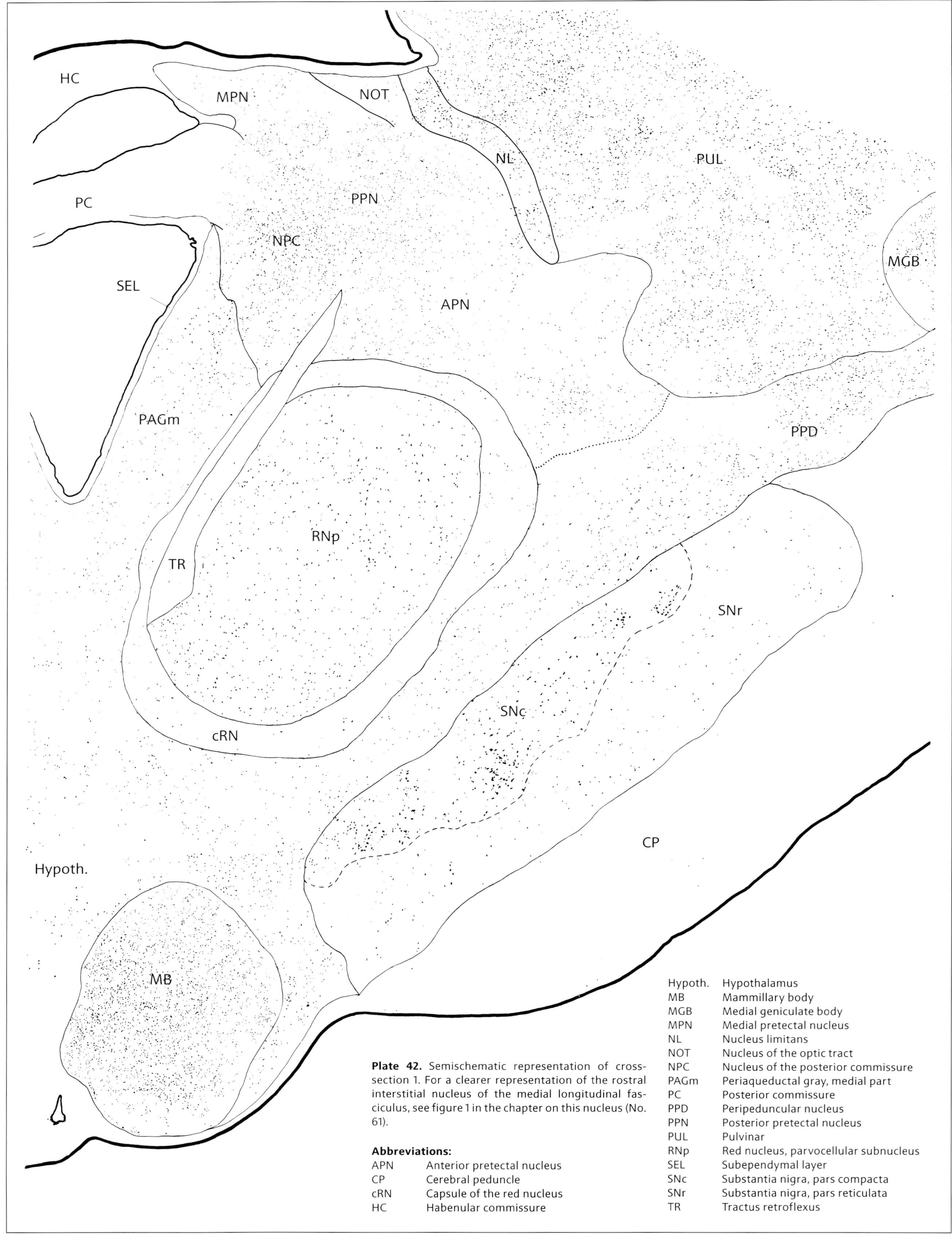

Plate 42. Semischematic representation of cross-section 1. For a clearer representation of the rostral interstitial nucleus of the medial longitudinal fasciculus, see figure 1 in the chapter on this nucleus (No. 61).

Abbreviations:

APN	Anterior pretectal nucleus
CP	Cerebral peduncle
cRN	Capsule of the red nucleus
HC	Habenular commissure
Hypoth.	Hypothalamus
MB	Mammillary body
MGB	Medial geniculate body
MPN	Medial pretectal nucleus
NL	Nucleus limitans
NOT	Nucleus of the optic tract
NPC	Nucleus of the posterior commissure
PAGm	Periaqueductal gray, medial part
PC	Posterior commissure
PPD	Peripeduncular nucleus
PPN	Posterior pretectal nucleus
PUL	Pulvinar
RNp	Red nucleus, parvocellular subnucleus
SEL	Subependymal layer
SNc	Substantia nigra, pars compacta
SNr	Substantia nigra, pars reticulata
TR	Tractus retroflexus

Chapters of Individual Nuclei

1 Gracile Nucleus (GR)

Original name:
Nucleus gracilis

Location and Cytoarchitecture – Original Text

The cells of this nucleus appear among the fibers of the gracile fasciculus in the caudal part of the medulla, and gradually replace it. The nucleus measures 13 mm in length, extending orally to the level of the caudal pole of the medial vestibular nucleus.

Caudal to the obex, the gracile nucleus (GR) appears on cross-section as a bulb-shaped structure with its narrow portion directed ventrally. This portion of the nucleus is related medially to the posterior median septum, laterally to the cuneate fasciculus, while ventrally its relations vary from section to section. Proceeding orally the nucleus gradually increases in size. Oral to the opening of the fourth ventricle the GR is displaced laterally and appears as an oblong cell group related medially to the nucleus of the solitary tract, ventrally to the oral pole of the medial cuneate nucleus and laterally to the lateral cuneate nucleus.

The cells of the GR exhibit considerable pleomorphism in regard to their size, shape and staining qualities, thus rendering description of a typical cell type difficult. The majority are of medium size, and may be elongated or plump and multipolar in form. The nucleus is centrally placed and although in most instances the Nissl granules are indistinct, the cytoplasm is stained intensely. Many similar but smaller cells are present, and occasionally a larger cell is observed in which the Nissl substance is peripherally arranged (fig. 1–8).

The cells composing the GR are widely scattered in the caudal portions of the nucleus but as one proceeds orally they become much more compactly arranged. Cells are more numerous in the dorsal than in the ventral part of the nucleus, and there is a tendency for the cells to congregate into small groups.

Posterior column [or dorsal column] fibers originating caudal to the sixth thoracic segment proceed orally in the gracile fasciculus and terminate by synapsing about the neurons of the GR [...].

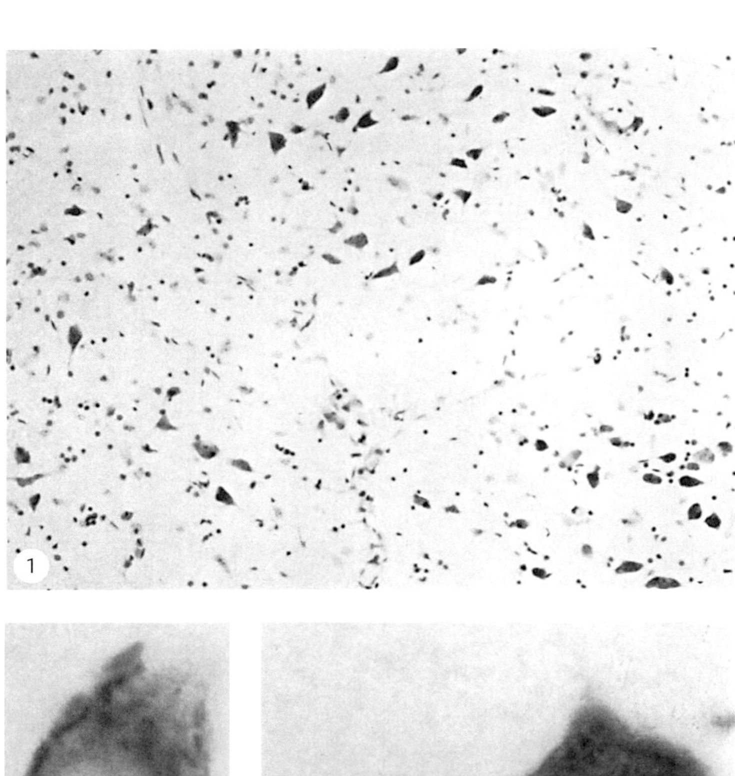

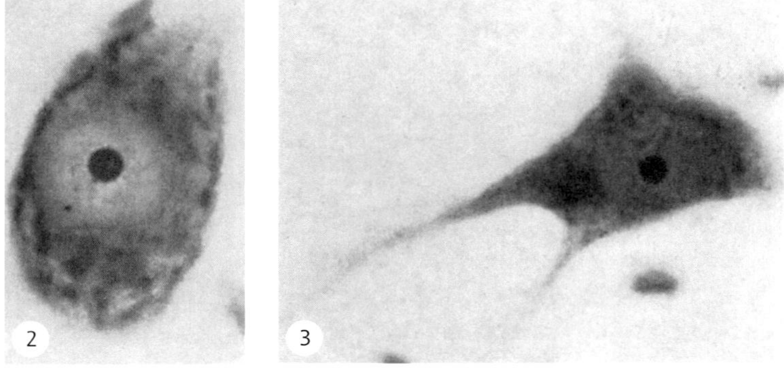

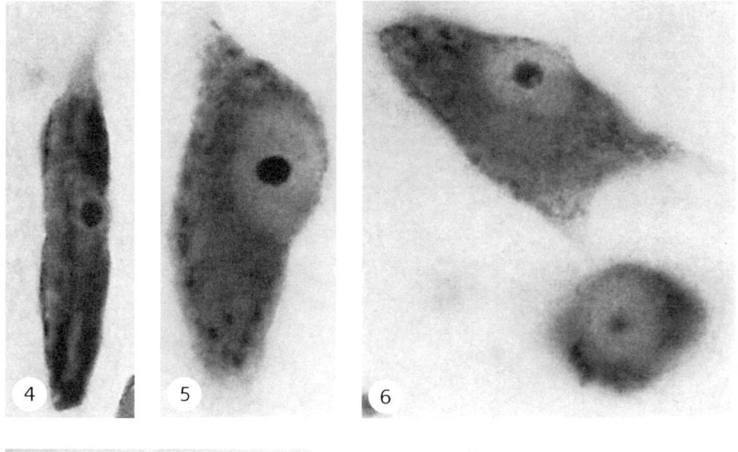

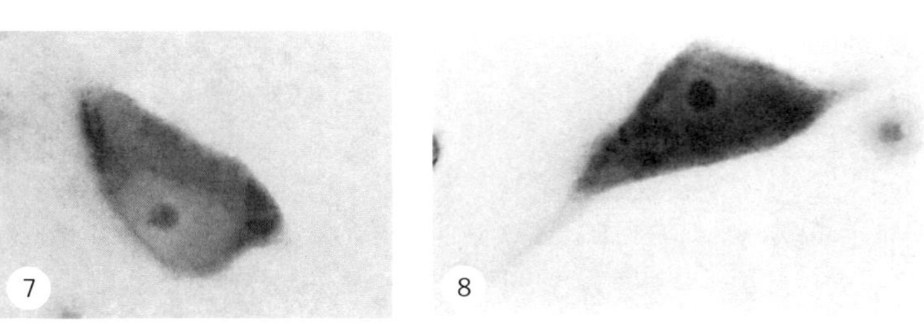

Fig. 1. Gracile nucleus. Magnification ×150.
Fig. 2–8. Different cell types from the gracile nucleus. Magnification ×1,000.

Functional Neuroanatomy

Function

The GR and cuneate nucleus, or dorsal column nuclei, are relays of the lemniscal pathway which carry tactile and proprioceptive information from the periphery to the somatosensory cerebral cortex. The lemniscal pathways are important for the conscious perception of these sensations; they maintain a strict topography, and provide the cortex with modality-specific inputs.

Whereas the cuneate nuclei receive afferents from the chest, upper limbs and neck (see nucleus No. 3), the GR receives afferents from the foot, leg, hip and abdomen via the dorsal columns of the spinal cord. The fibers encode mainly touch-pressure, vibration or proprioception (composed of inputs from muscle spindles, Golgi tendon organs and joint receptors). In addition, recent evidence suggests that the dorsal column pathways to the GR also play a role in the processing of pain, including pelvic visceral pain [Nauta et al., 1997; Nieuwenhuys, 2008; Kaas, 2012; Westlund and Willis, 2012].

Spinal lesions destroying the dorsal column input to the GR (as in Brown-Séquard syndrome) lead to a reduction in the discrimination of touch, pressure, vibration and positional sensations in the lower extremities, and may cause ataxia. However, the studies of dorsal column lesions on behavior are difficult to interpret, since the survival of very few fibers can have a substantial impact on cortical neurons, and furthermore the dorsal columns are not the only pathways carrying somatosensory information, for example the spinothalamic tract [Jain et al., 1997; Qi et al., 2011; Kaas, 2012].

Connections

Afferents. The main input to the GR arises from the gracile fasciculus which ascends medially in the dorsal columns of the spinal cord. Fibers from first-order sensory neurons in the dorsal root ganglia bifurcate into a short descending branch and a long ascending branch. The ascending fibers are segregated into tracts according to their specific sensory modality (touch, pain etc.), and the fibers that enter the ipsilateral gracile fasciculus encode mainly touch and proprioception from the sacral, lumbar and lower 6 thoracic dorsal roots [Al-Chaer et al., 1996]. The dorsal column fasciculi are organized topographically, and the pathways retain this organization throughout their relay in the GR, the thalamus and cortex [Nieuwenhuys, 2008]. At the level of the spinal cord, sacral fibers assume a medial position in the gracile fasciculus and the fibers from progressively more rostral levels are added laterally. After the level T_6, the fibers enter the cuneate fasciculus (see medial cuneate nucleus, No. 3). Some fibers ascending from levels below T_6 to the GR travel in the dorsolateral fasciculus [Gordon and Grant, 1982]. Most group I afferents from muscle spindles and Golgi tendon organs of the lower limbs travel at first in the gracile fasciculus up to the level of L_2, where they branch off, synapse in the dorsal nucleus of Clarke (Clarke's column), from where axons ascend in the dorsal spinocerebellar tract and send collaterals to nucleus Z. Since these proprioceptors are mainly slowly adapting receptors, the final ascending dorsal column input to the GR is predominantly from the rapidly adapting cutaneous receptors [Kaas, 2012].

The second-order somatosensory neurons of the GR which mainly receive the excitatory inputs from the fibers of the gracile fasciculus, occupy the medial and caudal part of the nucleus, and are cholinergic [Paxinos et al., 2012]. There is a strict topography and some separation of modalities [Dykes, 1983; Haines, 2006]. The outer shell area of the GR contains smaller cells than the mediocaudal parts, and the small cells are thought to exert a feedback inhibition on the projection neurons [Barbaresi et al., 1986].

Additional sensory information to the GR arises from postsynaptic dorsal column pathways, specifically from the nucleus proprius (lamina IV) in the dorsal horn: the projection was shown to provide nociceptive and viscerosomatic information to the dorsal column nuclei [Cliffer and Willis, 1994]. A specific component of the postsynaptic dorsal column pathways through the GR that is important in viscerosensory transmission is reviewed by Nauta et al. [1997]. Further afferents to modulate the activity of the GR neurons arise from the pontomedullary reticular formation [Fernandez de Sevilla et al., 2006], and from the primary somatosensory cortex [Jabbur and Towe, 1961; Bentivoglio and Rustioni, 1986].

Efferents. The central core of the GR provides the main outputs, which project through the medial lemniscus to the ventroposterior complex of the contralateral thalamus. The topography is such that the lower extremities are represented in the lateral part of the lateral ventroposterior nucleus and the upper body represented medially [Kaas, 2012]. The more diffusely organized outer shell of the GR projects to the cerebellum, the inferior olive, the tectum, the anterior nucleus of the pretectum, the red nucleus and provides a feedback pathway to the spinal cord [Nieuwenhuys, 2008].

References

Al-Chaer ED, Lawand NB, Westlund KN, Willis WD: Pelvic visceral input into the nucleus gracilis is largely mediated by the postsynaptic dorsal column pathway. J Neurophysiol 1996;76:2675–2690.

Barbaresi P, Spreafico R, Frassoni C, Rustioni A: GABAergic neurons are present in the dorsal column nuclei but not in the ventroposterior complex of rats. Brain Res 1986;382:305–326.

Bentivoglio M, Rustioni A: Corticospinal neurons with branching axons to the dorsal column nuclei in the monkey. J Comp Neurol 1986;253:260–276.

Cliffer KD, Willis WD: Distribution of the postsynaptic dorsal column projection in the cuneate nucleus of monkeys. J Comp Neurol 1994;345:84–93.

Dykes RW: Parallel processing of somatosensory information: a theory. Brain Res 1983;287:47–115.

Fernandez de Sevilla D, Rodrigo-Angulo M, Nunez A, Buno W: Cholinergic modulation of synaptic transmission and postsynaptic excitability in the rat gracilis dorsal column nucleus. J Neurosci 2006;26:4015–4025.

Gordon G, Grant G: Dorsolateral spinal afferents to some medullary sensory nuclei. An anatomical study in the cat. Exp Brain Res 1982;46:12–23.

Haines DE: Fundamental Neuroscience for Basic and Clinical Applications. Philadelphia, Churchill Livingstone Elsevier, 2006.

Jabbur SJ, Towe AL: Cortical excitation of neurons in dorsal column nuclei of cat, including an analysis of pathways. J Neurophysiol 1961;24:499–509.

Jain N, Catania KC, Kaas JH: Deactivation and reactivation of somatosensory cortex after dorsal spinal cord injury. Nature 1997;386:495–498.

Kaas JH: Somatosensory system; in Mai JK, Paxinos G (eds): The Human Nervous System, ed 3. Amsterdam, Academic Press, 2012.

Nauta HJ, Hewitt E, Westlund KN, Willis WD Jr: Surgical interruption of a midline dorsal column visceral pain pathway. Case report and review of the literature. J Neurosurg 1997;86:538–542.

Nieuwenhuys R: General sensory systems and taste; in Nieuwenhuys R, Voogd J, van Huijzen C (eds): The Human Central Nervous System. Berlin, Springer, 2008.

Paxinos G, Huang X-F, Sengul G, Watson C: Organization of the brainstem; in Mai JK, Paxinos G (eds): The Human Nervous System, ed 3. Amsterdam, Academic Press, 2012.

Qi HX, Chen LM, Kaas JH: Reorganization of somatosensory cortical areas 3b and 1 after unilateral section of dorsal columns of the spinal cord in squirrel monkeys. J Neurosci 2011;31:13662–13675.

Westlund KN, Willis WD: Pain system; in Mai JK, Paxinos G (eds): The Human Nervous System, ed 3. Amsterdam, Academic Press, 2012.

2 Nucleus Parvocellularis Compactus (PVC)

Original name:
Nucleus parvocellularis compactus

Associated area:
A significantly larger region than the nucleus parvocellularis compactus, but including it, has been called the parasolitary nucleus by Paxinos et al. [2012]; the relationship between the two nuclei is not clear (see chapter 19)

Location and Cytoarchitecture – Original Text

This nucleus is formed by a small cell group which appears between the nucleus of the solitary tract (SOL) and the gracile nucleus (GR) at the level of the area postrema. The orocaudal extent of the nucleus does not exceed 1 mm.

On cross-section, the nucleus is oval in outline and lies with its long axis parallel to the lateral border of the SOL.

The nucleus parvocellularis compactus (PVC) is composed of exceedingly small, closely packed, oval cells which possess a relatively large nucleus and scanty, palely stained cytoplasm (fig. 1, 3).

No function of this nucleus has yet been established, and it is difficult to decide on morphological grounds whether it should be grouped with the SOL or GR. The cells, although smaller, are very similar to those of the SOL but on myelin-stained sections the PVC is seen to lie within the heavily myelinated area of the GR rather than within the relatively unmyelinated area of the SOL (fig. 2).

Functional Neuroanatomy

Function
The function of the PVC is not known, but its connections indicate that it probably should be grouped with the somatosensory system, rather than the viscerosensory SOL (see above). Its location within the boundaries of the GR means that it could be the target for an unusual ascending pathway ending in the GR, and carrying signals concerning visceral pain from pelvic regions (see below).

Connections
This specialized region within the GR has not received much attention recently, but studies in the mouse show that the nucleus has individual acid phosphatase staining properties dif-

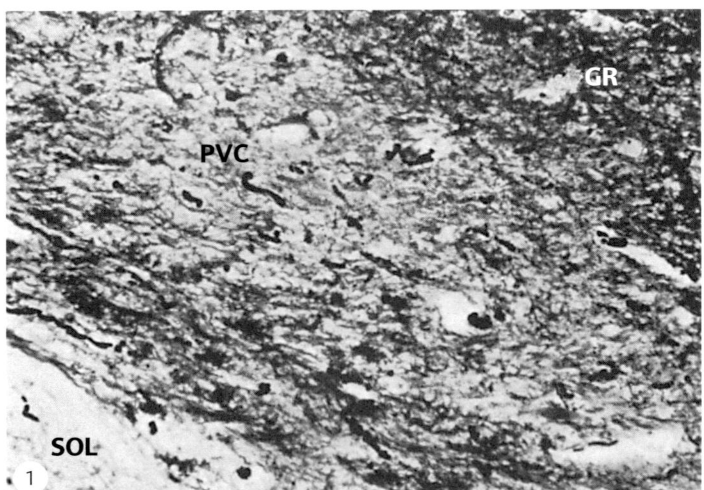

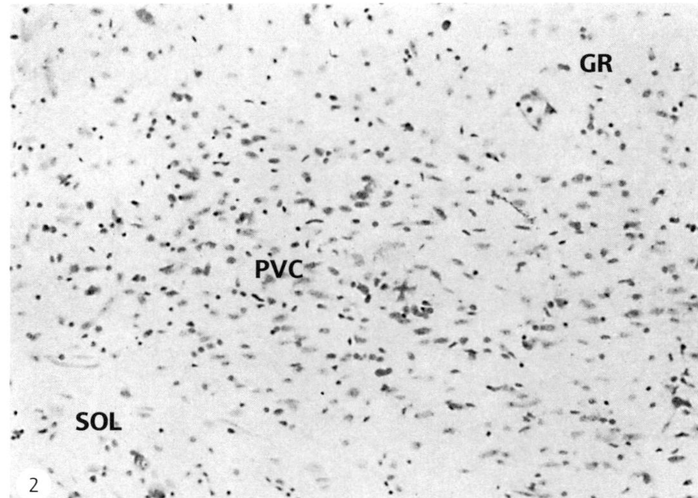

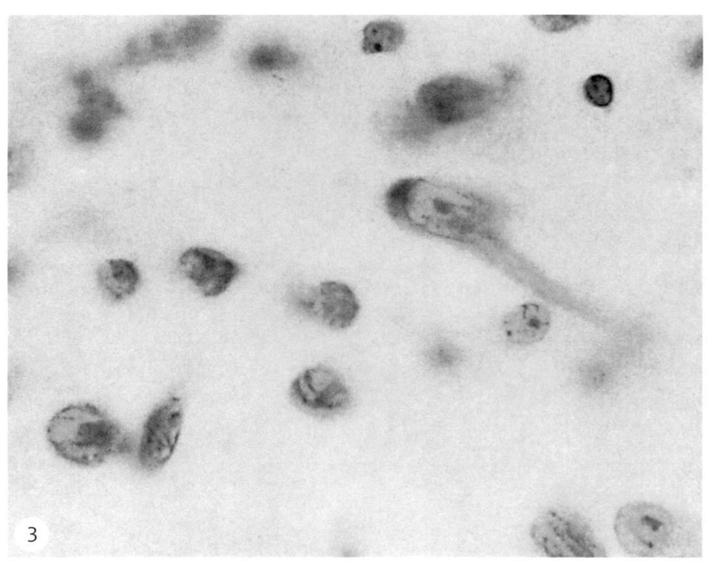

Fig. 1. Nucleus parvocellularis compactus (PVC). GR = Gracile nucleus; SOL = solitary nucleus. Heidenhain stain. Magnification ×150.
Fig. 2. Nucleus parvocellularis compactus (PVC). GR = Gracile nucleus; SOL = solitary nucleus. Magnification ×150.
Fig. 3. Cells from the nucleus parvocellularis compactus. Magnification ×1,000.

ferent from the GR [Sood and Mulchandani, 1977]. Furthermore, an anatomical analysis of the somatosensory cortex limb region efferents of the cat showed a descending projection to the PVC, as well as to the adjacent GR [Flindt-Egebak, 1979]. Importantly, the results imply that the PVC should be grouped with the somatosensory dorsal column nuclei.

The GR receives primary somatosensory afferents ascending from the lower limbs through the dorsal columns, but the GR also plays an important role in processing pelvic visceral input and relaying it to the ventral posterolateral nucleus of the thalamus [Al-Chaer et al., 1996b]. A specific component of the dorsal column fasciculus has been identified as postsynaptic; the fibers ascend in the spinal cord ipsilaterally to terminate in the medial GR [Al-Chaer et al., 1996a]. The pathway conveys pain sensation from pelvic regions, and although it has not been specifically reported, it is possible that this relay is associated with small-celled PVC.

References

Al-Chaer ED, Lawand NB, Westlund KN, Willis WD: Pelvic visceral input into the nucleus gracilis is largely mediated by the postsynaptic dorsal column pathway. J Neurophysiol 1996a;76:2675–2690.

Al-Chaer ED, Lawand NB, Westlund KN, Willis WD: Visceral nociceptive input into the ventral posterolateral nucleus of the thalamus: a new function for the dorsal column pathway. J Neurophysiol 1996b;76:2661–2674.

Paxinos G, Huang X-F, Sengul G, Watson C: Organization of the brainstem; in Mai JK, Paxinos G (eds): The Human Nervous System, ed 3. Amsterdam, Academic Press, 2012.

Flindt-Egebak P: An autoradiographical study of the projections from the feline sensorimotor cortex to the brain stem. J Hirnforsch 1979;20:375–390.

Sood PP, Mulchandani MM: A comparative study of histoenzymological mapping on the distribution of acid and alkaline phosphatases and succinic dehydrogenase in the spinal cord and medulla oblongata of mouse. Acta Histochem 1977; 60:180–203.

Sensory Systems – Somatosensory Nuclei (Plates 6–11)

3 Medial Cuneate Nucleus (MCU)

Original name:
Nucleus cuneatus medialis

Alternative name:
Cuneate nucleus, or main cuneate nucleus: in this terminology the laterally adjacent nucleus is usually called the external cuneate nucleus, but here it is referred to as the lateral cuneate nucleus (nucleus No. 4)

Location and Cytoarchitecture – Original Text

The medial cuneate nucleus (MCU) begins caudally at a slightly more oral level than does the gracile nucleus (GR) and extends to a point just oral to the opening of the fourth ventricle. The nucleus appears on cross-sections as a wedge-shaped area which projects upwards into the cuneate fasciculus. The base of the wedge extends from the GR medially, nearly to the spinal trigeminal complex laterally. Ventrally, the nucleus is related to the nucleus of the solitary tract and to the central nucleus of the medulla oblongata. Proceeding orally, the nucleus increases progressively in size until it occupies the greater part of the cuneate fasciculus. With the appearance of the lateral cuneate nucleus (LCU), the MCU rapidly diminishes in size and disappears just oral to the opening of the fourth ventricle.

The cells composing the MCU are round or oval, plump and possess short dendrites and excentric nuclei. The cytoplasm stains with moderate intensity and the Nissl granules are peripherally arranged. These cells vary in size from small to large, the larger elements being more numerous in the caudal part of the nucleus. The cells congregate about the dorsal margin of the nucleus in caudal sections, whereas at other levels they tend to gather into irregular cell clusters (fig. 1–7).

Posterior column [or dorsal column] fibers originating above the level of the sixth thoracic segment travel in the cuneate fasciculus and terminate about the cells of the MCU. Fibers originating in this nucleus then proceed to the nucleus ventralis posterior of the thalamus via the contralateral medial lemniscus.

Functional Neuroanatomy

Function

The cuneate nucleus and GR, also called the dorsal column nuclei, have many features in common. They are all relays of the lemniscal pathways which carry tactile and proprioceptive information from the periphery to the contralateral somatosensory cerebral cortex. The pathways are important for the conscious perception of these sensations. Whereas the GR receives afferents from the ipsilateral lower body (see nucleus No. 1), the MCU receives afferents from the ipsilateral upper body, from the chest, fingers, hands, upper limbs and neck, and the trigeminal complex, directly adjacent, receives the afferents from the face and mouth (see nuclei No. 6–12). The MCU afferents, ascending in the dorsal columns of the spinal cord, encode touch-pressure, and vibration, from rapidly adapting receptors. The upper limb proprioceptors, also ascending through the cuneate fasciculus, terminate in peripheral and rostral regions of the MCU and also supply the adjacent LCU (nucleus No. 4) [Hummelsheim et al., 1985; Leiras et al., 2010]. Recent evidence suggests that dorsal column inputs to the MCU may also play a role in the processing of pain [Nauta et al., 1997; Nieuwenhuys, 2008; Kaas, 2012; Westlund and Willis, 2012].

A comparison of the properties of sensory neurons in the lemniscal pathway suggests that the relay nuclei, like the MCU, strictly preserve somatotopy and modality, but modulate receptive fields by sharpening borders, suppressing inputs from hairy parts of the hand, and refining and amplifying inputs from the glabrous parts [Bjorkeland and Boivie, 1984; Xu and Wall, 1999].

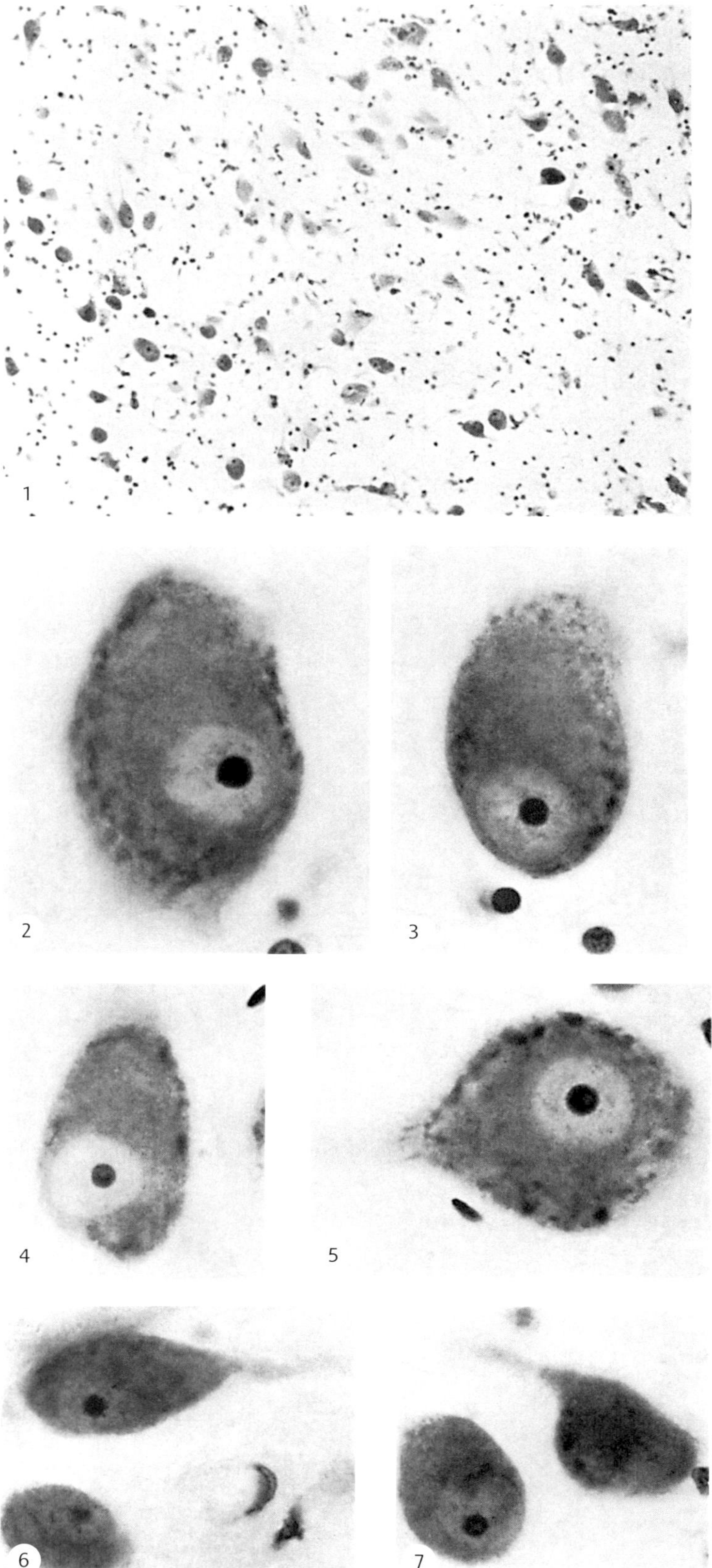

Fig. 1. Medial cuneate nucleus. Magnification ×150.
Fig. 2–7. Different cell types from the medial cuneate nucleus. Magnification ×1,000.

In man, lesions of the dorsal columns do not markedly disrupt simple tactile tasks, unless the spinothalamic tract is also damaged; however, they do severely affect the ability to detect speed or direction of moving stimuli or to interpret patterns drawn on the skin. The lesions also disrupt knowledge of movement and position, causing ataxia and clumsiness in the use of the hands. It is reported that lesions of the dorsal columns cause an increase in pain, tickle, warmth and cold, in keeping with the concept that the pathways through the dorsal column nuclei also process pain [Wall and Noordenbos, 1977; Nathan et al., 1986; Qi et al., 2011].

Connections

Afferents. The dorsal column fibers of the cuneate fasciculus lie lateral to the gracile fasciculus, and contain fibers from the ipsilateral dorsal roots above the level of T_6 and below C_{3-4}. They are topographically arranged, with lower regions represented medially. The neural connections of the MCU are very similar to those of the GR, except that they carry the tactile and proprioceptive signals from the ipsilateral upper body including the hands. The tactile representation of the hand in the MCU lies centrally, with the thumb lateral and the little finger medial, and occupies a much larger volume of neural tissue than that of the foot in the GR [Xu and Wall, 1996; Nieuwenhuys, 2008; Kaas, 2012]. The central core of the MCU is called pars rotunda and is composed of clusters of round, acetylcholine esterase-positive cells with bushy dendrites [Paxinos et al., 2012]. Here there is a very precise somatotopic map. In the periphery of the MCU, the somatotopic maps are less precise, the cells smaller, and shown to be to a large extent inhibitory [Barbaresi et al., 1986; Florence et al., 1989]. However, it is in the periphery of the MCU that the proprioceptive fibers terminate, supplying information from joints [Leiras et al., 2010].

A widespread inhibition in the MCU of the primate has been described, and is thought to arise from descending fibers of the sensorimotor cortex [Biedenbach et al., 1971; Leiras et al., 2010].

Efferents. The projection neurons of both the central core and the periphery of the MCU send axons across the midline, as the internal arcuate fibers, into the contralateral medial lemniscus, which in turn terminates in the lateral ventroposterior nucleus of the thalamus. Both cutaneous and proprioceptive modalities are transmitted [Rosén and Sjölund, 1973; Leiras et al., 2010]. The information is then relayed to Brodmann areas 3a, 3b, 1 and 2 of the primary somatosensory cortex [Nieuwenhuys, 2008]. The periphery of the MCU, mainly in the rostral region, also projects to the cerebellum, the inferior olive, the tectum and the cochlear nuclei, and is considered to support sensory and motor integration [Cheek et al., 1975; Cerminara et al., 2003; Leiras et al., 2010].

References

Barbaresi P, Spreafico R, Frassoni C, Rustioni A: GABAergic neurons are present in the dorsal column nuclei but not in the ventroposterior complex of rats. Brain Res 1986;382:305–326.

Biedenbach MA, Jabbur SJ, Towe AL: Afferent inhibition in the cuneate nucleus of the rhesus monkey. Brain Res 1971;27:179–183.

Bjorkeland M, Boivie J: The termination of spinomesencephalic fibers in cat. An experimental anatomical study. Anat Embryol 1984;170:265–277.

Cerminara NL, Makarabhirom K, Rawson JA: Somatosensory properties of cuneocerebellar neurones in the main cuneate nucleus of the rat. Cerebellum 2003; 2:131–145.

Cheek MD, Rustioni A, Trevino DL: Dorsal column nuclei projections to the cerebellar cortex in cats as revealed by the use of the retrograde transport of horseradish peroxidase. J Comp Neurol 1975;164:31–46.

Florence SL, Wall JT, Kaas JH: Somatotopic organization of inputs from the hand to the spinal gray and cuneate nucleus of monkeys with observations on the cuneate nucleus of humans. J Comp Neurol 1989;286:48–70.

Hummelsheim H, Wiesendanger R, Wiesendanger M, Bianchetti M: The projection of low-threshold muscle afferents of the forelimb to the main and external cuneate nuclei of the monkey. Neuroscience 1985;16:979–987.

Kaas JH: Somatosensory system; in Mai JK, Paxinos G (eds): The Human Nervous System, ed 3. Amsterdam, Academic Press, 2012.

Leiras R, Velo P, Martin-Cora F, Canedo A: Processing afferent proprioceptive information at the main cuneate nucleus of anesthetized cats. J Neurosci 2010; 30:15383–15399.
Nathan PW, Smith MC, Cook AW: Sensory effects in man of lesions of the posterior columns and of some other afferent pathways. Brain 1986;109:1003–1041.
Nauta HJ, Hewitt E, Westlund KN, Willis WD Jr: Surgical interruption of a midline dorsal column visceral pain pathway. Case report and review of the literature. J Neurosurg 1997;86:538–542.
Nieuwenhuys R: General sensory systems and taste; in Nieuwenhuys R, Voogd J, van Huijzen C (eds): The Human Central Nervous System. Berlin, Springer, 2008.
Paxinos G, Huang X-F, Sengul G, Watson C: Organization of the brainstem; in Mai JK, Paxinos G (eds): The Human Nervous System, ed 3. Amsterdam, Academic Press, 2012.
Qi HX, Chen LM, Kaas JH: Reorganization of somatosensory cortical areas 3b and 1 after unilateral section of dorsal columns of the spinal cord in squirrel monkeys. J Neurosci 2011;31:13662–13675.
Rosén I, Sjölund B: Organization of group I activated cells in the main and external cuneate nuclei of the cat: identification of muscle receptors. Exp Brain Res 1973;16:221–237.
Wall PD, Noordenbos W: Sensory functions which remain in man after complete transection of dorsal columns. Brain 1977;100:641–653.
Westlund KN, Willis WD: Pain system; in Mai JK, Paxinos G (eds): The Human Nervous System, ed 3. Amsterdam, Academic Press, 2012.
Xu J, Wall JT: Cutaneous representations of the hand and other body parts in the cuneate nucleus of a primate, and some relationships to previously described cortical representations. Somatosens Mot Res 1996;13:187–197.
Xu J, Wall JT: Functional organization of tactile inputs from the hand in the cuneate nucleus and its relationship to organization in the somatosensory cortex. J Comp Neurol 1999;411:369–389.

Sensory Systems – Somatosensory Nuclei (Plates 10–15)

4 Lateral Cuneate Nucleus (LCU)

Original name:
Nucleus cuneatus lateralis

Alternative names:
External cuneate nucleus
Accessory cuneate nucleus

Location and Cytoarchitecture – Original Text

This is an elongated nucleus lying close to the periphery of the brainstem in the dorsolateral part of the medulla. It appears on the dorsolateral aspect of the medial cuneate nucleus (MCU) just caudal to the obex, and extends orally for approximately 8 mm, terminating at the level of the caudal pole of the descending vestibular nucleus.

Caudal to the level of the obex, the lateral cuneate nucleus (LCU) appears on cross-section as a conspicuous, darkly stained, irregularly shaped cell group lying lateral to the MCU, dorsal to the central nucleus of the medulla oblongata and dorsomedial to the spinal trigeminal complex.

At the level of the opening of the fourth ventricle, the nucleus has increased considerably in size and lies lateral to the gracile nucleus (GR), dorsal to the oral pole of the MCU and to the spinal trigeminal complex. At this level, groups of cells similar to and usually continuous with the cell groups of the LCU are found interposed between the spinal trigeminal tract and the newly developed restiform body. These groups (fig. 2), designated the islands of lateral cuneate nuclei (in plate 10), are often in continuity ventrally with similar cell groups which lie ventral to the spinal trigeminal complex (lateral reticular nucleus, subtrigeminal part). (These islands of cells have also been referred to as the paratrigeminal nucleus [Chan-Palay, 1978].)

The LCU attains its maximum dimensions just oral to the level at which the GR disappears. Here, the nucleus lies lateral to the medial vestibular nucleus and to the nucleus of the solitary tract (SOL), dorsal to the spinal trigeminal complex and medial to the inferior cerebellar peduncle. The nucleus approaches its oral pole without much diminution in size and is directly continuous with the caudal pole of the descending vestibular nucleus.

The cells of the LCU (fig. 1) may easily be distinguished from those of the MCU. On the whole, they are much larger and more intensely stained. The most characteristic cells (fig. 2, 4–6) present are large, oval, plump neurons with short dendrites, excentric nuclei and peripherally distributed Nissl granules. Such cells bear a striking resemblance to those of Clarke's column of the spinal cord[1]. Scattered among these rather typical cells, and just as numerous, are medium-sized, triangular, multipolar cells with long dendrites, centrally placed nuclei and darkly stained, evenly distributed Nissl granules (fig. 7–10). The oral pole of the LCU is composed almost exclusively of the latter cell type, and this often renders the distinction between the oral pole of the LCU and the caudal pole of the descending vestibular nucleus (DVN) very difficult. The differentiation can usually be made, however, by the observation that the cells of the LCU are larger and more closely arranged than are those of the DVN. Further, glial nuclei are more numerous within the DVN than within the LCU.

Large fiber bundles traverse the LCU and break up its constituent cells into irregular groups or masses. This accounts for the characteristic 'lobulated' appearance of the nucleus on both Nissl- and myelin-stained sections. The superficially situated cells of the LCU stain more intensely than the more deeply situated cells, but this may be due to the effects of fixation.

At the level of the oral half of the LCU, groups of small, oval, lightly stained cells with excentric nuclei, and peripherally arranged Nissl granules may be distinguished. These groups lie ventromedial to the LCU, several of them between the SOL medially and spinal trigeminal complex laterally. The cells of these groups are, on average, intermediate in size and staining qualities between the cells of the MCU and LCU (fig. 3). Such a group may be clearly seen in plates 12 and 13 [Paxinos et al. [2012] have referred to this nucleus as the medial pericuneate nucleus.] [...].

[1] In the light of present knowledge, it is clear that the neurons of the LCU and Clarke's dorsal column in the spinal cord (T1–L4) have nearly identical functions: the LCU receiving first-order sensory afferents from upper limb muscles and projecting to the cerebellum, and Clarke's column receiving first-order sensory afferents from lower limb muscles and also projecting to the cerebellum. It is quite awe-inspiring, even for a neuroanatomist, to read here that Olszewski and Baxter predicted this functional similarity in 1952 on the basis of cytoarchitectural likeness alone.

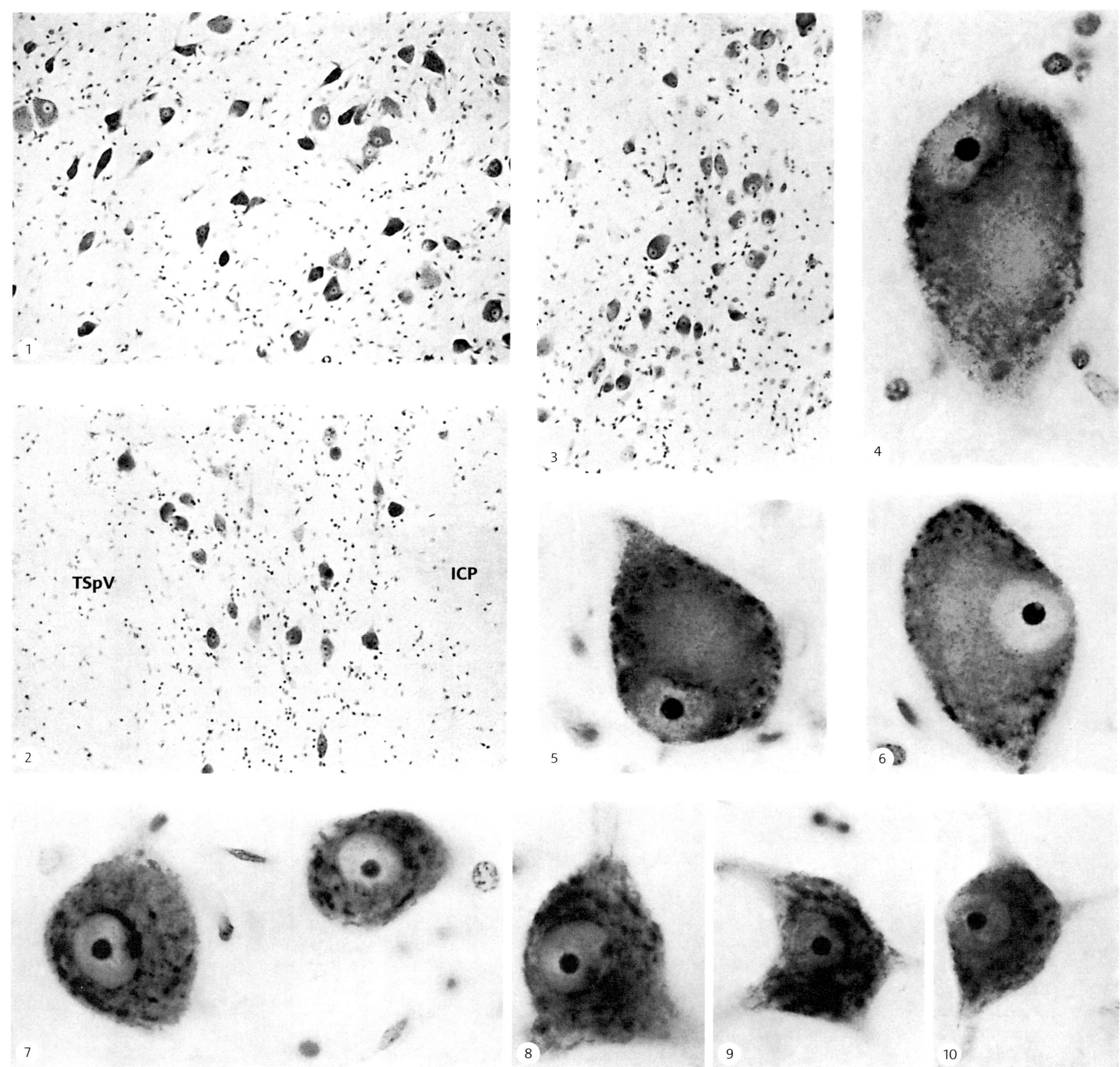

Fig. 1. Lateral cuneate nucleus. Magnification ×150.

Fig. 2. Islands of the lateral cuneate nucleus, which lie between the spinal trigeminal tract (TSpV) and the inferior cerebellar peduncle (ICP). Magnification ×150.

Fig. 3. Cell group lying between the solitary nucleus and the spinal trigeminal complex. Magnification ×150.

Fig. 4–6. Cells from the lateral cuneate nucleus with excentric nuclei and peripherally distributed Nissl granules. Magnification ×1,000.

Fig. 7–10. Medium-sized cells of the lateral cuneate nucleus with evenly distributed Nissl granules.

Functional Neuroanatomy

Function

The LCU is a dorsal column nucleus, like the GR and MCU. It receives predominantly muscle afferents from the forelimb and neck [Hummelsheim et al., 1985; Bakker et al., 1985; Witham and Baker, 2011], and in cats the output of the LCU enters the inferior cerebellar peduncle and terminates in the cerebellum [Cooke et al., 1971]. However, in primates and rats the signals of the LCU are also sent via the medial lemniscus to the thalamus [Boivie et al., 1975; Boivie and Boman, 1981; Mantle-St John and Tracey, 1987]. This difference may reflect the versatile use that primates and rats make of their glabrous hands compared with the more locomotive function of cat paws.

Clinically, a severe loss of neurons in the LCU is found in cases of spinocerebellar ataxia type 3; the loss may contribute to the ataxia typical of these patients [Rüb et al., 2002].

Connections

Afferents. In humans the LCU is the largest of the dorsal column nuclei. The major inputs to LCU neurons are axons from muscle receptors of the forelimb and neck, whose cell bodies are in the upper thoracic and lower cervical dorsal root ganglia. The fibers ascend ipsilaterally in the cuneate fasciculus. Like the other dorsal column nuclei, the representation of the muscles within the LCU is somatotopically organized [Bakker et al., 1985]. The descending afferent pathways to the LCU from the sensory and motor cortices are more extensive in primates than in cats [Bentivoglio and Rustioni, 1986].

Efferents. The LCU provides the proprioceptive component of the cuneocerebellar pathway, which enters the inferior cerebellar peduncle, innervates the cerebellar nuclei and the vermal and paravermal regions of the cerebellar cortex in the anterior lobe (lobules I–IVa) and the pyramis and uvula [Cooke et al., 1971; Cheek et al., 1975; Massopust et al., 1985; Zguczynski et al., 2012]. In the cat this is the major efferent pathway but in the rat and primate the LCU also projects through the medial lemniscus to the ventroposterior superior nucleus of the thalamus [Boivie et al., 1975; Boivie and Boman, 1981; Mantle-St John and Tracey, 1987; Kaas, 2012].

References

Bakker DA, Richmond FJ, Abrahams VC, Courville J: Patterns of primary afferent termination in the external cuneate nucleus from cervical axial muscles in the cat. J Comp Neurol 1985;241:467–479.

Bentivoglio M, Rustioni A: Corticospinal neurons with branching axons to the dorsal column nuclei in the monkey. J Comp Neurol 1986;253:260–276.

Boivie J, Boman K: Termination of a separate (proprioceptive?) cuneothalamic tract from external cuneate nucleus in monkey. Brain Res 1981;224:235–246.

Boivie J, Grant G, Albe-Fessard D, Levante A: Evidence for a projection to the thalamus from the external cuneate nucleus in the monkey. Neurosci Lett 1975;1: 3–8.

Chan-Palay V: The paratrigeminal nucleus. I. Neurons and synaptic organization. J Neurocytol 1978;7:405–418.

Cheek MD, Rustioni A, Trevino DL: Dorsal column nuclei projections to the cerebellar cortex in cats as revealed by the use of the retrograde transport of horseradish peroxidase. J Comp Neurol 1975;164:31–46.

Cooke JD, Larson B, Oscarsson O, Sjolund B: Origin and termination of cuneocerebellar tract. Exp Brain Res 1971;13:339–358.

Hummelsheim H, Wiesendanger R, Wiesendanger M, Bianchetti M: The projection of low-threshold muscle afferents of the forelimb to the main and external cuneate nuclei of the monkey. Neuroscience 1985;16:979–987.

Kaas JH: Somatosensory system; in Mai JK, Paxinos G (eds): The Human Nervous System, ed. 3. Amsterdam, Academic Press, 2012.

Mantle-St John LA, Tracey DJ: Somatosensory nuclei in the brainstem of the rat: independent projections to the thalamus and cerebellum. J Comp Neurol 1987; 255:259–271.

Massopust LC, Hauge DH, Ferneding JC, Doubek WG, Taylor JJ: Projection systems and terminal localization of dorsal column afferents: an autoradiographic and horseradish peroxidase study in the rat. J Comp Neurol 1985;237:533–544.

Paxinos G, Huang X-F, Sengul G, Watson C: Organization of the brainstem; in Mai JK, Paxinos G (eds): The Human Nervous System, ed 3. Amsterdam, Academic Press, 2012.

Rüb U, de Vos RAI, Schultz C, Brunt ER, Paulson H, Braak H: Spinocerebellar ataxia type 3 (Machado-Joseph disease): severe destruction of the lateral reticular nucleus. Brain 2002;125:2115–2124.

Witham CL, Baker SN: Modulation and transmission of peripheral inputs in monkey cuneate and external cuneate nuclei. J Neurophysiol 2011;106:2764–2775.

Zguczynski L, Bukowska D, Mierzejewska-Krzyzowska B: Dorsal column nuclei projection to the cerebellar caudal vermis in the rabbit revealed by the fluorescent double-labeling method. Cells Tissues Organs 2012;196:280–290.

5 Intercollicular Nucleus (ICOL)

Original name:
Nucleus intercollicularis

Alternative names:
Precuneiform area
The nucleus is not homologous with the avian intercollicular nucleus (see below)

Location and Cytoarchitecture – Original Text

The intercollicular nucleus (ICOL) forms the transitional zone which is interposed between the oral pole of the inferior colliculus (IC) and the caudal pole of the superior colliculus (SC).

This nucleus is composed of uniformly distributed, medium-sized, oval, triangular or fusiform cells which stain with light or medium intensity. Glial satellites are associated with the majority of these cells (fig. 1, 2, 4, 5): in the central portion of the nucleus one occasionally notes large, more darkly stained, multipolar neurons similar to those of the sixth layer of the SC (fig. 3).

The cytoarchitecture of the ICOL differs from that of the IC in that the former is less cellular and is composed predominantly of morphologically similar cells. The characteristic cell types and the laminated distribution of cells found in the SC allow this structure to be readily differentiated from the ICOL.

Functional Neuroanatomy

Function

The ICOL of the mesencephalic tectum lies between the visual SC and the auditory IC. Its neurons, along with some adjacent areas, receive topographically arranged somatosensory afferents, and appear to be the only purely somatosensory receiver, and the most direct somatosensory receiver, in the colliculi. Experimental results suggest that it forms a somesthetic entity: as a third sensory center in the tectal plate, it could contribute to the establishment of a multisensory map of the surroundings in the colliculus [Danielsson and Norrsell, 1986; Blomqvist et al., 1990].

Connections

The ICOL receives bilateral inputs from the spinal cord, dorsal column nuclei and the sensorimotor cortex (cat [Itoh et al., 1984; Blomqvist et al., 1990]; opossum [Robards et al., 1976]), also from the auditory cortex in the ferret [Bajo et al., 2007]. Some immediately adjacent regions, such as the IC external nucleus, the deep layers of the SC, and the dorsal periaqueductal gray (PAG), receive these somatosensory afferents too, and form part of an 'intercollicular terminal zone'.

The ICOL region projects to the lateral division of the posterior complex of the thalamus, a multimodal region that in turn projects to the retroinsular cortex surrounding the secondary sensory cortex [Itoh et al., 1984; Cooper and Dostrovsky, 1985; Nieuwenhuys et al., 2008].

Reach-related neurons with tactile properties have been found in deep layers of both caudal and rostral SC [Reyes-Puerta et al., 2011; Linzenbold and Himmelbach, 2012]. It is possible that the intercollicular zone, including the ICOL, forms an entity which contributes somatosensory information for the multisensory topographic 'map' in the SC of the surrounding world [Blomqvist et al., 1990; Stein et al., 2002; Krueger et al., 2009].

The ICOL should not be confused with the collicular commissural systems crossing the midline (sometimes referred to as intercollicular), nor with the avian intercollicular nucleus of the tectum, a region associated with vocalization and thought to be a 'homologue' of the dorsal PAG of mammals [Kingsbury et al., 2011].

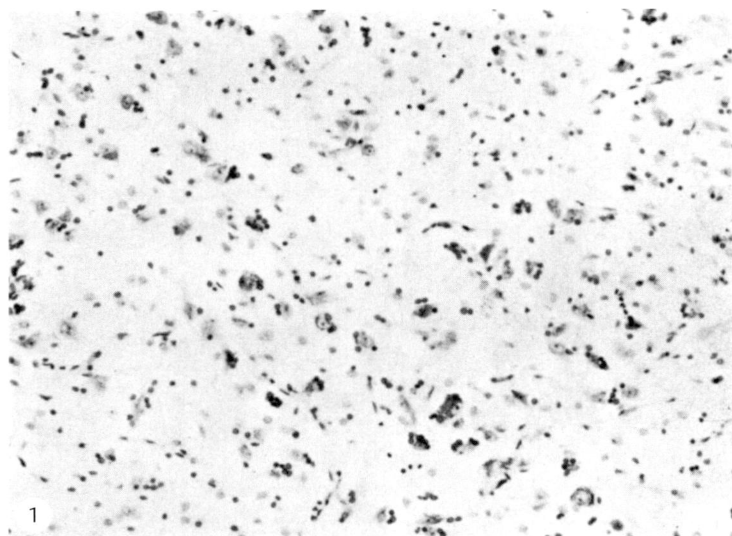

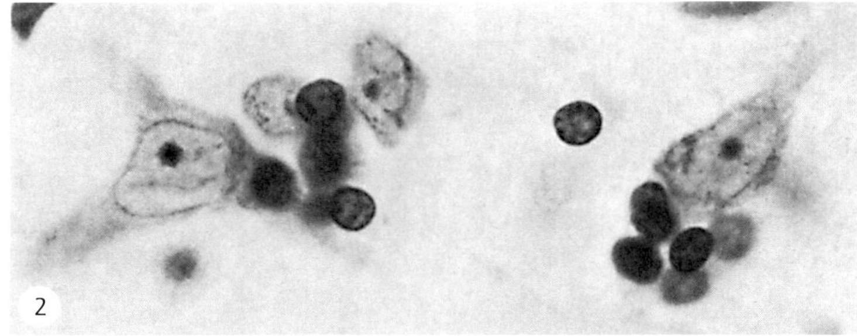

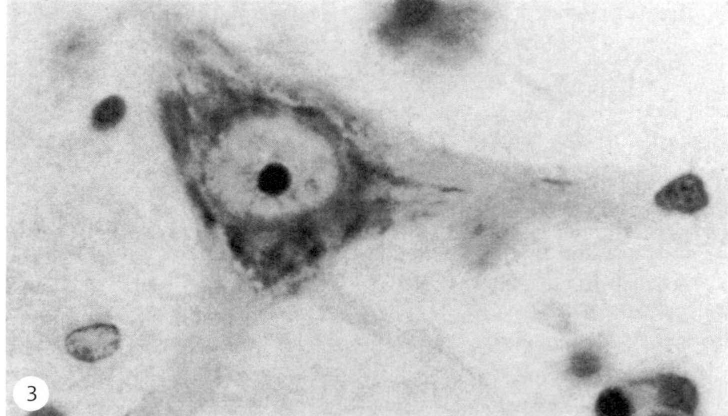

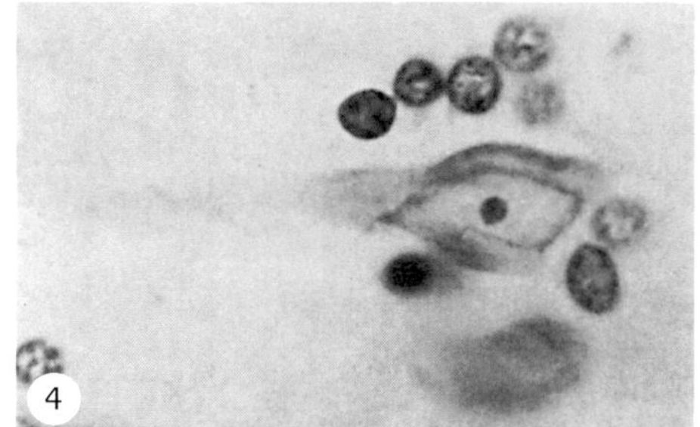

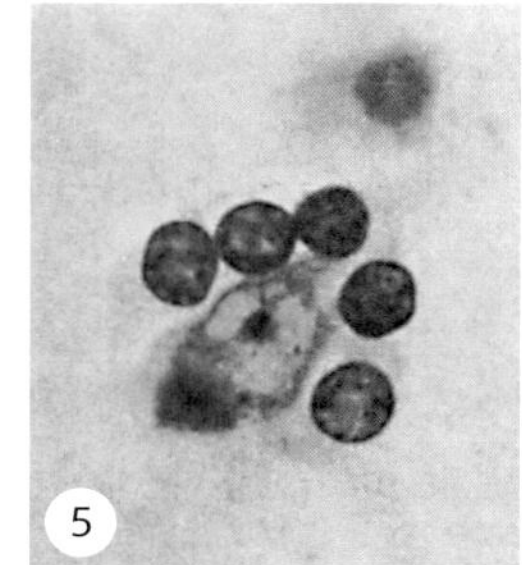

Fig. 1. Intercollicular nucleus. Magnification ×150.

Fig. 2–5. Different cells from the intercollicular nucleus. Note the glial satellitosis. Magnification ×1,000.

References

Bajo VM, Nodal FR, Bizley JK, Moore DR, King AJ: The ferret auditory cortex: descending projections to the inferior colliculus. Cereb Cortex 2007;17:475–491.

Blomqvist A, Danielsson I, Norrsell U: The somatosensory intercollicular nucleus of the cat's mesencephalon. J Physiol 1990;429:191–203.

Cooper LL, Dostrovsky JO: Projection from dorsal column nuclei to dorsal mesencephalon. J Neurophysiol 1985;53:183–200.

Danielsson I, Norrsell U: Somatosensory units in the cat's intercollicular region. Acta Physiol Scand 1986;128:579–586.

Itoh K, Kaneko T, Kudo M, Mizuno N: The intercollicular region in the cat: a possible relay in the parallel somatosensory pathways from the dorsal column nuclei to the posterior complex of the thalamus. Brain Res 1984;308:166–171.

Kingsbury MA, Kelly AM, Schrock SE, Goodson JL: Mammal-like organization of the avian midbrain central gray and a reappraisal of the intercollicular nucleus. PLoS One 2011;6:e20720.

Krueger J, Royal DW, Fister MC, Wallace MT: Spatial receptive field organization of multisensory neurons and its impact on multisensory interactions. Hear Res 2009;258:47–54.

Linzenbold W, Himmelbach M: Signals from the deep: reach-related activity in the human superior colliculus. J Neurosci 2012;32:13881–13888.

Nieuwenhuys R, Voogd J, van Huijzen C: Motor systems; in Nieuwenhuys R, Voogd J, van Huijzen C (eds): The Human Central Nervous System, ed 4. Berlin, Springer, 2008.

Reyes-Puerta V, Philipp R, Lindner W, Hoffmann KP: Neuronal activity in the superior colliculus related to saccade initiation during coordinated gaze-reach movements. Eur J Neurosci 2011;34:1966–1982.

Robards MJ, Watkins DW 3rd, Masterton RB: An anatomical study of some somesthetic afferents to the intercollicular terminal zone of the midbrain of the opossum. J Comp Neurol 1976;170:499–524.

Stein BE, Wallace MW, Stanford TR, Jiang W: Cortex governs multisensory integration in the midbrain. Neuroscientist 2002;8:306–314.

(Fig. 1, p. 18)

Overview

Some knowledge of the unusual arrangement of the trigeminal nerve (NV) is important for understanding the central organization of the trigeminal nuclei in the brainstem. The NV, or fifth cranial nerve, is a mixed nerve, with a large sensory portion (portio major) carrying touch, nociception, temperature and proprioception information from the face and intraoral regions, and a small motor component (portio minor) innervating the muscles of mastication. The NV has 3 branches: the ophthalmic (V_1), maxillary (V_2) and mandibular (V_3) nerves which supply the upper, medial and lower face regions, respectively, and join intracranially to build the trigeminal (semilunar) ganglion. The combined NV enters the brainstem in the lateral pons and is associated with 4 main cranial nerve nuclei: the spinal trigeminal nucleus (SpV), the principal sensory trigeminal nucleus (PrV), the motor trigeminal nucleus (MoV) and the mesencephalic trigeminal nucleus (MesV), whereby each nucleus processes signals of different modalities. The SpV is important for the transmission of nondiscriminative touch, nociception and thermal stimuli, the PrV for light touch, the MesV for proprioception and the MoV for motor control. The 4 trigeminal nuclei extend throughout the whole length of the brainstem (see 'Plates of Serial Sections through the Human Brainstem', fig. 1, p. 18).

Unlike the PrV and MoV, the SpV has an unusual morphology. It builds a long column of cells divided into 3 clear cytological subdivisions: SpV caudal part (SpVc), SpV interpolar part (SpVi) and the SpV oral part (SpVo). Its role in nociception is a result of the predominantly small-caliber trigeminal afferents (Aδ and C nerve fibers) that collect together and descend laterally to the SpV, down its length, as the spinal trigeminal tract (TSpV, see below). The TSpV supplies the SpV with sensory afferents, topographically. In a horizontal brainstem section (as seen in plates 4–22) through the SpV and its tract, there is a strict topographical organization in both structures, with the jaw represented dorsally and the forehead ventrally – chin up! Along the rostral-caudal axis of the SpV, the topographical organization of trigeminal afferents results in an 'onion skin pattern' of facial pain and temperature, clearly illustrated by Warren et al. [2006], the SpVc receiving from the lateral face (forehead, cheek and jaw), the SpVi receiving from the medial facial regions (around the eyes, medial cheek and point of the chin), and the SpVo near the obex receiving from the circumoral and internal structures of the nose and mouth.

The topographical organization of the SpV and TSpV has important diagnostic value for differentiating between central lesions of the trigeminal system in the medulla, or peripheral lesions of the trigeminal nerve: central lesions of the SpV result in loss of pain, but not touch, in a facial area crossing all 3 dermatomes (onion skin pattern), whereas a lesion of V_1, V_2 or V_3 results in hypesthesia of a cutaneous dermatome of the face.

Reference

Warren S, Yezierski RP, Capra NF: The somatosensory system II: touch, thermal sense, and pain; in Haines DE (ed): Fundamental Neuroscience for Basic and Clinical Applications. Philadelphia, Churchill Livingstone Elsevier, 2006.

Sensory Systems – Trigeminal Complex

(Plates 4–9)

6 Spinal Trigeminal Nucleus, Caudal Part (SpVc)

Original name:
Nucleus tractus spinalis nervi trigemini caudalis

Alternative names:
Nucleus spinalis nervi trigemini, pars caudalis
Spinal nucleus of the trigeminal nerve, caudal part
Trigeminal spinal nucleus

Location and Cytoarchitecture – Original Text

This nucleus is formed by an oral prolongation of the apical gray substance of the dorsal horn of the cervical spinal cord [see also Waite and Ashwell, 2004]. In the medulla, the nucleus extends orally some 13 mm and terminates at the obex or about the level at which the lateral cuneate nucleus (LCU) appears.

In the caudal part of the medulla, the nucleus is triangular in shape on cross-section, but as one proceeds orally it assumes a globular form and

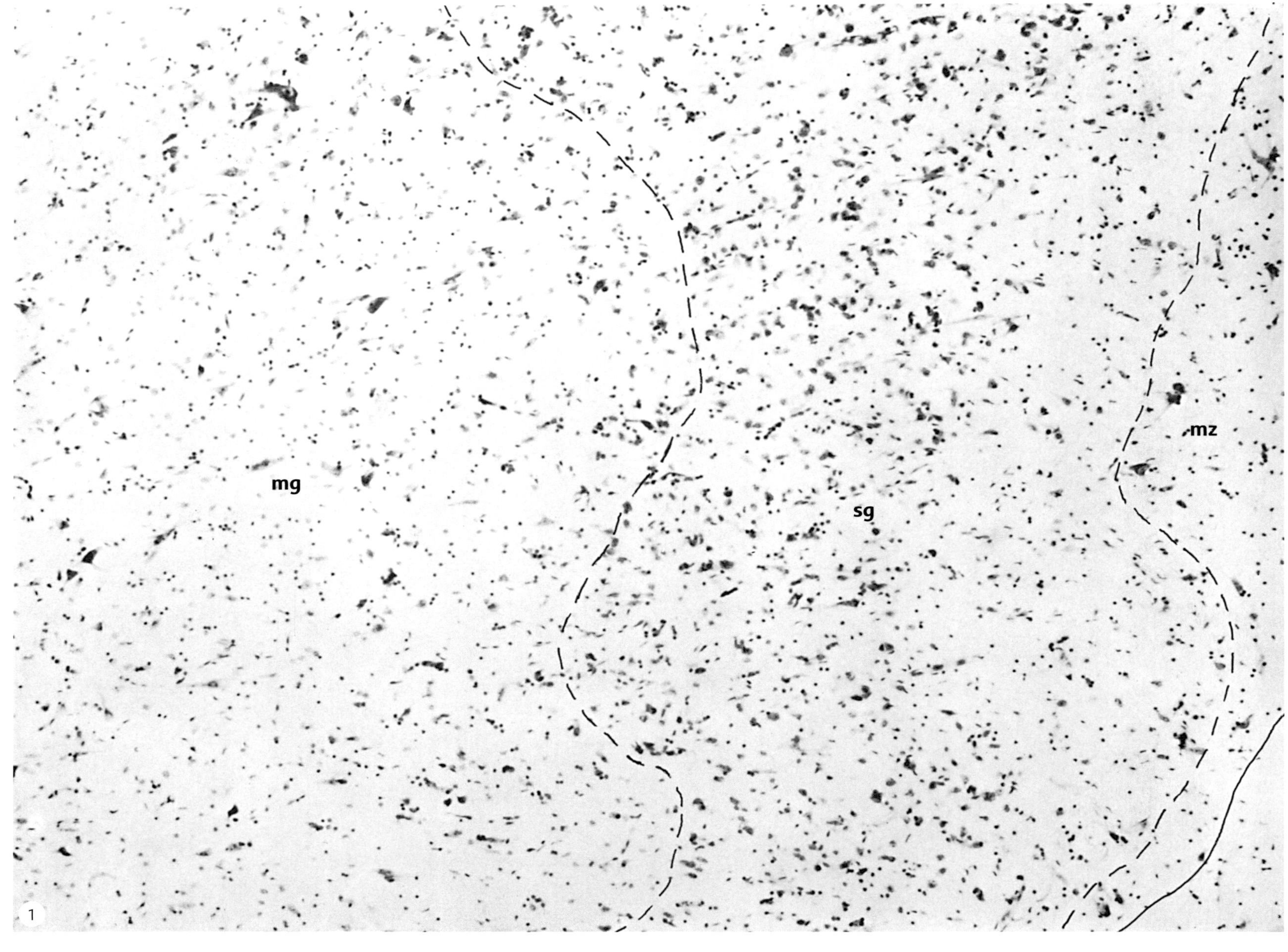

Fig. 1. Spinal trigeminal nucleus, caudal part. Note the subdivision into marginal zone (mz), substantia gelatinosa (sg) and magnocellular zone (mg). Magnification ×150.

increases considerably in size. In cross-sections through the pyramidal decussation, the nucleus lies in the dorsolateral part of the medulla; oral to the decussation it is displaced slightly ventrally due to the widening of the dorsal columns. Throughout its extent, the spinal trigeminal nucleus caudal part (SpVc) is related ventromedially to the central nucleus of the medulla oblongata, dorsomedially to the fasciculus cuneatus, dorsolaterally and ventrolaterally to the spinal trigeminal and the dorsal spinocerebellar tracts. Orally, it is directly continuous with the spinal trigeminal nucleus interpolar part (SpVi).

The cytoarchitecture of the SpVc is essentially the same as that of the apex of the dorsal horn of the spinal cord. Three subdivisions can be distinguished – marginal zone (mz), substantia gelatinosa (sg) and the magnocellular zone (mg) (fig. 1).

The mz corresponds to the marginal nucleus (Rexed lamina I) in the spinal cord and is composed of a few fusiform, multipolar, darkly stained cells scattered about the periphery of the sg. In general, the long axes of these cells lie parallel to the border of the zone (fig. 9–12).

The sg is the largest of these and forms the horseshoe-shaped area which is characteristic of the SpVc in both Nissl- and myelin-stained sections. This area is analogous to the 'substantia gelatinosa of Rolando' (Rexed lamina II) in the spinal cord. It is composed of small, closely packed, lightly stained, oval or spindle-shaped cells arranged, in general, with their long axes directed towards the hilus of the horseshoe (fig. 5–8).

The mg occupies the hilus of the horseshoe and corresponds to the nucleus proprius of the dorsal horn in the spinal cord (Rexed lamina III–IV). It is composed of small- and medium-sized, lightly stained cells, and of larger cells which contain abundant Nissl substance and are fusiform in shape. Occasionally, however, cells composing the mg are slender and thus stain darkly. The great majority of the cells resemble those in the LCU (fig. 2–4).

The oral limit of the SpVc is evident in both Nissl- and myelin-stained preparations. The most obvious feature is the breaking up of the sg into small islands which quickly disappear. The mz and mg end at the same level and the cytoarchitecture of the cell column changes completely into that of the SpVi.

Functional Neuroanatomy

Function

The SpVc is the rostral continuation of the dorsal horn in the spinal cord (laminae I–IV). The dorsal horn receives information from sensory structures of a spinal segment, the SpVc receives ipsilateral signals from the lateral face (forehead, cheek

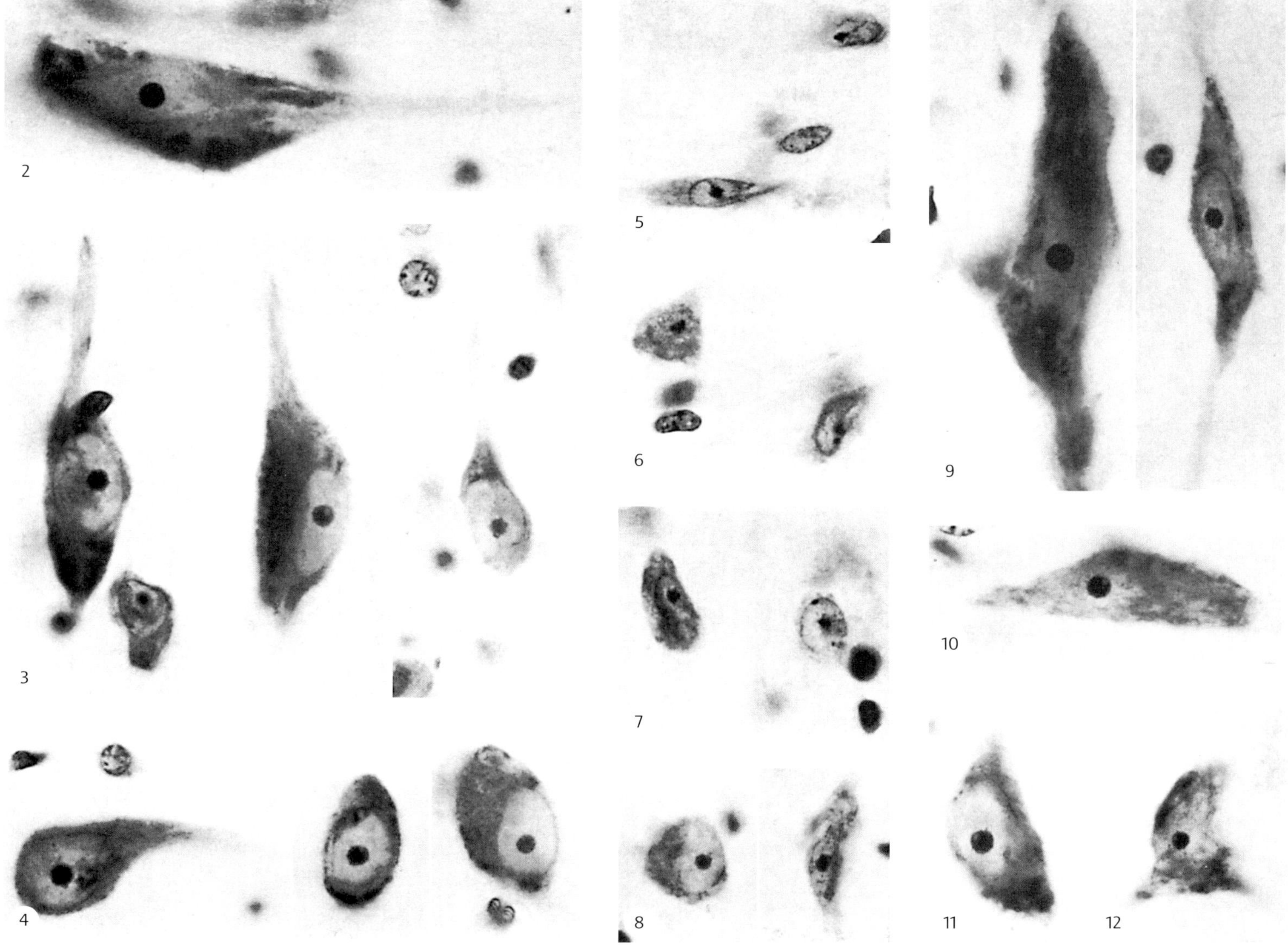

Fig. 2–4. Spinal trigeminal nucleus, caudal part: different cell types from the magnocellular zone. Magnification ×1,000.

Fig. 5–8. Cells from the substantia gelatinosa. Magnification ×1,000.
Fig. 9–12. Different cell types from the marginal zone. Magnification ×1,000.

and jaw) and from meningeal blood vessels. The modality of sensory signals in the SpVc is focussed only on nociceptive and thermal stimuli, and nondiscriminative touch similar to the laminae I–IV [Dubner and Bennett, 1983], and the representation of pain of the lateral face is strictly topographically organized in the SpVc – chin up (see 'Sensory Systems – Trigeminal Complex: Overview')!

Like the spinal laminae I and II, the SpVc mz and sg contain C fibers carrying the neurotransmitters such as acetylcholine, substance P, cholecystokinin, somatostatin, calcitonin gene-related substance and enkephalin [Waite and Ashwell, 2004].

Lesions in the SpVc diminish painful cutaneous stimuli of the lateral face (onion skin pattern, see 'Sensory Systems – Trigeminal Complex: Overview'), but they do not affect touch and 2-point discrimination, which is associated with the principal sensory trigeminal nucleus (PrV); corneal sensation and the corneal reflex also remain intact but may be slowed [Dubner and Bennett, 1983; Willis and Westlund, 2004]. The trigeminovascular system plays an important role in migraine [Mitsikostas and Sanchez del Rio, 2001].

Connections

Afferents. The afferents to the SpV through the spinal trigeminal tract arise from 3 sharply segregated facial dermatomes, with no overlap peripherally. A strict, but different, topography is preserved within the SpVc, see 'Sensory Systems – Trigeminal Complex: Overview' [Kerr, 1963]. The dorsomedial regions of the SpVc also receive afferent fibers from the glossopharyngeal vagus and facial nerves, which carry sensory inputs from the larynx, pharynx and small regions around and in the ear [Rhoton et al., 1966]. In addition the SpVc receives descending corticobulbar fibers from the contralateral somatosensory cortex, which may serve to suppress nociceptive information [Gojyo et al., 2002].

Efferents. The efferents of the SpVc play an important role in transmitting facial pain to higher centers, and form part of the ascending nociception and temperature sensitivity (protopathic) pathway of the trigeminal system. Experimental evidence confirms that SpVc efferents cross the midline in the caudal medulla oblongata, and ascend as the lateral trigeminothalamic tract, medial to the spinothalamic (anterolateral) pathways and dorsal to the medial lemniscus, to terminate in the contralateral posteromedial and posterolateral ventral nucleus of the thalamus, also in the intralaminar nuclei and the posterior group [Gauriau and Bernard, 2004; Nieuwenhuys et al., 2008]. The parabrachial nuclei are also a prominent target for ascending trigeminal collaterals, and may participate in the integration of sensory information with visceral sensations [Pinto et al., 2006]. In addition, axons from the SpVc project into an 'intranuclear pathway' which ascends to the PrV. The neurons of the mz and sg project to the solitary nucleus [Menétrey et al., 1992], while the mg laminae target the facial nucleus [Erzurumlu and Killackey, 1979].

References

Dubner R, Bennett GJ: Spinal and trigeminal mechanisms of nociception. Annu Rev Neurosci 1983;6:381–418.

Erzurumlu RS, Killackey HP: Efferent connections of the brainstem trigeminal complex with the facial nucleus of the rat. J Comp Neurol 1979;188:75–86.

Gauriau C, Bernard JF: Posterior triangular thalamic neurons convey nociceptive messages to the secondary somatosensory and insular cortices in the rat. J Neurosci 2004;24:752–761.

Gojyo F, Sugiyo S, Kuroda R, Kawabata A, Varathan V, Shigenaga Y, Takemura M: Effects of somatosensory cortical stimulation on expression of c-Fos in rat medullary dorsal horn in response to formalin-induced noxious stimulation. J Neurosci Res 2002;68:479–488.

Kerr FWL: The divisional organization of afferent fibers of the trigeminal nerve. Brain 1963;86:721–732.

Menétrey D, Depommery J, Baimbridge KG, Thomasset M: Calbindin-D28K (CaBP28k)-like immunoreactivity in ascending projections. I. Trigeminal nucleus caudalis and dorsal vagal complex projections. Eur J Neurosci 1992;4:61–69.

Mitsikostas DD, Sanchez Del Rio M: Receptor systems mediating c-fos expression within trigeminal nucleus caudalis in animal models of migraine. Brain Res Brain Res Rev 2001;35:20–35.

Nieuwenhuys R, Voogd J, van Huijzen C: The Human Central Nervous System. Berlin, Springer, 2008.

Pinto ML, de Cássia Machado R, Schoorlemmer GH, Colombari E, de Cássia Ribeiro da Silva Lapa R: Topographic organization of the projections from the interstitial system of the spinal trigeminal tract to the parabrachial nucleus in the rat. Brain Res 2006;1113:137–145.

Rhoton AL, O'Leary JL, Ferguson JP: The trigeminal, facial, vagal and glossopharyngeal nerves in the monkey. Arch Neurol 1966;14:530–540.

Waite PME, Ashwell KWS: Trigeminal sensory system; in Paxinos G, Mai JK (eds): The Human Nervous System, ed 2. Amsterdam, Elsevier Academic Press, 2004.

Warren S, Yezierski RP, Capra NF: The somatosensory system II: touch, thermal sense, and pain; in Haines DE (ed): Fundamental Neuroscience for Basic and Clinical Applications, ed 3. Philadelphia, Churchill Livingstone Elsevier, 2006.

Willis WD, Westlund KN: Pain system; in Paxinos G, Mai JK (eds): The Human Nervous System, ed 2. Amsterdam, Elsevier Academic Press, 2004.

7 Spinal Trigeminal Nucleus, Interpolar Part (SpVi)

Original name:
Nucleus spinalis nervi trigemini interpolaris

Alternative names:
Nucleus spinalis nervi trigemini, pars interpolaris
Spinal nucleus of the trigeminal nerve, interpolar part
Trigeminal spinal nucleus

Location and Cytoarchitecture – Original Text

The spinal trigeminal nucleus interpolar part (SpVi) lies in the lateral part of the medullary tegmentum. It extends from the junction of the caudal and middle thirds to the junction of the middle and oral thirds of the inferior olivary complex – a distance of 6 mm. Orally, the nucleus is directly continuous with the spinal trigeminal nucleus oral part (SpVo), and caudally with the spinal trigeminal nucleus caudal part (SpVc).

Caudally the SpVi is oblong in outline and lies with its long axis directed slightly dorsomedially. As one proceeds orally, it diminishes somewhat in size and in many sections is broken up into 2 or 3 islands of cells by fibers of the vagus and glossopharyngeal nerves. The nucleus is related to the lateral cuneate nucleus (LCU) dorsally, to the central nucleus of the medulla oblongata (CN) medially, and to the lateral reticular nucleus ventrally. That portion of the CN which comes into direct contact with the SpVi appears relatively acellular due to the accumulation of internal arcuate fibers dorsomedially, olivocerebellar fibers ventrally and concomitant fibers of the trigeminal nucleus medially. Fibers of the spinal trigeminal and dorsal spinocerebellar tracts separate the SpVi from the periphery of the medulla. The former tract increases progressively in size as one proceeds orally.

The SpVi is composed of diffusely arranged, lightly stained, small or medium-sized, oval, fusiform or triangular cells among which are scattered larger plump cells with excentric nuclei and peripherally placed Nissl granules (fig. 1–3). These latter cells are similar to those found in the LCU. This cytoarchitectural pattern of the SpVi is uniform so that no division into subnuclei is possible.

The oral pole of the nucleus is marked by a sudden change in the cell pattern to that of SpVo.

Functional Neuroanatomy

Function

The SpVi receives sensory inputs from the cutaneous and part of the medial face, and importantly, in rats, from the whisker region. The SpVi inputs process nociceptive and thermal stimuli, and nondiscriminative touch [Dubner and Bennett, 1983]. The SpV is the first relay of the trigeminal pain system, and it conveys the information to the cerebral cortex, parallel to the principal sensory trigeminal nucleus (PrV); for a review, see

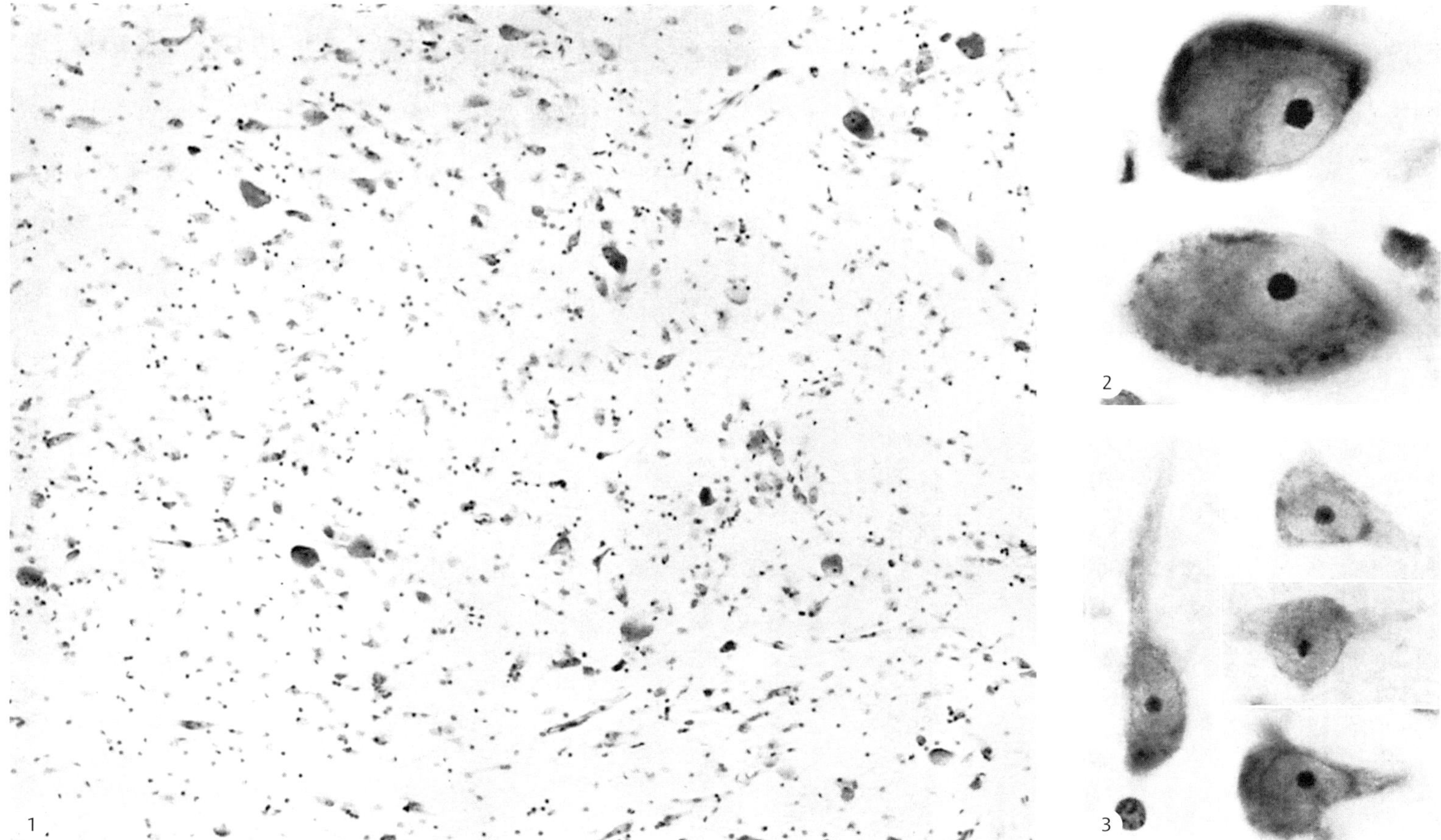

Fig. 1. Spinal trigeminal nucleus, interpolar part. Magnification ×150.
Fig. 2. Medium-sized cells, with excentric nuclei and peripherally arranged Nissl granules, from the spinal trigeminal nucleus, interpolar part. Magnification ×1,000.

Fig. 3. Small cells from the spinal trigeminal nucleus, interpolar part. Magnification ×1,000.

Waite and Ashwell [2004]. The SpVi is the central portion of the spinal trigeminal nuclear column with a very different structure, and presumably function, from the laminated SpVc (SpV, see 'Sensory Systems – Trigeminal Complex: Overview'). One special role for the SpVi in the trigeminal complex may be through its intranuclear connections (see below), indicating that it participates in intratrigeminal coordination. A second difference is that the SpVi has a strong association with the cerebellum, which is not shared by the other parts of the SpV. A probable function for this connection is related to the vital importance of sensory whisker input for most laboratory models. They terminate only in the SpVi. The whisker input in rats is essential for motor coordination; a rat appears ataxic and cannot walk along a narrow bar without whiskers. However, this does not appear relevant in humans.

Lesions in the SpVi diminish painful cutaneous stimuli of the medial face (onion skin pattern, see 'Sensory Systems – Trigeminal Complex: Overview'), but they do not affect touch and 2-point discrimination, which is associated with the PrV [Dubner and Bennett, 1983; Willis and Westlund, 2004]. Clinically it has been implicated in the development of persistent orofacial pain [Ren and Dubner, 2011].

Histochemistry. In transverse sections of the human brainstem, the SpVi stands out in acetylcholinesterase stains on account of its moderately high content of acetylcholine and the occasional intense patches corresponding to either the parvicellular regions or, ventrally, to a group of small compact neurons [Paxinos and Huang, 1995; Koutcherov et al., 2004]. In the rat many cells have been found to contain parvalbumin, calbindin, GABA or glutamate, while somatostatin and enkephalin have been reported in the ferret [Waite and Ashwell, 2004].

Connections

Afferents. Inputs to the SpVi enter via the spinal trigeminal tract (see 'Sensory Systems – Trigeminal Complex: Overview' and TSpV No. 9) and carry ipsilateral primary afferents from the main cutaneous part of the face, including the whisker area of rats and cats. Proprioceptive afferents from eye muscles were reported to terminate in the ventrolateral SpVi [Porter, 1986; Donaldson, 2000]. Both SpVi and SpVo receive extensive inputs from intraoral structures including tooth pulp. Afferents were reported to be less densely represented in the SpVi than in the oral part [Takemura et al., 1991].

Interestingly, afferents from the periaqueductal gray (PAG) [Hayashi et al., 1984], and from the pedunculopontine nucleus (PPT) to the SpVi have been reported to target the SpVi [Timofeeva et al., 2005]. Both of these inputs are able to modulate the transmission of pain through the SpVi to higher centers.

Efferents. Major targets for SpVi neurons are the cerebellum, via direct pathways in the inferior cerebellar peduncle, and indirect pathways through the inferior olive, the spinal cord and the superior colliculus [Huerta et al., 1981, 1983; Huerta and Harting, 1984]. These projections originate mainly from the

SpVi and not the other regions of the SpV, and their possible function is considered above. In addition there are both SpVi descending pathways to the spinal cord, and ascending paralemniscal output pathways to the same thalamic regions targeted by the SpVc: the medial part of the posterior group and the nucleus ventralis posterior medialis [Magnusson et al., 1987; Chiaia et al., 1991]. Finally the SpVi also has prominent local intranuclear connections to other parts of the trigeminal complex such as the PrV and SpVc which may be important for trigeminal sensory integration. This could be particularly relevant since neuromodulatory projections from the PAG and PPT to the SpVi have recently been demonstrated (see above) [Jacquin et al., 1989a, b].

References

Chiaia NL, Rhoades RW, Bennett-Clark CA, Fish SE, Killackey HP: Thalamic processing of vibrissal information in the rat. I. Afferent input to the medial ventral posterior and posterior nuclei. J Comp Neurol 1991;314:201–216.

Donaldson IML: The functions of the proprioceptors of the eye muscles. Phil Trans R Soc London B 2000;355:1685–1754.

Dubner R, Bennett GJ: Spinal and trigeminal mechanisms of nociception. Annu Rev Neurosci 1983;6:381–418.

Hayashi H, Sumino R, Sessle BJ: Functional organization of trigeminal subnucleus interpolaris: nociceptive and innocuous afferent inputs, projections to thalamus, cerebellum, and spinal cord, and descending modulation from periaqueductal gray. J Neurophysiol 1984;51:890–905.

Huerta MF, Frankfurter AJ, Harting JK: The trigeminocollicular projection in the cat: patch-like endings within the intermediate gray. Brain Res 1981;211:1–13.

Huerta MF, Frankfurter A, Harting JK: Studies of the principal sensory and spinal trigeminal nuclei of the rat: projections to the superior colliculus, inferior olive and cerebellum. J Comp Neurol 1983;220:147–167.

Huerta MF, Harting JK: The mammalian superior colliculus: studies of its morphology and connections; in Vanegas H (ed): Comparative Neurology of the Optic Tectum. New York, Plenum Publishing Corporation, 1984.

Jacquin MF, Barcia M, Rhoades RW: Structure function relationship in rat brainstem subnucleus interpolaris. IV. Projection neurons. J Comp Neurol 1989a;282:45–62.

Jacquin MF, Golden J, Rhoades RW: Structure function relationship in rat brainstem subnucleus interpolaris. III. Local circuit neurons. J Comp Neurol 1989b;282:24–44.

Koutcherov Y, Huang X-F, Halliday G, Paxinos G: Organization of human brain stem nuclei; in Paxinos G, Mai JK (eds): The Human Nervous System, ed 2. Amsterdam, Elsevier Academic Press, 2004.

Magnusson KR, Clements JR, Larson AA, Madl JE, Beitz AJ: Localization of glutamate in trigeminothalamic projection neurons: a combined retrograde transport-immunohistochemical and immunoradiochemical study. Somatosens Res 1987;4:177–190.

Paxinos G, Huang X-F: Atlas of the Human Brainstem. San Diego, Academic Press, 1995.

Porter JD: Brainstem terminations of extraocular muscle primary afferent neurons in the monkey. J Comp Neurol 1986;247:133–143.

Ren K, Dubner R: The role of trigeminal interpolaris-caudalis transition zone in persistent orofacial pain. Int Rev Neurobiol 2011;97:207–225.

Takemura M, Sugimoto T, Shigenaga Y: Difference in central projection of primary afferent innervating facial and intraoral structures in the rat. Exp Neurol 1991;111:324–331.

Timofeeva E, Dufresne C, Sik A, Zhang ZW, Deschenes M: Cholinergic modulation of vibrissal receptive fields in trigeminal nuclei. J Neurosci 2005;25:9135–9143.

Waite PME, Ashwell KWS: Trigeminal sensory system; in Paxinos G, Mai JK (eds): The Human Nervous System, ed 2. Amsterdam, Elsevier Academic Press, 2004.

Willis WD, Westlund KN: Pain system; in Paxinos G, Mai JK (eds): The Human Nervous System, ed 2. Amsterdam, Elsevier Academic Press, 2004.

8 Spinal Trigeminal Nucleus, Oral Part (SpVo)

Original names:
Nucleus spinalis nervi trigemini oralis
Nucleus spinalis nervi trigemini, pars oralis

Location and Cytoarchitecture – Original Text

The spinal trigeminal nucleus oral part (SpVo) lies in the lateral tegmentum of the oral portion of the medulla and the caudal part of the pons (fig. 1–4); it extends from the oral pole of the spinal trigeminal nucleus interpolar part (SpVi) to the caudal pole of the principal sensory trigeminal nucleus (PrV).

On cross-section, the SpVo is oval in outline, and increases in size progressively from its caudal to its oral pole. Within the medulla the nucleus is related dorsomedially to the nucleus of the solitary tract (SOL), medially and ventrally to the parvocellular reticular nucleus (PCR), laterally to the spinal trigeminal tract (TSpV) and dorsally to fibers of the trigeminal nerve and to the spinal vestibular nucleus (SpV). The dorsomedial, medial, and ventral borders of this portion of the nucleus are difficult to define due to the fact that the cells of the SOL and PCR are also small and lightly stained. The latter nucleus can usually be distinguished by the fact that its cells are larger and less compactly arranged than those of the SpVo or the SOL. Oral to the level at which the SOL disappears, the SpVo is related dorsally to the medial vestibular nucleus and dorsolaterally to fibers of the vestibular nerve. Within the pons, the nucleus is related laterally to the TSpV, ventrally to the facial nucleus, medially to the PCR and dorsally to the lateral vestibular nucleus and rootlets of the vestibular nerve.

The SpVo is composed of densely arranged, irregularly oriented, small and oral ends of the nucleus; the cells are larger and less compactly arranged [...].

The cells of the spinal trigeminal complex give rise to quintothalamic fibers which proceed to the ventral posteromedial nucleus of the contralateral thalamus [Waite and Ashwell, 2004]. In the monkey, according to Walker [1939], these fibers decussate and then proceed orally in the white matter between the inferior olivary complex and the pyramid. Farther orally they are said to be associated with the spinothalamic tracts along the lateral margin of the medial lemniscus [...].

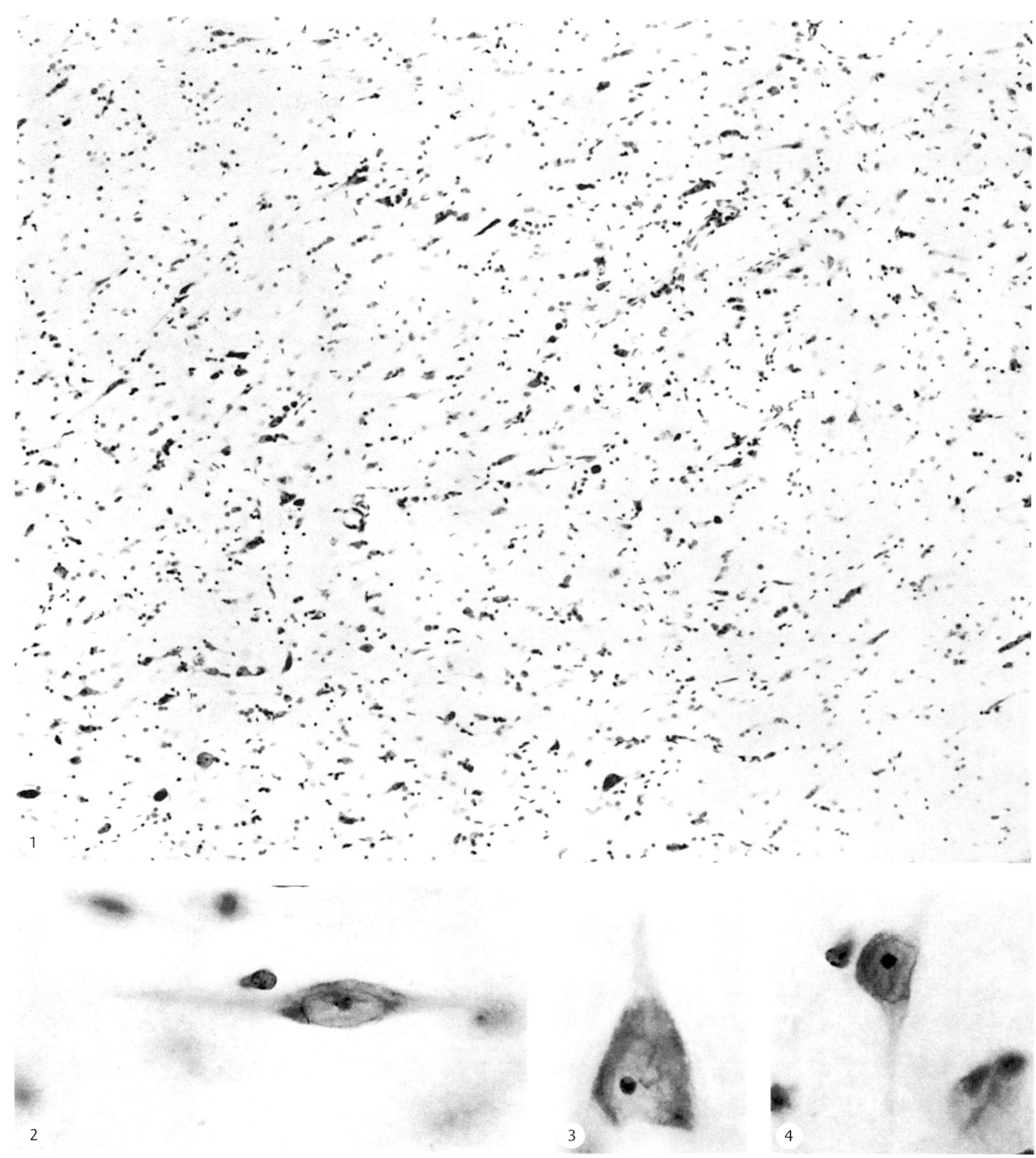

Fig. 1. Spinal trigeminal nucleus, oral part. Magnification ×150.
Fig. 2–4. Cells from the spinal trigeminal nucleus, oral part. Magnification ×1,000.

Functional Neuroanatomy

The SpVo participates in the transmission of painful stimuli from internal areas of the nose and mouth. It receives extensive inputs from intraoral structures including tooth pulp and periodontium [Jacquin and Rhoades, 1990; Waite and Ashwell, 2004], and in humans lesions of the SpVo are associated with the loss of oral sensation.

The spinal trigeminal nucleus caudal part (SpVc) is closely associated with processing nociceptive signals of the forehead, cheek and jaw. Its cytoarchitecture corresponds exactly with that of the apex of the dorsal horn of the spinal cord, whereas those of the SpVo and SpVi are entirely different, implying a functional difference, but no functional differences between 'SpVc' and 'SpVi and SpVo' are clear [Guy et al., 2005].

The specific projections of the SpVo have been studied in rats [Ruggiero et al., 1981], where strong pathways to the facial nucleus and spinal cord were demonstrated, with less dense projections to the cerebellum, thalamus and motor trigeminal nucleus.

A review of the chemoarchitecture of the SpVo shows it to have similar properties to those of the SpVi and the adjacent PrV [Waite and Ashwell, 2004].

References

Guy N, Chalus M, Dallel R, Voisin DL: Both oral and caudal parts of the spinal trigeminal nucleus project to the somatosensory thalamus in the rat. Eur J Neurosci 2005;21:741–754.

Jacquin MF, Rhoades RW: Cell structure and response properties in the trigeminal subnucleus oralis. Somatosens Motor Res 1990;7:265–288.

Ruggiero DA, Ross CA, Reis DJ: Projections from the spinal trigeminal nucleus to the entire length of the spinal cord in the rat. Brain Res 1981;225:225–233.

Waite PME, Ashwell KWS: Trigeminal sensory system; in Paxinos G, Mai JK (eds): The Human Nervous System, ed 2. Amsterdam, Elsevier Academic Press, 2004.

Walker AE: The origin, course and terminations of the secondary pathways of the trigeminal nerve in primates. Neurology 1939;71:59–89.

Sensory Systems – Trigeminal Complex (Plates 10–22)

9 Spinal Trigeminal Tract (TSpV)

Original name:
Tractus nervi trigemini spinalis

Alternative names:
Spinal tract of trigeminal nerve
Trigeminospinal tract

Functional Neuroanatomy

Fibers supplying all 3 nuclei of the spinal trigeminal complex enter the brainstem via the sensory divisions of the trigeminal nerve and to a much lesser extent via the facial, glossopharyngeal and vagus nerves (see 'Sensory Systems – Trigeminal Complex: Overview') [Brodal, 1947; Olszewski and Baxter, 1982; Waite and Ashwell, 2004; Nieuwenhuys, 2008]. Many of the trigeminal fibers bifurcate, giving rise to a short branch which passes to the main sensory trigeminal nucleus, and to a long branch which descends into the 'spinal tract of the trigeminal nerve' (TSpV). Finely myelinated or unmyelinated fibers which do not bifurcate also enter the TSpV. It is probable that the bifurcating fibers mediate the sensation of touch whereas the finer, nonbifurcating fibers mediate pain and temperature. Within the TSpV there exists an apparent lamination of the fibers from the 3 sensory divisions of trigeminal nerve, resulting in topography – with the chin up! Thus, the fibers from the ophthalmic division are ventrally placed, those from the mandibular division are dorsally placed and those from the maxillary division are intermediate in position. This results from a medial rotation of the trigeminal root as it enters the pons, and persists throughout the length of the TSpV [Warren et al., 2006]. Although collaterals from these fibers enter the spinal trigeminal complex at all levels, the terminals from all 3 divisions descend into the caudal part of the medulla, and some pass into the first cervical segment before terminating about the cells of the spinal trigeminal complex [Jacobs, 1970; Nieuwenhuys et al., 2008].

A group of small chromophobic interstitial neurons lies within the TSpV at the level of the caudal SpV interpolar part [Ramón y Cajal, 1909], and experimental studies indicate that they may receive projections from the tooth pulp and cornea [Waite and Ashwell, 2004; Shigenaga et al., 1988].

In the floor of the fourth ventricle there is a longitudinal protuberance on the dorsolateral surface of the medulla oblongata, caudal to the obex, called the trigeminal tubercle (tuberculum cinereum). It lies lateral to the cuneate tubercle, and provides a useful landmark for surgeons endeavoring to section the underlying TSpV to relieve ipsilateral facial pain (tractotomy) [Morita and Hosobuchi, 1992]. However, the results of trigeminal tractotomies in general indicate that nociceptive pathways are more complicated than originally thought [Dallel et al., 1988; Young, 1982].

References

Brodal A: Central course of afferent fibers for pain in fascial, glossopharyngeal and vagus nerves. Arch Neurol Psychiatry 1947;57:292–306.

Dallel R, Raboisson P, Auroy P, Woda A: The rostral part of the trigeminal sensory complex is involved in orofacial nociception. Brain Res 1988;448:7–19.

Jacobs MJ: The development of the human motor trigeminal complex and accessory facial nucleus and their topographic relations with the facial and abducens nuclei. J Comp Neurol 1970;138:161–194.

Morita M, Hosobuchi Y: Descending trigeminal tractotomy for trigeminal neuralgia after surgical failure. Stereotact Funct Neurosurg 1992;59:52–55.

Nieuwenhuys R: General sensory systems and taste; in Nieuwenhuys R, Voogd J, van Huijzen C (eds): The Human Central Nervous System. Berlin, Springer, 2008.

Nieuwenhuys R, Voogd J, Van Huijzen C: The Human Central Nervous System. Berlin, Springer, 2008.

Olszewski J, Baxter D: Cytoarchitecture of the Human Brain Stem. Basel, Karger, 1982.

Ramón y Cajal S: Histologie du système nerveux de l'homme et des vertébrés. Paris, Maloine, 1909.

Shigenaga Y, Mitsuhiro Y, Yoshida A, Cao CQ, Tsuru H: Morphology of single mesencephalic trigeminal neurons innervating masseter muscle of the cat. Brain Res 1988;445:392–399.

Waite PME, Ashwell KWS: Trigeminal sensory system; in Paxinos G, Mai JK (eds): The Human Nervous System, ed 2. Amsterdam, Elsevier Academic Press, 2004.

Warren S, Yezierski RP, Capra NF: The somatosensory system II: touch, thermal sense, and pain; in Haines DE (ed): Fundamental Neuroscience for Basic and Clinical Applications, ed 3. Philadelphia, Churchill Livingstone Elsevier, 2006.

Young RF: Effect of trigeminal tractotomy on dental sensation in humans. Neurosurg 1982;56:812–818.

10 Principal Sensory Trigeminal Nucleus (PrV)

Original name:
Nucleus nervi trigemini sensibilis principalis

Alternative names:
Main sensory trigeminal nucleus [Olszewski and Baxter, 1982]
Principal (superior) trigeminal sensory nucleus

Location and Cytoarchitecture – Original Text

This nucleus lies in the lateral portion of the midpontine tegmentum. Caudally, it is continuous with the spinal trigeminal nucleus oral part, and orally it terminates about 1 mm caudal to the oral pole of the motor trigeminal nucleus (MoV). It measures 5 mm in the orocaudal direction.

On cross-section the principal sensory trigeminal nucleus (PrV) is oblong or ovoid in outline. Its caudal portion is related medially to the parvocellular reticular nucleus, dorsally and laterally to fibers of the spinal trigeminal tract and ventrally to the lateral tegmental process of the pons. More orally the nucleus lies partially surrounded by afferent and efferent fibers of the trigeminal or fifth cranial nerve. These fibers separate the nucleus medially from the MoV and laterally from the brachium pontis. Dorsally this oral portion of the PrV is related to the superior vestibular nucleus.

The cells composing this nucleus are small, oval or round, and contain relatively large nuclei and diffusely stained Nissl substance. These cells tend to congregate in irregular groups or clusters (fig. 1–3). A detailed study of this nucleus in humans has focused on its morphometry and development [Hamano et al., 1997].

Functional Neuroanatomy

Function
The PrV receives primary sensory information concerning discriminative touch and proprioception of the ipsilateral face region, from all 3 trigeminal nerve branches, and it relays the information to the thalamus and higher centers [Kamitani et al., 2004; Waite and Ashwell, 2004; Nieuwenhuys et al., 2008]. The neurons of the PrV are thought to respond to a wide range of pressure stimuli without adaptation. The nucleus plays an important role in feeding behavior, and in some animals becomes enlarged with the pressure of specific feeding habits [Gutierrez-Ibanez et al., 2009].

Connections
The topography within the PrV is, like the spinal trigeminal nucleus, chin up (see 'Sensory Systems – Trigeminal Complex: Overview').
Afferents. A large proportion of the trigeminal afferents to the PrV arise from sensory fibers innervating the face, jaw and oral cavity. Apart from these, few other inputs to the nucleus have been described. However, some inputs to the PrV from the spinal trigeminal nucleus caudal part, and the cerebral cortex have been reported [Ganchrow, 1978; Künzle and Lotter, 1996].

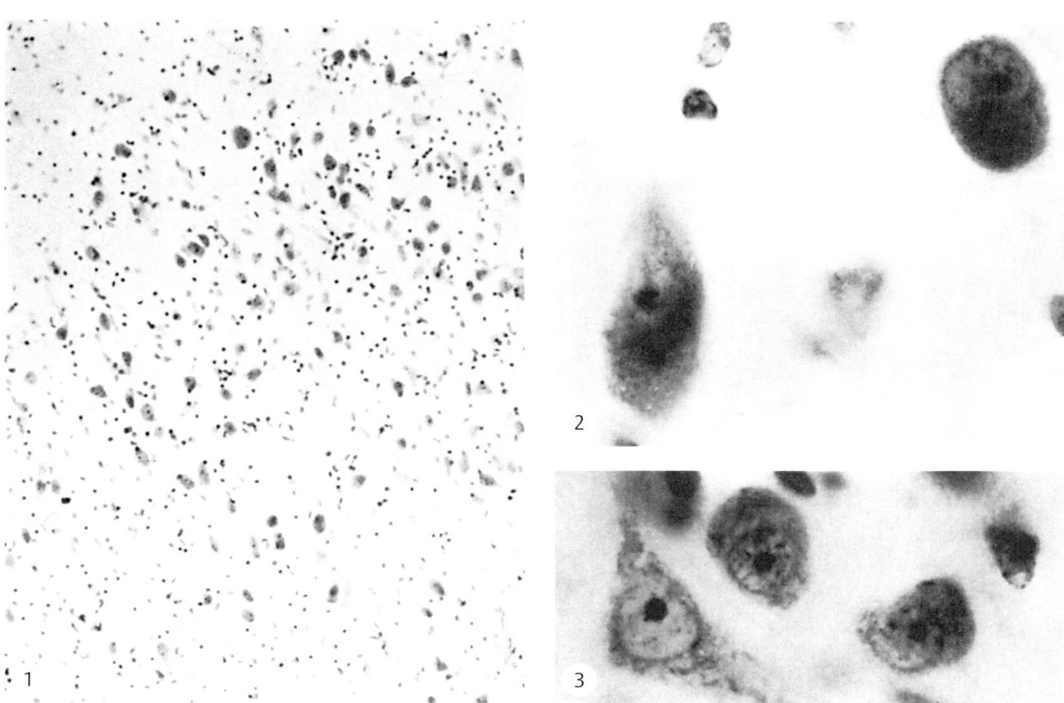

Fig. 1. Principal sensory trigeminal nucleus. Magnification ×150.
Fig. 2, 3. Cells from the principal sensory trigeminal nucleus. Magnification ×1,000.

Efferents. The PrV gives rise to 2 ascending pathways: the major crossed pathway which joins the contralateral medial lemniscus medially, terminates in the ventral posteromedial nucleus of the thalamus, and at the same time sends a significant projection to the superior colliculus [Huerta et al., 1983; Kivrak and Erzurumlu, 2012], and an uncrossed pathway, the dorsal trigeminothalamic tract, which arises from the dorsal part of the PrV and carries information from intraoral structures [Smith, 1975, 2004; Matsushita et al., 1982; Haines, 2006]. An area in the PrV and the adjacent oral spinal trigeminal nucleus send projections to motoneurons participating in the blink reflex, i.e. to motoneurons of the orbicularis oculi in the facial nucleus, the nictitating membrane motoneurons in the accessory abducens nucleus (in rabbits but not primates), and levator palpebrae motoneurons in the oculomotor nucleus. Presumably, the oculomotor input provides an inhibition to lower the upper eyelid [May, 1995; van Ham and Yeo, 1996]. These connections and those of the adjacent pontine reticular formation, called the 'pontine blink premotor area', coordinate the different (R1 and R2) components of the blink reflex (for a review, see Horn and Adamczyk [2012]).

References

Ganchrow D: Intratrigeminal and thalamic projections of nucleus caudalis in the squirrel monkey *(Saimiri sciureus)*: a degeneration and autoradiographic study. J Comp Neurol 1978;178:281–312.

Gutierrez-Ibanez C, Iwaniuk AN, Wylie DR: The independent evolution of the enlargement of the principal sensory nucleus of the trigeminal nerve in three different groups of birds. Brain Behav Evol 2009;74:280–294.

Haines DE: Fundamental Neuroscience for Basic and Clinical Applications. Philadelphia, Churchill Livingstone Elsevier, 2006.

Hamano S, Goto N, Nara T, Okada A, Maekawa K: Development of the human principal sensory trigeminal nucleus: a morphometric analysis. Early Hum Dev 1997;48:225–235.

Horn AK, Adamczyk C: Reticular formation: eye movements, gaze and blinks; in Mai JK, Paxinos G (eds): The Human Nervous System, ed 3. San Diego, Academic Press, 2012.

Huerta MF, Frankfurter A, Harting JK: Studies of the principal sensory and spinal trigeminal nuclei of the rat: projections to the superior colliculus, inferior olive and cerebellum. J Comp Neurol 1983;220:147–167.

Kamitani T, Kuroiwa Y, Hidaka M: Isolated hypesthesia in the right V2 and V3 dermatomes after a midpontine infarction localised at an ipsilateral principal sensory trigeminal nucleus. J Neurol Neurosurg Psychiatry 2004;75:1508–1509.

Kivrak BG, Erzurumlu RS: Development of the principal nucleus trigeminal lemniscal projections in the mouse. J Comp Neurol 2012;521:299–311.

Künzle H, Lotter G: Efferents from the lateral frontal cortex to spinomedullary target areas, trigeminal nuclei, and spinally projecting brainstem regions in the hedgehog tenrec. J Comp Neurol 1996;372:88–110.

Matsushita M, Ikeda M, Okado N: The cells of origin of the trigeminothalamic, trigeminospinal and trigeminocerebellar projections in the cat. Neuroscience 1982;7:1439–1454.

May PJ: Sources of premotor input directing primate eyelid motoneurons. Invest Ophthalmol Vis Sci 1995;37:154.

Nieuwenhuys R, Voogd J, van Huijzen C: The Human Central Nervous System. Berlin, Springer, 2008.

Olszewski J, Baxter D: Cytoarchitecture of the Human Brain Stem. Basel, Karger, 1982.

Smith RL: Axonal projections and connections of the principal sensory trigeminal nucleus in the monkey. J Comp Neurol 1975;163:347–374.

Smith RL: The ascending fiber projections from the principal sensory trigeminal nucleus in the rat. J Comp Neurol 2004;148:423–445.

van Ham JJ, Yeo CH: Trigeminal inputs to eyeblink motoneurons in the rabbit. Exp Neurol 1996;142:244–257.

Waite PME, Ashwell KWS: Trigeminal sensory system; in Paxinos G, Mai JK (eds): The Human Nervous System, ed 2. Amsterdam, Elsevier Academic Press, 2004.

Sensory Systems – Trigeminal Complex (Plates 24, 26–30, 32–34 and 36)

11 Mesencephalic Trigeminal Nucleus (MesV)

Original name:

Nucleus nervi trigemini mesencephalicus

Alternative names:

Nucleus mesencephalicus nervi trigemini
Mesencephalic nucleus of the trigeminal nerve
Mesencephalic nucleus of the fifth nerve

Location and Cytoarchitecture – Original Text

The cells composing this nucleus make their appearance at the level of the root entrance zone of the trigeminal, or fifth cranial, nerve (NV) and extend orally to levels through the oral part of the superior colliculus – a distance of 22 mm. In cross-sections through the oral part of the pons, the mesencephalic trigeminal nucleus (MesV) usually appears as a cluster of 2–15 cells lying medial to the fibers of the mesencephalic root and separated from the lateral angle of the fourth ventricle by the central gray matter. Laterally, the nucleus and its associated root are related to the medial parabrachial nucleus and the superior cerebellar peduncle. The ventral relations of the nucleus are inconstant due to the variability in its dorsoventral situation from section to section, except at levels in which the locus coeruleus (LC) is well developed. Here, the pigmented LC cells partially envelop the ventral pole of the MesV. Cells morphologically similar to those of the MesV are occasionally found lying in the anterior medullary velum.

In cross-sections through the midbrain, the cells of the MesV are less compactly arranged, less numerous and do not regularly appear in each cross-section. When present, the cells are invariably located along the outer border of the periaqueductal gray matter, dorsal or lateral to the cerebral aqueduct. From 1 to 15 such cells may appear in a single cross-section. The cells are most numerous at the level of the oculomotor (III) and trochlear nuclei (IV). Pearson has described cells similar to those of the MesV as appearing in the nuclei and along the intracranial course of the third [Houser et al., 1980] and fourth [Pearson, 1944] cranial nerves in human fetuses.

The morphological features of the individual cells of the MesV are quite distinctive. They are large, oval or round, and regular in outline. Fine granules of Nissl substance are regularly distributed throughout the cytoplasm, and larger, more darkly stained granules are located at the

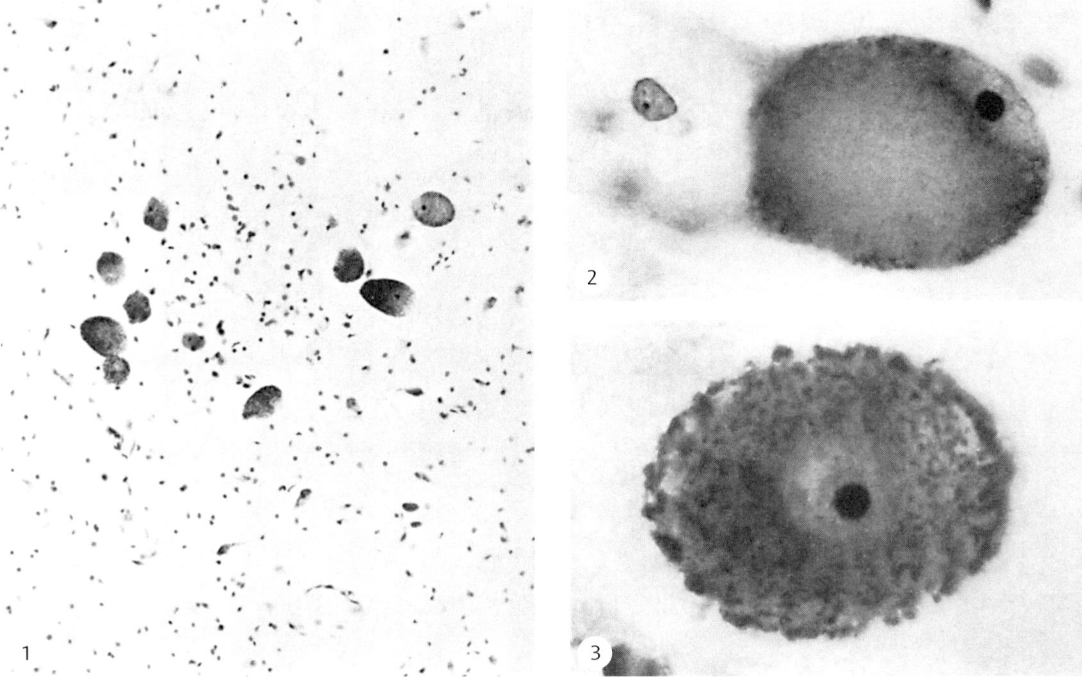

Fig. 1. Mesencephalic trigeminal nucleus. Magnification ×150.
Fig. 2, 3. Cells of the mesencephalic trigeminal nucleus. Note that the cell in figure 2 has at least 2 processes. Magnification ×1,000.

periphery of the cell (fig. 1–3). A few cells, particularly those closely associated with the cells of the LC may contain variable amounts of melanin. The nucleus, which is large, may be central or excentric. These cells are often described as unipolar in the adult human and are compared to cells of the dorsal root ganglia. Nevertheless, even in Nissl-stained preparations one may occasionally distinguish bipolar or multipolar cells (fig. 2).

The peripheral processes of the cells of the MesV are distributed in both the motor and sensory divisions of the NV [Pearson, 1949a], and possibly contribute to other cranial nerves as well. The trigeminal fibers are believed to conduct proprioceptive impulses from the muscles of mastication, the teeth and the hard palate. It is believed by many that proprioceptive impulses from the eye muscles, presumably travelling in the oculomotor and trochlear nerves, also reach cells which are morphologically similar to those of the mesencephalic trigeminal nucleus. Such cells may or may not be located within the usual boundaries of this nucleus [Pearson, 1949a, b).

Other processes derived from the cells of the MesV are believed to be distributed to the motor nuclei of III, IV, motor trigeminal nucleus (MoV) and facial cranial nerves and to the cerebellum.

Functional Neuroanatomy

Function

The main function of the MesV is the control of trigeminal motoneurons in the MoV during chewing and biting, as well as the modulation of oral-motor behavior, such as swallowing, feeding behavior and speech. The MesV consists of a separate population of primary sensory neurons, in addition to those located in the trigeminal ganglion (semilunar or Gasser's ganglion). Whereas the sensory neurons of the trigeminal ganglion innervate mechanopreceptors, thermoreceptors and nociceptors of the head region, those of the MesV carry proprioceptive information from muscle spindles, mainly from the muscles of mastication (80–90%), but also from extraocular muscles if muscle spindles are present in the species [Wang and May, 2008; Waite and Ashwell, 2012]. Ganglion cells of the sensory and kinoreceptors of the teeth also lie in the MesV. The sensory information is relayed through the centrally projecting processes of the primary neurons to the trigeminal motoneurons directly and indirectly through the surrounding trigeminal premotor areas, such as the supratrigeminal, juxtatrigeminal and intertrigeminal areas. For a review, see Lazarov [2007] or Waite and Ashwell [2012].

The MesV drives the jaw-closing motoneurons of the MoV mono- and multisynaptically for the 'masseter reflex', also called the 'jaw-jerk reflex'. A gentle tap on the point of the chin activates the muscle spindles in the jaw-closing muscles (temporal, masseter and pterygoid), which project through the MesV to the MoV, and leads to a quick reflex jaw closure. The reflex is highly useful clinically to localize the level of a brain dysfunction; a heightened reflex indicates the loss (or abnormality) of inputs from higher centers to the MoV.

Development

The pseudounipolar MesV neurons develop from the neural crest in the typical manner of ganglion cells, and they retain the characteristic cytoarchitecture of ganglia, but unlike other ganglion cells they remain within the neural tube. The MesV constitutes the only centrally lying group of ganglion cells that is well- documented.

The development of the MesV is dependent on the expression of Krox-20 [De et al., 2005], and suggests that some MesV components are derived from the hindbrain rather than the mesencephalon [De et al., 2005].

Connections

Afferents. The peripheral processes of the MesV neurons provide the major proprioceptive input to the nucleus. They travel ipsilaterally to the MesV, where periodontal cells are found caudally and jaw cells are scattered throughout the length of the nucleus [Luschei, 1987]. In addition to these large neurons, small GABAergic cells are also found in the MesV [Lazarov, 2007].

One important difference between MesV cells and typical ganglion cells is that MesV soma receive synaptic inputs, as originally proposed by Ramón y Cajal [Lazarov, 2007]. The axosomatic and dendritic contacts arise from monoaminergic and peptidergic sources in the brainstem, and are cabale of modulating the proprioceptive signals in the MesV (see also 'Neuromodulatory Systems: Overview'). Furthermore, the electrotonic coupling, or gap junctions, described between MesV neurons may enable the nucleus to operate as an integral unit. A hypothalamic input to the MesV has also been reported and suggested to be able to influence feeding behavior among other things [Nagy et al., 1986].

Efferents. Projections have been followed from the MesV to the lateral MoV, and also to the supratrigeminal and intertrigeminal areas, the parvocellular reticular region and to the Probst nucleus [Luschei, 1987; Shigenaga et al., 1988; Holstege, 1991; Nieuwenhuys et al., 2008] (see fig. 21.16 of Nieuwenhuys et al. [2008]). There is evidence that all of these relay nuclei are also involved in the control of mastication, swallowing and other orofacial reflexes, via connections to the MoV [Cunningham and Sawchenko, 2000]. It is interesting that in the primate the labeling from the MesV to the supratrigeminal area and a region immediately dorsolateral to the MoV were even stronger than the projections to the MoV itself.

The descending fibers from MesV axons run in the mesencephalic trigeminal tract, also called Probst's tract (see nucleus No. 12). Some fibers descend to the level of the cervical cord [Nomura and Mizuno, 1985]. Unlike the latter authors, other studies found no projections to the cerebellum from the MesV [Shigenaga et al., 1988]. Ascending projections from the MesV have been reported to terminate in the thalamus bilaterally [Pearson and Garfunkel, 1983].

References

Cunningham ET Jr, Sawchenko PE: Dorsal medullary pathways subserving oromotor reflexes in the rat: implications for the central neural control of swallowing. J Comp Neurol 2000;417:448–466.

De S, Nguyen AQ, Shuler CF, Turman JEJ: Mesencephalic trigeminal nucleus development is dependent on Krox-20 expression. Dev Neurosci 2005;27:49–58.

Holstege G: Descending motor pathways and the spinal motor system: limbic and non-limbic components. Prog Brain Res 1991;87:307–421.

Houser CR, Vaughn JE, Barber RP, Roberts E: GABA neurons are the major cell type of the nucleus reticularis thalami. Brain Res 1980;200:341–354.

Lazarov NE: Neurobiology of orofacial proprioception. Brain Res Rev 2007;56:362–383.

Luschei ES: Central projections of the mesencephalic nucleus of the fifth nerve: an autoradiographic study. J Comp Neurol 1987;263:137–145.

Nagy JI, Buss M, Daddona PE: On the innervation of trigeminal mesencephalic primary afferent neurons by adenosine deaminase-containing projections from the hypothalamus in the rat. Neuroscience 1986;17:141–156.

Nieuwenhuys R, Voogd J, van Huijzen C: The Human Central Nervous System. Berlin, Springer, 2008.

Nomura S, Mizuno N: Differential distribution of cell bodies and central axons of mesencephalic trigeminal nucleus neurons supplying the jaw-closing muscles and periodontal tissue: a transganglionic tracer study in the cat. Brain Res 1985;359:311–319.

Pearson AA: The oculomotor nucleus in the human fetus. J Comp Neurol 1944;80:47–63.

Pearson AA: Development and connections of the mesencephalic root of the trigeminal nerve in man. J Comp Neurol 1949a;90:1–46.

Pearson AA: Further observations on the mesencephalic root of the trigeminal nerve. J Comp Neurol 1949b;91:147–194.

Pearson JC, Garfunkel DA: Evidence for thalamic projections from external cuneate nucleus, cell groups Z and X, and the mesencephalic nucleus of the trigeminal nerve in squirrel monkey. Neurosci Lett 1983;41:41–47.

Shigenaga Y, Mitsuhiro Y, Yoshida A, Cao CQ, Tsuru H: Morphology of single mesencephalic trigeminal neurons innervating masseter muscle of the cat. Brain Res 1988;445:392–399.

Waite PME, Ashwell KWS: Trigeminal sensory system; in Mai JK, Paxinos G (eds): The Human Nervous System, ed 3. Amsterdam, Academic Press, 2012.

Wang N, May PJ: Peripheral muscle targets and central projections of the mesencephalic trigeminal nucleus in macaque monkeys. J Comp Neurol 2008;291:974–987.

Sensory Systems – Trigeminal Complex (Plates 24–29)

12 Mesencephalic Tract of the Trigeminal Nerve (TMesV)

Original name:
Tractus nervi trigemini mesencephalicus

Associated name:
Tract of Probst

Functional Neuroanatomy

The fibers from the jaw muscle stretch-receptors, as well as fibers conveying pressure and kinesthesis of the teeth, sensory receptors from the periodontium and joint receptors, all enter the pons via the trigeminal nerve and collect at the border of the tegmentum with the pontine central gray along with the sensory ganglion cells, the mesencephalic trigeminal nucleus (MesV). Some fibers arising from facial, lingual and extraocular muscles are also thought to use these pathways [Wang and May, 2008]. The ascending fibers build the mesencephalic trigeminal tract, and it is surrounded by the sensory ganglion cells from the caudal levels up to its rostral tip, which lies at the level of the superior colliculus, exactly at the border of the periaqueductal gray and the mesencephalic reticular formation. This is the only known example of sensory ganglion cells within the brainstem up to now, but other examples may well be found.

The descending fibers of the MesV follow the medial border of the sensory spinal trigeminal nucleus. At levels caudal to the trigeminal motor nucleus (MoV), the tract forms the so-called tract of Probst [1899] which can be followed to upper cervical levels in the cat [Holstege et al., 1990]. In the medulla oblongata, the tract of Probst is less compact; and lies subjacent to the solitary nucleus (not on plates). Collaterals of the tract terminate in the jaw-closing motoneuron subdivisions of the MoV, in the supra- and intertrigeminal regions, and at the level of the dorsal motor nucleus of the vagus nerve in the dorsolateral medullary reticular formation. These MesV-innervated areas provide the mono- and polysynaptic pathways for the classical myotatic jaw-closing reflex (jaw jerk), which is an invaluable neurological tool used to test the general integrity of neural structures rostral to the pontomedullary junction [Holstege et al., 1990; Nieuwenhuys et al., 2008]. Jaw-opening reflexes utilize different pathways [Lund and Olsson, 1983].

Cytologically no supratrigeminal area has been recognized in humans up to now [Usunoff et al., 1997] although Paxinos and Huang [1995] outline a peritrigeminal zone.

References

Holstege G, Blok BFM, ter Horst GJ: Brain stem systems involved in the blink reflex, feeding mechanisms, and micturition; in Paxinos G (ed): The Human Nervous System. San Diego, Academic Press, 1990.

Lund JP, Olsson KA: The importance of reflexes and their control during jaw movements. Trends Neurosci 1983;11:458–463.

Nieuwenhuys R, Voogd J, van Huijzen C: The Human Central Nervous System. Berlin, Springer, 2008.

Paxinos G, Huang X-F: Atlas of the Human Brainstem. San Diego, Academic Press, 1995.

Probst M: Über vom Vierhügel, von der Brücke und vom Kleinhirn absteigende Bahnen: Monakow'sches Bündel, Vierhügel-Vorderstrangbahn, Kleinhirn-Vorderstrangbahn, dorsales Längsbündel, cerebrale Trigeminuswurzel und andere motorische Haubenbündel. Dtsch Z Nervenheilkd 1899;15:192–221.

Usunoff KG, Marani E, Schoen JH: The trigeminal system in man. Adv Anat Embryol Cell Biol 1997;136:I–X, 1–126.

Wang N, May PJ: Peripheral muscle targets and central projections of the mesencephalic trigeminal nucleus in macaque monkeys. J Comp Neurol 2008;291: 974–987.

Sensory Systems – Vestibular Nuclei

(Fig. 1, p. 18)

Overview

The functions of the vestibular complex are well studied. The complex is divided into over 6 brainstem nuclei on the basis of cytoarchitecture, and these nuclei are identifiable in all vertebrates. However, it is difficult to associate specific functions with the individual nuclei, and even when this is possible, the functions are rarely confined to any cytoarchitectural borders [Büttner-Ennever, 1992]. Despite this complexity the subdivisions of the vestibular nuclei are of great importance for understanding, and ordering the vast amount of information that has been gathered. For these reasons a short overview will be presented here on the functional organization of the vestibular system, which is relevant for all parts of the complex described below. Several extensive reviews have covered the history, topography, cytochemistry and connections of the vestibular nuclei in detail [Neuhuber and Bankoul, 2000; Büttner-Ennever and Gerrits, 2004; Highstein and Holstein, 2006; Nieuwenhuys et al., 2008; Goldberg et al., 2011].

Vestibular signals together with visual and proprioceptive inputs are essential for stabilizing the body, head and eyes during movement in space. The vestibular nerve carries sensory afferents from the 3 semicircular canals of the labyrinth (the anterior, posterior and horizontal canals), each encoding the magnitude and direction of angular acceleration of the head in space, in one specific plane. The nerve also contains afferent fibers from the otoliths, the utriculus and sacculus, which detect linear acceleration and produce signals encoding horizontal and vertical head position, respectively. The bipolar ganglion cells of the vestibular fibers lie near their entrance to the internal auditory meatus, in Scarpa's ganglion, ordered in a strict topography [Gacek, 1969; Sando et al., 1972]. The vestibular nerve also carries *efferent* fibers arising from the group-e near the abducens nucleus [Goldberg et al., 2000].

The vestibular nerve joins with the cochlear and facial nerves, and together they enter the rostral medulla as the eighth cranial nerve. Each fiber from the labyrinth bifurcates into an ascending and a descending branch. The ascending branch proceeds to the superior vestibular nucleus (SVN) and the cerebellum, and the descending branch distributes to the medial and descending vestibular nuclei (MVN and DVN). In addition to vestibular information, the vestibular complex receives proprioceptive signals from the neck and spinal cord, visual activity from the accessory optic system [Gamlin, 2006], autonomic afferents as well as pathways from the cerebral cortex and cerebellum. The strong reciprocal connections between the cerebellum and the vestibular complex have led to the vestibular nuclei being sometimes described as 'an external cerebellar nucleus'. A striking feature of the vestibular nuclei is the massive commissural system that interconnects the two sides. As in the auditory system this plays a fundamental role in the improvement of sensory processing and compensatory mechanisms [Malinvaud et al., 2010]. The output of the vestibular complex influences eye movements (the vestibulo-ocular reflex), head and neck movements as well as posture, serving to stabilize the body and images on the retina [Büttner-Ennever, 2006; Shinoda et al., 2006]. It also conveys information to higher centers for the conscious perception of orientation [Guldin and Grüsser, 1998; Shiroyama et al., 1999], and in turn areas of the cerebral cortex modulate the vestibular complex [Akbarian et al., 1994]. Finally a close association has recently been emphasized between the caudal vestibular complex and sympathetic brainstem centers controlling respiration and blood pressure [Holstein et al., 2011].

References

Akbarian S, Grüsser OJ, Guldin WO: Corticofugal connections between the cerebral cortex and brainstem vestibular nuclei in the macaque monkey. J Comp Neurol 1994;339:421–437.

Büttner-Ennever JA: Patterns of connectivity in the vestibular nuclei. Ann NY Acad Sci 1992;656:363–378.

Büttner-Ennever JA: The extraocular motor nuclei: organization and functional neuroanatomy. Prog Brain Res 2006;151:95–125.

Büttner-Ennever JA, Gerrits NM: Vestibular system; in Paxinos G, Mai JK (eds): The Human Nervous System, ed 2. Amsterdam, Elsevier Academic Press, 2004.

Gacek RR: The course and central termination of first order neurons supplying vestibular end organs in the cat. Acta Otolaryngol 1969;254:1–66.

Gamlin PDR: The pretectum: connections and oculomotor-related roles. Prog Brain Res 2006;151:379–405.

Goldberg JM, Brichta AM, Wackym PA: Efferent vestibular system: anatomy, physiology, and neurochemistry; in Beitz AJ, Anderson JH (eds): Neurochemistry of the Vestibular System. Boca Raton, CRC Press, 2000.

Goldberg SJ, Wilson VJ, Cullen KE, Angelaki DE, Broussard DM, Büttner-Ennever J, Fukushima K, Minor LB: The Vestibular System: A Sixth Sense. New York, Oxford University Press, 2011.

Guldin W, Grüsser OJ: Is there a vestibular cortex? Trends Neurosci 1998;254–259.

Highstein SM, Holstein GR: The anatomy of the vestibular nuclei. Prog Brain Res 2006;151:157–203.

Holstein GR, Friedrich VLJ, Kang T, Kukielka E, Martinelli GP: Direct projections from the caudal vestibular nuclei to the ventrolateral medulla in the rat. Neuroscience 2011;175:104–117.

Malinvaud D, Vassias I, Reichenberger I, Rössert C, Straka H: Functional organization of vestibular commissural connections in frog. J Neurosci 2010;30:3310–3325.

Neuhuber WL, Bankoul S: Somatosensory influences on the vestibular system; in Beitz AJ, Anderson JH (eds): Neurochemistry of the Vestibular System. Boca Raton, CRC Press, 2000.

Nieuwenhuys R, Voogd J, van Huijzen C: The Human Central Nervous System. Berlin, Springer, 2008.

Sando I, Black FO, Hemenway WG: Spatial distribution of vestibular nerve in internal auditory canal. Ann Otolaryngol 1972;81:305–314.

Shinoda Y, Sugiuchi Y, Izawa Y, Hata Y: Long descending motor tract axons and their control of neck and axial muscles. Prog Brain Res 2006;151:527–563.

Shiroyama T, Kayahara T, Yasui Y, Nomura J, Nakano K: Projections of the vestibular nuclei to the thalamus in the rat: a Phaseolus vulgaris leucoagglutinin study. J Comp Neurol 1999;407:318–332.

Sensory Systems – Vestibular Nuclei (Plates 12–20)

13 Medial Vestibular Nucleus (MVN)

Original names:
Nucleus vestibularis medialis
Nucleus interpositus (now marginal zone of medial vestibular nucleus)

Alternative names:
Triangular nucleus (no longer in use)
Nucleus of Schwalbe (no longer in use)

Location and Cytoarchitecture – Original Text

This nucleus lies directly beneath the lateral part of the floor of the fourth ventricle in the upper medulla and lower pons. Caudo-orally, the nucleus measures 9 mm and extends from 1 mm caudal to the oral pole of the hypoglossal nucleus to the level of the caudal pole of the abducens nucleus. On cross-section, the nucleus is roughly triangular in outline with its apex directed medially.

The caudal part of the nucleus lies dorsolateral to the nucleus of the solitary tract (SOL) and medial to the oral pole of the lateral cuneate nucleus (LCU). The oral part of the nucleus shifts slightly medially, and its medial part, the 'marginal zone', is related medially to the prepositus nucleus (PrP), ventrally to the parvocellular reticular nucleus, and laterally to the descending vestibular (DVN) and the lateral vestibular nuclei (LVN) consecutively. With the appearance of the abducens nucleus (VI), the medial vestibular nucleus (MVN) rapidly diminishes in size and disappears.

The striking characteristic of the MVN is the great variety of cell forms found within the nucleus and the particularly dense arrangement of the glial nuclei. The cells are irregularly arranged (fig. 1). The majority are slender, oval, triangular or spindle-shaped, and are lightly or moderately darkly stained. In addition, there are scattered, larger elements of varying form, which contain dark rather large Nissl granules. Finally mainly lightly stained, small spindle-shaped or round cells may be found (fig. 2–11).

Functional Neuroanatomy

Function
The MVN has different functional regions. The rostral third contains the large-celled MVN magnocellular part (MVNm), which is an output zone for the region (fig. 12–17). It contains the oculomotor-projecting vestibular neurons essential for the vestibulo-ocular reflex (VOR) in all directions [McCrea et al., 1987a, b]. The VOR is a powerful, robust and very fast reflex, in which the eyes counter-rotate to compensate for a head movement. It enables eye movements to be tested even when a pa-

Fig. 1. Low-power photomicrograph of the 3 regions in the medial vestibular nucleus – magnocellular (MVNm), parvocellular (MVNp) and marginal zone (MZ) – and the lateral vestibular nucleus (LVN) as the structures lie beneath the floor of the fourth ventricle, in the oral medulla. The right broken line shows the border between MVN and LVN, whereby the medial region is often ascribed to the LVN. Here, it is considered as the magnocellular portion of the MVN, and it contains the secondary vestibulo-ocular neurons projecting to the oculomotor, trochlear and abducens nuclei. The marginal zone in the medial MVN contains nonsecondary vestibulo-ocular neurons. NVII = Facial nerve; PGiD = dorsal paragigantocellular nucleus; SEL = subependymal layer; SOLo = solitary nucleus, oral part. Magnification ×40.

Fig. 2. Medial vestibular nucleus, parvocellular region. Magnification ×150.

Fig. 3–11. Different cell types from the medial vestibular nucleus parvocellular region. Magnification ×1,000.

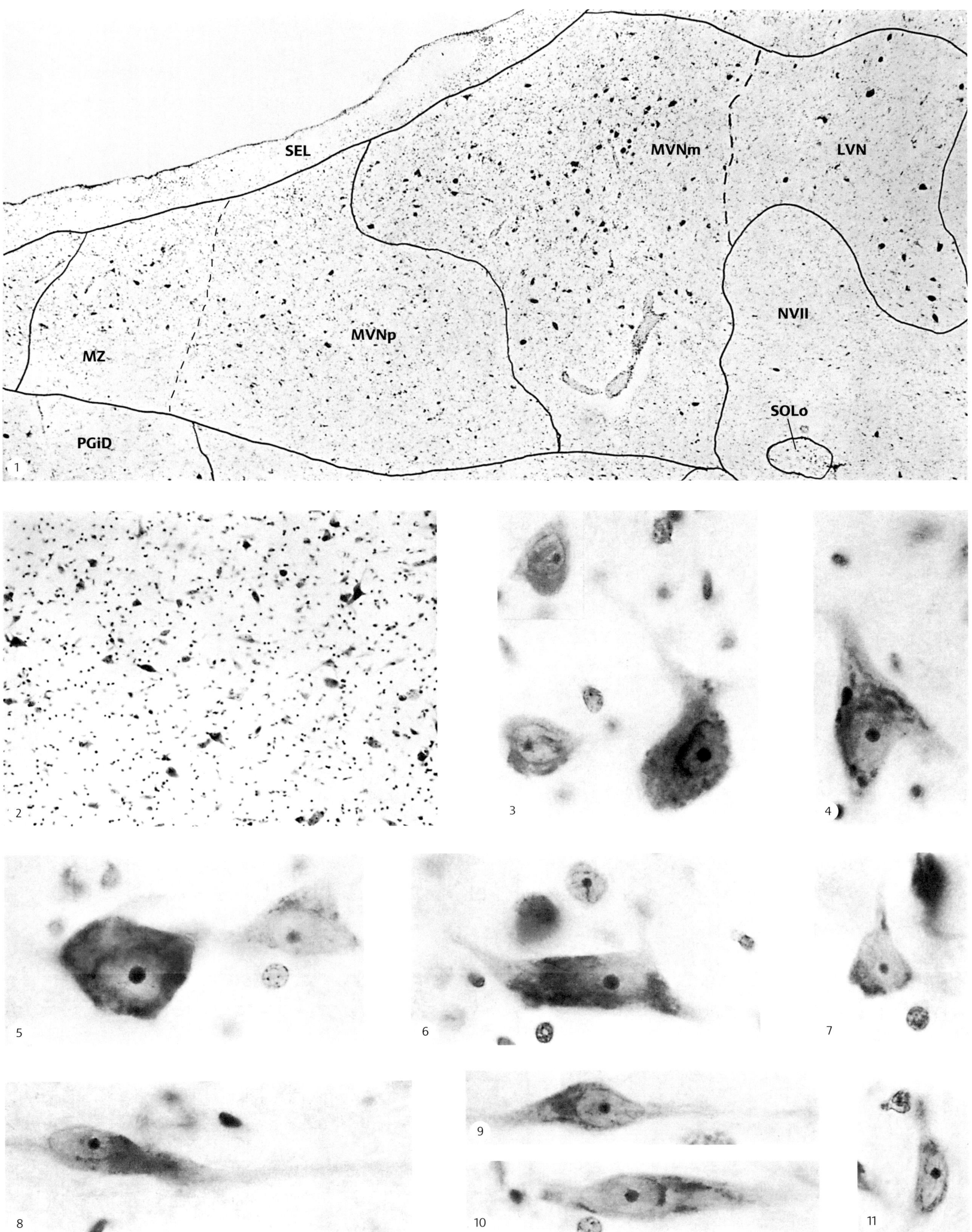
SEL
MVNm
LVN
NVII
MZ
MVNp
SOLo
PGiD
1
2
3
4
5
6
7
8
9
10
11

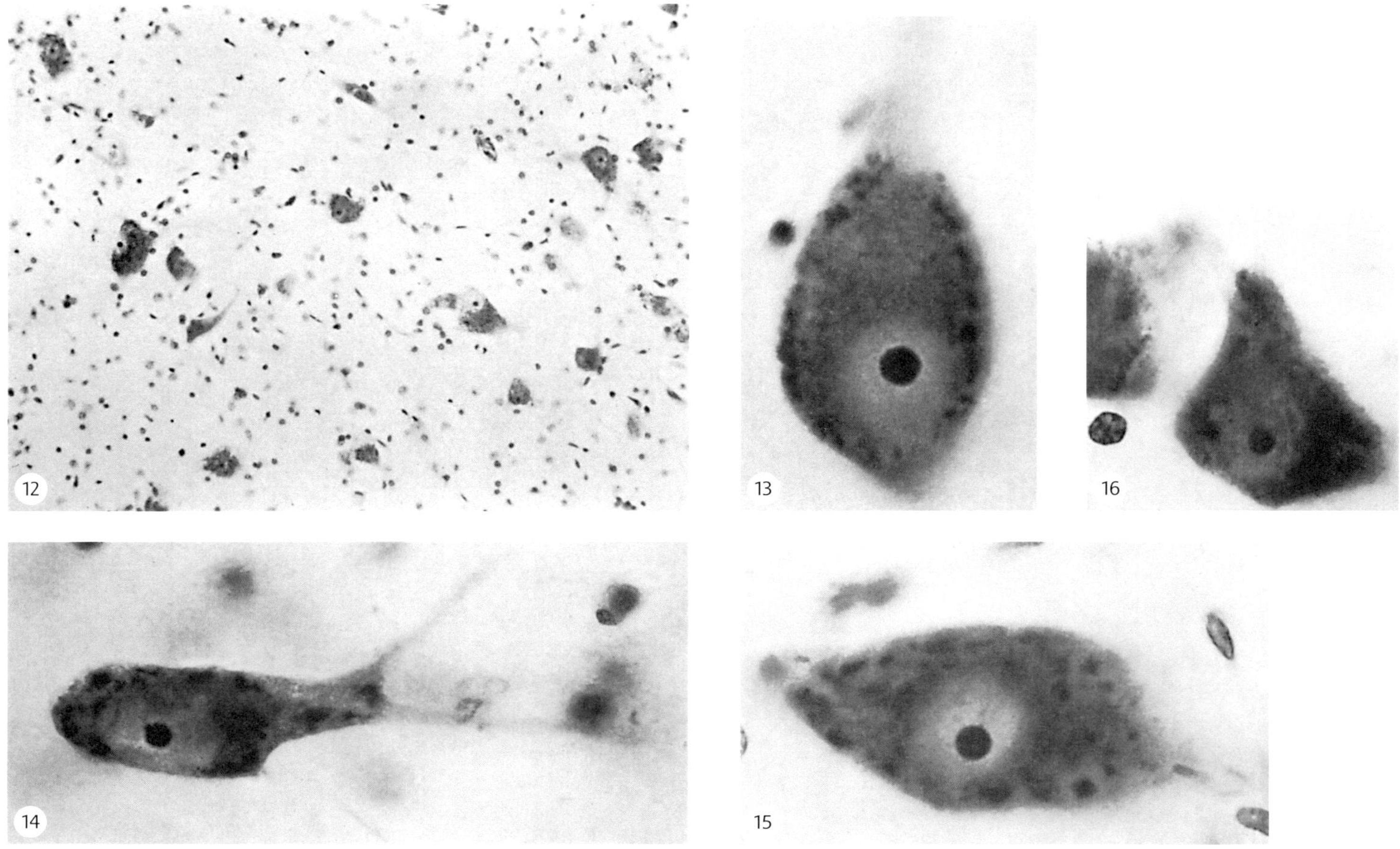

Fig. 12. Medial vestibular nucleus, magnocellular region. Magnification ×150.
Fig. 13–16. Examples of cells in the medial vestibular nucleus, magnocellular region. Magnification ×1,000.

tient is unconscious. The reciprocal connections between the MVNm and interstitial nucleus of Cajal (INC), and also between the MVNm and the cerebellum are important for the maintenance of vertical gaze and vertical head position. The MVN output zone is also the origin of the medial vestibulospinal tract, which brings neck and postural musculature under vestibular control. Despite its name, the ascending tract of Deiters also arises from neurons in the MVNm (see nucleus No. 15), and projects ipsilaterally to medial rectus motoneurons. It may participate in producing the viewing distance-related changes in the VOR [Chen-Huang and McCrea, 1998].

In contrast, the smaller-celled regions of the MVN parvocellular part (MVNp) around the MVN output zone have many intrinsic and commissural interconnections (fig. 2–11), which play an important role in vestibular compensation, while the marginal zone (MZ; fig. 18–23) projects almost exclusively to the abducens nucleus possibly controlling horizontal gaze [Langer et al., 1986]. Finally the caudal MVN, along with the adjacent DVN participates in the stabilization of blood pressure and respiratory processes during postural changes [Holstein et al., 2011b].

Connections

In the recent literature 4 parts of the medial vestibular nucleus are usually recognized: an MVNm, MVNp, caudal part and the MZ [Büttner-Ennever, 1992; Büttner-Ennever and Gerrits, 2004; Büttner-Ennever, 2006]. The MVNm lies ventrolaterally in the rostral third of the vestibular complex, encompassing part of the region Olszewski and Baxter allot to the medial LVN (fig. 1). The MVNm receives canal and otolith inputs, and contains the cell bodies of neurons projecting via the medial longitudinal fasciculus and the ascending tract of Deiters to the oculomotor nucleus and participating in the vertical and horizontal VOR reflex [McCrea et al., 1987a, b]. The MVNm has reciprocal connections with the INC, important for vertical gaze and head position. It is also the origin of fibers descending to the contralateral spinal cord in the medial vestibulospinal tract, to regulate the neck and axial musculature. The cerebellar connections with the MVNm primarily arise from the flocculus and ventral paraflocculus, which serve to modulate activity related to optokinetic space [Voogd and Barmack, 2006]. The MVNm is continuous with a similar magnocellular region in the central superior vestibular nucleus (SVN). Together these regions form a central core of the vestibular complex, which appears to be an output zone, fed by inputs from the surrounding parvocellular regions. These parvocellular parts of the vestibular nuclei are characterized by their prominent intrinsic and commissural interconnections [Epema et al., 1988].

The MVNp surrounds the MVNm dorsomedially, lying along the floor and the wall of the fourth ventricle and extending laterally to the SVN. The connectivity of the MVNp is different from that of the MVNm. The MVNp has reciprocal connections with the nodulus and uvula, the cerebellar regions that encode vestibular space, and it is interconnected with the contralateral MVNp and SVN by commissural fibers.

At the medial border of the MVNp, adjacent to the PrP, lies the MZ, where cells are intermediate in size, and densely packed [Langer et al., 1986]. This subdivision is an output region of the

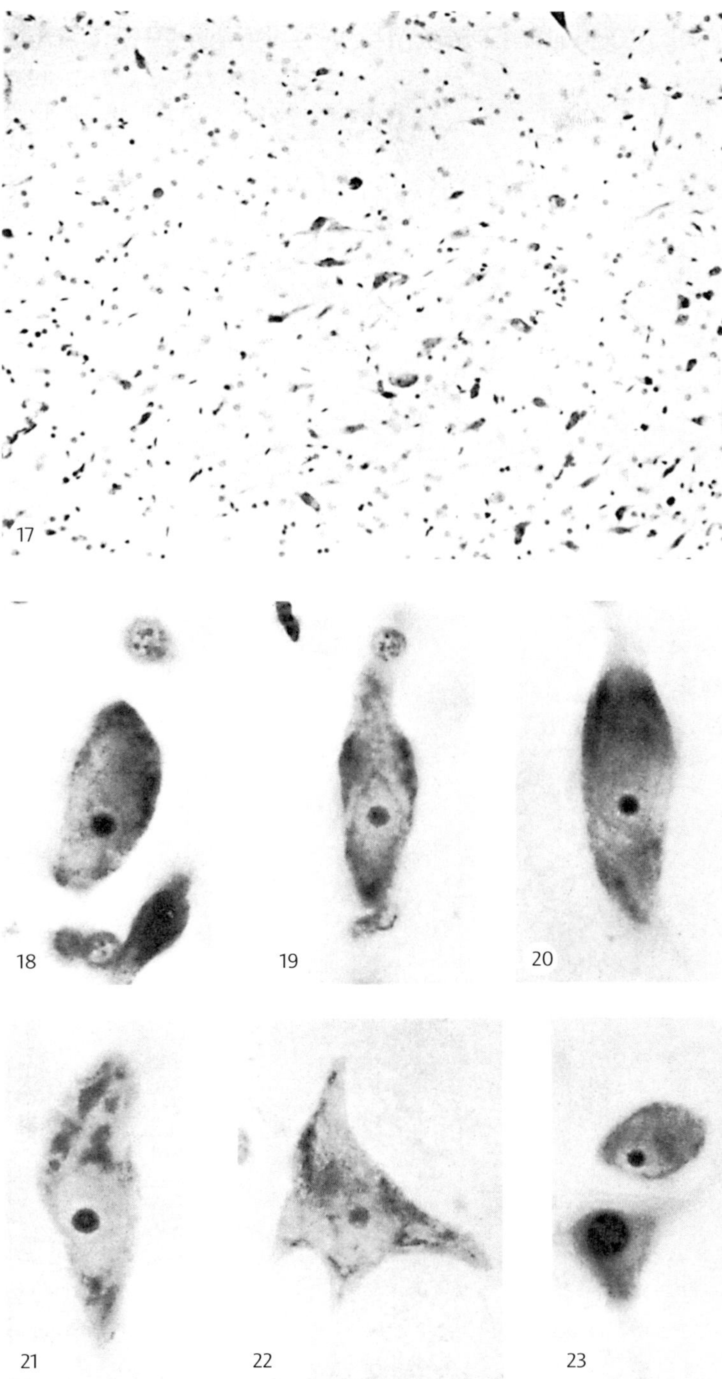

Fig. 17. Medial vestibular nucleus, marginal zone. Magnification ×150.
Fig. 18–23. Examples of cells in the medial vestibular nucleus, marginal zone. Magnification ×1,000.

MVN which contains nonsecondary vestibular neurons projecting to the abducens nucleus, whereby the rostral zone projects ipsilaterally and is excitatory, and the caudal MZ is inhibitory and projects to the contralateral VI. Interestingly, the MZ contains numerous strongly nitric oxide (NO)-sensitive neurons, but is devoid of NO-releasing neurons i.e. NO synthase neurons. These however are plentiful in the adjacent PrP. Pharmacological studies in cats demonstrate that the balanced production of NO in these structures (MZ and PrP) is necessary for the production of correct eye movements [Moreno-Lopez et al., 2001].

The functions associated with the caudal part of the MVN are not clear; it is not associated with the variety of vestibular-oculomotor-related activity found in the rostral third of the vestibular complex (MVNm, MZ and MVNp) in behaving primates. Canal and spinal afferents terminate in parts, and some regions give rise to descending fibers [Büttner-Ennever, 1992]. The caudal MVN also receives otolith afferents (see below), and a small region in the caudalmost tip of the vestibular complex, called the parasolitary nucleus (PSol), provides a GABAergic pathway to the inferior olive, which is the only source of otolith input to the olive [Barmack, 2006]. The PSol forms the caudal pole of the MVN, and in humans the area has not yet been defined. It may be included in the larger nucleus defined as PSol by Paxinos et al. [2012] (see nucleus No. 2).

It is clear that projections from the caudal MVN, and from the DVN, contribute to the stabilization of blood pressure and respiratory processes during postural changes. Direct vestibular pathways running in the solitary tract (TSOL) to the ventrolateral medulla, the solitary and dorsal vagal motor nuclei, play a role in autonomic effector responses [Balaban and Yates, 2004; Nisimaru, 2004; Holstein et al., 2011a, b]. Indirect vestibular influences from both rostral and caudal vestibular nuclei may be affected via their projections to the parabrachial complex, which in turn sends descending fibers to the same medullary autonomic centers.

References

Balaban CD, Yates BJ: Vestibuloautonomic interactions: a teleologic perspective; in Highstein SM, Fay RR, Popper AN (eds): The Vestibular System. New York, Springer, 2004.

Barmack NH: Inferior olive and oculomotor system. Prog Brain Res 2006;151:269–291.

Büttner-Ennever JA: Patterns of connectivity in the vestibular nuclei. Ann NY Acad Sci 1992;656:363–378.

Büttner-Ennever JA: The extraocular motor nuclei: organization and functional neuroanatomy. Prog Brain Res 2006;151:95–125.

Büttner-Ennever JA, Gerrits NM: Vestibular system; in Paxinos G, Mai JK (eds): Amsterdam, Elsevier Academic Press, 2004.

Chen-Huang C, McCrea RA: Viewing distance related sensory processing in the ascending tract of Deiters vestibulo-ocular reflex pathway. J Vestib Res 1998; 8:175–184.

Epema AH, Gerrits NM, Voogd J: Commissural and intrinsic connections of the vestibular nuclei in the rabbit: a retrograde labeling study. Exp Brain Res 1988; 71:129–146.

Holstein GR, Friedrich VLJ, Kang T, Kukielka E, Martinelli GP: Direct projections from the caudal vestibular nuclei to the ventrolateral medulla in the rat. Neurosci 2011a;175:104–117.

Holstein GR, Martinelli GP, Friedrich VLJ: Anatomical observations of the caudal vestibulo-sympathetic pathway. J Vestib Res 2011b;21:49–62.

Langer TP, Kaneko CR, Scudder CA, Fuchs AF: Afferents to the abducens nucleus in the monkey and cat. J Comp Neurol 1986;245:379–400.

McCrea RA, Strassman A, Highstein SM: Anatomical and physiological characteristics of vestibular neurons mediating the vertical vestibulo-ocular reflex of the squirrel monkey. J Comp Neurol 1987a;264:571–594.

McCrea RA, Strassman E, May E, Highstein SM: Anatomical and physiological characteristics of vestibular neurons mediating the horizontal vestibulo-ocular reflex of the squirrel monkey. J Comp Neurol 1987b;264:547–570.

Moreno-Lopez B, Escudero M, de Vente J, Estrada C: Morphological identification of nitric oxide sources and targets in the cat oculomotor system. J Comp Neurol 2001;435:311–324.

Nisimaru N: Cardiovascular modules in the cerebellum. Jpn J Physiol 2004;54: 431–448.

Paxinos G, Huang X-F, Sengul G, Watson C: Organization of the brainstem; in Mai JK, Paxinos G (eds): The Human Nervous System, ed 3. Amsterdam, Academic Press, 2012.

Voogd J, Barmack NH: Oculomotor cerebellum. Prog Brain Res 2006;151:231–268.

14 Superior Vestibular Nucleus (SVN)

Original name:
Nucleus vestibularis superior

Alternative name:
Angular nucleus of Bechterew

Location and Cytoarchitecture – Original Text

This nucleus, the most oral representative of the vestibular complex, appears caudally at the level of the oral pole of the facial nucleus and extends orally for a distance of 4 mm.

On cross-section, the nucleus is irregularly oblong in outline with its long axis directed dorsomedially. It lies in the dorsolateral corner of the midpontine tegmentum, ventral to the superior cerebellar peduncle (SCP) and the mesencephalic trigeminal nucleus, medial to the middle cerebellar peduncle, dorsal to the motor and principal sensory trigeminal nuclei and lateral to the gray matter beneath the floor of the lateral part of the fourth ventricle.

The majority of cells composing this nucleus are of medium size, plump, round or oval, with short dendrites, centrally placed nuclei and large, darkly stained, evenly distributed Nissl granules. A few larger and many smaller but similar cells are also present (fig. 1–7). […]

Functional Neuroanatomy

Function
The superior vestibular nucleus (SVN) is associated with the vertical vestibulo-ocular reflex, as well as the modification of vestibular and optokinetic signals to structures controlling vertical gaze. The signal modification is carried out by reciprocal interconnections with the cerebellum, and provides at least in part the basis for smooth pursuit eye movements and plasticity [Blazquez et al., 2007].

Afferent sensory inputs to the SVN arise from structures processing vertical movement in space, i.e. the anterior and posterior canals, the vertical nuclei of the accessory optic system and the interstitial nucleus of Cajal (INC). The anterior and posterior canal signals that relay in SVN then ascend to the oculomotor and trochlear nuclei, and through excitatory and inhibitory connections to the different eye muscle motoneuron subgroups, the signals drive upward gaze [Goldberg et al., 2012].

The reciprocal connections between the vestibular nuclei (VN; including SVN) and INC, along with similar interconnections between VN and the cerebellum, are considered to build the neuroanatomical basis for neural integration of eye and head position to enable vertical gaze-holding [Fukushima and Kaneko, 1995]. However, why every single vestibular nerve fiber that enters the vestibular complex always sends a branch into the SVN is unclear (see 'Sensory Systems – Vestibular Nuclei: Overview', also for reviews).

Connections
The center of the SVN is filled with medium-sized cells. It is usually referred to as the magnocellular part (SVNm), and it extends medially into the magnocellular medial vestibular nucleus (MVNm). The magnocellular regions are considered to provide the main output pathways of the vestibular complex, and are the origin of the vestibulo-ocular projections. The central SVNm is surrounded on the other 3 sides by small cells, the parvocellular part (SVNp, see plate 22 and fig. 6.5 in Goldberg et al. [2012]). As in the MVN, the parvocellular regions provide the commissural and intrinsic connections within the vestibular complex. The commissural fibers from the SVNp cross the midline dorsally through the cerebellum and provide a particularly strong projection to the contralateral SVNp, parvocellular MVN and the descending vestibular nucleus. The SVNp also feeds input locally to the ipsilateral SVNm and MVNm output regions [Epema et al., 1988].

Afferents. All afferents from the vestibular end organs send terminals to the peripheral SVNp via the ascending branches of the vestibular nerve (see 'Overview'). Some branches continue on and also provide primary afferents to the cerebellar vermis [Maklad and Fritzsch, 2003]. Additional major inputs to the SVN are from other VN (intrinsic connections), the cerebellum and the accessory optic system [Giolli et al., 1985; Büttner-Ennever, 1992]. The Purkinje cells of the cerebellar flocculus, associated with processing optokinetic space, project to the SVNm, whereas the nodulus and uvula, which encode vestibular space, project to the SVNp [Carpenter and Cowie, 1985]. The cerebellar modulation of the so-called floccular target neurons in the SVNm (and Y group), which project to the oculomotor system, provide the basis for smooth pursuit eye movements and plasticity [Blazquez et al., 2007].

Efferents. The main efferent targets of SVN neurons are the oculomotor nuclei, the INC and the cerebellar floccular region, with a marked absence of spinal efferents. The vestibular inputs to oculomotor neurons always target muscle pairs (e.g. the superior oblique and inferior rectus muscles, or the superior rectus and the inferior oblique), and there is always an excitatory and an inhibitory input, producing a push-pull system [Goldberg et al., 2012]. Ipsilateral efferents of the SVNm to the oculomotor nuclei are inhibitory, and travel in the medial longitudinal fasciculus (MLF) to contact the pairs of upward and downward motoneuron subgroups in the oculomotor and trochlear nuclei. This projection mediates the inhibitory limb of the vertical vestibulo-ocular reflex. Parallel excitatory projections to the oculomotor nucleus ascend in the contralateral MLF. A second, separate excitatory pathway to the oculomotor nuclei is considered to arise mainly from the adjacent dorsal Y group, and only in part from the SVN; the ascending pathway travels outside the MLF following the SCP (brachium conjunctivum), or lying ventral to it in the 'crossing ventral tegmental

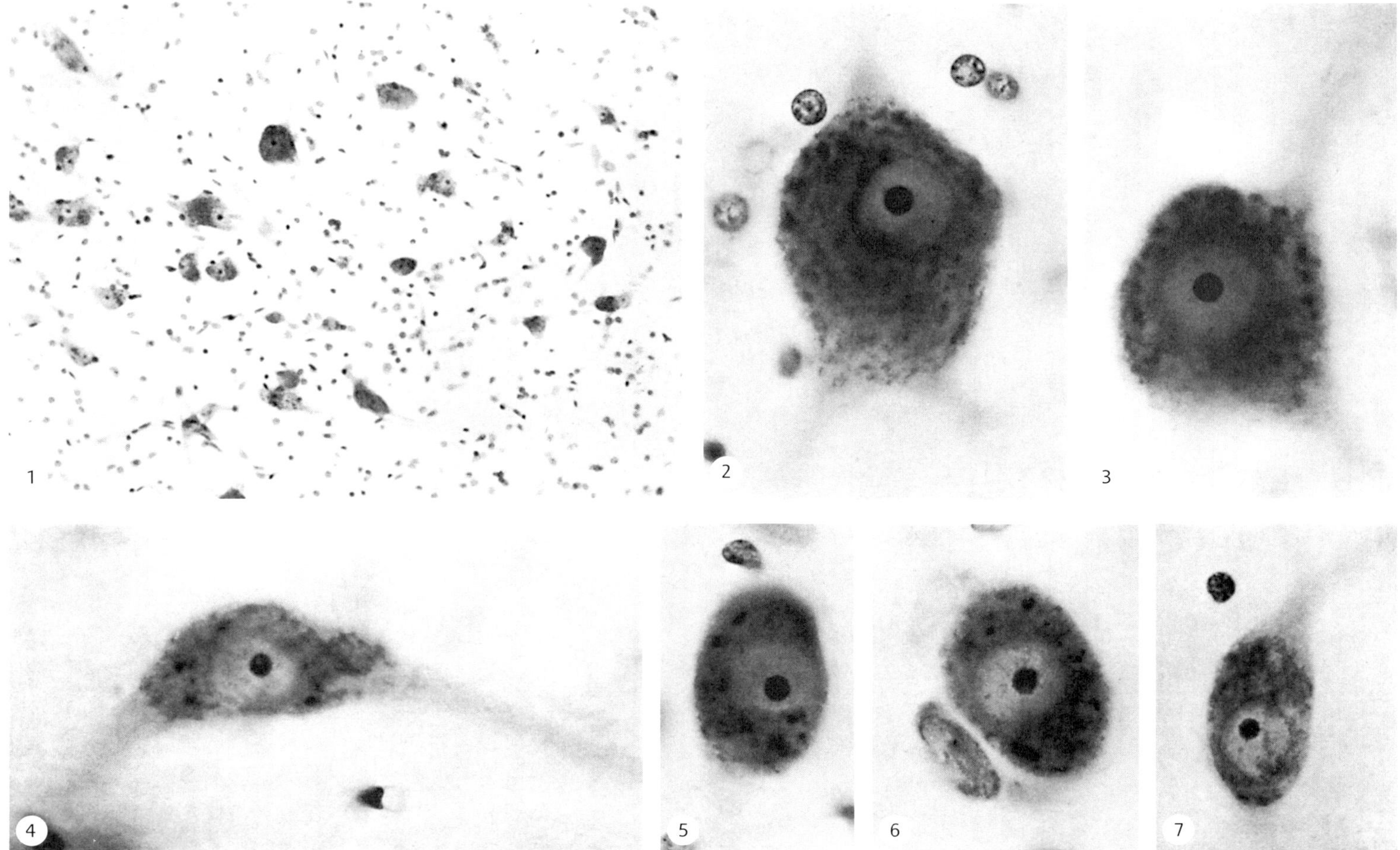

Fig. 1. Superior vestibular nucleus. Magnification ×150.
Fig. 2–7. Cells from the superior vestibular nucleus. Magnification ×1,000.

tract', which like the SCP crosses the midline in the midpons [Sato et al., 1984; Zwergal et al., 2008, 2009]. This pathway is thought to participate upward smooth pursuit eye movements (see Y group, chapter 18).

The connections between the INC and the VN are reciprocal. The descending pathway from the INC to the VN (SVN, MVN and the prepositus nucleus) runs in the MLF and is considered to participate in the neural integration of eye and head position to enable vertical gaze-holding [Spence and Saint-Cyr, 1988; Fukushima et al., 1992; Fukushima and Kaneko, 1995; Kokkoroyannis et al., 1996]. Lesions of the INC lead to a vertical gaze-holding deficit, torsional and vertical nystagmus, and head tilt [Büttner and Helmchen, 2002].

References

Blazquez PM, Davis-Lopez de Carrizosa MA, Heiney SA, Highstein SM: Neuronal substrates of motor learning in the velocity storage generated during optokinetic stimulation in the squirrel monkey. J Neurophysiol 2007;97:1114–1126.

Büttner-Ennever JA: Patterns of connectivity in the vestibular nuclei. Ann NY Acad Sci 1992;656:363–378.

Büttner U, Helmchen C: Eye movement deficits after unilateral mesencephalic lesions. Neuroophthalmology 2002;24:469–484.

Carpenter MB, Cowie RJ: Connections and oculomotor projections of the superior vestibular nucleus and cell group 'y'. Brain Res 1985;336:265–287.

Epema AH, Gerrits NM, Voogd J: Commissural and intrinsic connections of the vestibular nuclei in the rabbit: a retrograde labeling study. Exp Brain Res 1988; 71:129–146.

Fukushima K, Kaneko CR: Vestibular integrators in the oculomotor system. Neurosci Res 1995;22:249–258.

Fukushima K, Kaneko CRS, Fuchs AF: The neuronal substrate of integration in the oculomotor system. Prog Neurobiol 1992;39:609–639.

Giolli RA, Blanks RHI, Torigoe Y, Williams DD: Projections of medial terminal accessory optic nucleus, ventral tegmental nuclei, and substantia nigra of rabbit and rat as studied by retrograde axonal transport of horseradish peroxidase. J Comp Neurol 1985;232:99–116.

Goldberg JM, Wilson VJ, Cullen KE, Angelaki DE, Broussard DM, Büttner-Ennever JA, Fukushima K, Minor LB: The Vestibular System – A Sixth Sense. Oxford, Oxford University Press, 2012.

Kokkoroyannis T, Scudder CA, Balaban CD, Highstein SM, Moschovakis AK: Anatomy and physiology of the primate interstitial nucleus of Cajal. I. Efferent projections. J Neurophysiol 1996;75:725–739.

Maklad A, Fritzsch B: Partial segregation of posterior crista and saccular fibers to the nodulus and uvula of the cerebellum in mice, and its development. Brain Res Dev Brain Res 2003;140, 223–236.

Sato Y, Yamamoto F, Shojaku H, Kawasaki T: Neuronal pathway from floccular caudal zone contributing to vertical eye movements in cats – role of group y nucleus of vestibular nuclei. Brain Res 1984;294:375–380.

Spence SJ, Saint-Cyr JA: Mesodiencephalic projections to the vestibular complex in the cat. J Comp Neurol 1988;268:375–388.

Zwergal A, Büttner-Ennever J, Brandt T, Strupp M: An ipsilateral vestibulothalamic tract adjacent to the medial lemniscus in humans. Brain 2008;131:2928–2935.

Zwergal A, Strupp M, Brandt T, Büttner-Ennever J: Parallel ascending vestibular pathways: anatomical localization and functional specialization. Ann NY Acad Sci 2009;1164:51–59.

15 Lateral Vestibular Nucleus (LVN)

Original name:
Nucleus vestibularis lateralis

Alternative name:
Deiters' nucleus

Location and Cytoarchitecture – Original Text

This nucleus is located in the dorsolateral part of the tegmentum of the caudal pons. It commences caudally at the oral pole of the spinal vestibular nucleus (SpV) and extends to the level of the caudal pole of the motor trigeminal nucleus – a distance of 4 mm.

The nucleus is related dorsally to the lateral part of the floor of the fourth ventricle, medially to the medial vestibular nucleus (MVN), laterally to the inferior cerebellar peduncle, ventrally to the parvocellular reticular nucleus (PCR), the SpV, and to the fibers of the vestibular division of the eighth cranial nerve. With the appearance of the superior vestibular nucleus (SVN), the lateral vestibular nucleus (LVN) rapidly diminishes in size. Its oral pole appears as a small band of cells lying between the cells of the PCR ventromedially and the SVN dorsolaterally.

At the level of the LVN, small groups of cells are constantly found along the course of the afferent vestibular fibers through the middle cerebellar peduncle of the LVN, whereas those of others are distinctly smaller. [These groups have been designated to the interstitial nucleus of the vestibular nerve (IVN, No. 17)].

The spinal vestibular root traverses [through to a more cellular magnocellular area now called the medial vestibular nucleus (MVNm, No. 13, fig. 1)]. The majority of cells composing the MVNm are distinctly smaller and more varied in size and shape than are those of the LVN. Superficial examination of the MVNm may lead to the impression that it represents an area in which the cells of MVN and LVN are intermingled.

The cells of the LVN are large, plump or spindle shaped, and multipolar [and often referred to as Deiters' neurons] (fig. 1–4). They possess prominent, darkly stained Nissl granuli which tend to increase in size from the central portion to the periphery of the cell body. The nucleus may not be excentric. An occasional cell resembling those of, and probably belonging functionally to, the mesencephalic trigeminal nucleus of the fifth cranial nerve may also be found in the lateral subnucleus.

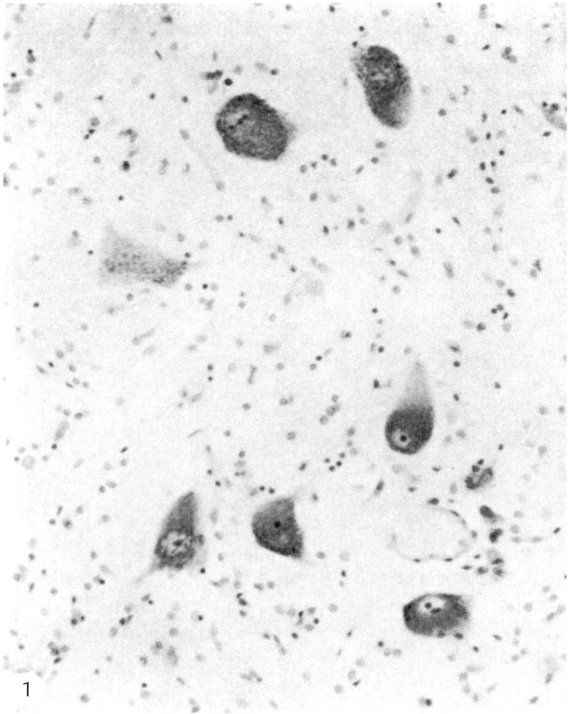

Fig. 1. Lateral vestibular nucleus neurons. Magnification ×150.

Functional Neuroanatomy

Function

The LVN participates in the control of posture through the lateral vestibulospinal tract (LVST), which exerts a facilitatory influence on spinal reflexes and on extensor muscle tone. The LVST is the main output of LVN, and the spinocerebellum provides the major input to the nucleus.

Connections

The border between LVN and MVN is confused, because the area described here as MVNm was included in the LVN as a medial subnucleus in the previous edition. Nowadays the area is usually allotted to the magnocellular region of the MVN [Epema et al., 1988; Büttner-Ennever and Gerrits, 2004; Highstein and Holstein, 2006; and see chapter 17, Nieuwenhuys et al., 2008]. The grounds for this change are based on functional neuroanatomical studies which show that the MVNm (unlike the LVN) receives a strong input from the vestibular nerve [Newlands and Perachio, 2003] and it is the source of secondary vestibular neurons projecting to the oculomotor nuclei and providing the basis for the vestibulo-ocular reflex. For a review, see Büttner-Ennever [1992]; however, see Goldberg et al. [2012] for a different opinion.

The area here taken as LVN is recognized by its lack of primary afferent input from the semicircular canals, a small sacculus input [Imagawa et al., 1998], as well as a by its strong GABAergic input from Purkinje cells of the anterior cerebellar vermis (B zone). The LVN is the source of origin of the LVST [Shinoda et al., 2006]. This is an excitatory uncrossed pathway which descends along the entire spinal cord at the junction of the ventral and lateral funiculi; it terminates ipsilaterally in the cervical cord exciting ipsilateral neck muscles and via commissural interneurons inhibits those on the contralateral side. But the majority of LVST terminals are in the lumbar spinal cord, where it excites extensor motoneurons and inhibits flexor motoneurons, via interneurons [Shinoda et al., 2006].

The ventral LVN is reported to give rise to the 'ascending tract of Deiters' which projects to the ipsilateral oculomotor nucleus (medial rectus subdivision); but the cells of origin lie in the MVNm region rather than in the LVN [Highstein and Hol-

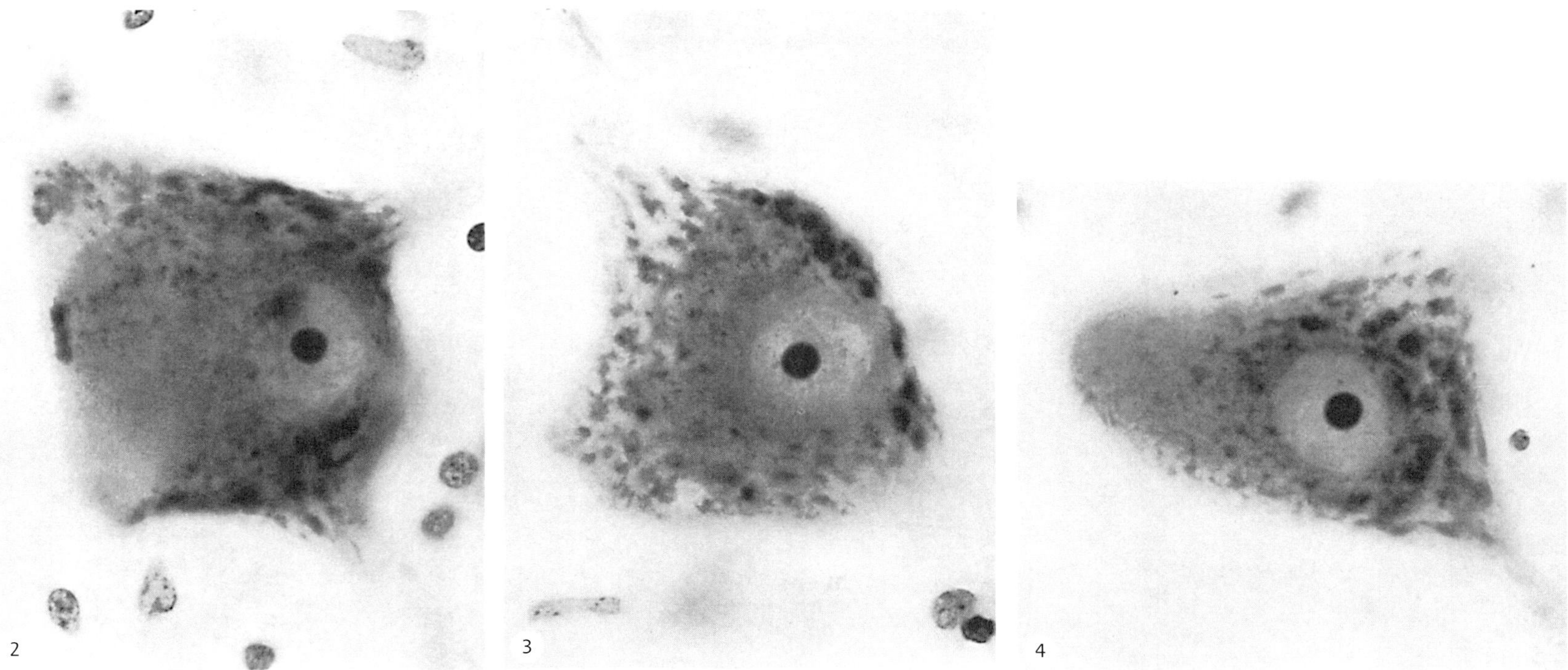

Fig. 2–4. Large cells from the lateral vestibular nucleus. Magnification ×1,000.

stein, 2006]. Likewise, inputs from the fastigial oculomotor region reported to terminate in the ventral LVN mainly project to the MVNm, and not the LVN defined here [Sugita and Noda, 1991]; nevertheless, descending fibers of the LVST were found to be monosynaptically activated from the fastigial nucleus [Akaike, 1983].

References

Akaike T: Neuronal organization of the vestibulospinal system in the cat. Brain Res 1983;259:217–227.

Büttner-Ennever JA: Patterns of connectivity in the vestibular nuclei. Ann NY Acad Sci 1992;656:363–378.

Büttner-Ennever JA, Gerrits NM: Vestibular system; in Paxinos G, Mai JK (eds): The Human Nervous System, ed 2. Amsterdam, Elsevier Academic Press, 2004.

Epema AH, Gerrits NM, Voogd J: Commissural and intrinsic connections of the vestibular nuclei in the rabbit: a retrograde labeling study. Exp Brain Res 1988;71:129–146.

Goldberg JM, Wilson VJ, Cullen KE, Angelaki DE, Broussard DM, Büttner-Ennever JA, Fukushima K, Minor LB: The Vestibular System – A Sixth Sense. Oxford, Oxford University Press, 2012.

Highstein SM, Holstein GR: The anatomy of the vestibular nuclei. Prog Brain Res 2006;151:157–203.

Imagawa M, Graf WM, Sato H, Suwa H, Isu N, Izumi R, Uchino Y: Morphology of single afferents of the saccular macula in cats. Neurosci Lett 1998;240:127–130.

Newlands SD, Perachio AA: Central projections of the vestibular nerve: a review and single fiber study in the Mongolian gerbil. Brain Res Bull 2003;60:475–495.

Nieuwenhuys R, Voogd J, van Huijzen C: The Human Central Nervous System. Berlin, Springer, 2008.

Shinoda Y, Sugiuchi Y, Izawa Y, Hata Y: Long descending motor tract axons and their control of neck and axial muscles. Prog Brain Res 2006;151:527–563.

Sugita S, Noda H: Pathways and terminations of axons arising in the fastigial oculomotor region of macaque monkeys. Neurosci Res 1991;10:118–136.

Sensory Systems – Vestibular Nuclei (Plates 16–18)

16 Descending Vestibular Nucleus (DVN)

Original name:
Nucleus vestibularis spinalis

Alternative names:
Spinal vestibular nucleus
Inferior vestibular nucleus

Location and Cytoarchitecture – Original Text

This nucleus is located in the dorsolateral part of the oral medulla and extends from the oral pole of the lateral cuneate nucleus to the level of the caudal pole of the facial nucleus (VII) a distance of 5 mm. Its cells lie, in the main, amongst and about the fibers of the spinal vestibular root. Caudally, the nucleus is irregularly quadrangular in outline, whereas more orally it assumes a curved outline along the dorsal and medial surfaces of

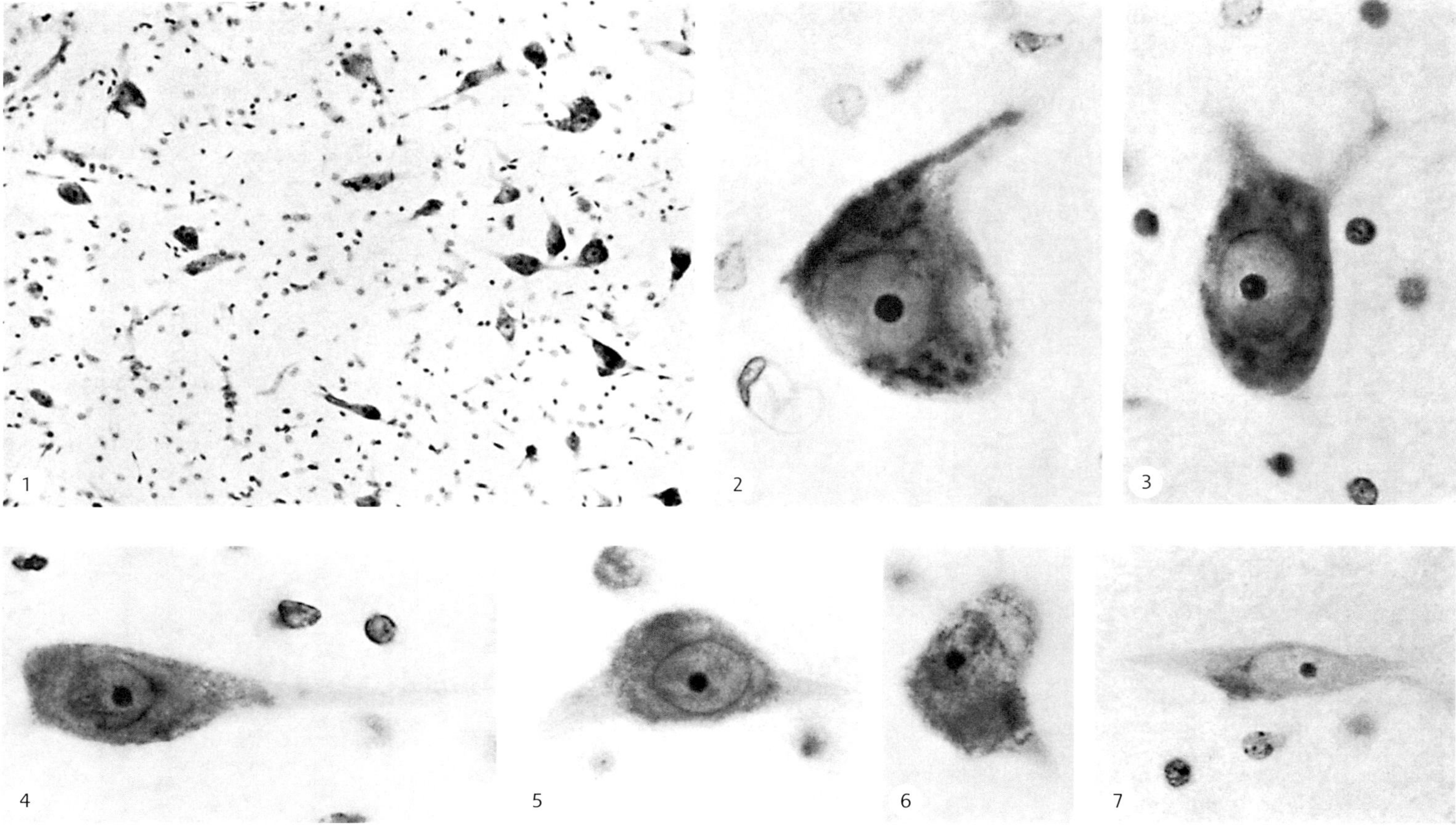

Fig. 1. Descending vestibular nucleus neurons. Magnification ×150.
Fig. 2–7. Different cell types from the descending vestibular nucleus. Magnification ×1,000.

the inferior cerebellar peduncle (ICP). At its caudal tip the nucleus blends into the parasolitary nucleus. Throughout its course the descending vestibular nucleus (DVN) is related medially to the medial vestibular nucleus (MVN), laterally to the ICP, and ventrolaterally to the fibers of the eighth and ninth cranial nerves. Dorsally the oral portion of the nucleus lies in the floor of the lateral recess of the fourth ventricle – the caudal portion is usually separated from the ventricular floor by cells of the supravestibular nucleus. Approaching the level of the caudal pole of the VII the cells of the DVN gradually diminish in number, those in the most medial part of the nucleus are the last to disappear.

Cells composing the DVN are larger and not as densely arranged as are the cells of the MVN (fig. 1). The majority are medium-sized, triangular or oval, multipolar cells with long dendrites, evenly distributed darkly stained Nissl granules and centrally placed nuclei (fig. 2, 3). Scattered among these are small elements of similar shape and Nissl arrangement (fig. 4–7). In general, the cells in the medial part of the nucleus are smaller than those in the lateral part.

Comment

A characteristic feature of the DVN seen in all mammals is its fasciculated appearance, evident only in fiber-stained material. These stains reveal the regular organization of small fiber bundles that run throughout the nucleus. The fiber bundles are composed of olivocerebellar climbing fibers, and efferents from the cerebellum such as the uncinate tract from the contralateral fastigial nucleus to the brainstem, and direct cerebellar corticovestibular fibers from the flocculus, nodulus and uvula. The fibers of the descending root of the vestibular nerve also lie in the DVN but between the prominent fiber bundles [Gerrits, 1990].

Functional Neuroanatomy

Function

The DVN has not been associated with a specific vestibular function, but it is clearly related to the cerebellum, and participates in vestibulomotor coordination. The rostral half of the DVN appears to resemble the MVN in its functional connectivity, but the caudal DVN is strongly related to a role in autonomic functions; for example controlling blood pressure and respiration, during postural changes [Balaban and Yates, 2004; Jian et al., 2005; Mori et al., 2005; Holstein et al., 2011a, b].

Connections

The rostrodorsal DVN connectivity is similar to that of its adjacent MVN, receiving canal afferents, and containing a subset of vestibular-ocular-collic neurons which project both to the oculomotor nuclei via the medial longitudinal fasciculus and to the spinal cord in the medial vestibulospinal tract [Büttner-Ennever, 1992; Shinoda et al., 2006]. Ascending inputs from the spinal cord carry proprioceptive inputs to terminate in the same area [Neuhuber and Bankoul, 2000]. Intrinsic connections within the vestibular complex interconnect the DVN with the superior vestibular nucleus (SVN) and the MVN of the same side, and commissural fibers connect the DVN with the contralateral DVN and MVN. The DVN projects into the MVN magnocellular part region, which is thought to provide the major output pathways for vestibular influences [Epema et al., 1988].

Inhibitory inputs to the DVN arise from the cerebellar Purkinje cells of the flocculus, ventral paraflocculus, nodulus and adjacent uvula, while mossy fibers from the DVN (also MVN and SVN) project back to similar regions, as well as the rostral anterior lobe and into the depth of the transverse fissure [Voogd et al., 1996].

The caudal DVN is not involved in oculomotor control, but it is strongly interconnected with autonomic structures, and has been shown to control cardiovascular and respiratory responses to postural changes [Jian et al., 2005; Holstein et al., 2011a, b].

References

Balaban CD, Yates BJ: Vestibuloautonomic interactions: a teleologic perspective; in Highstein SM, Fay RR, Popper AN (eds): The Vestibular System. New York, Springer, 2004.

Büttner-Ennever JA: Patterns of connectivity in the vestibular nuclei. Ann NY Acad Sci 1992;656:363–378.

Epema AH, Gerrits NM, Voogd J: Commissural and intrinsic connections of the vestibular nuclei in the rabbit: a retrograde labeling study. Exp Brain Res 1988; 71:129–146.

Gerrits NM: Vestibular nuclear complex; in Paxinos G (ed): The Human Nervous System. San Diego, Academic Press, 1990.

Holstein GR, Friedrich VLJ, Kang T, Kukielka E, Martinelli GP: Direct projections from the caudal vestibular nuclei to the ventrolateral medulla in the rat. Neuroscience 2011a;175:104–117.

Holstein GR, Martinelli GP, Friedrich VLJ: Anatomical observations of the caudal vestibulo-sympathetic pathway. J Vestib Res 2011b;21:49–62.

Jian BJ, Acernese AW, Lorenzo J, Card JP, Yates BJ: Afferent pathways to the region of the vestibular nuclei that participates in cardiovascular and respiratory control. Brain Res 2005;1044:241–250.

Mori RL, Cotter LA, Arendt HE, Olsheski CJ, Yates BJ: Effects of bilateral vestibular nucleus lesions on cardiovascular regulation in conscious cats. J Appl Physiol 2005;98:526–533.

Neuhuber WL, Bankoul S: Somatosensory influences on the vestibular system; in Beitz AJ, Anderson JH (eds): Neurochemistry of the Vestibular System. Boca Raton, CRC Press, 2000.

Shinoda Y, Sugiuchi Y, Izawa Y, Hata Y: Long descending motor tract axons and their control of neck and axial muscles. Prog Brain Res 2006;151:527–563.

Voogd J, Gerrits NM, Ruigrok TJH: Organization of the vestibulocerebellum. Ann NY Acad Sci 1996;781:553–579.

Sensory Systems – Vestibular Nuclei (Plate 20)

17 Interstitial Nucleus of the Vestibular Nerve (INV)

Original name:

β-Cell groups of the vestibular nerve

Location and Cytoarchitecture – Original Text

The vestibular nerve passes between the trigeminal spinal root and the inferior cerebellar peduncle to reach the vestibular nuclei. The cell groups of the interstitial nucleus of the vestibular nerve (INV) lie mainly within the section that curves around the ventrolateral border of the lateral vestibular nucleus (LVN) and spinal trigeminal nucleus, oral part (fig. 1).

The cell groups of the INV are comprised mainly of medium-sized fusiform cells lying along the axis of the vestibular nerve. In humans several very large neurons are evident, and these have led to suggestions that there are similarities between the INV and LVN. [The neurons in the INV are parvalbumin positive.]

Fig. 1. Cells of the interstitial nucleus of the vestibular nerve as it courses through the medial cerebellar peduncle (see plate 20). Magnification ×150.

Functional Neuroanatomy

Careful and exacting intracellular tracing studies show that vestibular horizontal canal afferents terminate in this group of cells [Sato et al., 1989], but no otolith afferents have yet been described. Efferent projections from the INV to the cerebellum, specifically the flocculus [Langer et al., 1985] have been reported, as well as a modest number of commissural connections between the INV of each side [Epema et al., 1988]. The cell group obviously operates within the vestibular system but its function is unknown.

References

Epema AH, Gerrits NM, Voogd J: Commissural and intrinsic connections of the vestibular nuclei in the rabbit: a retrograde labeling study. Exp Brain Res 1988; 71:129–146.

Langer TP, Fuchs AF, Chubb MC, Scudder CA, Lisberger SG: Floccular efferents in the rhesus macaque as revealed by autoradiography and horseradish peroxidase. J Comp Neurol 1985;235:26–37.

Sato F, Sasaki H, Ishizuka N, Sasaki S, Mannen H: Morphology of single primary vestibular afferents originating from the horizontal semicircular canal in the cat. J Comp Neurol 1989;290:423–439.

18 Y Group (Y)

Original name:
Not described

Alternative names:
Y group dorsal division is called the infracerebellar nucleus [Gacek, 1977]
Y group ventral division is called the Y group [Gacek, 1977]

Location and Cytoarchitecture

The Y group lies at the cerebellomedullary junction (fig. 1, 2). Rostrally it merges with the caudal border of the superior vestibular nucleus (SVN), and caudally its cells spread over the dorsal surface of the descending and caudal medial vestibular nuclei. The supravestibular nucleus lies medial to the caudal Y group, and may adjoin it. The Y group is bordered ventrally by the dorsal acoustic striae and the restiform body (the inferior cerebellar peduncle, ICP) and dorsally by the cerebellar white matter containing the floccular peduncle.

The Y group is well developed in primates. In most mammals it possesses 2 subdivisions: a small-celled, more densely packed, ventral part with fusiform neurons, which is continuous with the dorsal SVN; and a dorsal, more loosely scattered division, with larger, often multipolar cells, extending from the middle of the nucleus towards the dentate nucleus.

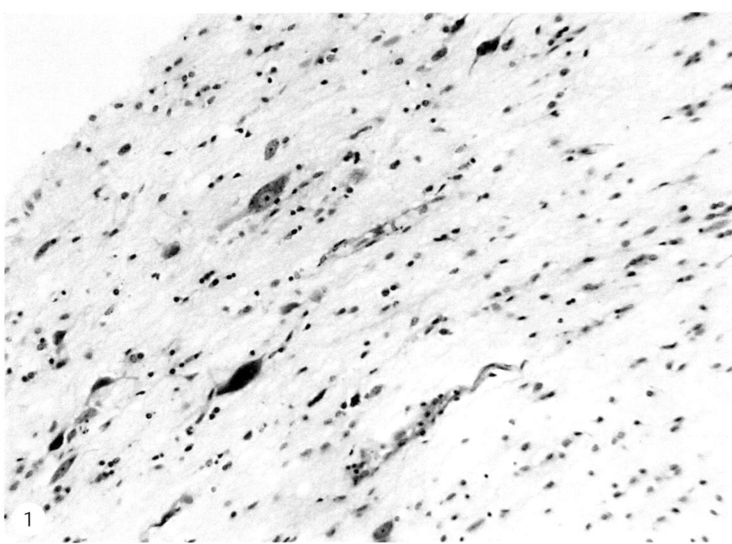

Fig. 1. Y group. Magnification ×150.

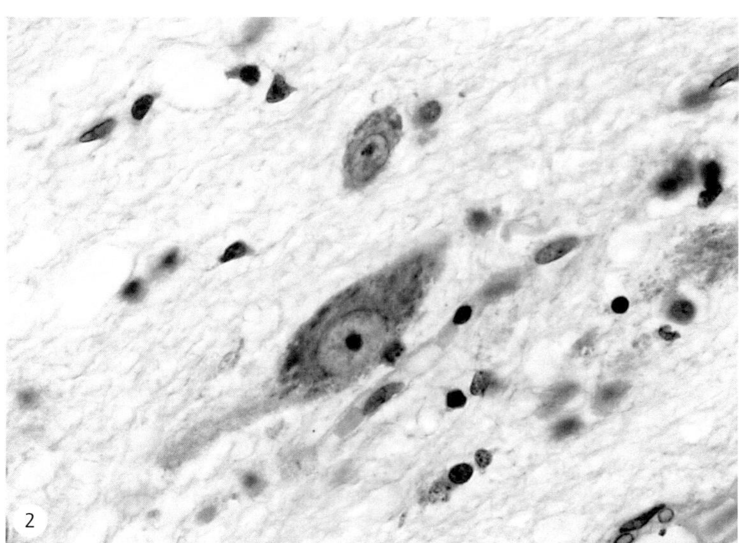

Fig. 2. Cells from the Y group. Magnification ×700.

Functional Neuroanatomy

Function

The dorsal Y group, also known as infracerebellar nucleus in the cat [Gacek, 1977], plays a role in the generation of upward smooth pursuit eye movements, and participates in the modulation of the vertical vestibulo-ocular reflex (i.e. motor-learning or plasticity), under cerebellar control [Blazquez et al., 2007]. The role of the ventral Y group is not clear.

Various clinical syndromes with upbeat nystagmus may be attributed to lesions in the pontine tegmentum which damage pathways from the dorsal Y group ascending to the oculomotor nuclei, and crossing in the ventral tegmentum near the brachium conjunctivum (the 'crossing ventral tegmental tract') [Ranalli and Sharpe, 1988; Pierrot-Deseilligny and Tilikete, 2008]. An additional ascending pathway (called the ipsilateral vestibulothalamic tract) arising from the Y group/SVN region was identified as the result of clinical studies. The patients had brainstem infarctions that led to an isolated deviation of the 'subjective-visual-vertical' [Zwergal et al., 2008]. The common effective lesioned area lay around the medial lemniscus, and was assumed to injure fibers of a vestibulothalamocortical pathway.

Connections

The ventral Y group receives saccular afferents [Carleton and Carpenter, 1984], projects to the flocculus [Langer et al., 1985b], and may have modest commissural connections [Carpenter and Cowie, 1985]. In contrast the main afferents to the dorsal Y group arise from the parvocellular SVN of both sides, Purkinje cells of the flocculus (in F2 and F4, the vertical optokinetic- and visual-related zones) and the ventral paraflocculus [Langer et al., 1985a; Blazquez et al., 2000].

The dorsal Y group is the source of 2 different efferent pathways that ascend to extraocular motoneurons to participate in upward gaze: first, strong monosynaptic connections to the downward-moving motoneurons of the trochlear and oculomotor nuclei (superior oblique and inferior rectus), which are thought to be inhibitory; and second, a pathway which projects to the upward-moving motoneurons in the oculomotor nucleus (III; inferior oblique and superior rectus) [Sato et al., 1984; Yamamoto et al., 1986; Sato and Kawasaki, 1987]. The pathways from the Y group to the III travel in the 'crossing ventral tegmental tract' near the superior cerebellar peduncle (or brachium conjunctivum), not in the medial longitudinal fasciculus. These findings support the current hypothesis that the dorsal Y group neurons are active during upward optokinetic and smooth pursuit eye movements [Chubb and Fuchs, 1982; Partsalis and Highstein, 1996]. Recent results show that the excitatory pathway to upward-moving motoneurons in the III contains calretinin, a calcium-binding protein which can be used to selectively label the cells of origin and the ascending pathway [Ahlfeld et al., 2011]. Since the calretinin cells are non-GABAergic, the GABAergic cells of the dorsal Y group probably give rise to the ascending inhibitory pathway to the downward-moving neurons of the III.

References

Ahlfeld J, Mustari MJ, Horn AKE: Sources of calretinin inputs to motoneurons of extraocular muscles involved in upgaze. Ann NY Acad Sci 2011;1233:91–99.

Blazquez PM, Davis-Lopez de Carrizosa MA, Heiney SA, Highstein SM: Neuronal substrates of motor learning in the velocity storage generated during optokinetic stimulation in the squirrel monkey. J Neurophysiol 2007;97;1114–1126.

Blazquez P, Partsalis A, Gerrits NM, Highstein SM: Input of anterior and posterior semicircular canal interneurons encoding head-velocity to the dorsal Y group of the vestibular nuclei. J Neurophysiol 2000;83:2891–2904.

Carleton SC, Carpenter MB: Distribution of primary vestibular fibers in the brainstem and cerebellum of the monkey. Brain Res 1984;294:281–298.

Carpenter MB, Cowie RJ: Connections and oculomotor projections of the superior vestibular nucleus and cell group 'y'. Brain Res 1985;336:265–287.

Chubb MC, Fuchs AF: Contribution of y group of vestibular nuclei and dentate nucleus of cerebellum to generation of vertical smooth eye movements. J Neurophysiol 1982;48:75–99.

Gacek RR: Location of brain stem neurons projecting to the oculomotor nucleus in the cat. Exp Neurol 1977;57:725–749.

Langer TP, Fuchs AF, Chubb MC, Scudder CA, Lisberger SG: Floccular efferents in the rhesus macaque as revealed by autoradiography and horseradish peroxidase. J Comp Neurol 1985a;235:26–37.

Langer TP, Fuchs AF, Scudder CA, Chubb MC: Afferents to the flocculus of the cerebellum in the rhesus macaque as revealed by retrograde transport of horseradish peroxidase. J Comp Neurol 1985b;235:1–25.

Partsalis AM, Highstein SM: Role of the Y group of the vestibular nuclei in motor learning or plasticity of the vestibulo-ocular reflex in the squirrel monkey. Ann NY Acad Sci 1996;781:513–524.

Pierrot-Deseilligny C, Tilikete C: New insights into the upward vestibulo-oculomotor pathways in the human brainstem. Prog Brain Res 2008;171:509–518.

Ranalli PJ, Sharpe JA: Upbeat nystagmus and the ventral tegmental pathway of the upward vestibulo-ocular reflex. Neurology 1988;38:1329–1330.

Sato Y, Kawasaki T: Target neurons of floccular caudal zone inhibition in Y-group nucleus of vestibular nuclear complex. J Neurophysiol 1987;57:460–480.

Sato Y, Yamamoto F, Shojaku H, Kawasaki T: Neuronal pathway from floccular caudal zone contributing to vertical eye movements in cats – role of group y nucleus of vestibular nuclei. Brain Res 1984;294:375–380.

Yamamoto F, Sato Y, Kawasaki T: The neuronal pathway from the flocculus to the oculomotor nucleus: an electrophysiological study of group y nucleus in cats. Brain Res 1986;371:350–354.

Zwergal A, Büttner-Ennever J, Brandt T, Strupp M: An ipsilateral vestibulothalamic tract adjacent to the medial lemniscus in humans. Brain 2008;131:2928–2935.

Sensory Systems – Vestibular Nuclei (Not on plates)

19 Cell Groups Associated with the Vestibular Complex

Original name:
Not described

List of vestibular complex cell groups:
Parasolitary nucleus (PSol)
Group e (E)
Group f (F)
Group l (L)
Group m (M)
Group x (X)
Group z (Z)
Islands of the lateral cuneate nucleus (ISL) [Olszewski and Baxter, 1982]

Location and Functional Neuroanatomy

The *parasolitary nucleus (PSol)* as defined by Barmack [2006] and colleagues is the caudal extreme of the medial vestibular nucleus (MVN), adjacent to the solitary tract close to plate 12. The cells in this region were shown to be GABAergic, receive primary afferents from vertical semicircular canals and utricular otoliths, and to relay the signals to the inferior olive (b-nucleus, and dorsomedial cell column). The olivary climbing fibers carry the vestibular information to the nodulus, which in turn projects back to the vestibular complex, as well as sending terminals to the cerebellar nuclei [for a review, see Barmack, 2006]. In the human the PSol has not yet been identified, but a large area which may include this vestibular cell group has been outlined by Koutcherov et al. [2004] and called PSol (see nucleus No. 2). It has small cells and lies at the junction of the solitary tract (TSOL), the MVN and the lateral cuneate nucleus (LCU).

Group e is a collection of multipolar neurons that lie between the abducens nucleus and the superior vestibular nucleus in higher mammals, dorsolateral to the facial genu. A possible example of this group e in humans is indicated in figure 1 of nucleus No. 41. The neurons have been recognized as the main origin of the efferent axons innervating the vestibular end organ, and hence the name 'group e' [Goldberg and Fernandez, 1980; Goldberg et al., 2000; Lysakowski and Goldberg, 2004]. Another smaller cluster of vestibular efferent neurons lies dorsomedial to the abducens nucleus in the monkey, and these tend to be more fusiform in appearance, while similar cells may also be found scattered in the caudal pontine reticular formation. In lower vertebrates the vestibular efferent neurons lie in or near the facial nucleus, associated with the facial branchial motoneurons. The efferent axons innervate the ipsilateral vestibular end organ, but a roughly equal number cross the midline to the contralateral labyrinth, passing over or through the spinal trigeminal tract.

In the human, cholinesterase stains outlined a region dorsolateral to the abducens nucleus and facial nerve, which was labeled the 'nucleus of origin of vestibular efferents' [Paxinos and Huang, 1995] or the paragenual nucleus in rats [Paxinos and Watson, 1986]. Group e neurons use acetylcholine as their neural transmitter to target nicotinic and muscarinic receptors in the hair cells.

The function of the vestibular efferent pathways is not clearly established. The fibers are modulated by vestibular and nonvestibular stimuli (e.g. skin pressure, limb movement or arousal). They play a role in modulating the activity of the peripheral end organs, and one suggestion proposed that they participate in the balance of activity between the vestibular organs of both sides, but that could not be verified experimentally. Another hypothesis suggests that they may serve to extend the dynamic range of the vestibular afferent. For a full review, see Goldberg et al. [2000, 2012, chapter 5].

Other cell groups associated with the vestibular complex include groups f, l, m, x and z, described in cats by Brodal and Pompeiano [1957], Sadjadpour and Brodal [1968] and by Suarez et al. [1997] in humans. None of these cell clusters are known to receive a significant vestibular-nerve input. They lie near the obex around the borders of the gracile, cuneate and spinal trigeminal nuclei at their junction with the caudal MVN and descending vestibular nucleus (DVN). Various authors have given these indistinct cell groups different names – a subject reviewed by Koutcherov et al. [2004] under the heading 'Pericuneate, Peritrigeminal, X and Paratrigeminal Nuclei'.

Group f is composed of large to medium ovoid cells which form clusters at the oral pole of the cuneate nucleus and inside the caudal pole of the vestibular nuclei (not shown in the plates). Brodal [1984] suggests that groups f and x have some similar features but group f lies further caudally. It is difficult to distinguish from the DVN, but has projections to the spinal cord and cerebellum: it also receives a direct input from cervical dorsal roots [Neuhuber and Bankoul, 2000; for a review, see Rubertone et al., 1995].

Group l is a variable, small-celled cluster of fusiform neurons at the lateral border of the lateral vestibular nucleus and medial to the restiform body, at a level where the vestibular nerve enters the vestibular nuclei (it is not shown in the plates but would lie close to plate 20). In the monkey, saccular afferents terminate on these cells [Newlands et al., 2003].

Group m lies in the caudal third of the MVN, a densely packed group of medium-sized cells that are closely associated with the TSOL. In man this region is often not identifiable (not shown in the plates), but in the monkey it seems well developed [Brodal, 1983], and in *Galago* group m was shown to project mainly to the paraflocculus [Rubertone and Haines, 1981].

Group x lies just rostral to the LCU, medial to the inferior cerebellar peduncle and lateral to the DVN (see dashed border on plate 17). In myelin-stained material this region is devoid of the typical bundles characterizing the DVN. Its cells receive inputs from the cervical spinal cord [Neuhuber and Bankoul, 2000], and the group is reciprocally connected to the cerebellum [Somana and Walberg, 1978; Brodal, 1984; Rubertone et al., 1995].

Group z is situated dorsal to the caudal DVN/MVN border (not shown in plates). The cells receive a rich innervation from the spinal cord, and they may provide a link between hindlimb afferents and the thalamus [Ostapoff et al., 1988].

Islands of the lateral cuneate nucleus (ISL) are cell groups outlined in the 2nd edition of this book, lateral and ventral to the LCU, and lying in the inferior cerebellar peduncle (plate 10). They are part of a series of cell groups surrounding the gracile, cuneate and peritrigeminal nuclei, which were considered by Olszewski and Baxter to be similar to the pontine nuclei [Braak, 1971; Koutcherov et al., 2004]. If these cell groups are homologues of the paratrigeminal nucleus, as suggested by Chan-Palay [1978], then they form a site for orofacial nociceptive sensory processing, with inputs from spinal, trigeminal, vagus and glossopharyngeal afferents. The paratrigeminal nucleus has efferent connections to brain structures associated with nociception and cardiorespiratory functions [Caous et al., 2008].

References

Barmack NH: Inferior olive and oculomotor system. Prog Brain Res 2006;151:269–291.

Braak H: Über die Kerngebiete des menschlichen Hirnstammes. IV. Der Nucleus reticularis lateralis und seine Satelliten. Z Zellforsch Mikrosk Anat 1971;122:145–159.

Brodal A: The perihypoglossal nuclei in the macaque monkey and the chimpanzee. J Comp Neurol 1983;218:257–269.

Brodal A: The vestibular nuclei in the macaque monkey. J Comp Neurol 1984;227:252–266.

Brodal A, Pompeiano O: The vestibular nuclei in cat. J Anat 1957;91:438–454.

Caous CA, Koepp J, Couture R, Balan AC, Lindsey CJ: The role of the paratrigeminal nucleus in the pressor response to sciatic nerve stimulation in the rat. Auton Neurosci 2008;140:72–79.

Chan-Palay V: The paratrigeminal nucleus. I. Neurons and synaptic organization. J Neurocytol 1978;7:405–418.

Goldberg JM, Brichta AM, Wackym PA: Efferent vestibular system: anatomy, physiology, and neurochemistry; in Beitz AJ, Anderson JH (eds): Neurochemistry of the Vestibular System. Boca Raton, CRC Press, 2000.

Goldberg JM, Fernandez C: Efferent vestibular system in the squirrel monkey: anatomical location and influence on afferent activity. J Neurophysiol 1980;43:986–1025.

Goldberg JM, Wilson VJ, Cullen KE, Angelaki DE, Broussard DM, Büttner-Ennever JA, Fukushima K, Minor LB: The Vestibular System – A Sixth Sense. Oxford, Oxford University Press, 2012.

Koutcherov Y, Huang X-F, Halliday G, Paxinos G: Organization of human brain stem nuclei; in Paxinos G, Mai JK (eds): The Human Nervous System, ed 2. Amsterdam, Elsevier Academic Press, 2004.

Lysakowski A, Goldberg JM: Morphophysiology of the vestibular periphery; in Highstein M, Fay RR, Popper AN (eds): The Vestibular System. New York, Berlin, Heidelberg, Springer, 2004.
Neuhuber WL, Bankoul S: Somatosensory influences on the vestibular system; in Beitz AJ, Anderson JH (eds): Neurochemistry of the Vestibular System. Boca Raton, CRC Press, 2000.
Newlands SD, Vrabec JT, Purcell IM, Stewart CM, Zimmerman BE, Perachio AA: Central projections of the saccular and utricular nerves in macaques. J Comp Neurol 2003;466:31–47.
Ostapoff EM, Johnson JI, Albright BC: Medullary sources of projections to the kinesthetic thalamus in raccoons: external and basal cuneate nuclei and cell groups x and z. J Comp Neurol 1988;267:231–252.
Paxinos G, Huang X-F: Atlas of the Human Brainstem. San Diego, Academic Press, 1995.
Paxinos G, Watson C: The Rat Brain in Stereotaxic Coordinates. Orlando, Academic Press, 1986.
Rubertone JA, Haines DE: Secondary vestibulocerebellar projections to flocculonodular lobe in a prosimian primate, *Galago senegalensis*. J Comp Neurol 1981; 200:255–272.
Rubertone JA, Mehler WR, Voogd J: The vestibular nuclear complex; in Paxinos G (ed): The Rat Nervous System. San Diego, Academic Press, 1995.
Sadjadpour K, Brodal A: The vestibular nuclei in man. A morphological study in the light of experimental findings in the cat. J Hirnforsch 1968;10:299–323.
Somana R, Walberg F: Cerebellar afferents from the paramedian reticular nucleus studied with retrograde transport of horseradish peroxidase. Anat Embryol 1978;154:353–368.
Suarez C, Diaz C, Tolivia J, Alvarez JC, Gonzalez del Rey C, Navarro A: Morphometric analysis of the human vestibular nuclei. Anat Rec 1997;247:271–288.

Sensory Systems – Vestibular Nuclei

(Plates 38, 39 and 41)

20 Interstitial Nucleus of Cajal (INC)

Original name:
Nucleus interstitialis (Cajal)

Alternative name:
Interstitial nucleus of the medial longitudinal fasciculus [Paxinos and Huang, 1995; Koutcherov et al., 2004]: it is unadvisable to use this name since it causes confusion with other nuclei (see chapters on nuclei 61 and 96)

Location and Cytoarchitecture – Original Text

The cells which form the interstitial nucleus of Cajal (INC) lie among the fibers of the medial longitudinal fasciculus (MLF) in the dorsomedial part of the rostral mesencephalic tegmentum. On cross-section, the nucleus is ovoid in outline and lies with its long axis directed dorsolaterally. It is related dorsomedially to the nucleus of Darkschewitsch (ND) and the central gray matter. Ventrolaterally, fibers of the capsule of the red nucleus (RN) and medial tegmental tract intervene between the INC and RN (see nucleus 101, fig. 9 and 10).

The cells composing the INC are compactly arranged, randomly oriented, medium sized, oval, round, fusiform or triangular in outline. They possess centrally placed, clear nuclei, distinct nucleoli and small to medium-sized, evenly distributed, darkly stained Nissl bodies. Occasional larger multipolar cells with long dendrites and large Nissl bodies may also be present (fig. 1–3).

Comment

The INC is grouped here with the vestibular complex because it has close connections with the system functionally. However, anatomically INC can be grouped with the reticular formation, like the rostral interstial nucleus of the MLF (nucleus No. 61).

Functional Neuroanatomy

Function

Numerous physiological and clinical studies indicate that the INC is essential for vertical gaze-holding and vertical eye-head coordination [Fukushima, 1987; Fukushima et al., 1992; Farshadmanesh et al., 2007; for reviews, see Leigh and Zee, 2006; Horn, 2006]. In order to maintain vertical gaze, the velocity-coded neural signals to the INC, from vestibular and oculomotor centers, must be transformed, or mathematically speaking *integrated*, to an eye position signal before they can drive the vertical ocular motoneurons. This is achieved through the reciprocal interconnections between the INC and vestibular nuclei, also involving the cerebellar flocculus [Fukushima et al., 1992; Fukushima and Kaneko, 1995]. Unilateral lesions of the INC lead to an ocular-tilt reaction, with contralateral head tilt, a tonic ocular torsion to the contralateral side, and skew deviation; in addition there is torsional nystagmus usually to the ipsilateral side [Helmchen et al., 2002]. Bilateral lesions of the INC result in upbeat nystagmus, neck retroflexion and a restriction of the vertical ocular range, which are also clinical signs characteristic of progressive supranuclear palsy [Fukushima-Kudo et al., 1987].

Connections

Three main efferent systems originate from the INC [Kokkoroyannis et al., 1996]: *ascending projections* to the (ipsilateral) mesencephalic reticular formation, including the rostral interstitial nucleus of the medial longitudinal fascicle (RIMLF) and zona incerta, weaker projections to the centromedian and parafascicular thalamic nuclei, mediodorsal, central medial and lateral nuclei of the thalamus; *descending projections*, which travel in the MLF and innervate the oculomotor (III) and trochlear nuclei (IV), the ipsilateral paramedian pontine reticu-

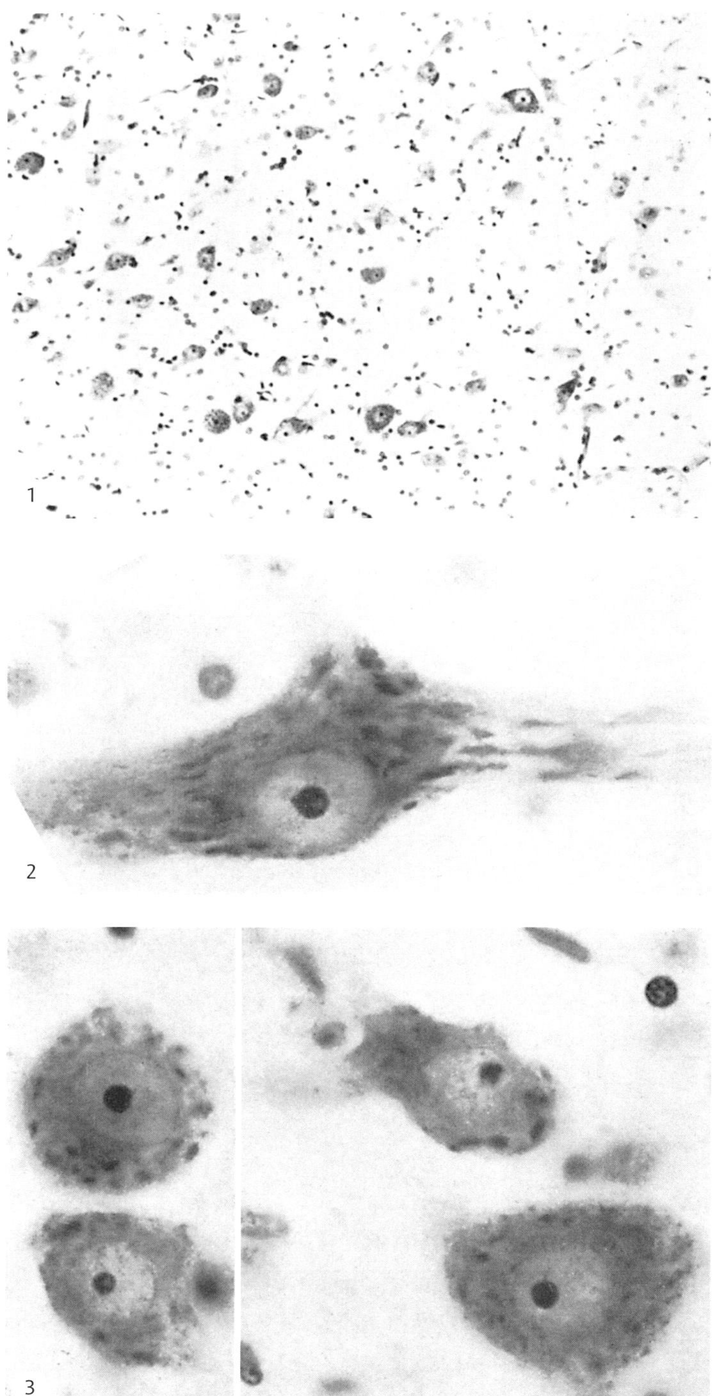

Fig. 1. Interstitial nucleus of Cajal. Magnification ×150.
Fig. 2, 3. Cells from the interstitial nucleus of Cajal. Magnification ×1,000.

lar formation, and the rostral cap of the abducens nucleus corresponding to one of the paramedian tract cell groups (nucleus No. 96), which are thought to contribute to gaze-holding. Additional descending projections mediating head movement control terminate in the vestibular nuclei, the prepositus nucleus, the gigantocellular nucleus, the inferior olive and the central horn of cervical segments C_1–C_4 [Holstege and Cowie, 1989; Cowie et al., 1994].

The commissural system from the INC projects through the posterior commissure (PC) to the nucleus of the PC, the contralateral INC, to the III and IV. At least some of these commissural fibers are GABAergic and target the motoneurons of exclusively vertically pulling extraocular eye muscles in the III and IV [Horn et al., 2003]. In addition the INC receives afferents from all premotor sources that encode eye or head velocity, e.g. secondary vestibulo-ocular neurons including the Y group, and the RIMLF in the mesencephalic reticular formation, which contains premotor burst neurons for vertical saccadic eye movements [Moschovakis et al., 1996].

Additional Features

The medium-sized neurons of the INC, which represent the premotor neurons of the III and IV express strong parvalbumin immunoreactivity and are ensheathed by perineuronal nets [Horn and Büttner-Ennever, 1998]. A population of large neurons in the INC are GABAergic and project to the contralateral III and IV via the PC [Horn and Büttner-Ennever, 1998; Horn et al., 2003]. These properties have been used to delineate the INC clearly from the dorsally adjacent ND, which contains much less parvalbumin, and from the RIMLF, which lacks GABAergic neurons and has a more reticulated appearance.

References

Cowie RJ, Smith MK, Robinson DL: Subcortical contributions to head movements in macaques. II. Connections of a medial pontomedullary head-movement region. J Neurophysiol 1994;72:2665–2682.
Farshadmanesh F, Klier EM, Chang P, Wang H, Crawford JD: Three-dimensional eye-head coordination after injection of muscimol into the interstitial nucleus of Cajal (INC). J Neurophysiol 2007;97:2322–2338.
Fukushima-Kudo J, Fukushima K, Tashiro K: Rigidity and dorsiflexion of the neck in progressive supranuclear palsy and the interstitial nucleus of Cajal. J Neurol Neurosurg Psychiatry 1987;50:1197–1203.
Fukushima K: The interstitial nucleus of Cajal and its role in the control of movements of head and eyes. Progr Neurobiol 1987;29:107–192.
Fukushima K, Kaneko CR: Vestibular integrators in the oculomotor system. Neurosci Res 1995;22:249–258.
Fukushima K, Kaneko CR, Fuchs AF: The neuronal substrate of integration in the oculomotor system. Prog Neurobiol 1992;39:609–639.
Helmchen C, Rambold HC, Kempermann U, Büttner-Ennever JA, Büttner U: Localizing value of torsional nystagmus in small midbrain lesions. Neurology 2002; 59:1956–1964.
Holstege G, Cowie RJ: Projections from the rostral mesencephalic reticular formation to the spinal cord. Exp Brain Res 1989;75:265–279.
Horn AKE: The reticular formation. Prog Brain Res 2006;151:127–155.
Horn AKE, Büttner-Ennever JA: Premotor neurons for vertical eye-movements in the rostral mesencephalon of monkey and man: the histological identification by parvalbumin immunostaining. J Comp Neurol 1998;392:413–427.
Horn AKE, Helmchen C, Wahle P: GABAergic neurons in the rostral mesencephalon of the macaque monkey that control vertical eye movements. Ann NY Acad Sci 2003;1004:19–28.
Kokkoroyannis T, Scudder CA, Balaban CD, Highstein SM, Moschovakis AK: Anatomy and physiology of the primate interstitial nucleus of Cajal. I. Efferent projections. J Neurophysiol 1996;75:725–739.
Koutcherov Y, Huang X-F, Halliday G, Paxinos G: Organization of human brain stem nuclei; in Paxinos G, Mai JK (eds): The Human Nervous System, ed 2. Amsterdam, Elsevier Academic Press, 2004.
Leigh RJ, Zee DS: The Neurology of Eye Movements. New York, Oxford University Press, 2006.
Moschovakis AK, Scudder CA, Highstein SM: The microscopic anatomy and physiology of the mammalian saccadic system. Prog Neurobiol 1996;50:133–133.
Paxinos G, Huang X-F: Atlas of the Human Brainstem. San Diego, Academic Press, 1995.

21 Ventral Cochlear Nucleus (VCN)

Original name:
Nucleus cochlearis ventralis

Alternative name:
Anterior cochlear nucleus

Location and Cytoarchitecture – Original Text

This nucleus is located on the ventrolateral aspect of the inferior cerebellar peduncle (ICP) at the junction of the medulla and pons. It is pear shaped in outline on both cross- and sagittal sections with its apex directed dorsomedially. Cells from the apex of the nucleus stream along the lateral border of the ICP to come into association with the cells of the dorsal cochlear nucleus (DCN). Medially and ventrally the nucleus is related to the afferent fibers of the cochlear nerve. Laterally, the caudal portion of the nucleus forms the free border of the medulla which, at this level, is separated from the cerebellum by the foramen of Luschka. The oral portion of the nucleus is related laterally to the white matter of the cerebellum. Caudo-orally, the ventral cochlear nucleus (VCN) measures 3 mm.

The internal structure of the VCN is quite characteristic. The cells are of medium size, oval or round, and possess very short processes (fig. 1–5). In some cells the nucleus may be excentrically placed. The Nissl granules are very fine, darkly stained and evenly distributed, except at the periphery of the perikaryon, where, in many cells, there is a paucity of granules. In a few cells the Nissl granules are arranged in a concentric pattern about the nucleus. The majority of cells lie with their long axes directed dorsomedially, giving the nucleus a very orderly appearance (fig. 1). [...]

The VCN and DCN receive afferent fibers concerned with the conduction of auditory impulses. These fibers have their cell bodies in the spiral ganglion of Corti and proceed to the central nervous system via the cochlear division of the eighth nerve. It has long been accepted that immediately after entering the brainstem these cochlear fibers bifurcate into an ascending branch which proceeds to the ventral cochlear nucleus and a descending branch which proceeds over the tuberculum acusticum, a prominence formed by the DCN, to terminate in that nucleus, the DCN. [...].

Functional Neuroanatomy

Function

The VCN and DCN are the first relay of the auditory nerve in the brainstem, and serve as a point of divergence in the representation of auditory information. Specific groups of neurons in the VCN distribute the information to different brainstem targets such as the superior olivary complex (SOC), the nuclei of the lateral lemniscus and the inferior colliculus (IC). In some cases the VCN relay modifies the information to reduce synaptic jitter and secure precise signal transfer, which is important for the comparison of timing of acoustic signals, the basis of sound localization.

Unilateral damage to the cochlear nuclei causes monaural deafness, whereas lesions to higher auditory centers are usually less pronounced due to the multiple commissural pathways of the auditory system in the brainstem [Strominger et al., 1977]. The use of auditory brain implants, targeting the cochlear nuclei, to alleviate hearing loss is the topic of much current research [Adams, 1986; McCreery and Otto, 2011/2012].

Connections

The auditory system is adapted to receive sound waves. Humans are sensitive to sound frequencies of about 20–20,000 Hz, with the most sensitive range at 1,000–4,000 Hz. Auditory processing already starts in the outer ear where sound waves are reflected into the external auditory meatus from different points of the complicated pinna surface, and form different sound streams with different latencies. The various sound streams are compared in the brainstem nuclei, and the information is used as cues for sound localization, and possibly speech analysis [Hofman et al., 1998]. In some animals active motor movements of the pinna orienting towards sound sources may be used to provide the central pathways with information for building up an auditory map of the surroundings.

Afferents. The auditory hair cells are innervated by primary afferents. Their bipolar ganglion cells lie in the spiral ganglion embedded in the bony modiolus: inner hair cells are supplied by type 1 ganglion neurons, and the outer hair cells by the smaller type 2. The central processes of the primary afferents join to form the cochlear nerve, whose fibers enter at the central region of the VCN; the ascending branches supply the anterior part of the VCN, the descending branches supply the posterior part and continue on to the DCN. This may apply only to type 1 cochlear afferents and not to the thin unmyelinated axons of type 2 [Hurd et al., 1999].

The auditory fibers in both the anterior and posterior parts of the VCN terminate in a strictly tonotopic fashion, with high frequencies generated at the base of the cochlear spiral represented dorsally, and low frequencies from the apex ventrally. The dorsalmost region of the VCN is comparatively amorphous, and covers the lower 2 parts like a cap: it receives very thin myelinated branches of the auditory nerve [Moore and Linthicum, 2004].

Further afferents to the VCN arise from the DCN, originating from small glycine- and GABA-positive neurons called either 'vertical', 'corn' or 'fan cells' [Oertel and Wu, 1989]. The cells provide a precise inhibition to the central VCN in the corresponding isofrequency planes. Also large glycine-positive cells have been shown to provide an inhibitory commissural system between the cochlear nuclei of both sides [Bledsoe et al., 2009].

Efferents. The VCN is characterized by the presence of anatomically and physiologically distinct cell groups, each with different output pathways: there are 'spherical bushy cells' and 'globular bushy cells', 'octopus cells' and 'multipolar cells' [Moore and Osen, 1979; Nieuwenhuys et al., 2008]. The cell types and their dendrites can be visualized by immunostaining for microtubule-associated protein-2 but are not seen clearly in Nissl stains; for a review, see Moore and Linthicum [2004].

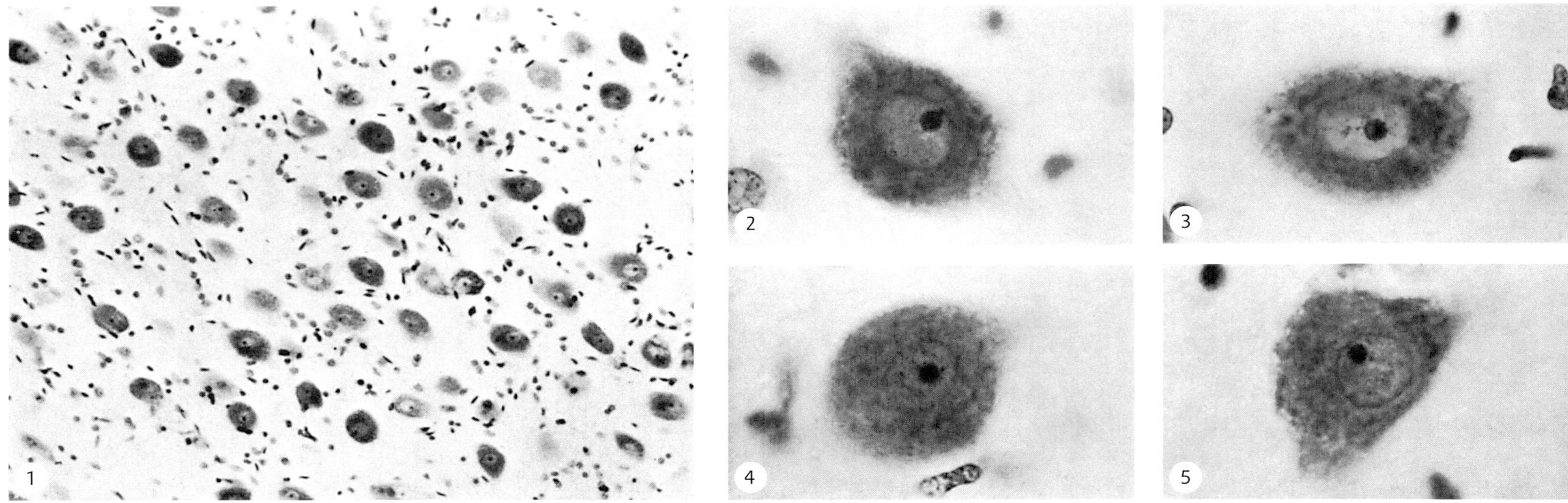

Fig. 1. Ventral cochlear nucleus. Magnification ×150.
Fig. 2–5. Cells from the ventral cochlear nucleus. Magnification ×1,000.

In the anterior VCN lie the groups of spherical bushy cells and globular bushy cells, which have almost a 1:1 relationship with cochlear nerve fibers and project to the trapezoid body, where they participate in processing binaural information and sound localization. Some of the synapses of the primary auditory nerve terminate on the spherical bushy cells, and form some of the largest synapses in the brain: they completely encircle the neuron with hundreds of contacts, and are called 'end bulbs of Held' [Rouiller et al., 1986]. The spherical bushy cells project directly to the SOC bilaterally, and the globular bushy cells relay in part of the contralateral SOC, the medial nucleus of the trapezoid body, also ending here in giant calyces [Ryugo and Spirou, 2009]. In the posterior VCN lie the group of octopus cells which form an oval region of larger cells. They project through the intermediate acoustic stria to the periolivary complex and the contralateral ventral nucleus of the lateral lemniscus, and travel on to the IC: they participate in the indirect monoaural pathway used amongst other things for speech analysis. The octopus cell area of the VCN sends axons to the DCN, and finally, the multipolar cells (also known as 'stellate cells') in the anterior VCN project directly and bilaterally to the inferior colliculus [Oliver, 1987]. For reviews of these pathways, see Nieuwenhuys et al. [2008] and Moore and Linthicum [2004].

The VCN sends efferents to the facial and trigeminal nuclei as part of a reflex arc to activate the stapedius and tensor tympani muscles, respectively, in the case of a loud disturbance [Nieuwenhuys et al., 2008, fig. 18.5B]. The middle ear muscle activity then reduces sound transmission as a protection mechanism: similar pathways may be in continuous use to reduce hearing sensitivity during self-generated speaking or shouting.

References

Adams JC: Neuronal morphology in the human cochlear nucleus. Arch Otolaryngol Head Neck Surg 1986;112:1253–1261.

Bledsoe SC Jr, Koehler S, Tucci DL, Zhou J, Le Prell C, Shore SE: Ventral cochlear nucleus responses to contralateral sound are mediated by commissural and olivocochlear pathways. J Neurophysiol 2009;102:886–900.

Hofman P, van Riswick J, van Opstal A: Relearning sound localization with new ears. Nat Neurosci 1998;1:417–421.

Hurd LB, Hudson KA, Morest DK: Cochlear nerve projections to the small cell shell of the cochlear nucleus: the neuroanatomy of extremely thin sensory axons. Synapse 1999;33:83–117.

McCreery DB, Otto SR: Cochlear nucleus auditory prostheses; in Zeng FG, Popper AN, Fay RR (eds): Auditory Prostheses: New Horizons. Springer Handbook of Auditory Research. New York, Springer, 2011/2012, vol 39.

Moore JK, Linthicum JR: Auditory System; in Paxinos G, Mai JK (eds): The Human Nervous System, ed 2. San Diego, Elsevier Academic Press, 2004.

Moore JK, Osen KK: The cochlear nuclei in man. Am J Anat 1979;154:393–418.

Nieuwenhuys R, Voogd J, van Huijzen C: The Human Central Nervous System. Berlin, Springer, 2008.

Oertel D, Wu SH: Morphology and physiology of cells in slice preparations of the dorsal cochlear nucleus of mice. J Comp Neurol 1989;283:228–247.

Oliver DL: Projections to the inferior colliculus from the anteroventral cochlear nucleus in the cat: possible substrates for binaural interaction. J Comp Neurol 1987;264:24–46.

Rouiller EM, Cronin-Schreiber R, Fekete DM, Ryugo DK: The central projections of intracellularly labeled auditory nerve fibers in cats: an analysis of terminal morphology. J Comp Neurol 1986;249:261–278.

Ryugo DK, Spirou GA: Auditory system: giant synaptic terminals, endbulbs and calyces; in Squire LR (ed): New Encyclopedia of Neuroscience. Oxford, Academic Press, 2009.

Strominger NL, Nelson LR, Dougherty WJ: Second order auditory pathways in the chimpanzee. J Comp Neurol 1977;172:349–365.

22 Dorsal Cochlear Nucleus (DCN)

Original name:
Nucleus cochlearis dorsalis

Alternative name:
Posterior cochlear nucleus

Location and Cytoarchitecture – Original Text

This nucleus is formed by cells which lie among the fibers of the cochlear nerve on the dorsolateral surface of the inferior cerebellar peduncle (ICP) at the junction of the medulla and pons. Laterally, this nucleus and the fibers which stream through it form the lateral border of the medulla and are separated from the cerebellum by the foramen of Luschka. The most dorsally situated cells lie in the floor of the lateral recess of the fourth ventricle; the most ventrally situated cells are intimately related to cells of the ventral cochlear nucleus (VCN). Caudo-orally the nucleus measures only 2–3 mm.

The dorsal cochlear nucleus (DCN) is much less conspicuous in man than in lower animals in which the nucleus is distinctly laminated and forms a prominent bulge on the surface of the ICP, usually described as the tuberculum acusticum.

The internal architecture of the DCN differs remarkably from that of the VCN. Its cells are also clearly distinguishable from the nucleus of the pontobulbar body (nucleus No. 97). The cells are medium sized, predominantly spindle shaped, and possess a centrally placed nucleus and long dendrites (fig. 2, 3). In the majority the Nissl granules are indistinct, but the perikaryon is darkly stained. Other smaller, but otherwise similar cells stain very lightly (fig. 4, 5). All of the cells are arranged with their long axes parallel to the dorsolateral surface of the ICP. Although usually no lamination similar to that seen in animals is distinguishable in man, one may occasionally see some indication of cell layering as illustrated in figure 1.

Functional Neuroanatomy

Function

The DCN has a unique position in the auditory system as a point of convergence of auditory and somatosensory information, as well as inputs from higher centers such as the cerebral cortex. This, combined with results from lesion and recording experiments, has led to numerous hypotheses of its function, mainly centering on a role in the localization of sound sources based on pinna cues, and in the extraction of novel components of sounds. However, neurons of the DCN have very complex firing patterns and their function is not clear [for reviews, see Young and Davis, 2001; Oertel and Young, 2004].

Some clinical studies have discussed the relevance of DCN activity to tinnitus [Sutherland et al., 1998; Levine, 1999; Young and Davis, 2001]. The DCN and the VCN are the targets for auditory brainstem implants, which can improve speech recognition to a limited extent [McCreery, 2008; McCreery and Otto, 2011/2012].

Connections

Afferents. All cochlear nerve fibers from type 1 spiral ganglion cells divide into an ascending and a descending branch as they enter the cochlear nuclear complex (nucleus No. 21). The fibers of the descending branch innervate first the caudal part of the VCN, then converge into a bundle and ascend to supply the DCN. In humans the fibers of the descending branch of the cochlear nerve run parallel to the surface of the DCN: this is different to other mammals where the fibers entering the DCN turn towards the brainstem surface and form a tonotopic plane at right angles to the nuclear surface [Moore and Linthicum, 2004]. Another difference in the primate DCN is the reorientation of the fusiform cells: in lower mammals with a clearly laminated DCN (and well-developed pinnae) the fusiform cells are oriented radially to the surface, with their dendrites at their poles stretching towards the surface or towards the deep layers, but in humans the fusiform neurons lie parallel to the nuclear surface. Studies of the DCN have equated several of the neural components found in lower animals with those in humans [for a review, see Moore and Linthicum, 2004].

The DCN receives inputs from auditory and somatosensory sources, including information about the pinna and head position. More specifically, it is targeted by the dorsal column nuclei, vestibular afferents, pontine nuclei, the 'octopus cell' area of the VCN, the inferior colliculus and the auditory cortex [for reviews, see Young et al., 1995; Oertel and Young, 2004; Kanold et al., 2011].

Efferents. Axons arising from fusiform cell bodies, also called 'pyramidal cells', course posteriorly over the restiform body, the ICP, as the dorsal acoustic stria [Adams and Warr, 1976]. The fibers decussate through the brainstem, and enter the lateral lemniscus where they ascend to terminate mainly in the central nucleus of the inferior colliculus [Osen, 1972; Ryugo and May, 1993].

Several different types of cells have been described in the DCN [Lorente de Nó, 1981; Wouterlood and Mugnaini, 1984]: 'granule cells' with parallel fiber efferents lying in the molecular layer; glycinergic 'cartwheel cells' in the deeper layers, as well as the glutaminergic 'fusiform' and 'giant cells', which are the principal cell type and provide the major efferent pathway from the DCN [for a review, see Oertel and Young, 2004]. Interestingly these complex intrinsic circuits in the DCN have been compared to the cerebellar circuitry, with which they have several features in common both cytoarchitecturally and developmentally [Rhode et al., 1983; Funfschilling and Reichardt, 2002; Tzounopoulos et al., 2004], albeit these features appear less striking in humans.

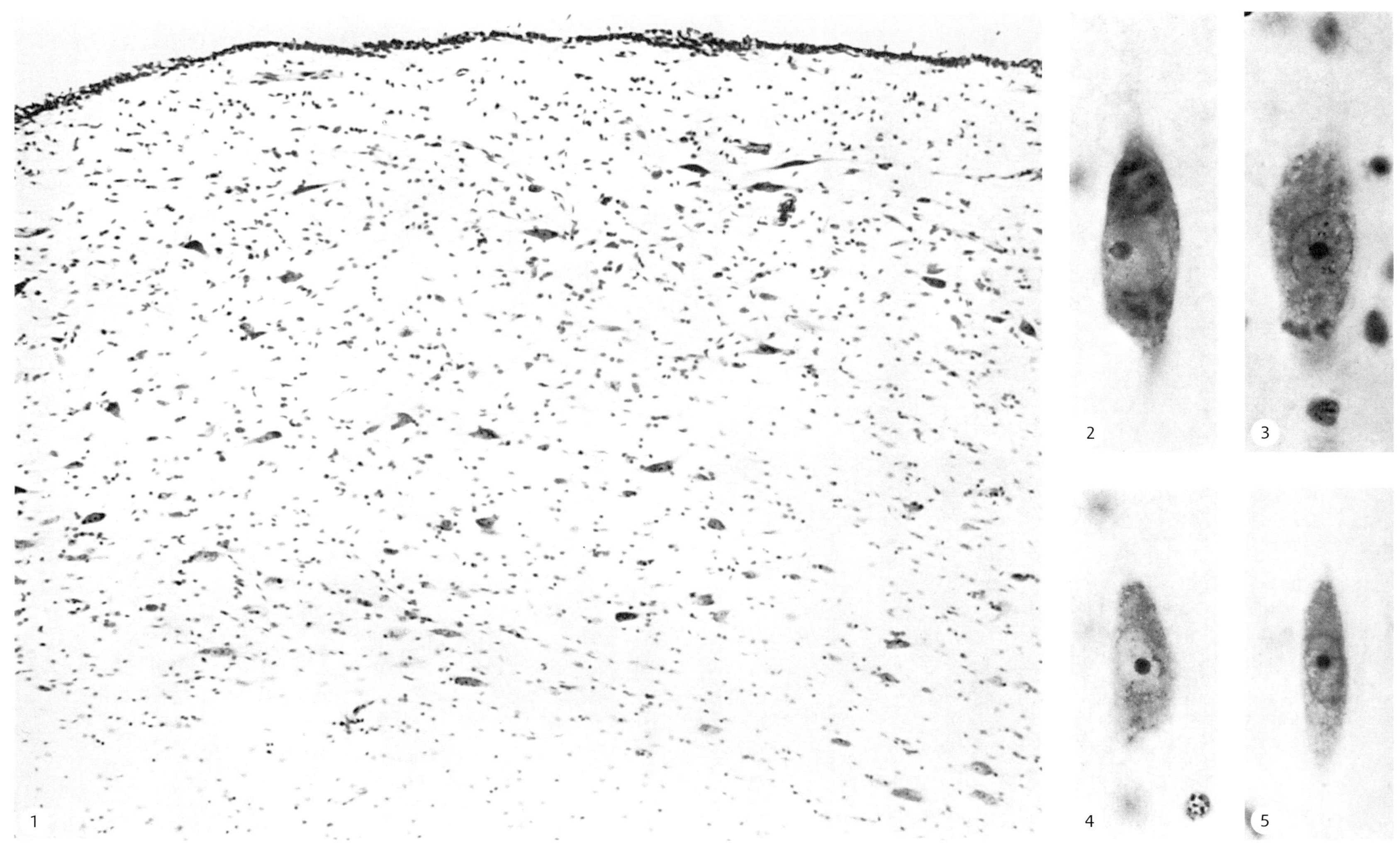

Fig. 1. Dorsal cochlear nucleus. Magnification ×150.
Fig. 2–5. Cells from the dorsal cochlear nucleus. Magnification ×1,000.

References

Adams JC, Warr WB: Origins of axons in the cat's acoustic striae determined by injection of horseradish peroxidase into severed tracts. J Comp Neurol 1976; 170:107–122.

Funfschilling U, Reichardt LF: Cre-mediated recombination in rhombic lip derivatives. Genesis 2002;33:160–169.

Kanold PO, Davis KA, Young ED: Somatosensory context alters auditory responses in the cochlear nucleus. J Neurophysiol 2011;105:1063–1070.

Levine RA: Somatic (craniocervical) tinnitus and the dorsal cochlear nucleus hypothesis. Am J Otolaryngol 1999;20:351–362.

Lorente de Nó R: The Primary Acoustic Nuclei. New York, Raven Press, 1981.

McCreery DB: Cochlear nucleus auditory prostheses. Hear Res 2008;242:64–73.

McCreery DB, Otto SR: Cochlear nucleus auditory prostheses; in Zeng FG, Popper AN, Fay RR (eds): Auditory Prostheses: New Horizons. Springer Handbook of Auditory Research. New York, Springer, 2011/2012, vol 39.

Moore JK, Linthicum JR: Auditory system; in Paxinos G, Mai JK (eds): The Human Nervous System, ed 2. San Diego, Elsevier Academic Press, 2004.

Oertel D, Young ED: What's a cerebellar circuit doing in the auditory system? Trends Neurosci 2004;27:104–110.

Osen KK: Projection of the cochlear nuclei on the inferior colliculus of the cat. J Comp Neurol 1972;144:355–372.

Rhode WS, Oertel D, Smith PH: Physiological response properties of cells labeled intracellularly with horseradish peroxidase in cat dorsal cochlear nucleus. J Comp Neurol 1983;213:426–447.

Ryugo DK, May SK: The projections of intracellularly labeled auditory nerve fibers to the dorsal cochlear nucleus of cats. J Comp Neurol 1993;329:20–35.

Sutherland DP, Masterton RB, Glendenning KK: Role of acoustic striae in hearing: reflexive responses to elevated sound-sources. Behav Brain Res 1998;97:1–12.

Tzounopoulos T, Kim Y, Oertel D, Trussell LO: Cell-specific, spike timing-dependent plasticities in the dorsal cochlear nucleus. Nat Neurosci 2004;7:719–725.

Wouterlood FG, Mugnaini E: Cartwheel neurons of the dorsal cochlear nucleus: a Golgi-electron microscopic study in rat. J Comp Neurol 1984;227:136–157.

Young ED, Davis KA: Circuitry and function of the dorsal cochlear nucleus; in Oertel D, Popper AN, Fay RR (eds): Integrative Functions of the Mammalian Auditory Pathway. New York, Springer, 2001.

Young ED, Nelken I, Conley RA: Somatosensory effects on neurons in dorsal cochlear nucleus. J Neurophysiol 1995;73:743–765.

23 Periolivary Complex (POC)

Original name:
Nucleus trapezoidalis or nucleus of the trapezoid body

Subdivisions:
No subnuclei are evident in the periolivary complex of humans [Moore, 1987]; however, in lower animals over 10 subnuclei have been defined (see below)

Location and Cytoarchitecture – Original Text

In man, the periolivary nuclei or complex (POC) is less well developed than in other mammals. The area is composed of small, poorly defined cell groups which lie among the fibers of the trapezoid body, lateral, ventral and medial to the complex of the superior olive and dorsal to the pontine nuclei (PN). The periolivary nuclei are best developed about the caudal half of the superior olivary complex (SOC).

The cells composing the POC are loosely arranged, small to medium sized, triangular or multipolar and possess long dendrites and large, darkly stained Nissl bodies (fig. 1–4). The distinction of these cells from those composing the lateral superior olive may be difficult. However, the cells of the latter are plumper, possess shorter dendrites and more discrete Nissl bodies. The cells of the adjacent PN are larger and more intensely stained than those of the POC. [...]

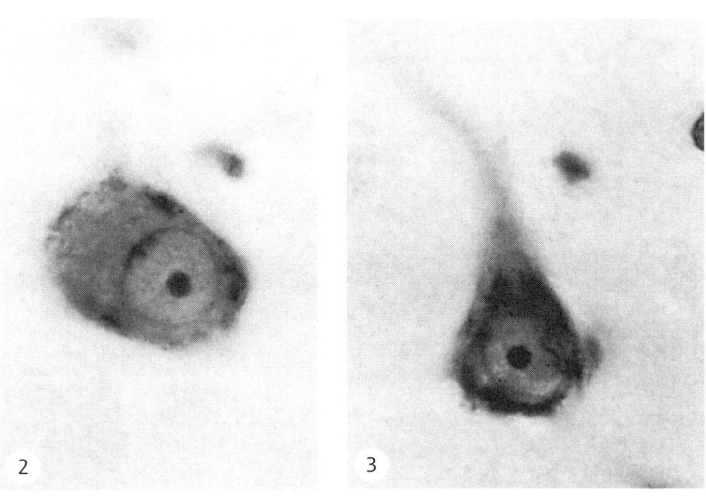

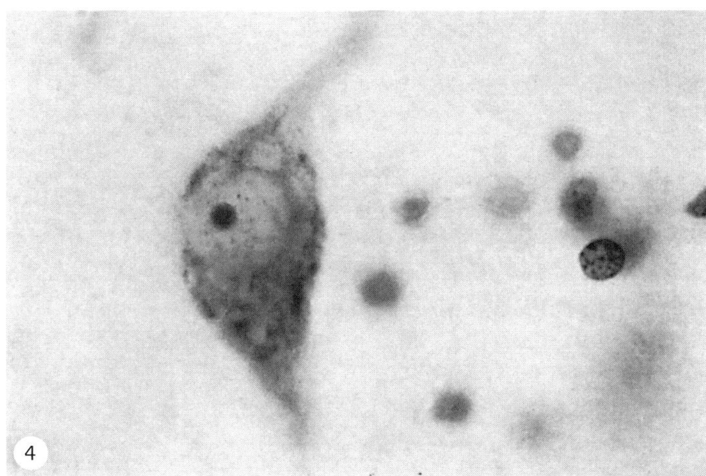

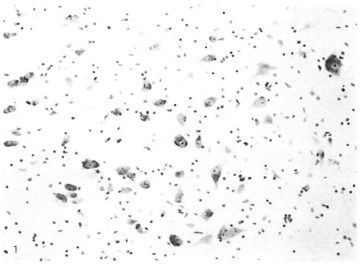

Fig. 1. Periolivary complex. Magnification ×150.
Fig. 2–4. Cells from the periolivary complex. Magnification ×1,000.

Functional Neuroanatomy

Function
The POC in humans forms a ring of loosely arranged cell groups surrounding the SOC. It contributes ascending and descending pathways to the auditory system, and is the origin of the efferent olivocochlear bundles (OCB). The OCB modulate the discharge rate of cochlear afferent fibers and may, in part, play a role in the olivocochlear reflexes preventing overstimulation of the peripheral neurons. Many different cell types have been identified in the POC of the cat and rat, but a general function has not been recognized [for a review, see Moore, 1987; Helfert and Aschoff, 1997; Schofield, 2005].

Connections
The principal connections of the POC are (a) inputs from the posterior part of the ventral cochlear nucleus (VCN), more specifically from the 'octopus cells', and (b) efferent projections to the central nucleus of the inferior colliculus (IC), mainly ipsilaterally [Adams, 1983; Schofield and Cant, 1992; Thompson and Schofield, 2000]. The rostral POC is associated with motor neurons of the trigeminal nucleus and supports the middle ear reflex of the tensor tympani muscle, along with the cochlear nuclei; similarly the caudal POC interacts with the facial motor neurons innervating the stapedius muscle [Lee et al., 2006; Jones et al., 2008].

Periolivary Nuclear Homologues

The periolivary nuclei are named according to their location around the superior olive [Koutcherov et al., 2004]; however, homologues of the nuclei seen in lower mammals are not evident in humans [Moore, 1987, Schmidt et al., 2010]. A region that can be called the medioventral periolivary nucleus in humans may in part represent the homologue of the medial nucleus of the trapezoid body (MNTB) in lower mammals. The MNTB is usually included in the term 'superior olivary complex' but in humans its location is not clear, and it appears to be integrated into the POC [Richter et al., 1983]. The MNTB contains glycinergic neurons enveloped by the calyces of Held (nucleus No. 24) [Kulesza, 2008]. These neurons are the 'principal cells' of the MNTB which send inhibitory signals to the lateral superior olivary nucleus (LSO), while the nonprincipal MNTB cells project to the IC [Schofield, 1994].

A medial region of the human POC is called the dorsomedial periolivary nucleus by Koutcherov et al. [2004]. It is equated with the superior paraolivary nucleus of the rat. The area receives a distinctly different projection from the cochlear nuclei to the medial superior olivary nucleus and LSO, namely inputs from 'multipolar' and 'octopus cell' groups of the contralateral VCN, and its GABAergic neurons provide a major projection to the ipsilateral IC [Schofield, 2005].

The POC is the source of the OCB, carrying efferents to the cochlea [Adams, 1983]. In general, the cochlea of adult mammals receives two types of efferent innervation: one division is called the medial OCB, which arises from large cells, usually lying in the medial POC, and projecting bilaterally to the type 2 outer hair cells; the second division is the lateral OCB, which arises from small cells and projects ipsilaterally to the type 1 inner hair cells. The OCB cells are strongly cholinergic and have been plotted in humans [Moore and Linthicum, 2004; Koutcherov et al., 2004]. The large 'medial OCB cells' lie scattered in several periolivary groups in humans, whereas the smaller 'lateral OCB cells' lie mainly dorsally. The efferent medial OCB may be the peripheral part of the descending auditory system, stretching from the cortex to the cochlea, which can enhance selective attention to auditory signals [for a review, see Moore and Linthicum, 2004; Cooper and Guinan, 2006]. The function of the lateral OCB is not known. Guinan [2006] reviews the efferent innervation of the cochlea and its association with otoacoustic emissions (low-level sounds produced in healthy cochleae, that can be measured noninvasively in humans with a sensitive microphone in the ear canal).

References

Adams JC: Cytology of periolivary cells and the organization of their projections in the cat. J Comp Neurol 1983;215:275–289.

Cooper NP, Guinan JJ Jr: Efferent-mediated control of basilar membrane motion. J Physiol 2006;576:49–54.

Guinan JJ Jr: Olivocochlear efferents: anatomy, physiology, function, and the measurement of efferent effects in humans. Ear Hear 2006;27:589–607.

Helfert RH, Aschoff A: Superior olivary complex and nuclei of the lateral lemniscus; in Ehret G, Romand R (eds): The Central Auditory System. New York, Oxford University Press, 1997.

Jones SE, Mason MJ, Sunkaraneni VS, Baguley DM: The effect of auditory stimulation on the tensor tympani in patients following stapedectomy. Acta Otolaryngol 2008;128:250–254.

Koutcherov Y, Huang X-F, Halliday G, Paxinos G: Organization of human brain stem nuclei; in Paxinos G, Mai JK (eds): The Human Nervous System, ed 2. Amsterdam, Elsevier Academic Press, 2004.

Kulesza RJ Jr: Cytoarchitecture of the human superior olivary complex: nuclei of the trapezoid body and posterior tier. Hear Res 2008;241:52–63.

Lee DJ, De Venecia RK, Guinan JJ Jr, Brown MC: Central auditory pathways mediating the rat middle ear muscle reflexes. Anat Rec A Discov Mol Cell Evol Biol 2006;288:358–369.

Moore J: The human auditory brain stem: a comparative view. Hear Res 1987;29: 1–32.

Moore JK, Linthicum JR: Auditory system; in Paxinos G, Mai JK (eds): The Human Nervous System, ed 2. San Diego, Elsevier Academic Press, 2004.

Richter EA, Norris BE, Fullerton BC, Levine RA, Kiang NY: Is there a medial nucleus of the trapezoid body in humans? Am J Anat 1983;168:157–166.

Schmidt E, Wolski TP Jr, Kulesza RJ Jr: Distribution of perineuronal nets in the human superior olivary complex. Hear Res 2010;265:15–24.

Schofield BR: Projections to the cochlear nuclei from principal cells in the medial nucleus of the trapezoid body in guinea pigs. J Comp Neurol 1994;344:83–100.

Schofield BR: Connections of the superior olivary complex and lateral lemniscus with the inferior colliculus; in Winer JA, Schreiner CE (eds): The Inferior Colliculus. New York, Springer, 2005.

Schofield BR, Cant NB: Organization of the superior olivary complex in the guinea pig. II. Patterns of projection from the periolivary nuclei to the inferior colliculus. J Comp Neurol 1992;317:438–455.

Thompson AM, Schofield BR: Afferent projections of the superior olivary complex. Microsc Res Tech 2000;51:330–354.

Sensory Systems – Auditory Nuclei

(Plate 20–22)

24 Superior Olivary Complex (SOC)

Original name:
Nucleus olivaris superior

Subdivisions:
Medial superior olivary nucleus (MSO)
Lateral superior olivary nucleus (LSO)
Medial nucleus of the trapezoid body (or nucleus of the trapezoid body, NTB) is included in the superior olive in lower mammals but it is considered in humans to lie in the periolivary complex, see below [Richter et al., 1983]

Location and Cytoarchitecture – Original Text

Whereas in lower mammals, the superior olive or superior olivary complex (SOC) is represented by 2 large conspicuous cell groups, in man only one of these groups, the medial superior olive (MSO), remains prominent. The other, the lateral superior olive (LSO), is reduced to a vestigial structure which is often difficult to delineate.

The SOC is situated in the ventrolateral corner of the caudal pontine tegmentum and extends from the caudal pole of the facial nucleus (VII) to the caudal pole of the motor trigeminal nucleus – a distance of 4–5 mm. Orally, the cells of the MSO are in direct continuity with the cells of the ventral nucleus of the lateral lemniscus.

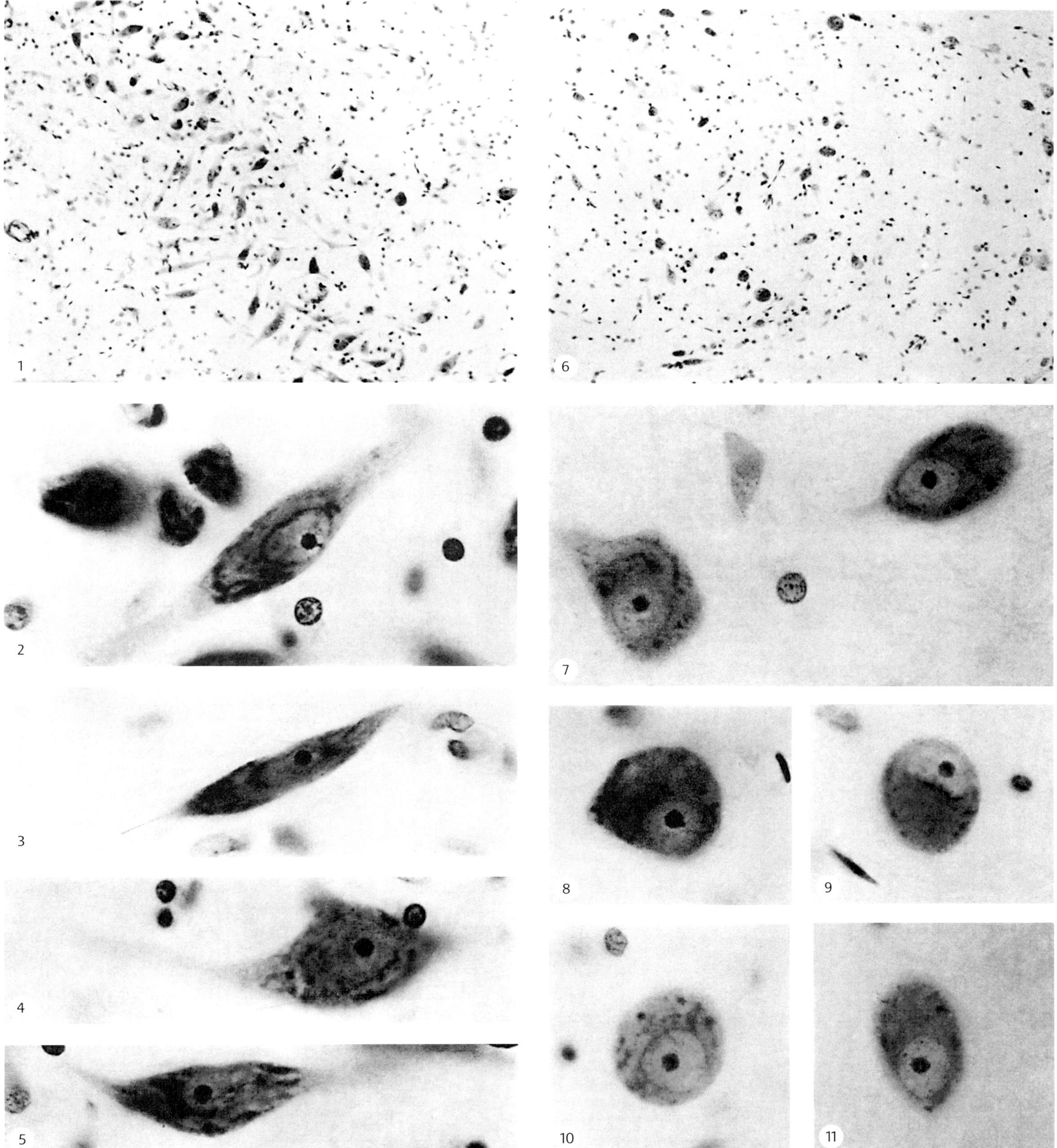

Fig. 1. Superior olivary complex, medial superior olive subnucleus. Magnification ×150.

Fig. 2–5. Cells from the superior olivary complex, medial superior olive subnucleus. Magnification ×1,000.

Fig. 6. Superior olivary complex, lateral superior olive subnucleus. Magnification ×150.

Fig. 7–11. Cells from the superior olivary complex, lateral superior olive subnucleus. Magnification ×1,000.

On cross-section, the SOC is roughly circular in outline. The caudal portion of the complex is related medially, ventrally and laterally to the nuclei of the periolivary complex (POC) and dorsally to the VII. The oral portion is bordered by fibers of the lateral lemniscus (LL) medially and ventrally, and by the POC laterally. Dorsally, this portion of the complex is related to the nucleus subcoeruleus.

The MSO is a crescent-shaped cell group which is surrounded by an oval, dense glial network. The concavity of the nucleus is directed laterally and this hilus is cell free. The cells which make up the MSO are slender and spindle shaped. They possess long dendrites, a centrally placed nucleus and darkly stained Nissl granules which tend to accumulate at the extremities or about the periphery of the cell (fig. 1–5). The long axes of these cells are directed at right angles to the long axis of the nucleus. [...]

The LSO is composed of several small cell groups congregated to form a curved band of cells which lies dorsal and dorsolateral to the MSO. One of these cell groups, situated dorsolateral to the MSO, is more conspicuous than the others due to the fact that its cells are more compactly arranged. The cells which compose the LSO are small to medium sized, round or plump in outline and stain with moderate intensity. They possess short dendrites, a centrally placed nucleus and finely granulated Nissl bodies (fig. 6–11). The distinction between these cells and those of the POC is very often difficult [Bazwinsky et al., 2003; Kulesza, 2007]. [...]

Functional Neuroanatomy

Function

The nuclei of the SOC are the first site of convergence of the cochlear input from the two ears. Their precise comparison of the binaural inputs is generally accepted as the basis for building an auditory map and locating sounds in the horizontal plane. The MSO specializes in interaural timing, and the LSO in intensity comparison, both used to locate sound sources [Webster, 1992].

Superior Olivary Complex Subdivisions

Medial Superior Nucleus. The MSO is the most conspicuous component of the SOC. The cells are surrounded by the POC, and are embedded in the crossing fibers of the trapezoid body, through which the MSO receives its ipsi- and contralateral afferents from the cochlear nuclei. The auditory signals in the MSO are tonotopically organized with lower frequencies represented dorsally and the higher frequencies ventrally [Oliver et al., 2003]. The MSO neurons are fusiform with the dendrites, at each pole, oriented horizontally: these receive afferents from the 'spherical bushy cells' of the ventral cochlear nucleus (VCN), the ipsilateral VCN innervates the dendrites of the lateral pole while the contralateral VCN innervates the medial dendrites. The contralateral VCN fibers plot a delay line along the neurons of the MSO, from rostral to caudal, with a different relative delay between the ipsi- and contralateral inputs to each cell, and in so doing produce a place map of the interaural time differences along the MSO nucleus [for reviews, see Smith et al., 1993; Joris et al., 1998]. The MSO map of the interaural time differences is projected onto the ipsilateral central nucleus of the inferior colliculus (IC) via the LL; it is also relayed to the dorsal nucleus of the LL (DNLL), which in turn sends GABAergic inhibitory projections to the contralateral IC and DNLL, as well as to the ipsilateral IC [for a review, see Nieuwenhuys et al., 2008].

Lateral Nucleus of the Superior Olive. The LSO is S-shaped in many mammals but not in humans [Kulesza, 2007]. Experiments show that it receives tonotopically organized projections from each side: excitatory afferents from the ipsilateral ear via spherical bushy cells of the VCN, as well as inhibitory afferents from the contralateral ear via the nucleus of the trapezoid body (NTB) [Nieuwenhuys et al., 2008]. These inhibitory signals are relayed from the contralateral VCN 'globular bushy cells' to the NTB part of the SOC, terminating as giant synapses, called calyces of Held. The recipient cells are glycinergic and project locally to the ipsilateral LSO. Experimental results indicate that the LSO participates like the MSO in the analysis of interaural differences, but does so in terms of responding to sound intensity, and so contributes to the localization of sound sources in the horizontal plane [Glendenning et al., 1991; Smith et al., 1998].

Nucleus of the Trapezoid Body. The homologue of the NTB, with its giant calyces of Held from VCN globular bushy cells, has not been clearly identified in humans, although it is prominent in other mammals [Richter et al., 1983; Moore, 2000]. In humans it consists of a few scattered pale oval cells lying along the ventral border of the MSO [Koutcherov et al., 2004]. Typically the cells are glycinergic [Bledsoe et al., 1990], and provide major inhibitory inputs to the LSO, as well as to other regions of the SOC [Smith et al., 1998]. Through these connections the NTB is considered to contribute to the localization of sound sources in the horizontal plane.

References

Bazwinsky I, Hilbig H, Bidmon HJ, Rubsamen R: Characterization of the human superior olivary complex by calcium binding proteins and neurofilament H (SMI-32). J Comp Neurol 2003;456:292–303.

Bledsoe SC Jr, Snead CR, Helfert RH, Prasad V, Wenthold RJ, Altschuler RA: Immunocytochemical and lesion studies support the hypothesis that the projection from the medial nucleus of the trapezoid body to the lateral superior olive is glycinergic. Brain Res 1990;517:189–194.

Glendenning KK, Masterton RB, Baker BN, Wenthold RJ: Acoustic chiasm. III. Nature, distribution, and sources of afferents to the lateral superior olive in the cat. J Comp Neurol 1991;310:377–400.

Joris PX, Smith PH, Yin TC: Coincidence detection in the auditory system: 50 years after Jeffress. Neuron 1998;21:1235–1238.

Koutcherov Y, Huang X-F, Halliday G, Paxinos G: Organization of human brain stem nuclei; in Paxinos G, Mai JK (eds): The Human Nervous System, ed 2. Amsterdam, Elsevier Academic Press, 2004.

Kulesza RJ Jr: Cytoarchitecture of the human superior olivary complex: medial and lateral superior olive. Hear Res 2007;225:80–90.

Moore JK: Organization of the human superior olivary complex. Microsc Res Tech 2000;51:403–412.

Nieuwenhuys R, Voogd J, van Huijzen C: The Human Central Nervous System. Berlin, Springer, 2008.

Oliver DL, Beckius GE, Bishop DC, Loftus WC, Batra R: Topography of interaural temporal disparity coding in projections of medial superior olive to inferior colliculus. J Neurosci 2003;23:7438–7449.

Richter EA, Norris BE, Fullerton BC, Levine RA, Kiang NY: Is there a medial nucleus of the trapezoid body in humans? Am J Anat 1983;168:157–166.

Smith PH, Joris PX, Yin TC: Projections of physiologically characterized spherical bushy cell axons from the cochlear nucleus of the cat: evidence for delay lines to the medial superior olive. J Comp Neurol 1993;331:245–260.

Smith PH, Joris PX, Yin TC: Anatomy and physiology of principal cells of the medial nucleus of the trapezoid body (MNTB) of the cat. J Neurophysiol 1998;79:3127–3142.

Webster DB: An overview of mammalian auditory pathways with an emphasis on humans; in Webster DB, Popper AN, Fay RR (eds): The Mammalian Auditory Pathway: Neuroanatomy. New York, Springer, 1992.

25 Ventral Nucleus of the Lateral Lemniscus (VLL)

Original name:
Nucleus lemnisci lateralis ventralis

Alternative name:
Anterior nucleus of the lateral lemniscus

Location and Cytoarchitecture – Original Text

This elongated nuclear mass lies in relation to the caudal two thirds of the lateral lemniscus (LL) in the ventrolateral part of the pons. The nucleus commences at the level of the caudal pole of the motor trigeminal nucleus (MoV) and extends orally for 8 mm. The caudal pole of the ventral nucleus of the lateral lemniscus (VLL) is directly continuous with the oral pole of the medial superior olive (and difficult to define in humans) [Moore, 1987; Koutcherov et al., 2004].

On cross-section, the caudal part of the VLL is round or oval in outline and is bordered on all aspects except dorsally by fibers of the LL. Dorsally, the nucleus is related to the nucleus subcoeruleus. At a point just oral to the MoV, the VLL migrates a short distance dorsolaterally and assumes an oblong outline with the long axis directed dorsoventrally. This oral portion of the nucleus is related to fibers of the LL in all directions except medially where it is approximated by either the nucleus reticularis pontis oralis or by the lateral parabrachial nucleus.

The VLL is composed of loosely arranged, small, round or oval cells which possess relatively large nuclei and nucleoli, and fine, evenly distributed, darkly stained Nissl granules (fig. 1–4). Particularly in the caudal portion of the nucleus, the glial nuclei are more densely arranged than in the adjacent nuclei.

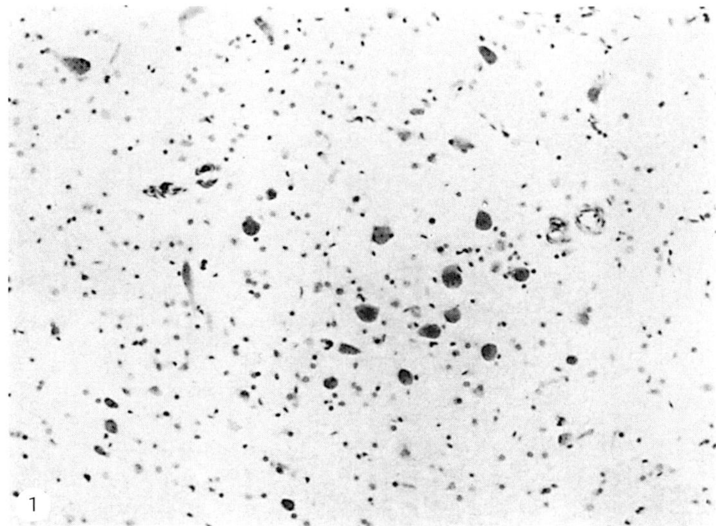

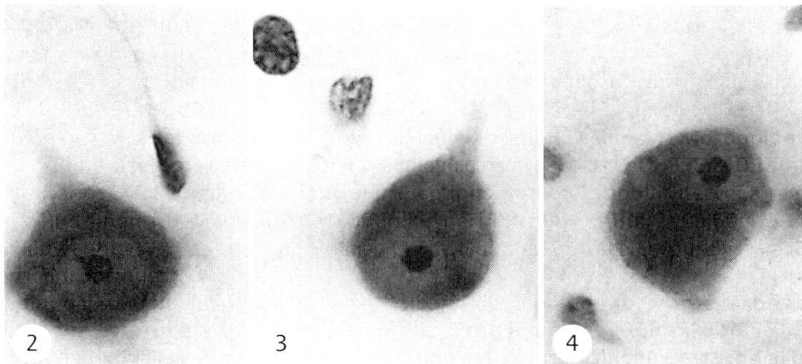

Fig. 1. Ventral nucleus of the lateral lemniscus. Magnification ×150.
Fig. 2–4. Cells from the ventral nucleus of the lateral lemniscus. Magnification ×1,000.

Functional Neuroanatomy

Function
The VLL relays auditory signals to the inferior colliculus (IC), and most current hypotheses consider it to participate in the temporal analysis of sounds [Batra and Fitzpatrick, 2002; Zhang and Kelly, 2006].

Connections
The VLL is a major source of ascending input to the ipsilateral IC; almost all cells in the VLL project to its central nucleus through the LL, some project to the external nucleus of the IC [Adams, 1979; Kudo, 1981]. Conversely the IC sends descending projections back to the VLL [Nieuwenhuys et al., 2008]. The majority of VLL neurons are inhibitory [Saint Marie et al., 1997; Riquelme et al., 2001].

The source of auditory inputs to the VLL comes mainly from the contralateral ventral cochlear nucleus (VCN) with a small projection from the ipsilateral VCN [Friauf and Ostwald, 1988; Smith et al., 2005]. In addition, the VLL receives afferents from the lateral superior olive (LSO) and the nucleus of the trapezoid body (NTB): all three of these nuclei (VLL, LSO and NTB) are comparatively small in humans, but very well developed in echolocating species such as bats [Moore and Linthicum, 2004]. In the bats a concentric tonotopic map of frequencies has been described in the VLL [Metzner and Radtke-Schuller, 1987], but in mammals there are several conflicting reports on the existence of a VLL tonotopic map: some authors confirm its presence (rat [Merchan and Berbel, 1996]; gerbil [Benson and Cant, 2008]; cat [Malmierca et al., 1998]), and others do not (rat [Zhang and Kelly, 2006; Nayagam et al., 2006]).

References

Adams JC: Ascending projections to the inferior colliculus. J Comp Neurol 1979;183:519–538.

Batra R, Fitzpatrick DC: Monaural and binaural processing in the ventral nucleus of the lateral lemniscus: a major source of inhibition to the inferior colliculus. Hear Res 2002;168:90–97.

Benson CG, Cant NB: The ventral nucleus of the lateral lemniscus of the gerbil *(Meriones unguiculatus)*: organization of connections with the cochlear nucleus and the inferior colliculus. J Comp Neurol 2008;510:673–690.
Friauf E, Ostwald J: Divergent projections of physiologically characterized rat ventral cochlear nucleus neurons as shown by intra-axonal injection of horseradish peroxidase. Exp Brain Res 1988;73:263–284.
Koutcherov Y, Huang X-F, Halliday G, Paxinos G: Organization of human brain stem nuclei; in Paxinos G, Mai JK (eds): The Human Nervous System, ed 2. Amsterdam, Elsevier Academic Press, 2004.
Kudo M: Projections of the nuclei of the lateral lemniscus in the cat: an autoradiographic study. Brain Res 1981;221:57–69.
Malmierca MS, Leergaard TB, Bajo VM, Bjaalie JG, Merchan MA: Anatomic evidence of a three-dimensional mosaic pattern of tonotopic organization in the ventral complex of the lateral lemniscus in cat. J Neurosci 1998;18:10603–10618.
Merchan MA, Berbel P: Anatomy of the ventral nucleus of the lateral lemniscus in rats: a nucleus with a concentric laminar organization. J Comp Neurol 1996; 372:245–263.
Metzner W, Radtke-Schuller S: The nuclei of the lateral lemniscus in the rufous horseshoe bat, *Rhinolophus rouxi*. A neurophysiological approach. J Comp Physiol A 1987;160:395–411.
Moore JK: The human auditory brain stem: a comparative view. Hear Res 1987;29: 1–32.
Moore JK, Linthicum JR: Auditory system; in Paxinos G, Mai JK (eds): The Human Nervous System, ed 2. San Diego, Elsevier Academic Press, 2004.
Nayagam DA, Clarey JC, Paolini AG: Intracellular responses and morphology of rat ventral complex of the lateral lemniscus neurons in vivo. J Comp Neurol 2006; 498:295–315.
Nieuwenhuys R, Voogd J, van Huijzen C: The Human Central Nervous System. Berlin, Springer, 2008.
Riquelme R, Saldana E, Osen KK, Ottersen OP, Merchan MA: Colocalization of GABA and glycine in the ventral nucleus of the lateral lemniscus in rat: an in situ hybridization and semiquantitative immunocytochemical study. J Comp Neurol 2001;432:409–424.
Saint Marie RL, Shneiderman A, Stanforth DA: Patterns of gamma-aminobutyric acid and glycine immunoreactivities reflect structural and functional differences of the cat lateral lemniscal nuclei. J Comp Neurol 1997;389:264–276.
Smith PH, Massie A, Joris PX: Acoustic stria: anatomy of physiologically characterized cells and their axonal projection patterns. J Comp Neurol 2005;482:349–371.
Zhang H, Kelly JB: Responses of neurons in the rat's ventral nucleus of the lateral lemniscus to amplitude-modulated tones. J Neurophysiol 2006;96:2905–2914.

Sensory Systems – Auditory Nuclei (Plate 28)

26 Dorsal Nucleus of the Lateral Lemniscus (DLL)

Original name:
Nucleus lemnisci lateralis dorsalis

Alternative name:
Dorsal lemniscal nucleus [Nieuwenhuys et al., 2008]

Location and Cytoarchitecture – Original Text

The dorsal nucleus of the lateral lemniscus (DLL) is a small nuclear mass which lies among the fibers of the orodorsal third of the lateral lemniscus (LL). It has an orocaudal length of 1–2 mm.

On cross-section the nucleus is irregularly oval in outline and is surrounded on all sides by fibers of the LL. These fibers separate the nucleus laterally from the sagulum nucleus, and medially from the lateral parabrachial nucleus.

The DLL is composed of medium-sized, loosely arranged cells, which are oval, triangular, or slender and multipolar. These cells possess long dendrites and relatively large, darkly stained Nissl granules (fig. 1–4). [...]

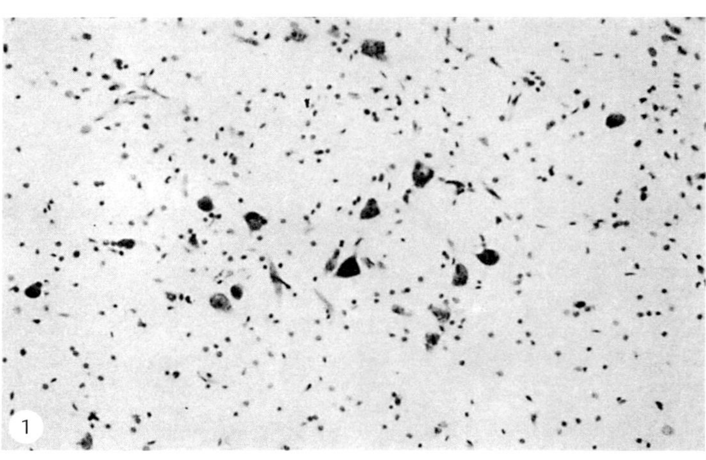

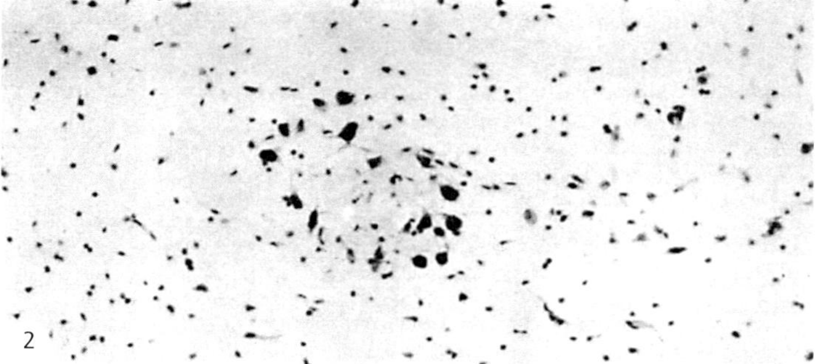

Fig. 1, 2. Dorsal nucleus of the lateral lemniscus. Note the difference in cell size and arrangement in these two examples. Magnification ×150.

Functional Neuroanatomy

Function

The DLL is an inhibitory and binaural nucleus in the chain of auditory pathways linking the cochlear nuclei and superior olivary nuclei (SOC) with the inferior colliculi (IC). It is presumed to play an important role in shaping the 'interaural time difference' responses in neurons of the IC, which in turn determine the balance of the auditory signals between one ear and the

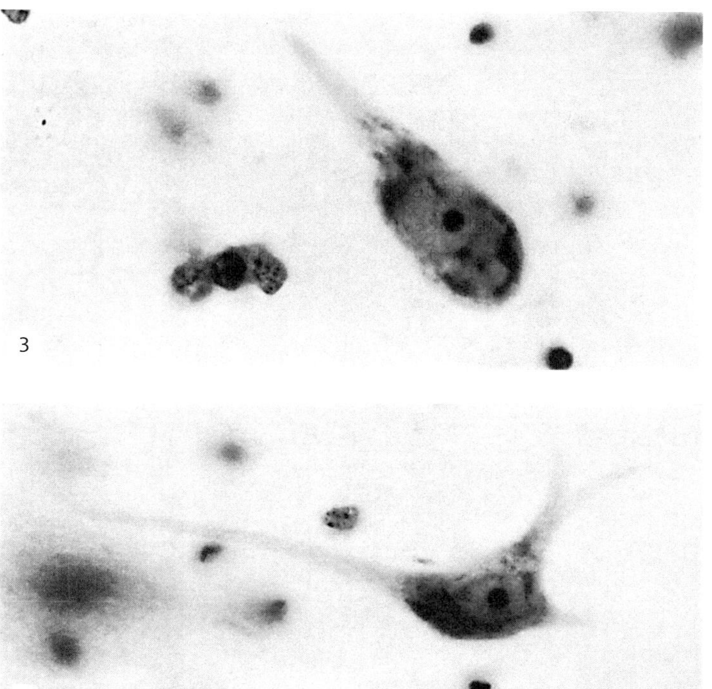

Fig. 3, 4. Cells from the dorsal nucleus of the lateral lemniscus. Magnification ×1,000.

other. Interaural differences in stimulus intensity and timing are major cues for sound localization [Shneiderman et al., 1988; Siveke et al., 2012].

Connections

Afferents. The DLL is phylogenetically a stable nucleus in terms of organization and size, and a tonotopic organization is evident in all the species studied, including humans [Moore, 1987]. The main afferents to the DLL arise from the ipsilateral SOC (medial and lateral superior olivary nuclei bilaterally), and weaker inputs originate from the cochlear nuclei [Glendenning et al., 1981; Nieuwenhuys et al., 2008]. The binaural inputs to the DLL ascend among the fibers of the LL, which terminate orthogonally on the horizontally oriented dendrites of the DLL neurons [Kane and Barone, 1980].

Efferents. The neurons of the DLL are mainly, if not all, GABAergic, that is inhibitory in nature [Adams and Mugnaini, 1984; Saint Marie et al., 1997]. Their efferent axons ascend a short distance to end strongly in the central nucleus of the IC. In order to supply the IC and DLL of the contralateral side, DLL efferents cross the midbrain in the dorsal commissure of the LL (alternatively called the posterior tegmental commissure or the commissure of Probst) [Kudo, 1981; Oliver and Shneiderman, 1989].

References

Adams JC, Mugnaini E: Dorsal nucleus of the lateral lemniscus: a nucleus of GABAergic projection neurons. Brain Res Bull 1984;13:585–590.

Glendenning KK, Brunso-Bechtold JK, Thompson GC, Masterton RB: Ascending auditory afferents to the nuclei of the lateral lemniscus. J Comp Neurol 1981; 197:673–703.

Kane ES, Barone LM: The dorsal nucleus of the lateral lemniscus in the cat: neuronal types and their distributions. J Comp Neurol 1980;192:797–826.

Kudo M: Projections of the nuclei of the lateral lemniscus in the cat: an autoradiographic study. Brain Res 1981;221:57–69.

Moore J: The human auditory brain stem: a comparative view. Hear Res 1987;29: 1–32.

Nieuwenhuys R, Voogd J, van Huijzen C: The Human Central Nervous System. Berlin, Springer, 2008.

Oliver DL, Shneiderman A: An EM study of the dorsal nucleus of the lateral lemniscus: inhibitory, commissural, synaptic connections between ascending auditory pathways. J Neurosci 1989;9:967–982.

Saint Marie RL, Shneiderman A, Stanforth DA: Patterns of gamma-aminobutyric acid and glycine immunoreactivities reflect structural and functional differences of the cat lateral lemniscal nuclei. J Comp Neurol 1997;389:264–276.

Shneiderman A, Oliver DL, Henkel CK: Connections of the dorsal nucleus of the lateral lemniscus: an inhibitory parallel pathway in the ascending auditory system? J Comp Neurol 1988;276:188–208.

Siveke I, Leibold C, Schiller E, Grothe B: Adaptation of binaural processing in the adult brainstem induced by ambient noise. J Neurosci 2012;32:462–473.

Sensory Systems – Auditory Nuclei (Plates 30 and 31)

27 Inferior Colliculus (IC)

Original name:
Nucleus colliculi inferioris

Subdivisions:
Central nucleus (ICc)
Dorsal cortex or pericentral nucleus (ICd)
Dorsal peripheral zone
Intercollicular area
External nucleus (ICx) or lateral peripheral zone or lateral peripheral cortex

Location and Cytoarchitecture – Original Text

The inferior colliculi (IC) form the two elevations which make up the caudal half of the mesencephalic tectal plate. They extend from the level of the oral extremity of the anterior medullary velum to the transitional zone which separates them from the superior colliculi (SC). Each IC contains a large nucleus which is oval on cross-section and which lies with its long axis directed dorsomedially. The fibers of the lateral lemniscus (LL) are so distributed that they form a capsule about the ventrolateral pole of the nucleus of the IC. These fibers allow sharp delineation of the ventromedial and ventrolateral borders of the nucleus and separate it from the periphery of the section and the sagulum nucleus (SAG) laterally, and from

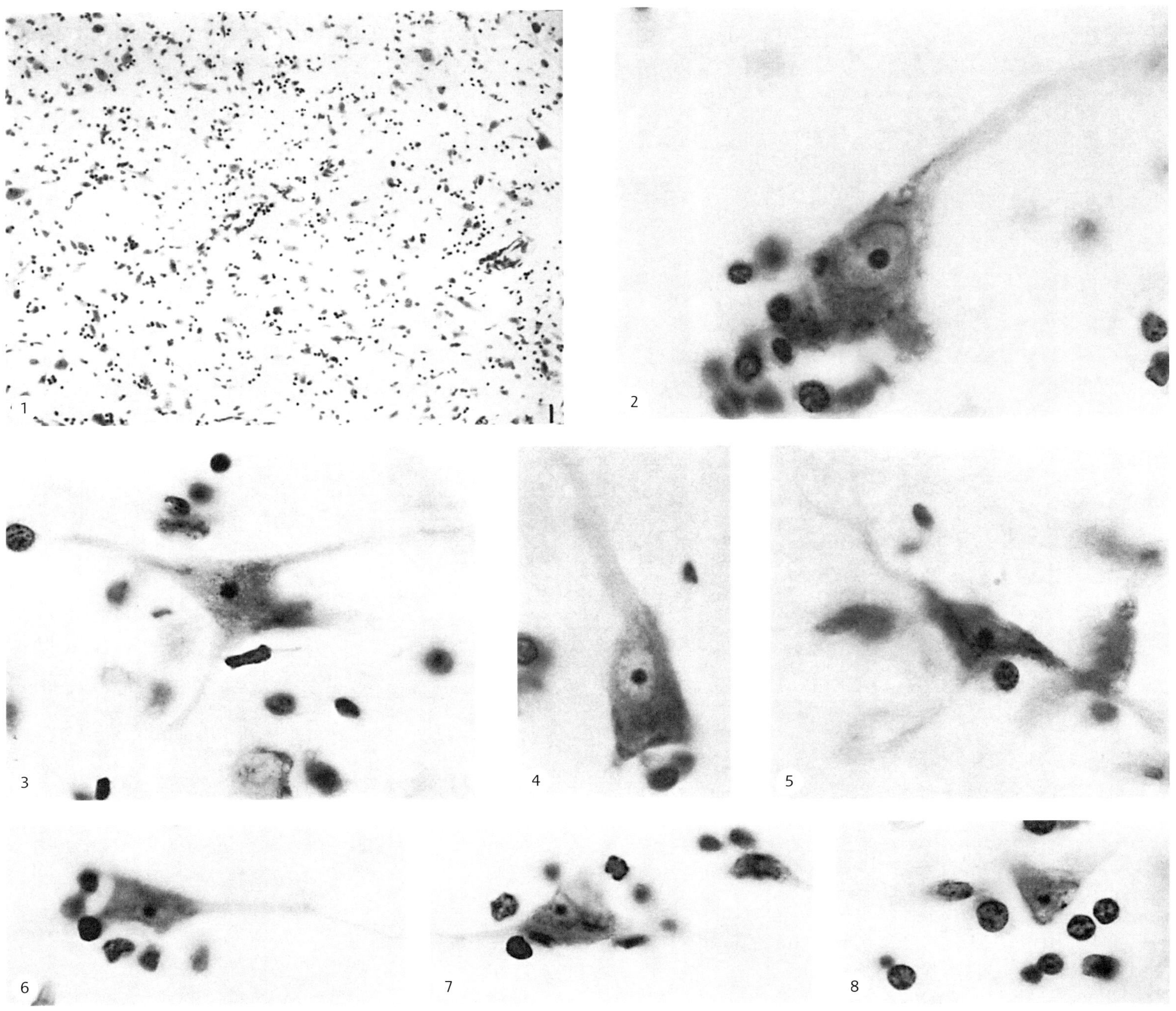

Fig. 1. Inferior colliculus. Magnification ×150.
Fig. 2. Medium-sized cell from the inferior colliculus. Note the presence of glial satellites. Magnification ×1,000.

Fig. 3–8. Small, variously shaped cell types from the inferior colliculus. Note the presence of glial satellites. Magnification ×1,000.

the cuneiform nucleus and the periaqueductal gray (PAG) medially. Dorsomedially, the nuclei of the two colliculi approach each other dorsal to the PAG. In this dorsal region the cells are very loosely arranged and the distinction between the PAG and the gray matter of the IC may be difficult. In sections through the oral portion of the IC this dorsal area is traversed by fibers of the commissure of the IC. This latter structure, however, attains its greatest development in sections through the intercollicular transition zone. At the level of the oral pole of the trochlear nucleus the cells of this transitional zone encroach upon and replace those of the nucleus of the IC. The ventrolateral portion of the nucleus of the IC is the last to disappear. The fibers of the inferior brachium form an easily recognizable structure which lies lateral to the oral pole of the IC.

The nucleus of the IC is characterized by a marked cellularity, an impression which is intensified by the presence of an abundance of glial elements (fig. 1). In the caudal portions of the nucleus the cells are arranged, in the main, with their long axes directed dorsomedially, parallel to the fibers of the LL. In the oral portion of the nucleus the fiber pattern becomes more complex and the cells appear to be arranged in a haphazard manner. Two main cell types may be distinguished:

(a) the majority of cells are small, oval, pyriform, spindle shaped or triangular; they possess a scanty, lightly stained cytoplasm and no distinct Nissl bodies; the nuclei and nucleoli are unremarkable (fig. 3–8);

(b) the remaining cells are medium sized, oval or triangular with a moderately dark cytoplasm, distinct, small Nissl granules and a nucleus which is usually centrally placed (fig. 2).

The majority of cells of the IC are associated with numerous glial satellites. [...]

Functional Neuroanatomy

Function

The IC is the major brainstem auditory relay, transmitting parallel pathways from all auditory brainstem nuclei to the ipsilateral thalamus and cerebral cortex. The IC can be divided into at least 3 regions, each with a different connectivity and function; however, the exact functions are still unclear. The *central nucleus* (ICc) carries finely tuned, tonotopically organized, binaural auditory signals to the thalamus, which relays on to the primary auditory cortex. This applies to the ventrolateral subdivision of the ICc but not the dorsomedial ICc which is the origin of the commissural fibers. The *external nucleus* (ICx) covers the rostral and lateral surface of the IC, receives auditory and somatosensory inputs, and is considered to participate in auditory-motor reflexes. The *dorsal cortex* (ICd) is a thin sheet of densely packed cells covering the dorsocaudal IC; due to its monaural inputs and projections to the secondary auditory cortex, it has been tentatively proposed as a link in controling auditory attention [Geniec and Morest, 1971; Aitkin et al., 1975; Aitkin, 1979; Webster and Garey, 1990; Nieuwenhuys et al., 2008; Aitkin et al., 2011]. The use of calcium-binding proteins on human IC highlights the subdivisions described here, and indicates that there are additional functional areas that have not as yet been defined [Tardif et al., 2003].

Connections

Afferents and Efferents of the Central Nucleus. The ICc receives the densest termination of the ascending LL. The fibers arise from the contralateral cochlear nuclei, bilaterally from the lateral superior olive and ipsilaterally from the medial superior olive and the nuclei of the LL. They traverse the laminae of ICc cells from ventrolateral to dorsomedial, and terminate tonotopically, forming curved sheets of neurons with isofrequency planes, and with a regular progression of frequencies from low dorsally in the ICc to high frequencies ventrally [Webster and Garey, 1990].

The strict tonotopic organization is found in the ventrolateral ICc but not in the dorsomedial part of the ICc: here the lemniscal input is more diffuse, the cells are more densely packed and give rise to a true commissure that terminates in the corresponding dorsomedial subdivision of the ICc of the other side [Geniec and Morest, 1971; Moore et al., 1977]. The dorsomedial ICc receives much of its input from the central and external nuclei of the contralateral side. A further difference between the subdivisions of the ICc is that the dorsomedial ICc receives bilateral inputs from the auditory cortex, but the ventromedial ICc only from the ipsilateral auditory cortex [Fitzpatrick and Imig, 1978].

The efferent pathways from the ICc preserve tonotopicity and carry short-latency, sharply frequency-tuned responses. They exit laterally, and ascend in the brachium of the IC (BIC) to terminate in the ventral thalamic division of the medial geniculate body (MGB). A crossed projection to the contralateral MGB exits the ICc medially, and travels in the commissure of the IC. Descending pathways to the brainstem auditory nuclei are reviewed by Nieuwenhuys et al. [2008].

Afferents and Efferents of the External Nucleus. The ICx lies lateral to the ICc, and is traversed by both the IC input fibers of the LL, and the IC output fibers to the BIC ascending to the thalamus. The ICx is also sometimes called the intercollicular area since it is continuous with the bridge of neurons at the transition between IC and SC where the intercollicular nucleus is to be found (nucleus No. 5). Very few LL afferents terminate in the ICx, but auditory signals enter it from the ICc, the dorsal lemniscal nuclei and the contralateral dorsal cochlear nucleus. The cells of the ICx have a more broadly tuned spectrum compared to the sharp tuning of the ICc cells. The ICx is a multisensory region which also receives retinal, somatosensory, spinal trigeminal and dorsal column nuclear afferents, as well as inputs from the substantia nigra and the cerebral cortex [Coleman and Clerici, 1987; Olazabal and Moore, 1989; Spangler and Warr, 1991].

The deep layers of the SC in mammals contain maps of both the visual and auditory fields, which participate in eye and head orientation to visual and auditory stimuli [Harting and Van Lieshout, 2000; Garcia Del Cano et al., 2006; Leo et al., 2008]. There is ample evidence for projections from the IC to the deep layers of the SC [for a review, see May, 2006]. The auditory connections in mammals are thought to originate from two main sources: from the ICx and the nucleus of the BIC (nucleus No. 29); in the owl they come from the homologue of the ICx, the nucleus mesencephalicus lateralis dorsalis [Knudsen and Konishi, 1978]. In some animals, especially echolocating species, there is evidence for a topographic representation of auditory space in the ICx which could be used to create the auditory map in the SC, but such a topography is not yet certain in the ICx of mammals [Binns et al., 1992; Hyde and Knudsen, 2000; Maier and Groh, 2009].

The ascending efferents of the ICx to the thalamus project into the BIC, and terminate mainly in the medial division of the MGB, unlike the ICc (see above) [Calford and Aitkin, 1983]. The descending ICx efferents supply the periolivary nuclei ipsilaterally, and the lemniscal nuclei (dorsal and ventral) and dorsal cochlear nuclei bilaterally [Nieuwenhuys et al., 2008, fig. 18.4].

Afferents and Efferents of the Dorsal Cortex or Pericentral Nucleus. The ICd contains densely packed cells that are organized in a 4-layered cortex-like cytoarchitecture with larger cells in its deeper layers [Geniec and Morest, 1971]. The region receives some lemniscal afferents as well as inputs from the SAG and from the primary auditory cortex [Andersen et al., 1980; Henkel and Shneiderman, 1988]. The ICd cells have a broadly tuned spectrum responding to contralateral monaural inputs, and they project into the BIC to terminate in the diffuse auditory thalamocortical system, mainly in the dorsal division of the MGB [Calford and Aitkin, 1983].

References

Aitkin LM: The auditory midbrain. Trends Neurosci 1979;2:308–310.

Aitkin LM, Irvine DRF, Webster WR: Central neural mechanisms of hearing. 2011. http://www.comprehensivephysiology.com.

Aitkin LM, Webster WR, Veale JL, Crosby DC: Inferior colliculus. I. Comparison of response properties of neurons in central, pericentral, and external nuclei of adult cat. J Neurophysiol 1975;38:1196–1207.

Andersen RA, Roth GL, Aitkin LM, Merzenich MM: The efferent projections of the central nucleus and the pericentral nucleus of the inferior colliculus in the cat. J Comp Neurol 1980;194:649–662.

Binns KE, Grant S, Withington DJ, Keating MJ: A topographic representation of auditory space in the external nucleus of the inferior colliculus of the guinea-pig. Brain Res 1992;589:231–242.

Calford MB, Aitkin LM: Ascending projections to the medial geniculate body of the cat: evidence for multiple, parallel auditory pathways through thalamus. J Neurosci 1983;3:2365–2380.

Coleman JR, Clerici WJ: Sources of projections to subdivisions of the inferior colliculus in the rat. J Comp Neurol 1987;262:215–226.

Fitzpatrick KA, Imig TJ: Projections of auditory cortex upon the thalamus and midbrain in the owl monkey. J Comp Neurol 1978;177:573–555.

Garcia del Cano G, Gerrikagoitia I, Alonso-Cabria A, Martinez-Millan L: Organization and origin of the connection from the inferior to the superior colliculi in the rat. J Comp Neurol 2006;499:716–731.

Geniec P, Morest DK: The neuronal architecture of the human posterior colliculus: a study with the Golgi method. Acta Otolaryngol Suppl 1971;295:1–33.

Harting JK, van Lieshout DP: Projections from the rostral pole of the inferior colliculus to the cat superior colliculus. Brain Res 2000;881:244–247.

Henkel CK, Shneiderman A: Nucleus sagulum: projections of a lateral tegmental area to the inferior colliculus in the cat. J Comp Neurol 1988;271:577–588.

Hyde PS, Knudsen EI: Topographic projection from the optic tectum to the auditory space map in the inferior colliculus of the barn owl. J Comp Neurol 2000; 421:146–160.
Knudsen EI, Konishi M: A neural map of auditory space in the owl. Science 1978; 200:795–797.
Leo F, Bertini C, di Pellegrino G, Ladavas E: Multisensory integration for orienting responses in humans requires the activation of the superior colliculus. Exp Brain Res 2008;186:67–77.
Maier JX, Groh JM: Multisensory guidance of orienting behavior. Hear Res 2009; 258:106–112.
May PJ: The mammalian superior colliculus: laminar structure and connections. Prog Brain Res 2006;151:321–378.
Moore JK, Karapas F, Moore RY: Projections of the inferior colliculus in insectivores and primates. Brain Behav Evol 1977;14:301–327.
Nieuwenhuys R, Voogd J, van Huijzen C: The Human Central Nervous System. Berlin, Springer, 2008.
Olazabal UE, Moore JK: Nigrotectal projection to the inferior colliculus: horseradish peroxidase transport and tyrosine hydroxylase immunohistochemical studies in rats, cats, and bats. J Comp Neurol 1989;282:98–118.
Spangler KM, Warr WB: The descending auditory systems; in Altschuler RA, Bobbin RP, Clopton BM, Hofman DW (eds): Neurobiology of Hearing: The Central Auditory System. New York, Raven, 1991.
Tardif E, Chiry O, Probst A, Magistretti PJ, Clarke S: Patterns of calcium-binding proteins in human inferior colliculus: identification of subdivisions and evidence for putative parallel systems. Neuroscience 2003;116:1111–1121.
Webster WR, Garey LJ: Auditory system; in Paxinos G (ed): The Human Nervous System. San Diego, Academic Press, 1990.

28 Sagulum Nucleus (SAG)

Original name:
Nucleus sagulum

Alternative name:
Sagulum [Beneyto et al., 1998]

Location and Cytoarchitecture – Original Text

The sagulum nucleus (SAG) (of Ziehen) is formed by a flattened band of cells which intervenes between the lateral surface of the lateral lemniscus (LL) and the periphery of the oral pontine tegmentum. It extends from the level of the trochlear decussation to the level of the main trochlear nucleus – a distance of approximately 7 mm.

Caudally the nucleus appears in the ventrolateral pontine tegmentum in the angle formed by the LL and the pontine gray. More orally its cells sweep dorsolaterally between the LL and the periphery of the section. The oral pole of the nucleus lies in the dorsolateral pontine tegmentum, at the ventrolateral pole of the inferior colliculus (IC).

The SAG is composed of irregularly distributed, small, darkly stained, elongated triangular or fusiform cells which possess relatively long processes (fig. 1, 2).

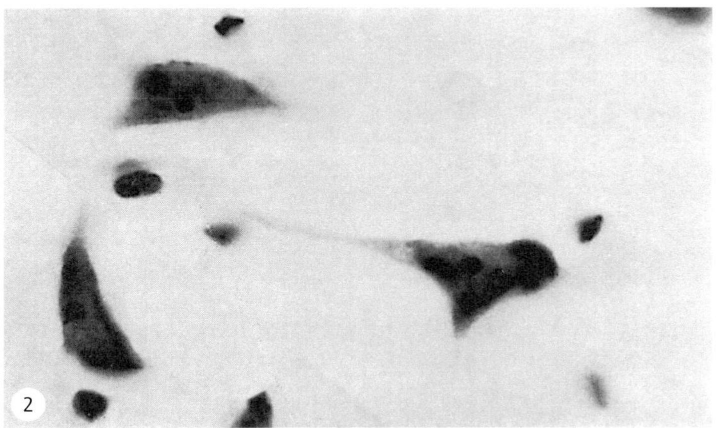

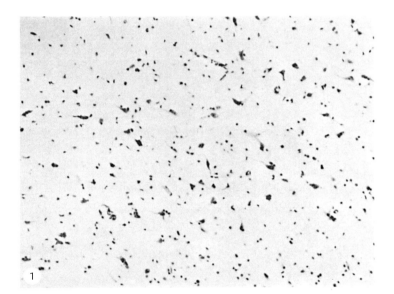

Fig. 1. Sagulum nucleus. Magnification ×150.
Fig. 2. Cells from the sagulum nucleus. Magnification ×1,000.

Functional Neuroanatomy

Function

The SAG is part of the ascending auditory pathways sending excitatory signals to the IC and thalamus, and receiving a major input from the cerebral cortex. Through these connections it has been suggested that it may play a role in motor adjustments to sounds [Beneyto et al., 1998]. For example, SAG neurons were shown to be active during vocalization, implying a role for them in communicative aspects of auditory behavior [Jurgens et al., 1996]; other authors proposed a role of the SAG in the acoustic startle response.

It has been suggested that the intermediate nucleus of the LL, which is unique to bats, may be the homologue of the SAG in mammals [Schweizer, 1981].

Connections

Efferents. The major projection of the SAG is to the caudal part of the dorsal cortex of the IC, where it terminates superficially in dense patches, ipsilaterally. The SAG also projects to the dorsal division of the medial geniculate body. This region of the thalamus projects to the nonprimary auditory cortex and subcortical forebrain regions. Interestingly the SAG projections to the thalamus bypass the IC, providing an auditory, nonlemniscal, parallel pathway to the thalamus. In this way the anatomical connections of the SAG are clearly distinct from other midbrain auditory nuclei [Calford and Aitkin, 1983; Henkel and Shneiderman, 1988; Beneyto et al., 1998]. Further SAG efferents target the superior colliculus, the pretectum, the mesencephalic reticular formation and a small paralemniscal zone just rostral to the ventral nucleus of the lateral lemniscus which projects to the pinna motoneurons; for references, see Beneyto et al. [1998].

Afferents. The major input to the SAG arises from the secondary auditory cortex, but it also receives significant cortical afferents from diverse polymodal association areas. In addition, projections to the SAG arise from the dorsal nucleus of the lateral lemniscus and the contralateral SAG [Beneyto et al., 1998].

Descending tectal fibers pass through the SAG in humans, tectal fibers in primates supply the parabigeminal pontine nuclei, but inputs to the SAG have not been reported [Harting, 1977; Naidich et al., 2009].

References

Beneyto M, Winer JA, Larue DT, Prieto JJ: Auditory connections and neurochemistry of the sagulum. J Comp Neurol 1998;401:329–351.

Calford MB, Aitkin LM: Ascending projections to the medial geniculate body of the cat: evidence for multiple, parallel auditory pathways through thalamus. J Neurosci 1983;3:2365–2380.

Harting JK: Descending pathways from the superior colliculus: an autoradiographic analysis in the rhesus monkey *(Macaca mulatta)*. J Comp Neurol 1977; 173:583–612.

Henkel CK, Shneiderman A: Nucleus sagulum: projections of a lateral tegmental area to the inferior colliculus in the cat. J Comp Neurol 1988;271:577–588.

Jurgens U, Lu CL, Quondamatteo F: C-fos expression during vocal mobbing in the new world monkey *Saguinus fuscicollis*. Eur J Neurosci 1996;8:2–10.

Naidich TP, Duvernoy HM, Delman BN, Sorensen AG, Kollias SS, Haake EM: Duvernoy's atlas of the human brain stem and cerebellum. Wien, Springer, 2009.

Schweizer H: The connections of the inferior colliculus and the organization of the brainstem auditory system in the greater horseshoe bat *(Rhinolophus ferrumequinum)*. J Comp Neurol 1981;201:25–49.

Sensory Systems – Auditory Nuclei (Plates 34 and 36–38)

29 Nucleus of the Brachium of the Inferior Colliculus (nBIC)

Original name:

Nucleus paralemniscalis (rostral two thirds): caudal part is now called parabigeminal nucleus (PBG, chapter No. 31)

Location and Cytoarchitecture – Original Text

The main body of the nucleus of the brachium of the inferior colliculus (nBIC) is formed by a sheet of cells which lies between the lateral surface of the medial lemniscus and the medial surface of the brachium of the inferior colliculus (BIC), in the oral two thirds of the mesencephalon. On cross-section, the nucleus is elongated in outline with its long axis directed dorsoventrally. The caudal pole of the nBIC adjoins [the rostral pole of the parabigeminal nucleus, and extends rostrally lateral to the mesencephalic reticular formation, medial to the BIC, and it lies dorsal to the lateral tip of the substantia nigra]. The oral pole of the nBIC lies medial to the medial geniculate body (MGB), dorsal to the peripeduncular nucleus and ventral to the brachium of the superior colliculus.

The nBIC is composed predominantly of irregularly oriented, small, triangular or fusiform cells with the dendrites. The Nissl substance of these cells is evenly distributed and in the majority of cases is darkly stained. Occasional very lightly stained cells are also present. The cytoarchitecture of this nucleus is very similar to that of the sagulum nucleus (fig. 1–4).

Functional Neuroanatomy

Function

Neurons in the nBIC respond to auditory signals with a clear preference for particular regions of space, within the contralateral hemifield [Morest and Oliver, 1984; Schnupp and King, 1997]. The nBIC is thought to provide the main auditory projection to the deep layers of the superior colliculus (SC), where spatially tuned receptive fields are arranged to form a map of auditory space. The nBIC-SC projection is topographically organized and may convey information about sound azimuth [King et al., 1998; Doubell et al., 2000].

Connections

Efferents. The main efferent pathway from the nBIC targets the ipsilateral SC, mainly the deep gray layer and also the optic layer [Edwards et al., 1979; Van Buskirk, 1983; Kudo et al., 1984; King et al., 1998]. The network demonstrates a topographical organization, whereby medial, or anterior, sound source positions are represented predominantly in the rostral nBIC and more peripheral sound source positions are represented in caudal regions of the nucleus.

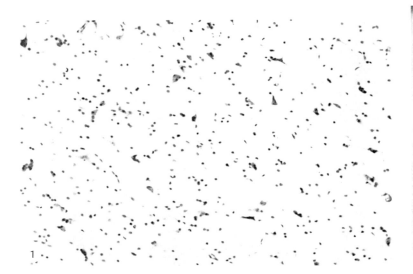

Fig. 1. Nucleus of the brachium of the inferior colliculus. Magnification ×150.

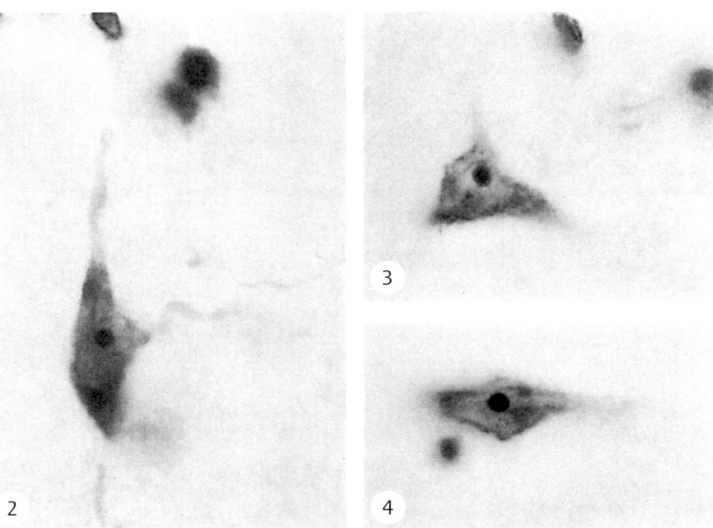

Fig. 2–4. Cells from the nucleus of the brachium of the inferior colliculus. Magnification ×1,000.

In addition to the strong nBIC-SC projection, ascending fibers from the nBIC travel along the medial border of the BIC and terminate in the dorsal and medial subdivisions of the MGB, also in the pretectum, periaqueductal gray, the pedunculopontine nucleus (PPT), thalamus [Kudo et al., 1984] and primary auditory cortex [Winer, 1984].

Afferents. The nBIC receives auditory-related afferents from the central and external nuclei of the inferior colliculus (central and external nuclei), mainly ipsilaterally [Kudo and Niimi, 1980]. Recordings in ferret nBIC reveal not only auditory units, but also visual units and bimodal auditory-visual units. The visual responses are at least in part driven by a topographically organized projection from visual neurons in the superficial layers of the SC to nBIC [Doubell et al., 2000].

Cholinergic inputs, probably from the PPT and laterodorsal tegmental nucleus, are thought to be important for coupling arousal with sound-processing [Habbicht and Vater, 1996]. Of all the auditory nuclei inspected, the nBIC, along with the ventral nucleus of the MGB, were found to contain the most cholinergic terminal-like endings. Thus, the modulation of nBIC signals by arousal may play an important role in the coding of auditory space. Furthermore, in the light of current ideas on nBIC function, these cholinergic pathways may participate in the influence of arousal on the auditory map in the SC [Wallace and Fredens, 1988; Hamada et al., 2010].

References

Doubell TP, Baron J, Skaliora I, King AJ: Topographical projection from the superior colliculus to the nucleus of the brachium of the inferior colliculus in the ferret: convergence of visual and auditory information. Eur J Neurosci 2000; 12:4290–4308.

Edwards SB, Ginsburgh CL, Henkel CK, Stein BE: Sources of subcortical projections to the superior colliculus in the cat. J Comp Neurol 1979;184:309–330.

Habbicht H, Vater M: A microiontophoretic study of acetylcholine effects in the inferior colliculus of horseshoe bats: implications for a modulatory role. Brain Res 1996;724:169–179.

Hamada S, Houtani T, Trifonov S, Kase M, Maruyama M, Shimizu J, Yamashita T, Tomoda K, Sugimoto T: Histological determination of the areas enriched in cholinergic terminals and M2 and M3 muscarinic receptors in the mouse central auditory system. Anat Rec (Hoboken) 2010;293:1393–1399.

King AJ, Jiang ZD, Moore DR: Auditory brainstem projections to the ferret superior colliculus: anatomical contribution to the neural coding of sound azimuth. J Comp Neurol 1998;390:342–365.

Kudo M, Niimi K: Ascending projections of the inferior colliculus in the cat: an autoradiographic study. J Comp Neurol 1980;191:545–556.

Kudo M, Tashiro T, Higo S, Matsuyama T, Kawamura S: Ascending projections from the nucleus of the brachium of the inferior colliculus in the cat. Exp Brain Res 1984;54:203–211.

Morest DK, Oliver DL: The neuronal architecture of the inferior colliculus in the cat: defining the functional anatomy of the auditory midbrain. J Comp Neurol 1984;222:209–236.

Schnupp JW, King AJ: Coding for auditory space in the nucleus of the brachium of the inferior colliculus in the ferret. J Neurophysiol 1997;78:2717–2731.

Van Buskirk RL: Subcortical auditory and somatosensory afferents to hamster superior colliculus. Brain Res Bull 1983;10:583–587.

Wallace MN, Fredens K: Origin of high acetylcholinesterase activity in the mouse superior colliculus. Exp Brain Res 1988;72:335–346.

Winer JA: Identification and structure of neurons in the medial geniculate body projecting to primary auditory cortex (AI) in the cat. Neuroscience 1984;13: 395–413.

30 Superior Colliculus (SC)

Original name:
Colliculus superior

Alternative name:
Optic tectum

Location and Cytoarchitecture – Original Text

The laminated gray substance of the superior colliculi (SC) forms the mound-like elevations of the oral half of the mesencephalic tectal plate. Each colliculus commences at the level of the caudal pole of the oculomotor nucleus and extends orally to the level of the posterior commissure – a distance of 6 mm. Caudally, the gray matter of the colliculus is continuous with the intercollicular nucleus; orally, it is continuous with the pretectal region.

On cross-section the dorsal and lateral margins of the SC form part of the free margin of the section. The ventral surface of the colliculus is related to the mesencephalic reticular formation (MRF) laterally, and to the periaqueductal gray matter (PAG) medially. Cells of the mesencephalic trigeminal nucleus may lie on the boundary line between the PAG and the colliculus. Dorsal to the PAG the medial extremities of the two colliculi meet and fuse in the midline. This median region is relatively acellular and is occupied by the myelinated fibers of the commissure of the SC.

The characteristic laminated arrangement of the SC can be distinguished on both myelin- and Nissl-stained sections. [...]

Current Nomenclature of Superior Colliculus Lamination

In the 'Original Text' the SC is divided into 8 layers on the basis of cytological criteria. Currently it is usual to distinguish 7 different layers (fig. 1, 2) [May, 2006]. The following description of the SC lamination also includes observations from the 'Original Text'.

Layer 1: Zonal Layer
This most superficial layer (stratum zonale, SZ) measures about 0.17 mm, and contains mainly glia cells which are usually found immediately beneath the pia. It was subdivided into 3 layers by Olszewski and Baxter in the 2nd edition. The lower region is characterized by the presence of sparsely arranged, very small, spindle-shaped or triangular, darkly stained nerve cells. These cells often exhibit a peculiar crumpled appearance (fig. 6).

Layer 2: Superficial Gray Layer
This stratum griseum superficiale (SGS) is a densely cellular area approximately 0.2 mm in thickness. The cells are slightly larger and stain more lightly than those above it in the SZ; otherwise their morphology is similar (fig. 7).

Layer 3: Optic Layer
This layer (stratum opticum, SO) is 0.8 mm thick. It is dominated by retinal afferents but nevertheless contains cells that are similar to, but in general larger than, those of the preceding layer.

Layer 4: Intermediate Gray Layer
In this layer (stratum griseum intermediale, SGI), occasional large cells similar to those of the overlying SO may be observed (fig. 8). In the lower regions of the SGI there appear collections of small, round or oval, densely arranged nerve cells. These cells possess a very scanty, lightly stained cytoplasm and a relatively large nucleus. In myelin-stained sections these 'glomeruli' appear as small, round, light areas devoid of myelinated fibers (fig. 3–5). The characteristic feature of the lower SGI is the presence of scattered large, plump, multipolar neurons possessing large, darkly stained, regular Nissl bodies, centrally placed nuclei and long dendrites (fig. 9).

Layer 5: Intermediate White Layer
The cellular elements of the stratum album intermediale (SAI) are identical with those of the lower SGI, but lie in an enriched myelin felt-work, which corresponds to the stratum lemnisci of the myelin-stained preparations.

Layer 6: Deep Gray Layer
This layer (stratum griseum profundum, SGP), approximately 0.6 mm thick, is more cellular than the SGI or SAI, and contains no large cells. The majority of its cells are small to medium-sized, triangular, irregularly oval or polygonal in shape (fig. 10).

Layer 7: Deep White Layer
This layer (stratum album profundum, SAP) measures 0.2 mm in thickness and is relatively acellular due to the dense accumulation of fibers. The cells which lie among these fibers are similar to those of the SGP.

This current nomenclature follows May [2006]: a fuller discussion of the nomenclatures, in particular those used in primate unit-recording, are to be found in the following articles [Ma et al., 1990; May and Porter, 1992]. Two additional points regarding the cytoarchitecture of the SC should be mentioned: (a) the medium-sized cells of the intermediate and deep layers exhibit a very intense glial satellitosis – as many as 10 glial nuclei may be closely associated with a single neuron; (b) the cells of the deep collicular layers are morphologically identical with those of the PAG. As well, in myelin-stained sections the background felt of these areas is similar. [...]

Functional Neuroanatomy

Function
The primary function of the SC is the orientation of the head and senses towards objects of interest. In humans the major input to the SC is from the eye which builds a retinotopic map in the superficial collicular layers. In rats the main collicular input comes from the trigeminal afferents innervating the whiskers. Visual, auditory and somatosensory signals are

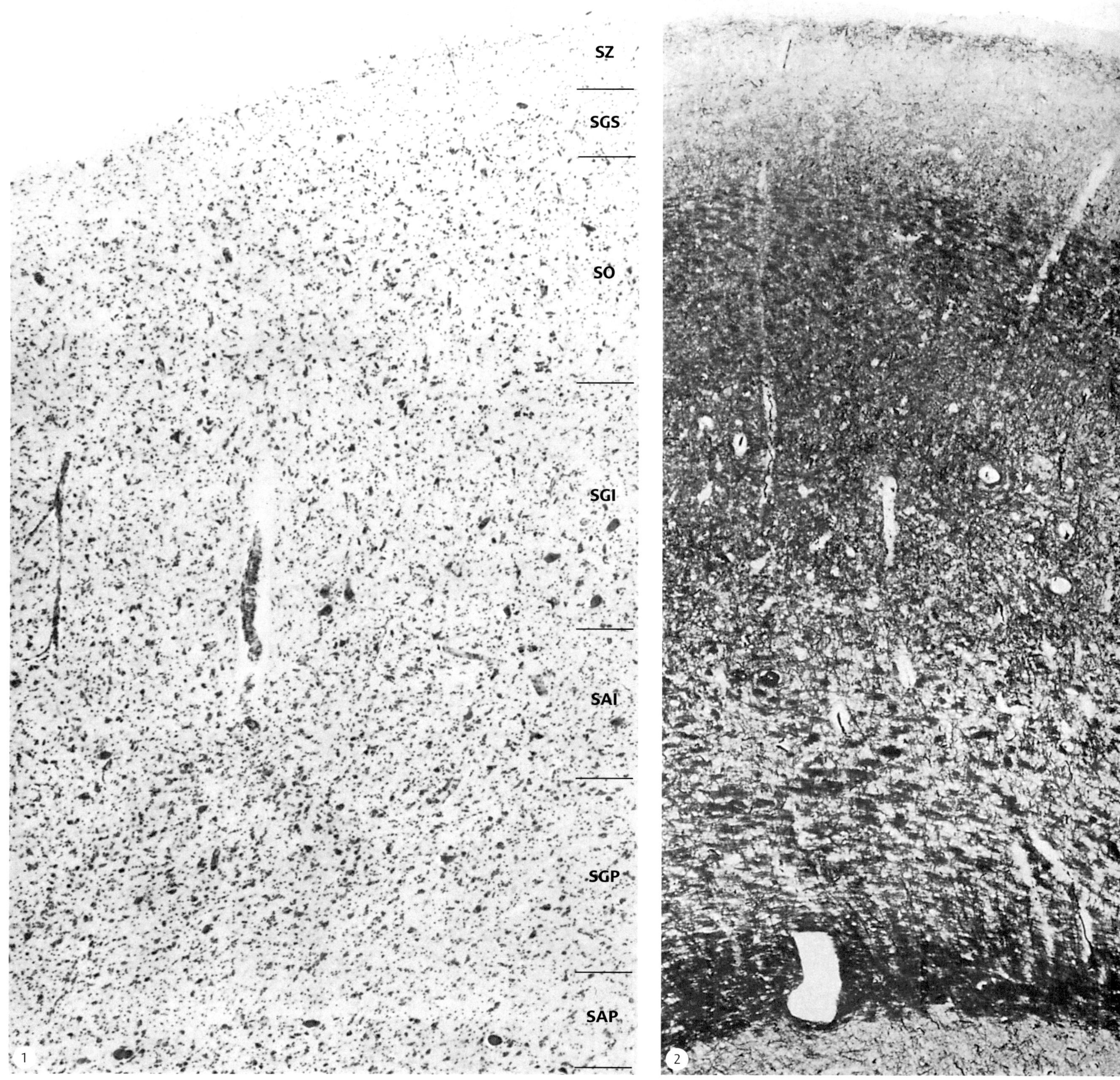

Fig. 1. Colliculus superior. See text for the description of the layers: SZ = stratum zonale; SGS = stratum griseum superficiale; SO = stratum opticum; SGI = stratum griseum intermediale; SAI = stratum album intermediale; SGP = stratum griseum profundum; SAP = stratum album profundum. Magnification ×150.

Fig. 2. Colliculus superior. Myelinated fibers stained by the Heidenhain method. Magnification ×150.

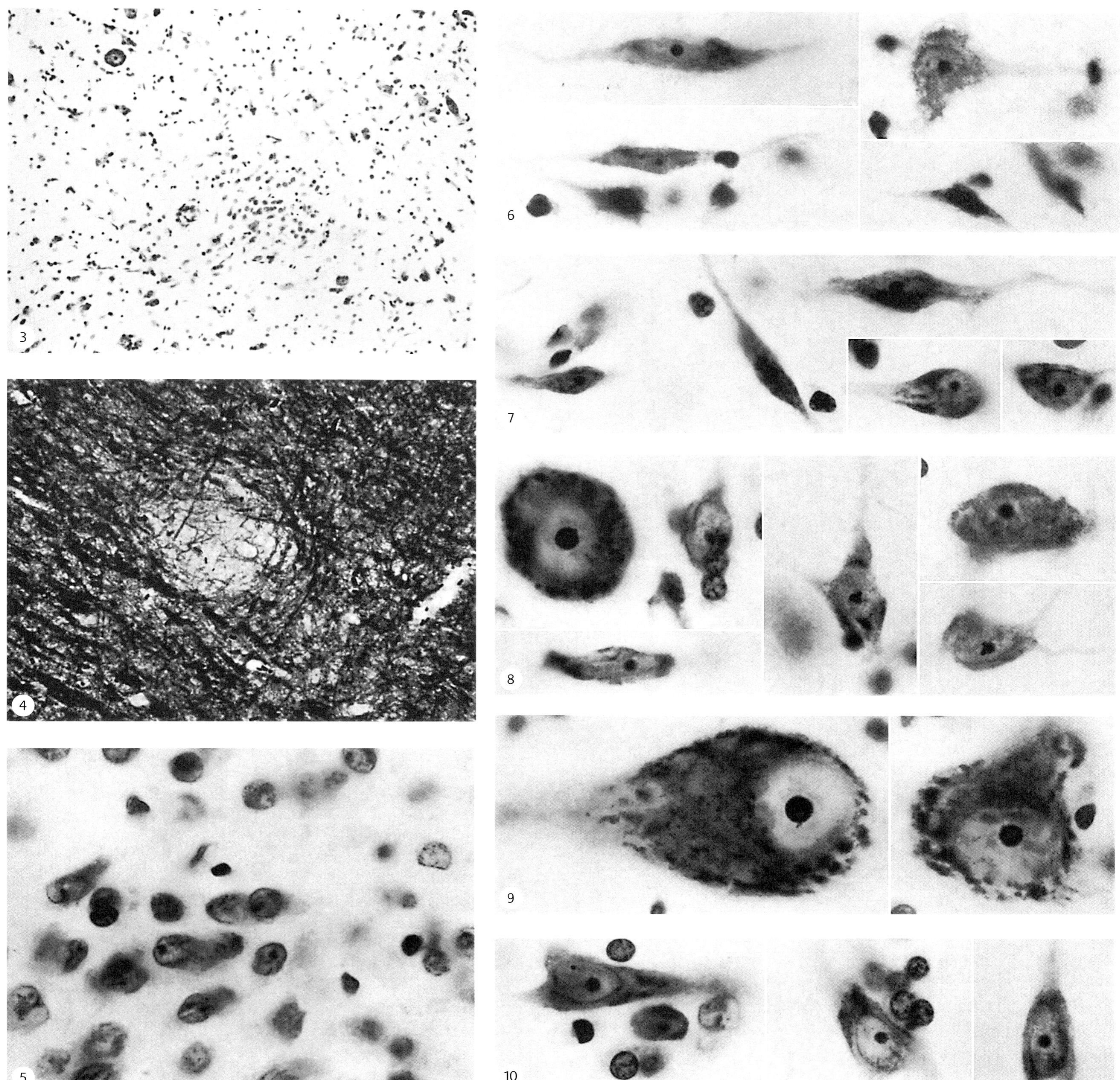

Fig. 3. Glomerulus of the SGI, superior colliculus. Magnification ×150.
Fig. 4. Glomerulus of the SGI, superior colliculus. Heidenhain fiber stain. Magnification ×150.
Fig. 5. Glomerulus of the SGI, superior colliculus. Magnification ×1,000.
Fig. 6. Cells from the SZ of the superior colliculus. Magnification ×1,000.
Fig. 7. Cells from the SGS of the superior colliculus. Magnification ×1,000.
Fig. 8. Cells from the SGI of the superior colliculus. Magnification ×1,000.
Fig. 9. Large cells from the SAI of the superior colliculus. Magnification ×1,000.
Fig. 10. Cells from the SGP and SAP of the superior colliculus. Magnification ×1,000.

channeled to the intermediate and deep layers of the SC, which provide premotor signals to direct the eyes, head, ears and possibly limbs (see chapter on nucleus No. 5). Ascending tectal pathways serve to direct attention to the object of interest. In the case of a threatening stimulus, separate neural pathways from the medial, as opposed to the lateral, colliculus are thought to participate in defense-related responses, such as orienting the head and body away from the threat [Comoli et al., 2012]. For reviews, see May [2006], Nieuwenhuys et al. [2008], and Lee and Groh [2012].

Connections

Afferents. Retinotectal input to the SC in primates arises from both retinae and enters via the brachium of the SC into the optic layer. Dense terminals are found in the SZ, and in the SGS there is a patchy distribution of contralateral and ipsilateral retinal afferents. The retinotopic map in the superficial layers of the SC represents the contralateral hemifield, with the fovea represented rostrolaterally, peripheral visual fields represented caudally, upper visual fields medially and lower visual fields laterally. This map coincides with the saccadic motor map in the deeper SC layers [Sparks and Hartwich-Young, 1989], and the auditory map arising from inputs of the nucleus of the brachium of the inferior colliculus and in part from the external nucleus of the inferior colliculus [Doubell et al., 2000; Harting and van Lieshout, 2000; May, 2006; Nieuwenhuys et al., 2008, fig. 19.12; May, 2006]. A somatosensory map of the body is also to be found in the deeper SC layers [Meredith et al., 1991].

Additional visual signals to the complete superficial layer of the SC arise from the adjacent nucleus of the optic tract (see 'Pretectum'). The information carried in these pathways favors large, slow-moving visual fields more suited to detecting self-motion, but some also respond to visual-jerk stimuli. The projection is GABAergic and may contribute to gaze fixation [Büttner-Ennever et al., 1996]. The projection from the parabigeminal nucleus also targets the superficial SC layers, and is reciprocal. Furthermore well-documented reciprocal projections exist between the SC and the subjacent MRF; the connection is topographic and concentrates around a region (central MRF) known to participate in horizontal saccades and head movements [Cohen and Büttner-Ennever, 1984; Warren et al., 2008; Wang et al., 2010].

Corticotectal projections from the visual cortex (V1 and V2) and medial temporal cortex (MT) terminate in the SC superficial layers but the deeper layers of the SC receive inputs from the frontal, supplementary and parietal eye fields as well as the MT.

An important projection to the SC which is inhibitory (GABAergic) originates in the substantia nigra, pars reticulata (SNr). A similar GABAergic input arises from the zona incerta. The nigrotectal pathway is controlled by inputs from the striatum, and thought to facilitate the visuomotor activity in the SC for acquiring a selected target [Harting and Updyke, 2006]. Careful tracing experiments have shown some of the complexity and exquisite order within these pathways. The SGI contains 3 tiers, or sublaminae, of nigrotectal projections (dorsal, middle and ventral), and terminals in each tier have a different site of origin in the SNr. The nigrotectal terminals form regular patches across the SGI within the different tiers, and the patches of the middle tier are interdigitated by similar patches from trigeminotectal afferents [Harting and van Lieshout, 1991]. These same experiments showed that the nigrotectal terminals of the middle SGI tier coincide with the afferent terminal patches of the pedunculopontine nucleus (PPT), a nucleus which modulates information in relation to attention, motivation and other cognitive functions. A further, independent nigrotectal projection arises from the lateral substantia nigra, but it covers all collicular layers and is sparse.

The SC has a well-developed commissural system consisting of excitatory and inhibitory connections. Their organization has been well demonstrated in recent experiments in the cat [Takahashi et al., 2010]. For details of intralaminar connections and other inputs, see May [2006].

Finally a comparison of the connectivity of 'medial colliculus deep layers' versus 'lateral colliculus deep layers' in the rat has shown that, surprisingly, they have very different afferent inputs, and there was little overlap found in their afferent control [Comoli et al., 2012]. Only a few structures targeted both regions (e.g. SNr, zona incerta, PPT, spinal trigeminal nucleus and locus coeruleus). More afferents were found to target the medial colliculus compared with lateral areas. These connectivity results support the hypothesis that the medial colliculus is associated with defense-related reactions, and the lateral colliculus with orienting responses.

Efferents. The SC output pathways contact several motor circuits. The efferents originate from large multipolar neurons in the intermediate and deep layers of the SC which give rise to the predorsal bundle. The descending predorsal bundle crosses the midline in the mesencephalon and sends a compact pathway to the premotor cell groups of the paramedian pontine reticular formation (nucleus No. 57) for the generation of horizontal saccadic eye movements [Büttner-Ennever et al., 1999; Izawa et al., 2007; Shinoda et al., 2008]. The collicular projection to the supraoculomotor area may influence the oculomotor system when saccades and vergence occur simultaneously [May, 2006]. Ascending branches of the predorsal bundle also target the vertical saccade generators in the mesencephalon, including the interstitial nucleus of Cajal (INC), rostral interstitial nucleus of the medial longitudinal fasciculus (RIMLF) as well as the mediodorsal thalamic nucleus and the intralaminar thalamic nuclei. These thalamic relay nuclei in turn contact not only the cerebral cortex, but also the basal ganglia, which influences the SNr and hence provides a feedback to the SC [Harting and Updyke, 2006]. The predorsal bundle terminates in premotor regions for head movements through projections to the medullary reticular formation and the spinal cord [Cowie et al., 1994; Quessy and Freedman, 2004].

Additional control over motor networks arises via SC interconnections with the cerebellum, through several precerebellar nuclei: there is a direct pathway to the caudal part of the contralateral medial accessory olive, to the dorsolateral pontine nuclei, and to the nucleus reticularis tegmenti pontis [May, 2006]. The cerebellar nuclei, including the fastigial nucleus, all project back to the SC [for a review, see May, 2006].

The output of the superficial layers of the SC projects to visual relay nuclei, such as the dorsal and ventral lateral geniculate nuclei, the pulvinar, pretectum and parabigeminal nucleus [May, 2006; Nieuwenhuys et al., 2008].

References

Büttner-Ennever JA, Cohen B, Horn AKE, Reisine H: Efferent pathways of the nucleus of the optic tract in monkey and their role in eye movements. J Comp Neurol 1996;373:90–107.

Büttner-Ennever JA, Horn AKE, Henn V, Cohen B: Projections from the superior colliculus motor map to omnipause neurons in monkey. J Comp Neurol 1999; 413:55–67.

Cohen B, Büttner-Ennever JA: Projections from the superior colliculus to a region of the central mesencephalic reticular formation (cMRF) associated with horizontal saccadic eye movements. Exp Brain Res 1984;57:167–176.

Comoli E, Das Neves Favaro P, Vautrelle N, Leriche M, Overton PG, Redgrave P: Segregated anatomical input to sub-regions of the rodent superior colliculus associated with approach and defense. Front Neuroanat 2012;6:9.

Cowie RJ, Smith MK, Robinson DL: Subcortical contributions to head movements in macaques. II. Connections of a medial pontomedullary head-movement region. J Neurophysiol 1994;72:2665–2682.

Doubell TP, Baron J, Skaliora I, King AJ: Topographical projection from the superior colliculus to the nucleus of the brachium of the inferior colliculus in the ferret: convergence of visual and auditory information. Eur J Neurosci 2000; 12:4290–4308.

Harting JK, Updyke BV: Oculomotor-related pathways of the basal ganglia. Prog Brain Res 2006;151:441–460.

Harting JK, van Lieshout DP: Spatial relationships of axons arising from the substantia nigra, spinal trigeminal nucleus, and pedunculopontine tegmental nucleus within the intermediate gray of the cat superior colliculus. J Comp Neurol 1991;305:543–558.

Harting JK, van Lieshout DP: Projections from the rostral pole of the inferior colliculus to the cat superior colliculus. Brain Res 2000;881:244–247.

Izawa Y, Sugiuchi Y, Shinoda Y: Neural organization of the pathways from the superior colliculus to trochlear motoneurons. J Neurophysiol 2007;97:3696–3712.

Lee J, Groh JM: Auditory signals evolve from hybrid- to eye-centered coordinates in the primate superior colliculus. J Neurophysiol 2012;108:227–242.

Ma TP, Cheng HW, Czech JA, Rafols JA: Intermediate and deep layers of the macaque superior colliculus: a Golgi study. J Comp Neurol 1990;295:92–110.

May PJ: The mammalian superior colliculus: laminar structure and connections. Prog Brain Res 2006;151:321–378.

May PJ, Porter JD: The laminar distribution of macaque tectobulbar and tectospinal neurons. Vis Neurosci 1992;8:257–276.

Meredith MA, Clemo HR, Stein BE: Somatotopic component of the multisensory map in the deep laminae of the cat superior colliculus. J Comp Neurol 1991;312: 353–370.

Nieuwenhuys R, Voogd J, van Huijzen C: The Human Central Nervous System. Berlin, Springer, 2008.

Quessy S, Freedman EG: Electrical stimulation of rhesus monkey nucleus reticularis gigantocellularis. I. Characteristics of evoked head movements. Exp Brain Res 2004;156:342–356.

Shinoda Y, Sugiuchi Y, Izawa Y, Takahashi M: Neural circuits for triggering saccades in the brainstem. Prog Brain Res 2008;171:79–85.

Sparks DL, Hartwich-Young R: The deep layers of the superior colliculus; in Wurtz RH, Goldberg ME (eds): The Neurobiology of Saccadic Eye Movements. Amsterdam, Elsevier, 1989.

Takahashi M, Sugiuchi Y, Shinoda Y: Topographic organization of excitatory and inhibitory commissural connections in the superior colliculi and their functional roles in saccade generation. J Neurophysiol 2010;104:3146–3167.

Wang N, Warren S, May PJ: The macaque midbrain reticular formation sends side-specific feedback to the superior colliculus. Exp Brain Res 2010;201:701–717.

Warren S, Waitzman DM, May PJ: Anatomical evidence for interconnections between the central mesencephalic reticular formation and cervical spinal cord in the cat and macaque. Anat Rec 2008;291:141–160.

Sensory Systems – Visual Nuclei (Plate 32)

31 Parabigeminal Nucleus (PBG)

Original name:
Not described; included in nucleus paralemniscalis [Olszewski and Baxter, 1982]

Alternative name:
Nucleus isthmi (in nonmammalian species [Diamond et al., 1992])

Location and Cytoarchitecture

The parabigeminal nucleus (PBG) lies on the lateral edge of the brainstem at the level of the ventral inferior colliculus (IC). It is subadjacent to the IC external nucleus, and bordered dorsomedially by the lateral lemniscal fibers, and dorsolaterally by the brachium of the IC. Nieuwenhuys et al. [2008] overestimate the rostral extent of the PBG in their figure 6.30: that area is termed here the nucleus of the brachium of the inferior colliculus (see plate 34). Medial to the PBG lies the mesencephalic reticular formation, but between these two structures is a small-celled region that is tentatively called the microcellular tegmental nucleus by Paxinos and colleagues [Koutcherov et al., 2004] or the periparabigeminal nucleus [May, 2006]. This small nucleus is not treated further here.

The PBG is inconspicuous in Nissl sections, but is readily highlighted in sections stained for acetylcholine revealing a small oval nucleus; the PBG is designated as the cholinergic cell group Ch8 [Mufson et al., 1986; Paxinos and Huang, 1995].

Functional Neuroanatomy

Function
The PBG is a topographically organized, subcortical, visual center with strong reciprocal connections to the superior colliculus (SC). It is often referred to as satellite nucleus of SC. The neurons respond consistently to both moving and stationary visual stimul [Cui and Malpeli, 2003]. It has been proposed that the PBG functions as a cholinergic neuromodulatory nucleus, which may modulate visual attention through connections to the SC, as well as contributing to an ascending visual pathway parallel to the striate system, involving koniocellular (W cell) geniculate layers. The function of this ascending pathway is unclear. The PBG connections to the amygdala nuclei and pretectum may enable a rapid response to visual threats, before the signals are analyzed consciously [Usunoff et al., 2007].

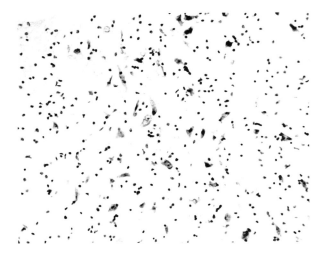

Fig. 1. Parabigeminal nucleus. Magnification ×150.

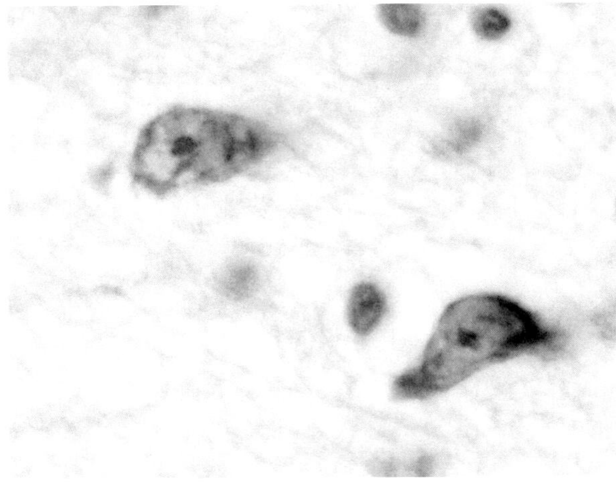

Fig. 2. Cells from the parabigeminal nucleus. Magnification ×1,000.

Connections

Afferents. There is a dense reciprocal interconnection between the PBG and the superficial layers of the SC. The SC efferent projections provide the main input to PBG neurons; they project ipsilaterally and are visuotopographically organized [Sherk, 1978; Baizer et al., 1991; Feig and Harting, 1992; Lee et al., 2001; May, 2006]. Some tectoparabigeminal projection fibers are substance P positive, providing the PBG with its sole source of substance P [Bennett-Clarke et al., 1989]. Additional inputs to the PBG originate from neurons in the prepositus nucleus (PrP), the ventral nucleus of the lateral geniculate body, locus coeruleus, the cuneiform nucleus, the periaqueductal gray and the dorsomedial hypothalamic area [Baleydier and Magnin, 1979]. The microcellular tegmental nucleus immediately medial to the PBG receives axon terminals from the deep SC and also PrP [May, 2006].

Efferents. Parabigeminal neurons provide the major cholinergic input to the superficial layers of the SC, bilaterally. Since acetylcholine has been shown, specifically in the superficial SC, to play an important role in the modulation of visual signals, it is probable that this is, at least in part, the role of the PBG [Binns and Salt, 2000]. The axon terminals from the PBG terminate on GABAergic interneurons in the SC, inducing an inhibition of the tectogeniculate projection neurons through GABA-B receptors [Lee et al., 2001]. Subgroups within the PBG have been described, with respect to its efferent targets: in the cat the rostral PBG projects to the contralateral SC and the caudal PBG to the ipsilateral SC [Graybiel, 1978; Edwards et al., 1979]. In contrast to the cat, ferret and monkey, the PBG in rats can be subdivided into a dorsal, middle and ventral subgroup, whereby the dorsal and ventral subgroups project to the ipsilateral SC, and the middle subgroup to the contralateral SC via the supraoptic commissure [Watanabe and Kawana, 1979].

Similar to the SC, the PBG projects heavily to the koniocellular layers of the thalamic lateral geniculate nucleus (LGN), that is to the small-celled layers associated with the W-cell retinogeniculocortical pathway. However, unlike the SC, the PBG also sends a smaller projection to the other magnocellular and parvocellular LGN layers [Harting et al., 1991; Bickford et al., 2000]. In some species connections to the striate-recipient pulvinar have been established [Diamond et al., 1992].

Comments

Three main cell groups make up the cholinergic midbrain area: the laterodorsal nucleus (LDT) lying medially in the central gray; the pedunculopontine tegmental nucleus (PPT), which lies lateral to it around the brachium conjunctivum (superior cerebellar peduncle) as it crosses the upper pons, and third the PBG on the lateral edge of the midbrain [Mufson et al., 1986; Paxinos and Huang, 1995; Oda and Nakanishi, 2000]. The neuromodulatory roles of the cholinergic networks from PPT and LDT are well documented [Saper et al., 2001; Butcher and Woolf, 2004] – see 'Neuromodulatory Systems: Overview' – but the role of PBG efferents in neuromodulation is not clear. The PBG, along with the medial habenular nucleus, is considered to form a separate cholinergic system from PPT and LDT by Geula and Mesulam [2012]. Differences in the PPT and PBG projections to the LGN are discussed by Bickford et al. [1993, 2000]. However, there is some evidence suggesting that the PBG plays a modulatory role in visual processing of the colliculus, and influences the W-cell koniocellular pathways [Lee et al., 2001; Goddard et al., 2007].

References

Baizer JS, Whitney JF, Bender DB: Bilateral projections from the parabigeminal nucleus to the superior colliculus in monkey. Exp Brain Res 1991;86:467–470.

Baleydier C, Magnin M: Afferent and efferent connections of the parabigeminal nucleus in cat revealed by retrograde axonal transport of horseradish peroxidase. Brain Res 1979;161:187–198.

Bennett-Clarke C, Mooney RD, Chiaia NL, Rhoades RW: A substance P projection from the superior colliculus to the parabigeminal nucleus in the rat and hamster. Brain Res 1989;500:1–11.

Bickford ME, Gunluk AE, Guido W, Sherman SM: Evidence that cholinergic axons from the parabrachial region of the brainstem are the exclusive source of nitric oxide in the lateral geniculate nucleus of the cat. J Comp Neurol 1993;334:410–430.

Bickford ME, Ramcharan E, Godwin DW, Erisir A, Gnadt JW, Sherman M: Neurotransmitters contained in the subcortical extraretinal inputs to the monkey lateral geniculate nucleus. J Comp Neurol 2000;424:701–717.

Binns KE, Salt TE: The functional influence of nicotinic cholinergic receptors on the visual responses of neurones in the superficial superior colliculus. Vis Neurosci 2000;17:283–289.

Butcher LL, Woolf NJ: Cholinergic neurons and networks revisited; in Paxinos G (ed): The Rat Central Nervous System. San Diego, Academic Press, 2004.

Cui H, Malpeli JG: Activity in the parabigeminal nucleus during eye movements directed at moving and stationary targets. J Neurophysiol 2003;89:3128–3142.

Diamond IT, Fitzpatrick D, Conley M: A projection from the parabigeminal nucleus to the pulvinar nucleus in *Galago*. J Comp Neurol 1992;316:375–382.

Edwards SB, Ginsburgh CL, Henkel CK, Stein BE: Sources of subcortical projections to the superior colliculus in the cat. J Comp Neurol 1979;184:309–330.
Feig S, Harting JK: Ultrastructural studies of the primate parabigeminal nucleus: electron microscopic autoradiographic analysis of the tectoparabigeminal projection in *Galago crassicaudatus*. Brain Res 1992;595:334–338.
Geula C, Mesulam MM: Brainstem cholinergic systems; in Mai JK, Paxinos G (eds): The Human Nervous System, ed 3. Amsterdam, Academic Press, 2012.
Goddard CA, Knudsen EI, Huguenard JR: Intrinsic excitability of cholinergic neurons in the rat parabigeminal nucleus. J Neurophysiol 2007;98:3486–3493.
Graybiel AM: A satellite system of the superior colliculus: the parabigeminal nucleus and its projections to the superficial collicular layers. Brain Res 1978;145: 365–374.
Harting JK, van Lieshout DP, Hashikawa T, Weber JT: The parabigeminogeniculate projection: connectional studies in eight mammals. J Comp Neurol 1991;305: 559–581.
Koutcherov Y, Huang X-F, Halliday G, Paxinos G: Organization of human brain stem nuclei; in Paxinos G, Mai JK (eds): The Human Nervous System, ed 2. Amsterdam, Elsevier Academic Press, 2004.
Lee PH, Schmidt M, Hall WC: Excitatory and inhibitory circuitry in the superficial gray layer of the superior colliculus. J Neurosci 2001;21:8145–8153.
May PJ: The mammalian superior colliculus: laminar structure and connections. Prog Brain Res 2006;151:321–378.
Mufson EJ, Martin TL, Mash DC, Wainer BH, Mesulam MM: Cholinergic projections from the parabigeminal nucleus (Ch8) to the superior colliculus in the mouse: a combined analysis of horseradish peroxidase transport and choline acetyltransferase immunohistochemistry. Brain Res 1986;370:144–148.
Nieuwenhuys R, Voogd J, van Huijzen C: The Human Central Nervous System. Berlin, Springer, 2008.
Oda Y, Nakanishi I: The distribution of cholinergic neurons in the human central nervous system. Histol Histopathol 2000;15:825–834.
Olszewski J, Baxter D: Cytoarchitecture of the Human Brain Stem. Basel, Karger, 1982.
Paxinos G, Huang X-F: Atlas of the Human Brainstem. San Diego, Academic Press, 1995.
Saper CB, Chou TC, Scammell TE: The sleep switch: hypothalamic control of sleep and wakefulness. Trends Neurosci 2001;24:726–731.
Sherk H: Visual response properties and visual field topography in the cat's parabigeminal nucleus. Brain Res 1978;145:375–379.
Usunoff KG, Schmitt O, Itzev DE, Rolfs A, Wree A: Efferent connections of the parabigeminal nucleus to the amygdala and the superior colliculus in the rat: a double-labeling fluorescent retrograde tracing study. Brain Res 2007;1133:87–91.
Watanabe K, Kawana E: Efferent projections of the parabigeminal nucleus in rats: a horseradish peroxidase (HRP) study. Brain Res 1979;168:1–11.

Sensory Systems – Visual Nuclei (Plates 36 and 37)

32 Lateral Terminal Nucleus (LTN)

Original name:
Not described

Alternative name:
Nucleus terminalis lateralis

Location and Cytoarchitecture

The lateral terminal nucleus (LTN) lies on the lateral surface of the mesencephalic brainstem, with the main part of the nucleus in the angle between the dorsal border of the cerebral peduncle and lateral substantia nigra. It is bounded dorsomedially by the brachium of the inferior colliculus and its nucleus, or further rostrally by the medial geniculate body (not shown). The inconspicuous cell groups extend from the LTN dorsally toward the colliculus, lying between the fiber pathways on the brain surface. Similar cell groups can be found ventrally around the cerebral peduncle, close to the surface of the brainstem. This population of neurons is called the interstitial nucleus of the superior fascicle, posterior fibers (ITN), and is part of a continuum with the LTN (see below). The cells of the LTN are small, lightly staining and loosely arranged between the afferent fibers of the accessory optic tract (fig. 1–3). Finger-like extensions of the LTN/ITN interdigitate with the substantia nigra, and strands continue into the medial mesencephalon, following the fascicles of the accessory optic tract and blood vessels towards the medial terminal nuclei [Fredericks et al., 1988; Giolli et al., 2006]. The LTN is more compact in monkeys than in humans, and forms a small bean-like nucleus on the lateral border of the brachium of the inferior colliculus [Baleydier et al., 1990; Cooper et al., 1990]. The LTN is an accessory optic cell group; and accessory optic pathways can be clearly visualized with parvalbumin (fig. 2) and cytochrome oxidase staining techniques (pers. observation) or using neuropeptide Y immunoreactivity [Borostyankoi-Baldauf and Herczeg, 2002].

Functional Neuroanatomy

Function

The LTN is part of the accessory optic system which receives a direct retinal input, from ganglion cells encoding information about slow, global (wide-field) visual shifts. The LTN neurons encode the vertical component of motion of the visual surround. Their information, together with that of other accessory optic nuclei encoding different directions of motion, contributes to the generation of optokinetic eye movements, through vestibular-oculomotor pathways. Optokinetic eye movements stabilize the visual background on the retina during self-motion through space, and they complement the vestibulo-ocular reflex (VOR) at low frequencies, where the VOR gain is low [Büttner and Büttner-Ennever, 2006; Giolli et al., 2006; see chapter 19, 'Visual System', in Nieuwenhuys et al., 2008].

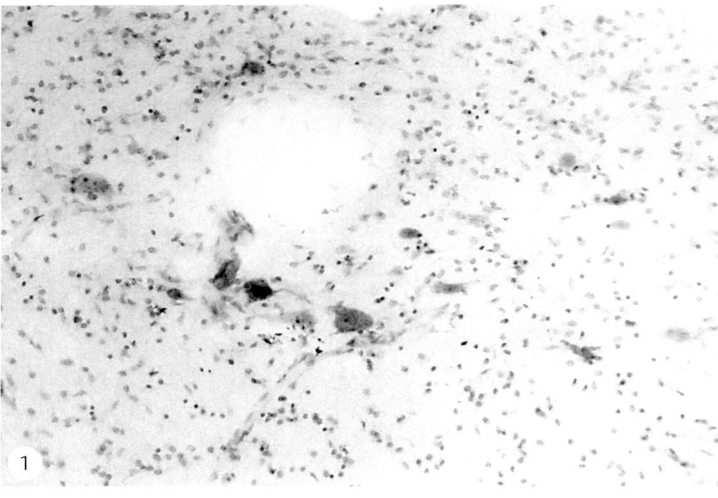

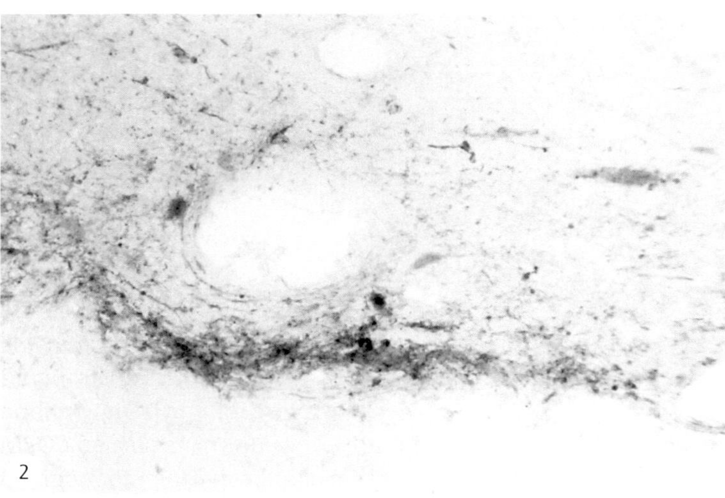

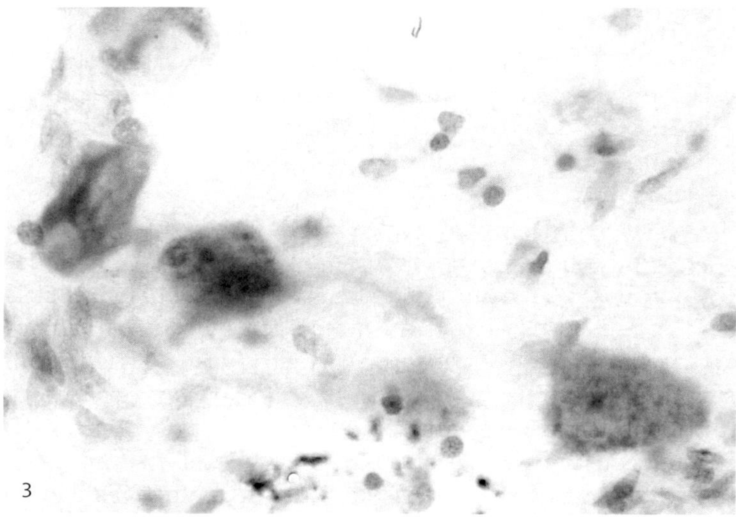

Fig. 1. Lateral terminal nucleus neurons at the lateral edge of the pons. Typically the lateral terminal nucleus cells follow along routes taken by blood vessels. Note the numerous blood vessels around the nucleus. Nissl stain. Magnification ×150.

Fig. 2. The same region as figure 1 showing the lateral terminal nucleus at the lateral edge of the pons. The afferent fibers of the optic nerve and their terminals are stained with parvalbumin in order to locate the inconspicuous lateral terminal nucleus. Parvalbumin stain. Magnification ×150.

Fig. 3. Cells of the lateral terminal nucleus. Nissl stain. Magnification ×1,000.

Connections

The direct retinal input to the LTN arises from crossed fibers of the optic tract called the 'superior fascicle of the accessory optic tract'. The fascicle branches off from the brachium of the colliculus superior, turns ventrally, and can be seen with the naked eye coursing across the cerebral peduncle on the ventral surface of the mesencephalon. These fibers were first given the name 'transpeduncular tracts' by von Gudden. They terminate in the accessory optic nuclei: the dorsal, lateral, medial and interstitial terminal nuclei. In primates the medial terminal nucleus has a dorsal and a ventral division [Giolli et al., 2006]. The LTN is the only nucleus of all the accessory terminal nuclei visible in the plates of this atlas: but all of them lie between plates 36 and 38.

Apart from retinal afferents, the accessory optic nuclei, including the LTN, receive information from the cerebral cortical visual areas [Lui et al., 1995]. The nuclei relay signals to the vestibular nuclei and the cerebellum, via efferent projections to the inferior olive, the prepositus nucleus and the pontine nuclei [Giolli et al., 2006]. In addition the accessory optic nuclei are tightly interconnected with reciprocal connections to each other, and to areas of the adjacent pretectum, specifically the nucleus of the optic tract (NOT) and the pretectal olivary nucleus [Cooper et al., 1990]. Many of these pathways are GABAergic [van der Want et al., 1992]. The NOT is also considered important for the transfer of accessory optic signals to optokinetic pathways, as is an area called the ventral tegmental relay zone, lying between the rootlets of the oculomotor nucleus in the ventral tegmental area. Many of these connections have been established in the rat and rabbit, and not in humans; for comprehensive reviews, see Giolli et al. [2006] and Gamlin [2006].

References

Baleydier C, Magnin M, Cooper HM: Macaque accessory optic system. II. Connections with the pretectum. J Comp Neurol 1990;302:405–416.

Borostyankoi-Baldauf Z, Herczeg L: Parcellation of the human pretectal complex: a chemoarchitectonic reappraisal. Neuroscience 2002;110:527–540.

Büttner U, Büttner-Ennever JA: Present concepts of oculomotor organization. Prog Brain Res 2006;151:1–42.

Cooper HM, Baleydier C, Magnin M: Macaque accessory optic system. I. Definition of the medial terminal nucleus. J Comp Neurol 1990;302:394–404.

Fredericks CA, Giolli RA, Blanks RHI, Sadun AA: The human accessory optic system. Brain Res 1988;454:116–122.

Gamlin PDR: The pretectum: connections and oculomotor-related roles. Prog Brain Res 2006;151:379–405.

Giolli RA, Blanks RHI, Lui F: The accessory optic system: basic organization with an update on connectivity, neurochemistry, and function. Prog Brain Res 2006; 151:407–440.

Lui F, Gregory KM, Blanks RHI, Giolli RA: Projections from visual areas of the cerebral cortex to pretectal nuclear complex, terminal accessory optic nuclei, and superior colliculus in macaque monkey. J Comp Neurol 1995;363:439–460.

Nieuwenhuys R, Voogd J, van Huijzen C: The Human Central Nervous System. Berlin, Springer, 2008.

Van der Want JJL, Nunes Cardozo JJ, van der Togt C: GABAergic neurons and circuits in the pretectal nuclei and the accessory optic system of mammals; in Mize RR, Marc R, Silito A (eds): Progress in Brain Research. Amsterdam, Elsevier, 1992.

33 Pretectum

Original name:
Regio pretectalis

Individual nuclei:
Nucleus of the optic tract (NOT)
Pretectal olivary nucleus (PON)
Posterior pretectal nucleus (PPN)
Medial pretectal nucleus (MPN)
Anterior pretectal nucleus (APN)

Location and Cytoarchitecture

The pretectum lies between the superior colliculus (SC) and the caudal thalamus. Although the pretectum has recently been shown to develop from the caudal synencephalon (prosomere 1), it is still usually considered with the mesencephalon [Nieuwenhuys et al., 2008]. The pretectum is subdivided into 5 subnuclei [Simpson et al., 1988; Borostyankoi-Baldauf and Herczeg, 2002; Gamlin, 2006]: the posterior pretectal nucleus (PPN), the nucleus of the optic tract (NOT), the pretectal olivary nucleus (PON), the medial pretectal nucleus (MPN) and finally the anterior pretectal nucleus (APN), which can be divided into a reticular and compact subnucleus [Borostyankoi-Baldauf and Herczeg, 2002].

The neurons of the NOT lie scattered amongst the fibers of the brachium of the superior colliculus, with the PPN lying ventrally and bordering on the nucleus of the posterior commissure. The neurons of the PON form a compact spherical structure embedded within the NOT in primates, but not in lower mammals [Büttner-Ennever et al., 1996; Gamlin, 2006]. The PON has a complex internal structure containing glomeruli [Klooster and Vrensen, 1997]. The MPN is the smallest nucleus and lies directly caudal to the habenular nuclei on the floor of the third ventricle. The compact subnucleus of the APN provides the dividing border to the rostral thalamic structures, i.e. nucleus limitans. For reviews of the development of various nomenclatures of the pretectum, and in particular the NOT, see Weber [1985] or Simpson et al. [1988]; these reviews also explain the frequent confusion of the NOT with the nucleus limitans. For details of pretectal cytoarchitecture, see the study of Hutchins and Weber [1985].

Functional Neuroanatomy

Function

The pretectum is a heterogeneous region, and known to be an important subcortical center for processing visual signals. It receives direct retinal afferents, and the main retinorecipient nuclei are the NOT and PON [Hutchins and Weber, 1985]. Recent research has concentrated primarily on these two regions [Gamlin, 2006]. The NOT is intimately interconnected with the accessory optic system, in particular with the dorsal terminal nucleus, and plays an important role in optokinetic responses to slowly moving wide-field visual stimuli, particularly in the horizontal direction. However, the NOT is not a homogeneous structure it also contains 'jerk neurons' which respond to very fast-moving large visual fields moving in any direction [Sudkamp and Schmidt, 1995]. It participates also in short latency ocular following, smooth pursuit eye movements and the adaptation of the gain of the horizontal vestibular ocular reflex [Yakushin et al., 2000; Mustari et al., 2009]. In contrast the PON has been shown to play a critical role in the pupillary light reflex, light-evoked blinks, and rapid eye movement sleep triggering. As a recipient of intrinsically photosensitive retinal ganglion cells, the PON also plays a role in the modulation of subcortical nuclei in circadian rhythms [Hattar et al., 2006; Allen et al., 2011].

The MPN and PPN have also been attributed visuomotor functions, but not the APN. The rostral part of the APN was thought to be part of a descending inhibitory pathway with antinociceptive influences on spinal cord inputs. This hypothesis stands in sharp contrast to possible visuomotor functions of the caudal APN [Foster et al., 1989; Rees and Roberts, 1993].

Connections

Nucleus of the Optic Tract. There are two main inputs to the NOT, from the retina and cerebral cortex. The retinal input is contralateral in lower animals but extensively bilateral in humans [Hutchins and Weber, 1985]. The descending connections from the cerebral cortex arise mainly from visual motion-related areas such as the middle temporal area/middle superior temporal area [Distler et al., 2002].

The NOT is not an accessory optic nucleus, but it has reciprocal connections to all 3 accessory optic nuclei [dorsal, lateral (LTN) and medial terminal nuclei]. They may participate in refining the directional tuning of the neurons to optokinetic stimuli [Giolli et al., 2006].

With respect to efferents, in all species the NOT projects to the SC, which lies immediately adjacent to it. The fibers are GABAergic and terminate mainly in the stratum griseum superficiale of the rostral SC, predominantly ipsilaterally. Given that the SC participates in the visual-grasp reflex, this inhibitory project has been suggested to play a role in gaze fixation [Büttner-Ennever et al., 1996]. The projections from the NOT to the vestibular nuclei and prepositus nucleus are thought to form the anatomical basis of the slow, or delayed, increase in optokinetic eye movements during vestibular stimulation [Büttner-Ennever et al., 1996; Gamlin, 2006; Giolli et al., 2006]. Additional projections have been demonstrated to the visual thalamic relay nuclei, locus coeruleus and the pulvinar, precerebellar structures including the inferior olive (IO), and preoculomotor nuclei; all the structures confirm the visuomotor role of the NOT, of which several aspects are still unclear [for a review, see Gamlin, 2006].

Pretectal Olivary Nucleus. The main PON afferent arises from the retina including intrinsically photosensitive retinal ganglion cells [Klooster and Vrensen, 1997; Allen et al., 2011]. Other inputs originate in the cortex, including the striate and extrastriate visual areas, the frontal and supplementary eye fields, as well as the ventral thalamus, the superior colliculus and the accessory optic nucleus LTN [Gamlin, 2006]. Efferent fibers of the PON cross in the posterior commissure and supply the bilateral (indirect) pathways to the Edinger-Westphal nuclei, which provide the basis for the pupillary constriction [Gamlin, 2006]. The PON also projects to structures controlling the circadian clock such as the suprachiasmatic nucleus [Tritto et al., 2009] and the ventral thalamus [Fernandez et al., 1988].

Posterior Pretectal Nucleus. Like the APN, the PPN receives somatosensory inputs from the lateral cervical nucleus, dorsal column nuclei and the somatosensory cortex [Wiberg et al., 1987; Eatock et al., 2008]. Descending pathways target the SC, pontine nuclei and IO. A compact projection from the PPN to the caudal dorsal accessory olive was reported [Goldberg and Cullen, 2011].

Medial Pretectal Nucleus. Few inputs to the MPN have been reported up to now: in primates a bilateral input from the retina was shown, and there are reports of afferents from the zona incerta, and the ventrolateral geniculate nucleus [Desai et al., 2005]. One controversial projection of the MPN is whether or not it is the source of retinopetal fibers [Labandeira-Garcia, 1988; Baizer and Broussard, 2010]. The existence of retinopetal fibers is generally accepted, but results concerning specific nuclei as origins are inconsistent and not well understood.

Anterior Pretectal Nucleus. Afferent inputs from the spinal cord, trigeminal and dorsal column nuclei to the rostral APN are prominent [Yoshida et al., 1992], and descending projections from the somatosensory and motor cortex combine to underline its role in sensory or sensorimotor integration [Berman, 1977; Foster et al., 1989]. The APN projects to the posterior thalamic nucleus, the zona incerta and IO [Murray et al., 2010]. Although the APN has no direct projection to the spinal cord, its antinociceptive effects are thought to operate through projections to the lateral midbrain and medulla, which descend in the dorsolateral funiculus and modulate lamina I spinal cord cells [Rees and Roberts, 1993].

References

Allen AE, Brown TM, Lucas RJ: A distinct contribution of short-wavelength-sensitive cones to light-evoked activity in the mouse pretectal olivary nucleus. J Neurosci 2011;31:16833–16843.

Baizer JS, Broussard DM: Expression of calcium-binding proteins and nNOS in the human vestibular and precerebellar brainstem. J Comp Neurol 2010;518:872–895.

Berman N: Connections of the pretectum in the cat. J Comp Neurol 1977;174:227–254.

Borostyankoi-Baldauf Z, Herczeg L: Parcellation of the human pretectal complex: a chemoarchitectonic reappraisal. Neuroscience 2002;110:527–540.

Büttner-Ennever JA, Cohen B, Horn AKE, Reisine H: Efferent pathways of the nucleus of the optic tract in monkey and their role in eye movements. J Comp Neurol 1996;373:90–107.

Desai SS, Ali H, Lysakowski A: Comparative morphology of rodent vestibular periphery. II. Cristae ampullares. J Neurophysiol 2005;93:267–280.

Distler C, Mustari MJ, Hoffmann KP: Cortical projections to the nucleus of the optic tract and dorsal terminal nucleus and to the dorsolateral pontine nucleus in macaques: a dual retrograde tracing study. J Comp Neurol 2002;444:144–158.

Eatock RA, Xue J, Kalluri R: Ion channels in mammalian vestibular afferents may set regularity of firing. J Exp Biol 2008;211:1764–1774.

Fernandez C, Baird RA, Goldberg JM: The vestibular nerve of the chinchilla. I. Peripheral innervation patterns in the horizontal and superior semicircular canals. J Neurophysiol 1988;60:167–181.

Foster GA, Sizer AR, Rees H, Roberts MH: Afferent projections to the rostral anterior pretectal nucleus of the rat: a possible role in the processing of noxious stimuli. Neuroscience 1989;29:685–694.

Gamlin PDR: The pretectum: connections and oculomotor-related roles. Prog Brain Res 2006;151:379–405.

Giolli RA, Blanks RHI, Lui F: The accessory optic system: basic organization with an update on connectivity, neurochemistry, and function. Prog Brain Res 2006; 151:407–440.

Goldberg JM, Cullen KE: Vestibular control of the head: possible functions of the vestibulocollic reflex. Exp Brain Res 2011;210:331–345.

Hattar S, Kumar M, Park A, Tong P, Tung J, Yau KW, Berson DM: Central projections of melanopsin-expressing retinal ganglion cells in the mouse. J Comp Neurol 2006;497:326–349.

Hutchins B, Weber JT: The pretectal complex of the monkey: a reinvestigation of the morphology and retinal terminations. J Comp Neurol 1985;232:425–442.

Klooster J, Vrensen GFJM: The ultrastructure of the olivary pretectal nucleus in rats. A tracing and GABA immunohistochemical study. Exp Brain Res 1997;114: 51–62.

Labandeira-Garcia JL: The retinopetal system in the rat. Neurosci Res 1988;6:88–95.

Murray PD, Masri R, Keller A: Abnormal anterior pretectal nucleus activity contributes to central pain syndrome. J Neurophysiol 2010;103:3044–3053.

Mustari MJ, Ono S, Das VE: Signal processing and distribution in cortical-brainstem pathways for smooth pursuit eye movements. Ann NY Acad Sci 2009; 1164:147–154.

Nieuwenhuys R, Voogd J, van Huijzen C: Development; in Nieuwenhuys R, Voogd J, van Huijzen C (eds): The Human Central Nervous System. Berlin, Springer, 2008.

Rees H, Roberts MH: The anterior pretectal nucleus: a proposed role in sensory processing. Pain 1993;53:121–135.

Simpson JI, Giolli RA, Blanks RHI: The pretectal nuclear complex and the accessory optic system; in Büttner-Ennever JA (ed): Neuroanatomy of the Oculomotor System. Amsterdam, Elsevier, 1988.

Sudkamp S, Schmidt M: Physiological characterization of pretectal neurons projecting to the lateral posterior-pulvinar complex in the cat. Eur J Neurosci 1995; 7:881–888.

Tritto S, Botta L, Zampini V, Zucca G, Valli P, Masetto S: Calyx and dimorphic neurons of mouse Scarpa's ganglion express histamine H_3 receptors. BMC Neurosci 2009;10:70.

Weber JT: Pretectal complex and accessory optic system of primates. Brain Behav Evol 1985;26:117–140.

Wiberg M, Westman J, Blomqvist A: Somatosensory projection to the mesencephalon: an anatomical study in the monkey. J Comp Neurol 1987;264:92–117.

Yakushin SB, Reisine H, Büttner-Ennever J, Raphan T, Cohen B: Functions of the nucleus of the optic tract (NOT). I. Adaptation of the gain of the horizontal vestibulo-ocular reflex. Exp Brain Res 2000;131:416–432.

Yoshida A, Sessle BJ, Dostrovsky JO, Chiang CY: Trigeminal and dorsal column nuclei projections to the anterior pretectal nucleus in the rat. Brain Res 1992;590: 81–94.

34 Solitary Nucleus (SOL)

Original name:
Nucleus tractus solitarii

Alternative name:
Nucleus of the solitary tract

Associated names:
Solitary tract (TSOL) or fasciculus solitarius
Oval nucleus of the solitary tract (rostral tip of SOL, SOLo)
Oval fasciculus of the solitary tract (rostral tip of TSOL)
Solitary nucleus, gelatinous subnucleus (SOLg)
Nucleus of the gelatinous substance of the solitary bundle [Pritchard and Norgren, 2004]

Location and Cytoarchitecture – Original Text

The solitary nucleus (SOL) appears in the closed portion of the medulla oblongata at the level of the caudal pole of the dorsal motor nucleus (DMX). It extends orally to the level of the caudal pole of the facial motor nucleus (VII), a distance of approximately 16 mm.

In the caudal portion of the medulla, the nucleus lies dorsolateral to the central canal, is oblong in outline, and its long axis is directed dorsomedially. Dorsally the nuclei of each side approximate each other above the central canal so that together they assume the form of an inverted V. This region dorsal to the central canal and on either side of the median raphe has been referred to as the 'commissural nucleus', but the cells of each side are separated by a narrow glial zone which extends dorsally from the central canal, and hence speaks against considering the commissural subnucleus as 'unpaired'. The SOL is related ventromedially to the DMX while the nuclei gracilis (GR) and cuneatus (LCU) lie along its dorsolateral border. The solitary tract (TSOL) is first discernable about 4 mm oral to the caudal extremity of the nucleus. In Nissl-stained sections, the tract appears as a round area almost devoid of cells, situated in the inferior pole, and surrounded by cells of the SOL. In the most caudal sections, the TSOL is small in diameter but increases rapidly in size as one proceeds orally.

In the periventricular portion of the medulla, the SOL lies in the floor of the fourth ventricle and assumes a more lateral position. It has increased somewhat in size, and has elongated so that its ventral pole extends farther ventrally than the ventral pole of the DMX. Medially, the main body of the nucleus maintains its relationship to the DMX, while laterally it lies in relation at first to the GR and more orally to the medial vestibular nucleus (MVN). The inferior pole of the nucleus is now bordered laterally by the LCU, ventrally by the central nucleus of the medulla oblongata and the parvocellular reticular nucleus (PCR).

Just prior to their disappearance, the SOL and TSOL, both somewhat diminished in size, lie dorsomedial to the spinal trigeminal complex, ventral to the MVN and dorsolateral to the PCR. Due to the fact that small cells are found in 3 of the structures mentioned above, it is often difficult to delineate the borders of the SOL at this level.

Although grossly the structure of the SOL appears uniform, a more detailed examination reveals slight differences in cell structure and arrangement in 3 areas (fig. 1).

That portion of the nucleus lying dorsal to the TSOL, and often referred to as the dorsal sensory nucleus of the vagus [perhaps equivalent to the dorsolateral subnucleus of Paxinos], is composed, in the main of small, fusiform or oval cells which contain a very scanty cytoplasm and lightly stained Nissl substance (fig. 3, 4). These cells are loosely arranged and the majority of them lie with their long axes directed dorsomedially.

That portion of the nucleus which surrounds the TSOL and is often termed the ventral sensory nucleus of the vagus [perhaps equivalent to the ventral subnucleus of Paxinos], is composed of cells which are similar in shape to the foregoing, but which are definitely larger and stain more intensely. In general, these cells lie with their long axis parallel to the circumference of the tract. A few nerve cells are always found scattered between the fibers of the TSOL.

At levels in which the DMX is a prominent structure, cells similar to those composing this nucleus, and often containing pigment, are found scattered among the smaller, more lightly stained cells of the SOL. Usually 3–5 such cells appear on each cross-section.

The gelatinosus subnucleus is located in the dorsolateral corner of the SOL at the level of the area postrema (AP). It appears as 1–3 small, round areas characterized in Nissl stain by few glial elements and by very few, pale, small, triangular or elongated nerve cells (fig. 1, 5, 6). This subnucleus is also easily distinguishable in myelin-stained sections. Here, it is characterized by a peculiar gelatinous appearance of the ground substance and by the complete lack of myelinated fibers (fig. 2). [...]

Oval Nucleus – The Rostral Tip of the Solitary Nucleus

The oval nucleus (SOLo, now known to be the rostral tip of the SOL; plates 18 and 20) can be seen as a small, elongated, inconspicuous structure located in the dorsolateral tegmentum of the oral medulla and the caudal pons. The nucleus lies along the dorsal or dorsolateral border of the oral part of the spinal trigeminal nucleus (SpV), and extends from the oral pole of the DMX to the caudal pole of the principal trigeminal nucleus (PrV) a distance of 5 mm. Throughout its extent, the nucleus is surrounded by longitudinally coursing myelinated fibers, which, when traced caudally, appear to be directly continuous with the orality of the TSOL.

In each cross-section the oval nucleus appears as only 5–20 cells arranged in either 1 or 2 small groups. Thus, in Nissl-stained preparations the nucleus may be easily overlooked, whereas in myelin-stained sections the oval nucleus is usually clearly delineated by the surrounding ring of myelinated fibers (fig. 7). These fibers are more conspicuous about the caudal than about the oral part of the nucleus.

The cells composing the oval nucleus are small, plump, round or oval, with regular outline and short dendrites. They possess relatively large nuclei and their cytoplasm is stained darkly and diffusely (fig. 8, 9). These cells are quite similar to those of the PrV, and once this latter structure appears, one is unable to trace the oval nucleus farther orally. [...]

Functional Neuroanatomy

Function

The SOL is the principal sensory nucleus for receiving visceral inputs to the brainstem. Sensory fibers from taste buds, as well as receptors for taste and general sensation of the pharynx and larynx, all synapse in the SOL; in addition fibers carrying gastrointestinal, cardiovascular, baroreceptor and respiratory information from peripheral receptors terminate in the SOL. Their central branches enter the TSOL to terminate topographically within the nucleus. Here the representation of different modalities remains relatively separate: taste afferents from the tongue project to the rostral SOL, a region often called the

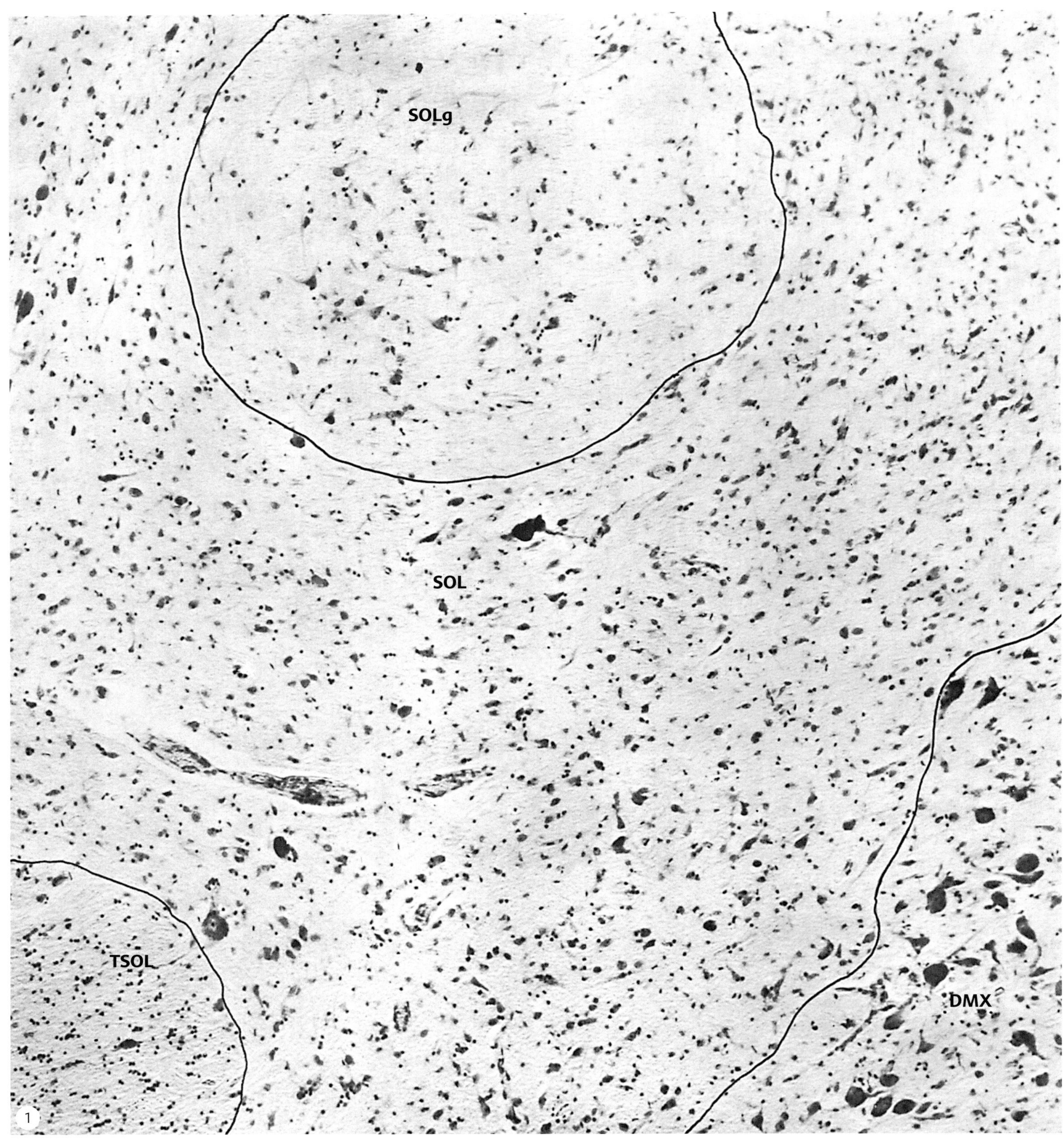

Fig. 1. Solitary nucleus. Note the position of the subnucleus gelatinosus, the solitary tract and the dorsal motor nucleus of the vagus. DMX = Dorsal motor nucleus of the vagus; SOLg = solitary nucleus, gelatinous subnucleus; SOL = solitary nucleus; TSOL = solitary tract. Magnification ×150.

'gustatory nucleus'. Gastrointestinal afferents synapse further caudally in the intermediate part of the SOL, adjacent to the AP. Cardiovascular, baro- and chemoreceptors are confined to the caudal half of the SOL, with respiratory inputs represented more laterally [for a review, see Saper, 1995]. The SOL is a complex integration center relaying sensory information that controls gustatory sensations, respiration, swallowing and cardiovascular functions. Neurons in the SOL, or close beside, are thought to contain the 'dorsal swallowing generator' [Jean, 2001]. Local central projections from the SOL to centers for respiration, swallowing and cardiovascular regulation in the medulla oblongata are vital for survival.

The sensory visceral inputs to the SOL are also the afferent limb for basic reflexes such as the gag reflex, cough reflex, and vomiting reflex, also for baroreflexes, chemoreflexes, respiratory reflexes and gastrointestinal reflexes.

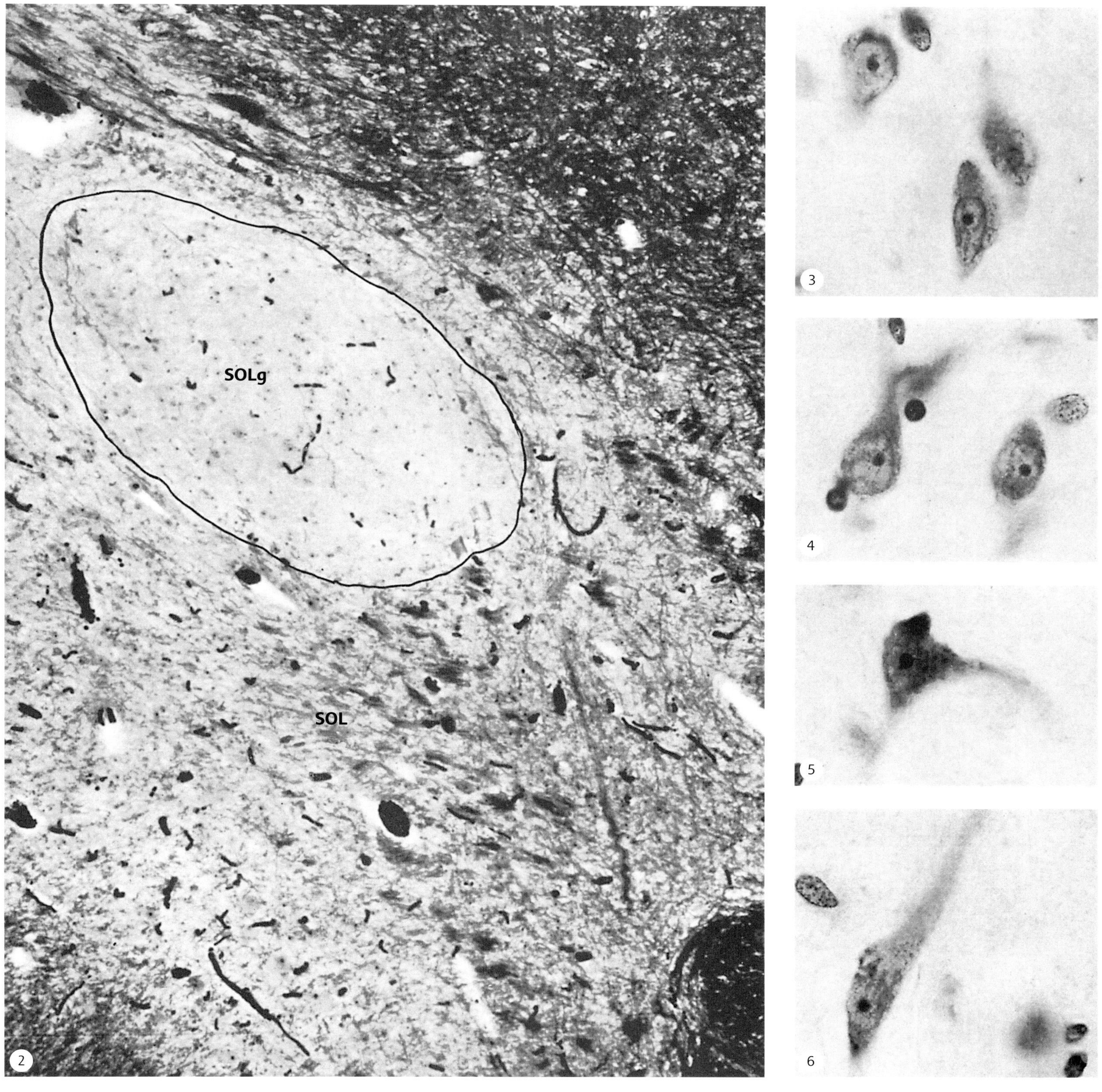

Fig. 2. Solitary nucleus. Note the homogeneous appearance of the background of the gelatinous subnucleus. SOLg = Solitary nucleus, gelatinous subnucleus; SOL = solitary nucleus. Heidenhain stain. Magnification ×150.

Fig. 3, 4. Small cells of the solitary nucleus. Magnification ×1,000.
Fig. 5, 6. Cells from the gelatinous subnucleus. Magnification ×1,000.

Organization of the Nucleus

The SOL has been divided into several subnuclei on the basis of cytoarchitectural and histochemical studies; this is comprehensively reviewed by Saper [1995] and Norgren [1995; see also Koutcherov et al., 2004; King, 2007]. The subnuclei are not so obvious in human material; however, the human SOL has been subdivided in a similar fashion [Paxinos and Huang, 1995; Koutcherov et al., 2004]. According to McRitchie and Tork [1993], the internal architecture of the human SOL is similar to that of rodents and carnivores, but with some notable species differences.

One such difference in the SOL of primates is the appearance of the so-called oval nucleus: in primates and cats the rostral tip of the SOL and TSOL extends further forward than in other species, i.e. beyond the level of the VII. It can be seen adjacent to the oral part of the SpV and even in the PrV, as an

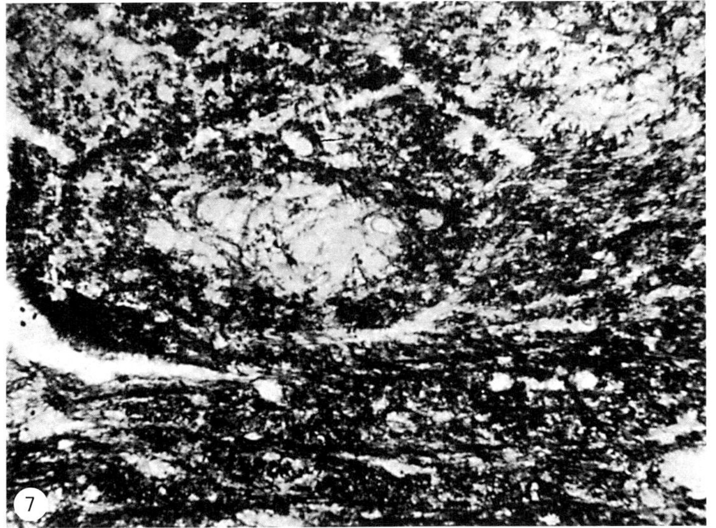

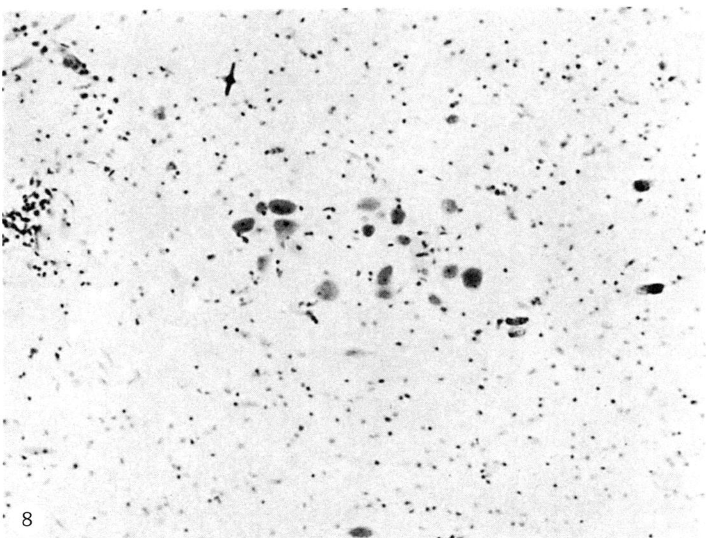

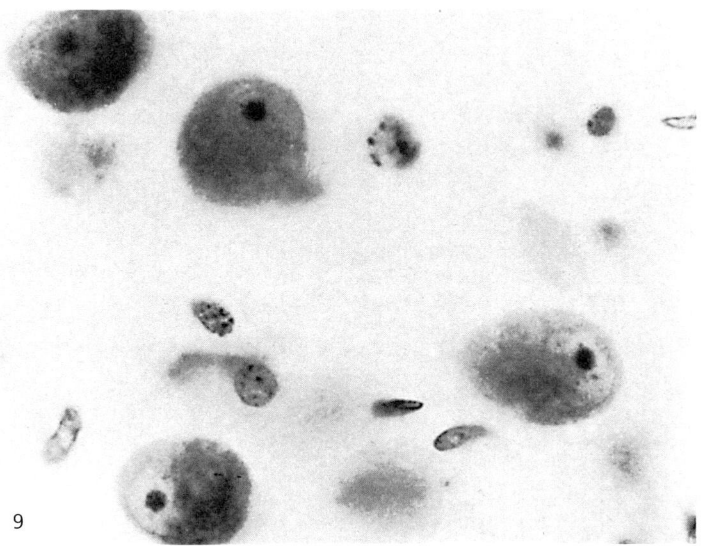

Fig. 7. Solitary nucleus, rostral tip. Haidenhain stain. Magnification ×150.
Fig. 8. Solitary nucleus, rostral tip. Magnification ×150.
Fig. 9. Cells from the solitary nucleus, rostral tip. Magnification ×1,000.

isolated group of plump cells surrounded by a fiber capsule (plates 18 and 20; fig. 7–9). These structures were called the 'oval nucleus (nucleus ovalis) and oval fascicle (fasciculus ovalis)' in earlier studies, and they were not recognized as the rostral pole of the SOL and TSOL. Pearson [1946] suggested a special visceral role for the oval nucleus, and the name is still used in some atlases, for instance in Nieuwenhuys et al. [2008, chapter 16 and fig. 6.25]. The oval nucleus is now shown in plates 18 and 20 as the rostral tip of the SOL (SOLo). It receives the gustatory fibers of the facial nerve (NVII) [Rhoton, 1968; Beckstead and Norgren, 1979; Pritchard and Norgren, 2004].

Many different neural transmitters, neuropeptides and neuropeptide receptors have been reported in the SOL, and the same neuropeptides can be found associated with relay nuclei carrying gustatory and visceral information to higher centers. The largest number of cell bodies containing neuropeptides and catecholamines, including the noradrenergic A2 and adrenergic C2 cell groups, are located around the SOL, DMX and PCR [for reviews, see Koutcherov et al., 2004; Bradley and King, 2007].

Connections

Afferents. Sensory visceral fibers project through the vagal, glossopharyngeal and facial nerves, leaving their ganglion cells in the inferior or geniculate ganglia, and their central branches enter the TSOL to branch off and terminate in the SOL.

The SOL has a close association with the AP, a sensory circumventricular organ which lies along the dorsal surface of the caudal SOL, its parvocellular subnucleus (nucleus No. 35). The AP has intimate contact with the liquor and vascular compartments, and major neuronal projections into the SOL. In addition some SOL neurons extend dendrites up into the AP [Herbert and Saper, 1990]. The results indicate a close relationship between these two nuclei, in which the AP can be considered as a chemosensory component of the solitary-vagal complex.

Efferents. The efferent projections of the SOL have been summarized by Saper [1995] into 3 major classes. First, the projections to the parasympathetic output neurons such as the DMX, or the salivatory nuclei, or the long descending pathways that target the preganglionic sympathetic system of the thoracic spinal cord, controlling cardiovascular structures [Torvik, 1957].

The second class of efferents is the local projections to the medullary reticular formation, targeting the specific areas that control vital functions, such as respiration and cardiovascular responses, swallowing or vomiting [for a review, see Nieuwenhuys et al., 2008, fig. 22.4 and 16.10]. The local projections to the rostral and caudal 'ventrolateral superficial medullary reticular formation', including the A1 noradrenaline-synthesizing neurons of the PCR, control respiration and cardiovascular responses, and possibly nausea and vomiting [Blessing, 2004]. The A5 group near the superior olive also appears to be part of this network, since it receives efferents from the SOL (as well as from the hypothalamus and parabrachial nuclei) and projects to the cervical and thoracic cord from where it could influence respiration vasoconstriction [Byrum and Guyenet, 1987]. There are also the direct projections from the SOL central subnucleus to the nucleus ambiguus controlling esophagoesophageal reflexes [Blessing, 2004]. Sensory outputs from the SOL play a critical role in swallowing, particularly those from the superior laryngeal nerve terminating as rosettes in the interstitial subnucleus of the SOL [for reviews, see Mrini and Jean, 1995; Jean, 2001; Feldman et al., 2003]. The information is relayed to neurons in, or close to, the SOL, which are considered to be generator neurons for triggering and shaping a rhythmic swallowing pattern [Jean, 2001]. The region is referred to as the 'dorsal swallowing generator'. This network drives a 'ventral swallowing generator' in the adjacent PCR, which distributes the swallowing drive to the various pools of motoneurons such as the ambiguus, retrofacial and hypoglossal nuclei involved in swallowing. The cell group referred to as the dorsal medullary reticular column in the studies on swallowing by Cunningham and Sawchenko [2000] lies around the SOL in the same region as the swallowing trigger network, at the dorsomedial tip of the caudal PCR, and near the A2 and C2 catecholaminergic cell groups.

The third group of SOL efferents are the ascending pathways. The largest of these efferent projections carries taste, general visceral and respiratory signals to the ipsilateral parabrachial nuclei. Like the SOL, the parabrachial nuclei also project to the thalamic ventroposterior medial nucleus and several limbic forebrain sites. The thalamic pathways relay gustatory information to the insula cortex and operculum [Pritchard and Norgren, 2004].

References

Beckstead RM, Norgren R: An autoradiographic examination of the central distribution of the trigeminal, facial, glossopharyngeal, and vagal nerves in the monkey. J Comp Neurol 1979;184:455–472.

Blessing WW: Lower brain stem regulation of visceral, cardiovascular, and respiratory function; in Paxinos G, Mai JK (eds): The Human Nervous System. Amsterdam, Elsevier Academic Press, 2004.

Bradley RM, King MS: The role of the nucleus of the solitary tract in gustatory processing; in Bradley RM (ed): The Role of the Nucleus of the Solitary Tract in Gustatory Processing. Boca Raton, Taylor & Francis, 2007.

Byrum CE, Guyenet PG: Afferent and efferent connections of the A5 noradrenergic cell group in the rat. J Comp Neurol 1987;261:529–542.

Cunningham ETJ, Sawchenko PE: Dorsal medullary pathways subserving oromotor reflexes in the rat: implications for the central neural control of swallowing. J Comp Neurol 2000;417:448–466.

Feldman JL, Mitchell GS, Nattie EE: Breathing: rhythmicity, plasticity, chemosensitivity. Annu Rev Neurosci 2003;26:239–266.

Herbert H, Saper CB: Cholecystokinin-, galanin-, and corticotropin-releasing factor-like immunoreactive projections from the nucleus of the solitary tract to the parabrachial nucleus in the rat. J Comp Neurol 1990;293:581–598.

Jean A: Brain stem control of swallowing: neuronal network and cellular mechanisms. Physiol Rev 2001;81:929–969.

King MS: Anatomy of the rostral nucleus of the solitary tract; in Bradley RM (ed): The Role of the Nucleus of the Solitary Tract in Gustatory Processing. Boca Raton, CRC Press, 2007.

Koutcherov Y, Huang X-F, Halliday G, Paxinos G: Organization of human brain stem nuclei; in Paxinos G, Mai JK (eds): The Human Nervous System, ed 2. Amsterdam, Elsevier Academic Press, 2004.

McRitchie DA, Tork I: The internal organization of the human solitary nucleus. Brain Res Bull 1993;31:171–193.

Mrini A, Jean A: Synaptic organization of the interstitial subdivision of the nucleus tractus solitarii and of its laryngeal afferents in the rat. J Comp Neurol 1995;355:221–236.

Nieuwenhuys R, Voogd J, van Huijzen C: The Human Central Nervous System. Berlin, Springer, 2008.

Norgren R: Gustatory system; in Paxinos G (ed): The Rat Nervous System, ed 2. San Diego, Academic Press, 1995.

Paxinos G, Huang X-F: Atlas of the Human Brainstem. San Diego, Academic Press, 1995.

Pearson AA: The development of the motor nuclei of the facial nerve in man. J Comp Neurol 1946;85:461–476.

Pritchard TC, Norgren R: Gustatory system; in Paxinos G, Mai JK (eds): The Human Nervous System, ed 2. Amsterdam, Elsevier Academic Press, 2004.

Rhoton AL Jr: Afferent connections of the facial nerve. J Comp Neurol 1968;133: 89–100.

Saper CB: Central autonomic system; in Paxinos G (ed): The Rat Nervous System, ed 2. San Diego, Academic Press, 1995.

Torvik A: The spinal projection from the nucleus of the solitary tract: an experimental study in the cat. J Anat 1957;91:314–322.

Sensory Systems – Viscerosensory Nuclei

(Plates 10 and 11)

35 Area Postrema (AP)

Original name:
Area postrema

Alternative names:
None

Location and Cytoarchitecture – Original Text

This structure lies in the most caudal part of the fourth ventricle, on the border between the medulla oblongata and the spinal cord. It has an approximate length of 1.5–2 mm. It may be visualized as a V-shaped cell group with its apex directed caudally and the two wings orally. The unpaired caudal portion spans the obex and lies dorsal to the dilated, most oral portion of the central canal. The paired, oral portions of the area postrema (AP) lie in the lateral wall of the caudal portion of the fourth ventricle. Immediately caudal to the obex, the AP appears on cross-section as an unpaired structure lying dorsal and dorsolateral to the dilated central canal, and bordered laterally by the solitary nucleus (SOL). Oral to the obex, the paired portions of the AP form mound-like elevations, one on each lateral wall of the fourth ventricle. On each side the area maintains its medial relationship to the SOL. In the floor of the fourth ventricle the AP is separated from the more medially situated area cinerea by a glistening band of neuroglial fibers known as the funiculus separans. These neuroglial fibers extend ventrally into the medulla and partially separate the ventromedial and inferior surfaces of the AP from the SOL.

The AP contains neurons embedded in a meshwork of astroglia, a few oligodendrocytes, and a strikingly rich bed of capillaries from the posterior inferior cerebellar artery [Dempsey, 1973]. In Nissl-stained sections the glia and neurons are often difficult to distinguish, and the appearance of the cells composing the AP varies depending on the age of the individual. In the child (fig. 2) the AP is composed of darkly stained, closely packed, round or oval nuclei, some of which contain nucleoli. No cytoplasm can be distinguished around the majority of these nuclei. However, a few large, clear, oval nuclei possess a small amount of palely stained, homogeneous cytoplasm. Such cells are usually spindle shaped.

In the young adult (fig. 1, 3), one can distinguish cytoplasm around many more cells than in the child, and further, dark, blue, intracytoplasmic granular masses may be seen. The nuclei of such cells tend to lie close

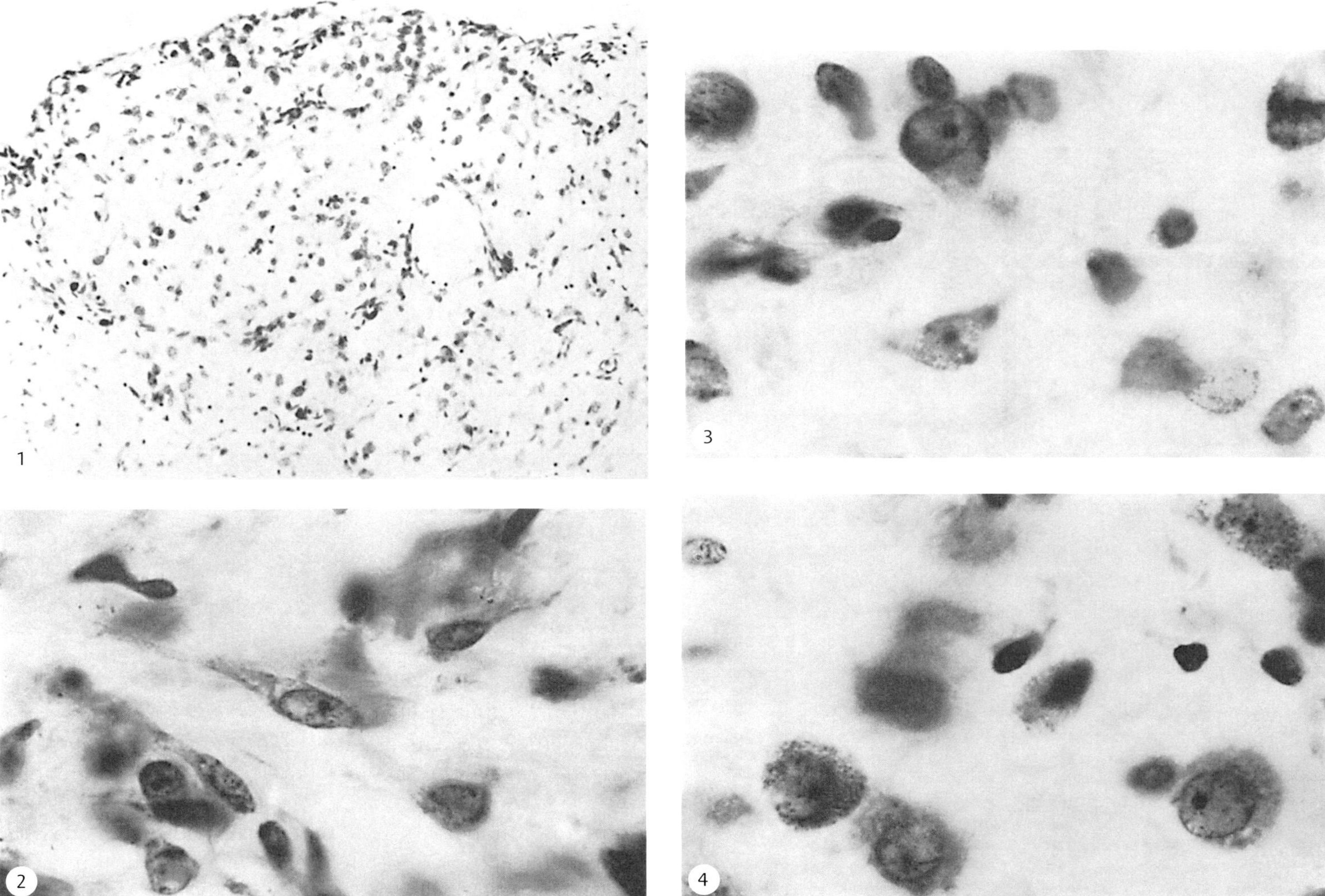

Fig. 1. Area postrema. Magnification ×150.
Fig. 2. Cells from the area postrema of an infant. Magnification ×1,000.
Fig. 3. Cells from the area postrema of a young adult. Magnification ×1,000.
Fig. 4. Cells from the area postrema of a middle-aged adult. Magnification ×1,000.

to the periphery of the perikaryon so that one aspect of the nucleus is related to a very scanty cytoplasm [King, 1937]. These cells are small, and may be round, oval or spindle shaped in outline, and possess distinct nucleoli. Other similar nuclei, unassociated with cytoplasm which stains with the Nissl technique, lie scattered among these cells. Clusters of indisputably glial nuclei are found at the medial and lateral extremities of the AP.

After the second decade of life (fig. 4), fine, greenish brown pigment granules tend to collect in the cytoplasm of many cells of the AP and, as a consequence, these cells are quite conspicuous. According to Cammermeyer [1949], such pigment granules are found only in the human species.

One of the most striking microscopic features of the AP at all ages is its marked vascularity – a feature which can be appreciated even in Nissl-stained preparations.

Functional Neuroanatomy

Function

The AP is a sensory circumventricular organ (CVO), which participates in the transfer of information between the blood, CVO neurons, and the cerebrospinal fluid. The CVOs regulate autonomic functions by the release of, or by sensing, hormones without the disruption of the blood-brain barrier. The best-established function of the AP is that it elicits nausea and vomiting in response to emetic or toxic chemical substances in the circulation, i.e. the AP is considered to act as a chemoreceptor trigger zone for the vomiting reflex [for reviews, see Borison, 1989; Cottrell and Ferguson, 2004; McKinley et al., 2004; Price et al., 2008]. Recent evidence shows that 5-hydroxtryptamine receptors in the AP may play a role in vomiting and nausea, and blocking the response of these receptors reduces the nausea experienced under chemotherapy [Miller and Leslie, 1994].

There is also evidence for additional functional roles of the AP in the maintenance of water and energy balance of the body, cardiovascular regulation [Cottrell and Ferguson, 2004], as well as acting as an immunological interface between the brain and the immune system [Cottrell and Ferguson, 2004; Goehler et al., 2006; Wuchert et al., 2008].

Circumventricular Organs

CVOs are specialized regions of the central nervous system within the walls of the brain ventricles, which have access to the cerebrospinal fluid and to the blood supply, due to the absence of local blood-brain barriers [for reviews, see Cottrell and Ferguson, 2004; Price et al., 2008]. The AP is the most caudal of the CVOs in the mammalian brain. According to a recent review there are 8 classically defined CVOs, 3 sensory, 4 secretory and 1 that is of indiscriminate character. The sensory CVOs include the AP, organum vasculosum of the lamina terminalis, and the subfornical organ. The secretory CVOs include the neurohypophysis, median eminence, intermediate lobe of the pituitary gland, and the pineal gland. Like all CVOs, the AP is characterized by a number of morphological features: it is highly vascularized with unusual vascular arrangements; the blood-brain barrier is absent due to fenestrations of the capillary walls; the blood vessels are surrounded by extensive perivascular spaces; the columnar ciliated cells of the normal ventricular walls are replaced by irregular and flattened ependymal cells; tight junctions between the ependymal cells form a barrier between the cerebrospinal fluid and the CVO cells. The specialized features of the CVO enable them to play a role in transducing information between the blood, CVO neurons and the cerebrospinal fluid, and thereby regulate autonomic functions by the release of, or by sensing, hormones without the disruption of the blood-brain barrier.

Connections

Afferents. The major afferents to the AP carry visceral sensory information; they arise from the SOL, the lateral parabrachial nuclei and the hypothalamus, specifically the paraventricular nucleus and adjacent areas [Shapiro and Miselis, 1985]. Other regions such as the dorsal motor vagal nucleus, ambiguus nucleus, spinal trigeminal nucleus and the ventrolateral medulla also contribute to AP inputs [McKinley et al., 2004]. Since the AP is outside the blood-brain barrier, hormones in the blood also influence its neuronal activity. Recent studies show the effects of hormones such as prolactin or angiotensin which bind to receptors in the AP and influence osmoregulation, blood pressure or heart rate [Cottrell and Ferguson, 2004].
Efferents. The AP has reciprocal efferent neural connections to the neighboring solitary-vagal nuclear complex and to the lateral parabrachial nuclei, according to tracing experiments in rats [Shapiro and Miselis, 1985]. Pathways from the AP have also been reported to project to the noradrenergic cell groups in the caudal ventrolateral medulla, the deep cerebellar nuclei and the dorsal tegmental nucleus. Most of these connections have been shown to be reciprocal [Shapiro and Miselis, 1985].

References

Borison HL: Area postrema: chemoreceptor circumventricular organ of the medulla oblongata. Prog Neurobiol 1989;32:351–390.

Cammermeyer J: Histochemistry of the mammalian area postrema. J Comp Neurol 1949;90:121–149.

Cottrell GT, Ferguson AV: Sensory circumventricular organs: central roles in integrated autonomic regulation. Regul Peptides 2004;117:11–23.

Dempsey EW: Neural and vascular ultrastructure of the area postrema in the rat. J Comp Neurol 1973;150:177–200.

Goehler LE, Erisir A, Gaykema RP: Neural-immune interface in the rat area postrema. Neuroscience 2006;140:1415–1434.

King LS: Cellular morphology in the area postrema. J Comp Neurol 1937;66:1–22.

McKinley MJ, Clarke IJ, Oldfield BJ: Circumventricular organs; in Paxinos G, Mai JK (eds): The Human Nervous System, ed 2. Amsterdam, Elsevier Academic Press, 2004.

Miller AD, Leslie RA: The area postrema and vomiting. Front Neuroendocrinol 1994;15:301–320.

Price CJ, Hoyda TD, Ferguson AV: The area postrema: a brain monitor and integrator of systemic autonomic state. Neuroscientist 2008;14:182–194.

Shapiro RE, Miselis RR: The central neural connections of the area postrema of the rat. J Comp Neurol 1985;234:344–364.

Wuchert F, Ott D, Murgott J, Rafalzik S, Hitzel N, Roth J, Gerstberger R: Rat area postrema microglial cells act as sensors for the toll-like receptor-4 agonist lipopolysaccharide. J Neuroimmunol 2008;204:66–74.

36 Hypoglossal Nucleus (XII)

Original name:
Nucleus nervi hypoglossi

Alternative names:
Twelfth cranial nucleus
Nucleus of the hypoglossal nerve

Location and Cytoarchitecture – Original Text

The hypoglossal nucleus (XII) appears caudally at the level of the caudal pole of the inferior olivary complex (IO) and extends orally to the level of the caudal pole of the prepositus nucleus. The nucleus is shaped like a cigar, is tapered at its caudal end and is of uniform diameter throughout the remainder of its extent (fig. 7). Orally, it terminates without much diminution in size. In the brainstem which we studied, the XII measured from 8 to 10 mm in length and not 18–20 mm, as is cited in some references. [...]

The caudal pole of the nucleus lies ventrolateral to the central canal, ventromedial to the dorsal motor nucleus of the vagus (DMX), dorsal and dorsolateral to the medial longitudinal fasciculus. Caudal to the obex, the small cells of the intercalated nucleus (INSt) appear dorsolateral and lateral to the XII and separate it from the DMX. At the level of the opening of the fourth ventricle, the XII becomes associated ventromedially with the nucleus of Roller. The cells of the XII together with those of the INSt form an elevation in the floor of the fourth ventricle known as the trigonum hypoglossi.

The XII is composed of large, multipolar, motor type neurons which contain conspicuous, deeply stained, geometrically shaped Nissl granules (fig. 8, 9).

Definite morphological subdivisions of the nucleus could be differentiated within most of the preparations studied. Caudally, the nucleus is represented by a few neurons which gradually increase in number to form the compact, round medially placed group A (fig. 1). Further orally, 2 small cell groups (B and C) appear and lie respectively dorsolateral and dorsal to group A (fig. 2). Group B which appears at a slightly more caudal level than group C maintains its original size, whereas group C enlarges progressively to displace group A, first in a ventral and then in a ventrolateral direction. In this latter site, group A is somewhat diminished in size and assumes an oblong outline on cross-section (fig. 3). Still further orally the cell group B disappears by merging with the dorsal extremity of group A; the latter then forms a prominent group lying lateral and ventrolateral to group C (fig. 4, 5). A few millimeters caudal to the oral pole of the nucleus, group C diminishes in size and disappears, leaving group A to form the oral pole (fig. 6).

The XII gives rise to somatic efferent fibers which supply the motor innervation to all the intrinsic and extrinsic muscles of the tongue with the exception of the palatoglossus muscle. Attempts have been made to map somatotopic areas within the XII (see below).

It is probable that the hypoglossal nerve (NXII) carries proprioceptive fibers from the tongue but where these fibers terminate is not yet clear. Pearson [1945] suggests that small sensory-type cells, demonstrably intermingled with and adjacent to the fibers of the NXII as they pass through the IO, may be the cell bodies of such sensory fibers. We examined our Nissl-stained section for evidence of such cells but were unable to locate them definitely. We did note groups of small, darkly stained cells lying between the inferior aspect of the IO and pyramid, some of which lie in relation to the fibers of the NXII. These may represent the cells described by Pearson, but undoubtedly comprise the conterminal nucleus (nucleus No. 103).

Functional Neuroanatomy

Function

The XII lies in the medial brainstem as part of the general somatic efferent zone considered to be the rostral continuation of the anterior horn of the spinal cord. It contains the motoneurons of tongue muscles which innervate the tongue through the ipsilateral NXII. The complex coordination of the XII with other sensorimotor systems by brainstem premotor networks enables the tongue to play a vital role in the mastication of food, swallowing, vocalization and human speech, licking, grooming and the maintenance of dilated airways in respiration [Popratiloff et al., 2001]. In early postnatal stages the tongue also plays an important role in suckling. Rhythmic oscillations in XII motoneurons of neonates are sustained by an electrical coupling, or gap junctions, between the cells, a feature which may be important for XII motoneuron coordination, particularly in neonatal development [Sharifullina et al., 2005].

Organization of the Nucleus

The tongue contains 8 muscles: 4 intrinsic muscles (the superior longitudinal, inferior longitudinal, the traverse and vertical muscles) that act to change the shape of the tongue and are not attached to any bone, and 4 extrinsic muscles (genioglossus, styloglossus, hyoglossus and palatoglossus) that change the position of the tongue and are anchored to bone. They are all innervated by motoneurons in the XII, with the exception of the palatoglossus muscle whose motoneurons lie in the ambiguus nucleus and travel in the vagal nerve. The geniohyoid muscle also contributes to tongue movements, specifically tongue retraction; and it is innervated by the ventral XII in monkeys [Uemura-Sumi et al., 1981]: this may be equivalent to the motoneurons lying ventral to group A in figures 1–5.

In general, the functional grouping of the motoneuron subnuclei is based on the principle of muscle action: in the XII the tongue retractors lie together (dorsally in rats, dorsolaterally in primates), and the tongue protruders lie together (ventrally in rats, ventrally and medially in primates). In addition the intrinsic muscle motoneurons were found to lie medially in the XII, and the extrinsic ones more laterally [Travers, 1995; McClung and Goldberg, 1999]. The XII is composed of longitudinal columns of motoneurons. The myotopic organization within the columns has been well documented in the rat, cat and monkey [Krammer et al., 1979; Uemura et al., 1979; Uemura-Sumi et al., 1981; Sokoloff and Deacon, 1992; Travers, 1995]. The organization in each species is slightly different, but according to Koutcherov et al. [2004] the pattern in humans is still unclear. In nonhuman primates the lateral/dorsolateral hypoglossal subnucleus contains the motoneurons of the styloglos-

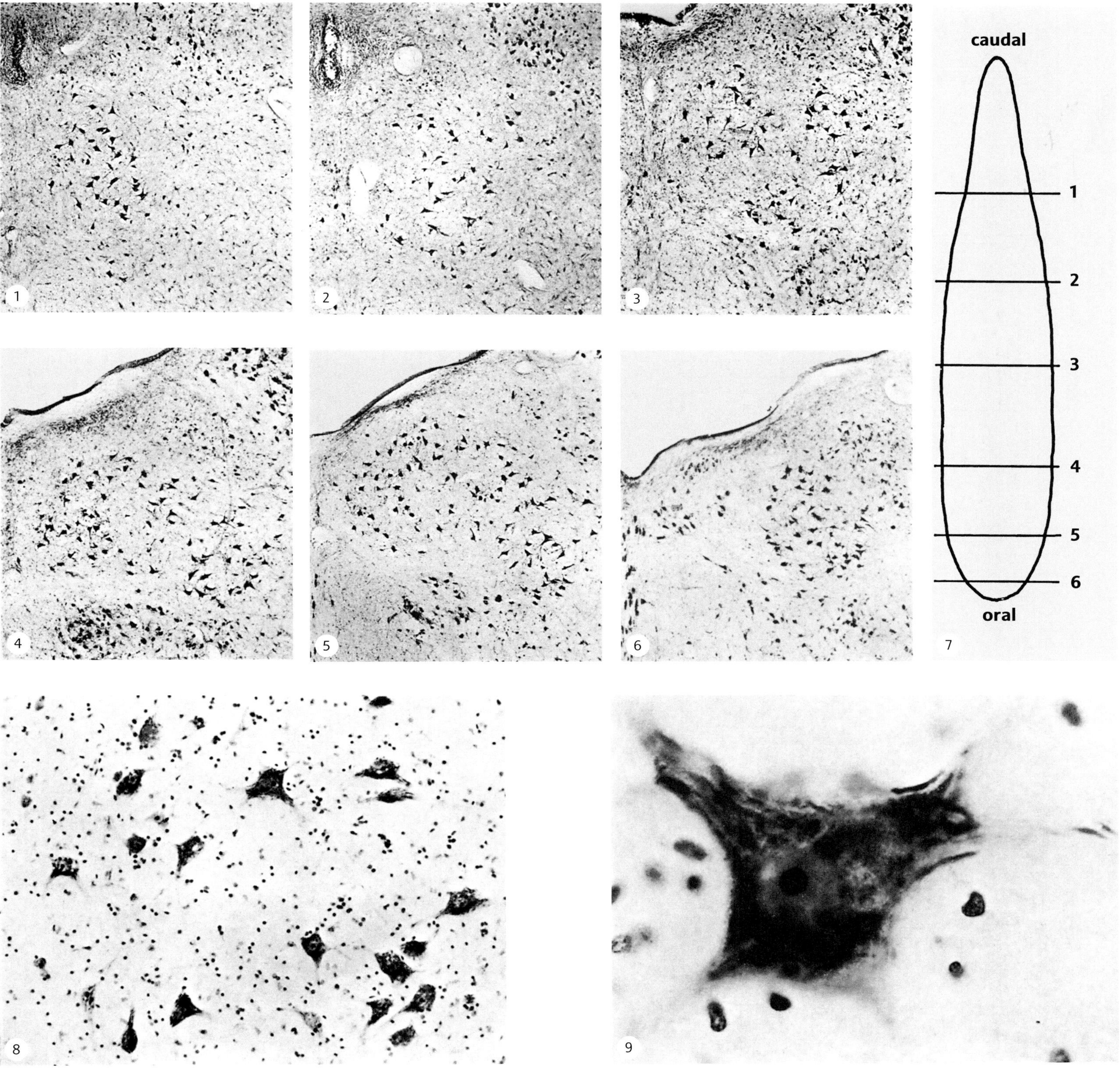

Fig. 1–6. Cross-sections through different levels of the hypoglossal nucleus to show the arrangement of cell groups. For motoneuron groups A, B and C, see 'Original Text'. Magnification ×25.

Fig. 7. Schematic representation of the longitudinal outline of the hypoglossal nucleus to show the levels from which the sections illustrated in figures 1–6 were taken.

Fig. 8. Hypoglossal nucleus. Magnification ×150.

Fig. 9. Cell from the hypoglossal nucleus. Magnification ×1,000.

sus and hyoglossus muscles. This subgroup is innervated by the lateral branch of the NXII, and it is probably homologous with group B in figures 2 and 3. The other motoneuron subgroups in the XII comprise the medioventral regions, and these are supplied by the medial branch of the NXII. The genioglossus subgroup occupies the dorsomedial regions, lying within group C in figures 2–5.

Connections

Efferents. The major projection of the XII is to the muscles of the tongue; however the nucleus also contains interneurons which have been studied with histochemical methods, and show that many are GABAergic, that is inhibitory [Travers, 1995]. Other studies in rats show that small cells in the nucleus project to the facial nucleus [Popratiloff et al., 2001], others demonstrate projections to the motoneurons of the vibrissae in the trigeminal motor nucleus and to the mesencephalic trigeminal nucleus containing sensory ganglion cells of the masseter muscle spindles, indicating a role for the XII in the sensorimotor control of orofacial systems [Mameli et al., 2008].

Afferents. As with other motor nuclei of the brainstem, the majority of premotor inputs to the XII arise from the surrounding medullary reticular formation [Ugolini, 1995]. Specifically there are afferents from: the retroambiguus nucleus, a premotor cell group controlling vocalization and respiration [Holstege, 1991]; the dorsal medial reticular formation close to the solitary nucleus contains a network for swallowing [Cunningham and Sawchenko, 2000]; the Kölliker-Fuse nucleus and the intertrigeminal region carry respiratory signals to the hypoglossal motoneurons [Roda et al., 2004], and serotonergic afferents to the XII mostly arise from the nucleus raphe pallidus and obscurus which would modulate the level but not the pattern of motoneuron activity [Manaker and Tischler, 1993]. Many of the premotor inputs to the XII arise from inhibitory (GABAergic) networks in the supratrigeminal region of the pontine or medullary lateral tegmental field, which modulate XII and other orofacial motoneuron pools [Li et al., 1997]. The cerebral cortex motor tongue area (M1) in lower primates projects to the premotor reticular formation of the parvocellular reticular nucleus, with no direct projections to the XII [Jürgens and Alipour, 2002]. Only in higher primates is a direct projection onto the tongue motoneurons seen. This implies an evolutionary trend towards an increasingly complex role of the tongue in human vocal behavior in contrast to nonhuman vocalization.

Proprioceptive afferents from muscle spindles in the tongue travel first in the NXII, then either to the spinal cord in the ansa cervicalis with cell bodies in C2 and C3, or to the spinal trigeminal complex or solitary nucleus, with their sensory cell bodies in the trigeminal ganglion, or the ganglia of the vagal nerve, respectively [Travers, 1995]. The proprioceptive afferents do not provide a monosynaptic reflex loop by terminating on motoneurons within the XII itself.

References

Cunningham ETJ, Sawchenko PE: Dorsal medullary pathways subserving oromotor reflexes in the rat: implications for the central neural control of swallowing. J Comp Neurol 2000;417:448–466.

Holstege G: Descending motor pathways and the spinal motor system: limbic and non-limbic components. Prog Brain Res 1991;87:307–421.

Jürgens U, Alipour M: A comparative study on the cortico-hypoglossal connections in primates, using biotin dextranamine. Neurosci Lett 2002;328:245–248.

Koutcherov Y, Huang X-F, Halliday G, Paxinos G: Organization of human brain stem nuclei; in Paxinos G, Mai JK (eds): The Human Nervous System, ed 2. Amsterdam, Elsevier Academic Press, 2004.

Krammer EB, Rath T, Lischka MF: Somatotopic organization of the hypoglossal nucleus: a HRP study in the rat. Brain Res 1979;170:533–537.

Li Y-Q, Takada M, Kaneko T, Mizuno N: Distribution of GABAergic and glycinergic premotor neurons projecting to the facial and hypoglossal nuclei in the rat. J Comp Neurol 1997;378:283–294.

Mameli O, Stanzani S, Russo A, Romeo R, Pellitteri R, Spatuzza M, Caria MA, De Riu PL: Hypoglossal nuclei participation in rat mystacial pad control. Pflugers Arch 2008;456:1189–1198.

Manaker S, Tischler LJ: Origin of serotoninergic afferents to the hypoglossal nucleus in the rat. J Comp Neurol 1993;334:466–476.

McClung JR, Goldberg SJ: Organization of motoneurons in the dorsal hypoglossal nucleus that innervate the retrusor muscles of the tongue in the rat. Anat Rec 1999;254:222–230.

Pearson AA: Further observations on the intramedullary sensory type neurones along the hypoglossal nerve. J Comp Neurol 1945;82:93–100.

Popratiloff AS, Streppel M, Gruart A, Guntinas-Lichius O, Angelov DN, Stennert E, Delgado-Garcia JM, Neiss WF: Hypoglossal and reticular interneurons involved in oro-facial coordination in the rat. J Comp Neurol 2001;433:364–379.

Roda F, Pio J, Bianchi AL, Gestreau C: Effects of anesthetics on hypoglossal nerve discharge and c-Fos expression in brainstem hypoglossal premotor neurons. J Comp Neurol 2004;468:571–586.

Sharifullina E, Ostroumov K, Nistri A: Metabotropic glutamate receptor activity induces a novel oscillatory pattern in neonatal rat hypoglossal motoneurones. J Physiol 2005;563:139–159.

Sokoloff AJ, Deacon TW: Musculotopic organization of the hypoglossal nucleus in the cynomolgus monkey, *Macaca fascicularis*. J Comp Neurol 1992;324:81–93.

Travers JB: Oromotor nuclei; in Paxinos G (ed): The Rat Nervous System, ed 2. San Diego, Academic Press, 1995.

Uemura-Sumi M, Mizuno N, Nomura S, Iwahori N, Takeuchi Y, Matsushima R: Topographical representation of the hypoglossal nerve branches and tongue muscles in the hypoglossal nucleus of macaque monkeys. Neurosci Lett 1981; 22:31–35.

Uemura M, Matsuda K, Kume M, Takeuchi Y, Matsushima R, Mizuno N: Topographical arrangement of hypoglossal motoneurons: an HRP study in the cat. Neurosci Lett 1979;13:99–104.

Ugolini G: Specificity of rabies virus as a transneuronal tracer of motor networks: transfer from hypoglossal motoneurons to connected second-order and higher order central nervous system cell groups. J Comp Neurol 1995;356:457–480.

Motor Systems – Somatomotor Nuclei (Plates 4–7)

37 Supraspinal Nucleus (SSp)

Original name:
Nucleus supraspinalis

Alternative names:
Spinal accessory nucleus
Spinal nucleus of the accessory nerve

Location and Cytoarchitecture – Original Text

This nucleus is represented by an extension of the anterior horn cells of the cervical cord into the medulla. In sections through the caudal medulla the supraspinal cells lie in the medial part of that portion of the ventral gray matter isolated by decussating pyramidal fibers. Oral to the decussation, at the level of the caudal pole of the inferior olivary complex, the cells of the supraspinal nucleus (SSp) migrate dorsally in the direction of the central canal and merge with the most caudal cells of the hypoglossal nucleus.

The SSp is composed of typical multipolar, motor type neurons among which are scattered, lightly stained, multipolar cells (fig. 1, 2). On most sections ventromedial and dorsolateral subdivisions of the nucleus may be distinguished. The relationship of these subgroups is inconstant – in many sections they appear to fuse to form a single group.

It has been customary to describe only the ventromedial group as the SSp – said to supply the motor elements of the first cervical nerve (see below). The dorsolateral cell group is usually described as the oral portion of the 'spinal accessory nucleus' the caudal part of which extends into the spinal cord where it occupies a lateral position in the ventral horn [...] and is thought to give rise to motor fibers proceeding to the trapezius and sternocleidomastoid muscles as the spinal component of the accessory nerve, the eleventh cranial nerve (NXI; see below).

Further comment is required regarding the cranial component of the NXI. The so-called cranial accessory nerve is formed by efferent fibers of the dorsal motor nucleus and the ambiguus nucleus, as well as afferent fibers which proceed to the nucleus of the solitary tract. These emerge through the jugular foramen as a part of the NXI but almost immediately branch off to join the vagus. It seems less confusing to regard these fibers as part of the vagal complex and not as part of the NXI. Consequently, we have omitted from our description the concept of 'cranial accessory nuclei and fibers'.

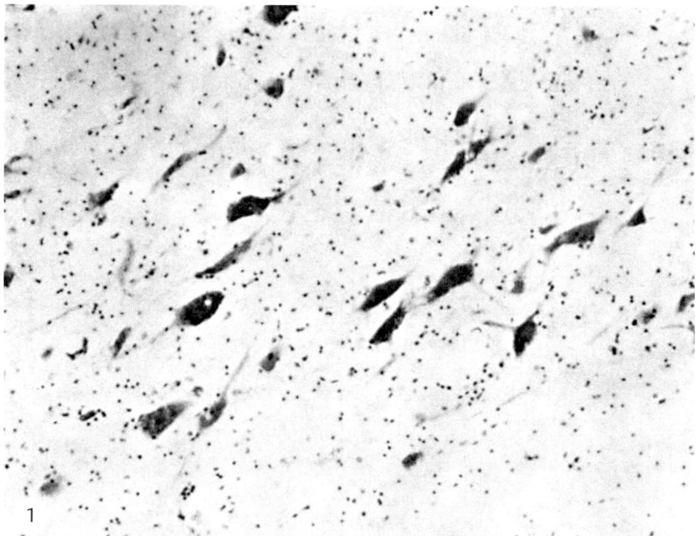

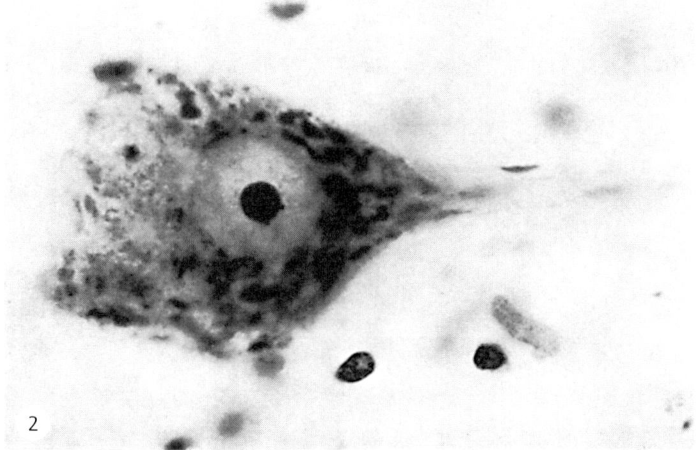

Fig. 1. Supraspinal nucleus. Magnification ×150.
Fig. 2. Cell from the supraspinal nucleus. Magnification ×1,000.

Functional Neuroanatomy

Function

The SSp is a long column of somatotopically organized motoneurons. Those in the oral part of the SSp near the caudal hypoglossal nucleus supply the thyrohyoid and other infrahyoid muscles, concerned with elevation of the larynx, or depressing the hyoid bone as in swallowing. The motoneurons in the caudal SSp, often referred to as the spinal accessory nucleus, innervate the ipsilateral sternocleidomastoideus and trapezoid muscles, responsible for turning the head and shoulder elevation, respectively. The axons exit from the spinal cord (C_{1-5}), and those of the caudal SSp form the spinal accessory nerve and the NXI (see above). Most anatomists classify spinal cord efferents as 'general somatic efferent'; however, the muscles supplied by the NXI are predominantly branchiometric in origin, and consistent with this view their efferent axons should be classified as 'special visceral efferents'.

Connections

The axons of the oral motoneurons of the SSp supplying the thyrohyoid and other infrahyoid muscles exit ventrolaterally from the spinal cord through the first cervical root and join the hypoglossal nerve [Kitamura et al., 1983, 1986].

The motoneurons lying dorsolateral to the SSp, in humans, form an independent column of motoneurons which extends caudally into the cervical cord. It is this set of motoneurons that gives rise to the spinal component of the accessory nerve (NXI) innervating the sternocleidomastoid and trapezius muscles, and it is conventionally referred to as the 'spinal accessory nucleus'. At the level of the lower medulla the dorsolateral motoneuron column innervates only the sternocleidomastoid muscle and extends caudally to about C_3. At the level of C_2 a second motoneuron column develops continuing caudally to C_5, and after a short interruption continues again between C_8 and S_3: this extensive column was considered to innervate the trapezoid muscle [Routal and Pal, 2000].

Afferent inputs to the SSp arise from the superior colliculus [Huerta and Harting, 1982], the medial vestibular nucleus but not the inferior or superior vestibular nuclei [Carleton and Carpenter, 1983]. The SSp projects to the caudal pharyngeal muscles involved in swallowing [Hudson, 1986], and muscles involved in head movement [Tellegen et al., 2001], as well as premotor head movement areas such as the gigantocellular nucleus [Cowie et al., 1994]. These connections imply a multifunctional role in head and pharynx control.

References

Carleton SC, Carpenter MB: Afferent and efferent connections of the medial, inferior and lateral vestibular nuclei in the cat and monkey. Brain Res 1983;278: 29–51.

Cowie RJ, Smith MK, Robinson DL: Subcortical contributions to head movements in macaques. II. Connections of a medial pontomedullary head-movement region. J Neurophysiol 1994;72:2665–2682.

Hudson LC: The origins of innervation of the canine caudal pharyngeal muscles: an HRP study. Brain Res 1986;374:413–418.

Huerta MF, Harting JK: Projections of the superior colliculus to the supraspinal nucleus and the cervical spinal cord gray of the cat. Brain Res 1982;242:326–331.

Kitamura S, Nishiguchi T, Okubo J, Chen KL, Sakai A: An HRP study of the motoneurons supplying the rat hypobranchial muscles: central localization, peripheral axon course and soma size. Anat Rec 1986;216:73–81.

Kitamura S, Nishiguchi T, Sakai A: Location of cell somata and the peripheral course of axons of the geniohyoid and thyrohyoid motoneurons: a horseradish peroxidase study in the rat. Exp Neurol 1983;79:87–96.

Routal RV, Pal GP: Location of the spinal nucleus of the accessory nerve in the human spinal cord. J Anat 2000;196:263–268.

Tellegen AJ, Arends JJ, Dubbeldam JL: The vestibular nuclei and vestibuloreticular connections in the mallard (*Anas platyrhynchos* L.). An anterograde and retrograde tracing study. Brain Behav Evol 2001;58:205–217.

38 Ambiguus Nucleus (AMB)

Original name:
Nucleus ambiguus

Location and Cytoarchitecture – Original Text

This nucleus extends as a long, slender column of cells from a point just caudal to the inferior olivary complex to the caudal pole of the facial nucleus – a distance of 16 mm.

The caudal portion of the nucleus lies among the cells of the lateral part of the central nucleus of the medulla oblongata (CN), ventromedial to the spinal trigeminal nucleus, and dorsal to the lateral reticular nucleus. With the change in the pattern of the nuclei which takes place in the upper medulla, the ambiguus nucleus (AMB) appears to migrate dorsally. The oral extremity of the nucleus thus lies among the cells of the parvocellular reticular nucleus (PCR), lateral to the gigantocellular nucleus and medial to the spinal trigeminal nucleus, oral part.

The caudal portion of the AMB is not a continuous cell column but is rather composed of irregularly arranged, small cell groups between which are scattered the cells of the CN. Thus, on cross-section through this portion of the nucleus one may see either a fairly discrete cell group, or only 1 or 2 cells. This fact makes accurate delineation of the caudal pole of the AMB difficult. The cells composing the intermediate and oral portions of the nucleus are more compactly arranged – 12 to 15 cells may usually be seen on each cross-section.

With the exception of its oral pole, the AMB is composed of large, somewhat elongated multipolar neurons containing discrete geometrically shaped Nissl granules (fig. 1, 2). The oral extremity of the nucleus, i.e. that portion which lies among the cells of the PCR, is made up of cells which, although similar to, are smaller and plumper than the cells of the remainder of the nucleus. [...]

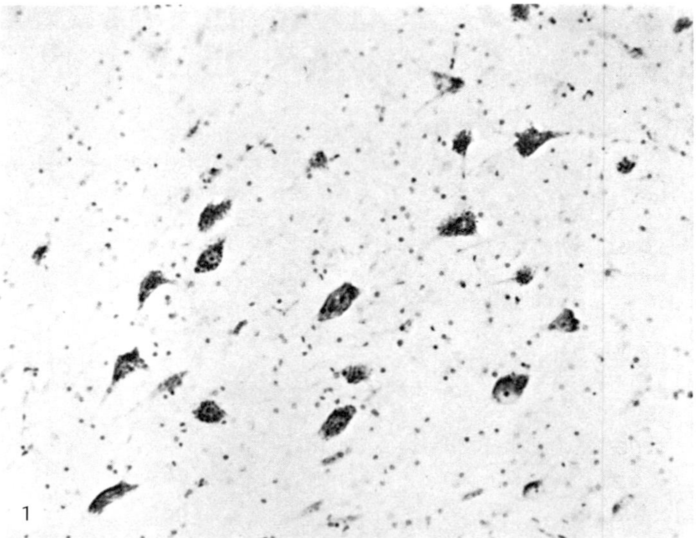
1

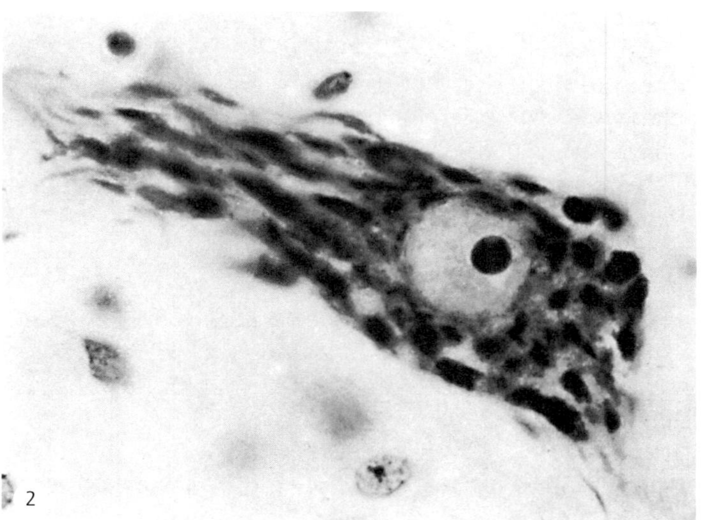
2

Fig. 1. Ambiguus nucleus. Magnification ×150.
Fig. 2. Cell from the ambiguus nucleus. Magnification ×1,000.

Functional Neuroanatomy

Function

The AMB gives rise to efferent fibers which provide motor innervation to the striated muscles of the larynx, pharynx, esophagus and the soft palate. Together the muscles are coordinated to effect swallowing, vocalization and assist in respiration (see PCR, nucleus No. 54) [Broussard et al., 1998; Cunningham and Sawchenko, 2000; Hannig and Jürgens, 2006]. The AMB motoneurons provide the efferent limb of the gag reflex: the reflex is evoked by painful stimulation of the pharynx, generating sensory signals that are relayed to the caudal part of the spinal trigeminal nucleus, and then to the AMB. The motoneurons also participate in coughing and vomiting reflexes. The nucleus is surrounded by an outer region of the AMB called the 'external formation' which contains neurons that participate in respiratory control, and a vagally mediated bradycardia [Nosaka et al., 1979]. The external formation includes what is now recognized as the pre-Bötzinger complex and the rostral ventrolateral respiratory group [Rekling and Feldman, 1998].

The AMB innervates muscles of the larynx through the recurrent laryngeal nerve of the vagus. If this is damaged, as in thyroid surgery, or by an aneurysm of the aorta, there is a resulting hoarseness of the voice. Injury to the lateral medulla, by an infarct of the posterior inferior cerebellar artery as in Wallenberg's syndrome, results in lesions of the AMB, and the patients have a resultant difficulty in swallowing (dysphagia), as well as speech difficulties (dysarthria and dysphonia) among other symptoms.

Organization of the Nucleus

The AMB is part of a continuous column of motoneurons starting with the trigeminal motor nucleus rostrally, and caudally including the facial and retrofacial nuclei (RFN) and the AMB [Jacobs, 1970]. The 'rostral AMB' innervates the stylopharyngeus and part of the cricothyroid muscle [Nieuwenhuys et al., 2008b]. However, here this region is called the retrofacial nucleus and considered under RFN (nucleus No. 39).

The AMB motoneurons show a common pattern of somatotopic organization that is seen across species, including primates [Yoshida et al., 1984, 1985; Bieger and Hopkins, 1987; Satoda et al., 1990, 1996; Nieuwenhuys et al., 2008a, chapter 21]. Four subdivisions in the AMB can be distinguished: a 'compact core' which lies dorsally and is distinguishable in Nissl stains, an adjacent 'semicompact region', a 'loose region', and the 'external formation' which lies outside the strict boundaries of the AMB [Broussard and Altschuler, 2000]. The 'compact core' contains motoneurons for all levels of the esophagus (upper esophagus rostrally, and lower esophagus caudally) as well as those of the stylopharyngeus, which lie most rostrally, dorsal to the facial nucleus in a region often referred to as RFN, or the retrofacial part of the AMB (see nucleus No. 39) [Holstege, 1991]. The motoneurons of the larynx lie more caudally in the 'loose subdivision' of the nucleus, and form the recurrent laryngeal nerve of the vagus. The motoneurons of the pharynx and the cricothyroid muscle lie in the 'semicompact subdivision' of the AMB.

The 'external formation' surrounds the motoneurons in the ventrolateral medulla like a sheath, and is part of a column which contains many cell groups important for the neural control of respiration. The pre-Bötzinger complex is thought to be a respiratory pacemaker: a group of cholinergic preganglionic parasympathetic neurons, which slow the heart [Nosaka et al., 1979]. The reticular area around the AMB conforms clearly to the boundaries of the intermediate reticular zone of Paxinos and colleagues (see 'Reticular Formation: Overview', nucleus No. 54).

The dendrites of the 'compact and semicompact division' of the AMB form bundles that run throughout the length of the nucleus and form an extensive intranuclear plexus. In addition the 'semicompact region' develops dendritic bundles which extend into the surrounding reticular formation, where premotor neurons active during swallowing have been identified [Broussard and Altschuler, 2000].

Connections

Efferents. The AMB efferents are classified as 'special visceral efferents' since they supply skeletal muscles which are derived from the branchial, or pharyngeal, arches, and not from the head mesoderm. The nerve fibers exit either through the vagus, which nowadays is taken to include the cranial part of the accessory nerve (see nucleus No. 37), or the glossopharyngeal nerve. Since the only group of fibers to join the glossopharyngeal nerve arise from the rostral AMB, they are considered here as RFN (nucleus No. 39); under these definitions, it can be said that all AMB axons exit in the vagus nerve.

Afferents. Afferent pathways to the AMB arise from the solitary nucleus, and from the surrounding reticular formation which serves as an important premotor area; these areas participate in the complex central control mechanism of swallowing, in addition to respiration and vocalization [Broussard et al., 1998; Cunningham and Sawchenko, 2000; Hannig and Jürgens, 2006] (see also nucleus No. 54). The periaqueductal gray (PAG) has both direct and indirect projections to the AMB. One pathway arises from an area involved in vocalization, which projects directly to the AMB, and also gives a strong projection to the retroambiguus nucleus (RAm) and the medullary reticular formation caudal to the AMB [Holstege, 1989]. The RAm and the reticular area also project onto the AMB. In summary these studies support the view that the RAm and medullary reticular formation caudal to the AMB mediate the drive from limbic areas such as the PAG or hypothalamus onto the AMB [Holstege, 1991].

Direct and indirect connections linking cortical areas with laryngeal motor neurons bilaterally have been postulated, and the laryngeal representation in the motor cortex has been well established in primates, including man [Jurgens and Ehrenreich, 2007; Brown et al., 2008]. The input to the AMB from several regions of the cerebral cortex supports the act of voluntary swallowing, and has been highlighted recently with transsynaptic tracing techniques [van Daele and Cassell, 2009]. These experiments also confirmed the afferent pathways to the AMB described above, and showed others.

References

Bieger D, Hopkins DA: Viscerotopic representation of the upper alimentary tract in the medulla oblongata in the rat: the nucleus ambiguus. J Comp Neurol 1987;262:546–562.

Broussard D, Altschuler S: Brainstem viscerotopic organization of afferents and efferents involved in the control of swallowing. Am J Med 2000;108(suppl 4a):79S–86S.

Broussard D, Lynn R, Wiedner E, Altschuler S: Solitarial premotor neuron projections to the rat esophagus and pharynx: implications for control of swallowing. Gastroenterology 1998;114:1268–1275.

Brown S, Ngan E, Liotti M: A larynx area in the human motor cortex. Cereb Cortex 2008;18:837–845.

Cunningham ETJ, Sawchenko PE: Dorsal medullary pathways subserving oromotor reflexes in the rat: implications for the central neural control of swallowing. J Comp Neurol 2000;417:448–466.

Hannig S, Jürgens U: Projections of the ventrolateral pontine vocalization area in the squirrel monkey. Exp Brain Res 2006;16992–105.

Holstege G: Anatomical study of the final common pathway for vocalization in the cat. J Comp Neurol 1989;284:242–252.

Holstege G: Descending motor pathways and the spinal motor system: limbic and non-limbic components. Prog Brain Res 1991;87:307–421.

Jacobs MJ: The development of the human motor trigeminal complex and accessory facial nucleus and their topographic relations with the facial and abducens nuclei. J Comp Neurol 1970;138:161–194.

Jurgens U, Ehrenreich L: The descending motorcortical pathway to the laryngeal motoneurons in the squirrel monkey. Brain Res 2007;1148:90–95.

Nieuwenhuys R, Voogd J, van Huijzen C: The Human Central Nervous System. Berlin, Springer, 2008a.

Nieuwenhuys R, Voogd J, van Huijzen C: Motor systems; in Nieuwenhuys R, Voogd J, van Huijzen C (eds): The Human Central Nervous System, ed 4. Berlin, Springer, 2008b.

Nosaka S, Yamamoto T, Yasunaga K: Localization of vagal cardioinhibitory preganglionic neurons with rat brain stem. J Comp Neurol 1979;186:79–92.

Rekling JC, Feldman JL: Pre-Bötzinger complex and pacemaker neurons: hypothesized site and kernel for respiratory rhythm generation. Annu Rev Physiol 1998;60:385–405.

Satoda T, Takahashi O, Murakami C, Uchida T, Mizuno N: The sites of origin and termination of afferent and efferent components in the lingual and pharyngeal branches of the glossopharyngeal nerve in the Japanese monkey *(Macaca fuscata)*. Neurosci Res 1996;24:385–392.

Satoda T, Uemura-Sumi M, Tashird T, Takahashi O, Matsushima R, Mizuno N: Localization of motoneurons innervating the stylohyoid muscle in the monkey, cat, rabbit, rat and shrew. J Hirnforsch 1990;31:731–737.

Van Daele DJ, Cassell MD: Multiple forebrain systems converge on motor neurons innervating the thyroarytenoid muscle. Neuroscience 2009;162:501–524.

Yoshida Y, Mitsumasu T, Hirano M, Kanaseki T: Somatotopic representation of the laryngeal motoneurons in the medulla of monkeys. Acta Otolaryngol 1985;100:299–303.

Yoshida Y, Mitsumasu T, Miyazaki T, Hirano M, Kanaseki T: Distribution of motoneurons in the brain stem of monkeys, innervating the larynx. Brain Res Bull 1984;13:413–419.

39 Retrofacial Nucleus (RFN)

Original name:
Nucleus retrofacialis

Alternative name:
Rostral ambiguus nucleus

Location and Cytoarchitecture – Original Text

This is a very small nucleus, only about 1 mm in length, found in the oral part of the medulla just caudal of the plane in which the facial nucleus (VII) appears. It lies centrolateral to the oral part of the ambiguus nucleus (AMB) between the lateral paragigantocellular nucleus ventrally and the parvocellular reticular nucleus dorsally. Thus, the retrofacial nucleus (RFN) occupies the same position in which the AMB was situated in more caudal sections, and in the VII appears about 2 mm more orally.

The nucleus is composed of a small group of loosely arranged, elongated, multipolar motor type cells (fig. 1, 2). Due to the fact that the cells are not compactly arranged and that the whole structure does not exceed 1 mm in length, the nucleus may be easily overlooked. [...]

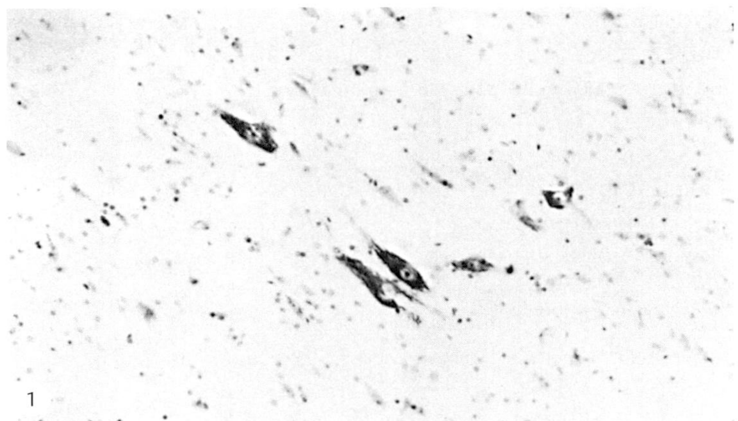

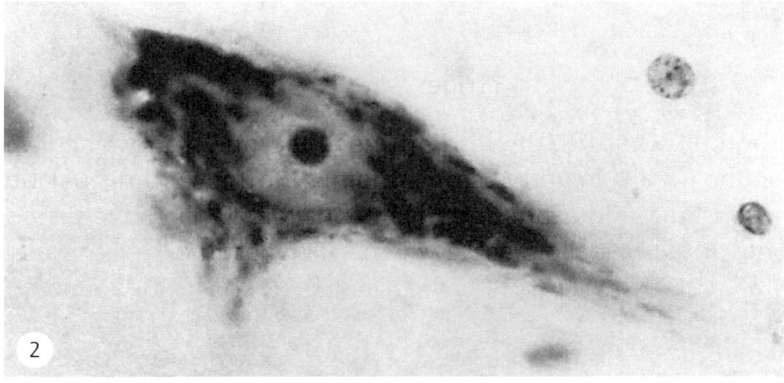

Fig. 1. Retrofacial nucleus. Magnification ×150.
Fig. 2. Cells from the retrofacial nucleus. Magnification ×1,000.

Functional Neuroanatomy

Function
The motoneurons of the RFN innervate the stylopharyngeus muscle and soft palate, which are important for swallowing [Yoshida et al., 1980; Holstege et al., 1983; van Loveren et al., 1985; Strutz et al., 1988]. Further caudally towards the AMB, the RFN motoneurons supply the cricothyroid muscle, important for phonation by tensing the vocal cords, and other RFN motoneurons innervate the upper portion of the esophagus.

The less compact RFN area surrounding the motoneurons is associated with many functions such as respiratory control, cardiovascular responses, chemosensitivity, coughing and vomiting, control of orofacial activity as well as neuroendocrine regulation and homeostasis [Bulloch and Moore, 1981; St. John et al., 1989; Miselis et al., 1989; Holstege, 1991; Blessing, 2004; Jakus et al., 2004, 2008]. The region includes the Bötzinger complex [Rekling and Feldman, 1998], which contains expiratory neurons with reciprocal local connections and efferents to the phrenic motoneurons [Anders et al., 1991; Blessing and Benarroch, 2012].

Organization of the Nucleus
The RFN is part of a continuous column of motoneurons starting with the trigeminal motor nucleus rostrally, and caudally including the VII, RFN and AMB [Jacobs, 1970]. The RFN innervates the stylopharyngeus and part of the cricothyroid muscle [Nieuwenhuys et al., 2008]. However, this region is often included in the AMB, constituting the 'rostral AMB' (nucleus No. 38). The motor column from the RFN is surrounded caudally with a sheath of reticular neurons that are a complex premotor region crucial for the neural control of respiration. The network is closely connected with the dorsal vagal complex and has immediate access to the motoneurons controlling respiration, swallowing, coughing and vocalization. This region fits closely with the cytoarchitectural boundaries of the intermediate reticular nucleus or zone, defined by Paxinos and colleagues (see 'Reticular Formation: Overview'). The RFN is associated with neurons containing noradrenaline, adrenaline, catecholaminergic, glutamatergic and GABA cell groups [Koutcherov et al., 2004].

Connections
The RFN receives extensive afferent projections from all parts of the caudal medulla, specifically the solitary nucleus, and other portions of the ventral respiratory group and ventrolateral medulla. It receives projections from limbic structures such as the parabrachial nuclei and the periaqueductal gray [Carrive et al., 1988; Holstege, 1991]. Efferents from RFN neurons, from cells other than the motoneurons, arborize within the medulla; they are strongly interconnected with the vagal nuclear complex, and the retroambiguus nucleus [Gerrits and Holstege, 1996].

References

Anders K, Ballantyne D, Bischoff AM, Lalley PM, Richter DW: Inhibition of caudal medullary expiratory neurones by retrofacial inspiratory neurones in the cat. J Physiol 1991;437:1–25.

Blessing WW: Lower brain stem regulation of visceral, cardiovascular, and respiratory function; in Paxinos G, Mai JK (eds): The Human Nervous System, ed 2. Amsterdam, Elsevier Academic Press, 2004.

Blessing WW, Benarroch EE: Lower brain stem regulation of visceral, cardiovascular, and respiratory function; in Mai JK, Paxinos G (eds): The Human Nervous System, ed 3. Amsterdam, Academic Press, 2012.

Bulloch K, Moore RY: Innervation of the thymus gland by brain stem and spinal cord in mouse and rat. Am J Anat 1981;162:157–166.

Carrive P, Bandler R, Dampney RA: Anatomical evidence that hypertension associated with the defence reaction in the cat is mediated by a direct projection from a restricted portion of the midbrain periaqueductal grey to the subretrofacial nucleus of the medulla. Brain Res 1988;460:339–345.

Gerrits PO, Holstege G: Pontine and medullary projections to the nucleus retroambiguus: a wheat germ agglutinin-horseradish peroxidase and autoradiographic tracing study in the cat. J Comp Neurol 1996;373:173–185.

Holstege G: Descending motor pathways and the spinal motor system: limbic and non-limbic components. Prog Brain Res 1991;87:307–421.

Holstege G, Graveland GA, Bijker-Biemond C, Schuddeboom I: Location of motoneurons innervating soft palate, pharynx and upper esophagus. Anatomical evidence for a possible swallowing center in the pontine reticular formation: an HRP and autoradiographical tracing study. Brain Behav Evol 1983;23:47–62.

Jacobs MJ: The development of the human motor trigeminal complex and accessory facial nucleus and their topographic relations with the facial and abducens nuclei. J Comp Neurol 1970;138:161–194.

Jakus J, Halasova E, Poliacek I, Tomori Z, Stransky A: Brainstem areas involved in the aspiration reflex: c-Fos study in anesthetized cats. Physiol Res 2004;53: 703–717.

Jakus J, Poliacek I, Halasova E, Murin P, Knocikova J, Tomori Z, Bolser DC: Brainstem circuitry of tracheal-bronchial cough: c-fos study in anesthetized cats. Respir Physiol Neurobiol 2008;160:289–300.

Koutcherov Y, Huang X-F, Halliday G, Paxinos G: Organization of human brain stem nuclei; in Paxinos G, Mai JK (eds): The Human Nervous System, ed 2. Amsterdam, Elsevier Academic Press, 2004.

Miselis RR, Rogers WT, Schwaber JS, Spyer KM: Localization of cardiomotor neurones in the anaesthetized rat: choleratoxin HRP conjugate and pseudorabies labeling. J Physiol (London) 1989;416:63P.

Nieuwenhuys R, Voogd J, van Huijzen C: Motor systems; in Nieuwenhuys R, Voogd J, van Huijzen C (eds): The Human Central Nervous System, ed 4. Berlin, Springer, 2008.

Rekling JC, Feldman JL: Pre-Bötzinger complex and pacemaker neurons: hypothesized site and kernel for respiratory rhythm generation. Annu Rev Physiol 1998;60:385–405.

St John WM, Hwang Q, Nattie EE, Zhou D: Functions of the retrofacial nucleus in chemosensitivity and ventilatory neurogenesis. Respir Physiol 1989;76:159–171.

Strutz J, Hammerich T, Amedee R: The motor innervation of the soft palate. An anatomical study in guinea pigs and monkeys. Arch Otorhinolaryngol 1988; 245:180–184.

Van Loveren H, Saunders MC, Cassini P, Keller JT: Localization of motoneurons innervating the stylopharyngeus muscle in the cat. Neurosci Lett 1985;58:251–255.

Yoshida Y, Miyazaki T, Hirano M, Shin T, Totoki T, Kanaseki T: Location of motoneurons supplying the cricopharyngeal muscle in the cat studied by means of the horseradish peroxidase method. Neurosci Lett 1980;18:1–4.

Motor System – Somatomotor Nuclei

(Plates 18, 20 and 21)

40 Facial Nucleus (VII)

Original name:
Nucleus nervi facialis

Alternative names:
Seventh cranial nerve nucleus
Motor nucleus of the facial nerve
Nucleus of the facial nerve

Location and Cytoarchitecture – Original Text

The facial nucleus (VII) first appears at the junction of the medulla and pons immediately rostral to the oral pole of the inferior olive. It extends orally for a distance of 3–4 mm and terminates just oral to the level of the caudal pole of the abducens nucleus.

The caudal extremity of the nucleus lies in the ventrolateral area of the tegmentum, close to the surface of the brainstem, between the parvocellular reticular nucleus dorsally and the lateral paragigantocellular nucleus ventrally. More rostrally, as the medial cerebellar peduncle forms, the nucleus is displaced from the lateral aspect of the brainstem and becomes related ventrally to the newly formed complex of the superior olive (SOC). Throughout its entire length, the VII lies ventromedial to the spinal tract of the trigeminal nerve. On cross-section the long axis of the nucleus is directed dorsolaterally.

The VII is composed of large, typically multipolar, motor type neurons (fig. 2, 3). These cells tend to congregate into more or less distinct groups. In line with Pearson's observations [1946], we distinguish the following subdivisions: (a) dorsal group, (b) intermediate group, (c) medial group, (d) ventral group, (e) ventromedial group, (f) ventrolateral group.

The relationship of these various groups, shown in figure 1, is fairly constant throughout the extent of the nucleus. Slight differences in size and arrangement of cells characterize some of the groups. Thus, the ventral group is characterized by the presence of relatively small cells, the dorsal group by the presence of relatively large cells, and the medial group by the loose arrangement of its constituent cells.

On examination of sagittal section(s) one finds that although the cells of the VII become closely associated with the cells of both the motor nucleus of the trigeminal nerve (MoV), retrofacial nucleus and the ambiguus nucleus (AMB), there is never any continuity of structure. [...]

Functional Neuroanatomy

Function

The VII contains the motoneurons of muscles for facial expression, and motoneurons of the stapedius muscle, which participates in the protection of the ear from loud noises by the stapedius reflex. The VII activity is integrated with the motor signals of the trigeminal, AMB, retrofacial and hypoglossal nuclei for feeding and vocalization. Central lesions to the cortical inputs of the VII cause a contralateral paralysis of the lower, but not upper face (see below). Lesions of the facial nerve (NVII) result in complete facial paralysis on the same side [for a review, see Nieuwenhuys et al., 2008].

Interestingly, the facial muscles are under both voluntary and emotional control. A dissociation between voluntary and emotional facial expressions may occur in patients with supra-

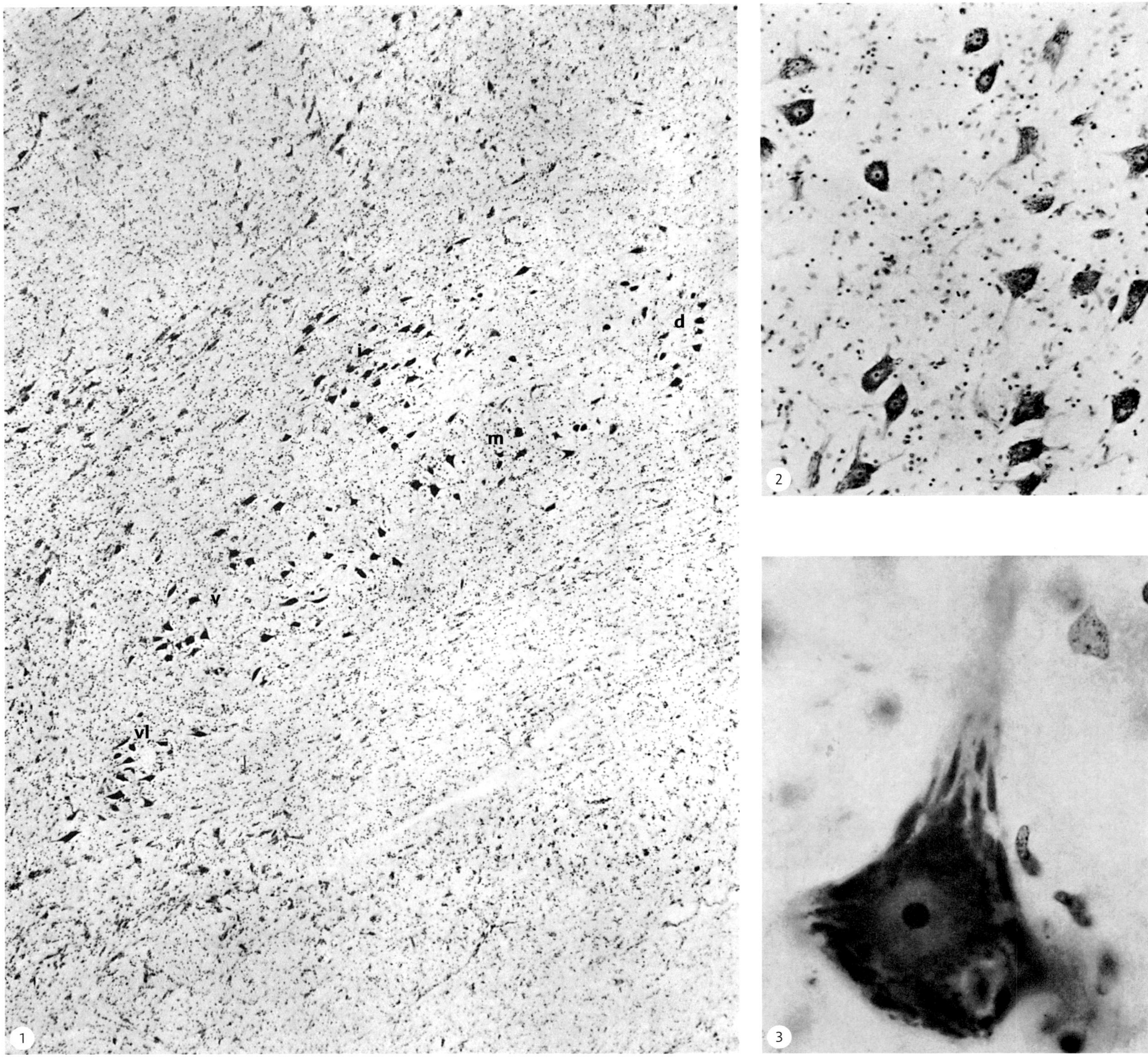

Fig. 1. Facial nucleus. Note the individual motoneuron subgroups innervating specific muscles or muscle groups, and described in the text: d = dorsal; i = intermediate; m = medial; v = ventral; vl = ventrolateral. Magnification ×40.

Fig. 2. Facial nucleus. Magnification ×150.
Fig. 3. Cells from facial nucleus. Magnification ×1,000.

nuclear lesions; for example, the patients cannot smile on command, but can smile at good jokes. This observation demonstrates that the voluntary and emotional pathways are separate [Morecraft et al., 2001; Holstege et al., 2004].

Topography of Motoneurons

The basic organization of VII motoneurons in primates into 4 or 5 subgroups has been verified by tract-tracing experiments in monkeys [Satoda et al., 1987; Porter et al., 1989; Welt and Abbs, 1990; Van der Werf et al., 1998]. Each subgroup subserves a specific muscle group of the ipsilateral side (fig. 1). In general the *lateral* subgroup innervates the perioral and nasolabial muscles, and is particularly prominent in man (ventrolateral group); the *medial* group is small in humans and supplies the posterior auricular and neck muscles; the *dorsal* and *intermediate* groups project to the orbicularis oculi, frontalis and scalp muscles, while the ventral subgroups supply the platysma and anterior neck muscles. There are still some discrepancies between both the older and more recent studies concerning details, such as the presence of a small contralateral component. Earlier studies found evidence for NVII fibers crossing beneath the floor of the fourth ventricle. The VII subgroups are labeled slightly differently by Paxinos and colleagues [Paxinos and Huang, 1995; Paxinos et al., 2012]; they divide the ventral, ventromedial and ventrolateral groups into a lateral and a ventrolateral group, which presumably innervate the perioral muscles and the platysma, shown in figure 1. The motoneurons for specific muscles are also organized in longitudinally oriented cell columns [Welt and Abbs, 1990]. The surprisingly large volume of the VII in great apes and humans is thought to reflect increased differentiation of the facial muscles of expression [Sherwood, 2005]. In this analysis the number of neurons in the human VII was estimated at about 6,000.

The facial muscles are derived from the branchial (pharyngeal) arches, and referred to as branchial motoneurons, and generally classified as part of the special visceral efferent cell column. Embryologically, the branchial motoneurons form a continuous cell column in the parasagittal plane composed of the VII, motor trigeminal, retrofacial nuclei and AMB [Jacobs, 1970]. Several small groups of motoneurons can be found scattered in this plane, for example the accessory VII innervating the posterior belly of the digastric muscle (nucleus No. 41). Motoneurons of the stapedius muscle of the middle ear lie scattered at the interface between VII and SOC [Thompson et al., 1985]. Also part of this column are the motoneurons of the posterior trigeminal nucleus supplying the anterior belly of the digastric muscle (nucleus No. 43), and in some species an accessory abducens nucleus can be found, which innervates muscles of the retractor bulbi and the nictitating membrane. No evidence of an accessory abducens nucleus has been found in humans, where these muscles are absent [Hutson et al., 1979].

Connections

Efferents. The vast majority of cells within the VII are α-motoneurons, but there may be a small fraction of neurons projecting to the cerebellar flocculus [Langer et al., 1985]. Only few small γ-motoneurons may be present. This coincides with the very low abundance of muscle spindles in superficial facial muscles [Sherwood, 2005].

Afferents. Premotor projections to the VII include the lateral tegmental field, AMB, nucleus retroambiguus, sensory trigeminal complex, the parabrachial nuclei (lateral and medial), the red nucleus and the cervical cord [Holstege, 1991; Nieuwenhuys et al., 2008]. The medullary reticular formation is the largest source of afferents to orofacial motoneurons including the VII, and contains both excitatory and inhibitory cell groups to coordinate swallowing and mastication through projections to the VII, MoV, AMB and hypoglossal nuclei. The two 'premotor blink areas' in the reticular formation of the pons and medulla project to the periorbital motor subgroups of the VII. This network is important for the control of reflex blinks to tactile, visual and auditory stimuli [Holstege, 1991; Sibony and Evinger, 1998; Horn and Adamczyk, 2012]. In addition, structures involved in oculomotor and head control, such as the superior colliculus (SC) and the pontine reticular formation, project directly to the VII [Horn, 2006; May, 2006].

In primates, the corticonuclear projections to the VII terminate both directly on motoneurons and on reticular formation interneurons around the nucleus: these projections are bilateral [Morecraft et al., 2001]. Different cortical areas were found to project with different intensities to different subgroups; but it was found that all cortical face representation areas innervate all nuclear subdivisions, to some degree. The face area of the primary motor cortex, the caudal cingulate cortex, the ventral and lateral premotor cortices projected strongly to the contralateral perioral muscles, whereas the supplementary motor cortex supplied the medial (ear) subgroups bilaterally, and the rostral cingulate cortex the orbicularis oculi motoneurons bilaterally. On the basis of these results the authors suggest that the classical clinical observation on sparing the upper facial muscles in a supranuclear palsy may be explained by the bilateral projection of the cingulate areas spared by the insult, rather than a bilateral projection of the primary cortex to the VII.

The neurons of the VII receive catecholaminergic afferents from the locus coeruleus and lateral parabrachial nucleus, as well as cholinergic afferents from an area ventral to the solitary nucleus [Fort et al., 1989]. A strong supply of serotonergic afferents arise from the nucleus raphe magnus, raphe obscurus and raphe pallidus nuclei, which are thought to adjust the excitability of facial motoneurons (see 'Neuromodulatory Systems: Overview') [Rasmussen and Aghajanian, 1990].

References

Fort P, Sakai K, Luppi P-H, Salvert D, Jouvet M: Monoaminergic, peptidergic, and cholinergic afferents to the cat facial nucleus as evidenced by a double immunostaining method with unconjugated cholera toxin as a retrograde tracer. J Comp Neurol 1989;283:285–302.

Holstege G: Descending motor pathways and the spinal motor system: limbic and non-limbic components. Prog Brain Res 1991;87:307–421.

Holstege GG, Mouton LJ, Gerrits NM: Emotional motor system; in Paxinos G, Mai JK (eds): The Human Nervous System, ed 2. Amsterdam, Elsevier Academic Press, 2004.

Horn AK, Adamczyk C: Reticular formation: eye movements, gaze and blinks; in Mai JK, Paxinos G (eds): The Human Nervous System, ed 3. San Diego, Academic Press, 2012.

Horn AKE: The reticular formation. Prog Brain Res 2006;151:127–155.

Hutson KA, Glendenning KK, Masterton RB: Accessory abducens nucleus and its relationship to the accessory facial and posterior trigeminal nuclei in cat. J Comp Neurol 1979;188:1–16.

Jacobs MJ: The development of the human motor trigeminal complex and accessory facial nucleus and their topographic relations with the facial and abducens nuclei. J Comp Neurol 1970;138:161–194.

Langer TP, Fuchs AF, Scudder CA, Chubb MC: Afferents to the flocculus of the cerebellum in the rhesus macaque as revealed by retrograde transport of horseradish peroxidase. J Comp Neurol 1985;235:1–25.

May PJ: The mammalian superior colliculus: laminar structure and connections. Prog Brain Res 2006;151:321–378.

Morecraft RJ, Louie J, Herrick JL, Stilwell-Morecraft KS: Cortical innervation of the facial nucleus in the non-human primate: a new interpretation of the effects of stroke and related subtotal brain trauma on the muscles of facial expression. Brain 2001;124:176–208.

Nieuwenhuys R, Voogd J, van Huijzen C: Motor systems; in Nieuwenhuys R, Voogd J, van Huijzen C (eds): The Human Central Nervous System, ed 4. Berlin, Springer, 2008.

Paxinos G, Huang X-F: Atlas of the Human Brainstem. San Diego, Academic Press, 1995.

Paxinos G, Huang X-F, Sengul G, Watson C: Organization of the brainstem; in Mai JK, Paxinos G (eds): The Human Nervous System, ed 3. Amsterdam, Academic Press, 2012.

Pearson AA: Facial motor nuclei. J Comp Neurol 1946;85:461–476.

Porter JD, Burns LA, May PJ: Morphological substrate for eyelid movements: innervation and structure of primate levator palpebrae superioris and orbicularis oculi muscles. J Comp Neurol 1989;287:64–81.

Rasmussen K, Aghajanian GK: Serotonin excitation of facial motoneurons: receptor subtype characterization. Synapse 1990;5:324–332.

Satoda T, Takahashi O, Tashiro T, Matsushima R, Uemura-Sumi M, Mizuno N: Representation of the main branches of the facial nerve within the facial nucleus of the Japanese monkey *(Macaca fuscata).* Neurosci Lett 1987;78:283–287.

Sherwood CC: Comparative anatomy of the facial motor nucleus in mammals, with an analysis of neuron numbers in primates. Anat Rec 2005;287A:1067–1079.

Sibony PA, Evinger C: Normal and abnormal eyelid function; in Miller NR, Newman NH (eds): Clinical Neuro-Ophthalmology. Baltimore, Williams & Wilkins, 1998.

Thompson GC, Igarashi M, Stach BA: Identification of stapedius muscle motoneurons in squirrel monkey and bush baby. J Comp Neurol 1985;231:270–279.

Van der Werf F, Aramideh M, Otto JA, Ongerboer de Visser BW: Retrograde tracing studies of subdivisions of the orbicularis oculi muscle in the rhesus monkey. Exp Brain Res 1998;121:433–441.

Welt C, Abbs JH: Musculotopic organisation of the facial motor nucleus in *Macaca fascicularis:* a morphometric and retrograde tracing study with cholera toxin B-HRP. J Comp Neurol 1990;291:621–636.

41 Accessory Facial Nucleus (VIIac)

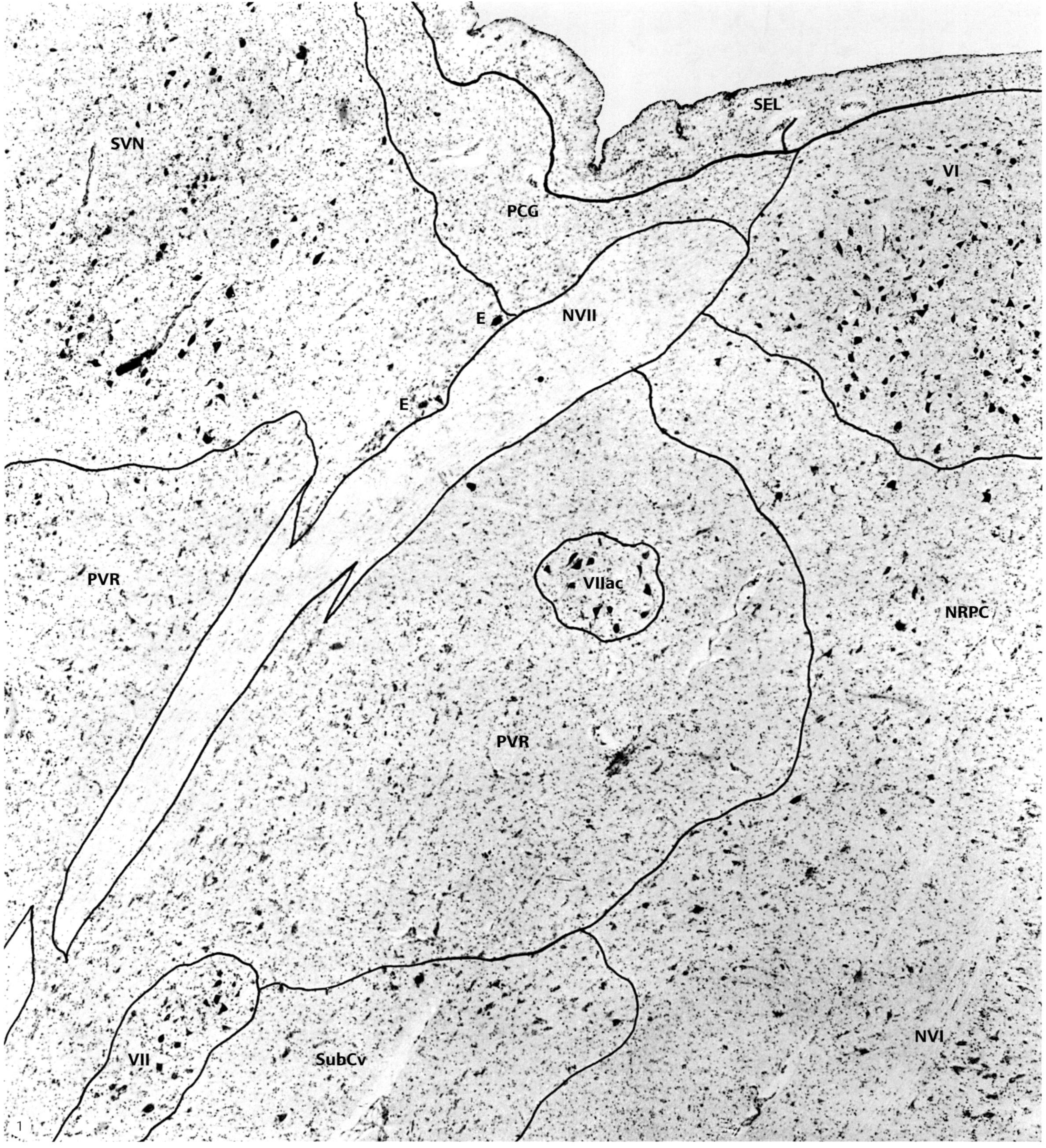

Fig. 1. Accessory facial nucleus (VIIac). E = Possibly the E group of the vestibular complex; NRPC = nucleus reticularis pontis caudalis; NVII = facial nerve; PCG = pontine central gray; PVR = parvocellular reticular nucleus; SEL = subependymal layer; SubCv = nucleus subcoeruleus, ventral part; SVN = superior vestibular nucleus; VI = abducens nucleus; VII = facial nucleus. Magnification ×40.

Original name:
Nucleus nervi facialis accessorius

Alternative names:
Nucleus nervi facialis, cellulae accessoriae
Accessory nucleus of the facial nerve

Location and Cytoarchitecture – Original Text

This nucleus is composed of small groups of motor-type cells resembling those of the facial (VII) and other somatic and branchiomeric motor nuclei, and situated within the parvocellular reticular nucleus (PCR) about midway between the oral pole of the VII ventrolaterally and the abducens nucleus (VI) dorsomedially (fig. 1). Occasionally, cells of the accessory facial nucleus (VIIac) may lie slightly more lateral in direct relationship to the descending intramedullary facial fibers.

On average the VIIac contains 131 cells [Van Buskirk, 1945]. These cells are arranged in several small inconstant groups. Many sections through the oral part of the main VII fail to reveal any accessory cells, and it is rare for more than one cell group of the VIIac to appear in a single cross-section. Since the VIIac is not represented in the serial cross-sections, a low-power photomicrograph is shown in figure 1.

Functional Neuroanatomy

Function

Szentágothai [1948] proposed many years ago that the VIIac was a small group of motoneurons innervating the posterior belly of the digastric muscle via the facial nerve; this has since been confirmed [Jacobs, 1970]. The activation of the muscle elevates the hyoid bone and assists in mouth opening. The nucleus falls into the parasagittal column of branchial motoneurons stretching from the VII to the trigeminal motor nucleus. In some mammals it lies close to the accessory VI which innervates the retractor bulbi muscles and those of the nictitating membrane, but no accessory VI has been found in humans [Hutson et al., 1979].

Comments

The VIIac motoneurons are cholinergic, and in immunohistochemical stains of human material they should be carefully distinguished from the cholinergic vestibular efferent neurons lying nearby in the E group (nucleus No. 19).

Fibers from the VIIac join the facial nerve fibers (NVII) passing dorsally and medially from the main VII towards the VI to form the facial genu. In those instances in which the VIIac is more laterally situated, its fibers may contribute directly to the descending bundle of the NVII.

References

Hutson KA, Glendenning KK, Masterton RB: Accessory abducens nucleus and its relationship to the accessory facial and posterior trigeminal nuclei in cat. J Comp Neurol 1979;188:1–16.

Jacobs MJ: The development of the human motor trigeminal complex and accessory facial nucleus and their topographic relations with the facial and abducens nuclei. J Comp Neurol 1970;138:161–194.

Szentágothai J: The representation of facial and scalp muscles in the facial nucleus. J Comp Neurol 1948;88:207–220.

Van Buskirk C: Seventh nerve complex. J Comp Neurol 1945;82:303–333.

Motor Systems – Somatomotor Nuclei

(Plates 24 and 25)

42 Motor Trigeminal Nucleus (MoV)

Original name:
Nucleus nervi trigemini motorius

Alternative names:
Nucleus motorius nervi trigemini
Motor nucleus of the trigeminal nerve
Fifth cranial nerve motor nucleus

Location and Cytoarchitecture – Original Text

This nucleus is located in the lateral part of the midpontine tegmentum at the level of the entrance zone of the motor and sensory roots of the trigeminal or fifth cranial nerve. The caudal pole of the nucleus appears about 2 mm rostral to the oral pole of the facial nucleus (VII), at a level in which the cells of the abducens nucleus are still present. From this point the motor trigeminal nucleus (MoV) extends orally for a further 4 mm and terminates at or about the level in which the pigmented cells of the locus coeruleus first appear.

On cross-section, as well as on sagittal section, the MoV is ovoid in outline. Its caudal portion is related ventrally and medially to the cells of the nucleus reticularis pontis caudalis. Laterally the root fibers of the fifth cranial nerve separate the motor from the principal sensory trigeminal nucleus, and dorsally the transversely coursing fibers of the seventh cranial nerve intervene between the MoV and the superior vestibular nucleus. The oral portion of the MoV is related laterally to the nucleus parabrachialis medialis (PBM); all other surfaces are bounded by the nucleus subcoeruleus.

The cells of the MoV are typical, large, multipolar, darkly stained motor-type neurons similar to those found in the VII, supraspinal and hypoglossal nuclei (XII) (fig. 1, 2).

As is the case in many other motor nuclei, one can often distinguish a definite tendency for the cells of this nucleus to congregate into discrete groups. In contrast, however, to the VII and XII, neither the size nor the

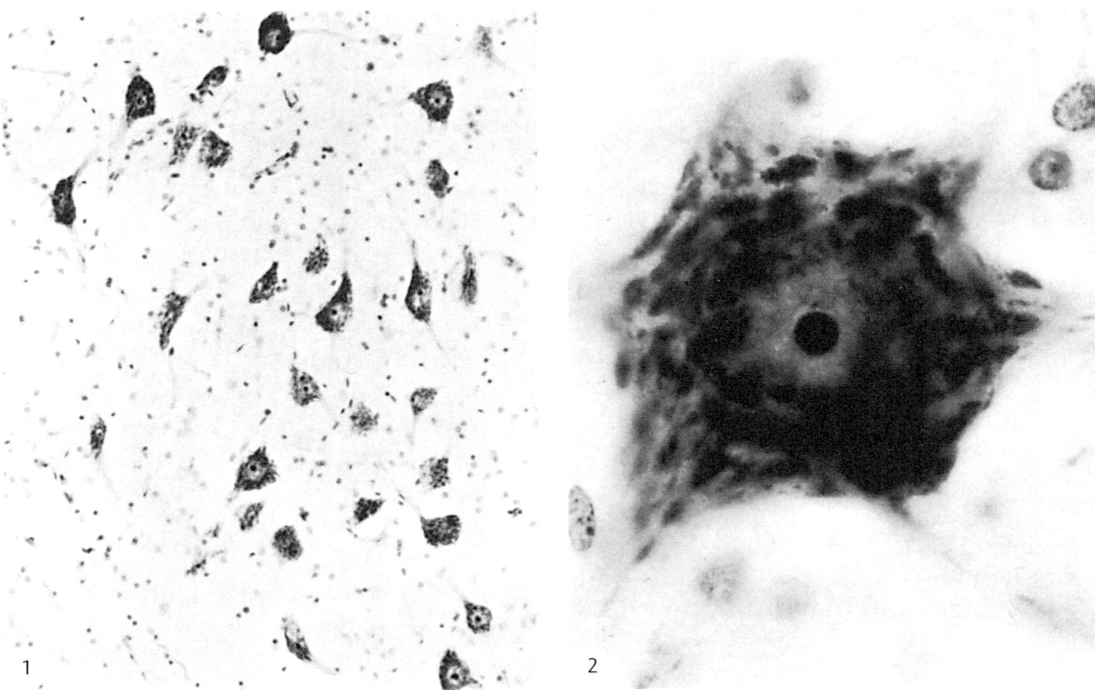

Fig. 1. Motor trigeminal nucleus. Magnification ×150.
Fig. 2. Cells from the motor trigeminal nucleus. Magnification ×1,000.

shape of such groups within the MoV is consistent, and occasionally no grouping at all can be distinguished.

The MoV gives rise to special visceral efferent fibers which proceed via the mandibular division of the fifth cranial nerve to supply the muscles of mastication, the anterior belly of the digastric (see posterior trigeminal nucleus below), the tensor tympani and the tensor palati muscle. [...]

Functional Neuroanatomy

Function

The MoV innervates the muscles of mastication and swallowing via the motor branch of the trigeminal nerve (portio minor). In addition the MoV innervates the tensor veli palatini and the tensor tympani muscles [Friauf and Baker, 1985]. The tensor veli palatini elevates the soft palate during swallowing, and in so doing assists in closing off the nasal cavities, and simultaneously assists in opening the auditory tube and equating the pressure of the middle ear with the oral cavity. The tensor tympani dampens the movement of the ossicles by tensing the tympanic membrane, and so reduces the transmission of sounds to the cochlea: it can do this in anticipation of loud sounds such as talking or chewing, and acts as a protection from loud noises. An important reflex associated with the MoV is the 'masseter or jaw reflex', where a light tap on the chin leads to jaw-closing through the activation of proprioceptive afferents in the mesencephalic trigeminal nucleus (MesV) onto MoV motoneurons. The reflex is useful to localize brain dysfunction; a heightened reflex indicates abnormalities of inputs to the MoV from higher centers (in clinical terms, from the 'upper motoneurons').

Connections

Efferents. Without tract-tracing methods, it is difficult to recognize the subgroups of motoneurons for individual muscles in the MoV. A summary of the localization of motoneurons innervating different muscles using tracers is illustrated by Voogd et al. [1998] in their figure 22.29a–c [Mizuno et al., 1981; Nieuwenhuys et al., 2008]. The motoneurons can be divided into a medial and a lateral group. Subgroups of the medial group supply motoneurons of jaw openers; the lateral group supplies jaw closers [Nieuwenhuys et al., 2008]. The smaller motoneurons of tensor veli palatini and tensor tympani muscles lie ventromedially [Friauf and Baker, 1985].

Afferents through Premotor Areas. In accordance with older reports, tract-tracing studies show that there are few *direct* cortical inputs to the trigeminal motor nucleus. Direct corticobulbar pathways terminate in the lateral tegmental area *surrounding* the MoV, a premotor region which is important for the coordination of the trigeminal motoneuron activity both ipsilaterally and contralaterally [Holstege, 1991; Li et al., 2002; Nieuwenhuys et al., 2008; see also Hage and Jürgens, 2006]. The premotor area around the MoV is divided into 3 regions: supratrigeminal, intertrigeminal and juxtatrigeminal areas (the latter including the PBM and Koelliker-Fuse nucleus). The premotor trigeminal regions receive excitatory afferents directly from several areas of the cerebral cortex, and through indirect pathways they receive inhibitory influences [Hatanaka et al., 2005]. Further afferents to the supratrigeminal area come from the marginal cells of the spinal trigeminal nucleus, caudal part [Li et al., 2005].

Direct Afferents. The motoneurons of the lateral MoV group, innervating jaw closers, receive ipsilateral collaterals from the MesV both directly and indirectly via the supratrigeminal area and the dorsolateral medullary reticular formation [Fay and Norgren, 1997; Cunningham and Sawchenko, 2000]. A network of local connections involving several premotor regions in the dorsomedial and dorsolateral medullary reticular formation, and the nucleus of Probst enable the MesV, carrying the muscle spindle activity of the muscles of mastication, to control jaw-closing muscles on both sides [Cunningham and Sawchenko, 2000; Nieuwenhuys et al., 2008, fig. 21.12]. This is the neuroanatomical basis of the efferent limb of the 'jaw-jerk' or 'masseter reflex'. In contrast to jaw-closing muscles, the jaw-opening muscles have few muscle spindles in the rat, and their motoneurons do not have prominent MesV afferents [Fay and Norgren, 1997].

Direct afferents to the medial MoV, the jaw-opening motoneurons, arise from the sensory trigeminal nuclei innervating lingual and oral mucous membranes and from the retroambiguus nucleus [Holstege, 1991; Shigenaga et al., 1988].

References

Cunningham ET Jr, Sawchenko PE: Dorsal medullary pathways subserving oromotor reflexes in the rat: implications for the central neural control of swallowing. J Comp Neurol 2000;417:448–466.

Fay RA, Norgren R: Identification of rat brainstem multisynaptic connections to the oral motor nuclei using pseudorabies virus. I. Masticatory muscle motor systems. Brain Res Rev Brain Res 1997;25:255–275.

Friauf E, Baker R: An intracellular HRP-study of cat tensor tympani motoneurons. Exp Brain Res 1985;57:499–511.

Hage SR, Jürgens U: Localization of a vocal pattern generator in the pontine brainstem of the squirrel monkey. Eur J Neurosci 2006;23:840–844.

Hatanaka N, Tokuno H, Nambu A, Inoue T, Takada M: Input-output organization of jaw movement-related areas in monkey frontal cortex. J Comp Neurol 2005;492:401–425.

Holstege G: Descending motor pathways and the spinal motor system: limbic and non-limbic components. Prog Brain Res 1991;87:307–421.

Li JL, Wu SX, Tomioka R, Okamoto K, Nakamura K, Kaneko T, Mizuno N: Efferent and afferent connections of GABAergic neurons in the supratrigeminal and the intertrigeminal regions. An immunohistochemical tract-tracing study in the GAD67-GFP knock-in mouse. Neurosci Res 2005;51:81–91.

Li YQ, Tao FS, Okamoto K, Nomura S, Kaneko T, Mizuno N: The supratrigeminal region of the rat sends GABA/glycine-cocontaining axon terminals to the motor trigeminal nucleus on the contralateral side. Neurosci Lett 2002;330:13–16.

Mizuno N, Matsuda K, Iwahori N, Uemura-Sumi M, Kume M, Matsushima R: Representation of the masticatory muscles in the motor trigeminal nucleus of the macaque monkey. Neurosci Lett 1981;21:19–22.

Nieuwenhuys R, Voogd J, van Huijzen C: The Human Central Nervous System. Berlin, Springer, 2008.

Shigenaga Y, Yoshida A, Mitsuhiro Y, Tsuru K, Doe K: Morphological and functional properties of trigeminal nucleus oralis neurons projecting to the trigeminal motor nucleus of the cat. Brain Res 1988;461:143–149.

Voogd J, Nieuwenhuys R, van Dongen PAM, ten Donkelaar HJ: Mammals; in Nieuwenhuys R, ten Donkelaar HJ, Nicholson C (eds): The Central Nervous System of Vertebrates. Berlin, Springer, 1998.

Motor Systems – Somatomotor Nuclei (Not on plates)

43 Posterior Trigeminal Nucleus (PoV)

Original name:
Nucleus retrotrigeminalis

Alternative names:
Accessory trigeminal nucleus
Retrotrigeminal nucleus
Nucleus motorius nervi trigemini accessorius

Location and Cytoarchitecture – Original Text

[Since the posterior trigeminal nucleus (PoV) is often continuous with or even overlaps the caudal border of the main trigeminal motor nucleus (MoV), the name 'posterior trigeminal nucleus', seems more appropriate than 'retrotrigeminal nucleus' used in the previous edition, and therefore will be used here [Jacobs, 1970].]

This is a small nucleus, less than 1 mm in length, located immediately caudal to the MoV and rostral to the facial nucleus (VII), in the dorsolateral part of the midpontine tegmentum. On cross-section the nucleus is oblong in outline with its long axis directed dorsoventrally. It is related medially and ventrally to the parvocellular reticular nucleus, dorsally and dorsolaterally to the fibers of the facial nerve which separates the nucleus from the central gray matter of the pons and from the superior vestibular nucleus (fig. 1). The caudal pole of the MoV appears in a more ventral and lateral position than that occupied by the PoV. The nucleus lies close, but rostral to the accessory facial nucleus (VIIac). The PoV is composed of typical motor-type cells similar to those of the adjacent VII and MoV. As a rule 10–20 cells appear in each cross-section through the nucleus.

Functional Neuroanatomy

The PoV innervates the anterior belly of the digastric muscle [Jacobs, 1970]. The efferent nerves exit with the trigeminal motor nerve (portio minor), pass through the trigeminal ganglion into the mandibular nerve and are active in opening the mouth and swallowing [Holstege et al., 1983]. This nucleus, along with the VIIac, which innervates the posterior belly of the digastric muscle, is part of a parasagittal column of branchial motoneurons stretching from the VII to MoV [Hutson et al., 1979].

References

Holstege G, Graveland GA, Bijker-Biemond C, Schuddeboom I: Location of motoneurons innervating soft palate, pharynx and upper esophagus. Anatomical evidence for a possible swallowing center in the pontine reticular formation: an HRP and autoradiographical tracing study. Brain Behav Evol 1983;23:47–62.

Hutson KA, Glendenning KK, Masterton RB: Accessory abducens nucleus and its relationship to the accessory facial and posterior trigeminal nuclei in cat. J Comp Neurol 1979;188:1–16.

Jacobs MJ: The development of the human motor trigeminal complex and accessory facial nucleus and their topographic relations with the facial and abducens nuclei. J Comp Neurol 1970;138:161–194.

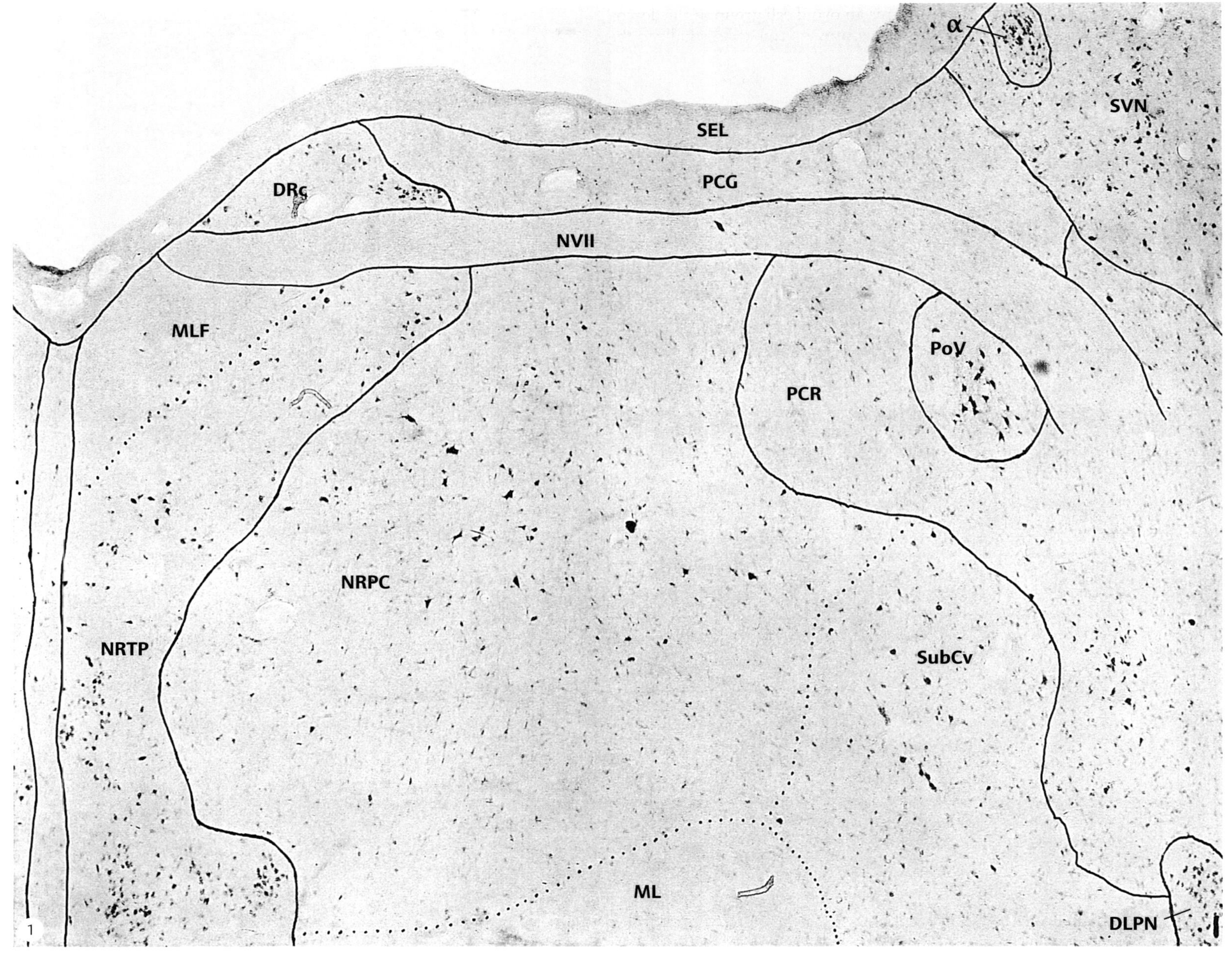

Fig. 1. Posterior trigeminal nucleus. See plates 22 and 24. DLPN = Dorsolateral pontine nuclei; DRc = dorsal raphe nucleus, caudal part; ML = medial lemniscus; MLF = medial longitudinal fasciculus; NRPC = nucleus reticularis pontis caudalis; NRTP = nucleus reticularis tegmenti pontis; NVII = facial nerve; PCG = pontine central gray; PCR = parvocellular reticular nucleus; PoV = posterior trigeminal nucleus; SEL = subependymal layer; SubCv = nucleus subcoeruleus, ventral part; SVN = superior vestibular nucleus; α = pontine nuclei, possible subgroups. Magnification ×40.

Motor Systems – Somatomotor Nuclei

(Plates 22 and 23)

44 Abducens Nucleus (VI)

Original name:
Nucleus nervi abducentis

Alternative names:
Abducens nerve nucleus
Sixth cranial nerve nucleus

Location and Cytoarchitecture – Original Text

The abducens nucleus (VI) is located beneath the facial colliculus in the floor of the midpontine portion of the fourth ventricle. It appears at the level of the oral pole of the facial nucleus (VII), extends orally for approximately 3 mm, and terminates at, or just oral to, the level in which the most caudal cells of the motor trigeminal nucleus appear.

On cross-section, the nucleus forms an ovoid cell group in the dorsomedial portion of the midpontine tegmentum. It is related medially to the medial longitudinal fasciculus (MLF), ventrally to the nucleus reticularis pontis caudalis, and laterally to the parvocellular reticular nucleus. Dorsomedially the nucleus is related to the fibers of the facial genu, and the oral portion of the nucleus is separated from the floor of the fourth ventricle by horizontally coursing facial fibers. A few abducens cells lie among these latter fibers and between them and the ependyma of the ventricular floor. The caudal portion of the nucleus is related dorsally to the supragenual nucleus.

The cells of the VI, trochlear and principal oculomotor nuclei (III), although similar to those of other somatic motor nuclei, are, in general, more plump, smaller and more lightly stained. Further, their Nissl granules are smaller and less regular (fig. 1–5).

Functional Neuroanatomy

Function

The VI contains the motoneurons of the ipsilateral lateral rectus muscle (LR), which control the horizontal abduction of the eye. However, the nucleus is not a homogeneous cell population. First, there are at least 2 types of motoneurons, those controlling the twitch muscle fibers and those controlling the nontwitch muscle fibers. The neural activity of nontwitch motoneurons is probably highly tonic [Dieringer and Precht, 1986], but their function is still unclear [Büttner-Ennever, 2006]. One hypothesis suggests that they control the tonic properties of the muscles (e.g. length or stiffness), whereas the twitch fibers drive the eye movements (see below).

Apart from motoneurons, there are 2 other major neuronal populations within the VI; the abducens internuclear neurons (INT), and paramedian tract neurons (PMT cell groups). The INT project to the contralateral medial rectus (MR) subgroup in the III. Their function is to coordinate the activity of the contralateral MR and LR for all horizontal conjugate eye movements (including the vestibulo-ocular reflex, saccades and smooth pursuit). Lesions to the internuclear neurons or their axons in the MLF cause paralysis of the contralateral MR for all horizontal conjugate eye movements, but vergence movements of the MR remain possible.

Finally the cytological boundaries of the rostral VI enclose cell groups which project to the cerebellar flocculus bilaterally. They are part of a continuum of cells called the PMT cell groups (nucleus No. 96), and may be important in the cerebellar control of gaze-holding in the horizontal or vertical plane [Dean and Porrill, 2008]. Experimentally, lesions to the different PMT cell groups close to the VI result in the lack of gaze-holding to the ipsilateral side, or to downbeat nystagmus [Cheron et al., 1992; Nakamagoe et al., 2000; Leigh and Zee, 2006].

Histochemistry

The nontwitch motoneurons innervating the orbital muscle layer lie mostly around the boundaries of the VI, medially and dorsally [Büttner-Ennever et al., 2001; Ugolini et al., 2001]. The two motoneuron populations, twitch and nontwitch, can be differentiated using histochemical techniques: both types are cholinergic but only the twitch motoneurons have perineuronal nets [Eberhorn et al., 2005]. The INT are morphologically similar to motoneurons [McCrea et al., 1986], possess perineuronal nets, but they are not cholinergic, they use glutamate and aspartate as transmitter [Nguyen et al., 1999; Nguyen and Spencer, 1999].

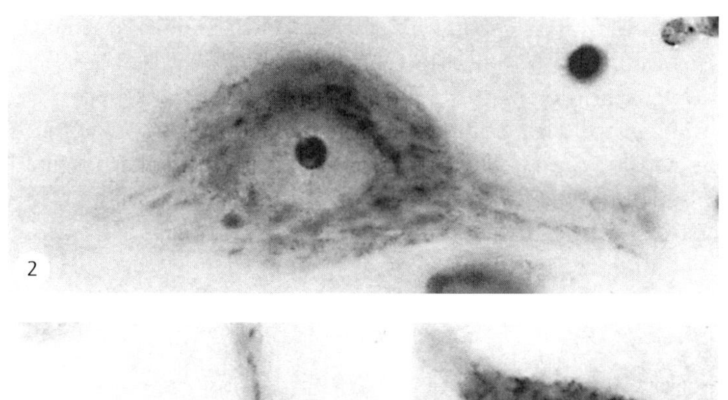

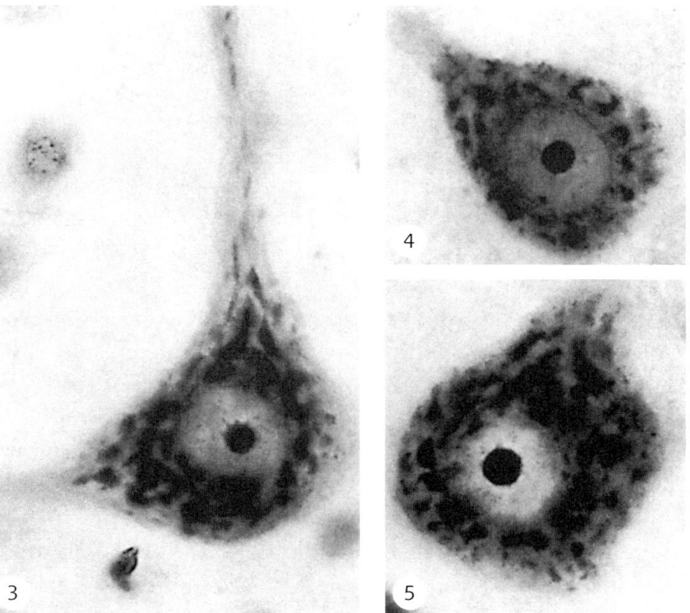

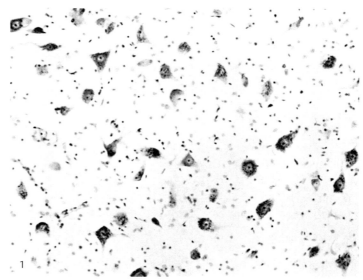

Fig. 1. Abducens nucleus. Magnification ×150.
Fig. 2–5. Cells from the abducens nucleus. Magnification ×1,000.

Development

The site of generation of the VI is in the ventral pons (rhombomere 5), whereas the origin of the VII lies rostrally and dorsally in rhombomere 4. At an early embryological stage the VI migrates dorsally to its adult position in the floor of the fourth ventricle, while the VII migrates medially to the VI, and caudolaterally to rhombomere 6 [Jacobs, 1970; Nieuwenhuys et al., 2008]. It has been suggested that the complex route of the facial nerve and the facial genu result from these migrations. In some animals the migration does not take place and the VI remains in the ventral pons (e.g. goldfish).

Connections

Efferents. The VI is the largest of the extraocular motor nuclei, with two major efferent systems: one through the abducens nerve (NVI) to the lateral rectus muscle, and the second projecting into the contralateral MLF, crossing the midline at the level of the VI and ascending in the MLF to the contralateral III.

The NVI carries efferents to the ipsilateral LR for abduction of the eye [Sylvestre and Cullen, 2002]; the fibers arise from the twitch motoneurons within the VI, and from the nontwitch motoneurons, whose distribution in the VI is still unclear. Recently the exciting possibility has been raised that the putative sensory receptors of eye muscles, called palisade endings, send their axons centrally in the NVI, and their sensory ganglion cells lie close to the nontwitch motoneurons of the VI, and similar results were obtained for the III and oculomotor nerve (NIII) [Lienbacher et al., 2011a, b; Zimmermann et al., 2011]. These results imply that the NVI (and NIII) also carry a sensory component.

The second major efferent pathway from the VI originates from abducens INT which compose about 25–30% of the VI population [Steiger and Büttner-Ennever, 1978]. In monkeys, the neurons lie rostrolaterally, but in humans their location is not clear. The INT efferents cross the midline at the level of the VI and ascend in the MLF to the contralateral III, terminating on the MR subgroups, thereby forming the anatomical basis for conjugate eye movements.

A third and less well-known group of efferents associated with the VI comes from the PMT cell groups, which project to the flocculus bilaterally (see nucleus No. 96). They lie around the abducens rostral cap of the VI merging between the fascicles of the MLF [Buresch, 2005; Büttner-Ennever, 2006]. The efferent fibers travel laterally in the internal arcuate bundles to join the inferior cerebellar peduncle and terminate in the flocculus.

Afferents. The motoneurons and internuclear neurons receive afferents from secondary vestibulo-ocular neurons in the magnocellular region of the medial vestibular nucleus; these are essential for the vestibulo-ocular reflex. For a review on this and subsequent connections see Büttner-Ennever [2006] and Horn and Adamczyk [2012]. Nonsecondary vestibular neurons in the marginal zone and the adjacent parvocellular medial vestibular nucleus also project to the VI: a rostral group targeting the ipsilateral VI and a caudal glycinergic group targeting the contralateral VI [Langer et al., 1986; McFarland et al., 1992]. Less is known about these nonsecondary vestibular inputs. In contrast, the monosynaptic inputs from the pontine reticular formation (nucleus No. 56) and dorsal paragigantocellular nucleus (nucleus No. 55) where the excitatory and inhibitory burst neurons for saccades lie, respectively, have been closely studied [Horn et al., 1995, 1996; Zhou and King, 1998]. Afferents to the VI arise from the prepositus nucleus, from internuclear neurons of the III and the central mesencephalic reticular formation [Langer et al., 1986; Spencer et al., 1989; Lahjouji et al., 1995; Büttner-Ennever, 2006; Ugolini et al., 2006]. Finally the superior colliculus has also been shown to project directly to the VI [Izawa et al., 1999].

The twitch and nontwitch motoneurons receive different afferent inputs indicating that they have different functions [Wasicky et al., 2004]. The nontwitch motoneurons of the VI receive monosynaptic inputs from the prepositus nucleus, parvocellular medial vestibular nucleus and the supraoculomotor area along with its adjacent mesencephalic reticular formation.

The PMT cell groups receive collaterals, from all known premotor inputs supplying the abducens motoneurons [Büttner-Ennever et al., 1989; Dean and Porrill, 2008]. However, some PMT groups encode horizontal eye signals and others vertical ones, and the inputs from premotor structures reflect this segregation. Thus, it is rather surprising but understandable, in the nonhuman primate, to find afferent projections from the interstitial nucleus of Cajal and the rostral interstitial nucleus of the MLF terminating in the rostral VI on a vertical PMT cell group [Büttner-Ennever and Horn, 1996].

References

Buresch N: Neuroanatomische Charakterisierung blickstabilisierender Neurone an der Hirnstammmittellinie der Primaten, einschliesslich des Menschen; dissertation, LMU München, 2005.

Büttner-Ennever JA: The extraocular motor nuclei: organization and functional neuroanatomy. Prog Brain Res 2006;151:95–125.

Büttner-Ennever JA, Horn AKE: Pathways from cell groups of the paramedian tracts to the floccular region. Ann NY Acad Sci 1996;781:532–540.

Büttner-Ennever JA, Horn AKE, Scherberger H, D'Ascanio P: Motoneurons of twitch and nontwitch extraocular muscle fibers in the abducens, trochlear, and oculomotor nuclei of monkeys. J Comp Neurol 2001;438:318–335.

Büttner-Ennever JA, Horn AKE, Schmidtke K: Cell groups of the medial longitudinal fasciculus and paramedian tracts. Rev Neurol (Paris) 1989;145:533–539.

Cheron G, Mettens P, Godaux E: Gaze holding defect induced by injections of ketamine in the cat brainstem. Neuroreport 1992;3:97–100.

Dean P, Porrill J: Oculomotor anatomy and the motor-error problem: the role of the paramedian tract nuclei. Prog Brain Res 2008;171:177–186.

Dieringer N, Precht W: Functional organization of eye velocity and eye position signals in abducens motoneurons of the frog. J Comp Physiol 1986;158:179–194.

Eberhorn AC, Ardelenanu P, Büttner-Ennever JA, Horn AKE: Histochemical differences between motoneurons supplying multiply and singly innervated extraocular muscle fibers. J Comp Neurol 2005;491:352–366.

Horn AK, Adamczyk C: Reticular formation: eye movements, gaze and blinks; in Mai JK, Paxinos G (eds): The Human Nervous System, ed 3. San Diego, Academic Press, 2012.

Horn AK, Büttner-Ennever JA, Suzuki Y, Henn V: Histological identification of premotor neurons for horizontal saccades in monkey and man by parvalbumin immunostaining. J Comp Neurol 1995;359:350–363.

Horn AKE, Büttner-Ennever JA, Büttner U: Saccadic premotor neurons in the brainstem: functional neuroanatomy and clinical implications. Neuro-Ophthalmology 1996;16:229–240.

Izawa Y, Sugiuchi Y, Shinoda Y: Neural organization from the superior colliculus to motoneurons in the horizontal oculomotor system of the cat. J Neurophysiol 1999;81:2597–2611.

Jacobs MJ: The development of the human motor trigeminal complex and accessory facial nucleus and their topographic relations with the facial and abducens nuclei. J Comp Neurol 1970;138:161–194.

Lahjouji F, Bras H, Barbe A, Chazal G: GABAergic innervation of rat abducens motoneurons retrogradely labelled with HRP: quantitative ultrastructural analysis of cell bodies and proximal dendrites. J Neurocytol 1995;24:29–44.

Langer TP, Kaneko CR, Scudder CA, Fuchs AF: Afferents to the abducens nucleus in the monkey and cat. J Comp Neurol 1986;245:379–400.

Leigh RJ, Zee DS: The Neurology of Eye Movements. New York, Oxford University Press, 2006.

Lienbacher K, Mustari M, Hess B, Büttner-Ennever J, Horn AK: Is there any sense in the palisade endings of eye muscles? Ann NY Acad Sci 2011a;1233:1–7.

Lienbacher K, Mustari M, Ying HS, Büttner-Ennever JA, Horn AK: Do palisade endings in extraocular muscles arise from neurons in the motor nuclei? Invest Ophthalmol Vis Sci 2011b;52:2510–2519.

McCrea RA, Strassman A, Highstein SM: Morphology and physiology of abducens motoneurons and internuclear neurons intracellulary injected with horseradish peroxidase in alert squirrel monkey. J Comp Neurol 1986;243:291–308.

McFarland JL, Fuchs AF, Kaneko CR: The nucleus prepositus and nearby medial vestibular nucleus and the control of simian eye movements; in Shinoda Y, Shimazu H (eds): Vestibular and Brain Stem Control of Eye, Head and Body Movements. Tokyo, Japan Acientific Societies Press, 1992.

Nakamagoe K, Iwamoto Y, Yoshida K: Evidence for brainstem structures participating in oculomotor integration. Science 2000;288:857–859.

Nguyen LT, Baker R, Spencer RF: Abducens internuclear and ascending tract of Deiters inputs to medial rectus motoneurons in the cat oculomotor nucleus: synaptic organization. J Comp Neurol 1999;405:141–159.

Nguyen LT, Spencer RF: Abducens internuclear and ascending tract of Deiters inputs to medial rectus motoneurons in the cat oculomotor nucleus: neurotransmitters. J Comp Neurol 1999;411:73–86.

Nieuwenhuys R, Voogd J, van Huijzen C: Development; in Nieuwenhuys R, Voogd J, van Huijzen C (eds): The Human Central Nervous System. Berlin, Springer, 2008.

Spencer RF, Wenthold RJ, Baker R: Evidence for glycine as an inhibitory neurotransmitter of vestibular, reticular, and prepositus hypoglossi neurons that project to the cat abducens nucleus. J Neurosci 1989;9:2718–2736.
Steiger HJ, Büttner-Ennever JA: Relationship between motoneurons and internuclear neurons in the abducens nucleus: a double retrograde tracer study in the cat. Brain Res 1978;148:181–188.
Sylvestre PA, Cullen KE: Dynamics of abducens nucleus neuron discharges during disjunctive saccades. J Neurophysiol 2002;88:3452–3468.
Ugolini G, Klam F, Doldan Dans M, Dubayle D, Brandi A-M, Büttner-Ennever JA, Graf W: Horizontal eye movement networks in primates as revealed by retrograde transneuronal transfer of rabies virus: differences in monosynaptic input to 'slow' and 'fast' abducens motoneurons. J Comp Neurol 2006;498:762–785.
Wasicky R, Horn AKE, Büttner-Ennever JA: Twitch and non-twitch motoneuron subgroups of the medial rectus muscle in the oculomotor nucleus of monkeys receive different afferent projections. J Comp Neurol 2004;479:117–129.
Zhou W, King WM: Premotor commands encode monocular eye movements. Nature 1998;393:692–695.
Zimmermann L, May PJ, Pastor AM, Streicher J, Blumer R: Evidence that the extraocular motor nuclei innervate monkey palisade endings. Neurosci Lett 2011;489:89–93.

Motor Systems – Somatomotor Nuclei (Plates 32 and 33)

45 Trochlear Nucleus (IV)

Original name:
Nucleus nervi trochlearis

Location and Cytoarchitecture – Original Text

The trochlear nucleus (IV) is located in the tegmentum of the midbrain at the level of the inferior colliculus and consists of 1 large oral cell group and 1 or 2 smaller, more caudally situated groups which lie partially embedded in the dorsal part of the medial longitudinal fasciculus (MLF). Orally, the large cell group is separated from the caudal pole of the oculomotor nucleus (III) by a very narrow cell-free band. All of these features can be clearly demonstrated on sagittal sections (fig. 1).

On cross-section the nucleus is round in outline at all levels and is usually composed of 3–40 cells, depending upon whether components of the large oral or the smaller caudal groups appear. As these latter groups are not symmetrically arranged, one often finds the IV represented on only one side of the cross-section. Cross-sections in the areas intervening between the component groups of the nucleus will contain no motor neurons. Dorsally and dorsolaterally the IV and the MLF are related to the dorsal raphe nucleus.

The distinction between the oral pole of the IV and the caudal pole of the III may at times be difficult, particularly on cross-sections. [...] The differentiation can usually be made on the basis of the following observation: (a) the IV lies in the MLF; the caudal pole of the III lies on the MLF; (b) trochlear cells do not infiltrate the surrounding MLF; oculomotor cells may be found scattered among and ventral to the fibers of the MLF.

The IV is composed of irregularly arranged plump, multipolar motor-type neurons similar to those of the III and abducens nucleus (fig. 1–4). Pearson [1943] has observed cells of the type found in the mesencephalic trigeminal nucleus [sensory ganglion-like neurons] along the course of trochlear nerve (NIV) fibers and within the IV.

Functional Neuroanatomy

Function

The IV gives rise to somatic efferent fibers which provide the motor innervation for the contralateral superior oblique (SO) muscle. The pulling direction of the SO, from the resting position, is downward with an outward rotational component. Lesions of either the IV, or NIV, cause a trochlear palsy, which is quite common. They result in a hypertropia of the affected eye, producing a diplopia which is largest when there is attempted gaze in the 'down and medial' quadrant of vision [for a comprehensive review, see Leigh and Zee, 2006]. The contribution of the SO motor units to convergence and counter-rolling during static tilt is not well understood [Mays et al., 1991; Sasaki et al., 1991].

Trochlear motoneurons, like all extraocular motoneuron groups, can be divided into two categories, those that innervate twitch motoneurons and those that innervate nontwitch motoneurons. The function of the nontwitch motoneurons in the dorsal cap of the IV is not clear, but they may participate more in the tonic properties of the SO, in contrast to the twitch motoneurons which are essential for movement of the eye [Büttner-Ennever, 2006].

Connections

Afferents. The IV motoneurons receive a strong input from the magnocellular regions of the vestibular nuclei via the MLF, which mediate the vertical component of vestibulo-ocular eye movements [Büttner-Ennever, 2006]. The excitatory pathway arises from the magnocellular part of the contralateral medial vestibular nucleus, the inhibitory pathway from the magnocellular part of the superior vestibular nucleus [Goldberg et al., 2012].

The premotor inputs to the IV for vertical and torsional saccadic eye movements arise from the rostral interstitial nucleus of the medial longitudinal fascicle (RIMLF) and the interstitial nucleus of Cajal (INC), whereby the latter is more involved in vertical gaze-holding [Fukushima et al., 1992]. The afferents from the RIMLF are excitatory and predominantly ipsilateral, those from the INC project bilaterally to the IV with the fibers crossing within the posterior commissure. The afferents from the ipsilateral INC are excitatory, and at least a portion of the contralateral afferents are GABAergic [Horn et al., 2003]. Interestingly, the motoneurons of all vertical-pulling eye muscles in the III and IV receive a strong GABAergic input from secondary

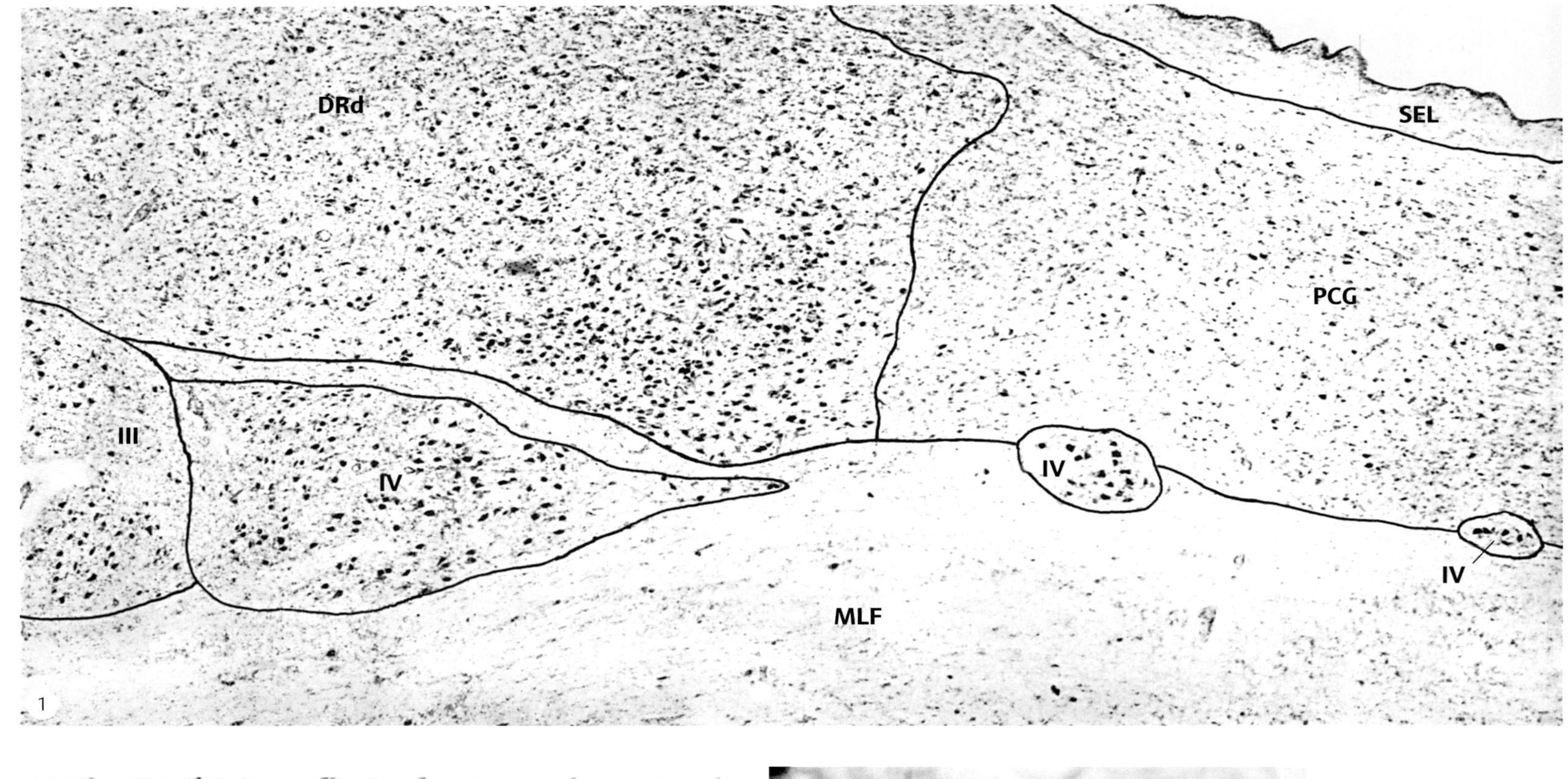

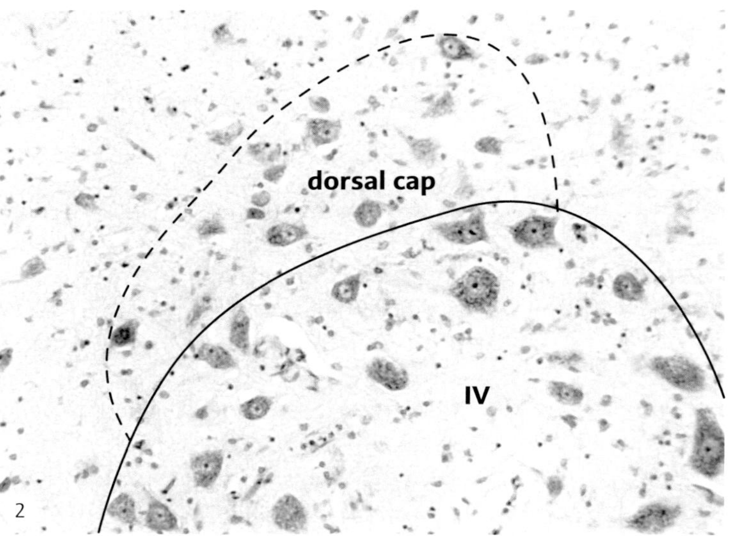

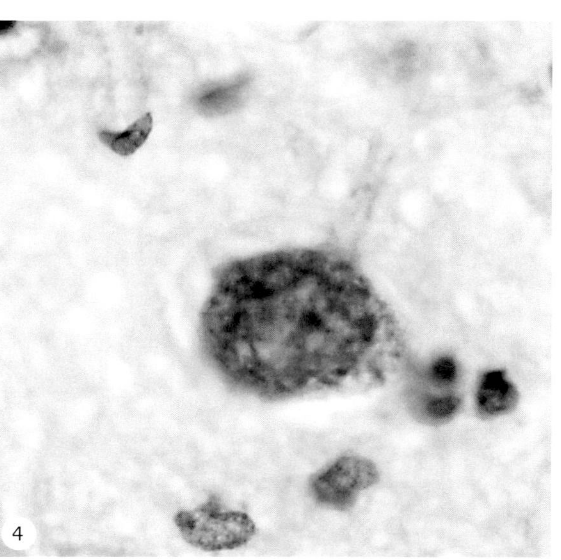

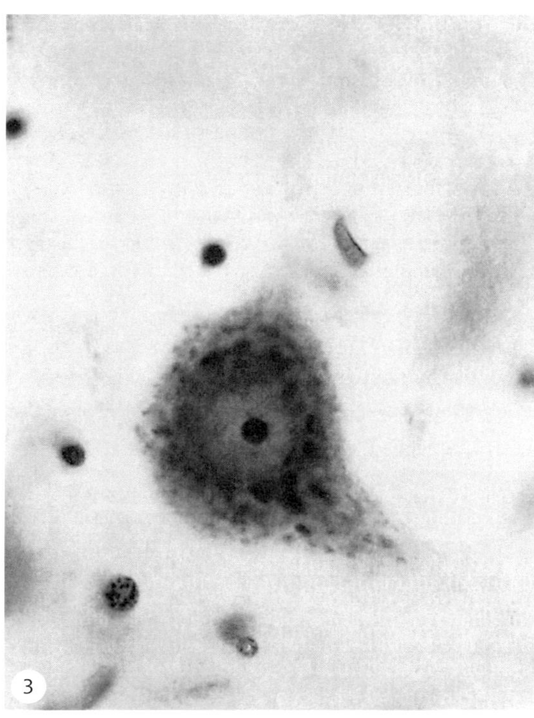

Fig. 1. Sagittal section through the region of the trochlear nucleus. Note that the trochlear nucleus is composed of three separate cell groups. III = Oculomotor nucleus; IV = trochlear nucleus; MLF = medial longitudinal fasciculus; PCG = pontine central gray; SEL = subependymal layer; DRd = dorsal raphe nucleus, dorsal part. Magnification ×25.

Fig. 2. Trochlear nucleus. Motoneurons of the main body of the trochlear nucleus lie ventrally: these innervate singly innervated muscle fibers (or twitch muscle fibers). Note the trochlear 'dorsal cap' containing motoneurons that innervate multiply innervated muscle fibers (or nontwitch muscle fibers). Magnification ×150.

Fig. 3. Singly innervated (twitch) motoneuron in the trochlear nucleus. Magnification ×1,000.

Fig. 4. Multiply innervated (nontwitch) motoneuron in the trochlear nucleus. Magnification ×1,000.

vestibulo-ocular neurons and the INC, whereas horizontal motoneurons receive glycinergic inhibition from the secondary vestibulo-ocular neurons [Spencer and Baker, 1992].

There are two categories of trochlear motoneurons: those that innervate twitch muscle fibers (or singly innervated muscle fibers), and those that innervate nontwitch muscle fibers (or multiply innervated muscle fibers; fig. 2–4). The two different types of motoneurons are known to receive different afferent inputs [Wasicky et al., 2004]. The two groups in humans can be distinguished on the basis of their histochemical properties: only the twitch motoneurons, which reside within the nucleus boundaries, contain nonphosphorylated neurofilaments and are ensheathed by perineuronal nets, but the nontwitch motoneurons in the dorsal cap lack both staining properties [Büttner-Ennever, 2006; Horn and Adamcyzk, 2011]. Both motoneuron groups are cholinergic and parvalbumin positive.

Additional Features

At about embryonic week 4–5, trochlear motoneurons in humans can be recognized within the neural tube in the second mesomere [Bruska et al., 1998; Straka et al., 1998].

References

Bruska M, Markowski M, Szyszka-Mroz J, Ulatowska-Blaszyk K, Wozniak W: Segmental pattern and nuclei in the human embryonic brain at stage 13. Folia Morphol (Warsz) 1998;57:321–330.

Büttner-Ennever JA: The extraocular motor nuclei: organization and functional neuroanatomy. Prog Brain Res 2006;151:95–125.

Fukushima K, Kaneko CR, Fuchs AF: The neuronal substrate of integration in the oculomotor system. Prog Neurobiol 1992;39:609–639.

Goldberg JM, Wilson VJ, Cullen KE, Angelaki DE, Broussard DM, Büttner-Ennever JA, Fukushima K, Minor LB: The Vestibular System – A Sixth Sense. Oxford, Oxford University Press, 2012.

Horn AKE, Adamcyzk C: Reticular formation: eye movements, gaze and blinks; in Mai JK, Paxinos G (eds): The Human Nervous System, ed 3. San Diego, Academic Press, 2012.

Horn AKE, Helmchen C, Wahle P: GABAergic neurons in the rostral mesencephalon of the macaque monkey that control vertical eye movements. Ann NY Acad Sci 2003;1004:19–28.

Leigh RJ, Zee DS: The Neurology of Eye Movements. New York, Oxford University Press, 2006.

Mays LE, Zhang Y, Thorstad MH, Gamlin PDR: Trochlear unit activity during ocular convergence. J Neurophysiol 1991;65:1484–1491.

Pearson AA: Trochlear nerve in human fetuses. J Comp Neurol 1943;78:29–43.

Sasaki M, Hiranuma K, Isu N, Uchino Y: Is there a three neuron arc in the cat utriculo-trochlear pathway? Exp Brain Res 1991;86:421–425.

Spencer RF, Baker R: GABA and glycine as inhibitory neurotransmitters in the vestibulo-ocular reflex. Ann NY Acad Sci 1992;656:602–611.

Straka H, Gilland E, Baker R: Rhombomeric organization of brainstem motor neurons in larval frogs. Biol Bull 1998;195:220–222.

Wasicky R, Horn AKE, Büttner-Ennever JA: Twitch and non-twitch motoneuron subgroups of the medial rectus muscle in the oculomotor nucleus of monkeys receive different afferent projections. J Comp Neurol 2004;479:117–129.

Motor Systems – Oculomotor Complex

(Plates 34, 36, 38 and 39)

46 Oculomotor Nucleus (III)

Original name:
Nucleus oculomotorius principalis

Alternative name:
Third nerve nucleus

Location and Cytoarchitecture – Original Text

This nucleus constitutes the major component of the oculomotor complex. It is located in the tegmentum of the midbrain, ventral to the gray matter and the cerebral aqueduct (PAG) at the level of the superior colliculus (fig. 1). It extends from the oral pole of the trochlear nucleus (IV) to the unpaired portion of the nucleus of Edinger-Westphal (EW), a distance of 5 mm.

On cross-section, the caudal pole of the nucleus is round or oval in outline and lies immediately dorsal and dorsomedial to the medial longitudinal fasciculus (MLF), ventral to the oral part of the dorsal raphe nucleus. Medially, the caudal central nucleus of Tsuchida (CCN) separates this portion of the oculomotor nucleus (III) from its opposite fellow (fig. 2).

The midportion of the III is triangular on cross-section with the base directed dorsally and the apex ventrally. It is separated from the corresponding nucleus of the opposite side by the nucleus of Perlia and is related dorsally and dorsomedially to components of the EW and PAG, laterally and ventrolaterally to the MLF (fig. 1, 3–5).

Orally, the nucleus diminishes gradually in size, becomes oblong rather than triangular in outline and approaches its opposite fellow in the midline. Components of the unpaired anterior portion of the EW lie dorsal, oral and ventral to the oral portion of the III (fig. 1, 5, 6).

Oculomotor cells infiltrate and lie irregularly arranged among the fibers of the MLF at all levels but particularly in levels through the caudal third of the nucleus (fig. 2). Such cells tend, on the whole, to be smaller than those found elsewhere in the nucleus.

A small dorsal and a larger ventral subdivision of the III may be distinguished at all levels with the exception of the caudal and oral poles where only the ventral subdivision persists.

The nucleus is composed of plump, multipolar motor type neurons similar to those found in the IV and abducens nucleus (VI). These cells are irregularly oriented, and are less compactly arranged in the caudal third than in the remainder of the nucleus (fig. 8–12). Very occasional cells resembling those of the mesencephalic trigeminal nucleus may also be present. [...] It is possible that the cells resembling those of the mesencephalic trigeminal nucleus receive afferent proprioceptive fibers from the eye muscles. [...]

Functional Neuroanatomy

Function

The III contains discrete subgroups of motoneurons supplying 4 extraocular eye muscles, e.g. the ipsilateral medial rectus (MR), inferior rectus (IR) and inferior oblique (IO), and the contralateral superior rectus (SR), see figure 7. Their axons travel

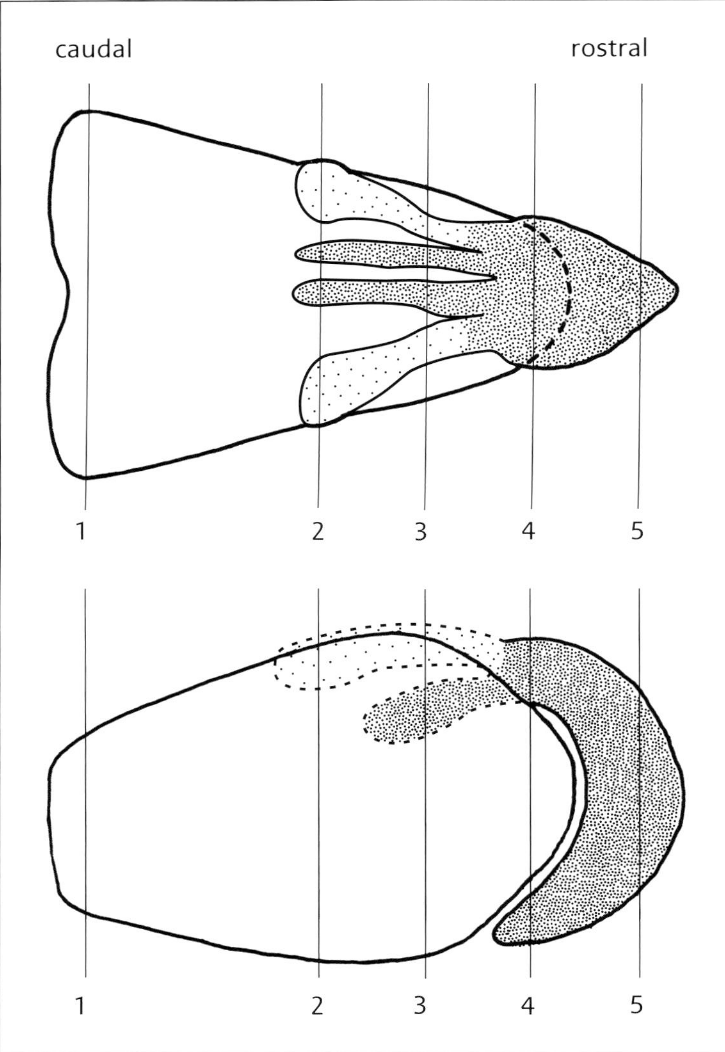

Fig. 1. Schematic drawings of a dorsal and a sagittal view of the oculomotor nucleus showing its relationship to the cell groups composing the Edinger-Westphal nucleus. The dotted area shows the extent of the cortically projecting part of the Edinger-Westphal nucleus (EWcp, see nucleus No. 49), with the sparsely dotted region highlighting its lateral subdivision seen in figures 2 and 3. The vertical lines represent the levels of cross-sections portrayed in figures 2–6.

to the orbit in the oculomotor nerve (NIII). The CCN is often included in the term 'III', because it lies at the caudal pole of the III, and innervates the levator muscle of the eyelid via the NIII, too; but the CCN is treated here separately (see nucleus No. 47). The NIII also contains parasympathetic fibers supplying the pupillary sphincter muscle and the ciliary muscle for lens accommodation, which originates in the EW. Thus, lesions of the NIII are accompanied by a 'down and out' resting position of the affected eye, double vision, a dilated pupil and ptosis. There are several comprehensive reviews on this topic [Büttner-Ennever, 2006;Leigh and Zee, 2006]. The function of different types of motoneurons within the III is discussed below in 'Organization of Motoneuron Subgroups'.

Connections

Efferents. Apart from the motoneuron efferents in the NIII, there is a significant population of internuclear neurons within the III, which project to the contralateral VI [Clendaniel and Mays, 1994; Ugolini et al., 2006]. These 'oculomotor internuclear neurons' appear to carry an excitatory signal to the contralateral lateral rectus motoneurons appropriate for horizontal conjugate gaze during saccades. The axons descend in the MLF, and cross the midline at the level of the VI. Their loss may contribute to the nystagmus seen on abduction in internuclear ophthalmoplegia, a useful diagnostic sign [Leigh and Zee, 2006].

Afferents. The main afferents to oculomotor neurons arise from the VI (nucleus No. 44), the vestibular nuclei (nuclei No. 13, 14, 16 and 18), premotor saccadic neurons of the rostral interstitial nucleus of the MLF (RIMLF; nucleus No. 61) and the interstitial nucleus of Cajal (INC; nucleus No. 20) [Büttner-Ennever, 2006]. The MR subgroups in the III receive afferents from the internuclear neurons of the contralateral VI via the MLF [Büttner-Ennever and Akert, 1981]. The pathway coordinates the activity of MR and LR motoneurons in horizontal conjugate gaze, and damage to the MLF causes a paresis of the MR for all eye movement types except convergence, where the activity of the MR remains intact. The premotor inputs for vergence arise in part from neurons in the supraoculomotor area above the III and the laterally adjoining mesencephalic reticular formation [Zhang et al., 1991; Graf et al., 2002]. An additional input to the MR subgroups arises from the ipsilateral ascending tract of Deiters, a vestibular pathway presumably involved in the control of vergence eye movements during translational head movements [Chen-Huang and McCrea, 1998].

The vertical components of vestibulo-ocular eye movements ascend from the magnocellular regions of the vestibular nuclei via the MLF, and from the Y group through crossing ventral tegmental tracts, including the brachium conjunctivum [for reviews, see Büttner-Ennever, 2006; Zwergal et al., 2009; Ahlfeld et al., 2011]. Tract-tracing, physiological and clinical studies bear out the rule that the afferent inputs to extraocular motoneurons always innervate conjugate 'yoke muscle pairs', one from each eye [Moschovakis et al., 1990].

The premotor inputs to the III for vertical and torsional saccadic eye movements arise from the RIMLF. Additional selective inputs to the motoneurons of upward moving eye muscles, e.g. SR and IO, originate in the adjacent M group (see nucleus No. 61) that targets in addition the motoneurons of the forehead frontalis muscle and the levator palpebrae muscle in the upper eyelids. The afferents from the INC have a function closely related to the RIMLF, and serve for vertical gaze-holding. The INC signals are integrated through neuronal networks involving the vestibular nuclei and cerebellar flocculus, in order to produce tonic signals for the muscles that maintain a stable eye and head position, during vertical gaze-holding [Fukushima et al., 1992]. Some of the fibers from the INC cross to the contralateral III in the posterior commissure [Horn et al., 2003].

Histochemistry

The excitatory vestibular afferents to motoneurons in the III always cross the midline at the level of the vestibular nuclei and ascend in the MLF; they probably use glutamate as their transmitter [Raymond et al., 1988]. The excitatory input to MR motoneurons from the internuclear neurons of the contralateral VI is mediated by glutamate and aspartate, whereas the afferents from the ascending tract of Deiters use only glutamate as a transmitter [Nguyen and Spencer, 1999].

The inhibitory secondary vestibulo-oculomotor afferents to the III always ascend ipsilaterally in the MLF. They were

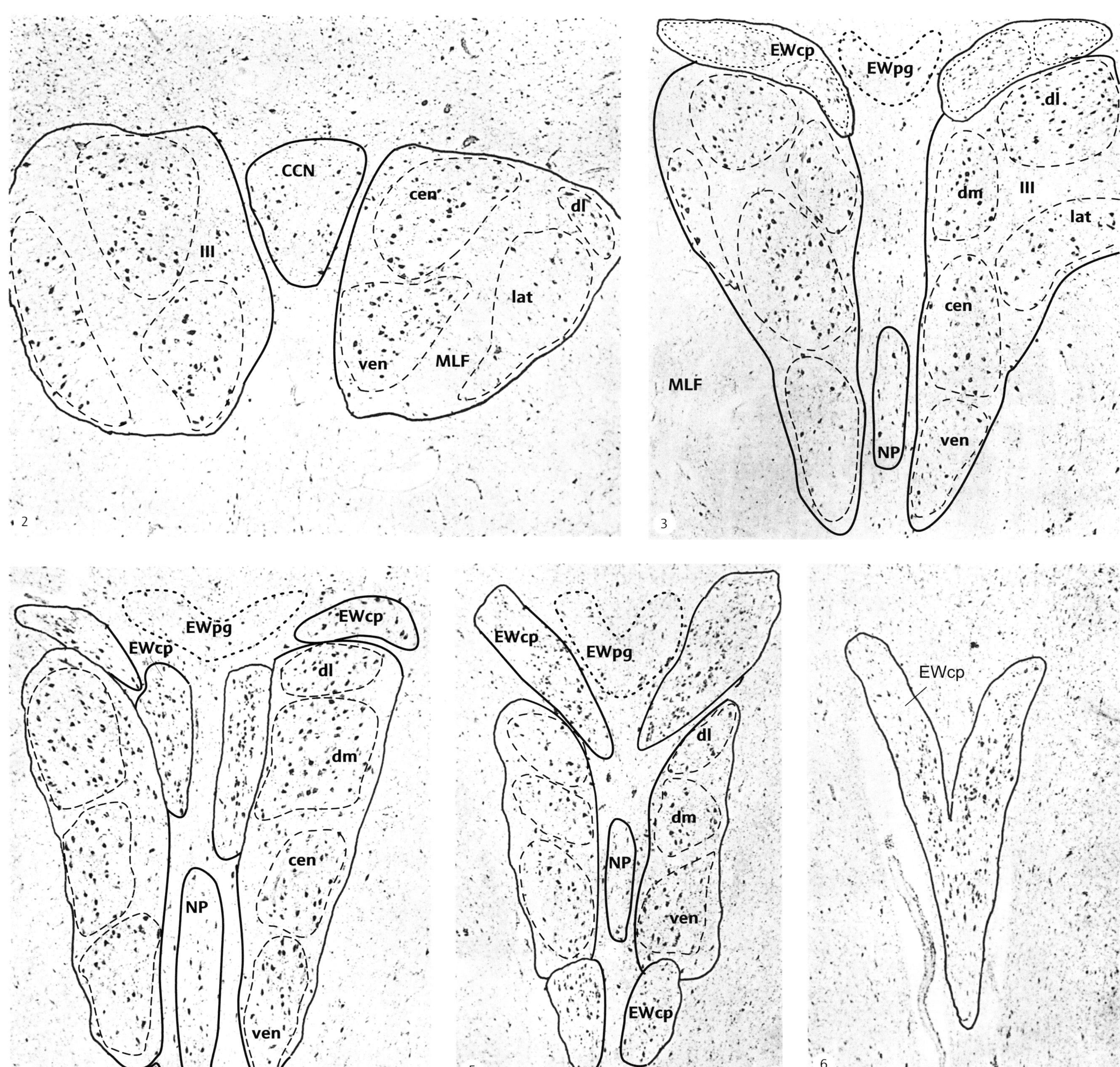

Fig. 2–6. Cross-sections through different levels of the oculomotor complex. The approximate level of each figure may be determined by reference to figure 1. CCN = Central caudal nucleus; cen = central subnucleus of III; dl = dorsolateral subnucleus of III; dm = dorsomedial subnucleus of III; EWcp = Edinger-Westphal nucleus, centrally projecting (urocortin) neurons; EWpg = Edinger-Westphal nucleus, preganglionic neurons; III = oculomotor nucleus; lat = lateral subnucleus of III; MLF = medial longitudinal fasciculus; NP = nucleus of Perlia; ven = ventral subnucleus of III. Magnification ×30.

shown to utilize different inhibitory transmitters in pathways mediating horizontal and vertical eye movements. The motoneurons of vertically pulling eye muscles in both III and IV receive a strong GABAergic, but a rather weak glycinergic input; conversely, the opposite pattern was found for the horizontal motoneuronal subgroups of MR and LR muscles in the VI [de la Cruz et al., 1992; Spencer and Baker, 1992]. The results indicate that inhibition in horizontal eye movement pathways is provided by glycine, while those for vertical eye movement pathways utilize GABA.

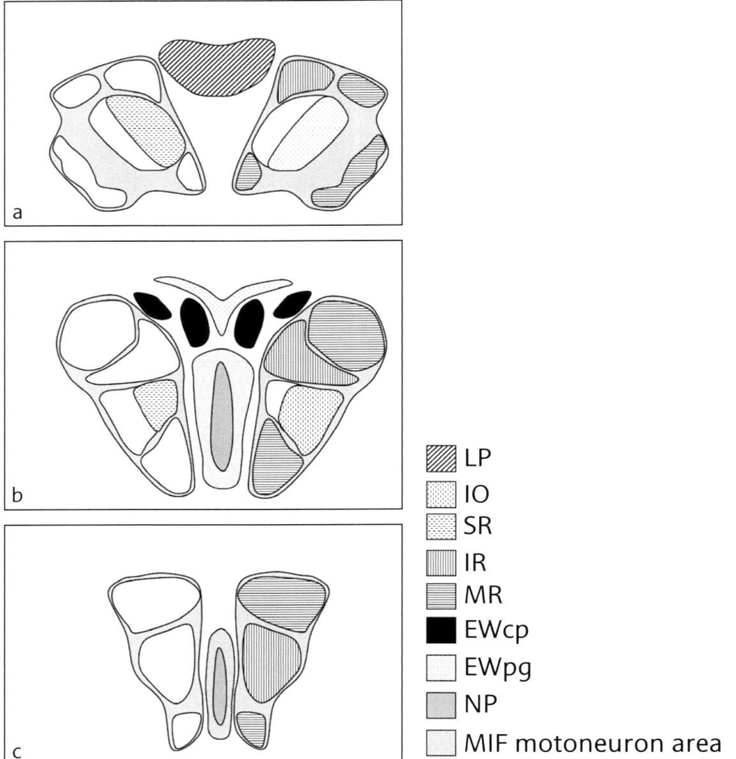

Fig. 7. a–c Schematic drawings of sections through the oculomotor nucleus similar to the photomicrographs shown in figures 2–6, and identifying some of the functional subgroups within the oculomotor complex: these are the singly innervated muscle fiber motoneuron subgroups (LP, IO, SR, IR and MR; see text), and the multiply innervated muscle fiber (MIF) motoneuron area, all innervating extraocular muscles, also the nucleus of Perlia (NP) and the Edinger-Westphal cell groups – preganglionic (EWpg) and the urocortin centrally projecting groups (EWcp). LP = Levator palpebrae muscle; IO = inferior oblique; SR = superior rectus; IR = inferior rectus; MR = medial rectus.

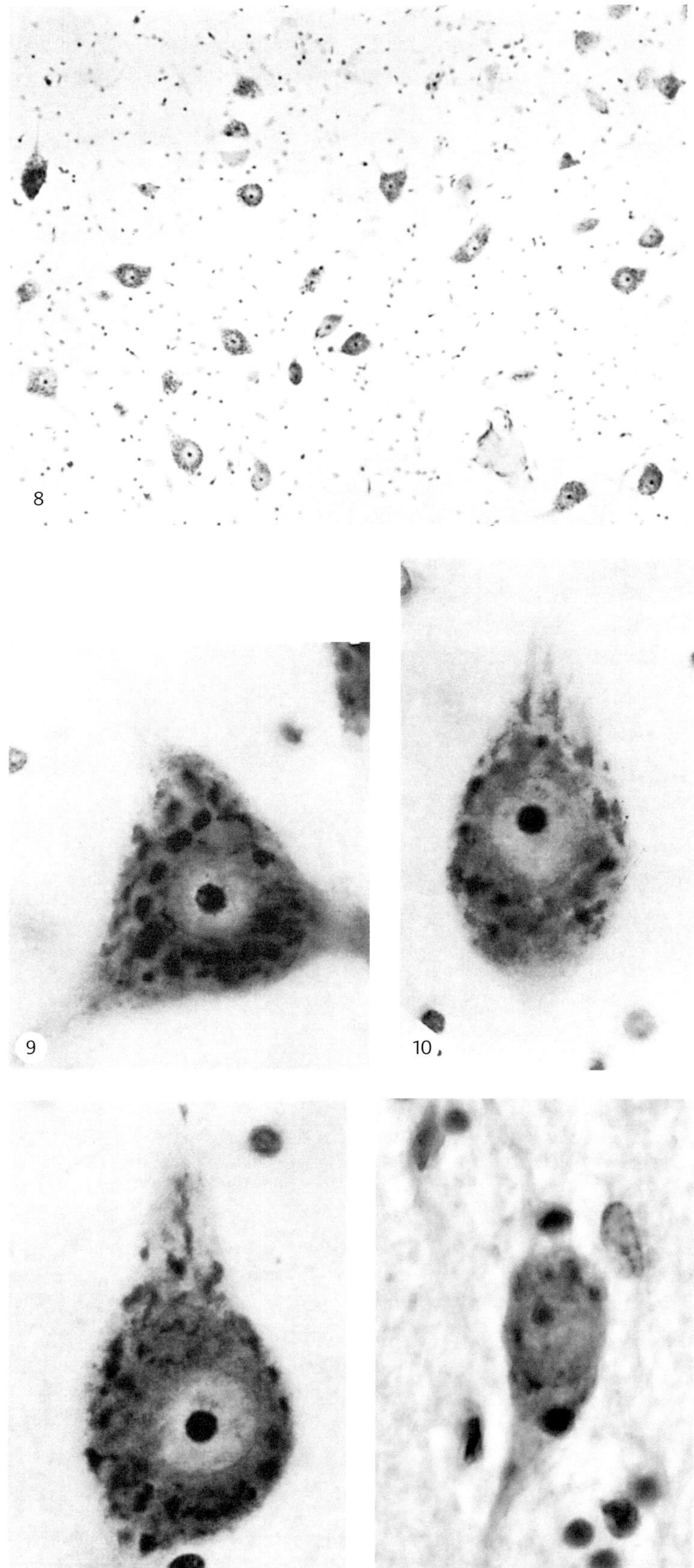

Fig. 8. Oculomotor nucleus. Magnification ×150.
Fig. 9–11. Singly innervated muscle fiber motoneurons in the oculomotor nucleus. Magnification ×1,000.
Fig. 12. Multiply innervated muscle fiber motoneuron in the oculomotor nucleus. Magnification ×1,000.

Furthermore pathways from the Y group to upward moving muscles were found to be calretinin positive [Ahlfeld et al., 2011]. These histochemical differences of afferents to individual III subgroups along with comparative studies in nonhuman primates have enabled the delineation of individual subgroups in the human III after many years of conflicting hypotheses, see figure 7 [Horn et al., in preparation]. The previous hypotheses of III topography have been fully reviewed by Warwick [1953] and Olszewski and Baxter [1982].

Organization of Motoneuron Subgroups

The motoneuron subgroups within the III form a topographic map, which has been well studied in many species [Büttner-Ennever, 2006]. There is a prominent change in the arrangement of MR motoneurons in primates, where 3 separate clusters have been identified: an A, B and C group [Büttner-Ennever and Akert, 1981; Büttner-Ennever et al., 2001]. There are cytological differences between the A and B groups, but the functional significance of this is unknown. In accordance with the abundance of muscle fiber types in extraocular muscles, two basic types of motoneurons can be distinguished, which differ in their afferent input [Wasicky et al., 2004] and histochemical properties [Eberhorn et al., 2005]. One group is formed by the majority of neurons within the III that supply singly innervated muscle fibers (SIF), also called twitch muscle fibers. The other group represents motoneurons that activate multiply innervated muscle fibers (MIF), also called nontwitch muscle fibers, which are unique to extraocular muscles. In general they are located around the periphery of the main nucleus, separate from the classical motoneurons, and they tend to be smaller (fig. 12). In nonhuman primates, the MIF motoneurons of MR and IR muscles are located in the 'C group' dorsomedial to the III, those of the SR and IO muscles lie in the 'S group' between the III [Büttner-Ennever, 2006]. A compact accumulation of MIF motoneurons in a 'C and a S group' is not obvious in humans, but the MIF motoneurons have been identified by their histochemical characteristics [Horn et al., 2008]. Current evidence supports the hypothesis that the MIF motoneurons may participate in determining the resting tension in the extraocular muscle, as in gaze-holding, vergence and eye alignment [Büttner-Ennever, 2006]. In contrast, the SIF motoneurons in the classical nuclei may provide the drive to move the eyes.

Interestingly putative-sensory endings at the myotendinous junctions of eye muscles, called palisade endings, also have their cell bodies around the III in the area of the 'C group' [Lienbacher et al., 2011]. Such cell bodies would be expected to have the morphology of sensory ganglia, like the mesencephalic trigeminal nucleus. Olszewski and Baxter [1982] described such cells in the original text of this book (see above), commenting that they are a rare but a recognized feature of III cytoarchitecture.

References

Ahlfeld J, Mustari MJ, Horn AKE: Sources of calretinin inputs to motoneurons of extraocular muscles involved in upgaze. Ann NY Acad Sci 2011;1233:91–99.

Büttner-Ennever JA: The extraocular motor nuclei: organization and functional neuroanatomy. Prog Brain Res 2006;151:95–125.

Büttner-Ennever JA, Akert K: Medial rectus subgroups of the oculomotor nucleus and their abducens internuclear input in the monkey. J Comp Neurol 1981;197: 17–27.

Büttner-Ennever JA, Horn AKE, Scherberger H, d'Ascanio P: Motoneurons of twitch and nontwitch extraocular muscle fibers in the abducens, trochlear, and oculomotor nuclei of monkeys. J Comp Neurol 2001;438:318–335.

Chen-Huang C, McCrea RA: Viewing distance related sensory processing in the ascending tract of deiters vestibulo-ocular reflex pathway. J Vestib Res 1998; 8:175–184.

Clendaniel RA, Mays LE: Characteristics of antidromically identified oculomotor internuclear neurons during vergence and versional eye movements. J Neurophysiol 1994;71:1111–1127.

de la Cruz RR, Pastor AM, Martinez-Guijarro FJ, Lopez-Garcia C, Delgado-Garcia JM: Role of GABA in the extraocular motor nuclei of the cat: a postembedding immunocytochemical study. Neuroscience 1992;51:911–929.

Eberhorn AC, Ardelenanu P, Büttner-Ennever JA, Horn AKE: Histochemical differences between motoneurons supplying multiply and singly innervated extraocular muscle fibers. J Comp Neurol 2005;491:352–366.

Fukushima K, Kaneko CR, Fuchs AF: The neuronal substrate of integration in the oculomotor system. Prog Neurobiol 1992;39:609–639.

Graf W, Gerrits N, Yatim-Dhiba N, Ugolini G: Mapping the oculomotor system: the power of transneuronal labelling with rabies virus. Eur J Neurosci 2002;15: 1557–1562.

Horn AK, Eberhorn A, Härtig W, Ardelenanu P, Messoudi A, Büttner-Ennever JA: Perioculomotor cell groups in monkey and man defined by their histochemical and functional properties: reappraisal of the Edinger-Westphal nucleus. J Comp Neurol 2008;507:1317–1335.

Horn AKE, Che Ngwa E: Identification of motoneurons innervating individual extraocular muscles within the oculomotor nucleus in human. In preparation.

Horn AKE, Helmchen C, Wahle P: GABAergic neurons in the rostral mesencephalon of the macaque monkey that control vertical eye movements. Ann NY Acad Sci 2003;1004:19–28.

Leigh RJ, Zee DS: The Neurology of Eye Movements. New York, Oxford University Press, 2006.

Lienbacher K, Mustari M, Hess B, Büttner-Ennever J, Horn AK: Is there any sense in the palisade endings of eye muscles? Ann NY Acad Sci 2011;1233:1–7.

Moschovakis AK, Scudder CA, Highstein SM: A structural basis for Hering's law: projections to extraocular motoneurons. Science 1990;248:1118–1119.

Nguyen LT, Spencer RF: Abducens internuclear and ascending tract of Deiters inputs to medial rectus motoneurons in the cat oculomotor nucleus: neurotransmitters. J Comp Neurol 1999;411:73–86.

Olszewski J, Baxter D: Cytoarchitecture of the Human Brain Stem. Basel, Karger, 1982.

Raymond J, Dememes D, Nieoullon A: Neurotransmitters in vestibular pathways. Prog Brain Res 1988;76:29–43.

Spencer RF, Baker R: GABA and glycine as inhibitory neurotransmitters in the vestibuloocular reflex. Ann NY Acad Sci 1992;656:602–611.

Ugolini G, Klam F, Doldan Dans M, Dubayle D, Brandi A-M, Büttner-Ennever JA, Graf W: Horizontal eye movement networks in primates as revealed by retrograde transneuronal transfer of rabies virus: differences in monosynaptic input to 'slow' and 'fast' abducens motoneurons. J Comp Neurol 2006;498:762–785.

Warwick R: Representation of the extraocular muscles in the oculomotor nuclei of the monkey. J Comp Neurol 1953;98:449–495.

Wasicky R, Horn AKE, Büttner-Ennever JA: Twitch and non-twitch motoneuron subgroups of the medial rectus muscle in the oculomotor nucleus of monkeys receive different afferent projections. J Comp Neurol 2004;479:117–129.

Zhang Y, Gamlin PDR, Mays LE: Antidromic identification of midbrain near response cells projecting to the oculomotor nucleus. Exp Brain Res 1991;84:525–528.

Zwergal A, Strupp M, Brandt T, Büttner-Ennever J: Parallel ascending vestibular pathways: anatomical localization and functional specialization. Ann NY Acad Sci 2009;1164:51–59.

Motor Systems – Oculomotor Complex (Plate 34)

47 Central Caudal Nucleus (CCN)

Original name:

Nucleus oculomotorius caudalis centralis (of Tsuchida)

Location and Cytoarchitecture – Original Text

This nucleus is composed of a small group of cells lying in the midline of the mesencephalon, directly caudal to the nucleus of Perlia (NP) and intervening between the dorsal parts of the caudal thirds of the principal oculomotor nuclei (III) (see nucleus No. 46, fig. 2). On cross-section the nucleus is wedge shaped with its narrow portion directed ventrally.

The cells composing the nucleus caudalis centralis (CCN) are similar to, but smaller than, those of the principal oculomotor nuclei (fig. 1, 2). The elongated neurons characteristic of the NP are not present. [...]

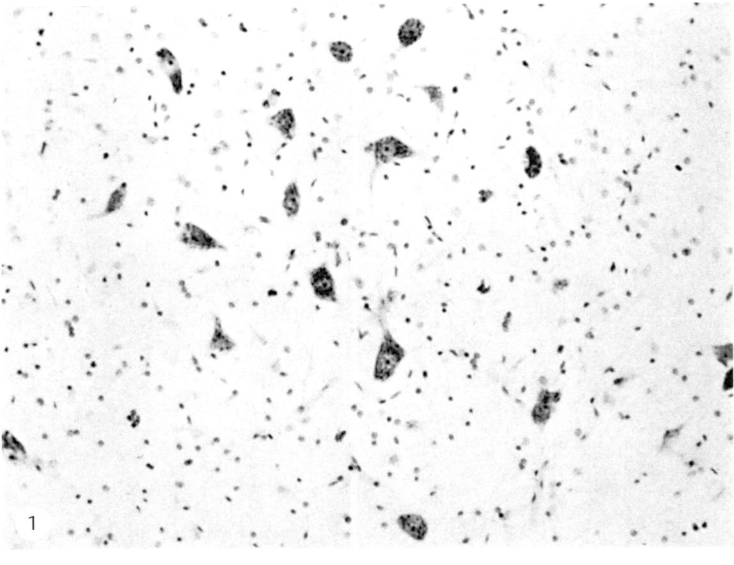

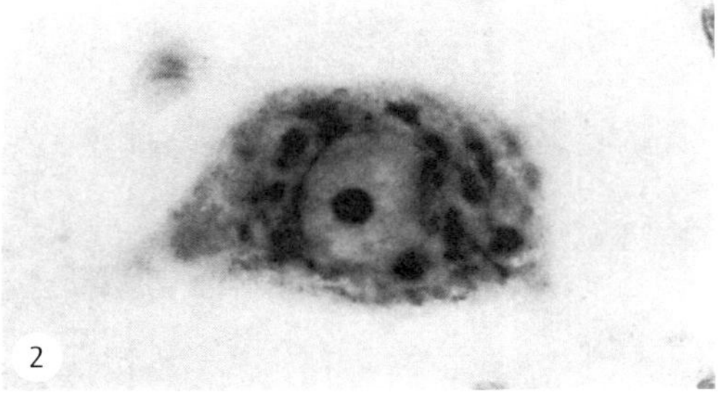

Fig. 1. Central caudal nucleus. Magnification ×150.
Fig. 2. Cell from the central caudal nucleus. Magnification ×1,000.

Functional Neuroanatomy

Function

The CCN contains the motoneurons innervating the levator palpebrae superioris muscle (LP), which raises the eyelid. In the awake state it is essential for LP motoneurons to maintain a high level of tonic excitatory activity to keep the eyelid open: this activity may be supplied through orexin-A afferents to the CCN [Schreyer et al., 2009]. The orexin afferents arise from the lateral hypothalamic area, a region which stabilizes the sharp transitions between sleep and wakefulness [Saper et al., 2001].

During reflex blinks triggered by activation of trigeminal afferents, LP motoneurons are inhibited and allow the orbicularis oculi muscle, controlled by the facial nucleus, to close the eye [May et al., 2012]. Lesions of the CCN usually cause bilateral ptosis (i.e. hanging eyelids), and damage to the oculomotor nerve causes ipsilateral ptosis [Averbuch-Heller, 1997; Rucker, 2011].

Connections

Efferents. In lateral-eyed species, such as the rabbit or rat, the LP motoneuron subgroups lie laterally in the oculomotor nucleus, and contralateral to the muscle innervated. In primates, the motoneurons of both eyelids appear intermixed in the unpaired CCN, but with a predominance contralaterally [Sun and May, 1993; Porter et al., 1989]. The axons of LP motoneurons cross through the CCN to join the oculomotor rootlets of the contralateral oculomotor nerve. The dendrites of LP motoneurons extend bilaterally into the supraoculomotor area (SOA) above the CCN [May et al., 2012]. It is unclear whether the motoneurons innervate only one eyelid or possibly both. For a general review of the CCN, see Horn and Adamczyk [2012].

Afferents. Several sources of afferents to the CCN motoneurons have been identified: First, excitatory afferents originate in the M group in the central gray immediately adjacent to the rostral interstitial nucleus of the medial longitudinal fascicle (nucleus No. 61) and the interstitial nucleus of Cajal (nucleus No. 20), which serve to coordinate the eyelid with vertical eye movements [Horn et al., 2000; Chen and May, 2007]. Second, inhibitory, presumably GABAergic inputs to the CCN and the SOA arise from a region at the rostral ventral border of (but outside) the principal sensory trigeminal nucleus [May et al., 2012]. This region was also shown to receive supraorbital trigeminal afferents, thereby providing an anatomical basis for the trigeminal blink reflex. It has been called the 'pontine blink premotor area' by Holstege and colleagues [for a review, see Horn and Adamczyk, 2012]. Another inhibitory input to the CCN via glycinergic afferents presumably originates from saccadic omnipause neurons in the nucleus raphe interpositus of the nucleus reticularis pontis caudalis (nucleus No. 56), and it may contribute to pathways involved in blink-saccade interaction [Zee et al., 1983; Horn and Büttner-Ennever, 2008]. Finally neurons of the lateral hypothalamus area, which synthesize orexin-A and help to maintain wakefulness through excitatory projections to nuclei involved in arousal, also project to the CCN, but not the other classical motoneuron subgroups of the III. This input may contribute to the tonic activation of the LP motoneurons during wakefulness, and its inactivation during sleepiness may result in passive lid lowering by the lack of LP motoneuron activity [Schreyer et al., 2009].

References

Averbuch-Heller L: Neurology of the eyelids. Curr Opin Ophthalmol 1997;8:27–34.

Chen B, May PJ: Premotor circuits controlling eyelid movements in conjunction with vertical saccades in the cat. II. Interstitial nucleus of Cajal. J Comp Neurol 2007;500:676–692.

Horn AK, Adamczyk C: Reticular formation: eye movements, gaze and blinks; in Mai JK, Paxinos G (eds): The Human Nervous System, ed 3. San Diego, Academic Press, 2012.

Horn AK, Büttner-Ennever JA: Brainstem circuits controlling lid-eye coordination in monkey. Prog Brain Res 2008;171:87–95.

Horn AKE, Büttner-Ennever JA, Gayde M, Messoudi A: Neuroanatomical identification of mesencephalic premotor neurons coordinating eyelid with upgaze in the monkey and man. J Comp Neurol 2000;420:19–34.

May PJ, Vidal PP, Baker H, Baker R: Physiological and anatomical evidence for an inhibitory trigemino-oculomotor pathway in the cat. J Comp Neurol 2012;520:2218–2240.

Porter JD, Burns LA, May PJ: Morphological substrate for eyelid movements: innervation and structure of primate levator palpebrae superioris and orbicularis oculi muscles. J Comp Neurol 1989;287:64–81.

Rucker JC: Normal and abnormal lid function; in Kennard C, Leigh JR (eds): Neuro-Ophthalmology, ed 3. Edinburgh, Elsevier, 2011.

Saper CB, Chou TC, Scammell TE: The sleep switch: hypothalamic control of sleep and wakefulness. Trends Neurosci 2001;24:726–731.

Schreyer S, Büttner-Ennever JA, Tang X, Mustari MJ, Horn AKE: Orexin-A inputs onto visuomotor cell groups in the monkey brainstem. Neuroscience 2009;164:629–640.

Sun WS, May PJ: Organization of the extraocular and preganglionic motoneurons supplying the orbit in the lesser galago. Anat Rec 1993;237:89–103.

Zee DS, Chu FC, Leigh RJ, Savino PJ, Schatz NJ, Reingold DB, Cogan DG: Blink-saccade synkinesis. Neurology 1983;33:1233–1236.

48 Nucleus of Perlia (NP)

Original name:
Nucleus of Perlia

Alternative names:
Interoculomotor nucleus [Paxinos and Huang, 1995]
Perioculomotor group of Perlia [Horn et al., 2008]

Location and Cytoarchitecture – Original Text

This is an unpaired, centrally situated cell group which lies between the middle thirds of the oculomotor nuclei (III). On cross-section the nucleus is elongated and ellipsoid in outline (see nucleus No. 46, fig. 3–5).

The cells composing the nucleus of Perlia (NP) are loosely arranged, and in general possess characteristics similar to those of the III. [...] Multipolar cells similar to the components of the III are present (fig. 1, 2). The majority of other cells are smaller medium-sized, spindle-shaped neurons which lie with their long axes directed dorsoventrally and contain darkly stained elongated Nissl granules (fig. 1, 2)[1].

According to Tsuchida [1906], this nucleus is present in 80% of humans and its degree of development varies from individual to individual. The nucleus is usually considered to be concerned with the control of convergence, although as Adler [1933] points out, the only evidence which suggests this function is 'the time of its appearance in both the species and the embryo which coincides with positioning of the eyes in the frontal plane where convergence becomes possible'.

Functional Neuroanatomy

Function

There is no current evidence for assuming that the NP is involved in convergence [Warwick, 1955]. The function of the NP is not known, but on the basis of morphological and histochemical properties the NP neurons appear to be twitch motoneurons and may represent superior rectus motoneurons that are separated from the main subgroup by dorsoventrally traveling nerve fibers [Horn et al., 2008].

Histochemistry

The cells of the NP share many histochemical properties with those of the twitch-type motoneurons within the III: they are cholinergic, they stain for nonphosphorylated neurofilaments, show a strong cytochrome oxidase content, and they are ensheathed by prominent perineuronal nets [Horn et al., 2008]. The latter 3 properties are not present in nontwitch motoneurons which are located between the NP and III as vertically oriented elongated spindle-shaped neurons (fig. 1) [Büttner-Ennever, 2006; Horn et al., 2008].

[1] The spindle-shaped neurons mentioned here are nontwitch multiply innervated motoneurons innervating the multiply innervated muscle fibers in extraocular muscles and therefore not part of the NP (fig. 1 and nucleus No. 46).

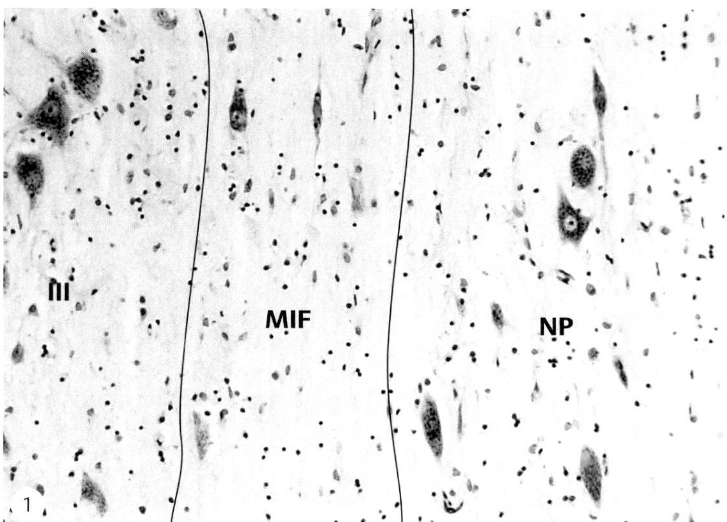

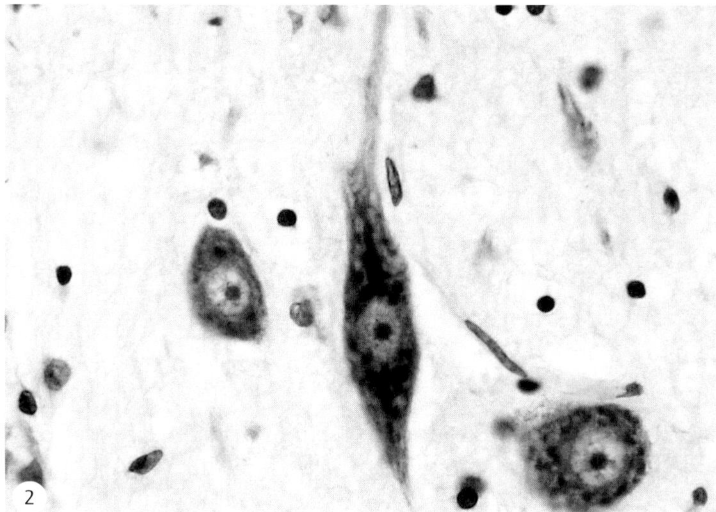

Fig. 1. Section showing both the oculomotor nucleus (III), the nucleus of Perlia (NP) and the zone between them that contains the multiply innervated muscle fiber (MIF) motoneurons of the S group. Note that III and NP motoneurons display similar features. These are different from the typical MIF motoneurons (see nucleus No. 46, fig. 12). Nissl stain. Magnification ×150.
Fig. 2. Cells from the nucleus of Perlia. Magnification ×1,000.

References

Adler A: Zur Lokalisation des Konvergenzzentrums und der Kerne der glatten Augenmuskeln. Z Ges Neurol Psychiatrie 1933;145:186–207.
Büttner-Ennever JA: The extraocular motor nuclei: organization and functional neuroanatomy. Prog Brain Res 2006;151:95–125.
Horn AK, Eberhorn A, Härtig W, Ardelenanu P, Messoudi A, Büttner-Ennever JA: Perioculomotor cell groups in monkey and man defined by their histochemical and functional properties: reappraisal of the Edinger-Westphal nucleus. J Comp Neurol 2008;507:1317–1335.
Paxinos G, Huang X-F: Atlas of the Human Brainstem. San Diego, Academic Press, 1995.
Tsuchida U: Ueber die Ursprungskerne der Augenbewegungs-Nerven im Mittel- und Zwischenhirn. Arb Hirnanat Inst Zurich 1906;1–2:1–205.
Warwick R: The so-called nucleus of convergence. Brain 1955;78:92–114.

49 Edinger-Westphal Nucleus (EW)

Original name:
Nucleus Edinger-Westphal

Alternative name:
Accessory oculomotor nucleus

Location and Cytoarchitecture – Original Text

The nucleus of Edinger-Westphal (EW) is an elongated slender structure which overlies the main oculomotor nuclei orodorsally. It lies ventral to the central gray matter of the cerebral aqueduct and extends from the level of the caudal pole of the red nucleus to the region of the interstitial nucleus of Cajal, a distance of 5 mm.

For purposes of description the EW may be subdivided into 3 portions: (a) an unpaired oral portion [the anteromedian nucleus], (b) a paired dorsomedial portion and (c) a paired caudolateral portion. [...]

An understanding of the complicated morphology of this nucleus is best achieved by supplementing the study of cross-sections with that of sagittal and schematic drawing sections (see nucleus No. 46, fig. 1–7, and the section below on 'Function').

The EW is composed of densely arranged, small and medium-sized, fusiform, oval or triangular cells whose long axes in general lie parallel to the long axis of the nucleus on cross-section. These cells have a prominent nucleus and the Nissl substance is arranged in clumps near the periphery of the cell. They resemble the cells of the dorsal motor nucleus of the vagus, with the exception that their edges are not as ragged. [A medial and lateral subnucleus can be distinguished in the EW (fig. 1); the medial subnucleus possesses round cells (fig. 2, 3), in contrast to the multipolar and fusiform cells of the lateral group (fig. 4, 5).]

The EW nuclei are believed to be the site of origin of impulses which bring about constriction of the pupils and relaxation of the ocular lenses [...] (see 'Function' below).

The evidence upon which the above 'statements' are based is surprisingly scanty. Stimulation of the EW results in constriction of the pupils [Benjamin, 1939; Szentágothai, 1942], and removal of the intraorbital contents or intracranial section of the oculomotor nerve results in chromatolysis of some cells in both the ipsilateral and the contralateral EW [Crouch, 1936]. However, axons of the EW cells have never been traced directly into the oculomotor nerve, and these cells show surprisingly little evidence of chromatolysis after extirpation of the ciliary ganglion [Crouch, 1936]. [...]

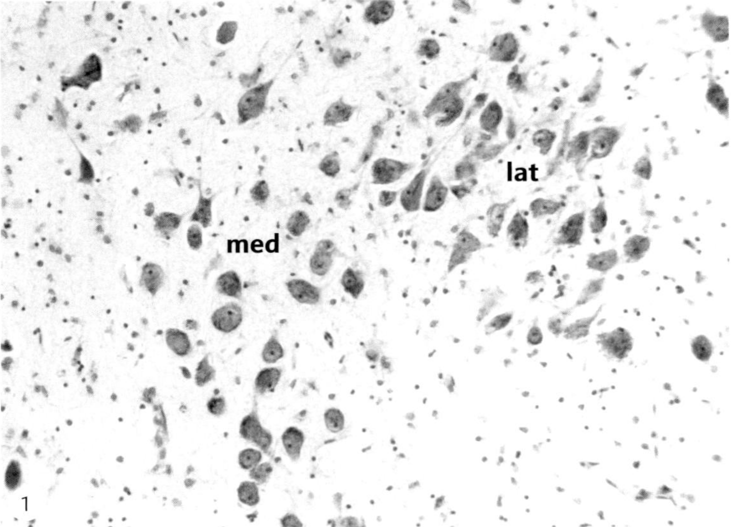

Fig. 1. Nucleus Edinger-Westphal, central projecting subdivision. Note the difference between a medial subgroup with round cells (med), and the lateral subgroup with multipolar and fusiform cells (lat), within the EWcp (see nucleus No. 46, fig. 3–5). Magnification ×150.

Functional Neuroanatomy

Function

The EW is traditionally considered as the location of parasympathetic preganglionic neurons of the ciliary ganglion, but tract-tracing studies in animals revealed a considerable variation in the location of preganglionic neurons across several species [Büttner-Ennever, 2006; May et al., 2008]. The neurons of the cytoarchitecturally circumscribed EW in humans described above have recently been shown to represent noncholinergic neurons indicating that they are not parasympathetic preganglionic neurons [Horn et al., 2008]. In fact this nucleus contains neurons expressing the neuropeptide urocortin-1, and is involved in modulatory functions related to stress adaptation [Kozicz, 2007], or regulation of food and alcohol intake [Ryabinin and Weitemier, 2006]. Since these peptidergic neurons represent centrally projecting (cp) neurons targeting neurons within the brain, the EW in humans, described in the 'Original Text' above and shown in nucleus No. 46, fig. 1, is now termed EWcp and is distinguished from the preganglionic neurons (EWpg) [Kozicz et al., 2011]. The EWcp may represent a group of modulator neurons (see 'Neuromodulatory Systems: Overview'). In the light of this recent finding it is interesting to see that in the 'Original Text' above Olszewski and Baxter state that the evidence for the role of their EW in the pupillary constriction and relaxation of the lens is scanty.

The preganglionic cells of the EWpg lie dorsally to the EWcp and form an inconspicuous cell group of cholinergic neurons that are scattered in the supraoculomotor area (fig. 6, 7; nucleus No. 46, fig. 3–5). These neurons carry signals to the ciliary ganglion and mediate pupillary constriction and accommodation of the lens. In cats the rostral EWpg, also called anteromedian nucleus, was shown to contain primarily preganglionic neurons related to pupillary control, whereas the caudal EWpg controls lens accommodation [Erichsen and May, 2002].

In clinical disorders, changes in accommodation and pupillary constriction or convergence can now be correlated with the appropriate functional cell groups. It appears that changes in the modulator cells of the EWcp in Alzheimer's disease have already been reported [Scinto et al., 2001], although in this study the cells were erroneously assumed to be preganglionic cells.

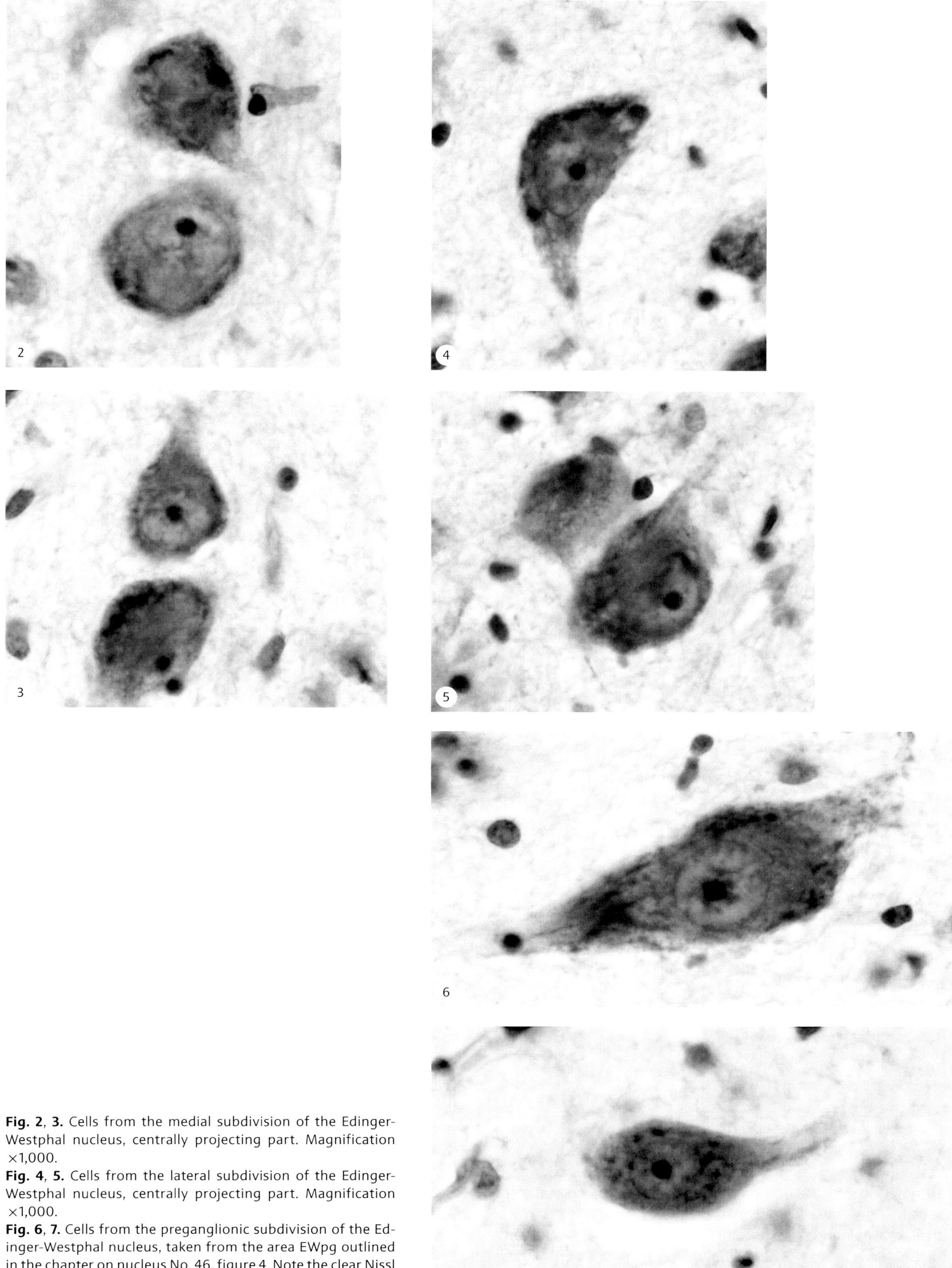

Fig. 2, 3. Cells from the medial subdivision of the Edinger-Westphal nucleus, centrally projecting part. Magnification ×1,000.
Fig. 4, 5. Cells from the lateral subdivision of the Edinger-Westphal nucleus, centrally projecting part. Magnification ×1,000.
Fig. 6, 7. Cells from the preganglionic subdivision of the Edinger-Westphal nucleus, taken from the area EWpg outlined in the chapter on nucleus No. 46, figure 4. Note the clear Nissl granules and the relative isolation of the individual neurons. Magnification ×1,000.

Connections

Efferents. The preganglionic cholinergic neurons of the EWpg project via the oculomotor nerve to the ciliary ganglion [Burde and Williams, 1989; Ishikawa et al., 1990; Sun and May, 1993]. From there cholinergic postganglionic fibers arise and proceed in the short ciliary nerves to approach the eyeball from the posterior side and innervate the smooth muscles, sphincter pupillae and the ciliary muscle [Kardon, 2005].

The EWcp is defined by noncholinergic centrally projecting neurons. A descending pathway from the EWcp targets the spinal cord, the inferior olive, the parabrachial nuclei, the facial nucleus, the spinal trigeminal nucleus and the C1 adrenergic neurons in the lateral reticular nucleus [Loewy and Saper, 1978]. Predominantly ipsilateral projections to the fastigial and interposed nucleus of the cerebellum were mainly found arising from the anterior part of the EWcp [May et al., 1992]. The ascending pathway from the EWcp to the lateral septum may be involved in alcohol consumption [Bachtell et al., 2003].

Afferents. One main input to the EWpg arises from the pretectal olivary nucleus, which receives retinal afferents and mediates pupillary constriction [Gamlin, 2006]. Additional inputs are described from the fastigial and interposed nucleus of the cerebellum [May et al., 1992]. The preganglionic neurons in the monkey receive a strong supply of orexin-A-positive nerve endings [Schreyer et al., 2009]. These projections arise from the lateral hypothalamus area and are involved in the maintenance of wakefulness [Sakurai, 2007]. The sources of afferents to preganglionic neurons in the EWpg mediating lens accommodation are less clear and involve the near-response region in the vicinity of the oculomotor nucleus [Mays and Gamlin, 1995].

A variety of further inputs was originally attributed to the preganglionic neurons, but in the light of the recent reinterpretation of the EW, they must now be considered to target the nonpreganglionic EWcp [for a review, see Koczic et al., 2011]. Sources of these inputs involve the solitary nucleus, the medullary reticular formation, the vestibular nuclei, the abducens nucleus, the raphe nuclei, the trigeminal nuclei, the locus coeruleus, the parabrachial nuclei, the cerebellar nuclei, the ventral tegmental area, the substantia nigra and the paraventricular nucleus of the thalamus [for a review, see Kozicz et al., 2011].

Histochemistry

The preganglionic neurons of the EWpg are cholinergic and express immunoreactivity for cytochrome oxidase and nonphosphorylated neurofilaments [Horn et al., 2008; May et al., 2008]. The neurons of the EWcp contain various neuropeptides, which include urocortin-1, cholecystokinin, substance P, neuropeptide B, cocaine- and amphetamine-regulated transcripts and nesfatin, the latter being involved in the pathobiology of anxiety and depressive-like behavior [Bloem et al., 2012; for a review, see Kozicz et al., 2011].

References

Bachtell RK, Weitemier AZ, Galvan-Rosas A, Tsivkovskaia NO, Risinger FO, Phillips TJ, Grahame NJ, Ryabinin AE: The Edinger-Westphal-lateral septum urocortin pathway and its relation to alcohol-induced hypothermia. J Neurosci 2003;23: 2477–2487.

Benjamin JW: The nucleus of the oculomotor nerve with special reference to innervation of the pupils and fibers from the pretectal region. J Nerv Mental Dis 1939;89:294–310.

Bloem B, Xu L, Morava E, Faludi G, Palkovits M, Roubos EW, Kozicz T: Sex-specific differences in the dynamics of cocaine- and amphetamine-regulated transcript and nesfatin-1 expressions in the midbrain of depressed suicide victims vs controls. Neuropharmacology 2012;62:297–303.

Burde RM, Williams F: Parasympathetic nuclei. Brain Res 1989;498:371–375.

Büttner-Ennever JA: The extraocular motor nuclei: organization and functional neuroanatomy. Prog Brain Res 2006;151:95–125.

Crouch RL: The efferent fibers of the Edinger-Westphal nucleus. J Comp Neurol 1936;64:365–373.

Erichsen JT, May PJ: The pupillary and ciliary components of the cat Edinger-Westphal nucleus: a transsynaptic transport investigation. Vis Neurosci 2002;19: 15–29.

Gamlin PDR: The pretectum: connections and oculomotor-related roles. Prog Brain Res 2006;151:379–405.

Horn AK, Eberhorn A, Härtig W, Ardelenanu P, Messoudi A, Büttner-Ennever JA: Perioculomotor cell groups in monkey and man defined by their histochemical and functional properties: reappraisal of the Edinger-Westphal nucleus. J Comp Neurol 2008;507:1317–1335.

Ishikawa S, Sekiya H, Kondo Y: The center for controlling the near reflex in the midbrain of the monkey: a double labelling study. Brain Res 1990;519:217–222.

Kardon R: Anatomy and physiology of the autonomic nervous system; in Miller NR, Newman NJ (eds): Walsh and Hoyt's Clinical Neuro-Ophthalmology, ed 6. Philadelphia, Lippincott, Williams & Wilkins, 2005.

Kozicz T: On the role of urocortin 1 in the non-preganglionic Edinger-Westphal nucleus in stress adaptation. Gen Comp Endocrinol 2007;153:235–240.

Kozicz T, Bittencourt JC, May PJ, Reiner A, Gamlin PDR, Palkovits M, Horn AKE, Toledo CAB, Ryabinin AE: The Edinger-Westphal nucleus: a historical, structural, and functional perspective on a dichotomous terminology. J Comp Neurol 2011;519:1413–1434.

Loewy AD, Saper CB: Edinger-Westphal nucleus: projections to the brain stem and spinal cord in the cat. Brain Res 1978;150:1–27.

May PJ, Porter JD, Gamlin PDR: Interconnections between the primate cerebellum and midbrain near-response regions. J Comp Neurol 1992;315:98–116.

May PJ, Reiner AJ, Ryabinin AE: Comparison of the distributions of urocortin-containing and cholinergic neurons in the perioculomotor midbrain of the cat and macaque. J Comp Neurol 2008;507:1300–1316.

Mays LE, Gamlin PDR: Neuronal circuitry controlling the near response. Curr Opin Neurobiol 1995;5:763–768.

Ryabinin AE, Weitemier AZ: The urocortin 1 neurocircuit: ethanol-sensitivity and potential involvement in alcohol consumption. Brain Res Rev 2006;52:368–380.

Sakurai T: The neural circuit of orexin (hypocretin): maintaining sleep and wakefulness. Nat Rev Neurosci 2007;8:171–181.

Schreyer S, Büttner-Ennever JA, Tang X, Mustari MJ, Horn AKE: Orexin-A inputs onto visuomotor cell groups in the monkey brainstem. Neuroscience 2009;164: 629–640.

Scinto LFM, Frosch M, Wu CK, Daffner KR, Gedi N, Geula C: Selective cell loss in Edinger-Westphal in asymptomatic elders and Alzheimer's patients. Neurobiol Aging 2001;22:729–736.

Sun W, May PJ: Organization of the extraocular and preganglionic motoneurons supplying the orbit in the lessor galago. Anat Rec 1993;237:89–103.

Szentágothai J: Die innere Gliederung des Oculomotoriuskernes. Arch Psychiatrie 1942;115:127–135.

50 Dorsal Motor Nucleus of the Vagal Nerve (DMX)

Original name:
Nucleus dorsalis motorius nervi vagi (X)

Alternative names:
Tenth cranial nerve nucleus
Dorsal vagal nucleus

Location and Cytoarchitecture – Original Text

The dorsal motor nucleus of the vagus (DMX; fig. 1) first appears in cross-sections through the medulla just caudal to the spinal end of the hypoglossal nucleus (XII). Its caudal pole lies lateral and dorsolateral to the central canal, dorsolateral to the XII and ventromedial to the nucleus of the solitary tract (SOL). More orally, the nucleus intercalatus (INSt) intervenes between DMX and XII. At the level of the opening of the floor of the fourth ventricle, the increasing size of the XII and INSt serves to displace the DMX into a more lateral position. Here, it lies directly beneath the floor of the fourth ventricle where it forms an elevation lateral to the trigonum hypoglossi, known as the ala cinerea or the trigonum vagi. Still more orally the DMX lies lateral to the prepositus nucleus. The oral pole of the nucleus is displaced ventrally and laterally away from the floor of the fourth ventricle by the development of the medial vestibular nucleus.

On cross-section the DMX is oblong in outline and lies with its long axis directed dorsomedially. It attains its maximal dimensions just oral to the level of the obex.

The DMX is composed of 3 cell types:

(a) medium-sized, predominantly fusiform cells with evenly distributed Nissl granules and centrally placed nuclei (fig. 5);

(b) cells of similar size and shape which contain large amounts of melanin-pigmented cells (fig. 6);

(c) larger, medium-sized, multipolar, fork-shaped or fusiform cells. In these cells the Nissl substance is distributed as large irregular masses located at the periphery of the perikaryon; the central area, adjacent to the nucleus, is lightly stained and finely granulated; the edges of the cell body are characteristically irregular and ragged and the nucleus is often excentric (fig. 3, 4).

These 3 cell types are not uniformly distributed throughout the extent of the nucleus. The medium-sized, fusiform cells (type a) are located predominantly in the oral, caudal and ventral portions of the nucleus. The larger cells with ragged edges (type c) are most abundant in the dorsal part of the intermediate portion. Finally, the pigmented cells (type b), which are not numerous, appear among the unpigmented medium-sized cells in the caudal and ventral portions of the nucleus.

The oral pole of the DMX is often associated dorsally with a small group of round or oval, darkly stained, plump cells (fig. 2). Due to their distinctive morphology, these cells do probably not form an integral part of the DMX. [...]

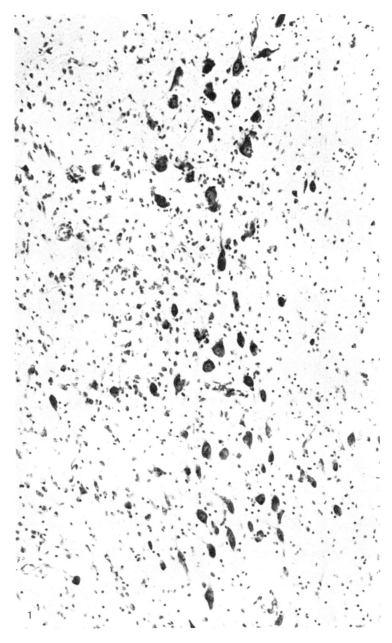

Fig. 1. Dorsal motor nucleus of the vagus. Note that the cells in the dorsal portion of the nucleus are larger than those of the ventral portion. Magnification ×150.

Functional Neuroanatomy

Function
The DMX is a 'general visceral efferent' nucleus, which gives rise to preganglionic parasympathetic fibers terminating in the viscera of the thoracic and abdominal cavities. The efferent fibers form the vagus nerve, the stimulation of which causes, classically, constriction of the bronchioles, a reduction in heart rate and in the gut an increase in blood flow, peristalsis and secretions. However, targets of the DMX may receive several independent vagal inputs, each mediating a different effect; thus, the responses to vagal stimulation are highly diverse [for reviews, see Jänig, 2008; Gibbins, 2012].

The DMX is considered by Nieuwenhuys et al. [2008] to be a caudal component of the 'core of neuraxis' of the limbic sys-

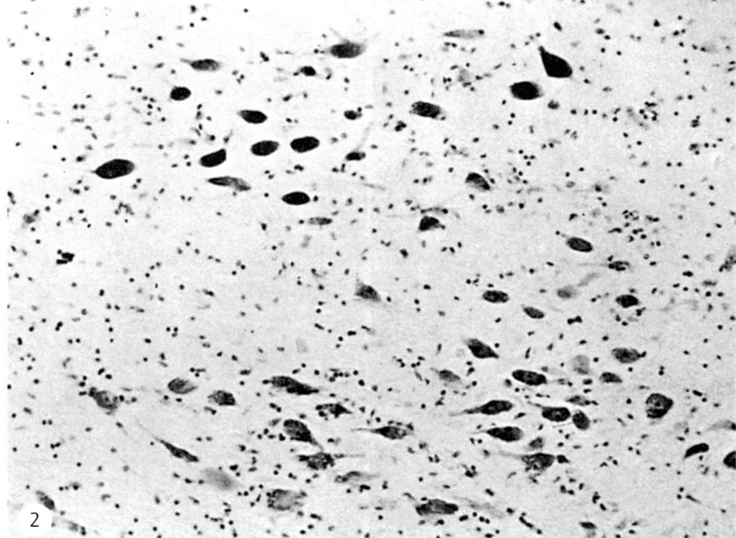

Fig. 2. The oral pole of the dorsal motor nucleus of the vagus is often related dorsally to a small group of plump, oval, darkly stained cells. This latter cell group is seen in the upper portion of the figure. Magnification ×150.

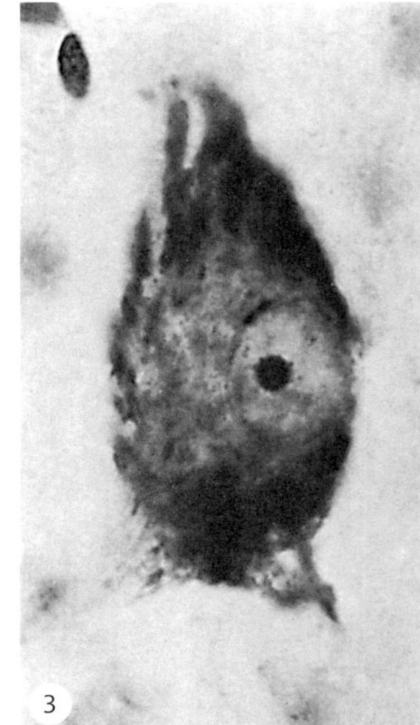

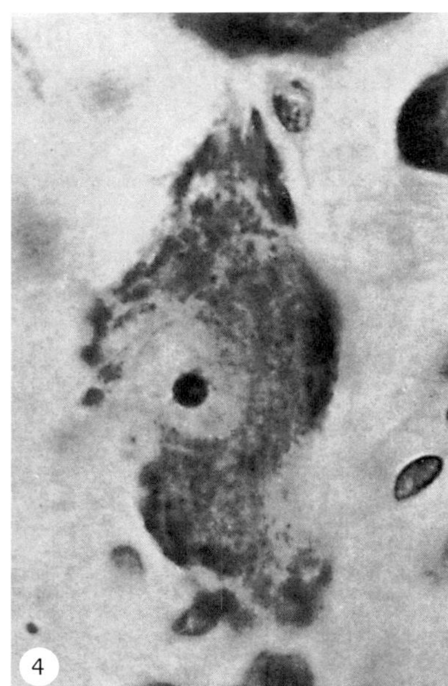

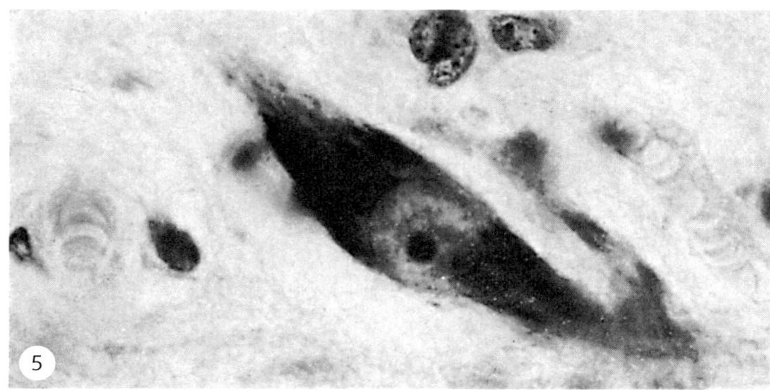

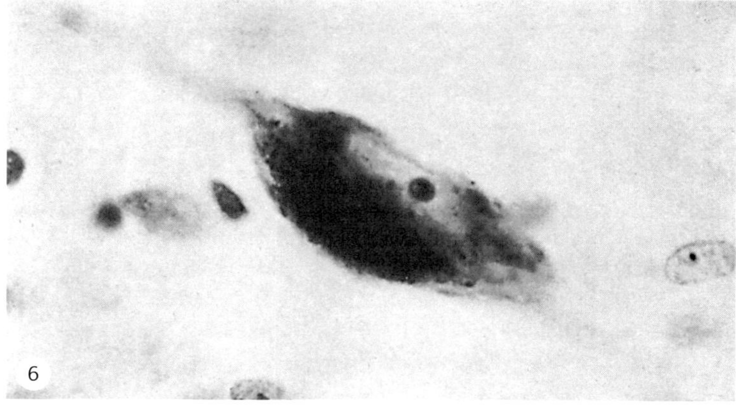

Fig. 3, 4. Medium-sized cells, with peripherally distributed Nissl substance and excentric nuclei, from the dorsal motor nucleus of the vagus. Magnification ×1,000.

Fig. 5. Medium-sized, fusiform cell with evenly distributed Nissl granules, from the dorsal motor nucleus of the vagus. Magnification ×1,000.

Fig. 6. Melanin-containing cell of the dorsal motor nucleus of the vagus. Magnification ×1,000.

tem. The DMX is controlled by a 'central autonomic network (CAN)', a functional group of interconnected nuclei, which influence visceromotor, neuroendocrine, pain and behavioral responses essential for survival. The CAN includes insular cortex, amygdala, hypothalamus, periaqueductal gray matter, parabrachial complex, ventrolateral medulla and SOL (nucleus No. 34) [Benarroch, 1993]. Through these pathways the DMX plays a role in eating behavior.

A unilateral vagal nerve lesion leaves no long-lasting symptoms specifically related to visceromotor dysfunction, but a bilateral lesion of the vagal output, as for example in a brainstem lesion intruding on the DMX, may cause aphagia, dyspnea and inspiratory stridor, which can be life threatening, although this is rare.

Connections

The DMX is organized roughly topographically in longitudinal columns; the columns stretch rostrocaudally, with the cecum represented laterally [Fox and Powley, 1985; Altschuler et al., 1991; Fox and Powley, 1992; Powley, 2000]. Each column has a similar cytoarchitecture, and the dendrites of DMX neurons remain mostly within these columns, but some range across borders and others supply the adjacent SOL [Jarvinen and Powley, 1999]. There is a particularly tight and reciprocal relationship between DMX and SOL so that the visceral afferents to the SOL can immediately affect the visceral efferents of the DMX.

Afferents. A major input to the DMX, as described above, is from the overlying SOL, and it is through the SOL that the signals from other areas, such as the area postrema – the chemical sensor of the vagal complex – mainly reach the DMX [Cunningham et al., 1994]. Limbic forebrain structures project directly to the DMX: for example the medial prefrontal orbital cortex (the visceral motor cortex), the central nucleus of the amygdala complex, the bed nuclei of the stria terminalis and the hypothalamus, specifically the paraventricular nucleus of the lateral hypothalamus and the medial hypothalamus [Neafsey, 1990; Holstege, 1991; Saper, 1995; Carrasco et al., 2001]. Neurochemical studies of DMX cells have shown that they have a dense concentration of binding sites for metabolic and hypothalamic peptides and hormones [for a review, see Powley, 2000].

Inputs onto the DMX from the medullary raphe nuclei (e.g. 5-hydroxytryptamine, substance P and thyrotropin-releasing factor) are found to contribute to vagal regulation of gastric mucosa function, and its response to injury [Kaneko and Tache, 1995; Tache, 2012].

Efferents. The main efferents of the DMX descend in the vagus nerve. The term 'vagus' is derived from the Latin word meaning 'wandering or rambling', and reflects the meandering path of the nerve. The axons terminate on the intramural ganglia in the heart, respiratory organs, pancreas and gastrointestinal tract.

Ascending DMX fibers that project to the SOL and locus coeruleus have been implicated in the mediation of memory storage induced by stressful situations that increase peripheral adrenaline [Chen and Williams, 2012].

References

Altschuler SM, Ferenci DA, Lynn RB, Miselis RR: Representation of the cecum in the lateral dorsal motor nucleus of the vagus nerve and commissural subnucleus of the nucleus tractus solitarii in rat. J Comp Neurol 1991;304:261–274.

Benarroch EE: The central autonomic network: functional organization, dysfunction, and perspective. Mayo Clin Proc 1993;68:988–1001.

Carrasco M, Portillo F, Larsen PJ, Vallo JJ: Insulin and glucose administration stimulates Fos expression in neurones of the paraventricular nucleus that project to autonomic preganglionic structures. J Neuroendocrinol 2001;13:339–346.

Chen CC, Williams CL: Interactions between epinephrine, ascending vagal fibers, and central noradrenergic systems in modulating memory for emotionally arousing events. Front Behav Neurosci 2012;6:35.

Cunningham ET Jr, Miselis RR, Sawchenko PE: The relationship of efferent projections from the area postrema to vagal motor and brain stem catecholamine-containing cell groups: an axonal transport and immunohistochemical study in the rat. Neuroscience 1994;58:635–648.

Fox EA, Powley TL: Longitudinal columnar organization within the dorsal motor nucleus represents separate branches of the abdominal vagus. Brain Res 1985;341:269–282.

Fox EA, Powley TL: Morphology of identified preganglionic neurons in the dorsal motor nucleus of the vagus. J Comp Neurol 1992;322:79–98.

Gibbins I: Peripheral autonomic pathways; in Mai JK, Paxinos G (eds): The Human Nervous System, ed 3. Amsterdam, Academic Press, 2012.

Holstege G: Descending motor pathways and the spinal motor system: limbic and non-limbic components. Prog Brain Res 1991;87:307–421.

Jänig W: Integrative Action of the Autonomic Nervous System: Neurobiology of Homeostasis. Cambridge, Cambridge University Press, 2008.

Jarvinen MK, Powley TL: Dorsal motor nucleus of the vagus neurons: a multivariate taxonomy. J Comp Neurol 1999;403:359–377.

Kaneko H, Tache Y: TRH in the dorsal motor nucleus of vagus is involved in gastric erosion induced by excitation of raphe pallidus in rats. Brain Res 1999;699:97–102.

Neafsey EJ: Prefrontal cortical control of the autonomic nervous system: anatomical and physiological observations. Prog Brain Res 1990;85:147–165, discussion 165–166.

Nieuwenhuys R, Voogd J, van Huijzen C: The Human Central Nervous System. Berlin, Springer, 2008.

Powley TL: Vagal circuitry mediating cephalic-phase responses to food. Appetite 2000;34:184–188.

Saper CB: Central autonomic system; in Paxinos G (ed): The Rat Nervous System, ed 2. San Diego, Academic Press, 1995.

Tache Y: Brainstem neuropeptides and vagal protection of the gastric mucosa against injury: role of prostaglandins, nitric oxide and calcitonin-gene related peptide in capsaicin afferents. Curr Med Chem 2012;19:35–42.

Overview

The reticular formation is cytoarchitecturally an inconspicuous, longitudinal column of neurons in the brainstem. It stretches from the rostral mesencephalon down to the caudal medulla and continues on into the spinal cord as the intermediate zone (Rexed laminae V–VIII). In the 1950s, Olszewski and Baxter carefully documented these differences and divided the reticular formation into about 10 nuclei, which are described in detail below. The multipolar reticular neurons receive countless afferents, which are coordinated to drive the specific activities of the brainstem nuclei usually embedded in the reticular formation, or close by. For example, the specific activities coordinated by the reticular formation around the facial and trigeminal nuclei include sneezing, swallowing, coughing, blinking etc., and the reticular areas around the abducens nucleus coordinate horizontal eye and head movements, whereas those around the oculomotor nucleus process vertical eye-head coordination (see 'Plates of Serial Sections through the Human Brainstem', fig. 1 'Organization of the Brainstem Nuclei'). The general functional role of the reticular formation can be summarized as 'coordination', and it is now clear that it does not play a role in arousal or attention. Functions such as arousal or attention are carried out by the diffusely projecting neuromodulatory nuclei lying just outside or around the borders of the reticular formation, such as the locus coeruleus, the pedunculopontine nucleus and the raphe nuclei (see 'Neuromodulatory Systems: Overview'). It is these cell groups that form the neuroanatomical basis of the 'ascending reticular activating system' for arousal, and not the 10 or more reticular nuclei defined by Olszewski and Baxter (see 'Introduction: The Reticular Formation 60 Years Later').

The 10 or more subdivisions of the reticular formation by Olszewski and Baxter are in general accepted, and have been verified in most vertebrates [Taber, 1961; Nieuwenhuys et al., 1998]. However, Paxinos and colleagues in their studies have outlined several other reticular areas, and one of these is particularly worthy of mention. It is called the 'intermediate reticular nucleus (or zone)' (IRt) and lies in the medulla oblongata [Paxinos et al., 2012]. The IRt cuts across the borders of 3 nuclei defined by Olszewski and Baxter, namely the central nucleus of the medulla oblongata, the lateral paragigantocellular nucleus, and the parvocellular reticular nucleus. The IRt is the junctional region of the alar and basal plates, and it contains the motoneurons of the ambiguus, retrofacial and facial nuclei, and the retroambiguus nucleus. The IRt around this motoneuron column includes the Bötzinger complex, the pre-Bötzinger complex and the ventral respiratory group, among others. These groups form a functional complex column controlling respiration, cardiovascular and other vital functions [Dobbins and Feldman, 1994; Blessing and Benarroch, 2012]. Thus, the IRt appears to represent a natural and functional subdivision of the reticular formation, and a study of its correlation with the functional columns of the ventrolateral medulla would be highly interesting. The IRt can be identified in several species, including the human, and it seems advisable in the next edition of this atlas to incorporate it into the scheme of the reticular formation [Jones, 1995; Paxinos and Huang, 1995; Paxinos et al., 2012]. In the descriptions of the individual reticular nuclei here, we have adhered to the original outlines of Olszewski and Baxter, but recognize the usefulness of the term IRt.

The reticular formation and its nuclei are usually divided into 2 functional zones: (1) 'the medial tegmental fields', which are mainly large-celled nuclei, with descending reticulospinal pathways which coordinate posture, eye and head movements, and (2) the 'lateral tegmental fields', containing smaller and fewer cells, with shorter local projections that coordinate automatic or limbic functions such as micturition, swallowing, mastication and vocalization; see 'Introduction: The Reticular Formation 60 Years Later' and figure 22.1 in Nieuwenhuys et al. [2008; Brodal, 1957; Holstege et al., 2004].

References

Blessing WW, Benarroch EE: Lower brain stem regulation of visceral, cardiovascular, and respiratory function; in Mai JK, Paxinos G (eds): The Human Nervous System, ed 3. Amsterdam, Elsevier Academic Press, 2012.

Brodal A: The Reticular Formation of the Brain Stem. Anatomical Aspects and Functional Correlations. London, William Ramsay Henderson Trust, 1957.

Dobbins EG, Feldman JL: Brain stem network controlling descending drive to phrenic motoneurons in rat. J Comp Neurol 1994;347:64–86.

Holstege GG, Mouton LJ, Gerrits NM: Emotional motor system; in Paxinos G, Mai JK (eds): The Human Nervous System, ed 2. Amsterdam, Elsevier Academic Press, 2004.

Jones BE: Reticular formation: cytoarchitecture, transmitters and projections; in Paxinos G (ed): The Rat Nervous System, ed 2. San Diego, Academic Press, 1995.

Nieuwenhuys R, Ten Donkelaar HJ, Nicholson C: The Central Nervous System of Vertebrates. Berlin, Springer, 1998.

Nieuwenhuys R, Voogd J, van Huijzen C: The Human Central Nervous System. Berlin, Springer, 2008.

Paxinos G, Huang X-F: Atlas of the Human Brainstem. San Diego, Academic Press, 1995.

Paxinos G, Huang X-F, Sengul G, Watson C: Organization of the brainstem; in Mai JK, Paxinos G (eds): The Human Nervous System, ed 3. Amsterdam, Academic Press, 2012.

Taber E: The cytoarchitecture of the brain stem of the cat. I. Brain stem nuclei of the cat. J Comp Neurol 1961;116:27–70.

51 Central Nucleus of the Medulla Oblongata (CN)

Original names:
Nucleus medullae oblongatae centralis, subnucleus dorsalis
Nucleus medullae oblongatae centralis, subnucleus ventralis

Alternative names:
Central nucleus of the medulla oblongata, lateral and medial subnucleus [Nieuwenhuys et al., 2008]
Medullary reticular nucleus, dorsal and ventral part
Dorsal, ventral and medial reticular nucleus, and intermediate reticular zone of the medulla [Paxinos et al., 1990; Koutcherov et al., 2004]

Location and Cytoarchitecture – Original Text

This nuclear group occupies the central portion of the medullary gray matter from the level of the junction of the spinal cord and medulla to the junction of the middle and oral thirds of the inferior olivary complex, a distance of about 18 mm.

The most caudally situated cells of this nucleus lie enmeshed among the decussating pyramidal fibers and due to the paucity of cellular elements at this level, the boundary line between the gray matter of the spinal cord and the central nucleus of the medulla oblongata (CN) is rather arbitrary.

Immediately oral to the pyramidal decussation, the CN is roughly quadrangular in outline and lies ventral to the solitary nucleus (SOL), medial cuneate nucleus (MCU) and spinal trigeminal complex, dorsolateral to the nucleus supraspinalis and fibers of the pyramid, and dorsomedial to the lateral white column of the medulla. The retroambiguus nucleus lies embedded in the lateral aspects of the CN in levels caudal to the inferior olivary complex (IO).

In cross-section through the caudal half of the IO, the CN lies dorsomedial to the lateral reticular nucleus (LRN), dorsal to the IO, lateral to the medial lemniscus and the medial longitudinal fasciculus, ventrolateral to the hypoglossal nucleus (XII), the dorsal motor nucleus (DMX) and SOL, ventral to the MCU, and ventromedial to the spinal trigeminal complex.

The oral portion of the CN lies dorsomedial to the LRN subtrigeminal part, dorsal to the IO, lateral to the interfascicular hypoglossal nucleus, and ventral to the XII, DMX, SOL and the lateral cuneate nucleus. The cells of the nucleus ambiguus lie embedded in the CN at all levels oral to the caudal pole of the IO.

The CN is composed of randomly oriented cells which vary from small to medium in size. The smallest cells are spindle shaped and contain a relatively large nucleus and indistinct palely stained Nissl granules (fig. 7, 8). The largest cells are multipolar and possess long dendrites and darkly stained, evenly distributed Nissl granules (fig. 3). Many cells are intermediate in histological appearance between these two extremes (fig. 4–6, 9).

Although both small and medium-sized cells are found throughout the nucleus, it is possible to distinguish a dorsal subnucleus (CNd) and a ventral subnucleus (CNv) on the basis of cell density and distribution. The CNd (fig. 2) is composed predominantly of densely arranged, small, spindle-shaped neurons whereas the CNv (fig. 1) is less cellular and contains relatively more numerous darkly stained, medium-sized cells.

Functional Neuroanatomy

Function
The CN is the most caudal part of the brainstem reticular formation at its transition to the spinal cord. The CNv is the rostral continuation of the ventral horn and intermediate zone of the cord, and as such is part of the 'medial tegmental field'. Nuclei in the medial tegmental fields are in general associated with somatic motor control, and in keeping with this the CNv was shown to participate in the premotor control of the head movements [Robinson et al., 1994]. The CNd belongs to the 'lateral tegmental field', associated with autonomic visceral functions (see 'Reticular Formation: Overview'). The CNd contains neural networks for respiratory and cardiovascular control. The networks are parts of functional centers that extend further rostrally into adjacent reticular nuclei [Blessing, 2004; Nieuwenhuys et al., 2008].

Connections
Large adrenergic neurons of the C2 group and noradrenergic cells of the A2 cell group lie along the medial border of the CN, between the XII, SOL, and the intercalated nucleus. Ventrolaterally in the area around the CNv-CNd border, the A1 and C1 catecholamine cell groups form part of a center for cardiovascular and respiratory control [Blessing, 2004; Nieuwenhuys et al., 2008]. The CN has been reported to send projections to the forebrain, brainstem and spinal cord, where many of the fibers ascending to the hypothalamus are catecholaminergic [Martin et al., 1990]. The majority of the CN output comes from the border region of the CNv with CNd, which includes the intermediate reticular zone; see 'Reticular Formation: Overview' [Paxinos et al., 2012].

Cardiovascular Centers. The CNd extends to the caudal ventrolateral medulla (CVLM) which contains GABAergic neurons and an inhibitory vasomotor center. Further rostrally, around A1 and the caudal part of C1, there is a second cardiovascular region containing sympathoexcitatory premotor circuits, called the rostral ventrolateral medulla (RVLM). The neurons in the CNd respond to internal bodily stress, for example an increase in arterial pressure: in such a case, the sensory baroreceptor information enters the brainstem through the glossopharyngeal or vagus nerve, to reach the caudal SOL. The SOL efferents relay the stress signals to the CVLM which inhibits the RVLM. Efferents from the RVLM project to the spinal cord and control the spinal preganglionic neurons of the sympathetic system to alleviate the pressure increase [Blessing, 2004; Nieuwenhuys et al., 2008].

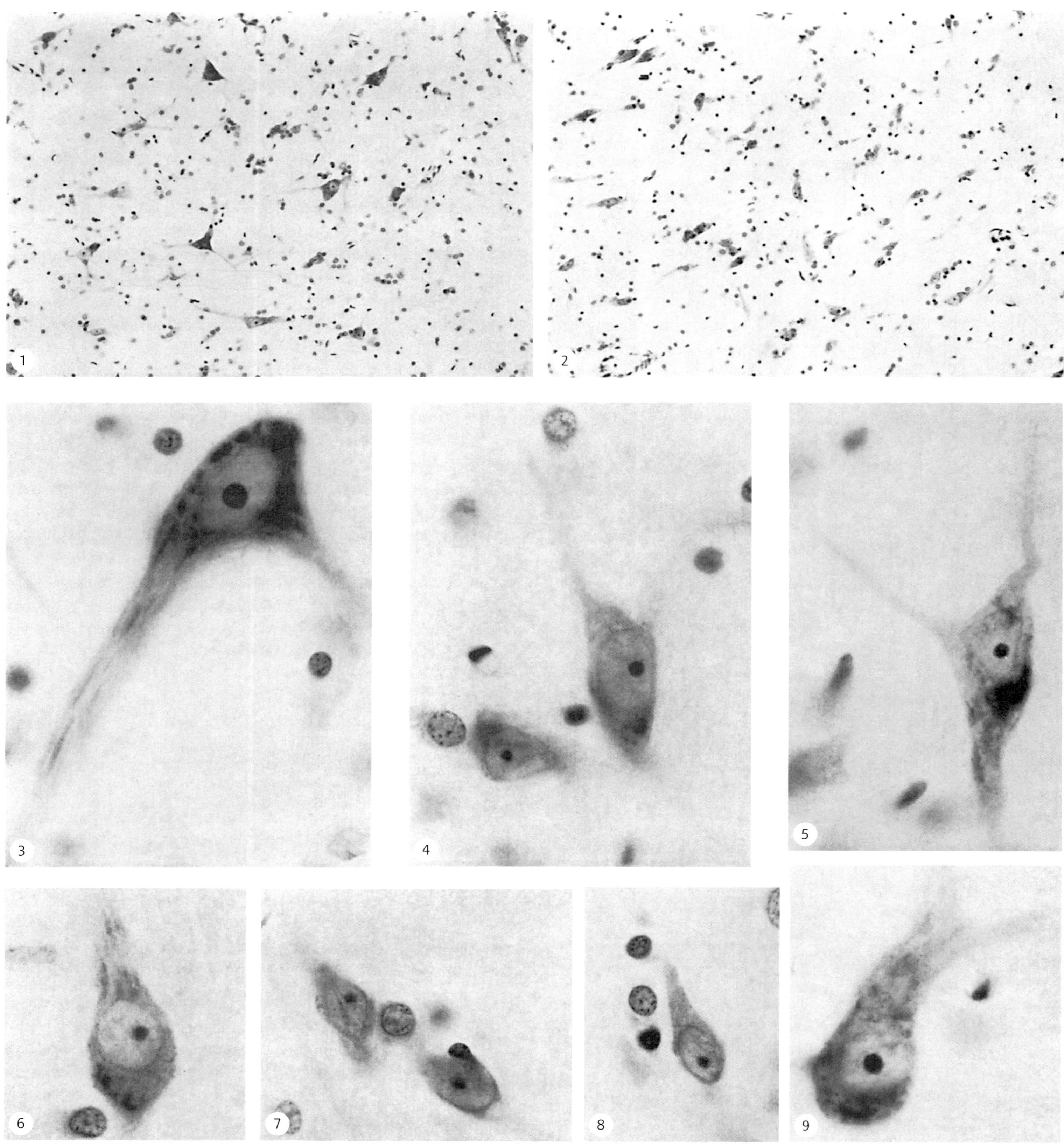

Fig. 1. Central nucleus of the medulla oblongata, ventral subnucleus. Magnification ×150.

Fig. 2. Central nucleus of the medulla oblongata, dorsal subnucleus. Magnification ×150.

Fig. 3–9. Different cell types from the central nucleus of the medulla oblongata. Magnification ×1,000.

These CNd cells were found to project to the hypothalamus, where their activity promotes the release of antidiuretic and adrenocorticotrophic hormones into the portal circulation, in response to stress stimuli [Blessing, 1997].

Respiratory Centers. Close to the cardiovascular centers, around the CNv-CNd border, and near the ambiguus nucleus, lie the rostral and caudal ventral respiratory groups [Blessing, 2004; Nieuwenhuys et al., 2008]. Expiratory premotor neurons are found most caudally, further rostrally the inspiratory premotor neurons. This column of respiratory neurons in the ventrolateral medulla extends further rostrally than the CN into the lateral paragigantocellular nucleus and the parvocellular reticular nucleus where the pacemaker cells of the pre-Bötzinger group and the expiratory neurons of the Bötzinger complex are found.

References

Blessing WW: The Lower Brain Stem and Bodily Homeostasis. New York, Oxford University Press, 1997.

Blessing WW: Lower brain stem regulation of visceral, cardiovascular, and respiratory function; in Paxinos G, Mai JK (eds): The Human Nervous System, ed 2. Amsterdam, Elsevier Academic Press, 2004.

Koutcherov Y, Huang X-F, Halliday G, Paxinos G: Organization of human brain stem nuclei; in Paxinos G, Mai JK (eds): The Human Nervous System, ed 2. Amsterdam, Elsevier Academic Press, 2004.

Martin GF, Holstege G, Mehler WR: Reticular formation of the pons and medulla; in Paxinos G (ed): The Human Nervous System. San Diego, Academic Press, 1990.

Nieuwenhuys R, Voogd J, van Huijzen C: The Human Central Nervous System. Berlin, Springer, 2008.

Paxinos G, Törk I, Halliday G, Mehler WR: Human homologs to brainstem nuclei identified in other animals as revealed by acetylcholinesterase activity; in Paxinos G (ed): The Human Nervous System. San Diego, Academic Press, 1990.

Paxinos G, Huang X-F, Sengul G, Watson C: Organization of the brainstem; in Mai JK, Paxinos G (eds): The Human Nervous System, ed 3. Amsterdam, Academic Press, 2012.

Robinson FR, Phillips JO, Fuchs AF: Coordination of gaze shifts in primates: brainstem inputs to neck and extraocular motoneuron pools. J Comp Neurol 1994; 346:43–62.

Reticular Formation

(Plates 14–23)

52 Gigantocellular Nucleus (Gi, Giv and Giα)

Original name:
Nucleus gigantocellularis

Alternative names:
Gigantocellular reticular nucleus

Subnuclei:

The gigantocellular nucleus (Gi) has at least 3 subnuclei: Gi proper refers to the area in the dorsal Gi containing the giant neurons; the Gi caudal subnucleus, nowadays referred to as the ventral Gi (Giv); the rostral ventral Gi around the border of the nucleus raphe magnus, called Gi pars α (Giα) [Martin et al., 1990]

The term 'magnocellular reticular nucleus' corresponds to the Giv together with the Giα [Berman, 1968; Cowie et al., 1994; Jones, 1995]

Location and Cytoarchitecture – Original Text

This nucleus comprises the ventromedial part of the tegmentum of the oral third of the medulla and the caudal half of the pons. It commences caudally at the level of the junction of the oral and middle thirds of the inferior olivary complex (IO) and extends orally to the level of the caudal pole of the motor trigeminal nucleus (MoV), a distance of 10–12 mm. The nucleus is bounded caudally by the central nucleus of the medulla oblongata and orally by the nucleus reticularis pontis caudalis (NRPC).

On cross-section the gigantocellular nucleus (Gi) is quadrangular in outline. In the medulla, it lies ventral to the dorsal part of the paragigantocellular nucleus, ventromedial to the parvocellular reticular nucleus (PCR), medial to the nucleus paragigantocellularis lateralis and dorsal to the IO. Medially, the nucleus is related to fibers of the medial lemniscus and to the nucleus raphe magnus (RMg).

In the caudal part of the pons the Gi retains its quadrangular outline and lies dorsal to the fibers of the medial lemniscus, lateral to the median raphe nucleus, ventral to the NRPC, and medial to the PCR and the nucleus of the trapezoid body. Approaching its oral pole, the outline of the Gi becomes progressively narrower in the dorsoventral direction.

The majority of cells composing the Gi are characteristically large, darkly stained and multipolar. Three main cell types may be distinguished:

(a) large, plump, multipolar cells which possess ragged edges, an excentric nucleus and darkly stained Nissl granules which tend to congregate in irregular clusters about the periphery of the cell; that surface of the excentric nucleus which faces the bulk of the cytoplasm is often covered by a cap of darkly stained Nissl substance (fig. 8, 9); the concentration of Nissl substance towards the center of the cell is less dense and here the granules are often arranged in the form of concentric lamellae about the nucleus; those nerve cells which possess this laminated pattern of Nissl granules have been referred to as 'onion skin cells' by Meessen and Olszewski [1949] (fig. 7);

(b) large, slender, multipolar neurons which possess regular edges, a centrally placed nucleus, long dendrites and darkly stained, evenly distributed Nissl granules; these cells are usually smaller than the first cell type and bear a marked resemblance to the neurons of the somatic and branchiomeric motor nuclei (fig. 4–6, 10, 11);

(c) small, spindle-shaped or triangular cells with lightly stained cytoplasm (fig. 12, 13).

On the basis of the distribution of these cell types the Gi may be divided into a ventral and an oral subnucleus.

The ventral subnucleus (Giv; fig. 1) forms that portion of the Gi which lies caudal to the level of the facial nucleus (VII). It is composed of cells of all 3 types, with those of type a predominating.

The oral subnucleus (Gi proper; fig. 2) comprises the remainder of the nucleus and is composed of cells of types b and c. In the oral part of the oral subnucleus, the cells of type b diminish slightly in size, become more elongated and characteristically lie with their long axes directed horizontally (fig. 3, 10, 11). [...]

Functional Neuroanatomy

Function

The Gi is part of the 'medial tegmental field' giving rise to reticulospinal pathways. Stimulation and tracer studies in the primate confirm that the Gi, and adjacent areas, play a role in the generation of movement patterns associated with head orientation – these include head, jaw, face and tongue movements – but not so clearly with eye movements [Cowie and Robinson, 1994]. In particular, Gi neurons encode head position, and may serve as a 'head integrator' [Robinson et al., 1994].

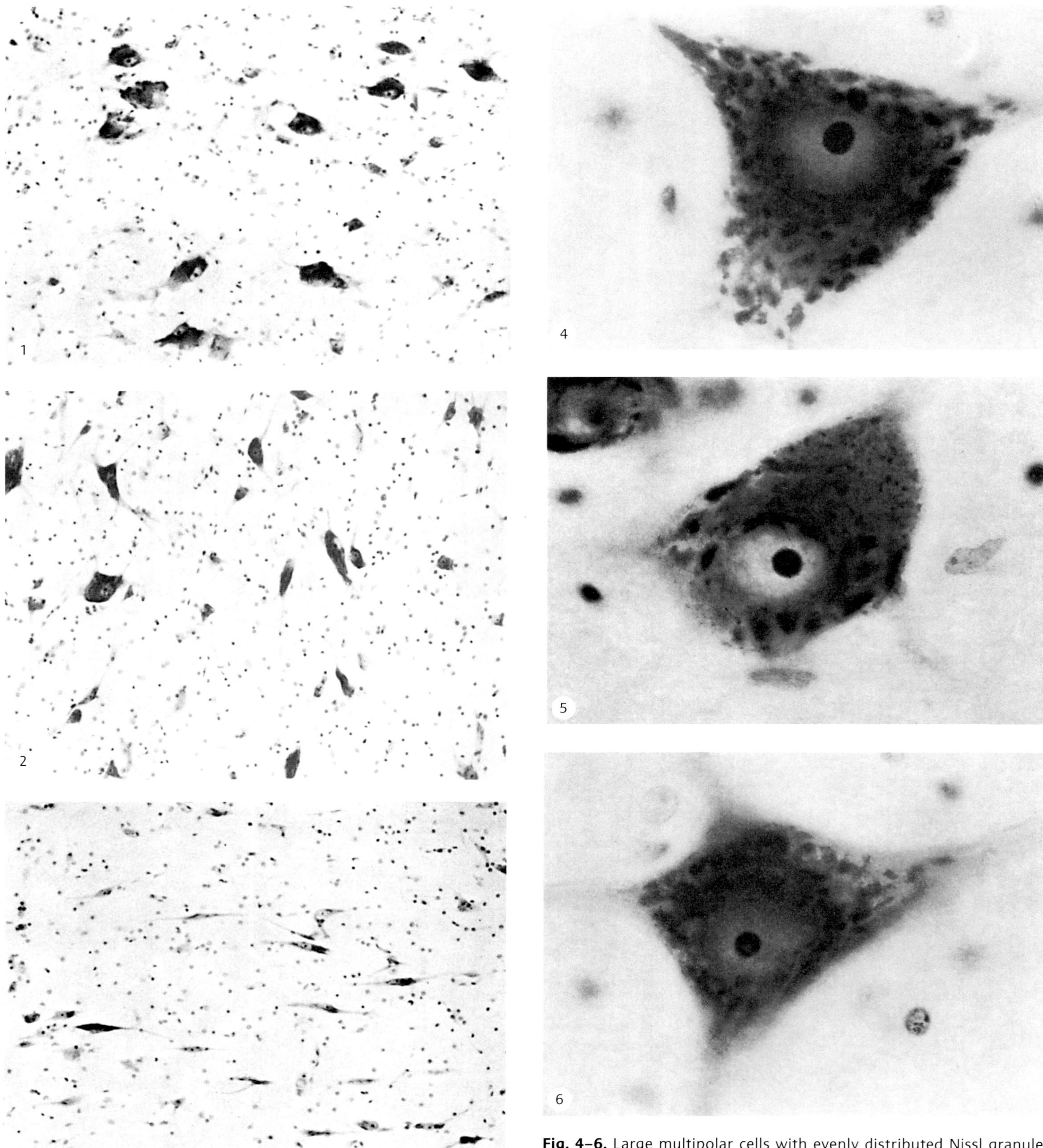

Fig. 1. Nucleus gigantocellularis, subnucleus caudalis, ventral part. Magnification ×150.
Fig. 2. Nucleus gigantocellularis proper. Magnification ×150.
Fig. 3. Nucleus gigantocellularis, pars α. Magnification ×150.

Fig. 4–6. Large multipolar cells with evenly distributed Nissl granules (type b) from the nucleus gigantocellularis. Magnification ×1,000.

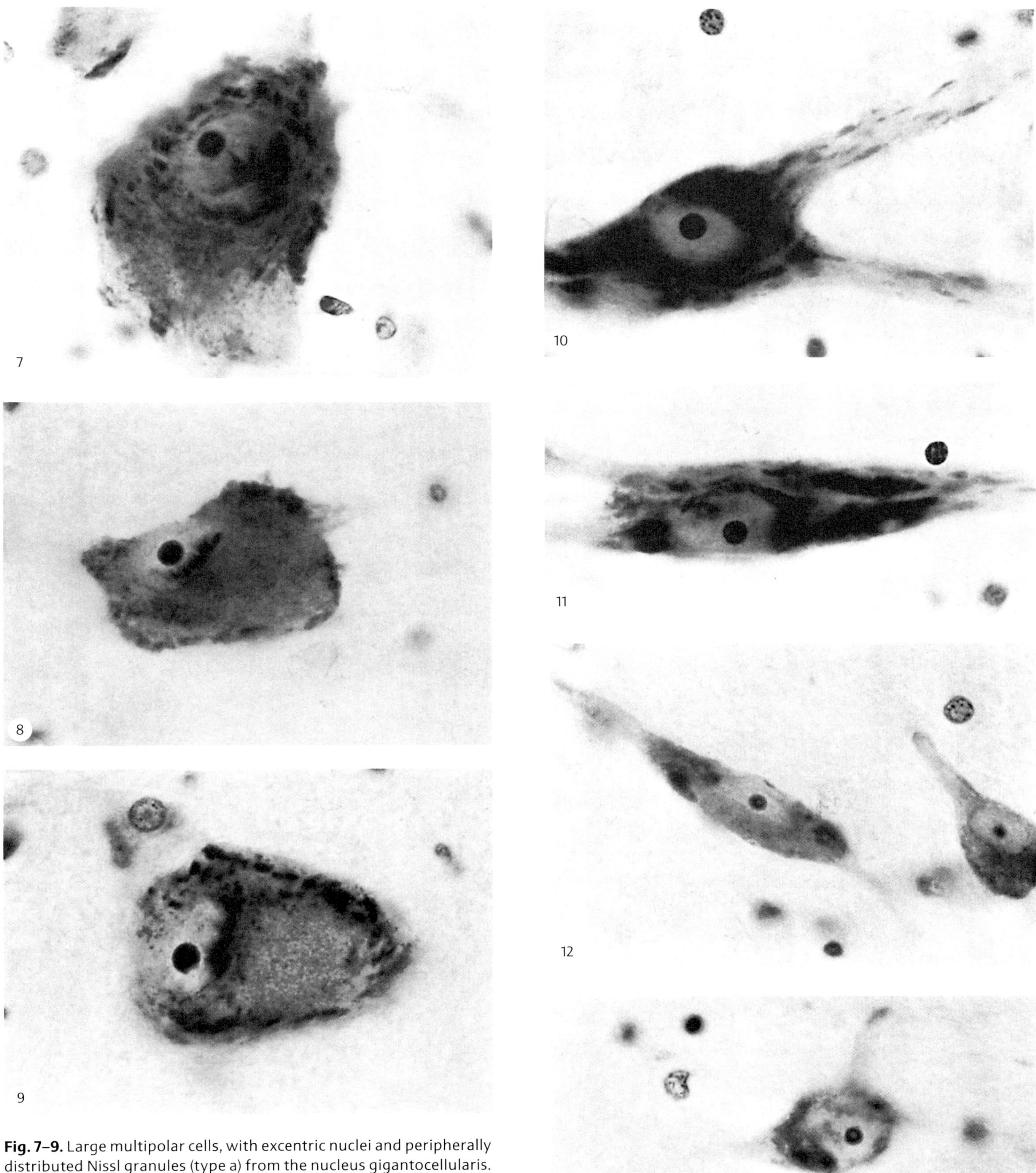

Fig. 7–9. Large multipolar cells, with excentric nuclei and peripherally distributed Nissl granules (type a) from the nucleus gigantocellularis. Magnification ×1,000.

Fig. 10, 11. Large, elongated, multipolar cells with evenly distributed Nissl granules (type b) from the oral portion of the subnucleus oralis of the nucleus gigantocellularis. Magnification ×1,000.

Fig. 12, 13. Small cells (type c) from the nucleus gigantocellularis. Magnification ×1,000.

The Gi pars α (Giα) area around the RMg is the source of neuromodulatory descending pathways to the dorsal horn, which participate in the control of pain through the spinal cord and the caudal spinal trigeminal nucleus (see fig. 16.6 in Nieuwenhuys et al. [2008] and RMg nucleus No. 66) [Willis and Westlund, 2004; Mason, 2005].

Histochemistry and Subdivisions

Cytoarchitectural descriptions of the Gi have been supplemented by many chemoarchitectural and connectivity studies which highlight specific regions or subnuclei within it [Paxinos et al., 2012]. The Gi contains a large population of serotonergic neurons which are concentrated rostrally around the RMg [Martin et al., 1990; Baker et al., 1991; Nieuwenhuys et al., 2008]. The serotonergic cell group RMg has been designated as B3 [Dahlström and Fuxe, 1964], but the 'wings' of B3 extend laterally into the Gi, an area which Meessen and Olszewski [1949] suggested be called Giα, in the rabbit. The name Giα is now generally accepted in all species for the serotonergic subdivision of the rostral Gi adjacent to the RMg. The term does not include the more caudal serotonergic region of the Gi adjacent to the nucleus raphe pallidus, often called Giv [Jones, 1995; Nieuwenhuys et al., 2008]. As a simplification several authors recognize only 1 subdivision in the Gi, Giα + Giv, and call it the magnocellular reticular nucleus or the 'magnocellular tegmental field' [Berman, 1968; Cowie et al., 1994].

Connections

In general, descending anatomical pathways to the somatic motor system can be divided into a medial and a lateral component [Holstege, 1991, 2009]. The Gi is part of the 'medial component of the somatic motor system' which controls posture through descending reticulospinal pathways. In contrast, the 'lateral component of the somatic motor system' controls refined and complex motor movements such as independent finger movements and speech, and utilizes the lateral corticospinal pathways. The spinal pathways of the 'medial motor system' are phylogenetically older than the lateral system, its fibers project to premotor interneurons in laminae VII and VIII, and usually bilaterally [Shinoda et al., 2006].

Efferents. In keeping with its role in the 'medial somatic motor system', the Gi contains many large reticulospinal cells whose axons descend mainly ipsilaterally, but with some bilaterality, to terminate close to motoneurons of multiple neck muscles [Mitani et al., 1988; Cowie et al., 1994; Sasaki, 1999; Shinoda et al., 2006]. Projections to the MoV, VII and hypoglossal nucleus were also found but none to the oculomotor nuclei, although ascending pathways from the Gi have been described which targeted premotor cell groups of the oculomotor system [Horn, 2006].

Both serotonergic and nonserotonergic pathways descend from the Gi to the spinal cord [Braz and Basbaum, 2008]. The serotonergic descending pathways from the Giα and RMg target the dorsal horn of the spinal cord (laminae I–V) bilaterally, throughout its whole length; slightly more caudal areas terminate on the intermediolateral cell columns as well. These pathways exert a general modulatory control of the sensory and autonomic activity (see 'Neuromodulatory Systems: Overview', and fig. 16.6 in Nieuwenhuys et al. [2008] as well as Holstege [1991]). These Giα pathways have been implied in pain control of the spinal cord and of the caudal spinal trigeminal nucleus [Willis and Westlund, 2004; Mason, 2005; Nieuwenhuys et al., 2008, fig. 16.6].

Afferents. A major input to the Gi arises from the superior colliculus projecting directly to the reticulospinal neurons. Other areas carrying activity related to head control also target the Gi, for example the rostral mesencephalic reticular formation, the interstitial nucleus of Cajal, the vestibular nuclear complex, the motor and postarcuate cortex [Cowie et al., 1994; Shinoda et al., 2006]. The limbic control of the Gi is carried out through descending pathways from the periaqueductal gray [Holstege, 1991; Holstege et al., 2004].

References

Baker KG, Halliday GM, Halasz P, Hornung JP, Geffen LB, Cotton RGH, Tork I: Cytoarchitecture of serotonin-synthesizing neurons in the pontine tegmentum of the human brain. Synapse 1991;7:301–320.

Berman AL: The Brain Stem of the Cat: A Cytoarchitectonic Atlas with Stereotaxic Coordinates. Madison, University of Wisconsin Press, 1968.

Braz JM, Basbaum AI: Genetically expressed transneuronal tracer reveals direct and indirect serotonergic descending control circuits. J Comp Neurol 2008; 507:1990–2003.

Cowie RJ, Robinson DL: Subcortical contributions to head movements in macaques. I. Contrasting effects of electrical stimulation of a medial pontomedullary region and the superior colliculus. J Neurophysiol 1994;72:2648–2664.

Cowie RJ, Smith MK, Robinson DL: Subcortical contributions to head movements in macaques. II. Connections of a medial pontomedullary head-movement region. J Neurophysiol 1994;72:2665–2682.

Dahlström A, Fuxe K: Evidence for the existence of monoamine neurons in the central nervous system. I. Demonstration of monoamines in the cell bodies of brain stem neurons. Acta Physiol Scand 1964;62:1–55.

Holstege G: Descending motor pathways and the spinal motor system: limbic and non-limbic components. Prog Brain Res 1991;87:307–421.

Holstege G: The mesopontine rostromedial tegmental nucleus and the emotional motor system: role in basic survival behavior. J Comp Neurol 2009;513:559–565.

Holstege G, Mouton LJ, Gerrits NM: Emotional motor system; in Paxinos G, Mai JK (eds): The Human Nervous System, ed 2. Amsterdam, Elsevier Academic Press, 2004.

Horn AKE: The reticular formation. Prog Brain Res 2006;151:127–155.

Jones BE: Reticular formation: cytoarchitecture, transmitters and projections; in Paxinos G (ed): The Rat Nervous System, ed 2. San Diego, Academic Press, 1995.

Martin GF, Holstege G, Mehler WR: Reticular formation of the pons and medulla; in Paxinos G (ed): The Human Nervous System. San Diego, Academic Press, 1990.

Mason P: Ventromedial medulla: pain modulation and beyond. J Comp Neurol 2005;493:2–8.

Meessen H, Olszewski J: A Cytoarchitectonic Atlas of the Rhombencephalon of the Rabbit. Basel, Karger, 1949.

Mitani A, Ito K, Mitani Y, McCarley RW: Descending projections from the gigantocellular tegmental field in the cat: cells of origin and their brainstem and spinal cord trajectories. J Comp Neurol1988;268:546–566.

Nieuwenhuys R, Voogd J, van Huijzen C: The Human Central Nervous System. Berlin, Springer, 2008.

Paxinos G, Huang X-F, Sengul G, Watson C: Organization of the brainstem; in Mai JK, Paxinos G (eds): The Human Nervous System, ed 3. Amsterdam, Academic Press, 2012.

Robinson FR, Phillips JO, Fuchs AF: Coordination of gaze shifts in primates: Brainstem inputs to neck and extraocular motoneuron pools. J Comp Neurol 1994; 346:43–62.

Sasaki S: Direct connection of the nucleus reticularis gigantocellularis neurons with neck motoneurons in cats. Exp Brain Res 1999;128:527–530.

Shinoda Y, Sugiuchi Y, Izawa Y, Hata Y: Long descending motor tract axons and their control of neck and axial muscles. Prog Brain Res 2006;151:527–563.

Willis WD, Westlund KN: Pain system; in Paxinos G, Mai JK (eds): The Human Nervous System, ed 2. Amsterdam, Elsevier Academic Press, 2004.

53 Lateral Paragigantocellular Nucleus (PGiL)

Original name:
Nucleus paragigantocellularis lateralis

Alternative name:
Subretrofacial vasomotor nucleus [Dampney, 1994]

Location and Cytoarchitecture – Original Text

This nucleus lies in the ventrolateral portion of the medullary tegmentum (fig. 1). Its orocaudal limits correspond to those of the caudal subnucleus of the gigantocellular nucleus (Gi; now ventral part, Giv), and those of the dorsal paragigantocellular nucleus. Caudally, the nucleus is bordered by the lateral reticular nucleus subtrigeminal part, orally by the nucleus of the trapezoid body and the superior olivary complex.

On cross-section the nucleus is irregularly quadrangular in outline and is related ventrally to the inferior olivary complex (IO), medially to the Giv, dorsally to the parvocellular reticular nucleus (PCR) and the spinal trigeminal tract and laterally to the fibers of the spinothalamic and other tracts, which course close to the periphery of the medulla.

The characteristic feature of the lateral paragigantocellular nucleus (PGiL) is the marked pleomorphism of its nerve cells. The majority of cells are medium sized, multipolar, usually slightly elongated, and possess dark Nissl granules. The arrangement of the latter and the position of the nucleus vary considerably (fig. 3–6). A few larger multipolar cells similar to type b cells of the Gi are seen (fig. 2). An occasional 'onion skin' cell may also be present. The distribution of cells within the PGiL is comparatively uniform and the majority of cells lie with their long axes directed dorsomedially.

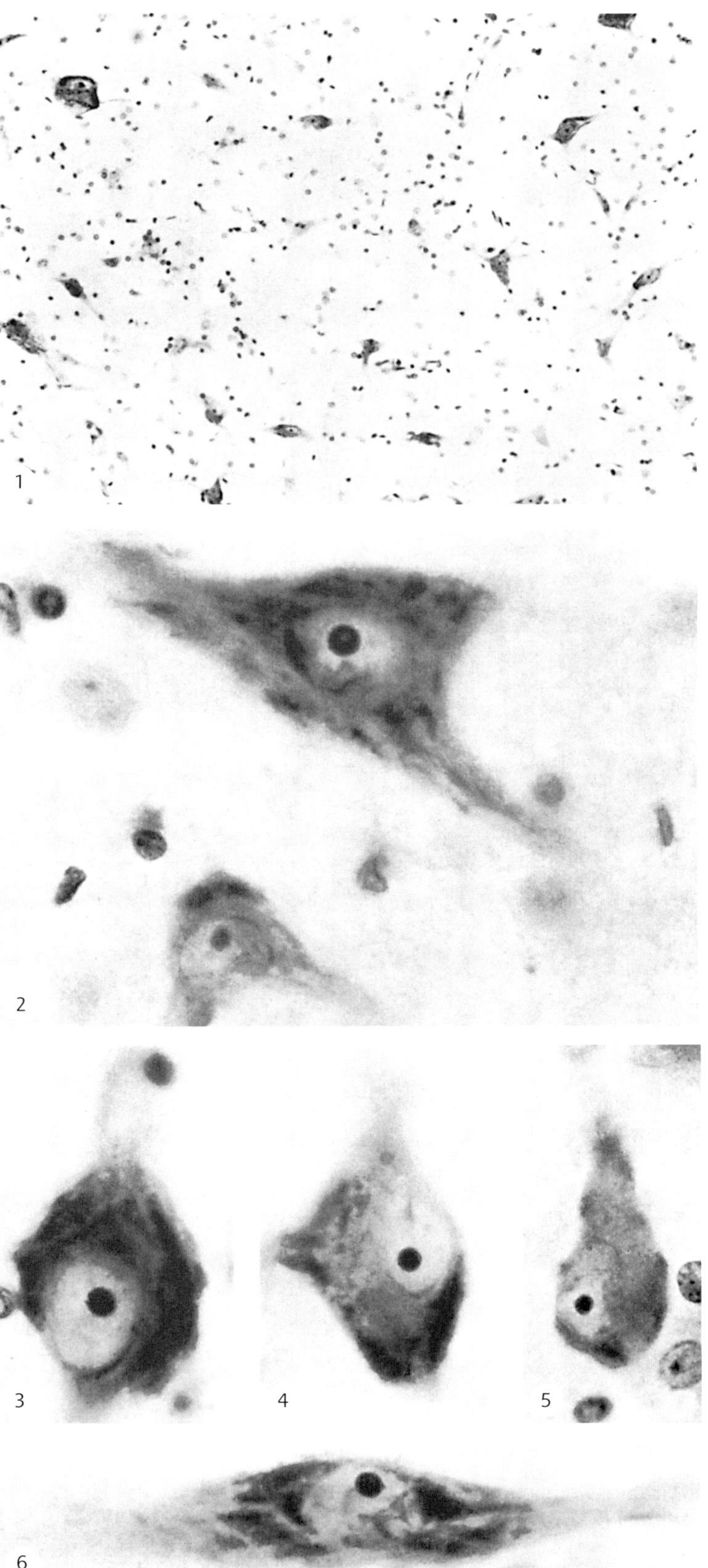

Fig. 1. Lateral paragigantocellular nucleus. Magnification ×150.
Fig. 2–6. Different cell types from the lateral paragigantocellular nucleus. Magnification ×1,000.

Functional Neuroanatomy

Function
The PGiL is part of the 'lateral tegmental field' which contains premotor networks subserving basic autonomic and limbic motor functions; see 'Reticular Formation: Overview' [Holstege, 1991; Holstege et al., 2004]. The networks controlling *cardiovascular* and *respiratory* functions lie in the rostral ventrolateral medullary region of the PGiL, central nucleus of the medulla oblongata (CN) and PCR [Nieuwenhuys et al., 2008; Blessing and Benarroch, 2012]. The area is important in the maintenance of normal arterial pressure and heart rate. Damage to this area, but not further rostrally, causes a sudden fall in arterial pressure to levels observed in spinal shock [Dampney, 1994; Blessing and Benarroch, 2012]. A neural network controlling *swallowing* has also been identified in or close to the PGiL and caudal nucleus parvocellularis compactus [Blessing and Benarroch, 2012].

Connections
Efferents. The PGiL encompasses rostrally part of the lateral noradrenergic A5 cell group of Dahlström and Fuxe [1964], as well as at least the rostral part of A1 and the adrenergic C1 group in its caudal part [Nieuwenhuys et al., 2008, fig. 22.2].

The A5 cell group supplies autonomic centers of the brainstem, the IO and cerebellum; the cholinergic cell groups of the dorsolateral pontine reticular formation associated with the induction of REM sleep, the paraventricular nucleus of the hypothalamus and A5 have modest projections to the spinal cord including the intermediolateral nucleus. Although several groups of neurons in the region of A5 provide potent excitatory amino acid inputs to the locus coeruleus (LC), constituting one of LC's main afferent inputs, A5 itself does not contribute to these projections [Aston-Jones et al., 1991; Camp and Wijesinghe, 2009].

Afferents. Afferents to the PGiL arise from a widespread collection of sources, mainly nuclei with autonomic, visceral and sensory-related functions, like the dorsal vagus nucleus, the parabrachial and solitary nuclei (SOL), the area of Kölliker-Fuse and paraventricular nucleus of the hypothalamus. For a review and further references, see chapter 22 in Nieuwenhuys et al. [2008].

The PGiL, including A5, receives descending control from limbic structures, specifically from the periaqueductal gray, bed nucleus of the stria terminalis and the central nucleus of the amygdala [Bandler et al., 1991; Holstege, 1991].

Functional Centers in the Ventrolateral Medulla

Cardiovascular. The Gi and the PGiL region contain the rostral part premotor networks controlling cardiovascular reflexes called the 'rostral ventrolateral medulla' [Blessing and Benarroch, 2012; Li et al., 2012]. The region lies close to the nucleus ambiguus and projects to the preganglionic neurons of the sympathetic system in the thoracic cord; see the chapter on the CN, No. 51 [Dobbins and Feldman, 1994].

Swallowing. The neural network for swallowing has two centers: a dorsal swallowing group close to the SOL and hypoglossal nucleus, and a ventral swallowing group just above the nucleus ambiguus and close to the PGiL (in the intermediate reticular nucleus, IRt, see 'Reticular Formation: Overview'). The ventral area distributes the swallowing drive to various brainstem motoneuronal pools [Amirali et al., 2001; Jean, 2001; Zerari-Mailly et al., 2001]; see also the chapter on the PCR, No. 54.

Respiration. The long column of respiratory neurons contains several functionally distinct cell groups [Nieuwenhuys et al., 2008; Blessing and Benarroch, 2012]. For example most caudally in the dorsal subnucleus of the CN lies a group of third-order expiratory neurons; just rostral to them and extending into the PGiL lies a group of second-order excitatory inspiratory premotor neurons, which in turn are influenced by a more rostral cluster of pacemaker neurons called the pre-Bötzinger group; further rostrally again, also in the PGiL, and just below the spinal trigeminal nucleus, is the Bötzinger cell group, whose neurons actively inhibit the phrenic motoneurons of the diaphragm in C_5–C_6, during expiration. The term 'Bötzinger' comes from the name of a German vineyard near Freiburg, where brainstem specialists were discussing possible nomenclatures for neuronal cell groups over a glass of wine, after a scientific meeting in 1980 on the SOL.

The PGiL contains part of the region called the IRt. The column of functional centers described above appears to be more related to this region than the PGiL; however, this suggestion must still be investigated (see 'Reticular Formation: Overview').

References

Amirali A, Tsai G, Schrader N, Weisz D, Sanders I: Mapping of brain stem neuronal circuitry active during swallowing. Ann Otol Rhinol Laryngol 2001;110:502–513.

Aston-Jones G, Shipley MT, Chouvet G, Ennis M, van Bockstaele E, Pieribone V, Shiekhattar R, Akaoka H, Drolet G, Astier B, et al: Afferent regulation of locus coeruleus neurons: anatomy, physiology and pharmacology. Prog Brain Res 1991;88:47–75.

Bandler R, Carrive P, Zhang SP: Integration of somatic and autonomic reactions within the midbrain periaqueductal grey: viscerotopic, somatotopic and functional organization. Prog Brain Res 1991;87:269–305.

Blessing WW, Benarroch EE: Lower brain stem regulation of visceral, cardiovascular, and respiratory function; in Mai JK, Paxinos G (eds): The Human Nervous System, ed 3. Amsterdam, Academic Press, 2012.

Camp AJ, Wijesinghe R: Calretinin: modulator of neuronal excitability. Int J Biochem Cell Biol 2009;41:2118–2121.

Dahlström A, Fuxe K: Evidence for the existence of monoamine neurons in the central nervous system. I. Demonstration of monoamines in the cell bodies of brain stem neurons. Acta Physiol Scand 1964;62:1–55.

Dampney RA: The subretrofacial vasomotor nucleus: anatomical, chemical and pharmacological properties and role in cardiovascular regulation. Prog Neurobiol 1994;42:197–227.

Dobbins EG, Feldman JL: Brain stem network controlling descending drive to phrenic motoneurons in rat. J Comp Neurol 1994;347:64–86.

Holstege G: Descending motor pathways and the spinal motor system: limbic and non-limbic components. Prog Brain Res 1991;87:307–421.

Holstege GG, Mouton LJ, Gerrits NM: Emotional motor system; in Paxinos G, Mai JK (eds): The Human Nervous System, ed 2. Amsterdam, Elsevier Academic Press, 2004.

Jean A: Brain stem control of swallowing: neuronal network and cellular mechanisms. Physiol Rev 2001;81:929–969.

Li FCH, Yen J-C, Chan SHH, Chang AYW: Bioenergetics failure and oxidative stress in brain stem mediates cardiovascular collapse associated with fatal methamphetamine intoxication. PLoS One 2012;7:e30589.

Nieuwenhuys R, Voogd J, van Huijzen C: The Human Central Nervous System. Berlin, Springer, 2008.

Zerari-Mailly F, Pinganaud G, Dauvergne C, Buisseret P, Buisseret-Delmas C: Trigemino-reticulo-facial and trigemino-reticulo-hypoglossal pathways in the rat. J Comp Neurol 2001;429:80–93.

54 Parvocellular Reticular Nucleus (PCR)

Original name:
Nucleus parvocellularis

Alternative name:
Parvocellular nucleus

Location and Cytoarchitecture – Original Text

This nucleus occupies the dorsolateral portion of the tegmentum of the medulla and pons from the level of the junction of the oral and middle thirds of the inferior olivary complex to the level of the caudal pole of the motor trigeminal nucleus (MoV).

On cross-section, the nucleus is kidney shaped in outline and lies with its hilus directed dorsolaterally. In the medulla this hilus partially encloses the spinal trigeminal nucleus, oral part (SpVo). The dorsal surface of the nucleus is related to the solitary nucleus (SOL), and the curved medioventral surface is related in a dorsoventral direction to the dorsal paragigantocellular, the gigantocellular (Gi) and the lateral paragigantocellular nuclei (PGiL), respectively. [The border region of these nuclei is called the intermediate reticular zone by Paxinos et al. [2012], see 'Reticular Formation: Overview'.]

At the junction of the pons and medulla the most orally situated cells of the ambiguus nucleus (AMB) lie embedded in the parvocellular reticular nucleus (PCR), medial to the spinal trigeminal complex.

In the caudal pons the PCR is related laterally to the SpVo and the facial nucleus (VII), ventrally to the periolivary complex and the superior olivary complex, medially to the Gi and the nucleus reticularis pontis caudalis (NRPC), and dorsally to the medial and lateral vestibular (LVN) nuclei. At the level of the abducens nucleus (VI), the PCR increases in volume, becomes quadrangular in outline and is traversed by fibers of the facial nerve. This oral pole of the nucleus is related laterally to the main sensory trigeminal nucleus, medially to the NRPC and ventrally to the caudal portion of the nucleus subcoeruleus. Dorsally, this portion of the nucleus reaches the floor of the fourth ventricle and intervenes between the VI dorsomedially and the oral pole of the LVN dorsolaterally.

The PCR is composed of loosely distributed, irregularly oriented, medium-sized or small, triangular or fusiform cells (fig. 1). The medium-sized cells stain with moderate intensity and possess Nissl substance in the form of distinct granules which occasionally clump together to form larger masses (fig. 2–4). Many of these cells are surrounded by glial satellites. The small cells possess indistinct Nissl granules and their cytoplasm stains very faintly (fig. 5). The cells composing the oral part of the nucleus are, on the average, less numerous and slightly larger.

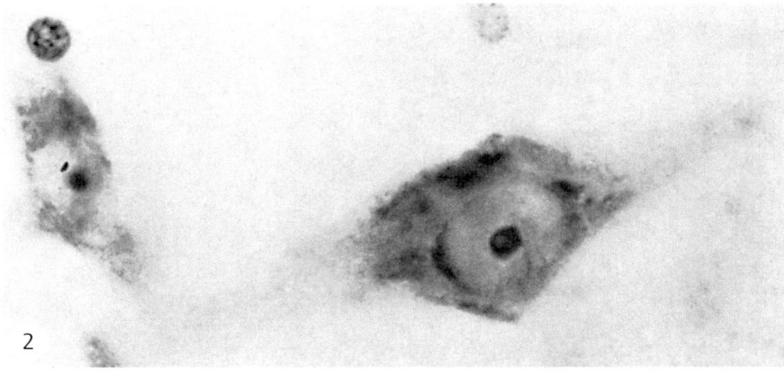

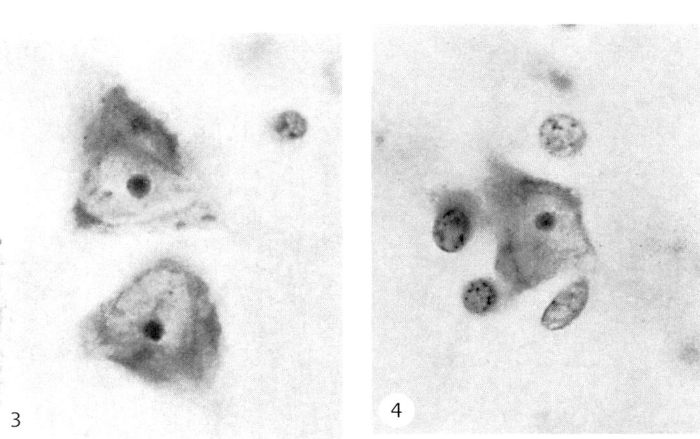

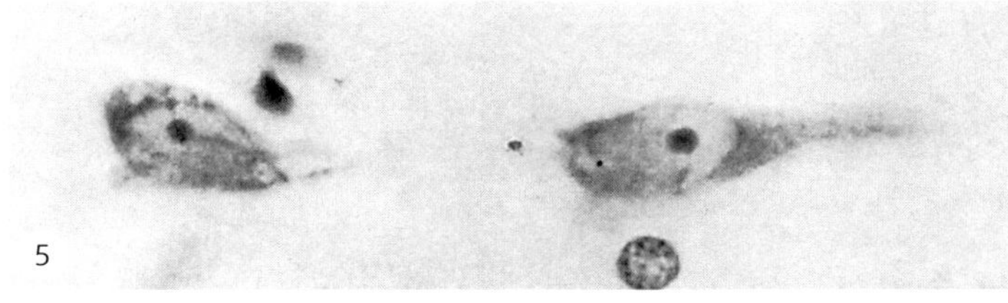

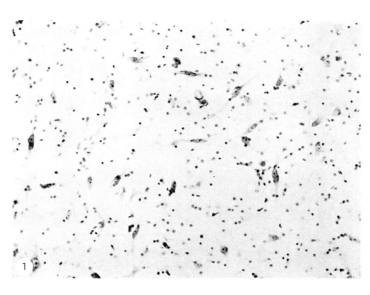

Fig. 1. Parvocellular reticular nucleus. Magnification ×150.
Fig. 2–5. Cells from the parvocellular reticular nucleus. Magnification ×1,000.

Functional Neuroanatomy

Function

The PCR extends from the caudal pons into the caudal medulla, and is part of the lateral tegmental fields, that is it contains interneurons which control the motoneurons of the brainstem and spinal cord [Holstege, 1991]. The PCR and adjacent areas contain numerous premotor networks that coordinate the motor activity in adjacent nuclei, such as the VII, AMB, trigeminal and hypoglossal (XII) nuclei. The PCR networks are under the descending influence of the limbic system, and generate basic movements such as micturition, blinking, salivation, mastication, vocalization, swallowing, and breathing [Holstege, 1991]. The PCR area also controls cardiovascular and visceral homeostasis [Blessing and Benarroch, 2012]. Several of the networks lie around the ventral border of the PCR, in the region called the intermediate reticular zone (see 'Reticular Formation: Overview') [Jones, 1995; Paxinos et al., 2012].

Unilateral damage to the PCR may not result in clinical symptoms, but bilateral damage is life threatening [Blessing and Benarroch, 2012].

Connections

In general, the afferents to the PCR arise from limbic structures such as the bed nucleus of the stria terminalis (lateral part), the amygdala and the lateral hypothalamus and the periaqueductal gray [Holstege, 1991]. There are also reciprocal connections with the SOL and the dorsal vagal complex. The local efferents of PCR networks target the MoV, VII, AMB and XII motor nuclei [Holstege et al., 1977]. An ascending pathway from the PCR terminates in the adrenergic A5 cells of the PGiL, and other fibers continue to the hypothalamus to the paraventricular nucleus, which secretes vasopressin. Descending PCR pathways terminate on effector neurons such as the phrenic, intercostal, abdominal and pelvic motoneurons, and the preganglionic sympathetic cells in the intermediolateral column of the cord.

The PCR is not functionally homogeneous, and its networks do not have clear cytoarchitectural borders. The neural centers associated with the PCR, or perhaps more closely to the intermediate reticular zone (see 'Reticular Formation: Overview'), are summarized below, roughly from rostral to caudal [Holstege et al., 2004; Nieuwenhuys et al., 2008; Blessing and Benarroch, 2012].

Micturition. The most rostral part of the PCR in the pons provides the supraspinal control for micturition. It is located in the dorsolateral tegmental field at approximately the level of the VI and has 2 centers: the M region of Holstege, or Barrington's nucleus, just medial and caudal to the locus coeruleus, and the L region further caudal and lateral [Holstege et al., 1986; Holstege, 1991]. The M region promotes micturition via bilateral projections to the sacral cord, terminating on parasympathetic motoneurons innervating the bladder, and on the commissural nucleus which inhibits motoneurons of the urethral sphincter in Onuf's nucleus. Stimulation of the L region supports urine retention via excitatory connections to Onuf's nucleus activating the urethral sphincter motoneurons. Studies with PET scans have verified that similar neural control centers are present in humans, although surprisingly it was reported that the cortical and pontine micturition sites are more active on the right than on the left side, in right-handed subjects [Blok et al., 1997, 1998].

Premotor Blink Center. At pontine levels a premotor blink center has been identified in the PCR, within the ventrolateral tegmental field just dorsal to the VII [Holstege, 1991; Büttner-Ennever and Horn, 2004]. The efferents of the pontine blink center project to the orbicularis oculi motoneurons in the VII and retractor bulbi motoneurons further rostrally.

Salivation. The lateral tegmental field of the pontomedullary junction contains the neural circuitry for salivation. Traditionally the parasympathetic salivary motoneurons are divided into the superior and inferior salivary subnuclei, whose axons were thought to travel to the periphery in the facial or glossopharyngeal nerve, respectively [Ter Horst et al., 1991; Blessing, 2004]. However, more recent experimental data show that there is virtually no difference in the distribution of the salivary neurons of either nerve: they lie intermingled in the ventrolateral PCR, dorsal and lateral to the VII in the caudal pons; for an excellent review, see Blessing [2004]. These neurons may be cholinergic, and stain strongly for NADPH and nitric oxide synthetase also in man [Ramos and Puerto, 1988; Gai and Blessing, 1996]. Presumably some of the surrounding cells subserve similar functions such as lacrimation and nasal mucosal secretions. The network probably regulates the extra- and intracranial vasculature as well, since the fluid in saliva is rapidly filtered from the blood, and its production is accompanied by a rise in blood flow to the salivary gland.

Mastication. Premotor networks regulating vocalization have been located in the rostral PCR close to the trigeminal nuclei of the mid pons. Other nearby areas also support mastication [Hannig and Jürgens, 2006]. The jaw-opening motoneurons receive their strongest projections from levels caudal to the obex in the nucleus retroambiguus, whereas jaw-closing motoneurons have their premotor network further rostrally at PCR levels near the XII, in Probst's nucleus, and up to the supra- and intertrigeminal nuclei, and continuing into the lower pons to the region of the parabrachial nuclei [Mizuno et al., 1981; Holstege, 1991; Minkels et al., 1995; Bourque and Kolta, 2001; Li et al., 2005]; for review diagrams, see figures 21.12 and 21.15 in Nieuwenhuys et al. [2008].

Swallowing. The swallowing network in the PCR lies further caudally in the medulla oblongata, and involves a central pattern generator for swallowing movements near the AMB. The cell groups in this area distribute the swallowing drive to the brainstem motoneuronal pools involved in the act, and are themselves triggered by inputs from a separate set of neurons near, or even within, the SOL [Jean, 2001; Feldman et al., 2003]. The cell group referred to as the dorsal medullary reticular column in the studies on swallowing by Cunningham and Sawchenko [2000] lies at the dorsomedial tip of the caudal PCR in the same region as the swallowing trigger network, and near the A2 and C2 catecholaminergic cell groups around the SOL, and the nucleus of Probst [Nieuwenhuys et al., 2008, fig. 21.16]. The experimental results on the control of swallowing are supported to some extent by clinical studies [Prosiegel et al., 2005].

Respiration. Lying close to the cardiovascular cell columns is another column of neurons controlling respiration: the 'caudal ventral respiratory group' contains expiratory neurons innervating abdominal muscles; immediately rostral to this is the 'rostral ventral respiratory group' containing inspiratory neurons projecting to motoneurons of the phrenic nerve; further rostrally lies the 'pre-Bötzinger cell group', near the border of the PGiL and ventral PCR, which acts as a pacemaker for respiration. The most rostral group is the 'Bötzinger complex', a group of expiratory neurons in the ventral PCR with local reciprocal connections, and projections to the phrenic motoneurons [Blessing, 2004; Nieuwenhuys et al., 2008, fig. 22.4]; see also nucleus No. 53.

Cardiovascular Control. The 'ventrolateral superficial medullary reticular formation' is recognized as the location of neural circuits regulating some aspects of cardiovascular function, including the baroreflex [Nieuwenhuys et al., 2008]. Other research groups use the name of the 'rostral ventrolateral medulla oblongata' for the same area [Blessing, 2004]. It contains the noradrenergic and adrenergic cell groups A1 and C1, and the neurons build a longitudinal column of cell groups stretching up to the pons, interconnected with the SOL, and projecting up to the A5 group near the superior olive. Several studies have established details of the neural circuitry of the cardiovascular cell groups [for reviews, see Blessing, 2004; Nieuwenhuys et al., 2008, fig. 22.4].

References

Blessing WW: Lower brain stem regulation of visceral, cardiovascular, and respiratory function; in Paxinos G, Mai JK (eds): The Human Nervous System, ed 2. Amsterdam, Elsevier Academic Press, 2004.

Blessing WW, Benarroch EE: Lower brain stem regulation of visceral, cardiovascular, and respiratory function; in Mai JK, Paxinos G (eds): The Human Nervous System, ed 3. Amsterdam, Elsevier Academic Press, 2012.

Blok BF, Sturms LM, Holstege G: Brain activation during micturition in women. Brain 1998;121:2033–2042.

Blok BF, Willemsen AT, Holstege G: A PET study on brain control of micturition in humans. Brain 1997;120:111–121.

Bourque MJ, Kolta A: Properties and interconnections of trigeminal interneurons of the lateral pontine reticular formation in the rat. J Neurophysiol 2001;86: 2583–2596.

Büttner-Ennever JA, Horn AKE: Reticular formation: eye movements, gaze and blinks; in Paxinos G, Mai JK (eds): The Human Nervous System, ed 2. Amsterdam, Elsevier Academic Press, 2004.

Cunningham ETJ, Sawchenko PE: Dorsal medullary pathways subserving oromotor reflexes in the rat: implications for the central neural control of swallowing. J Comp Neurol 2000;417:448–466.

Feldman JL, Mitchell GS, Nattie EE: Breathing: rhythmicity, plasticity, chemosensitivity. Annu Rev Neurosci 2003;26:239–266.

Gai WP, Blessing WW: Human brainstem preganglionic parasympathetic neurons localized by markers for nitric oxide synthesis. Brain 1996;119:1145–1152.

Hannig S, Jürgens U: Projections of the ventrolateral pontine vocalization area in the squirrel monkey. Exp Brain Res 2006;169:92–105.

Holstege G: Descending motor pathways and the spinal motor system: limbic and non-limbic components. Prog Brain Res 1991;87:307–421.

Holstege G, Griffiths D, de Wall H, Dalm E: Anatomical and physiological observations on supraspinal control of bladder and urethral sphincter muscles in the cat. J Comp Neurol 1986;250:449–461.

Holstege G, Kuypers HG, Dekker JJ: The organization of the bulbar fibre connections to the trigeminal, facial and hypoglossal motor nuclei. II. An autoradiographic tracing study in cat. Brain 1977;100:265–286.

Holstege GG, Mouton LJ, Gerrits NM: Emotional motor system; in Paxinos G, Mai JK (eds): The Human Nervous System, ed 2. Amsterdam, Elsevier Academic Press, 2004.

Jean A: Brain stem control of swallowing: neuronal network and cellular mechanisms. Physiol Rev 2001;81:929–969.

Jones BE: Reticular formation: cytoarchitecture, transmitters and projections; in Paxinos G (ed): The Rat Nervous System, ed 2. San Diego, Academic Press, 1995.

Li JL, Wu SX, Tomioka R, Okamoto K, Nakamura K, Kaneko T, Mizuno N: Efferent and afferent connections of GABAergic neurons in the supratrigeminal and the intertrigeminal regions. An immunohistochemical tract-tracing study in the GAD67-GFP knock-in mouse. Neurosci Res 2005;51:81–91.

Minkels RF, Juch PJ, van Willigen JD: Interneurones of the supratrigeminal area mediating reflex inhibition of trigeminal and facial motoneurones in the rat. Arch Oral Biol 1995;40:275–284.

Mizuno N, Matsuda K, Iwahori N, Uemura-Sumi M, Kume M, Matsushima R: Representation of the masticatory muscles in the motor trigeminal nucleus of the macaque monkey. Neurosci Lett 1981;21:19–22.

Nieuwenhuys R, Voogd J, van Huijzen C: The Human Central Nervous System. Berlin, Springer, 2008.

Paxinos G, Huang X-F, Sengul G, Watson C: Organization of the brainstem; in Mai JK, Paxinos G (eds): The Human Nervous System, ed 3. Amsterdam, Academic Press, 2012.

Prosiegel M, Holing R, Heintze M, Wagner-Sonntag E, Wiseman K: The localization of central pattern generators for swallowing in humans – a clinical-anatomical study on patients with unilateral paresis of the vagal nerve, Avellis' syndrome, Wallenberg's syndrome, posterior fossa tumours and cerebellar hemorrhage. Acta Neurochir Suppl 2005;93:85–88.

Ramos JM, Puerto A: The nucleus parvocellularis reticularis regulates submandibular-sublingual salivary secretion in the rat: a pharmacological study. J Auton Nerv Syst 1988;23:221–228.

Ter Horst GJ, Copray JC, Liem RS, van Willigen JD: Projections from the rostral parvocellular reticular formation to pontine and medullary nuclei in the rat: involvement in autonomic regulation and orofacial motor control. Neuroscience 1991;40:735–758.

Reticular Formation

(Plates 14–21)

55 Dorsal Paragigantocellular Nucleus (PGiD)

Original name:
Nucleus paragigantocellularis dorsalis

Alternative name:
Nucleus supragigantocellularis [Langer et al., 1986]

Location and Cytoarchitecture – Original Text

This nucleus comprises the dorsomedial part of the medullary tegmentum from the level of the junction, the oral and middle thirds of the inferior olivary complex to the level of the caudal pole of the facial nucleus. It is bounded caudally by the central nucleus of the medulla oblongata and orally by the nucleus reticularis pontis caudalis (NRPC).

On cross-section, the dorsal paragigantocellular nucleus (PGiD) is quadrangular in outline and lies, at all levels, ventral to the nucleus prepositus (PrP), medial to the parvocellular reticular nucleus, dorsal to the caudal subnucleus of the gigantocellular nucleus (Gi) and lateral to the medial longitudinal fasciculus (MLF) and the tectospinal tract. Towards the oral boundary of the medulla, the two latter structures become infiltrated with occasional cells which are indistinguishable from those of the PGiD. At such levels the medial border of the nucleus has been indicated with an interrupted line (plate 16).

The PGiD is less cellular than the nuclei which border it, and its constituent cells are loosely and irregularly arranged (fig. 1). The majority of cells vary from small to medium in size. The small cells are of variable form and possess distinct nuclei and nucleoli and a palely stained cytoplasm (fig. 4). The medium-sized cells are slender and multipolar, with long dendrites and darkly stained, evenly distributed Nissl granules (fig. 5). Many cells exhibit characteristics intermediate between those typical of the small- and medium-sized cells (fig. 6). Occasional large cells identical with either the 'motor' or 'onion skin' cells of the Gi may also be present (fig. 2, 3).

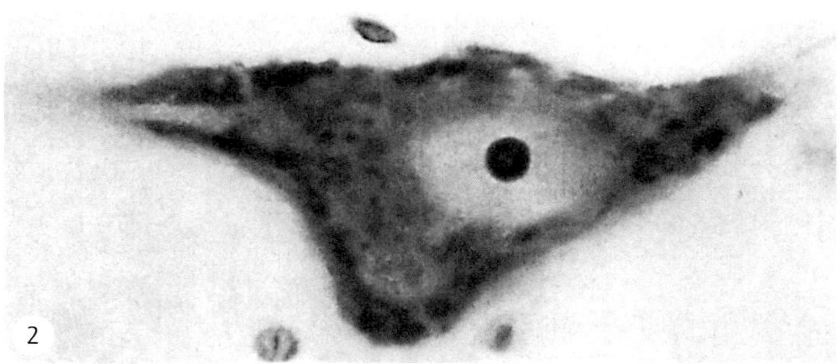

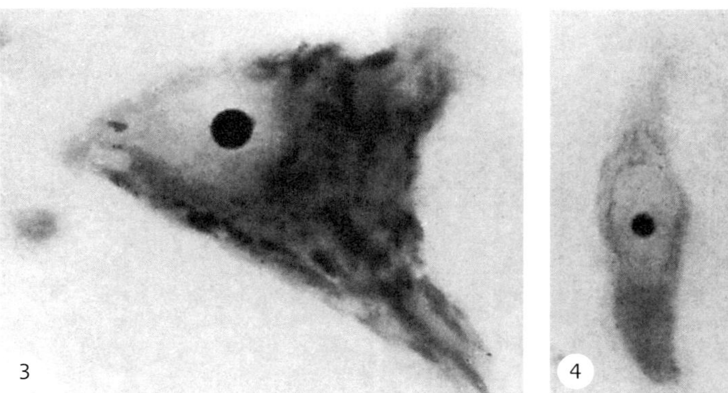

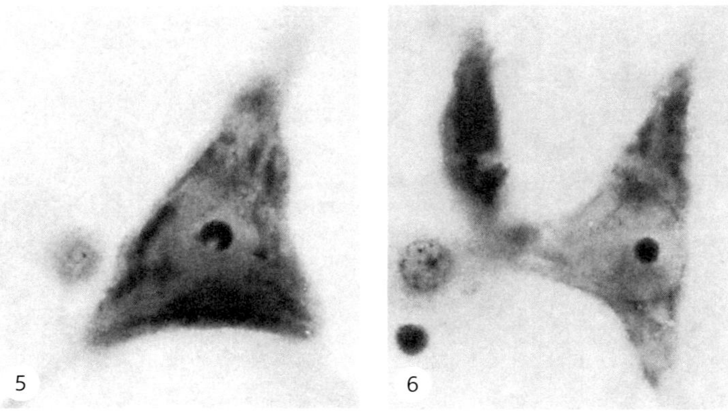

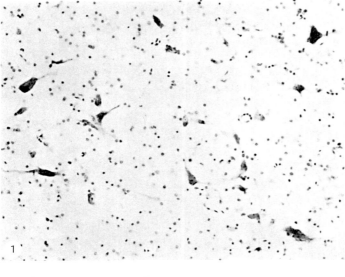

Fig. 1. Dorsal paragigantocellular nucleus. Magnification ×150.
Fig. 2–6. Different cell types from the dorsal paragigantocellular nucleus. Magnification ×1,000.

Functional Neuroanatomy

Function

The area belongs to the 'medial tegmental field', and participates in the coordination of eye and head movements. Through its tectal afferents, the PGiD is associated with the control of gaze, rather than head movements alone [Shinoda et al., 2006], although head movements can be evoked from the PGiD by electrical stimulation [Cowie et al., 1994].

Three groups of neurons in the PGiD are thought to be important for this function: premotor inhibitory burst neurons (IBN), burster-driving neurons (BDN), and reticulospinal neurons (RSN). The IBN lie medially near the MLF, and are an important part of the network for horizontal saccadic eye movements. During a saccade to one side the IBN of the same side inhibit the antagonistic extraocular motoneuron pools [Scudder and Kaneko, 2002; Büttner-Ennever and Horn, 2004]. The BDN also participate in saccade generation; their pattern of cell firing correlates with the quick-phases of vestibular nystagmus. They have been detected in the PrP and the reticular formation below it, that is the PGiD [Ohki et al., 1988]. Finally the RSN participate in the tectoreticulospinal activation of neck muscles in orienting responses [Shinoda et al., 2006]. The RSN can be divided into two groups, one centered on the NRPC and the other more caudally in the Gi. The cells in the PGiD belong to the more rostral group of RSN, which are driven mainly from the superior colliculus rather than the motor cortex.

Histochemistry

The IBN use glycine as the transmitter [Spencer et al., 1989], and they belong to the medium-sized cell population that expresses parvalbumin immunoreactivity: these properties have helped to outline the homologous region in humans as well [Horn et al., 1995, 1996].

Connections

Efferents. The IBN in the medial PGiD send their axons across the midline, to terminate in the contralateral abducens nucleus; they monosynaptically inhibit the antagonistic lateral rectus muscle motoneurons and the abducens internuclear neurons, during horizontal conjugate saccadic eye movements to the contralateral side [Strassman et al., 1986; Scudder et al., 1988; Horn et al., 1996]. The IBN axons also terminate in the pontine reticular formation and cell groups of the paramedian tracts (nucleus No. 96), which project to the cerebellum [Strassman et al., 1986; Ohgaki et al., 1989; Büttner-Ennever and Horn, 2004]. The descending axons of the RSN project ipsilaterally in the medial funiculus to the cervical cord and continue down to lumbar levels [Iwamoto and Sasaki, 1990; May and Porter, 1992; Shinoda et al., 2006].

Afferents. The cells in the PGiD receive tectal afferents from the contralateral superior colliculus, and inputs from the NRPC, more specifically from omnipause neurons and the excitatory burst neurons [Sugiuchi et al., 2005; Shinoda et al., 2008].

References

Büttner-Ennever JA, Horn AKE: Reticular formation: eye movements, gaze and blinks; in Paxinos G, Mai JK (eds): The Human Nervous System, ed 2. Amsterdam, Elsevier Academic Press, 2004.

Cowie RJ, Smith MK, Robinson DL: Subcortical contributions to head movements in macaques. II. Connections of a medial pontomedullary head-movement region. J Neurophysiol 1994;72:2665–2682.

Horn AKE, Büttner-Ennever JA, Büttner U: Saccadic premotor neurons in the brainstem: functional neuroanatomy and clinical implications. Neuro-Ophthalmology 1996;16:229–240.

Horn AKE, Büttner-Ennever JA, Suzuki Y, Henn V: Histological identification of premotor neurons for horizontal saccades in monkey and man by parvalbumin immunostaining. J Comp Neurol 1995;359:350–363.

Iwamoto Y, Sasaki S: Monosynaptic excitatory connexions of reticulospinal neurones in the nucleus reticularis pontis caudalis with dorsal neck motoneurones in the cat. Exp Brain Res 1990;80:277–289.

Langer TP, Kaneko CR, Scudder CA, Fuchs AF: Afferents to the abducens nucleus in the monkey and cat. J Comp Neurol 1986;245:379–400.

May PJ, Porter JD: The laminar distribution of macaque tectobulbar and tectospinal neurons. Vis Neurosci 1992;8:257–276.

Ohgaki T, Markham CH, Schneider JS, Curthoys IS: Anatomical evidence of the projection of pontine omnipause neurons to midbrain regions controlling vertical eye movements. J Comp Neurol 1989;289:610–625.

Ohki Y, Shimazu H, Suzuki I: Excitatory input to burst neurons from the labyrinth and its mediating pathway in the cat: location and functional characteristics of burster-driving neurons. Exp Brain Res 1988;72:457–472.

Scudder CA, Fuchs AF, Langer TP: Characteristics and functional identification of saccadic inhibitory burst neurons in the alert monkey. J Neurophysiol 1988;59:1430–1454.

Scudder CA, Kaneko CRS: The brainstem burst generator for saccadic eye movements. Exp Brain Res 2002;142:439–462.

Shinoda Y, Sugiuchi Y, Izawa Y, Hata Y: Long descending motor tract axons and their control of neck and axial muscles. Prog Brain Res 2006;151:527–563.

Shinoda Y, Sugiuchi Y, Izawa Y, Takahashi M: Neural circuits for triggering saccades in the brainstem. Prog Brain Res 2008;171:79–85.

Spencer RF, Wenthold RJ, Baker R: Evidence for glycine as an inhibitory neurotransmitter of vestibular, reticular, and prepositus hypoglossi neurons that project to the cat abducens nucleus. J Neurosci 1989;9:2718–2736.

Strassman A, Highstein SM, McCrea RA: Anatomy and physiology of saccadic burst neurons in the alert squirrel monkey. II. Inhibitory burst neurons. J Comp Neurol 1986;249:358–380.

Sugiuchi Y, Izawa Y, Takahashi M, Na J, Shinoda Y: Physiological characterization of synaptic inputs to inhibitory burst neurons from the rostral and caudal superior colliculus. J Neurophysiol 2005;93:697–712.

Reticular Formation (Plates 22–25)

56 Nucleus Reticularis Pontis Caudalis (NRPC)

Original name:
Nucleus pontis centralis caudalis

Associated name:
Paramedian pontine reticular formation

Location and Cytoarchitecture – Original Text

This nucleus occupies a major portion of the tegmentum of the caudal half of the pons. It commences at the level of the caudal pole of the facial nucleus (VII) and extends to the level of the midpoint of the motor trigeminal nucleus (MoV). It is bounded caudally by the dorsal paragigantocellular nucleus (PGiD), and orally by the nucleus reticularis pontis oralis (NRPO).

Caudally, the nucleus reticularis pontis caudalis (NRPC) appears on cross-sections as a quadrilateral area in the dorsomedial part of the pontine tegmentum. It is related dorsally to the prepositus nucleus and the medial vestibular nucleus, laterally to the parvocellular reticular nucleus (PCR), ventrally to the gigantocellular nucleus (Gi) and medially to the median raphe and the medial longitudinal fasciculus (MLF). Proceeding orally, the nucleus increases in size at the expense of the Gi and comes to lie ventral to the abducens nucleus. At this level the NRPC is traversed by the vertically coursing fibers of the sixth cranial nerve. Oral to the disappearance of the Gi and PCR, the NRPC occupies almost the entire pontine tegmentum medial to the trigeminal complex. The principal relations of' this oral portion of the nucleus are the median raphe and MLF medially, the fibers of the seventh cranial nerve and the central gray matter of the pons dorsally, the MoV laterally and the fibers of the medial lemniscus ventrally.

The majority of cells which compose the NRPC are small to medium sized, fusiform or triangular in outline and possess long dendrites and Nissl substance which is congregated into irregularly shaped small clumps and stains with moderate intensity (fig. 1, 3–5; see 'excitatory burst neurons' below). These cells are irregularly oriented and form a background upon which are scattered a few large, multipolar neurons, which possess small, evenly distributed, darkly stained Nissl granules and very long dendrites (fig. 2, see 'reticulospinal neurons' below). These cells superficially resemble those found in the somatic motor nuclei but on closer examination are found to possess distinctly smaller Nissl granules and longer dendrites.

At the level of the rootlets of the sixth cranial nerve (plates 22 and 23), the medial part of the NRPC possesses distinctive morphological characteristics. This portion of the nucleus is composed almost exclusively of numerous, medium-sized, elongated neurons which possess long dendrites and darkly stained, evenly distributed Nissl granules (fig. 6). The majority of these cells lie with their long axes perpendicular to the median raphe (see 'omnipause neurons' below, and fig. 2).

Functional Neuroanatomy

Function
This region of the reticular formation, the NRPC, contains premotor networks for the generation of the horizontal component of conjugate eye and head movements to the ipsilateral side. The cells discharge before all types of saccades, and quick-phases of nystagmus, with a horizontal component, or before a head movement [Robinson et al., 1994]. The saccadic networks are separate from those coordinating vestibular, optokinetic, smooth pursuit or gaze-holding eye movements. Clinical and experimental studies show that lesions in the NRPC produce ipsilateral horizontal gaze palsy. The effective brainstem lesions in macaque monkeys were centered on the

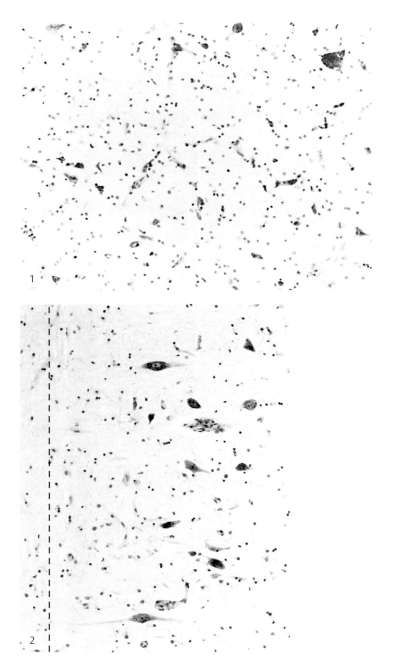

Fig. 1. Nucleus pontis centralis caudalis. Magnification ×150.
Fig. 2. Nucleus pontis centralis caudalis, medial part covering nucleus raphe interpositus which contains the horizontally oriented omnipause neurons. The dashed line marks the midline. Magnification ×150.

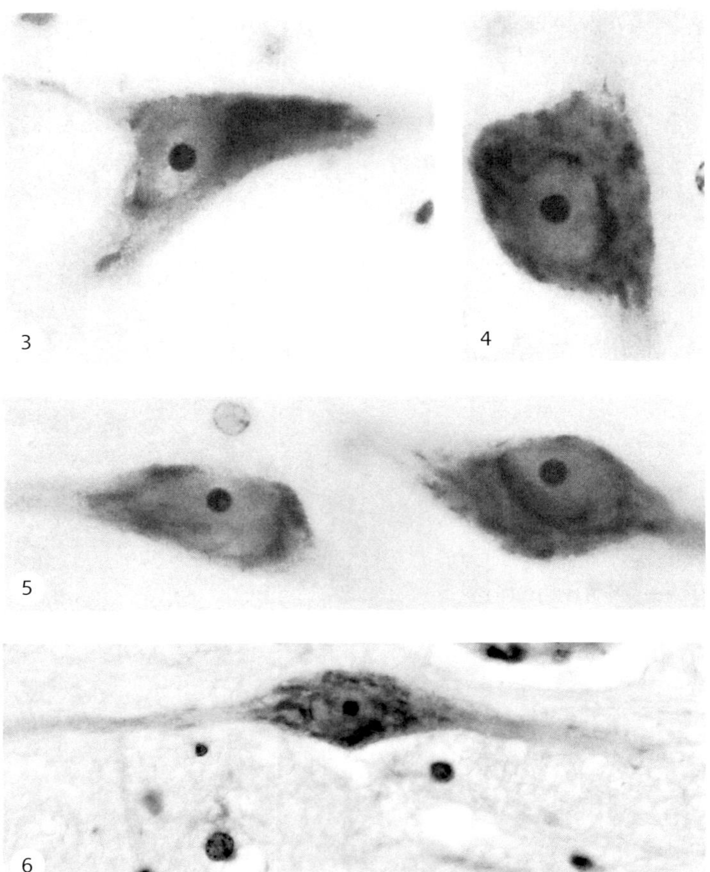

Fig. 3–5. Large multipolar neurons with evenly distributed Nissl granules and small and medium-sized cells from the nucleus pontis centralis caudalis. Magnification ×1,000.
Fig. 6. An identified omnipause neuron taken from the medial nucleus pontis centralis caudalis, in the nucleus raphe interpositus. Magnification ×1,000.

medial NRPC, and included parts of the NRPO; the term 'the paramedian pontine reticular formation (PPRF)' was coined to define this area [Cohen and Komatsuzaki, 1972]. For general reviews, see Büttner and Büttner-Ennever [2006], Horn [2006] and Leigh and Zee [2006].

Types of Neurons and Their Connections

The individual cell groups in the NRPC, which compose the networks for horizontal gaze, have been studied in great detail: they include the excitatory and inhibitory burst neurons (EBN and IBN), omnipause neurons (OPN) and saccadic long-lead burst neurons (LBN), all of which participate directly in the generation of horizontal saccadic eye movements; reticulospinal neurons (RSN) are involved in eye-head movements, and paramedian tract neurons (PMT) may play a role in gaze stabilization. Each of these groups and their connections are described shortly below.

The *EBN* are medium-sized, parvalbumin- and cytochrome-oxidase-positive immunostaining cells, which lie in a compact group (about 2.5 × 2 mm in humans) adjacent to the MLF in the dorsomedial NRPC [Horn et al., 1995; Horn, 2006; Ugolini et al., 2006]. They project ipsilaterally and directly to motoneurons and internuclear neurons in the abducens nucleus, and also to the IBN and PMT neurons [Langer et al., 1986; Strassman et al., 1986a]. They have been shown to be specifically targeted in some disorders that are characterized by slow saccades, for example in spinocerebellar ataxia type 2 (Wadia type) [Geiner et al., 2008].

The *LBN* fire before EBN and may serve to synchronize the onset of the saccade. They are found not only in the NRPC, but also further rostrally in the NRPO, in the nucleus reticularis tegmenti pontis (nucleus No. 99) and in the central mesencephalic reticular formation (nucleus No. 59). For a review of the properties of LBN, see Fukushima et al. [1995] and Scudder et al. [2002].

The *IBN* lie in the caudalmost PPRF, not in the NRPC but in the posteriorly adjacent PGiD (nucleus No. 55), where they are discussed more fully. Like the EBN they are medium sized and parvalbumin positive, but unlike the EBN they use glycine as their transmitter [Spencer et al., 1989]. The IBN efferents cross the midline and inhibit motoneurons and interneurons in the contralateral abducens nucleus [Strassman et al., 1986b]. The cells are driven by connections from the superior colliculus, EBN, OPN and the contralateral IBN during horizontal saccades; their discharges facilitate the inhibition of the antagonist abducens muscle [Sugiuchi et al., 2005]. IBNs may also participate in the termination of saccades [Quaia et al., 1999; Rucker et al., 2011].

The *OPN* form a distinctive group of cells called nucleus raphe interpositus (RIP; fig. 2, 6) [Büttner-Ennever et al., 1988; Horn et al., 1995]. The neurons lie medially in the NRPC, arranged either side of the midline, at the level of the abducens rootlets: the latter serve as a reliable landmark for OPN location in vertebrates (see 'Original Text' above). In terms of location the RIP conforms to the nomenclature of the raphe nuclei, but unlike the neuromodulatory classical raphe nuclei with widespread projections, the projections of the RIP are highly specific to saccadic networks. The majority of OPN axons cross the midline, and their axons innervate the horizontal saccadic EBN in the NRPC, IBN in the PGiD and (in the cat) reticulospinal neurons; an ascending branch targets the vertical and torsional saccadic EBN in the rostral interstitial nucleus of the MLF (nucleus No. 61) [Ohgaki et al., 1987, 1989; Strassman et al., 1987; Horn et al., 1994; Grantyn et al., 2010]. The OPN act as a trigger which synchronizes the saccadic network – and in some species the accompanying head movement. The cells are glycinergic [Horn et al., 1994], and discharge continuously to inhibit the saccadic EBN. The OPN pause before saccades in any direction (hence the name), and shortly before blinks [Hepp et al., 1989; Schultz et al., 2010]. Powerful inputs controlling the OPN arise from the superior colliculus, IBN and 'latch' neurons [Missal and Keller, 2002; Shinoda et al., 2008]. Several studies have considered how OPN disorders could lead to the slowing of saccadic eye movements [Leigh and Zee, 2006].

The *RSN* in the NRPC include the largest neurons of the region. Here, the location of the RSN population overlaps with that of the EBN in monkeys, confirming the hypothesis that the region participates in eye and head movement [Robinson et al., 1994]. In cats, neurons in the NRPC fire in relation to both neck muscle activity and saccadic eye movements [Grantyn et al., 1987], whereas in primates the situation is somewhat different, and the neurons discharge for either eye or head movements associated with gaze shifts [Robinson et al., 1994; Scudder et al., 1996; Gaymard et al., 2000]. In cats, the RSNs also project to the motoneurons of the pinna in the facial nucleus, indicating that the RSN control gaze and ear position [Holstege et al., 1977]. The axons of the RSN descend in the medial reticulospinal tract, just lateral and ventral to the MLF, and provide a major projection to the motoneurons of axial musculature, and laminae VII–VIII, in the cervical cord [Shinoda et al., 2006].

The *cell groups of the paramedian tracts* (PMT cell groups) are a more recently identified set of cell groups scattered throughout the caudal brainstem (nucleus No. 96). Some of the cell clusters lie around the fibers of the MLF: two lie within the NRPC and one within the abducens nucleus. All PMT cell groups receive inputs from afferents targeting extraocular motoneurons, and project bilaterally to the cerebellar flocculus. They may carry a feedback signal which plays a role in the cerebellar stabilization of gaze [Dean and Porrill, 2008].

References

Büttner-Ennever JA, Cohen B, Pause M, Fries W: Raphe nucleus of the pons containing omnipause neurons of the oculomotor system in the monkey, and its homologue in man. J Comp Neurol 1988;267:307–321.

Büttner U, Büttner-Ennever JA: Present concepts of oculomotor organization. Prog Brain Res 2006;151:1–42.

Cohen B, Komatsuzaki A: Eye movements induced by stimulation of the pontine reticular formation: evidence for integration in oculomotor pathways. Exp Neurol 1972;36:101–117.

Dean P, Porrill J: Oculomotor anatomy and the motor-error problem: the role of the paramedian tract nuclei. Prog Brain Res 2008;171:177–186.

Fukushima K, Ohashi T, Fukushima J, Kaneko CR: Discharge characteristics of vestibular and saccade neurons in the rostral midbrain of alert cats. J Neurophysiol 1995;73:2129–2143.

Gaymard B, Siegler I, Rivaud-Pechoux S, Israel I, Pierrot-Deseilligny C, Berthoz A: A common mechanism for the control of eye and head movements in humans. Ann Neurol 2000;47:819–822.

Geiner S, Horn AK, Wadia NH, Sakai H, Büttner-Ennever JA: The neuroanatomical basis of slow saccades in spinocerebellar ataxia type 2 (Wadia-subtype). Prog Brain Res 2008;171:575–581.

Grantyn A, Kuze B, Brandi AM, Thomas MA, Quenech'du N: Direct projections of omnipause neurons to reticulospinal neurons: a double-labeling light microscopic study in the cat. J Comp Neurol 2010;518:4792–4812.

Grantyn A, Ong Meang Jacques V, Berthoz A: Reticulo-spinal neurons participating in the control of synergic eye and head movements during orienting in the cat. II. Morphological properties as revealed by intra-axonal injections of horseradish peroxidase. Exp Brain Res 1987;66:355–377.

Hepp K, Henn V, Vilis T, Cohen B: Brainstem regions related to saccade generation; in Wurtz RH, Goldberg ME (eds): Reviews in Oculomotor Research. Amsterdam, Elsevier, 1989.

Holstege G, Kuypers HGJM, Dekker JJ: The organization of the bulbar fibre connections to the trigeminal, facial and hypoglossal motor nuclei. II. An autoradiographic tracing study in cat. Brain 1977;100:265–286.

Horn AKE: The reticular formation. Prog Brain Res 2006;151:127–155.

Horn AKE, Büttner-Ennever JA, Suzuki Y, Henn V: Histological identification of premotor neurons for horizontal saccades in monkey and man by parvalbumin immunostaining. J Comp Neurol 1995;359:350–363.

Horn AKE, Büttner-Ennever JA, Wahle P, Reichenberger I: Neurotransmitter profile of saccadic omnipause neurons in nucleus raphe interpositus. J Neurosci 1994; 14:2032–2046.

Langer TP, Kaneko CR, Scudder CA, Fuchs AF: Afferents to the abducens nucleus in the monkey and cat. J Comp Neurol 1986;245:379–400.

Leigh RJ, Zee DS: The Neurology of Eye Movements. New York, Oxford University Press, 2006.

Missal M, Keller EL: Common inhibitory mechanism for saccades and smooth-pursuit eye movements. J Neurophysiol 2002;88:1880–1892.

Ohgaki T, Curthoys IS, Markham CH: Anatomy of physiologically identified eye-movement-related pause neurons in the cat: pontomedullary region. J Comp Neurol 1987;266:56–72.

Ohgaki T, Markham CH, Schneider JS, Curthoys IS: Anatomical evidence of the projection of pontine omnipause neurons to midbrain regions controlling vertical eye movements. J Comp Neurol 1989;289:610–625.

Quaia C, Lefevre P, Optican LM: Model of the control of saccades by superior colliculus and cerebellum. J Neurophysiol 1999;82:999–1018.

Robinson FR, Phillips JO, Fuchs AF: Coordination of gaze shifts in primates: brainstem inputs to neck and extraocular motoneuron pools. J Comp Neurol 1994; 346:43–62.

Rucker JC, Ying SH, Moore W, Optican LM, Büttner-Ennever J, Keller EL, Shapiro BE, Leigh RJ: Do brainstem omnipause neurons terminate saccades? Ann NY Acad Sci 2011;1233:48–57.

Schultz KP, Williams CR, Busettini C: Macaque pontine omnipause neurons play no direct role in the generation of eye blinks. J Neurophysiol 2010;103:2255–2274.

Scudder CA, Kaneko CRS, Fuchs AF: The brainstem burst generator for saccadic eye movements: a modern synthesis. Exp Brain Res 2002;142:439–462.

Scudder CA, Moschovakis AK, Karabelas AB, Highstein SM: Anatomy and physiology of saccadic long-lead burst neurons recorded in the alert squirrel monkey. II. Pontine neurons. J Neurophysiol 1996;76:353–370.

Shinoda Y, Sugiuchi Y, Izawa Y, Hata Y: Long descending motor tract axons and their control of neck and axial muscles. Prog Brain Res 2006;151:527–563.

Shinoda Y, Sugiuchi Y, Izawa Y, Takahashi M: Neural circuits for triggering saccades in the brainstem. Prog Brain Res 2008;171:79–85.

Spencer RF, Wenthold RJ, Baker R: Evidence for glycine as an inhibitory neurotransmitter of vestibular, reticular, and prepositus hypoglossi neurons that project to the cat abducens nucleus. J Neurosci 1989;9:2718–2736.

Strassman A, Evinger C, McCrea RA, Baker RG, Highstein SM: Anatomy and physiology of intracellularly labelled omnipause neurons in the cat and squirrel monkey. Exp Brain Res 1987;67:436–440.

Strassman A, Highstein SM, McCrea RA: Anatomy and physiology of saccadic burst neurons in the alert squirrel monkey. I. Excitatory burst neurons. J Comp Neurol 1986a;249:337–357.

Strassman A, Highstein SM, McCrea RA: Anatomy and physiology of saccadic burst neurons in the alert squirrel monkey. II. Inhibitory burst neurons. J Comp Neurol 1986b;249:358–380.

Sugiuchi Y, Izawa Y, Takahashi M, Na J, Shinoda Y: Physiological characterization of synaptic inputs to inhibitory burst neurons from the rostral and caudal superior colliculus. J Neurophysiol 2005;93:697–712.

Ugolini G, Klam F, Doldan Dans M, Dubayle D, Brandi A-M, Büttner-Ennever JA, Graf W: Horizontal eye movement networks in primates as revealed by retrograde transneuronal transfer of rabies virus: differences in monosynaptic input to 'slow' and 'fast' abducens motoneurons. J Comp Neurol 2006;498:762–785.

57 Nucleus Reticularis Pontis Oralis (NRPO)

Original name:
Nucleus pontis centralis oralis

Location and Cytoarchitecture – Original Text

This nucleus forms a major portion of the oral pontine tegmentum. It extends from the level of the midpoint of the motor trigeminal nucleus to the level of the caudal border of the inferior colliculus. Caudally, the nucleus is bordered by the nucleus reticularis pontis caudalis (NRPC), orally by the decussation of the superior cerebellar peduncle (dSCP).

On cross-section, the outline of the nucleus reticularis pontis oralis (NRPO) and the structures to which it is related vary considerably from one level to another. Proceeding caudo-orally, the nucleus is related medially to the nucleus reticularis tegmenti pontis and median raphe nucleus (MnR), dorsally to the central gray matter of the pons and the locus coeruleus (LC), laterally to the nucleus subcoeruleus, the medial parabrachial nucleus and the fibers of the SCP, ventrally to the supralemniscal process of the pontine nuclei and to the decussating fibers of the SCP.

The NRPO is composed of cells which vary in size from small to large. The small and medium-sized cells are triangular or irregularly oval in outline and contain indistinct Nissl granules which stain with moderate intensity (fig. 1, 5, 6). In general, the small cells are more lightly stained than the medium-sized cells. These irregularly distributed, small and medium-sized cells form the background of the NRPO which differs from that of the NRPC in that it is more cellular and the cells are, on the whole, more darkly stained.

Scattered on this background one finds large, plump, darkly stained cells with short dendrites, ragged edges, an excentric nucleus and Nissl substance which is, in the main, peripherally distributed (fig. 2, 3). These large cells are very similar to both the large cells of the MnR and the unpigmented cells of the LC. A few of them may contain moderate amounts of melanin pigment (fig. 4).

Functional Neuroanatomy

The NRPO contains some components of the saccadic eye and head premotor networks, including long-lead burst neurons [Scudder et al., 1996], but the center of this network lies further caudally in the NRPC. The hallmark of the NRPO is the plump, large cells lying medially. These contribute to a minor extent to the reticulospinal projections described in the NRPC, which

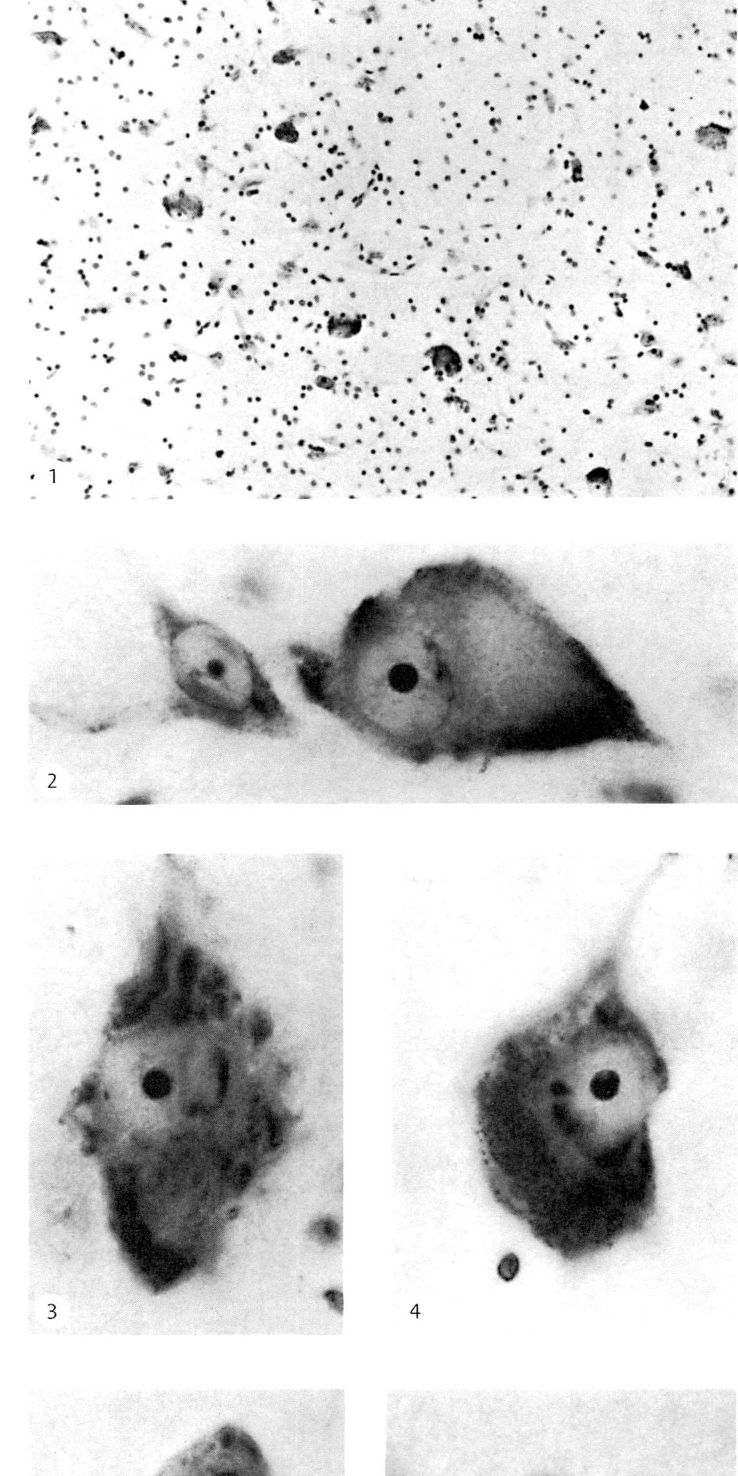

Fig. 1. Nucleus pontis centralis oralis. Magnification ×150.
Fig. 2, 3. Large cells with peripherally arranged Nissl substance from the nucleus pontis centralis oralis. Magnification ×1,000.
Fig. 4. Pigmented cell from the nucleus pontis centralis oralis. Magnification ×1,000.
Fig. 5, 6. A medium-sized and a small cell from the nucleus pontis centralis oralis. Magnification ×1,000.

innervate the cervical cord, but (unlike several reticulospinal regions) the NRPO cells have, in addition, projections to regions of the spinal cord further caudally, suggesting that the pontine reticular formation innervates not only the neck, but also the axial musculature [for reviews, see Holstege, 1988; Shinoda et al., 2006].

References

Holstege G: Brainstem-spinal cord projections in the cat, related to control of head and axial movements; in Büttner-Ennever JA (ed): Neuroanatomy of the Oculomotor System. Reviews in Oculomotor Research. Amsterdam, Elsevier, 1988, vol 2.

Scudder CA, Moschovakis AK, Karabelas AB, Highstein SM: Anatomy and physiology of saccadic long-lead burst neurons recorded in the alert squirrel monkey. II. Pontine neurons. J Neurophysiol 1996;76:353–370.

Shinoda Y, Sugiuchi Y, Izawa Y, Hata Y: Long descending motor tract axons and their control of neck and axial muscles. Prog Brain Res 2006;151:527–563.

58 Cuneiform Nucleus (CNF)

Original name:
Nucleus cuneiformis (caudal part)

Location and Cytoarchitecture – Comment

The name 'nucleus cuneiformis' was used by Olszewski and Baxter in the second edition to cover the whole mesencephalic reticular formation, but in this edition, and on the basis of arguments set out in the review by Jones [1995], we have conformed to the nomenclature of Paxinos and colleagues. The term 'cuneiform nucleus (CNF)' is used here, selectively, for the small caudal wedge-shaped region of the reticular formation at the pontomesencephalic border, which lies beneath the inferior colliculus and above the pedunculopontine nucleus (PPT), shown in plates 30 and 31. The term 'mesencephalic reticular formation (MRF)' is used here for the reticular larger area further rostrally.

The cells in the CNF are small and more closely packed than in the MRF, and resemble a dorsolaterally displaced region of the lateral tegmental field at the pontomesencephalic border (see 'Reticular Formation: Overview').

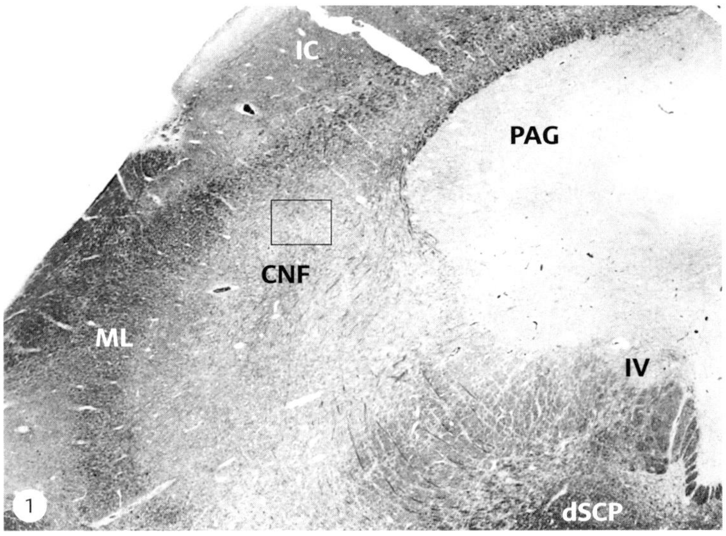

Fig. 1. Gallyas stain of the caudal mesencephalon to show fiber pathways and identify more easily the location of the cuneiform nucleus (CNF) illustrated in figure 2 (rectangle). dSCP = Decussation of the superior cerebellar peduncle; IC = inferior colliculus; IV = trochlear nucleus; ML = medial lemniscus; PAG = periaqueductal gray. Magnification ×8.

Functional Neuroanatomy

Function
The CNF (as newly defined above) is part of the limbic system, closely associated with the adjacent periaqueductal gray (PAG), and it has been included in the medial component of the emotional motor system [Holstege, 1991]. Its descending projections terminate in the raphe nuclei, which modulate locomotion and nociception control [Beitz, 1982]. This connection may be the neuroanatomical basis for the association of the CNF with the 'mesopontine tegmental anesthesia area' [Sukhotinsky et al., 2006], and the 'mesencephalic locomotion region'; see figure 21.17 in Nieuwenhuys et al. [2008] and Le Ray et al. [2011].

The pontomesencephalic border region is thought to contain a pattern generator, essential for initiating and controlling locomotor movements. It was initially identified in cats, and subsequently found in all vertebrates. The area is called the 'mesencephalic locomotion region' and the CNF is often included in it, although the subjacent PPT (nucleus No. 72) is currently considered to be the focus of the functional region. There is some evidence suggesting that neither PPT nor CNF are involved in locomotion [Allen et al., 1996].

Connections
The CNF receives afferents from several limbic structures, the bed nuclei of the stria terminalis, the amygdala nuclei, and the lateral hypothalamus [Holstege, 1991]. These pathways also terminate in the adjacent ventrolateral PAG with which the CNF has reciprocal connections. The PAG-CNF region, in turn, projects to the raphe nuclei, which provide extensive seroto-

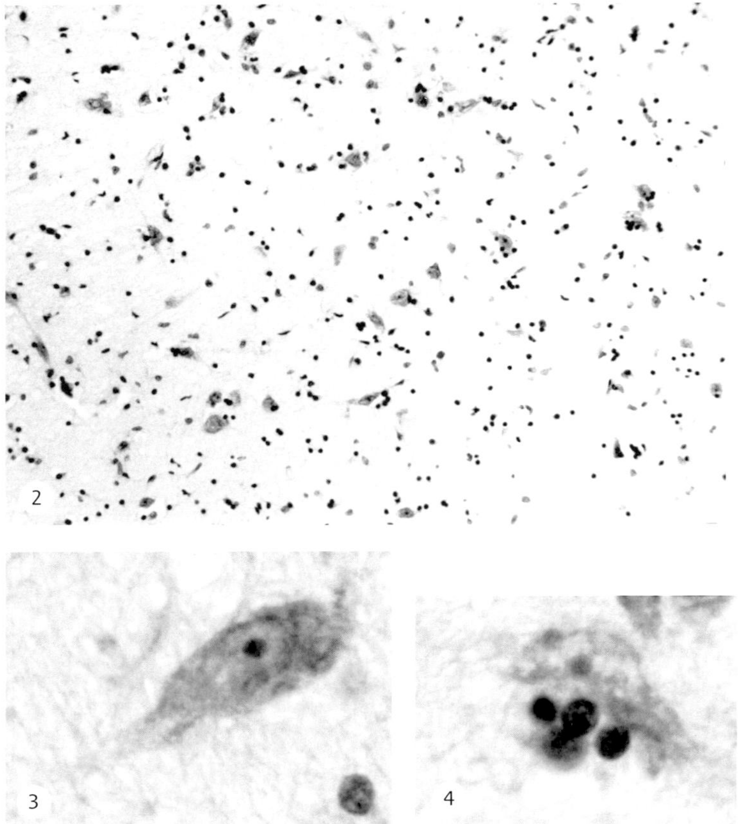

Fig. 2. Cuneiform nucleus. Nissl stain. Magnification ×150
Fig. 3, 4. Cells of the cuneiform nucleus. Magnification ×1,000.

nergic projections to the full length of the spinal cord, targeting both the sensory dorsal horn and the motor ventral horn [Beitz, 1982]. The projections are known to modulate the sensory and motor signals of the cord (see 'Neuromodulatory Systems: Overview'). In keeping with these functions the CNF contains high levels of substance P, at least in rats [Halliday et al., 1995], and moderate acetylcholine esterase levels in humans [Saper, 1990].

References

Allen LF, Inglis WL, Winn P: Is the cuneiform nucleus a critical component of the mesencephalic locomotor region? An examination of the effects of excitotoxic lesions of the cuneiform nucleus on spontaneous and nucleus accumbens induced locomotion. Brain Res Bull 1996;41:201–210.

Beitz AJ: The nuclei of origin of brain stem enkephalin and substance P projections to the rodent nucleus raphe magnus. Neuroscience 1982;7:2753–2768.

Halliday G, Harding A, Paxinos G: Serotonin and tachykinin systems; in Paxinos G (ed): The Rat Nervous System, ed 2. San Diego, Academic Press, 1995.

Holstege G: Descending motor pathways and the spinal motor system: limbic and non-limbic components. Prog Brain Res 1991;87:307–421.

Jones BE: Reticular formation: Cytoarchitecture, transmitters, and projections; in Paxinos G (ed): The Rat Nervous System, ed 2. San Diego, Academic Press, 1995.

Le Ray D, Juvin L, Ryczko D, Dubuc R: Supraspinal control of locomotion: the mesencephalic locomotor region. Prog Brain Res 2011;188:51–70.

Nieuwenhuys R, Voogd J, van Huijzen C: The Human Central Nervous System. Berlin, Springer, 2008.

Saper CB: Cholinergic system; in Paxinos G (ed): The Human Nervous System, ed 1. San Diego, Academic Press, 1990.

Sukhotinsky I, Reiner K, Govrin-Lippmann R, Belenky M, Lu J, Hopkins DA, Saper CB, Devor M: Projections from the mesopontine tegmental anesthesia area to regions involved in pain modulation. J Chem Neuroanat 2006;32:159–178.

Reticular Formation (Plates 32, 34, 36–39 and 41)

59 Mesencephalic Reticular Formation (MRF)

Original name:
The mesencephalic reticular formation includes the former nucleus cuneiformis (rostral part), the nucleus intracuneiformis and the nucleus subcuneiformis

Alternative name:
Central tegmental fields [Berman, 1968]

Nomenclature:
The 'newly' defined mesencephalic reticular formation (MRF) combines 3 nuclei of the 2nd edition. The use of the term 'MRF' conforms with current terminology [Brodal, 1957; Paxinos and Huang, 1995], and was decided on the basis of arguments set out in the review by Jones [1995]. The intracuneiform nucleus (nucleus No. 60) is enclosed in the rostral MRF, and the 'subcuneiform nucleus' is now called ventral MRF. The term 'cuneiform' (meaning wedge-shaped) is reserved for the caudal part of the former nucleus cuneiformis. It is treated here separately under the name cuneiform nucleus (nucleus No. 58).

Location and Cytoarchitecture – Original Text

[The mesencephalic reticular formation (MRF)] occupies an area ventral to the inferior and superior quadrigeminal bodies in the dorsolateral part of the mesencephalic tegmentum. It extends from the level of the caudal border of the inferior colliculus (IC) to the level of the oral border of the superior colliculus (SC) and increases in size progressively from its caudal to its oral pole. Orally it is bounded by the pretectal region.

On cross-section, the MRF is triangular or quadrangular in outline. Its principal relations are the periaqueductal gray (PAG) medially, the IC dorsally, the medial lemniscus and the nucleus of the brachium of the inferior colliculus laterally, and ventrally the pedunculopontine nucleus (PPT) and central tegmental tract. The nucleus intracuneiformis lies embedded in the oral part of the MRF.

The MRF is mainly composed of small to medium-sized, triangular, fusiform or oval cells which possess short processes, a centrally placed nucleus and scanty Nissl substance which stains with moderate intensity. The majority of cells are associated with numerous glial satellites, and lie with their long axes directed dorsomedially (fig. 1–5).

[The ventral part of the MRF (MRFv), formerly called 'nucleus subcuneiformis'] is composed predominantly of small to medium-sized, elongated, triangular or fusiform cells, the majority of which lie with their long axes directed dorsomedially (fig. 6). These cells contain a centrally placed nucleus and Nissl substance which stains with moderate intensity (fig. 9–11). Scattered among these cells are a few larger neurons similar to those which compose the PPT. Such cells are medium sized or large and either possess a centrally placed nucleus and evenly distributed Nissl granules (fig. 7), or an eccentric nucleus and peripherally arranged Nissl granules (fig. 8). The presence of these larger cells and the lesser degree of cellularity and of glial satellitosis distinguish the cytoarchitectonic picture of the MRFv from that of the overlying MRF.

Functional Neuroanatomy

Function

The close association of the MRF with the SC and the PAG are reflected in the results of behavioral studies of this area. In earlier studies electrical stimulation evoked emotional responses related to defense, vocalization, attacking or aggressive behavior, reminiscent of PAG responses. More recent studies have emphasized its role in gaze control, reminiscent of the SC function; however, when interpreting MRF studies, it should be remembered that the efferent pathways of the SC pass through the MRF [Chen and May, 2000; Horn, 2006].

Two general parts of the MRF can be recognized functionally and cytoarchitecturally: first the more dorsal and rostral MRF which resembles the lateral parvicellular reticular zone (or lateral tegmental fields), typically a sensory integrative zone [Jones, 1995], but exact functions have not been associated with this area; the second part is the large-celled MRFv which lies in the caudal half of the MRF, extending medially to the oculomotor nucleus, the interstitial nucleus of Cajal (INC) and the rostral interstitial nucleus of the medial longitudinal fasciculus (RIMLF). The MRFv represents the 'medial tegmental fields' or output zone of the MRF (see 'Reticular Formation: Overview'). It is the origin of the mesencephalic-cervical projections involved in head movements and head position [Holstege and Cowie, 1989; Robinson et al., 1994] and eye movements, i.e. gaze [Leigh and Zee, 2006]; for reviews, see Holstege [1988] and Horn [2006]. The MRFv may well be synonymous with the 'central mesencephalic reticular formation' (cMRF), a physiologically defined area associated with gaze control, and discussed at the end of this chapter.

Discrete lesions of the MRFv, centered in the cMRF region, can cause specific eye movement deficits [Leigh and Zee, 2006]. Long-lead burst neurons, which fire before and during saccadic eye movements, have been recorded in this area, and reversible lesions here cause hyper- or hypometria of saccades, and a disruption of gaze fixation [Waitzman et al., 2000a, b; Cromer and Waitzman, 2007]. Large lesions of the MRF cause changes in forebrain wakefulness due to damage to fibers of passage [Watson et al., 1974; Saper et al., 2001].

Subdivisions and Associated Structures

Apart from the rostrolateral MRF and MRFv described above, there are several other regions of the MRF that have been given specific names, or are associated with a specific function.

The most rostral parts of the MRF, the fields of Forel, lie beneath the thalamus around the zona incerta (ZI). Forel [1877] first named these 'Haubenfelder' (tegmental fields) with the H_1 field lying above the ZI , the H_2 field lying below the ZI, and the H (or H_3) field was the medial prerubral area where H_1 and H_2 joined (see nucleus No. 61, fig. 1). Part of the medial H field (or H_3) has been identified as the premotor nucleus for vertical and torsional saccadic eye movements, called RIMLF: this region is technically part of the reticular formation but due to its specialized function is treated separately in the chapter on nucleus No. 61. Medial to the caudal half of the RIMLF lies the M group, a small premotor nucleus responsible for eyelid coordination during upgaze (nucleus No. 61, fig. 1). The INC lies immediately caudal to the RIMLF, and is likewise part of the mesencephalic reticular fields but treated separately (nucleus No. 20). The cMRF is a functionally defined region within the MRF; it can be divided into 2 subregions, a rostral area subtending vertical and oblique saccades, and a caudal region corresponding to the MRFv for horizontal saccades [Waitzman et al., 2000b].

Connections

Efferents. The MRF has strong and reciprocal interconnections with its two neighboring structures SC and PAG [May, 2006]. The efferent projections of the MRF have been clearly analyzed using autoradiographic tracing techniques, which eliminate problematic false-positive results due to involvement of fibers of passage [Edwards and De Olmos, 1975]. Descending pathways, including those associated with cMRF control of gaze, terminate in the medial pontine and medullary reticular formation, nucleus raphe magnus, the omnipause neurons of the nucleus raphe interpositus and in the nucleus reticularis tegmenti pontis, of both sides. Ipsilateral projections to the locus coeruleus and the lateral reticular nucleus have been reported, as well as contralateral connections to the facial nucleus and the inferior olive. Some fibers innervate the spinal cord and even extend down to lumbosacral levels [Holstege, 1988; Yezierski, 1988]. The contralateral pathways from the MRF cross ventral to the oculomotor nucleus in the ventral tegmental decussation of Meynert, and descend close to the midline as the predorsal bundle. The ascending connections of the MRF target the pretectum, the nucleus of the posterior commissure, the posterior hypothalamus, the midline and intralaminar thalamic nuclei. No MRF efferents to the neostriatum or the specific thalamic nuclei were found.

Afferents. Apart from the strong reciprocal connections to the SC and PAG, there are reports of limbic afferents from the amygdala and hypothalamus to the MRF, but these target the rostrolateral MRF, regions different from the gaze-related cMRF described above, probably the retrorubral fields; see plates 36–38 [Holstege, 1987]. In contrast, the cMRF and SC receive direct projections from the frontal eye fields [Huerta et al., 1986; Stanton et al., 1988] and the supplementary eye fields [Huerta and Kaas, 1990; Shook et al., 1990]. In general it is found that descending afferents from the limbic structures avoid the oculomotor-related regions of the reticular formation [Büttner-Ennever and Holstege, 1986].

Central Mesencephalic Reticular Formation. One region of the MRF, the cMRF, has been studied in detail in terms of function and connectivity. Its specific role in the control of gaze and saccades is still unclear [Cromer and Waitzman, 2007]; see reviews of Horn [2006], Leigh and Zee [2006], May [2006] and Graf and Ugolini [2006]. The studies of May and colleagues emphasize the important role played by inhibitory GABAergic pathways, and they also show the presence of reciprocal cMRF

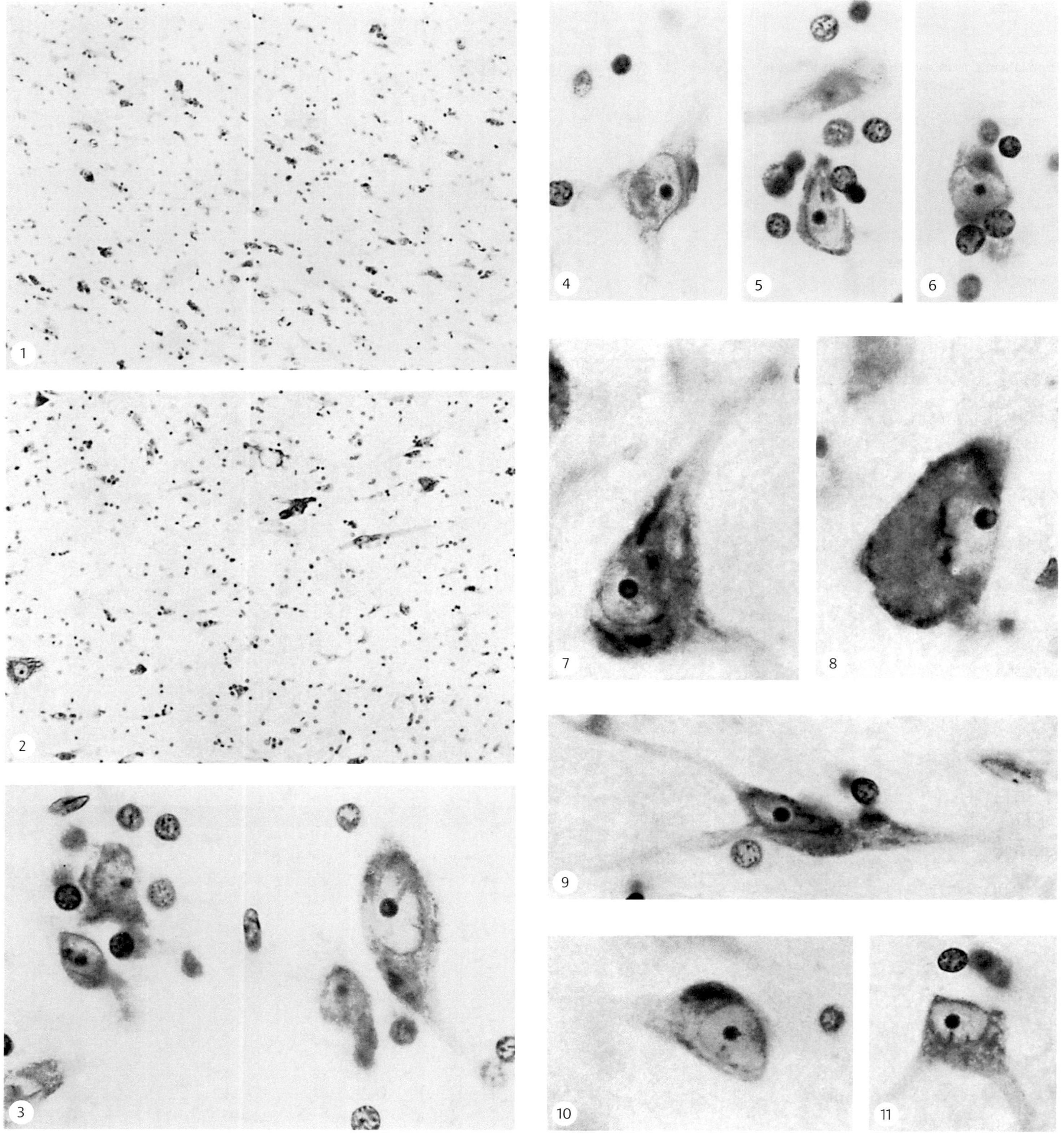

Fig. 1. Mesencephalic reticular formation. Magnification ×150.
Fig. 2. Mesencephalic reticular formation, ventral part. Magnification ×150.

Fig. 3–6. Cells from the mesencephalic reticular formation. Magnification ×1,000.
Fig. 7–11. Cells from the mesencephalic reticular formation, ventral part. Magnification ×1,000.

connections to the cervical cord, associated particularly with the medial cMRF [Zhou et al., 2008; Wang et al., 2010]. Finally projections from neurons in the medial part of the cMRF directly to motoneurons of extraocular nuclei were demonstrated using a transsynaptic tracer [Graf and Ugolini, 2006; Ugolini et al., 2006]. The cMRF input targeted nontwitch motoneurons whose function may be related to the resting tension in extraocular muscles, such as the near-response (see nucleus No. 46).

Several fiber tracts pass through the MRF. The pallidothalamic and cerebellothalamic ones travel in the H_1 fields of Forel, and pallidothalamic fibers pass through the H_2 fields of Forel. Ascending modulatory pathways associated with the ascending activating system also run through the MRF; they arise in the caudal brainstem from serotonergic, noradrenergic and cholinergic cell groups and project via catecholaminergic and cholinergic bundles into the medial forebrain bundle in the hypothalamus [Nieuwenhuys et al., 2008].

References

Berman AL: The Brain Stem of the Cat: A Cytoarchitectonic Atlas with Stereotaxic Coordinates. Madison, University Wisconsin Press, 1968.

Brodal A: The Reticular Formation of the Brain Stem: Anatomical Aspects and Functional Correlations. Henderson Trust Lectures. London, William Ramsay Henderson Trust, 1957, No 18.

Büttner-Ennever JA, Holstege G: Anatomy of premotor centers in the reticular formation controlling oculomotor, skeletomotor and autonomic motor systems. Prog Brain Res 1986;64:89–98.

Chen B, May PJ: The feedback circuit connecting the superior colliculus and central mesencephalic reticular formation: a direct morphological demonstration. Exp Brain Res 2000;131:10–21.

Cromer JA, Waitzman DM: Comparison of saccade-associated neuronal activity in the primate central mesencephalic and paramedian pontine reticular formations. J Neurophysiol 2007;98:835–850.

Edwards SB, de Olmos JS: Autoradiographic studies of the projections of the midbrain reticular formation: ascending projections of nucleus cuneiformis. J Comp Neurol 1975;165:417–432.

Forel A: Untersuchungen über die Haubenregion und ihre oberen Verknüpfungen im Gehirn des Menschen und einiger Säugetiere, mit Beiträgen zu den Methoden der Gehirnuntersuchungen. Arch Psychiatr 1877;7:1–495.

Graf WM, Ugolini G: The central mesencephalic reticular formation: its role in space-time coordinated saccadic eye movements. J Physiol 2006;570:433–434.

Holstege G: Some anatomical observations on the projections from the hypothalamus to brainstem and spinal cord: an HRP and autoradiographic tracing study in the cat. J Comp Neurol 1987;260:98–126.

Holstege G: Brainstem-spinal cord projections in the cat, related to control of head and axial movements; in Büttner-Ennever JA (ed): Neuroanatomy of the Oculomotor System. Reviews in Oculomotor Research. Amsterdam, Elsevier, 1988, vol 2.

Holstege G, Cowie RJ: Projections from the rostral mesencephalic reticular formation to the spinal cord. Exp Brain Res 1989;75:265–279.

Horn AKE: The reticular formation. Prog Brain Res 2006;151:127–155.

Huerta MF, Kaas JH: Supplementary eye field as defined by intracortical microstimulation: connections in macaques. J Comp Neurol 1990;293:299–330.

Huerta MF, Krubitzer LA, Kaas JH: Frontal eye field as defined by intracortical microstimulation in squirrel monkeys, owl monkeys, and macaque monkeys. I. Subcortical connections. J Comp Neurol 1986;253:415–439.

Jones BE: Reticular formation: cytoarchitecture, transmitters, and projections; in Paxinos G (ed): The Rat Nervous System, ed 2. San Diego, Academic Press, 1995.

Leigh RJ, Zee DS: The Neurology of Eye Movements. New York, Oxford University Press, 2006.

May PJ: The mammalian superior colliculus: laminar structure and connections. Prog Brain Res 2006;151:321–378.

Nieuwenhuys R, Voogd J, van Huijzen C: The Human Central Nervous System, Berlin, Springer, 2008.

Paxinos G, Huang X-F: Atlas of the Human Brainstem. San Diego, Academic Press, 1995.

Robinson FR, Phillips JO, Fuchs AF: Coordination of gaze shifts in primates: brainstem inputs to neck and extraocular motoneuron pools. J Comp Neurol 1994; 346:43–62.

Saper CB, Chou TC, Scammell TE: The sleep switch: hypothalamic control of sleep and wakefulness. Trends Neurosci 2001;24:726–731.

Shook BL, Schlag-Rey M, Schlag J: Primate supplementary eye field. I. Comparative aspects of mesencephalic and pontine connections. J Comp Neurol 1990;301: 618–642.

Stanton GB, Goldberg ME, Bruce CJ: Frontal eye field efferents in the macaque monkey. II. Topography of terminal fields in midbrain and pons. J Comp Neurol 1988;271:493–506.

Ugolini G, Klam F, Doldan Dans M, Dubayle D, Brandi A-M, Büttner-Ennever JA, Graf W: Horizontal eye movement networks in primates as revealed by retrograde transneuronal transfer of rabies virus: differences in monosynaptic input to 'slow' and 'fast' abducens motoneurons. J Comp Neurol 2006;498:762–785.

Waitzman DM, Silakov VL, Depalma-Bowles S, Ayers AS: Effects of reversible inactivation of the primate mesencephalic reticular formation. I. Hypermetric goal-directed saccades. J Neurophysiol 2000a;83:2260–2284.

Waitzman DM, Silakov VL, Depalma-Bowles S, Ayers AS: Effects of reversible inactivation of the primate mesencephalic reticular formation. II. Hypometric vertical saccades. J Neurophysiol 2000b;83:2285–2299.

Wang N, Warren S, May PJ: The macaque midbrain reticular formation sends side-specific feedback to the superior colliculus. Exp Brain Res 2010;201:701–717.

Watson RT, Heilman KM, Miller BD, King FA: Neglect after mesencephalic reticular formation lesions. Neurology 1974;24:294–298.

Yezierski RP: Spinomesencephalic tract; projections from the lumbosacral spinal cord of the rat, cat, and monkey. J Comp Neurol 1988;267:131–146.

Zhou L, Warren S, May PJ: The feedback circuit connecting the central mesencephalic reticular formation and the superior colliculus in the macaque monkey: tectal connections. Exp Brain Res 2008;189:485–496.

Reticular Formation (Plate 38)

60 Intracuneiform Nucleus (ICUN)

Original name:

Nucleus intracuneiformis

Location and Cytoarchitecture – Original Text

The intracuneiform nucleus (ICUN) is an irregularly oval cell group which lies surrounded by the cells of the mesencephalic reticular formation (MRF) in the oral portion of the mesencephalon, at the level of the oral pole of the principal oculomotor nuclei. Orocaudally the nucleus extends for a distance of only 1.5 mm.

This nucleus is composed of closely congregated randomly oriented, medium-sized, plump, triangular or fusiform neurons which possess a large, centrally placed nucleus and darkly stained, evenly distributed indistinct Nissl granules (fig. 1–3). Glial satellites are associated with some of the smaller cells. The cells of the ICUN differ from those of the surrounding MRF in that they are larger, more intensely stained, and more closely packed.

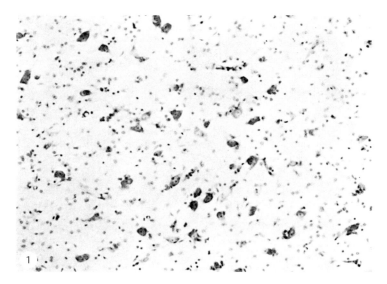

Fig. 1. Intracuneiform nucleus. Magnification ×150.

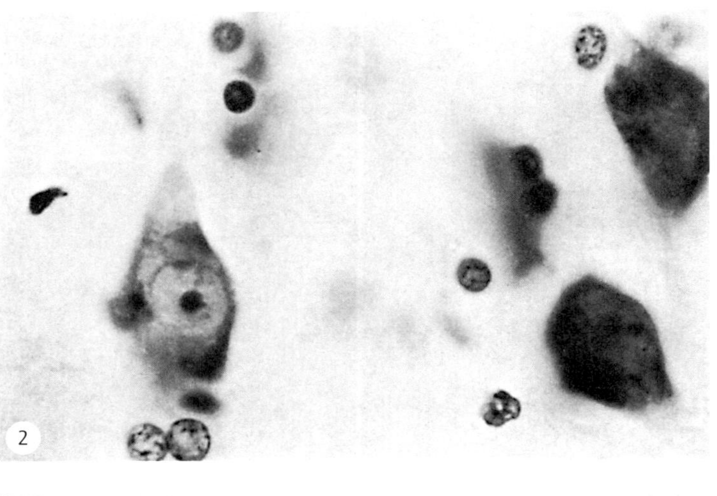

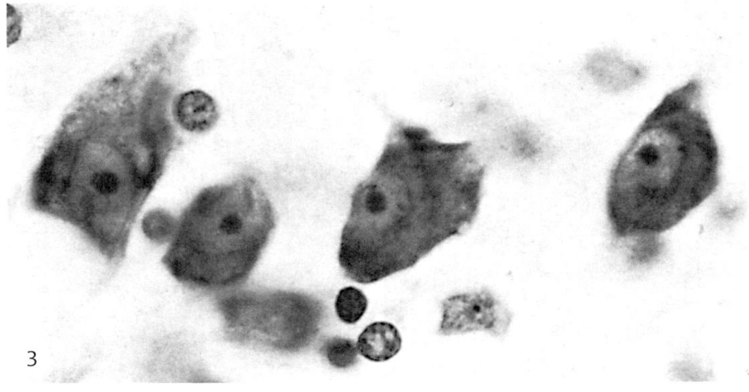

Fig. 2, 3. Cells from the intracuneiform nucleus. Magnification ×1,000.

Functional Neuroanatomy

No studies, up to now, have been able to relate this group of reticular cells to a specific function [Castiglioni et al., 1978]. However, in vascular studies of the region an area thought to be the ICUN is clearly visible due to its dense vascularization [Duvernoy, 1999]. In the behaving monkey, neurons have been recorded near this area, which play a role in the generation of vertical rapid eye movements [Waitzman et al., 2000], but it is unclear if they involve the ICUN specifically.

References

Castiglioni AJ, Gallaway MC, Coulter JD: Spinal projections from the midbrain in monkey. J Comp Neurol 1978;178:329–346.

Duvernoy HM: Human Brain Stem Vessels. Berlin, Springer, 1999.

Waitzman DM, Silakov VL, Depalma-Bowles S, Ayers AS: Effects of reversible inactivation of the primate mesencephalic reticular formation. II. Hypometric vertical saccades. J Neurophysiol 2000;83:2285–2299.

Reticular Formation

(Not on plates, see fig. 1)

61 Rostral Interstitial Nucleus of the Medial Longitudinal Fasciculus (RIMLF)

Original name:
Not described

Alternative name:
H field of Forel (medial part)

Location and Cytoarchitecture

The rostral interstitial nucleus of the medial longitudinal fasciculus (RIMLF; fig. 1, 2) lies at the most rostral tip of the mesencephalic reticular formation (MRF). The nucleus is situated at the medial border of the H fields of Forel, a name which is often used instead of RIMLF in lower species such as the cat. The term 'H fields of Forel' however designates a much larger region than the RIMLF. In transverse sections the RIMLF forms a wing-shaped nucleus, ventromedial to the third ventricle, and embedded in the rubral capsule surrounding the rostral parvocellular red nucleus. The posterior thalamosubthalamic paramedian artery serves as a helpful landmark, by bordering the dorsomedial margin of the RIMLF from the ventricular gray matter, like an eyebrow [Büttner-Ennever et al., 1982]. The ventrolateral border of the RIMLF is formed by the parvocellular portion of the red nucleus. The caudal limit of the RIMLF adjoins directly the rostral pole of the interstitial nucleus of Cajal (INC): this level is clearly marked by the traversing fibers of the tractus retroflexus. The rostral extent of the RIMLF is roughly demarcated by the traversing fibers of the mammillothalamic tract (MT) [Büttner and Büttner-Ennever, 1988].

On first inspection, the RIMLF is inconspicuous in humans; it has a typical reticulated appearance with a very meager neuronal population. More detailed inspection reveals several morphological cell types embedded between the rostralmost fibers of the MLF, including small to medium-sized multipolar

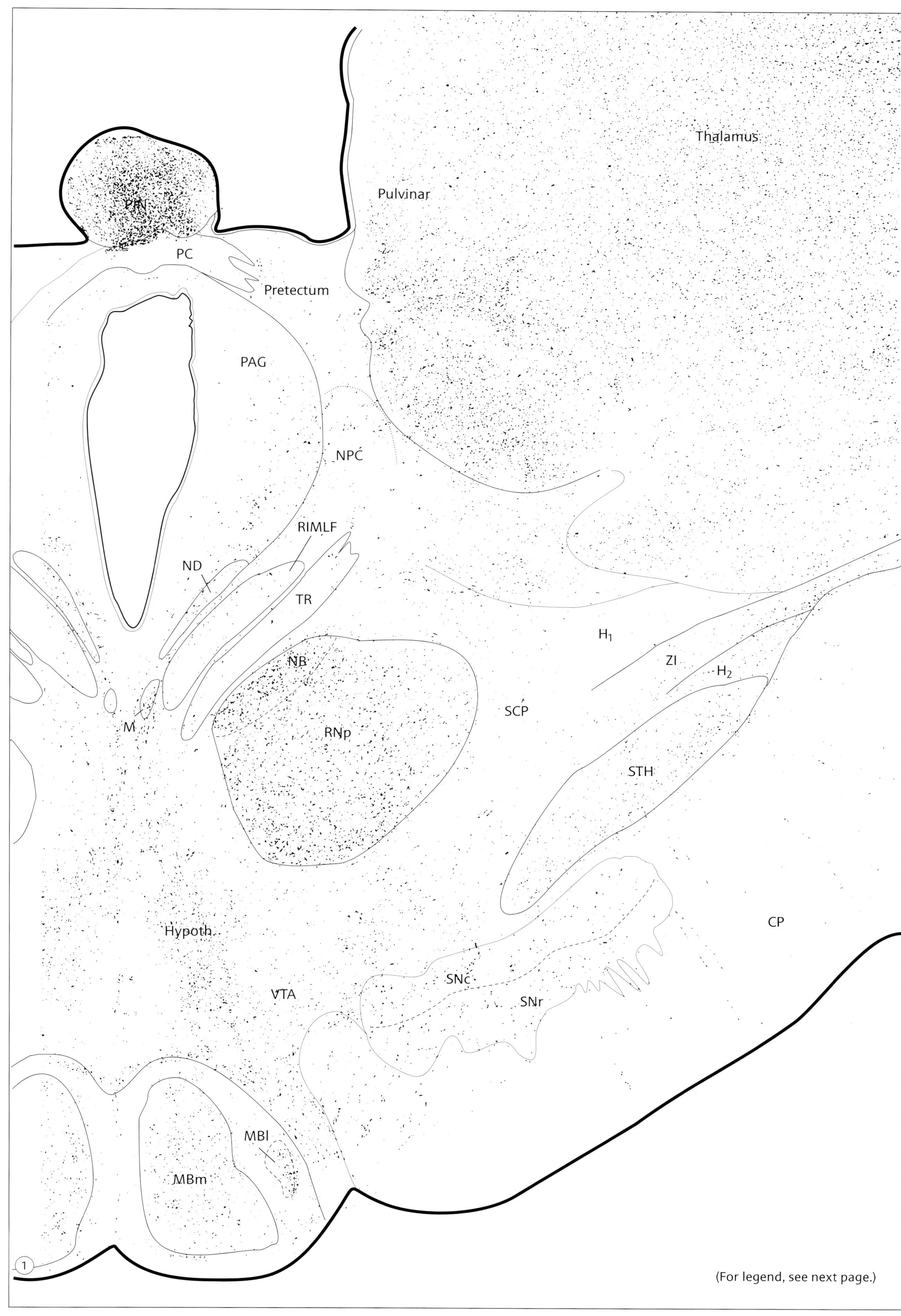

(For legend, see next page.)

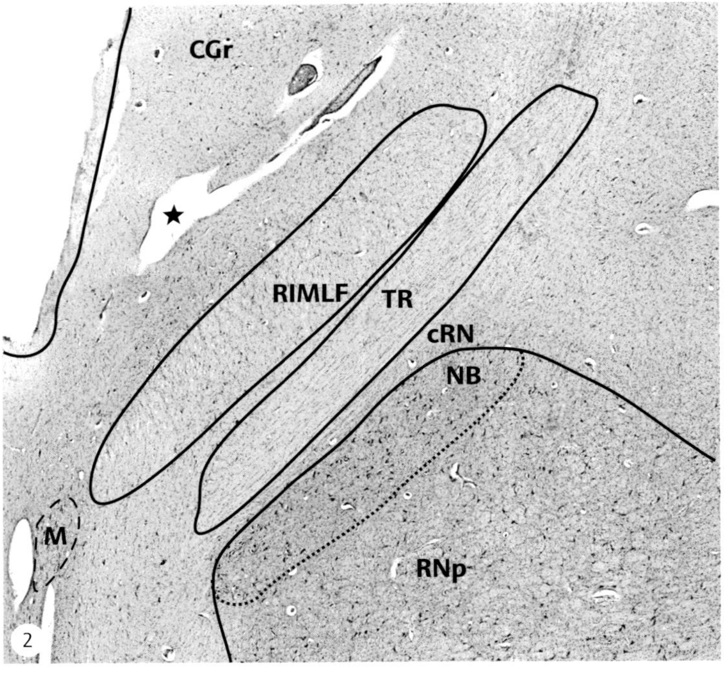

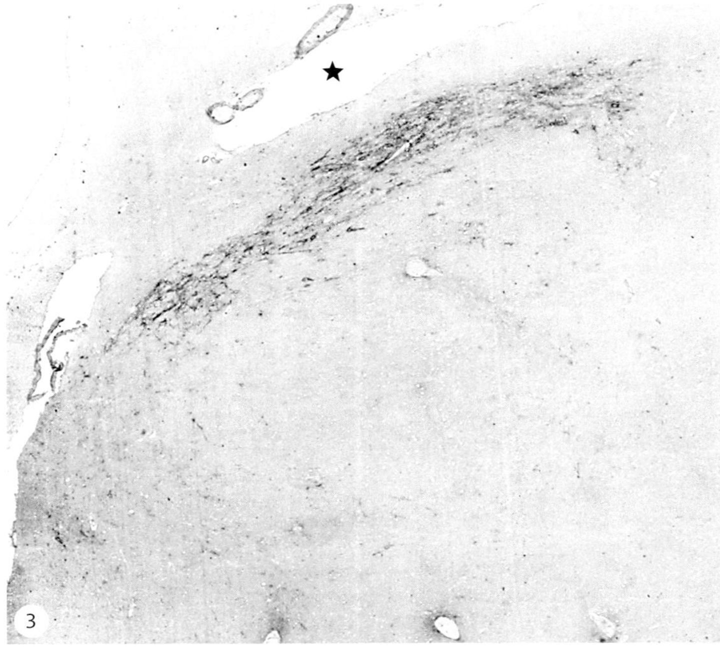

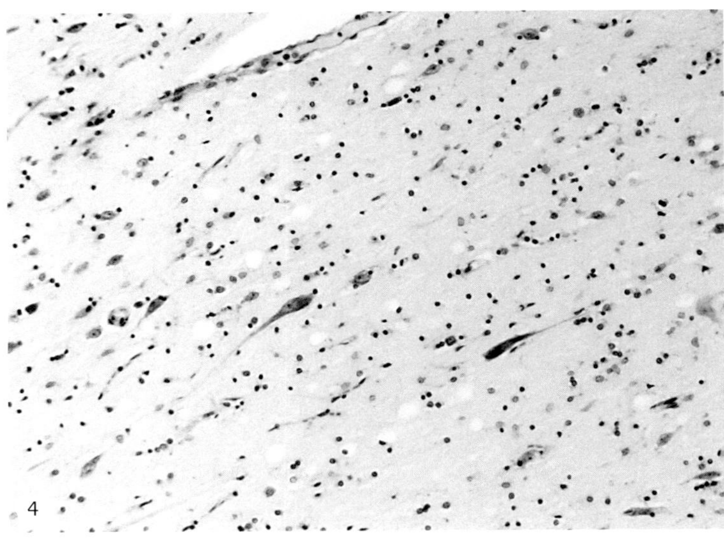

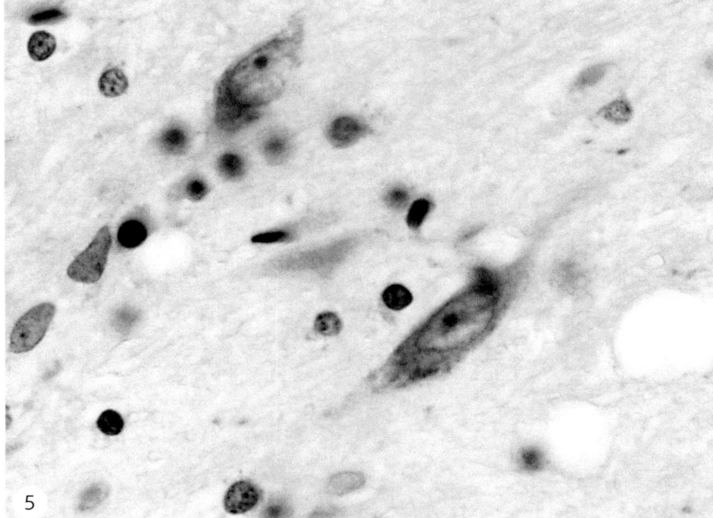

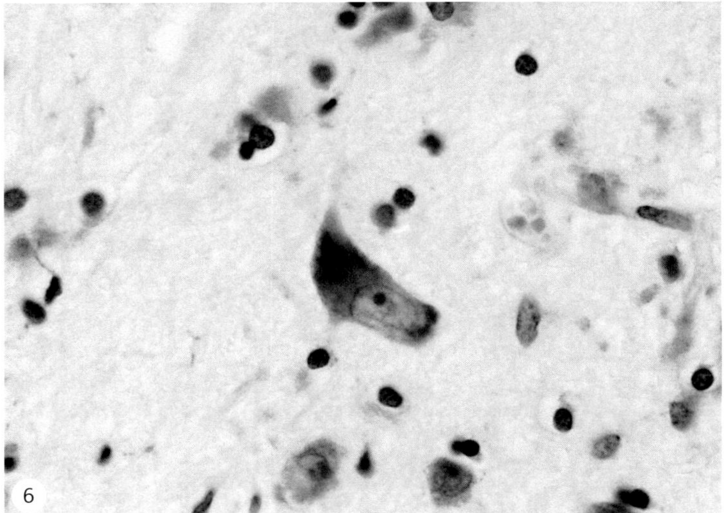

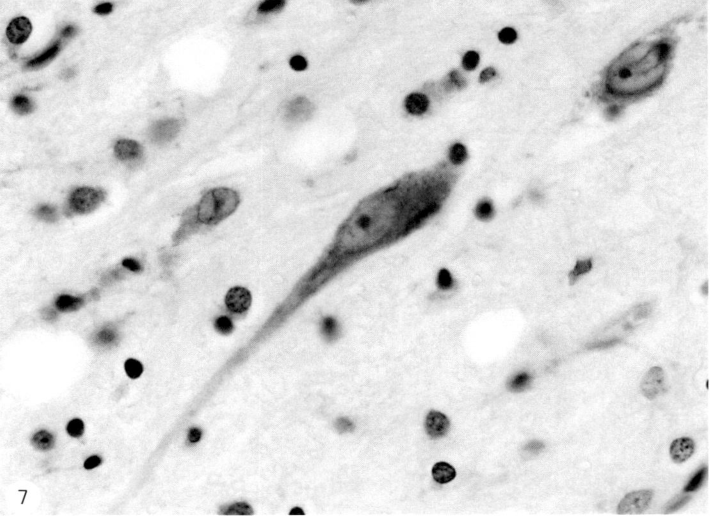

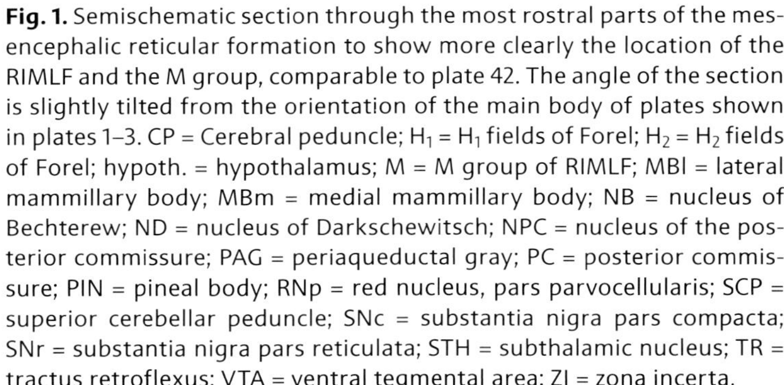

Fig. 1. Semischematic section through the most rostral parts of the mesencephalic reticular formation to show more clearly the location of the RIMLF and the M group, comparable to plate 42. The angle of the section is slightly tilted from the orientation of the main body of plates shown in plates 1–3. CP = Cerebral peduncle; H_1 = H_1 fields of Forel; H_2 = H_2 fields of Forel; hypoth. = hypothalamus; M = M group of RIMLF; MBl = lateral mammillary body; MBm = medial mammillary body; NB = nucleus of Bechterew; ND = nucleus of Darkschewitsch; NPC = nucleus of the posterior commissure; PAG = periaqueductal gray; PC = posterior commissure; PIN = pineal body; RNp = red nucleus, pars parvocellularis; SCP = superior cerebellar peduncle; SNc = substantia nigra pars compacta; SNr = substantia nigra pars reticulata; STH = subthalamic nucleus; TR = tractus retroflexus; VTA = ventral tegmental area; ZI = zona incerta.

Fig. 2. Photomicrograph to show details of the section used for figure 1. Note the posterior thalamosubthalamic paramedian artery (*) which serves as a useful landmark for the inconspicuous RIMLF. Magnification ×15. cRN = Perirubral capsule; CGr = central gray matter; for other abbreviations, see figure 1.

Fig. 3. Section through the RIMLF stained for perineuronal nets [Horn et al., 2003]. In this region the stain highlights only the premotor neurons of the RIMLF, which otherwise remains inconspicuous in Nissl stains. Note the location of the posterior thalamic subthalamic artery marked with an asterisk (*). Magnification ×18.

Fig. 4. RIMLF. Nissl-stained section. Magnification ×150.

Fig. 5–7. Cells from the RIMLF [Horn et al., 2003]. Magnification ×1,000.

and fusiform neurons [Crossland et al., 1994]. However, the RIMLF nucleus can be clearly demarcated by histochemical stains for cytochrome oxidase activity and parvalbumin expression [Horn and Büttner-Ennever, 1998; Horn et al., 2000]. An even clearer outline is obtained by stains that highlight perineuronal nets, such as *Wisteria floribunda* agglutinin-binding stains and chondroitin sulfate proteoglycan immunohistochemistry (fig. 3) [Horn et al., 2003a]. Immunostaining for the neurofilament protein SMI32 is less suitable for identifying the RIMLF: it is used in figure 34 of the atlas of Paxinos and Huang [1995], and labels the RIMLF along with the adjacent prerubral fields, surrounding the red nucleus.

Functional Neuroanatomy

Function

The RIMLF contains premotor burst neurons (fig. 4–7), which are essential for the generation of vertical and torsional saccades [Büttner-Ennever, 2006; Horn, 2006; Leigh and Zee, 2006]. A burst of activity in RIMLF on one side, evoked by electrical stimulation, evokes ipsitorsional conjugate eye movements (with the upper pole of ipsilateral eye rotating towards the activated side).

Unilateral lesions of the RIMLF have severe effects on torsional saccades and produce spontaneous torsional nystagmus; however, they produce little effect on vertical saccades [Suzuki et al., 1995; Helmchen et al., 1996, 2002]. Experimentally, bilateral lesions of the RIMLF cause the loss of both torsional and vertical saccades, but horizontal saccades, vertical gaze-holding, vestibular and smooth pursuit eye movements remain relatively unaffected [Suzuki et al., 1995]. In humans bilateral lesions lead to defective vertical saccades, sometimes affecting downward saccades more than upward ones. Finally medial lesions of the RIMLF area, which affect the M group (see below), may cause a dissociation of eye and eyelid movements, accompanied by lid lag or ptosis [Büttner-Ennever et al., 1996].

Connections

Efferents. The premotor neurons of the RIMLF are medium-sized burst neurons, which project monosynaptically to the motoneurons of the vertically pulling extraocular eye muscles in the oculomotor and trochlear nuclei. Upward-moving motoneurons of superior rectus and inferior oblique muscles are innervated bilaterally from the RIMLF, while downward motoneurons of inferior rectus and superior oblique muscles are innervated ipsilaterally [Moschovakis et al., 1991a, b, 1996]. Recently, the location of burst neurons in the RIMLF to the upward-moving eye muscles was identified using calretinin immunostaining, and they are shown to be evenly scattered throughout the nucleus [Ahlfeld et al., 2011]. Reticulospinal neurons are not present in the RIMLF of primates but are found in the adjacent MRF [Robinson et al., 1994]. In contrast, in the cat, reticulospinal neurons concerned with vertical head movements are found in the RIMLF [Isa and Sasaki, 1992]. Another difference between the cat and primate is the lack of GABAergic neurons in the primate RIMLF [Horn et al., 2003b]. Additional efferent targets of the descending RIMLF axons are the contralateral RIMLF, the paramedian tract neurons and the INC [Holstege and Cowie, 1989; Moschovakis et al., 1991a, b; Wang and Spencer, 1996]. The connections between RIMLF and INC are reciprocal, and play an important role in vertical gaze and vertical gaze-holding.

Afferents. A strong input to the burst neurons in the RIMLF arises from the inhibitory, glycinergic omnipause neurons within the pontine reticular formation in the nucleus reticularis pontis caudalis; they serve to synchronize the vertical and horizontal components of a saccade [Horn et al., 1994]. In addition, the RIMLF receives afferents from the deep layers of the superior colliculus, which may participate in the target selection [Nakao et al., 1990; Optican, 2005]. A careful study on primates revealed direct projections to the RIMLF from the frontal eye fields of the cerebral cortex [Yan et al., 2001; Lynch and Tian, 2006].

M Group. A cell group just medial to the RIMLF, called the 'M group', has recently been identified in humans [Horn et al., 2000]. It is interconnected with the cells of the RIMLF. In the monkey the M group neurons have been shown to project to the caudal oculomotor nucleus, terminating on upper eyelid and upward eye-moving motoneurons, then continuing caudally to supply motoneurons in the facial nucleus as well. Both the RIMLF and the M group participate in the control of vertical eye movements, but, unlike the RIMLF, the M group also coordinates the activity of additional muscles, such as the levator palpebrae and muscles of the forehead, with the vertical eye movements. For a review, see Horn and Adamczyk [2012].

References

Ahlfeld J, Mustari MJ, Horn AKE: Sources of calretinin inputs to motoneurons of extraocular muscles involved in upgaze. Ann NY Acad Sci 2011;1233:91–99.

Büttner-Ennever JA: The extraocular motor nuclei: organization and functional neuroanatomy. Prog Brain Res 2006;151:95–125.

Büttner-Ennever JA, Büttner U, Cohen B, Baumgartner G: Vertical gaze paralysis and the rostral interstitial nucleus of the medial longitudinal fasciculus. Brain 1982;105:125–149.

Büttner-Ennever JA, Jenkins C, Armin-Parsa H, Horn AKE, Elston JS: A neuroanatomical analysis of lid-eye coordination in cases of ptosis and downgaze paralysis. Clin Neuropathol 1996;15:313–318.

Büttner U, Büttner-Ennever JA: Present concepts of oculomotor organization; in Büttner-Ennever JA (ed): Neuroanatomy of the Oculomotor System. Reviews in Oculomotor Research. Amsterdam, Elsevier, 1988, vol 2.

Crossland WJ, Hu XJ, Rafols JA: Morphological study of the rostral interstitial nucleus of the medial longitudinal fasciculus in the monkey, *Macaca mulatta*, by Nissl, Golgi, and computer reconstruction and rotation methods. J Comp Neurol 1994;347:47–63.

Helmchen C, Glasauer S, Bartl K, Büttner U: Contralesionally beating torsional nystagmus in a unilateral rostral midbrain lesion. Neurology 1996;47:482–486.

Helmchen C, Rambold HC, Kempermann U, Büttner-Ennever JA, Büttner U: Localizing value of torsional nystagmus in small midbrain lesions. Neurology 2002; 59:1956–1964.

Holstege G, Cowie RJ: Projections from the rostral mesencephalic reticular formation to the spinal cord. Exp Brain Res 1989;75:265–279.

Horn AK, Adamczyk C: Reticular formation: eye movements, gaze and blinks; in Mai JK, Paxinos G (eds): The Human Nervous System, ed 3. San Diego, Academic Press, 2012.

Horn AKE: The reticular formation. Prog Brain Res 2006;151:127–155.

Horn AKE, Brückner G, Härtig W, Messoudi A: Saccadic omnipause and burst neurons in monkey and human are ensheathed by perineuronal nets but differ in their expression of calcium-binding proteins. J Comp Neurol 2003a;455:341–352.

Horn AKE, Büttner-Ennever JA: Premotor neurons for vertical eye-movements in the rostral mesencephalon of monkey and man: the histological identification by parvalbumin immunostaining. J Comp Neurol 1998;392:413–427.

Horn AKE, Büttner-Ennever JA, Gayde M, Messoudi A: Neuroanatomical identification of mesencephalic premotor neurons coordinating eyelid with upgaze in the monkey and man. J Comp Neurol 2000;420:19–34.

Horn AKE, Büttner-Ennever JA, Wahle P, Reichenberger I: Neurotransmitter profile of saccadic omnipause neurons in nucleus raphe interpositus. J Neurosci 1994; 14:2032–2046.

Horn AKE, Helmchen C, Wahle P: GABAergic neurons in the rostral mesencephalon of the macaque monkey that control vertical eye movements. Ann NY Acad Sci 2003b;1004:19–28.

Isa T, Sasaki S: Mono- and disynaptic pathways from Forel's field H to dorsal neck motoneurones in the cat. Exp Brain Res 1992;88:580–593.
Leigh RJ, Zee DS: The Neurology of Eye Movements. New York, Oxford University Press, 2006.
Lynch JC, Tian J-R: Cortico-cortical networks and cortico-subcortical loops for the higher control of eye movements. Prog Brain Res 2006;151:461–501.
Moschovakis AK, Scudder CA, Highstein SM: The structure of the primate oculomotor burst generator. I. Medium-lead burst neurons with upward on-directions. J Neurophysiol 1991a;65, 203–217.
Moschovakis AK, Scudder CA, Highstein SM: The microscopic anatomy and physiology of the mammalian saccadic system. Prog Neurobiol 1996;50:133–133.
Moschovakis AK, Scudder CA, Highstein SM, Warren JD: The structure of the primate oculomotor burst generator. II. Medium-lead burst neurons with downward on-directions. J Neurophysiol 1991b;65:218–229.
Nakao S, Shiraishi Y, Li WB, Oikawa T: Mono- and disynaptic excitatory inputs from the superior colliculus to vertical saccade-related neurons in the cat Forel's field H. Exp Brain Res 1990;82:222–226.
Optican L: Sensorimotor transformation for visually guided saccades. Ann NY Acad Sci 2005;1039:132–148.
Paxinos G, Huang X-F: Atlas of the Human Brainstem. San Diego, Academic Press, 1995.
Robinson FR, Phillips JO, Fuchs AF: Coordination of gaze shifts in primates: brainstem inputs to neck and extraocular motoneuron pools. J Comp Neurol 1994; 346:43–62.
Suzuki Y, Büttner-Ennever JA, Straumann D, Hepp K, Hess BJM, Henn V: Deficits in torsional and vertical rapid eye movements and shift of Listing's plane after uni- and bilateral lesions of the rostral interstitial nucleus of the medial longitudinal fasciculus. Exp Brain Res 1995;106:215–232.
Wang SF, Spencer RF: Spatial organization of premotor neurons related to vertical upward and downward saccadic eye movements in the rostral interstitial nucleus of the medial longitudinal fasciculus (riMLF) in the cat. J Comp Neurol 1996;366:163–180.
Yan Y, Cui D, Lynch JC: Overlap of saccadic and pursuit eye movement systems in the brain stem reticular formation. J Neurophysiol 2001;86:3056–3059.

Reticular Formation

(Plates 41 and 42)

62 Nucleus of the Posterior Commissure (NPC)

Original name:
Not described

Location and Cytoarchitecture

The nucleus of the posterior commissure (NPC; fig. 1–3) is usually considered with the mesencephalic tegmentum, although it is sometimes included in the descriptions of the pretectum and it has recently been shown to be derived from prosemere 1 [Kuhlenbeck and Miller, 1949; Nieuwenhuys et al., 2008]. The NPC lies medial to the caudal border of the pretectum, and lateral to the periaqueductal gray (PAG), immediately rostral to the level at which the ganglion cells of the mesencephalic trigeminal nucleus can be found. The deep layers of the rostral superior colliculus (SC) lie dorsal to the NPC, and ventrally it is continuous with the mesencephalic reticular formation (MRF).

At least 5 different cell groups have been identified in the monkey and human in the NPC (principal, magnocellular, rostral, subcommissural and infracommissural parts): these are based on the cytoarchitecture and relationship to the fibers of the posterior commissure (PC) [Carpenter and Peter, 1970]. Currently, the term NPC refers to the principal and magnocellular parts, since the 3 other subdivisions all lie within the PAG. The magnocellular part is composed mainly of medium and large cells interstitial to the commissural fibers, and it is surrounded ventrolaterally by the reticular-like principal part composed of small and medium-sized neurons [Bianchi and Gioia, 1993]. An area more ventral to the magnocellular NPC is called the 'interstitial tegmental NPC' by Kuhlenbeck and Miller [1949]. It contains scattered large cells, and lies within the rubral capsule, in the part of the MRF closely associated with the interstitial nucleus of Cajal (INC) [Holstege, 1988].

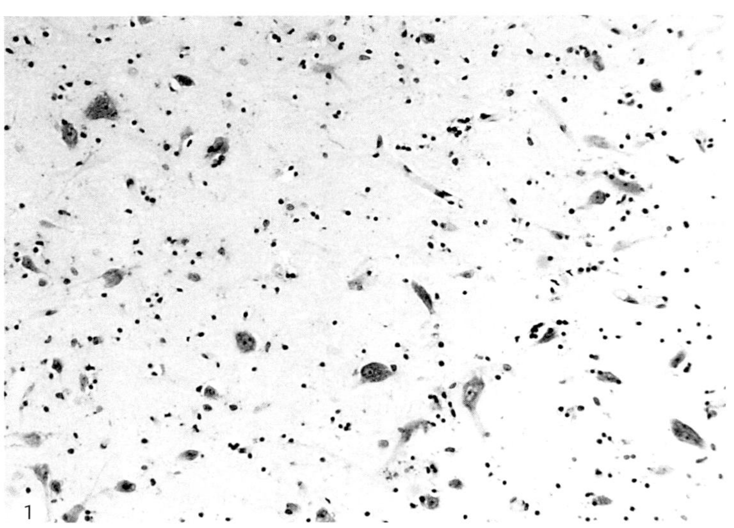

Fig. 1. Nucleus of the posterior commissure. Nissl stain. Magnification ×150.

The high levels of parvalbumin in the NPC can be used to delineate it from the rostral SC. Paxinos and Huang [Paxinos and Huang, 1995] use the name NPC for an area lying further dorsal and rostral to the NPC defined here; the area corresponds roughly to our medial pretectal nucleus (nucleus No. 33, plate 42).

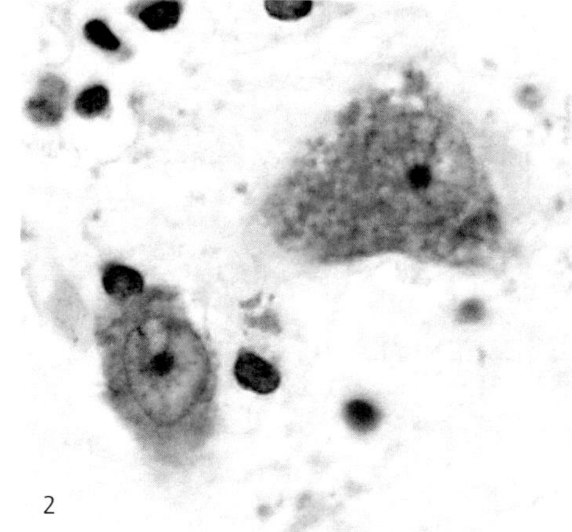

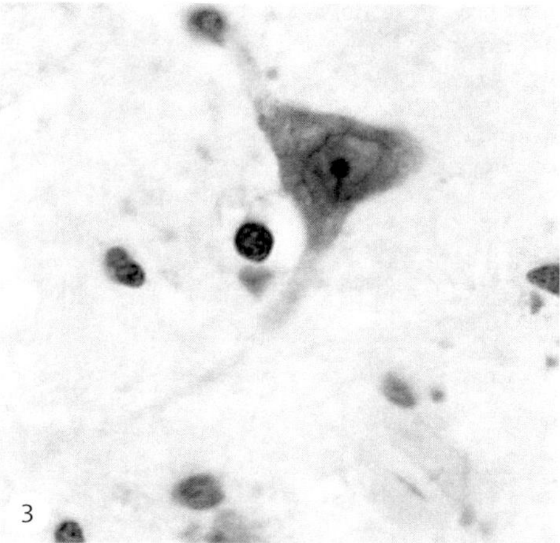

Fig. 2, 3. Cells of the nucleus of the posterior commissure. Magnification ×1,000.

Functional Neuroanatomy

Function

Chronic single-unit recordings in the macaque monkey revealed neurons in the NPC which fire with upward saccades. In contrast to the burst neurons in the rostral interstitial nucleus of the medial longitudinal fasciculus (RIMLF), the NPC saccade-related neurons were shown not to project to motoneurons of extraocular eye muscles, but to target neurons in the contralateral NPC, the INC, the RIMLF and intralaminar thalamic nuclei. The NPC was considered to play a role in modulating the vertical gaze integrator [Moschovakis et al., 1996].

On the basis of clinical observations the NPC is known to be involved in the generation of upward eye movements: lesions in patients can result in the impairment of all classes of vertical eye movements, especially upward gaze, and the loss of vertical gaze-holding (see 'neural integrator' function) [Leigh and Zee, 2006]. Additional symptoms such as upper eyelid retraction and 'convergence-retraction nystagmus' may also accompany these dorsal midbrain lesions. Experimental lesions of the NPC lead to less severe deficits: they may cause vertical gaze-evoked nystagmus [Partsalis et al., 1994]: larger lesions of the NPC including the PAG and PC impair the pupillary light reflex [Carpenter and Pierson, 1973].

Connections

Fibers of the NPC magnocellular part project through the ventral part of the PC to the contralateral INC, the magnocellular part of the NPC, RIMLF and the supraoculomotor area immediately dorsal to the oculomotor nucleus (III) [Büttner-Ennever and Horn, 2004]. The NPC also projects to a cell group just medial to the RIMLF, called the 'M group', which has recently been shown to project to the caudal III and terminate on the upper eyelid and upward eye moving motoneurons as well as facial motoneurons. The M group participates in the coordination of facial muscles with vertical eye movements [Horn et al., 2000]. Descending fiber pathways from the NPC terminate in the paramedian pontine reticular formation and spinal cord at cervical levels [Holstege, 1988; Holstege and Cowie, 1989]. The NPC has reciprocal connections with the SC, some of which are GABAergic [Appell and Behan, 1990; May, 2006]: it also receives a strong input from the frontal eye fields of the cortex and the dentate nucleus of the cerebellum.

References

Appell PP, Behan M: Sources of subcortical GABAergic projections to the superior colliculus in the cat. J Comp Neurol 1990;302:143–158.

Bianchi R, Gioia M: Accessory oculomotor nuclei of man. 3. The nuclear complex of the posterior commissure – a Nissl and Golgi study. Acta Anat 1993;146: 53–61.

Büttner-Ennever JA, Horn AKE: Reticular formation: eye movements, gaze and blinks; in Paxinos G, Mai JK (eds): The Human Nervous System, ed 2. Amsterdam, Elsevier Academic Press, 2004.

Carpenter MB, Peter P: Accessory oculomotor nuclei in the monkey. J Hirnforsch 1970;12:405–418.

Carpenter MB, Pierson RJ: Pretectal region and the pupillary light reflex. An anatomical analysis in the monkey. J Comp Neurol 1973;149:271–300.

Holstege G: Brainstem-spinal cord projections in the cat, related to control of head and axial movements; in Büttner-Ennever JA (ed): Neuroanatomy of the Oculomotor System. Reviews in Oculomotor Research. Amsterdam, Elsevier, 1988, vol 2.

Holstege G, Cowie RJ: Projections from the rostral mesencephalic reticular formation to the spinal cord. Exp Brain Res 1989;75:265–279.

Horn AKE, Büttner-Ennever JA, Gayde M, Messoudi A: Neuroanatomical identification of mesencephalic premotor neurons coordinating eyelid with upgaze in the monkey and man. J Comp Neurol 2000;420:19–34.

Kuhlenbeck H, Miller RN: The pretectal region of the human brain. J Comp Neurol 1949;91:369–407.

Leigh RJ, Zee DS: The Neurology of Eye Movements. New York, Oxford University Press, 2006.

May PJ: The mammalian superior colliculus: laminar structure and connections. Prog Brain Res 2006;151:321–378.

Moschovakis AK, Scudder CA, Highstein SM: The microscopic anatomy and physiology of the mammalian saccadic system. Prog Neurobiol 1996;50:133–254.

Nieuwenhuys R, Voogd J, van Huijzen C: Development; in Nieuwenhuys R, Voogd J, van Huijzen C (eds): The Human Central Nervous System. Berlin, Springer, 2008.

Partsalis AM, Highstein SM, Moschovakis AK: Lesions of the posterior commissure disable the vertical neural integrator of the primate oculomotor system. J Neurophysiol 1994;71:2582–2585.

Paxinos G, Huang X-F: Atlas of the Human Brainstem. San Diego, Academic Press, 1995.

63 Posterior Commissure (PC)

Original name:
Commissura posterior

Location

The compact crossing fibers of the posterior commissure (PC) mark the border between the mesencephalon and the diencephalon. The PC lies just rostral to the tectal commissure, and ventrocaudal to the habenular commissure, which interconnects the habenular nuclei of each side. In the early stages of myelination, during brain development, the PC emerges in the caudal pretectum as the first transversal tract [Keene, 1938; Ahsan et al., 2007; Ware and Schubert, 2011]. In many species, just ventral to the PC, in the roof of the aqueduct, the ependyma is specialized to form the 'subcommissural organ', a circumventricular organ, but, unlike other mammals, in humans it is only clearly evident in the fetus or newborns; see nucleus 85, figure 11–13, and figure 14.12 of Parent [1996; McKinley et al., 2004].

The line between the PC and the anterior commissure, taken in the midsagittal plane, is a useful horizontal referral line to align brain scans. It is based on the suggestion of Talairach and Toumoux [1988].

Functional Neuroanatomy

The PC contains numerous crossing pathways: for example those originating in the nucleus of the posterior commissure (NPC, nucleus No. 62) and interstitial nucleus of Cajal [Horn et al., 2003], those interconnecting the accessory optic nuclei [Giolli et al., 2006], also visual reflex pathways crossing the midline from cell groups of the pretectum to the contralateral Edinger-Westphal nucleus (EW) and oculomotor complex, which participate in the pupillary light reflex, accommodation and the near response [Carpenter and Pierson, 1973; Young and Lund, 1994; Wasicky et al., 2004; Gamlin, 2006; Nieuwenhuys et al., 2008]. However, lesions of the PC alone do not significantly modify the pupillary light reflex in spite of the (bilateral) interruption of the crossing fibers supplying the EW [Carpenter and Pierson, 1973]. The kinds of lesions that disrupt the pupillary light reflex are larger lesions, involving the PC, its nuclei and parts of the periaqueductal gray.

Interestingly, in a case of Niemann-Pick disease (a lipid storage disorder), the region of the PC (and the subcommissural organ) was found to have pathological features that could have intruded into the PC, and accounted for the patient's deficits in vertical eye movements [Solomon et al., 2005].

References

Ahsan M, Riley KL, Schubert FR: Molecular mechanisms in the formation of the medial longitudinal fascicle. J Anat 2007;211:177–187.
Carpenter MB, Pierson RJ: Pretectal region and the pupillary light reflex. An anatomical analysis in the monkey. J Comp Neurol 1973;149:271–300.
Gamlin PDR: The pretectum: connections and oculomotor-related roles. Prog Brain Res 2006;151:379–405.
Giolli RA, Blanks RHI, Lui F: The accessory optic system: basic organization with an update on connectivity, neurochemistry, and function. Prog Brain Res 2006;151:407–440.
Horn AKE, Helmchen C, Wahle P: GABAergic neurons in the rostral mesencephalon of the macaque monkey that control vertical eye movements. Ann NY Acad Sci 2003;1004:19–28.
Keene MF: The connexions of the posterior commissure. J Anat 1938;72:488–501.
McKinley MJ, Clarke IJ, Oldfield BJ: Circumventricular organs; in Paxinos G, Mai JK (eds): The Human Nervous System, ed 2. Amsterdam, Elsevier Academic Press, 2004.
Nieuwenhuys R, Voogd J, van Huijzen C: The Human Central Nervous System. Berlin, Springer, 2008.
Parent A: Carpenter's Human Neuroanatomy. Baltimore, Williams & Wilkins, 1996.
Solomon D, Winkelman C, Zee DS, Gray L, Büttner-Ennever JA: Niemann-Pick type C disease in two affected sisters: ocular motor recordings and brainstem neuropathology. Ann NY Acad Sci 2005;1039:436–445.
Talairach J, Toumoux P: Co-Planar Stereotaxic Atlas of the Human Brain. New York, Thieme, 1988.
Ware M, Schubert FR: Development of the early axon scaffold in the rostral brain of the chick embryo. J Anat 2011;219:203–216.
Wasicky R, Horn AKE, Büttner-Ennever JA: Twitch and non-twitch motoneuron subgroups of the medial rectus muscle in the oculomotor nucleus of monkeys receive different afferent projections. J Comp Neurol 2004;479:117–129.
Young MJ, Lund RD: The anatomical substrates subserving the pupillary light reflex in rats: origin of the consensual pupillary response. Neuroscience 1994;62:481–496.

Neuromodulatory Systems

Overview

The recognition of some neuronal networks as 'neuromodulators' is relatively recent. The neuromodulatory nuclei differ in several ways from those using classical fast synaptic transmission. Some synaptic transmitters, such as acetylcholine, can act as either a neurotransmitter or a neuromodulator depending on the receptors the chemical binds to. Neuromodulators do not participate in the transfer of information through excitatory or inhibitory pathways; they modulate the signal transfer by altering the cell properties, such as its excitability. That is they act much in the same way as the volume knob of an amplifier, which can make music louder, but the music signal itself is received through a separate channel. The term 'neurotransmission', or 'synaptic transmission', usually refers to fast ionotropic effects through ligand-gated ion channels, whereas the term 'neuromodulation' refers to the slower, longer-lasting effects at metabotropic 7-membrane-spanning G-protein-coupled receptors. They modulate functions such as arousal, sleep, learning, memory, cognition, locomotion and pain [Saper et al., 2001; Sillar et al., 2002; Fuxe et al., 2010]. Recent studies of neuromodulation, focusing on monoaminergic signal transfer, have led to the hypothesis that transmission travels not only by 'wire transmission' (i.e. through axons) but also through 'volume transmission' (using diffusion gradients in the intra- and extracellular space) [Hornung, 2004; Fuxe et al., 2010].

Neuromodulatory networks have several features in common; they possess extremely diffuse axonal projections covering immensely wide areas of the brain such as the whole cerebral cortex, basal ganglia, cerebellum or spinal cord, and in some cases, several of these. The neurons of origin lie in specific nuclei, or groups of nuclei, in the brainstem diencephalon or forebrain, with each characterized by a specific transmitter. The raphe nuclei (nucleus raphe pallidus, nucleus raphe obscurus, nucleus raphe magnus, median raphe nucleus, and dorsal raphe nucleus) use serotonin, dopamine is the transmitter of the neuromodulators in the ventral tegmental area (substantia nigra, caudal and rostral linear nuclei), and the pontine tegmental nuclei (pedunculopontine nucleus and laterodorsal tegmental nucleus) utilize acetylcholine. There are many further examples of neuromodulator networks in the diencephalon and telencephalon, but these are not considered here [Saper, 2006].

References

Fuxe K, Dahlstrom AB, Jonsson G, Marcellino D, Guescini M, Dam M, Manger P, Agnati L: The discovery of central monoamine neurons gave volume transmission to the wired brain. Prog Neurobiol 2010;90:82–100.

Hornung JP: Raphe nuclei; in Paxinos G, Mai JK (eds): The Human Nervous System, ed 2. Amsterdam, Elsevier Academic Press, 2004.

Saper CB: Staying awake for dinner: hypothalamic integration of sleep, feeding, and circadian rhythms. Prog Brain Res 2006;153:243–252.

Saper CB, Chou TC, Scammell TE: The sleep switch: hypothalamic control of sleep and wakefulness. Trends Neurosci 2001;24:726–731.

Sillar KT, Mclean DL, Fischer H, Merrywest SD: Fast inhibitory synapses: targets for neuromodulation and development of vertebrate motor behaviour. Brain Res Brain Res Rev 2002;40:130–140.

Neuromodulatory Systems – Serotonergic Nuclei

(Plates 10, 12 and 14)

64 Nucleus Raphe Pallidus (RPa)

Original name:
The nucleus raphe magnus was mistakenly called nucleus raphe pallidus in the 2nd edition [Braak, 1970; Paxinos et al., 2012]

Location and Cytoarchitecture

The cells of this nucleus are slender, elongated and with few dendrites (fig. 1, 2). They lie close to the midline, at the dorsal border of the pyramidal tract, extending from the level of the obex to the rostral medulla, parallel to the inferior olive. The neurons of the nucleus raphe pallidus (RPa) form a ventrocaudal continuation of the nucleus raphe magnus (RMg) [Braak, 1970], and like the RMg are serotonergic [Halliday et al., 1988; Baker et al., 1991]. The RPa is the smallest raphe nucleus in the brain, and roughly coincides with the B1 group of serotonergic neurons of Dahlström and Fuxe [1964]. Several serotonergic neurons lie lateral to the RPa, scattered in the ventromedial medullary reticular formation, and on the basis of connectivity studies they appear to cooperate with the RPa functionally.

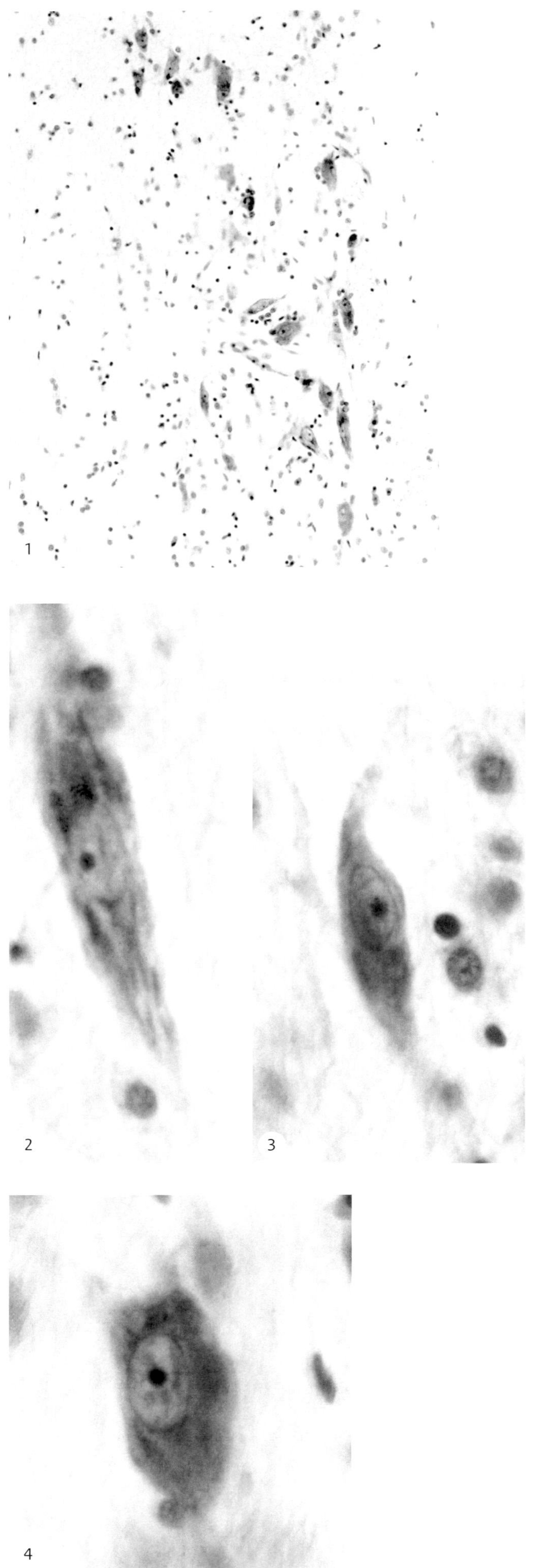

Fig. 1. Nucleus raphe pallidus: cells on the right of the midline in the raphe and medial to the descending fiber tracts. Magnification ×150.
Fig. 2–4. Cells from the nucleus raphe pallidus. Magnification ×1,000.

As in other raphe nuclei, the neurons of the RPa are not exclusively serotonergic, and may be considered as a multitransmitter system: nonserotonergic cells containing GABA or glycine have been reported, as well as several neuropeptides such as substance P, thyrotropin-releasing hormone or enkephalin [Johansson et al., 1981; Jones et al., 1991; Hornung, 2012]. Furthermore, many other substances that act as transmitters or neuromodulators have been located within serotonergic neurons and in their terminals [Nicholas et al., 1992].

Functional Neuroanatomy

Function

The RPa is a serotonergic raphe nucleus, and operates as a neuromodulator (see 'Neuromodulatory Systems: Overview'). Typically the efferent axons of neuromodulators are extremely widespread, and they set the level of activity of their neural targets, rather than transmitting specific information. The RPa targets mainly motoneurons in the ventral horn of the spinal cord, throughout its length, and in the brainstem. Through these pathways the RPa can control the vast majority of autonomic and somatic functions. The RPa can raise or lower the general excitability of motoneurons, and modulate respiratory, cardiovascular or somatic motor activity.

Clinically, serotonin is associated with many physiological and pathological functions including mood changes (depression with serotonin depletion), pain control, neurodegenerative disorders and schizophrenia [Hornung, 2004; Fuxe et al., 2009]. However, given the motor-specific projections of the RPa, a more applicable hypothesis is that the nucleus, particularly the caudal part, is concerned with 'shivering effects' related to hypothermia [Holstege, 1991]. The strong projections to the caudal RPa from the caudal hypothalamus, which control temperature regulation of the body, support this hypothesis. Experiments inducing motor fatigue reported reduced serotonin levels in the medullary raphe neurons [Fornal et al., 2006]. Finally, direct stimulation of the RPa in rats was shown to protect the gastric mucosa by activation of vagal pathways [Kaneko et al., 1998].

Connections

Efferents. The efferent pathways from the RPa project throughout the whole length of the spinal cord in a diffuse nontopographic fashion. They descend in the lateral funiculus, and terminals cover the intermediate zone and ventral horn where they contact sympathetic preganglionic neurons and somatic motoneurons. In addition the RPa innervates the brainstem motor nuclei: the trigeminal, facial, dorsal motor vagal and hypoglossal nuclei, but not the motoneurons controlling fine motor activity such as the distal limb or oculomotor muscles [Steinbusch, 1981; Holstege and Kuypers, 1982; Holstege, 1991].

Terminals from serotonergic raphe nuclei form 2 types of contacts: small or large varicosities. The small varicosities release the transmitter diffusely, without a synaptic specialization, and they arise from fine fibers; in contrast, the large varicosities terminate as synaptic contacts with membrane specializations, and arise from thicker fibers. The RPa and median raphe nucleus terminals tend to form large varicosities, whereas the RMg and the dorsal raphe nucleus form small varicosities [Mamounas et al., 1991; Descarries and Mechawar, 2000].

Afferents. Inputs into the RPa have been studied in the rat and cat [Holstege, 1991; Hermann et al., 1997]. They originate from medial limbic regions such as the anterolateral hypothalamus and the periaqueductal gray (PAG), dorsolateral and ventrolateral parts. These projections are of significance because the PAG does not project to the spinal cord itself. From the hypothalamus there appears to be a rough topography where the rostral hypothalamus projects to the rostral RPa, and the caudal hypothalamus to the caudal RPa, the latter being a strikingly strong pathway thought to participate in thermal control of the body [Hemingway, 1963; Holstege, 1991]. Small to moderate connections were demonstrated from the lateral preoptic area, bed nucleus of the stria terminalis, paraventricular hypothalamic nucleus, central nucleus of the amygdala, parabrachial nuclei, subcoeruleus area and parvocellular reticular nucleus, as well as from the insular and perirhinal cortices. The RPa neurons receive many catecholaminergic inputs [Tanaka et al., 1994].

References

Baker KG, Halliday GM, Halasz P, Hornung JP, Geffen LB, Cotton RGH, Törk I: Cytoarchitecture of serotonin-synthesizing neurons in the pontine tegmentum of the human brain. Synapse 1991;7:301–320.

Braak H: On the nuclei of the human brain stem. II. The raphe nuclei. Z Zellforsch Mikrosk Anat 1970;107:123–141.

Dahlström A, Fuxe K: Evidence for the existence of monoamine neurons in the central nervous system. I. Demonstration of monoamines in the cell bodies of brain stem neurons. Acta Physiol Scand 1964;62:1–55.

Descarries L, Mechawar N: Ultrastructural evidence for diffuse transmission by monoamine and acetylcholine neurons of the central nervous system. Prog Brain Res 2000;125:27–47.

Fornal CA, Martin-Cora FJ, Jacobs BL: 'Fatigue' of medullary but not mesencephalic raphe serotonergic neurons during locomotion in cats. Brain Res 2006; 1072:55–61.

Fuxe K, Marcellino D, Woods AS, Giuseppina L, Antonelli T, Ferraro L, Tanganelli S, Agnati LF: Integrated signaling in heterodimers and receptor mosaics of different types of GPCRs of the forebrain: relevance for schizophrenia. J Neural Transm 2009;116:923–939.

Halliday GM, Li YW, Joh TH, Cotton RGH, Howe PRC: Distribution of monoamine-synthesizing neurons in the human medulla oblongata. J Comp Neurol 1988; 273:301–317.

Hemingway A: Shivering. Physiol Rev 1963;43:397–422.

Hermann DM, Luppi PH, Peyron C, Hinckel P, Jouvet M: Afferent projections to the rat nuclei raphe magnus, raphe pallidus and reticularis gigantocellularis pars alpha demonstrated by iontophoretic application of choleratoxin (subunit b). J Chem Neuroanat 1997;13:1–21.

Holstege G: Descending motor pathways and the spinal motor system: limbic and non-limbic components. Prog Brain Res 1991;87:307–421.

Holstege G, Kuypers HGJM: The anatomy of brain stem pathways to the spinal cord in cat. A labeled amino acid tracing study. Brain Res 1982;57:145–175.

Hornung J-P: Raphe nuclei; in Mai JK, Paxinos G (eds): The Human Nervous System, ed 3. Amsterdam, Academic Press, 2012.

Hornung JP: Raphe nuclei; in Paxinos G, Mai JK (eds): The Human Nervous System, ed 2. Amsterdam, Elsevier Academic Press, 2004.

Johansson O, Hokfelt T, Pernow B, Jeffcoate SL, White N, Steinbusch HW, Verhofstad AA, Emson PC, Spindel E: Immunohistochemical support for three putative transmitters in one neuron: coexistence of 5-hydroxytryptamine, substance P- and thyrotropin releasing hormone-like immunoreactivity in medullary neurons projecting to the spinal cord. Neuroscience 1981;6:1857–1881.

Jones BE, Holmes CJ, Rodriguez-Veiga E, Mainville L: GABA-synthesizing neurons in the medulla: their relationship to serotonin-containing and spinally projecting neurons in the rat. J Comp Neurol 1991;313:349–367.

Kaneko H, Kaunitz J, Tache Y: Vagal mechanisms underlying gastric protection induced by chemical activation of raphe pallidus in rats. Am J Physiol 1998; 275:G1056–G1062.

Mamounas LA, Mullen CA, O'Hearn E, Molliver ME: Dual serotoninergic projections to forebrain in the rat: morphologically distinct 5-HT axon terminals exhibit differential vulnerability to neurotoxic amphetamine derivatives. J Comp Neurol 1991;314:558–586.

Nicholas AP, Pieribone VA, Arvidsson U, Hokfelt T: Serotonin-, substance P- and glutamate/aspartate-like immunoreactivities in medullo-spinal pathways of rat and primate. Neuroscience 1992;48:545–559.

Paxinos G, Huang X-F, Sengul G, Watson C: Organization of the brainstem; in Mai JK, Paxinos G (eds): The Human Nervous System, ed 3. Amsterdam, Academic Press, 2012.

Steinbusch HWM: Distribution of serotonin-immunoreactivity in the central nervous system of the rat-cell bodies and terminals. Neuroscience 1981;6:557–618.

Tanaka M, Okamura H, Tamada Y, Nagatsu I, Tanaka Y, Ibata Y: Catecholaminergic input to spinally projecting serotonin neurons in the rostral ventromedial medulla oblongata of the rat. Brain Res Bull 1994;35:23–30.

Neuromodulatory Systems – Serotonergic Nuclei

(Plates 14–16, 18 and 19)

65 Nucleus Raphe Obscurus (ROb)

Original name:

Nucleus raphe obscurus, subnuclei intraraphalis and extraraphalis

Location and Cytoarchitecture – Original Text

This nucleus lies in relation to the dorsal part of the raphe and to the medial longitudinal fasciculi (MLF) in the oral part of the medulla and the caudal part of the pons. It commences approximately 1 mm caudal to the oral pole of the hypoglossal nucleus (XII) and extends a distance of 5–6 mm to the level of the abducens nucleus and the caudal pole of the facial nucleus. [At this level the nucleus raphe obscurus (ROb) overlaps with the caudal part of the nucleus raphe magnus (RMg) lying ventrally.]

The ROb may be subdivided into a median, unpaired subnucleus intraraphalis (RObi), and a paired subnucleus extraraphalis (RObe) (fig. 1).

The subnucleus RObi extends throughout the entire length of the nucleus and is formed by an elongated, vertically directed cell group, the constituents of which lie among the fibers of the dorsal part of the raphe. At all levels this subnucleus is bordered laterally, by the MLF and tectospinal tracts and ventrally by the nucleus raphe pallidus (RPa).

The subnucleus RObe appears about 0.5 mm oral to the caudal pole of the ROb. It is formed by two wing-like processes which arise from the dorsal extremity of the RObi and arch laterally over the dorsal and dorsolateral surfaces of the MLF. Together with the RObi they give the entire nucleus a fountain-shaped outline on cross-section (fig. 1). Caudally, each RObe is related laterally to the oral pole of the XII, whereas at more oral levels they are related laterally to the prepositus nucleus. At levels in which they are present, elements of the nucleus paramedianus dorsalis lie between the RObe and the medial part of the floor of the fourth ventricle.

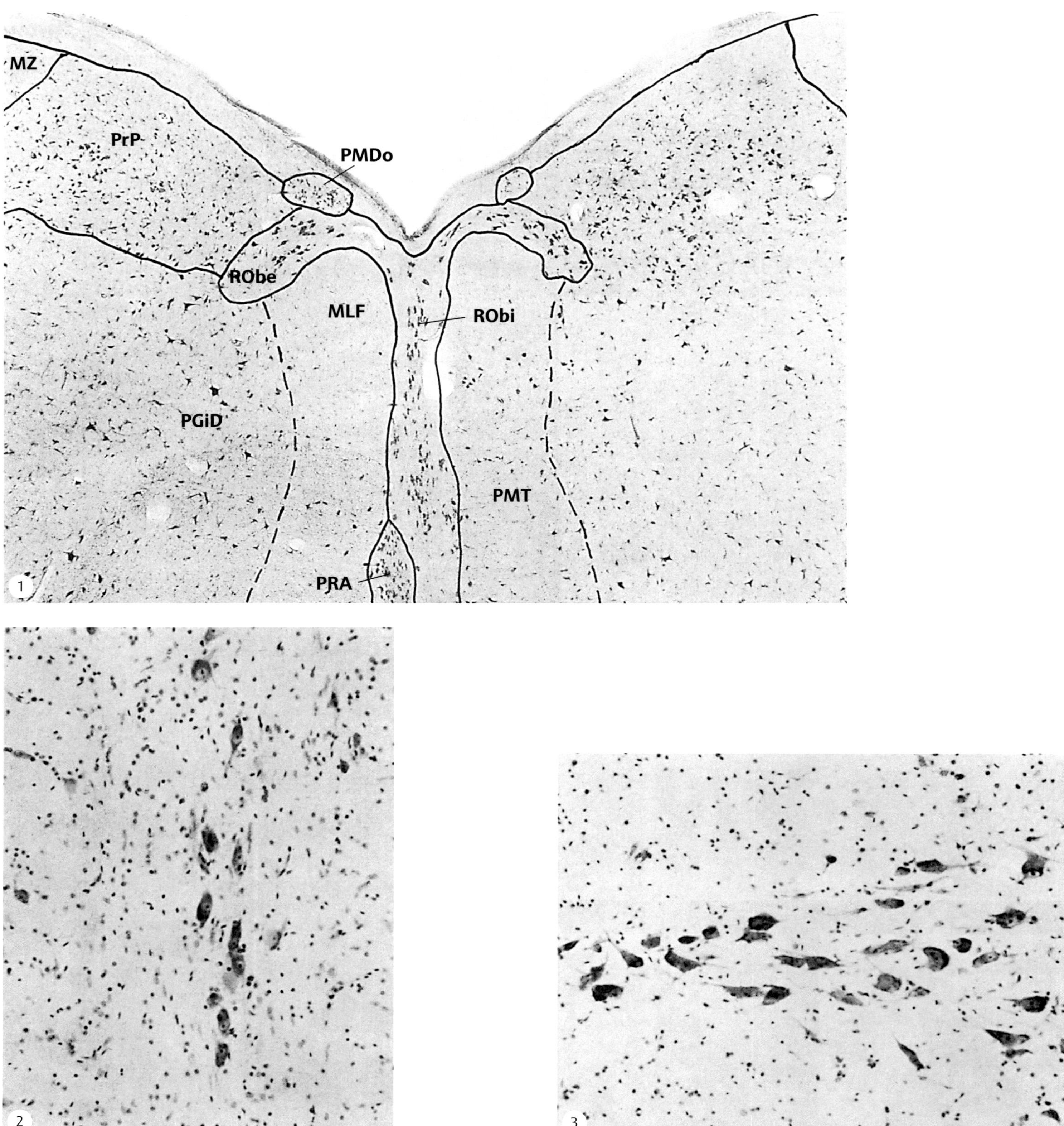

Fig. 1. Cross-section through the oral portion of the medullary tegmentum in the region of the nucleus raphe obscurus. MLF = Medial longitudinal fasciculus; MZ = marginal zone of the medial vestibular nucleus; PGiD = dorsal paragigantocellular nucleus; PMDo = nucleus paramedianus dorsalis, subnucleus oralis; PMT = paramedian tract; PRA = nuclei pararaphales; PrP = nucleus prepositus; RObe = nucleus raphe obscurus, subnucleus extraraphalis; RObi = nucleus raphe obscurus, subnucleus intraraphalis. Magnification ×30.

Fig. 2. Nucleus raphe obscurus, subnucleus intraraphalis. Magnification ×150.

Fig. 3. Nucleus raphe obscurus, subnucleus extraraphalis. Magnification ×150.

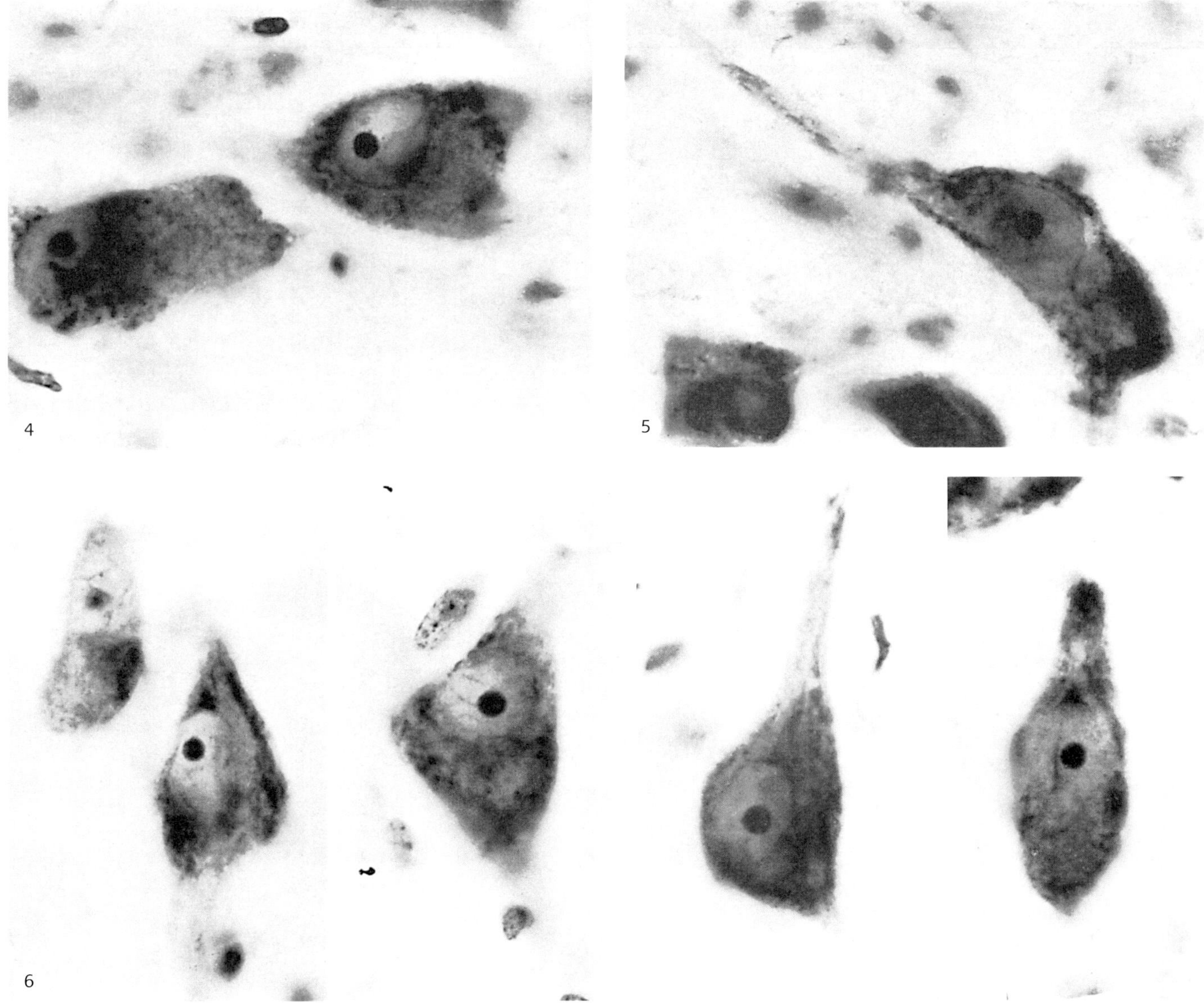

Fig. 4, 5. Cells from the nucleus raphe obscurus, subnucleus extraraphalis. Magnification ×1,000.

Fig. 6. Cells from the nucleus raphe obscurus, subnucleus intraraphalis. Magnification ×1,000.

Approaching the oral pole of the ROb, the cells of the RObe become less numerous and disappear at a more caudal level than do the cells of the RObi.

The cells of the RObi are large to medium sized, elongated and spindle shaped, and possess a centrally placed nucleus and large, distinct, darkly stained Nissl granules (fig. 2, 6). They lie with their long axes directed dorsoventrally, i.e. parallel to the long axis of the subnucleus. The cells of the RObe are elongated and multipolar rather than spindle shaped, possess longer dendrites than the RObi, and lie with their long axes directed parallel to the curved dorsal surface of the MLF (fig. 2–5).

Functional Neuroanatomy

Function

The ROb corresponds to the B2 group of serotonergic neurons of Dahlström and Fuxe [1964]. Like all raphe cell groups, the ROb has widespread axonal branches which *modulate* the level of ongoing neuronal activity of its targets, rather than causing the specific activity itself (see 'Neuromodulatory Systems: Overview'). Developmental criteria divide the raphe nuclei into two groups: a rostral group developing from rhombomeres 1–3, and consisting of the median raphe nucleus, raphe dorsalis and nucleus linearis, and a caudal group from rhombomeres 5–7, made up of the RMg, ROb and RPa [Alenina et al., 2006; Kiyasova and Gaspar, 2011].

Functionally the ROb is associated with the limbic system, being considered as the caudal part of the limbic system 'medial paracore' by Nieuwenhuys et al. [2008]. Holstege classifies the ROb as part of the emotional motor system, since the majority of ROb projections target autonomic and somatic motoneurons of the spinal cord [Taber-Pierce et al., 1977; Holstege et al., 2004].

More specifically, stimulation studies have shown that serotonergic projections, from the ROb to the caudal ventral respiratory group of the medulla, cause an inhibition of activity in phrenic motoneurons [Lalley et al., 1997].

Clinically serotonergic levels in the brain have been associated with disorders of mood, like depression [Hornung, 2012]. However, significant degenerative changes in the ROb have also been reported in neurodegenerative cases, such as Parkinson's disease [Halliday et al., 1990].

Connections

Efferents. The caudal group of raphe nuclei, ROb, RPa and RMg, is associated with pathways to the spinal cord and cerebellum [Holstege, 1991; Kerr and Bishop, 1991; Newman and Ginsberg, 1992]. Descending pathways from the ROb and RPa travel in the ventrolateral funiculus of the spinal cord and terminate mostly in the ventral horn and intermediate gray, where contacts are made with autonomic and somatosensory motoneurons. A significant proportion of the descending projections, labeled by injections into these raphe nuclei, arise from the cells lying laterally to the raphe nuclei in the adjacent medullary reticular formation [Braz et al., 2009]. Furthermore the terminals of the descending system contain many substances other than serotonin, including somatostatin, substance P, thyrotropin-releasing hormone and enkephalin, all of which influence signal transmission in the spinal cord [Bowker et al., 1982; Holstege, 1991]. Additional ROb projections affecting the autonomic dorsal vagal complex terminate in the nucleus of the solitary tract [Weissheimer and Machado, 2007].

Afferents. Even though the ROb is included in the emotional motor system [Holstege et al., 2004], inputs specifically to the ROb, rather than the caudal medullary raphe nuclei in general, are not often reported. The ventrolateral periaqueductal gray (PAG) and visceral nociceptors exert influences on ROb neurons [Snowball et al., 1997], and a direct projection to the ROb from the ventrolateral PAG has also been reported [Holstege, 1991].

References

Alenina N, Bashammakh S, Bader M: Specification and differentiation of serotonergic neurons. Stem Cell Rev 2006;2:5–10.

Bowker RM, Westlund KN, Sullivan MC, Wilber JF, Coulter JD: Transmitters of the raphe-spinal complex: immunocytochemical studies. Peptides 1982;3:291–298.

Braz JM, Enquist LW, Basbaum AI: Inputs to serotonergic neurons revealed by conditional viral transneuronal tracing. J Comp Neurol 2009;514:145–160.

Dahlström A, Fuxe K: Evidence for the existence of monoamine neurons in the central nervous system. I. Demonstration of monoamines in the cell bodies of brain stem neurons. Acta Physiol Scand 1964;62:1–55.

Halliday GM, Li YW, Blumberg PC, Joh TH, Cotton RG, Howe PR, Blessing WW, Geffen LB: Neuropathology of immunohistochemically identified brainstem neurons in Parkinson's disease. Ann Neurol 1990;27:373–385.

Holstege G: Descending motor pathways and the spinal motor system: limbic and non-limbic components. Prog Brain Res 1991;87:307–421.

Holstege GG, Mouton LJ, Gerrits NM: Emotional motor system; in Paxinos G, Mai JK (eds): The Human Nervous System, ed 2. Amsterdam, Elsevier Academic Press, 2004.

Hornung J-P: Raphe nuclei; in Mai JK, Paxinos G (eds): The Human Nervous System, ed 3. Amsterdam, Academic Press, 2012.

Kerr CWH, Bishop GA: Topographical organization in the origin of serotoninergic projections to different regions of the cat cerebellar cortex. J Comp Neurol 1991;304:502–515.

Kiyasova V, Gaspar P: Development of raphe serotonin neurons from specification to guidance. Eur J Neurosci 2011;34:1553–1562.

Lalley PM, Benacka R, Bischoff AM, Richter DW: Nucleus raphe obscurus evokes 5-HT-1A receptor-mediated modulation of respiratory neurons. Brain Res 1997; 747:156–159.

Newman DB, Ginsberg CY: Brainstem reticular nuclei that project to the cerebellum in rats: a retrograde tracer study. Brain Behav Evol 1992;39:24–68.

Nieuwenhuys R, Voogd J, van Huijzen C: The Human Central Nervous System. Berlin, Springer, 2008.

Snowball RK, Dampney RA, Lumb BM: Responses of neurones in the medullary raphe nuclei to inputs from visceral nociceptors and the ventrolateral periaqueductal grey in the rat. Exp Physiol 1997;82:485–500.

Taber-Pierce E, Hoddevik GH, Walberg F: The cerebellar projection from the raphe nuclei in the cat as studied with the method of retrograde transport of horseradish peroxidase. Anat Embryol 1977;152:73–87.

Weissheimer KV, Machado BH: Inhibitory modulation of chemoreflex bradycardia by stimulation of the nucleus raphe obscurus is mediated by 5-HT_3 receptors in the NTS of awake rats. Auton Neurosci 2007;132:27–36.

Neuromodulatory Systems – Serotonergic Nuclei (Plates 14, 16, 18 and 19)

66 Nucleus Raphe Magnus (RMg)

Original name:

Mistakenly called nucleus raphe pallidus in the 2nd edition

Location and Cytoarchitecture – Original Text from Nucleus Raphe Pallidus

This is an unpaired median nucleus formed by an accumulation of cells about the ventral part of the median raphe (midline) in the oral portion of the medulla and the caudal extremity of the pons. The cells lie among the fibers of the medial lemniscus (ML) and the median raphe and cause a rarefaction of these structures when viewed in myelin-stained sections [Riley, 1943, pp. 80–81, section T 9 660 B].

The caudal pole of the nucleus raphe magnus (RMg) is represented on cross-section by only a few cells scattered irregularly among the fibers of the ML. As one proceeds orally, the cell population increases rapidly and the nucleus attains its maximal development at the level of the oral pole of the inferior olivary complex. At this level, the RMg is irregularly quadrangular in outline. Throughout its extent the nucleus is related on each side to the pars α of the gigantocellular nucleus (Giα) [Martin et al., 1990], dorsally to the tectospinal tract and the nucleus raphe obscurus (ROb), and ventrally to the most ventral fibers of the ML. Orally the nucleus is contiguous to the [rostral] Gi. In those specimens in which the medullary striae and ventral external arcuate fibers are well developed, representative groups of the nuclei pararaphales may appear among the most medially situated cells of the RMg.

The cells of the RMg are extremely characteristic. They are large or medium sized, and possess a very palely stained perikaryon which contains only a few peripherally located Nissl bodies. The cell edges are ir-

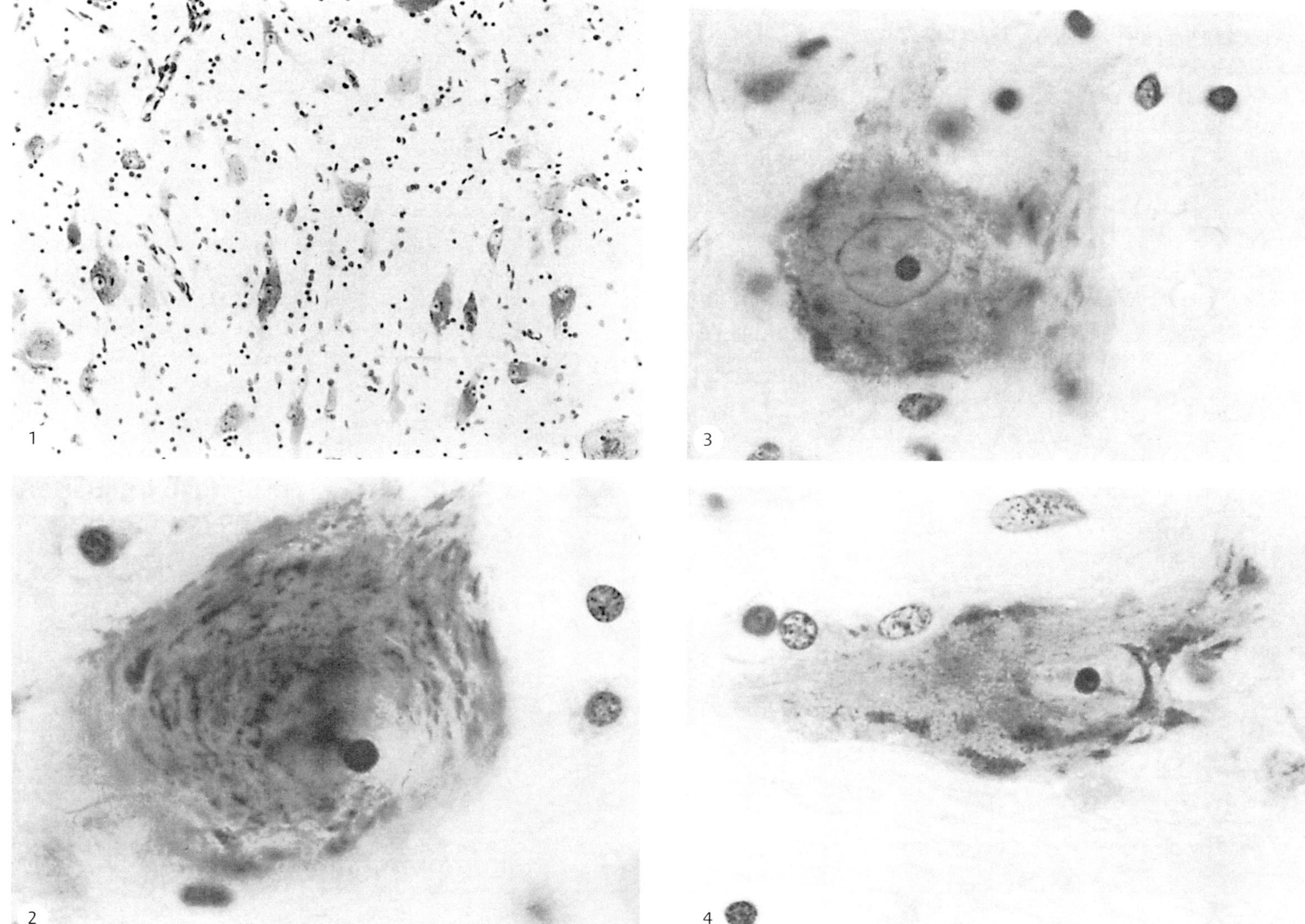

Fig. 1. Nucleus raphe magnus. Magnification ×150.
Fig. 2. 'Onion skin' cell from the nucleus raphe magnus. Magnification ×1,000.

Fig. 3, 4. Large palely stained cells from the nucleus raphe magnus. Magnification ×1,000.

regular or ragged and the nucleus is often excentric (fig. 1, 3, 4). In the vicinity of the raphe, the cells are elongated and their long axes are directed dorsoventrally. More laterally, the cells are triangular or multipolar in outline and are irregularly arranged [Dahlström and Fuxe, 1964]. Solitary 'onion skin cells', similar to those of the Gi, may occasionally be found among the more typical cells of the RMg (fig. 2).

Functional Neuroanatomy

Function

The majority of serotonergic neurons designated as B3 lie in the RMg, but the adjacent reticular formation also contains a significant serotonergic population, mixed amongst its reticular neurons [Dahlström and Fuxe, 1964; Baker et al., 1991]. There are several other nonserotonergic neurons in B3 and the adjacent reticular formation, but the major function of the RMg, in pain control, is most closely associated with the serotonergic efferent fibers [Hornung, 2012]. The descending RMg fibers target the dorsal horn throughout the whole length of the spinal cord, and through these can modulate the spinal transmission of sensory signals, in particular pain: the effects can operate as either nociceptive inhibitors or enhancers [Light et al., 1986; Holstege, 1988b, 1991; Fields et al., 1991; Braz and Basbaum, 2008; Zhang et al., 2009; Silveira et al., 2010; Todd, 2010]. Anatomical, electrophysiological and pharmacological evidence supports this role, but several other transmitters, not only serotonin, participate in the descending modulation of the pain [Antal et al., 1996; Hökfelt et al., 2000].

Connections

Efferents. The RMg, along with the nucleus raphe pallidus (RPa) and ROb, belong to the group of 'caudal raphe nuclei', which mainly project to the spinal cord. RMg efferents innervate the sensory neurons of the face in the spinal trigeminal nucleus, marginal zone, and continue caudally in the dorsolateral funiculus to terminate on the sensory neurons of the dorsal horn throughout the length of the cord. The pathways inhibit postsynaptic and presynaptic sensory relays, and autoradiographic studies show in detail that the caudal RMg terminates in laminae I, II and V, whereas the rostral RMg terminates in laminae II, III and IV [Basbaum et al., 1986; Holstege, 1988b, 1991]. The axon terminals of the RMg end as fine varicosities, as opposed to the large varicosities of the ventral horn, originating from the RPa and ROb [Ridet et al., 1994]. A significant proportion of the descending spinal projections from the RMg region arise from the adjacent reticular formation, specifically the Giα (the Giα together with the ventral Gi is often referred to as the magnocellular reticular nucleus). Many of these descending projections are from nonserotonergic sources using GABA or glycine as transmitters (40–50%) [Leanza et al., 1995]. The dorsal column nuclei are also innervated by RMg efferents [Basbaum et al., 1986].

While the major output of the RMg is the descending pathway regulating sensory input, there are also ascending and brainstem efferents from the RMg mainly to limbic structures. The RMg-hypothalamic axons to the preoptic area are mainly serotonergic [Leanza et al., 1995]. Other studies report projections to the lateral hypothalamus, parafascicular nucleus, ventral lateral periaqueductal gray (PAG), locus coeruleus, parabrachial nucleus (PB), A7, A5, and nucleus of the solitary tract [Sim and Joseph, 1992].

Afferents. Afferents to the RMg arise mostly from limbic structures such as the preoptic nuclei, the bed nuclei of the stria terminalis and PAG [Holstege, 1991; Holstege et al., 2004; Braz et al., 2009]. In addition the ventral PB, including the Kölliker-Fuse nucleus, project to the RMg and the adjacent tegmentum, suggesting that the inhibition of nociception induced by stimulation in the ventral PB may be based on the projections of this area to the RMg [Holstege, 1988a]. A spinal pathway ascending to the serotonergic neurons of the RMg has also recently been recognized, and it is presumed to arise from noci-responsive neurons of spinal cord laminae V–VIII; the pathway defines a route through which serotonergic neurons may act as a feedback pathway that can be activated by noxious stimuli [Braz et al., 2009].

References

Antal M, Petko M, Polgar E, Heizmann CW, Storm-Mathisen J: Direct evidence of an extensive GABAergic innervation of the spinal dorsal horn by fibres descending from the rostral ventromedial medulla. Neuroscience 1996;73:509–518.

Baker KG, Halliday GM, Halasz P, Hornung JP, Geffen LB, Cotton RGH, Törk I: Cytoarchitecture of serotonin-synthesizing neurons in the pontine tegmentum of the human brain. Synapse 1991;7:301–320.

Basbaum AI, Ralston DD, Ralston HJ 3rd: Bulbospinal projections in the primate: a light and electron microscopic study of a pain modulating system. J Comp Neurol 1986;250:311–323.

Braz JM, Basbaum AI: Genetically expressed transneuronal tracer reveals direct and indirect serotonergic descending control circuits. J Comp Neurol 2008; 507:1990–2003.

Braz JM, Enquist LW, Basbaum AI: Inputs to serotonergic neurons revealed by conditional viral transneuronal tracing. J Comp Neurol 2009;514:145–160.

Dahlström A, Fuxe K: Evidence for the existence of monoamine neurons in the central nervous system. I. Demonstration of monoamines in the cell bodies of brain stem neurons. Acta Physiol Scand 1964;62:1–55.

Fields HL, Heinricher MM, Mason P: Neurotransmitters in nociceptive modulatory circuits. Annu Rev Neurosci 1991;14:219–245.

Hökfelt T, Arvidson U, Cullheim S, Millhorn D, Nicholas AP, Pieribone VA, Seroogy K, Ulfhake B: Multiple messengers in descending serotonin neurons: localization and functional implications. J Chem Neuroanat 2000;18:75–86.

Holstege G: Anatomical evidence for a strong ventral parabrachial projection to nucleus raphe magnus and adjacent tegmental field. Brain Res 1988a;447:154–158.

Holstege G: Direct and indirect pathways to lamina I in the medulla oblongata and spinal cord of the cat. Prog Brain Res 1988b;77:47–94.

Holstege G: Descending motor pathways and the spinal motor system: limbic and non-limbic components. Prog Brain Res 1991;87:307–421.

Holstege GG, Mouton LJ, Gerrits NM: Emotional motor system; in Paxinos G, Mai JK: (eds): The Human Nervous System, ed 2. Amsterdam, Elsevier Academic Press, 2004.

Hornung J-P: Raphe nuclei; in Mai JK, Paxinos G (eds): The Human Nervous System, ed 3. Amsterdam, Academic Press, 2012.

Leanza G, Perez S, Pellitteri R, Russo A, Stanzani S: Branching serotonergic and non-serotonergic projections from caudal brainstem to the medial preoptic area and the lumbar spinal cord, in the rat. Neurosci Lett 1995;200:5–8.

Light AR, Casale EJ, Menetrey DM: The effects of focal stimulation in nucleus raphe magnus and periaqueductal gray on intracellularly recorded neurons in spinal laminae I and II. J Neurophysiol 1986;56:555–571.

Martin GF, Holstege G, Mehler WR: Reticular formation of the pons and medulla; in Paxinos G (ed): The Human Nervous System. San Diego, Academic Press, 1990.

Ridet JL, Geffard M, Privat A: Light and electron microscopic studies of the effects of *p*-chloroamphetamine on the monoaminergic innervation of the rat spinal cord. J Comp Neurol 1994;343:281–96.

Riley HA: An Atlas of the Basal Ganglia, Brain Stem and Spinal Cord. Baltimore, Williams & Wilkins Co, 1943, p 708.

Silveira JW, Dias QM, del Bel EA, Prado WA: Serotonin receptors are involved in the spinal mediation of descending facilitation of surgical incision-induced increase of Fos-like immunoreactivity in rats. Mol Pain 2010;6:17.

Sim LJ, Joseph SA: Efferent projections of the nucleus raphe magnus. Brain Res Bull 1992;28:679–682.

Todd AJ: Neuronal circuitry for pain processing in the dorsal horn. Nat Rev Neurosci 2010;11:823–836.

Zhang W, Gardell S, Zhang D, Xie JY, Agnes RS, Badghisi H, Hruby VJ, Rance N, Ossipov MH, Vanderah TW, Porreca F, Lai J: Neuropathic pain is maintained by brainstem neurons co-expressing opioid and cholecystokinin receptors. Brain 2009;132:778–787.

Neuromodulatory Systems – Serotonergic Nuclei (Plates 26–30 and 32)

67 Median Raphe Nucleus (MnR)

Original names:

In the previous edition this was called 'nucleus centralis superior'; it was subdivided into 4 parts, the medial, lateral, dorsal and ventral subnuclei; for each of these subdivisions, different names are more usual at present: *nucleus centralis superior, subnucleus medialis*, the main body of the nucleus is called now the median raphe nucleus (MnR) [Paxinos et al., 2012]; *nucleus centralis superior, subnucleus lateralis*, called now the paramedian raphe nucleus (PMR), see below [Paxinos et al., 2012]; *nucleus centralis superior, subnucleus dorsalis*, called now the dorsal raphe nucleus, pars caudalis (DRc, nucleus No. 68); *nucleus centralis superior, subnucleus ventralis*, or δ, used originally for the cells lying between the fibers of the superior cerebellar peduncle, called here MnR ventral subnucleus

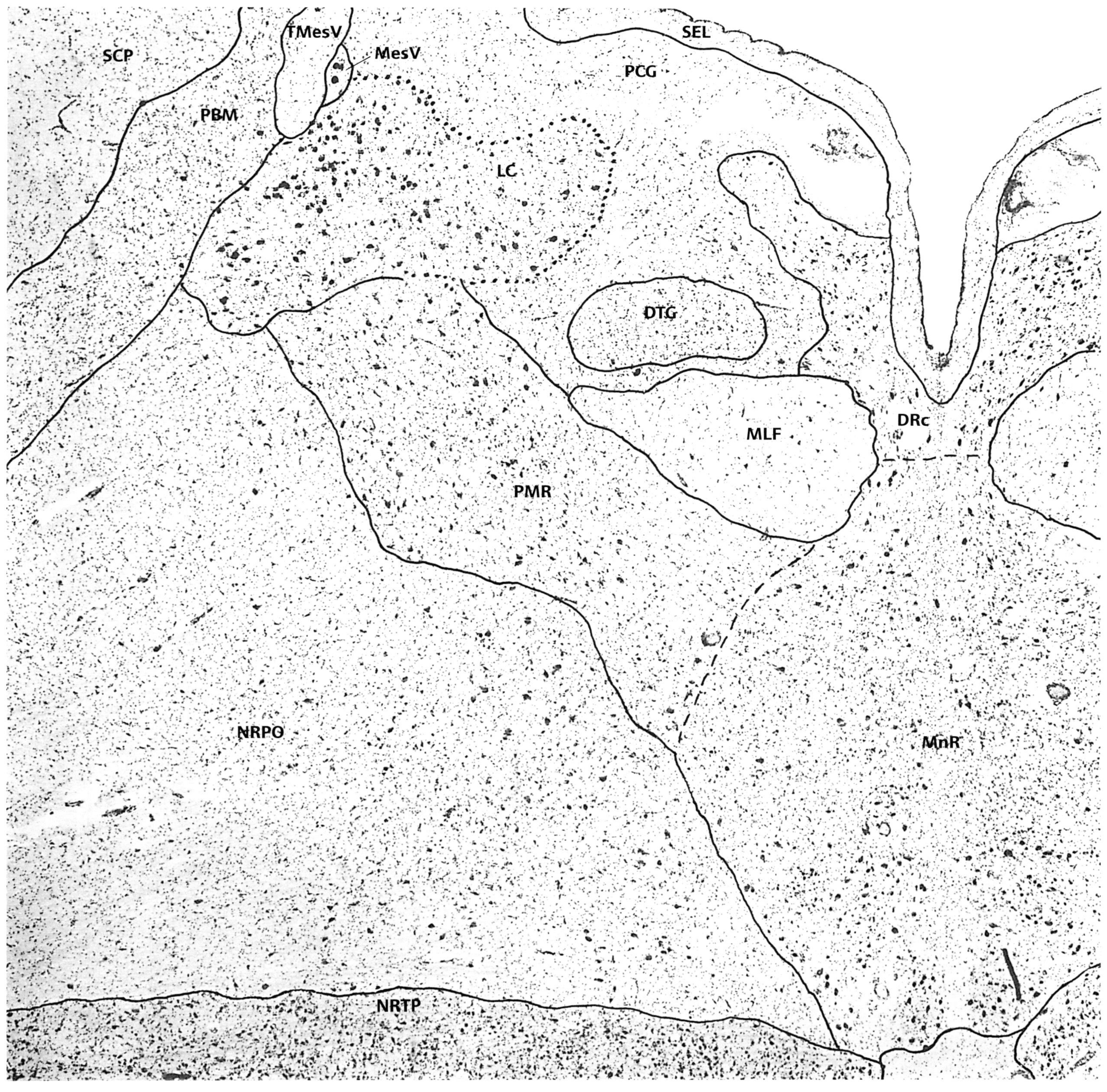

Fig. 1. Cross-section through the oral portion of the pontine tegmentum at the level of the median raphe nucleus (MnR). DRc = Dorsal raphe nucleus, caudal part; DTG = dorsal tegmental nucleus of Gudden; LC = locus coeruleus; MesV = mesencephalic trigeminal nucleus; MLF = medial longitudinal fasciculus; NRPO = nucleus reticularis pontis oralis; NRTP = nucleus reticularis tegmenti pontis; PBM = medial parabrachial nucleus; PCG = pontine central gray; PMR = paramedian raphe nucleus; SCP = superior cerebellar peduncle; SEL = subependymal layer; TMesV = mesencephalic trigeminal tract. Magnification ×40.

Location and Cytoarchitecture – Original Text from Nucleus Centralis Superior

This nucleus occupies the medial and dorsomedial portion of the oral pontine tegmentum and extends into the midbrain to the level of the main trochlear nucleus. Caudally the nucleus is directly contiguous to the oral pole of the nucleus reticularis tegmenti pontis.

The main body of the median raphe nucleus (MnR) occupies the central part of the pontine tegmentum on either side of the median raphe. It is related dorsally to the medial longitudinal fasciculi (MLF), laterally to the nuclei reticulares pontis oralis (NRPO) and ventrally to the nucleus reticularis tegmenti pontis (NRTP). The MnR is conspicuous on myelin-stained preparations as a lightly stained, lens-shaped area.

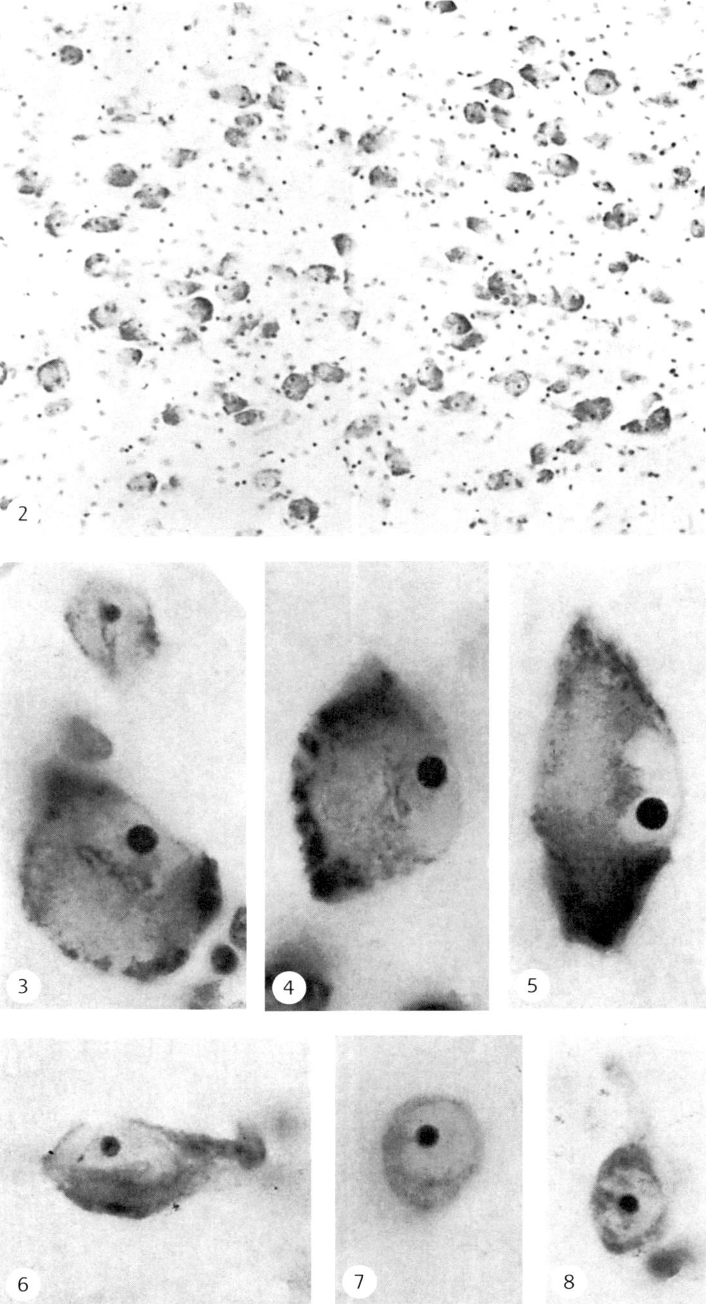

Fig. 2. Median raphe nucleus. Magnification ×150.
Fig. 3–5. Medium-sized cells with excentric nuclei and peripherally distributed Nissl granules from the median raphe nucleus. Magnification ×150.
Fig. 6–8. Small cells from the median raphe nucleus. Magnification ×1,000.

Cells similar to those of the MnR stream dorsally between the two MLF to form a V-shaped cell group beneath the central part of the floor of the fourth ventricle [and give rise to the dorsal raphe nucleus, pars caudalis (DRc, nucleus 68; fig. 8)]. [...]

From the dorsolateral aspect of the MnR a rather broad band of cells (subnucleus lateralis) sweeps across the dorsal pontine tegmentum [the name suggested by Mehler for this cell group was the paramedian raphe nucleus (PMR, fig. 1), which is used here [Paxinos et al., 2012]. However, recent literature has attempted to define a region including the PMR and its paramedian continuation into the caudal linear nucleus (CLi). This newly defined area is called the rostromedial tegmental nucleus [Goncalves et al., 2012]]. The PMR is present only in the caudal part of the MnR [...].

The oral pole of the MnR subnucleus ventralis has a very characteristic appearance in Nissl-stained preparations. It is formed by cells which lie scattered among the decussating fibers of the superior cerebellar peduncle in the ventral part of the mesencephalic tegmentum (plates 30 and 32) [and merge further rostrally into the CLi] [Törk and Hornung, 1990]. [...]

The MnR is composed predominantly of two cell types (fig. 2):

(a) medium-sized, oval or elongated neurons with short dendrites, nuclei which are usually excentric and darkly stained, peripherally arranged Nissl granules (fig. 3–5); those cells which lie close to the median raphe are elongated and their axes are directed dorsoventrally; the more laterally situated cells [of the PMR] tend to be plump rather than elongated and are randomly oriented; these medium-sized cells resemble very closely both the unpigmented cells of the locus coeruleus and the large cells found scattered throughout the NRPO;

(b) small, plump, spindle-shaped or oval cells with short dendrites, a relatively large nucleus and scanty cytoplasm which stains with medium intensity (fig. 6–8); these cells lie scattered among the medium-sized cells.

The cells of all 3 nuclei – MnR, PMR and DRc – are similar with the exception that those constituting the DRc may be smaller and more darkly stained.

Functional Neuroanatomy

Function

The MnR contains a large population of serotonergic neurons (B8) [Törk and Hornung, 1990]. Its ascending pathways target extensive areas of the basal forebrain, septal region, the hippocampus and cerebral cortex [Hornung, 2012]. Through these connections the MnR can modulate or drive the neural activity in vast regions of the brain (see 'Neuromodulatory Systems: Overview') [Hornung and Celio, 1992; Smiley and Goldman-Rakic, 1996; Hornung, 2012]. The serotonergic activation of the hippocampus results in a desynchronization of the hippocampal theta rhythm, which is critically involved in memory-processing functions, and has led to the hypothesis that the MnR may play a role in memory-processing [Vertes et al., 1999]. The pathways of the MnR also project to the suprachiasmatic nucleus and thereby participate in the control of the circadian rhythm [Morin and Allen, 2006].

Clinically serotonin is believed to be involved in mood disorders, addiction, sleep-wake cycles and motor activity [Hillegaart et al., 1989; Jacobs and Fornal, 1997; Gaspar et al., 2003; Hornung, 2012]. In addition neurodegenerative disorders such as Alzheimer's and Parkinson's disease have both been linked with a significant loss of MnR cells [Halliday et al., 1990, 1992].

Connections

Efferents. The MnR belongs to the rostral group of raphe nuclei, along with the dorsal raphe (DR) and CLi: these nuclei send their serotonergic efferents predominantly to the forebrain including the olfactory bulb [Hornung, 2012; Liu et al., 2012]. In general the fibers of the MnR innervate more medial structures than the DR, and the terminal fields of the two nuclei do not overlap [Vertes et al., 1999]. The ascending pathways from the MnR travel initially in two parallel pathways: a ventral bundle that passes through the ventral tegmental area, made up of relatively thick fibers (M fibers) that terminate in conventional synapses in the basal forebrain and medial cortex, and a second more dorsal bundle carrying thin fibers (D fibers) from the dorsolateral MnR and DR that run in the central gray matter to terminate as small varicosities in the lateral cortex [Molliver, 1987]. The two ascending bundles merge into the medial forebrain bundle. MnR fibers in the cerebral and hippocampal cortices may form characteristic 'baskets of terminals' around their targets. Further detailed studies on the hippocampal and neocortex inputs indicate that the serotonergic projections do not have a diffuse character, but terminate on a select population or class of interneurons [Molliver, 1987; Hornung and Celio, 1992; Smiley and Goldman-Rakic, 1996].

There is a topography in the ascending projections from the MnR [Imai et al., 1986]; rostral raphe nuclei of the midbrain are associated with the basal ganglia and related nuclei, intermediate raphe nuclei like the MnR target the amygdala nuclei and caudal nuclei project to the limbic system, whereas efferents from the PMR, lateral to the MnR, terminate in the brainstem and diencephalon. These connections are reported in detail by Vertes et al. [1999].

The scattered serotonergic neurons in the supralemniscal region ventrolateral to the MnR (sometimes referred to as B9), and the sparse cells of the nucleus raphe pontis (B5), can be considered to belong to the caudal pole of the MnR [Nieuwenhuys et al., 2008]. Serotonergic afferents to the cerebellum have been associated with this region in several studies, but at present the evidence suggests that this input originates predominantly from serotonergic neurons dispersed in various nuclei of the brainstem reticular formation [for a review, see Dieudonné, 2001]. Small varicosities arise from thin serotonergic fibers that target the deep cerebellar nuclei, and the cerebellar cortex, where they terminate mainly on the Lugano cells lying around the Purkinje cell layer. The input was found to modulate the activity of Purkinje cells by enhancing the activity of the Lugano cells.

Afferents. Unlike their efferents, the afferent inputs to the MnR and DR are similar; they arise from a large number of forebrain and hypothalamic regions, while relatively few brainstem regions project to these midbrain raphe nuclei [Arnsten and Goldman-Rakic, 1984; Behzadi et al., 1990]. Brainstem inputs to the MnR were reported from the nucleus raphe pallidus, nucleus raphe obscurus, nucleus of the solitary tract, prepositus nucleus, interpeduncular nucleus and the substantia nigra. Significant forebrain inputs to the MnR originate from the lateral habenula, hypothalamic and preoptic areas (like the suprachiasmatic nucleus) as well as the prefrontal cortex [for reviews, see Hornung, 2012; Nieuwenhuys et al., 2008, table 22.1].

References

Arnsten AF, Goldman-Rakic PS: Selective prefrontal cortical projections to the region of the locus coeruleus and raphe nuclei in the rhesus monkey. Brain Res 1984;306:9–18.

Behzadi G, Kalen P, Parvopassu F, Wiklund L: Afferents to the median raphe nucleus of the rat: retrograde cholera toxin and wheat germ conjugated horseradish peroxidase tracing, and selective D-[^{3}H]aspartate labelling of possible excitatory amino acid inputs. Neuroscience 1990;37:77–100.

Dieudonné S: Serotonergic neuromodulation in the cerebellar cortex: cellular, synaptic, and molecular basis. Neuroscientist 2001;7:207–219.

Gaspar P, Cases O, Maroteaux L: The developmental role of serotonin: news from mouse molecular genetics. Nat Rev Neurosci 2003;4:1002–1012.

Goncalves L, Sego C, Metzger M: Differential projections from the lateral habenula to the rostromedial tegmental nucleus and ventral tegmental area in the rat. J Comp Neurol 2012;520:1278–1300.

Halliday GM, Li YW, Blumberg PC, Joh TH, Cotton RG, Howe PR, Blessing WW, Geffen LB: Neuropathology of immunohistochemically identified brainstem neurons in Parkinson's disease. Ann Neurol 1990;27:373–385.

Halliday GM, McCann HL, Pamphlett R, Brooks WS, Creasey H, McCusker E, Cotton RG, Broe GA, Harper CG: Brain stem serotonin-synthesizing neurons in Alzheimer's disease: a clinicopathological correlation. Acta Neuropathol 1992;84:638–650.

Hillegaart V, Ahlenius S, Larsson K: Effects of local application of 5-HT into the median and dorsal raphe nuclei on male rat sexual and motor behavior. Behav Brain Res 1989;33:279–286.

Hornung J-P: Raphe nuclei; in Mai JK, Paxinos G (eds): The Human Nervous System, ed 3. Amsterdam, Academic Press, 2012.

Hornung JP, Celio MR: The selective innervation by serotoninergic axons of calbindin-containing interneurons in the neocortex and hippocampus of the marmoset. J Comp Neurol 1992;320:457–467.

Imai H, Steindler DA, Kitai ST: The organization of divergent axonal projections from the midbrain raphe nuclei in the rat. J Comp Neurol 1986;243:363–380.

Jacobs BL, Fornal CA: Serotonin and motor activity. Curr Opin Neurobiol 1997;7:820–825.

Liu S, Aungst JL, Puche AC, Shipley MT: Serotonin modulates the population activity profile of olfactory bulb external tufted cells. J Neurophysiol 2012;107:473–483.

Molliver ME: Serotonergic neuronal systems: what their anatomic organization tells us about function. J Clin Psychopharmacol 1987;7:3S–23S.

Morin LP, Allen CN: The circadian visual system, 2005. Brain Res Rev 2006;51:1–60.

Nieuwenhuys R, Voogd J, van Huijzen C: The Human Central Nervous System. Berlin, Springer, 2008.

Paxinos G, Huang X-F, Sengul G, Watson C: Organization of the brainstem; in Mai JK, Paxinos G (eds): The Human Nervous System, ed 3. Amsterdam, Academic Press, 2012.

Smiley JF, Goldman-Rakic PS: Serotonergic axons in monkey prefrontal cerebral cortex synapse predominantly on interneurons as demonstrated by serial section electron microscopy. J Comp Neurol 1996;367:431–443.

Törk I, Hornung JP: Raphe nuclei and the serotonergic system; in Paxinos G (ed): The Human Nervous System. San Diego, Academic Press, 1990.

Vertes RP, Fortin WJ, Crane AM: Projections of the median raphe nucleus in the rat. J Comp Neurol 1999;407:555–582.

Neuromodulatory Systems – Serotonergic Nuclei (Plates 26–30 and 32–34)

68 Dorsal Raphe Nucleus (DR)

Original name:
Nucleus supratrochlearis

Location and Cytoarchitecture – Original Text

This is a large, unpaired nuclear mass, the major portion of which lies dorsal to the medial longitudinal fasciculi (MLF) in the caudal part of the mesencephalon. It extends from the level of the caudal border of the inferior colliculus to a level through the caudal pole of the oculomotor complex (III), a distance of about 6 mm.

On cross-section, the characteristic form of the dorsal raphe nucleus (DR) is that of a fountain-shaped cell group composed of a medially situated central portion and two lateral wings. The central portion lies between the MLF of each side and extends dorsally to the base of the cerebral aqueduct, while the lateral wings arch dorsally to the MLF and the trochlear nuclei (IV). These lateral extensions are related dorsally to the central gray matter of the mesencephalon (PAG), and laterally to the ventral mesencephalic reticular formation (MRFv) and the fibers of the central tegmental tract (CTT).

Approaching the caudal border of the III, the centrally situated cells of the DR sweep oroventrally towards the interpeduncular fossa, a feature which is best visualized on sagittal sections (fig. 7). Thus, the oral pole of

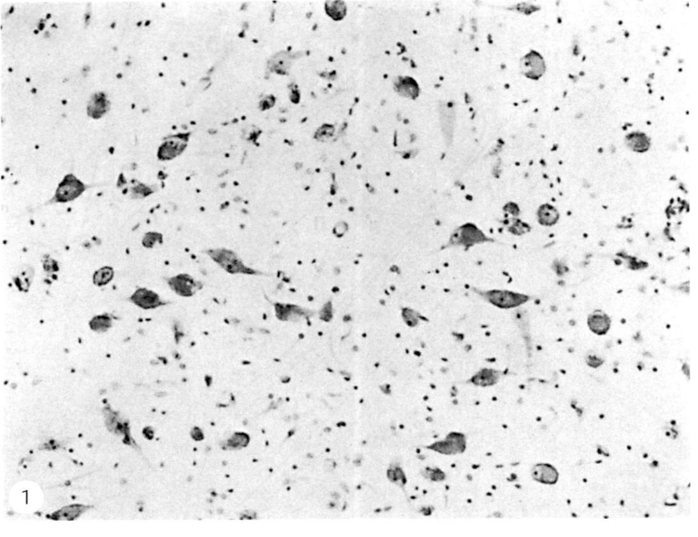

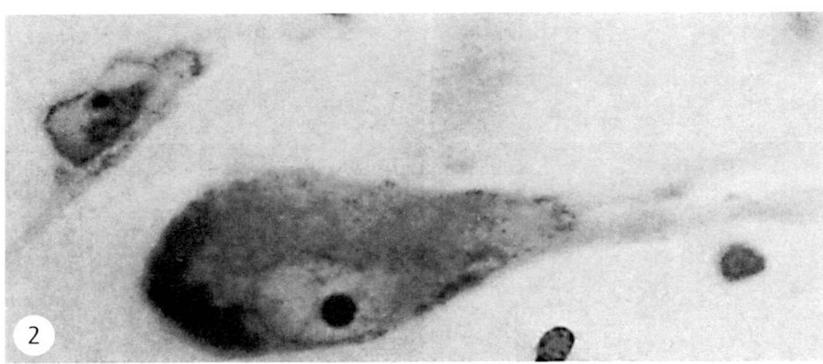

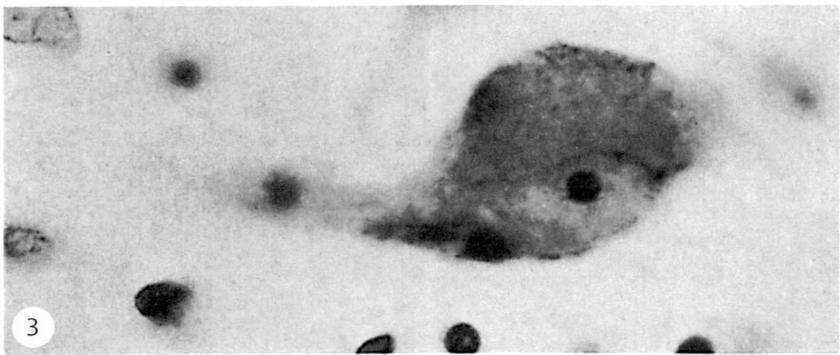

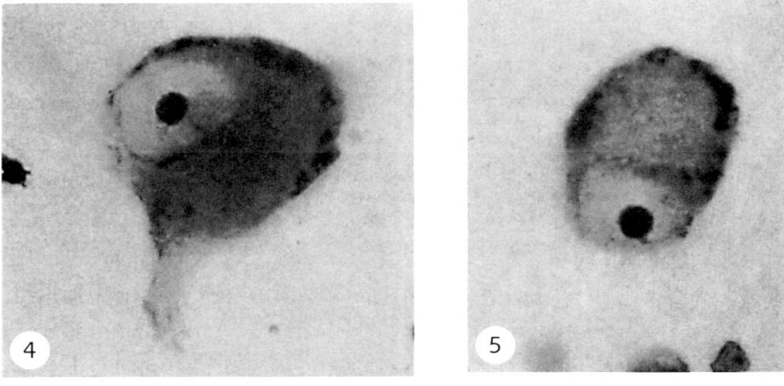

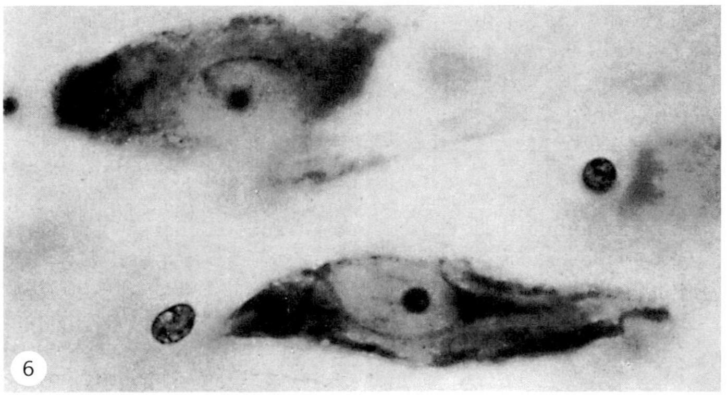

Fig. 1. Dorsal raphe nucleus. Magnification ×150.
Fig. 2–6. Cells from the dorsal raphe nucleus. Magnification ×1,000.

the DR is represented on cross-section as a median cell group lying between the two superior cerebellar peduncles, ventral to the caudal pole of the III and dorsal to the parabrachial pigmented nucleus (plates 34 and 35).

The DR is composed of irregularly oriented, darkly stained, medium-sized, plump, oval or fusiform cells, which possess conspicuous dendrites and nuclei which are often excentric. The Nissl substance appears as darkly stained, irregular clusters about the periphery or at the extremities of the cell, and as finer, more lightly stained granules in the central portion of the cytoplasm (fig. 1–6). The cell type is uniform throughout the major part of the DR, but there is considerable variation of cell size and of the density of cell distribution between different regions. Thus, the cells in the lateral portion of the nucleus tend to be larger and more densely arranged than those of the medial portion, and the caudal part of the nucleus is more cellular than the oral. Cells derived from the lateral portion of the DR infiltrate and lie among the fibers of the CTT.

Cells of the oroventral extension of the nucleus are spindle shaped and lie with their long axes directed dorsoventrally. A few of these cells possess darkly stained, evenly distributed Nissl granules.

Functional Neuroanatomy

Subdivisions

The human nucleus supratrochlearis of Olszewski and Baxter [1982] was shown to correspond to the DR in other mammals, on the basis of lipofuscin-stained material [Braak, 1970]. In most species the DR is subdivided into 5 regions, based on the differing morphology and cell density [Baker et al., 1990; Charara and Parent, 1998; Hornung, 2012]: the interfascicular nucleus is located at the midline between the trochlear nuclei and the MLF; the ventral subnucleus of the DR lies at the bottom of the central gray; the ventrolateral nucleus of the DR (DRvl) is located dorsolateral to the IV showing an intense Nissl staining, and its neurons merge with those of the dorsal subnucleus (DRd) immediately dorsally. The DRd lies dorsal to the caudal end of the III and IV, and is the largest subnucleus of the DR [Baker et al., 1990]. Caudal to the isthmus of the brainstem, the DR is reduced to a slim V-shaped group of neurons parallel to the ventricular surface and termed DR caudal subnucleus (DRc). Formerly the DRc was considered to be the dorsal part of the median raphe nucleus. Each wing of the DRc is related medially to the stratum gliosum and laterally to the MLF, the dorsal tegmental nucleus of Gudden and the central pontine gray matter (see fig. 1 of nucleus No. 67).

Transmitters

The DR contains the largest population of serotonergic cells in the human brainstem [Charara and Parent, 1998]. The neurons have diffuse widespread axonal branches that modulate an enormous region of the brain [Parent, 1996]. The DR neurons, rostral to the isthmus (border between the mesencephalon and the rhombencephalon), correspond to the B7 group of Dahlström and Fuxe [1964]; the DR neurons caudal to the isthmus correspond to the B6 group. Up to date, about 10 other transmitter-related substances have been reported in the DR, several coexisting in the same cells, although serotonin remains the major neurotransmitter, in 70% of the neurons [Michelsen et al., 2007]. The DR contains cell populations immunopositive for dopamine, glutamate, GABA, substance P, nitric oxide, vasoactive intestinal polypeptide, cholecystokinin and enkephalins; for reviews, see Charara and Parent [1998] and Sergeyev et al. [1999]. An extremely strong acetylcholine esterase activity is seen in the neuropil of the DRvl, and less in the medial DR [Paxinos et al., 2012].

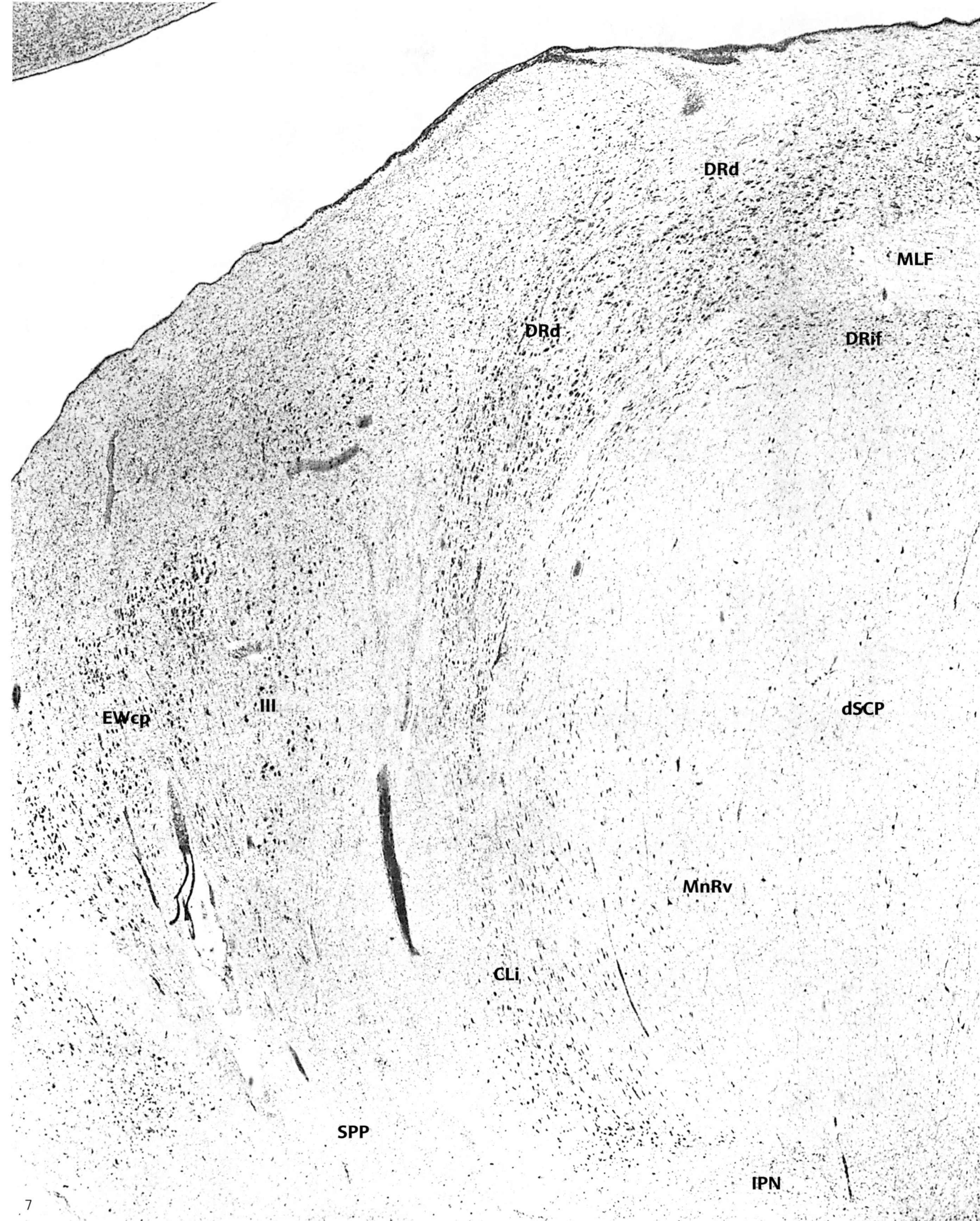

Fig. 7. Sagittal section to demonstrate the relations of the oral portion of the dorsal raphe nucleus. CLi = Caudal linear nucleus; DRd = dorsal raphe nucleus, dorsal part; DRif = dorsal raphe nucleus, interfascicular part; dSCP = decussation of the superior cerebellar peduncle; EWcp = Edinger-Westphal central projecting nucleus; III = oculomotor nucleus; IPN = interpeduncular nucleus; MLF = medial longitudinal fasciculus; MnRv = median raphe nucleus, ventral part; SPP = substantia perforata posterior. Magnification ×30.

Function

There are 2 types of serotonergic projection fibers, M and D fibers [Mamounas et al., 1991; Michelsen et al., 2007]. The M-type fibers are thick and exert their influence on their target structures through direct synaptic transmission (see nucleus No. 67). The D-type fibers are thin with small varicosities that modulate the level of activity in their targets, and exert a widespread influence through volume transmission (see 'Neuromodulatory Systems: Overview'). The fibers of the DR are of the D type and influence vast areas of the forebrain. In addition to this, axons of the DR penetrate into the ventricles early in development, and thereby have access to the cerebrospinal fluid [Jacobs and Azmitia, 1992]. The supraependymal axons, particularly from the DR, build a thick plexus of axons that cover the entire ventricular surface, and innervate circumventricular organs (see nucleus No. 35) [Azmitia and Gannon, 1982; Parent, 1996; Michelsen et al., 2007]. Some ventricular nerve fibers have been seen to turn and grow back into the brain, leading to the hypothesis that these fibers transmit trophic information about the cerebrospinal fluid back to neural structures.

The DR is thought to be involved in the regulation of learning, memory, sleep, anxiety-related functions and movement [Michelsen et al., 2007]. The serotonergic and nonserotonergic cells of the DR can be subdivided into different types on the basis of their firing patterns during sleep-wake cycles [Sakai and Crochet, 2001]. The nucleus is also associated with pain modulation and blood pressure control [Piper and Goadsby, 1985; Wang and Nakai, 1994]. Serotonin is the major transmitter of the DR, and it is considered to be a key factor in depression and other mood disorders. In depressed patients levels of serotonin and its metabolites are decreased, and a cell loss in the DR was reported [Baumann et al., 2002] – but there is conflicting evidence on this point.

Connections

Efferents. The ascending pathways from the raphe nuclei are mainly ipsilateral. There are two main ascending pathways from the DR, a medial pathway targeting the substantia nigra pars compacta (SNc), the caudate and putamen, and a ventral pathway which mainly innervates the hypothalamic and thalamic nuclei, the habenula, septum, central nucleus of the amygdala, hippocampus, interpeduncular nucleus (IPN), lateral geniculate nucleus and the olfactory bulb. The ventral pathway is shared with other raphe nuclei [Imai et al., 1986b; Michelsen et al., 2007; Hornung, 2012; Liu et al., 2012].

Several reports have emphasized a topography in the ascending efferents of the raphe nuclei [Imai et al., 1986b; Abrams et al., 2004]. This implies some functional segregation within the nuclei; and this has been verified at a single-cell level. In spite of the vast diffuse character of the serotonergic projection axons, the branches of individual raphe neurons send collaterals to functionally similar areas, for example to the SNc and caudate putamen [Imai et al., 1986a], or to the entorhinal cortex and the hippocampus [Kohler and Steinbusch, 1982], or the central nucleus of the amygdala and the paraventricular nucleus of the hypothalamus [Petrov et al., 1994].

Descending pathways from the DR are modest and terminate in the locus coeruleus (LC) and parabrachial nuclei [Bobillier et al., 1976].

Afferents. According to some authors the largest source of afferents to the raphe nuclei are the serotonergic interconnections amongst themselves [Jacobs and Azmitia, 1992]. In addition the important sources of inputs to the DR come from the limbic forebrain areas, such as the lateral habenula and IPN [Peyron et al., 1998], also the central nucleus of the amygdala, and the anterior hypothalamus [Holstege, 1991]. The brainstem afferents to the DR include the SNc, the ventral tegmental area, the PAG, LC and nucleus prepositus [Kalen et al., 1985; Jacobs and Azmitia, 1992; Poller et al., 2011].

References

Abrams JK, Johnson PL, Hollis JH, Lowry CA: Anatomic and functional topography of the dorsal raphe nucleus. Ann NY Acad Sci 2004;1018:46–57.

Azmitia EC, Gannon PJ: A light and electron microscopic analysis of the selective retrograde transport of (^{3}H)-5-hydroxytryptamine by serotonergic neurons. J Histochem Cytochem 1982;30:799–804.

Baker KG, Halliday GM, Törk I: Cytoarchitecture of the human dorsal raphe nucleus. J Comp Neurol 1990;301:147–161.

Baumann B, Bielau H, Krell D, Agelink MW, Diekmann S, Wurthmann C, Trubner K, Bernstein HG, Danos P, Bogerts B: Circumscribed numerical deficit of dorsal raphe neurons in mood disorders. Psychol Med 2002;32:93–103.

Bobillier P, Seguin S, Petitjean F, Salvert D, Touret M, Jouvet M: The raphe nuclei of the cat brain stem: a topographical atlas of their efferent projections as revealed by autoradiography. Brain Res 1976;113:449–486.

Braak H: On the nuclei of the human brain stem. II. The raphe nuclei. Z Zellforsch Mikrosk Anat 1970;107:123–141.

Charara A, Parent A: Chemoarchitecture of the primate dorsal raphe nucleus. J Chem Neuroanat 1998;72:111–127.

Dahlström A, Fuxe K: Evidence for the existence of monoamine neurons in the central nervous system. I. Demonstration of monoamines in the cell bodies of brain stem neurons. Acta Physiol Scand 1964;62:1–55.

Holstege G: Descending motor pathways and the spinal motor system: limbic and non-limbic components. Prog Brain Res 1991;87:307–421.

Hornung J-P: Raphe nuclei; in Mai JK, Paxinos G (eds): The Human Nervous System, ed 3. Amsterdam, Academic Press, 2012.

Imai H, Park MR, Steindler DA, Kitai ST: The morphology and divergent axonal organization of midbrain raphe projection neurons in the rat. Brain Dev 1986a;8: 343–354.

Imai H, Steindler DA, Kitai ST: The organization of divergent axonal projections from the midbrain raphe nuclei in the rat. J Comp Neurol 1986b;243:363–380.

Jacobs BL, Azmitia EC: Structure and function of the brain serotonin system. Physiol Rev 1992;72:165–229.

Kalen P, Karlson M, Wiklund L: Possible excitatory amino acid afferents to nucleus raphe dorsalis of the rat investigated with retrograde wheat germ agglutinin and D-[^{3}H]aspartate tracing. Brain Res 1985;360:285–297.

Kohler C, Steinbusch H: Identification of serotonin and non-serotonin-containing neurons of the mid-brain raphe projecting to the entorhinal area and the hippocampal formation. A combined immunohistochemical and fluorescent retrograde tracing study in the rat brain. Neuroscience 1982;7:951–975.

Liu S, Aungst JL, Puche AC, Shipley MT: Serotonin modulates the population activity profile of olfactory bulb external tufted cells. J Neurophysiol 2012;107:473–483.

Mamounas LA, Mullen CA, O'Hearn E, Molliver ME: Dual serotoninergic projections to forebrain in the rat: morphologically distinct 5-HT axon terminals exhibit differential vulnerability to neurotoxic amphetamine derivatives. J Comp Neurol 1991;314:558–586.

Michelsen KA, Schmitz C, Steinbusch HW: The dorsal raphe nucleus – from silver stainings to a role in depression. Brain Res Rev 2007;55:329–342.

Olszewski J, Baxter D: Cytoarchitecture of the Human Brain Stem. Basel, Karger, 1982.

Parent A: Carpenter's Human Neuroanatomy. Baltimore, Williams & Wilkins, 1996.

Paxinos G, Huang X-F, Sengul G, Watson C: Organization of the brainstem; in Mai JK, Paxinos G (eds): The Human Nervous System, ed 3. Amsterdam, Academic Press, 2012.

Petrov T, Krukoff TL, Jhamandas JH: Chemically defined collateral projections from the pons to the central nucleus of the amygdala and hypothalamic paraventricular nucleus in the rat. Cell Tissue Res 1994;277:289–295.

Peyron C, Petit JM, Rampon C, Jouvet M, Luppi PH: Forebrain afferents to the rat dorsal raphe nucleus demonstrated by retrograde and anterograde tracing methods. Neuroscience 1998;82:443–468.

Piper RD, Goadsby PJ: Pressor response to electrical and chemical stimulation of nucleus raphe dorsalis in the cat. Stroke 1985;16:307–312.

Poller WC, Bernard R, Derst C, Weiss T, Madai VI, Veh RW: Lateral habenular neurons projecting to reward-processing monoaminergic nuclei express hyperpolarization-activated cyclic nucleotid-gated cation channels. Neuroscience 2011;193:205–216.

Sakai K, Crochet S: Differentiation of presumed serotonergic dorsal raphe neurons in relation to behavior and wake-sleep states. Neuroscience 2001;104:1141–1155.

Sergeyev V, Hokfelt T, Hurd Y: Serotonin and substance P co-exist in dorsal raphe neurons of the human brain. Neuroreport 1999;10:3967–3970.

Wang QP, Nakai Y: The dorsal raphe: an important nucleus in pain modulation. Brain Res Bull 1994;34:575–585.

Neuromodulatory Systems – Noradrenergic Nuclei (Plates 26–30)

69 Locus Coeruleus (LC)

Original name:
Nucleus locus coeruleus

Location and Cytoarchitecture – Original Text

This nucleus lies ventral to the lateral angle of the fourth ventricle in the oral half of the pons. It commences caudally at a point just dorsal to the oral pole of the motor trigeminal nucleus and extends to the level of the caudal border of the inferior colliculus, a distance of approximately 12 mm. The nucleus is irregularly oval in outline and is related dorsomedially to the central gray matter, dorsolaterally to the mesencephalic trigeminal nucleus (MesV), ventrally and ventromedially to the nucleus reticularis pontis oralis (NRPO). Laterally it is bordered, as one proceeds caudo-orally, by the medial parabrachial nucleus (PBM), the superior cerebellar peduncle and the nucleus cuneiformis successively.

Caudally, the locus coeruleus (LC) is easily delineated as a uniform structure in both Nissl- and myelin-stained sections (fig. 1). At more oral levels its organization becomes more complicated due to the fact that its cells become more scattered and lie on either side of the line which delineates the central gray matter. This line is clearly visible on myelin-stained sections on which the central gray matter appears completely devoid of fiber bundles, whereas the adjacent part of the tegmentum is traversed by numerous fiber tracts. Reviewing the myelin-stained sections one gains the impression that the background of myelinated fibers in the LC is the same on both sides of this dividing line. The fiber bundles in the ventral part of the nucleus are probably 'fibers of passage', and it is very likely that they pass through this region without contributing to or receiving fibers from it. Similarly, the cells of the LC are identical on either side of this 'border of the central gray matter'. The diminished cellularity of the ventral portion of the nucleus is due to the presence of the large fiber bundles. It is thus probable that that part of the border of the central gray matter which traverses the LC is of no functional significance and has only topographical value (see 'pericoerulear region' below) [...].

Two main cell types may be distinguished in the nucleus of the LC:

(a) medium-sized, round, oval or multipolar cells with irregular edges, short dendrites, and nuclei which in the majority of these cells contain melanin pigment; the amount of this pigment varies from only a few small granules in the central portion of the cell to masses which completely obscure the cell cytoplasm; this pigment does not appear in the cells of the LC until about the fourth year of life, and even in the adult many unpigmented cells may be found (fig. 2, 3); (neuromelanin is a by-product of catecholamine synthesis [Double et al., 2008];) the Nissl substance of the pigmented cells is often obscured, but in those cells containing little or no pigment, it is found to be aggregated largely in irregular clusters about the periphery of the cell; in many respects, the unpigmented cells of the LC resemble the large cells of the median raphe nucleus and the NRPO;

(b) very small, oval or round, lightly stained cells similar to those which compose the central gray matter lie scattered between the larger cells of the LC (fig. 4–6).

Occasionally, cells from the ventral portion of the MesV are found among the cells of the LC. At various levels pigmented cells of the LC infiltrate the surrounding nuclei for a short distance. They may be found intermingled with the cells of the central gray matter, NRPO, PBM and MesV. [...]

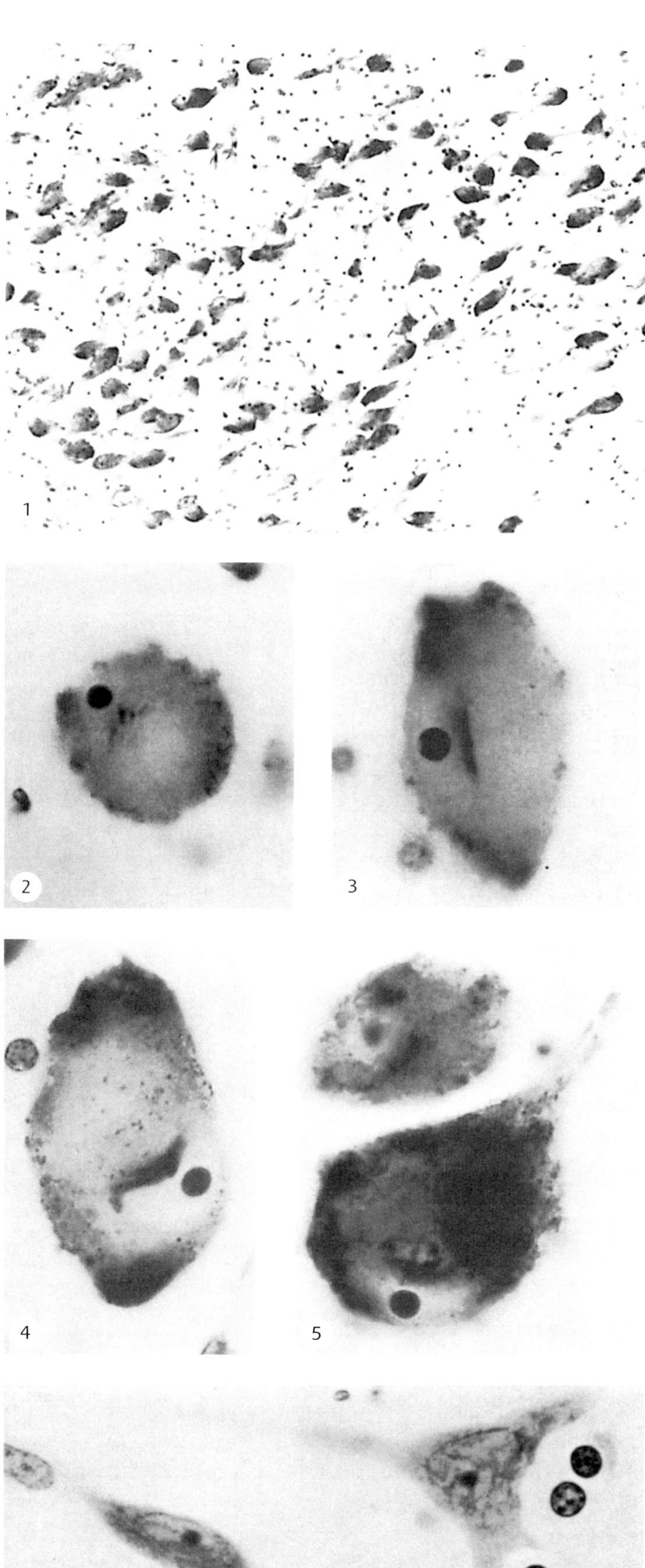

Fig. 1. Locus coeruleus. Magnification ×150.
Fig. 2, 3. Unpigmented cells from the locus coeruleus. Magnification ×1,000.
Fig. 4, 5. Cells from the locus coeruleus showing different degrees of pigmentation. Magnification ×1,000.
Fig. 6. Small, lightly stained cells from the locus coeruleus. Magnification ×1,000.

Functional Neuroanatomy

Function

The LC is a noradrenergic neuromodulator. Its neurons directly modulate the activity of immense areas of the brain during vigilance and arousal. In addition they control autonomic functions and furthermore are implicated in cognitive processes such as attention, memory and motivation [Sara, 2009]. The LC neurons themselves show a high level of tonic activity if the subject is stressed or excited, a slow irregular firing during quiet wakefulness, a more decreased firing rate during slow-wave sleep and virtually no activity during REM sleep [Aston-Jones et al., 1991a]. The LC neurons form the basic circuitry of the classical ascending activating system (see 'Introduction: The Reticular Formation 60 Years Later').

In old age there is a progressive loss of LC neurons (30–50%), predominantly from the rostral parts of the nucleus, which provide the forebrain with noradrenergic modulation. Some neurodegenerative diseases are also accompanied by a profound loss of LC neurons; it has been shown that in Alzheimer's and Parkinson's disease there is more loss of neurons in the LC than in the traditional nuclei targeted by the diseases, i.e. nucleus basalis or substantia nigra [Zarow et al., 2003].

Connections

Efferents. The LC cell group is designated as A6 according to the terminology of Dahlström and Fuxe [1964], and evidence of electrotonic coupling between its neurons suggests that it operates to some extent as a synchronized network [Van Bockstaele et al., 2004]. The LC is the largest source of noradrenaline in the central nervous system, supplying the entire neocortex, the olfactory bulb, the thalamus, limbic structures such as the amygdala and the hippocampus, the pallidum (not the striatum), sensory nuclei of the brainstem such as the spinal trigeminal nucleus, the solitary nucleus and cochlear nuclei, as well as the tectum, the periaqueductal gray, inferior olive, cerebellum and spinal cord. The gigantic axonal projection networks of the LC for all these regions arise from only 45,000 neurons. The first reports of the vast LC projections were received with disbelief but are now well established [Pickel et al., 1974; Jones, 1991]. The fine axon terminals synapse on dendrites and spines of neuronal perikarya in the target areas, but in these areas noradrenaline receptors are also found on axons, glia cells and blood vessels, which suggests a more global role of the LC in modulation during vigilance and arousal [Counts and Mufson, 2012].

Afferents. Some afferent inputs target the center of the LC and others the peripheral, or pericoerulear region. The two major inputs to the central LC arise from the rostral ventrolateral medulla, the lateral paragigantocellular nucleus (PGiL), providing excitation, and the prepositus nucleus (PrP) providing inhibition [Aston-Jones et al., 1991b]. The PGiL area processes autonomic visceral information and plays a role in cardiovascular control and pain [Van Bockstaele et al., 1989], and provides the LC with a variety of inputs (e.g. adrenaline, metenkephalin, GABA and corticotropin-releasing factor) [Aston-Jones, 2004]. The function of the inhibitory PrP input is not clear [Ennis and Aston-Jones, 1989].

The pericoerulear region of the LC contains dendrites of LC neurons, and this zone also receives afferent inputs from several areas which do not terminate in the central LC, for example from the medial prefrontal cortex, central amygdala, dorsomedial hypothalamus, periaqueductal gray, dorsal raphe and solitary nucleus [for a review, see Aston-Jones et al., 2004]. The peri-LC region has recently been shown to contain neurons which project into the central core of the LC, many of which are GABAergic [Aston-Jones et al., 2004]. It is hypothesized that this population of small GABAergic neurons in the peri-LC dendritic zone may provide interneuronal integration for LC noradrenergic neurons. The lateral hypothalamus supplies the LC (probably both peri-LC and LC center) with orexin inputs, which participate in the sleep/waking behavior [Espana et al., 2005].

References

Aston-Jones G: Locus coeruleus, A5 and A7 noradrenergic cell groups; in Paxinos G (ed): The Rat Nervous System. Amsterdam, Elsevier, 2004.

Aston-Jones G, Chiang C, Alexinsky T: Discharge of noradrenergic locus coeruleus neurons in behaving rats and monkeys suggests a role in vigilance. Prog Brain Res 1991a;88:501–520.

Aston-Jones G, Shipley MT, Chouvet G, Ennis M, van Bockstaele E, Pieribone V, Shiekhattar R, Akaoka H, Drolet G, Astier B, et al: Afferent regulation of locus coeruleus neurons: anatomy, physiology and pharmacology. Prog Brain Res 1991b;88:47–75.

Aston-Jones G, Zhu Y, Card JP: Numerous GABAergic afferents to locus ceruleus in the pericerulear dendritic zone: possible interneuronal pool. J Neurosci 2004;24:2313–2321.

Counts SE, Mufson EJ: Locus coeruleus; in Mai JK, Paxinos G (eds): The Human Nervous System, ed 3. Amsterdam, Academic Press, 2012.

Dahlström A, Fuxe K: Evidence for the existence of monoamine neurons in the central nervous system. I. Demonstration of monoamines in the cell bodies of brain stem neurons. Acta Physiol Scand 1964;62:1–55.

Double KL, Dedov VN, Fedorow H, Kettle E, Halliday GM, Garner B, Brunk UT: The comparative biology of neuromelanin and lipofuscin in the human brain. Cell Mol Life Sci 2008;65:1669–1682.

Ennis M, Aston-Jones G: GABA-mediated inhibition of locus coeruleus from the dorsomedial rostral medulla. J Neurosci 1989;9:2973–2981.

Espana RA, Reis KM, Valentino RJ, Berridge CW: Organization of hypocretin/orexin efferents to locus coeruleus and basal forebrain arousal-related structures. J Comp Neurol 2005;481:160–178.

Jones BE: Noradrenergic locus coeruleus neurons: their distant connections and their relationship to neighboring (including cholinergic and GABAergic) neurons of the central gray and reticular formation. Prog Brain Res 1991;88:15–30.

Pickel VM, Segal M, Bloom FE: A radioautographic study of the efferent pathways of the nucleus locus coeruleus. J Comp Neurol 1974;155:15–42.

Sara SJ: The locus coeruleus and noradrenergic modulation of cognition. Nat Rev Neurosci 2009;10:211–223.

Van Bockstaele EJ, Garcia-Hernandez F, Fox K, Alvarez VA, Williams JT: Expression of connexins during development and following manipulation of afferent input in the rat locus coeruleus. Neurochem Int 2004;45:421–428.

Van Bockstaele EJ, Pieribone VA, Aston-Jones G: Diverse afferents converge on the nucleus paragigantocellularis in the rat ventrolateral medulla: retrograde and anterograde tracing studies. J Comp Neurol 1989;290:561–584.

Zarow C, Lyness SA, Mortimer JA, Chui HC: Neuronal loss is greater in the locus coeruleus than nucleus basalis and substantia nigra in Alzheimer and Parkinson diseases. Arch Neurol 2003;60:337–341.

70 Nucleus Subcoeruleus (SubC)

Original name:
Nucleus subcoeruleus, dorsal and ventral subnuclei

Alternative name:
Locus coeruleus, pars α [Meessen and Olszewski, 1949]

Location and Cytoarchitecture – Original Text

This nucleus appears in the lateral part of the pontine tegmentum at the level of the oral pole of the superior olivary complex (SOC), and extends rostrally a further 7 mm to the level of the oral pole of the nucleus reticularis tegmenti pontis. It is characterized by the presence of numerous melanin-containing cells. The nucleus subcoeruleus (SubC) is roughly L shaped in outline with the horizontal limb directed orocaudally, and the vertical limb directed dorsally from the oral extremity of the horizontal limb. The ventral subnucleus (SubCv) is formed by the horizontal limb, the dorsal subnucleus (SubCd) by the vertical limb.

On cross-section, the SubCv is oblong in outline with its long axis directed dorsomedially. Caudally, it is related ventrally to the SOC and on all other aspects to the parvocellular reticular nucleus. More orally the SubCv is related ventrally to the ventral nucleus of the lateral lemniscus and to the nucleus reticularis pontis oralis (NRPO) medially, dorsally and laterally.

The SubCd appears dorsal to the SubCv at the level of the oral portion of the motor trigeminal nucleus (MoV, nucleus No. 42) and sweeps dorsomedially to reach, at a slightly more oral level, the locus coeruleus (LC). On cross-section, the SubCd is formed by a broad band of cells which extends from the LC dorsally to the SubCv ventrally, and which is related ventromedially to the NRPO and dorsolaterally to the MoV and the medial parabrachial nucleus (PBM).

The majority of cells composing the SubCv (fig. 1) are small, fusiform or triangular in outline and their cytoplasm stains with medium intensity (fig. 4). Scattered among these irregularly oriented small cells one finds larger, oval, melanin-containing cells similar to those of the LC (fig. 2, 3). The presence of these melanin-containing cells as well as the greater cellularity of the SubCv allow differentiation from the adjacent nuclei.

The SubCd (fig. 5) is composed of sparsely distributed, elongated, darkly stained, medium-sized to large cells which lie with their long axes directed dorsomedially. Melanin pigment is found within the cytoplasm of many of these cells. The Nissl granules of the pigmented cells are usually peripherally arranged whereas in the unpigmented cells they may be uniformly distributed (fig. 6–8).

Functional Neuroanatomy

Function

The SubC is closely associated with the limbic system, and is included by Nieuwenhuys et al. [2008] in the 'lateral paracore' of their classification of the 'greater limbic system'. In general it appears that the SubC is usually included in the LC [Camp and Wijesinghe, 2009], on the basis of the scattered noradrenergic neurons of A6 mainly in the LC, but spilling into the SubCd. However, this reduced concept of the SubC neglects the function of the large SubCv.

The SubC has been shown to be important in the generation of REM sleep, in the accompanying muscle atonia, and in the generation of pontogeniculo-occipital (PGO) waves [Simon et al., 2012]. PGO waves are prominent propagated field potentials generated by a phasic activation of a group of SubC neurons and adjacent structures. They are a cardinal sign of REM sleep [Jones, 1991; Datta et al., 1998; Jones, 2005]. Concerning atonia, the SubC appears to be part of a cholinoceptive trigger zone, which excites a medullary area in the vicinity of the gigantocellular nucleus (Gi): descending projections from the Gi then generate a motoneuron inhibition and muscle atonia during REM sleep [Chase and Morales, 1990]. These functions are associated with the Ch5 cholinergic cells of the SubC [Mesulam et al., 1984].

It is unclear as to how far the 'pontine pelvic organ-stimulating center' of Holstege, including the pontine micturition center, overlaps with the SubC, but the centers are not associated with the noradrenergic parts of the nucleus [Holstege, pers. commun.].

Finally, lesions of the SubC region lead to the loss of muscle atonia during REM sleep both experimentally in rats, and in humans [Karlsson et al., 2005; Xi and Luning, 2009].

Comments

The SubC can be usefully considered as the rostral extension of the pontine reticular formation segment defined as the 'intermediate reticular zone' (see 'Reticular Formation: Overview') [Clarkson et al., 2010; Paxinos et al., 2012]. It lies in the lateral tegmental field between the NRPO, PBM and the Kölliker-Fuse nucleus.

The SubC also contains the cholinergic neurons belonging to the Ch5 group, which continue rostrally into the pedunculopontine nucleus, pars compacta (PPT), and the laterodorsal tegmental nucleus [Mesulam et al., 1984]. The SubC is clearly identified as the acetylcholine esterase-positive area above the central tegmental tract [Paxinos et al., 2012]. There is some evidence of electrotonic coupling between SubC neurons, suggesting that it operates to some extent as a synchronized network [Ennis and Datta, 2007; Heister et al., 2007].

Connections

Efferents. The structures mainly targeted by the output fibers from the SubC are the occipital cortex, entorhinal cortex, piriform cortex, amygdala, hippocampus, thalamus and hypothalamus [Datta et al., 1998]. Furthermore SubC efferents project to the PPT, NRPO and Gi, which are brainstem nuclei that also participate in the generation of REM sleep, whereby the PPT appears to be the main center [Datta et al., 2009].

Defining the SubC projections to the spinal cord has been fraught with difficulties, since different strains of Sprague-Dawley rats were found to give different results: in 'Harlan'

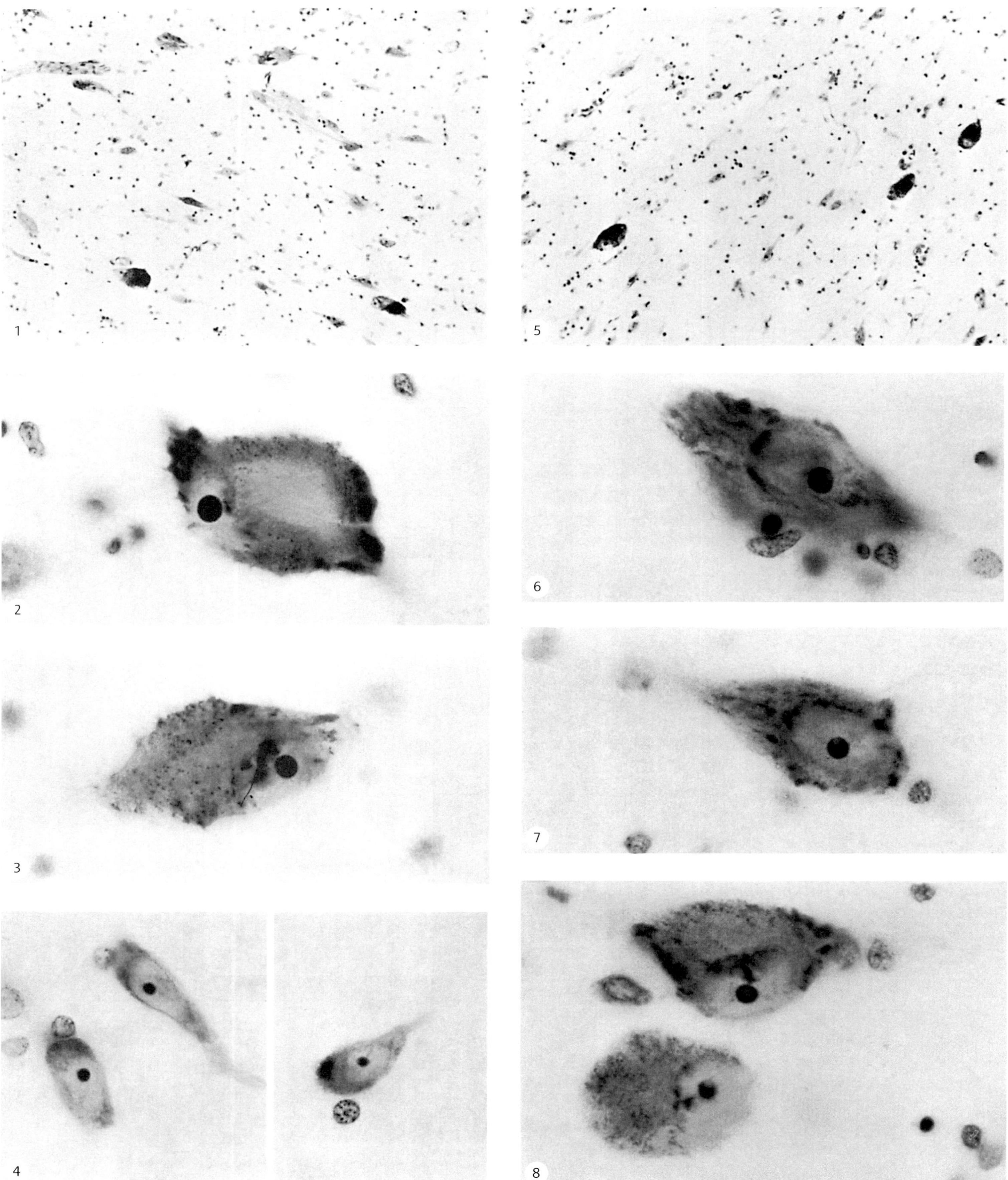

Fig. 1. Nucleus subcoeruleus, ventral subnucleus. Magnification ×150.
Fig. 2, **3.** Large, pigmented cells from the nucleus subcoeruleus, ventral subnucleus. Magnification ×1,000.
Fig. 4. Small, unpigmented cells from the nucleus subcoeruleus, ventral subnucleus. Magnification ×1,000.

Fig. 5. Nucleus subcoeruleus, dorsal subnucleus. Magnification ×150.
Fig. 6, **7.** Unpigmented cells of the nucleus subcoeruleus, dorsal subnucleus. Magnification ×1,000.
Fig. 8. Pigmented cells from the nucleus subcoeruleus, dorsal subnucleus. Magnification ×1,000.

rats the efferents targeted the dorsal horn, but in 'Sasco' rats they terminated in the intermediate gray and the ventral horn [Tracey, 1995].

Afferents. Autoradiographic studies have shown that the SubC receives a significant input from limbic structures such as the bed nuclei of the stria terminalis, anterior hypothalamus and preoptic area, the lateral hypothalamus and the periaqueductal gray [Holstege, 1991; Camp and Wijesinghe, 2009]. Tracer injections into the SubC, in a region which had been identified physiologically as a cholinoceptive 'PGO-wave-generating site', resulted in retrograde labeling of cell bodies in many parts of the brainstem: the majority of labeled cells were in the PPT, NRPO, parabrachial nucleus, vestibular nuclei, and Gi [Datta et al., 1999].

References

Camp AJ, Wijesinghe R: Calretinin: modulator of neuronal excitability. Int J Biochem Cell Biol 2009;41:2118–2121.

Chase MH, Morales FR: The atonia and myoclonia of active (REM) sleep. Annu Rev Psychol 1990;41:557–584.

Clarkson C, Juíz JM, Merchán MÁ: Long-term regulation in calretinin staining in the rat inferior colliculus after unilateral auditory cortical ablation. J Comp Neurol 2010;518:4261–4276.

Datta S, Patterson EH, Siwek DF: Brainstem afferents of the cholinoceptive pontine wave generation sites in the rat. Sleep Res Online 1999;2:79–82.

Datta S, Siwek DF, Patterson EH, Cipolloni PB: Localization of pontine PGO wave generation sites and their anatomical projections in the rat. Synapse 1998;30:409–423.

Datta S, Siwek DF, Stack EC: Identification of cholinergic and non-cholinergic neurons in the pons expressing phosphorylated cyclic adenosine monophosphate response element-binding protein as a function of rapid eye movement sleep. Neuroscience 2009;163:397–414.

Ennis M, Datta S: Electrotonic coupling in the nucleus subcoeruleus. Focus on 'evidence for electrical coupling in the subcoeruleus (SubC) nucleus'. J Neurophysiol 2007;97:2579.

Heister DS, Hayar A, Charlesworth A, Yates C, Zhou YH, Garcia-Rill E: Evidence for electrical coupling in the subcoeruleus (SubC) nucleus. J Neurophysiol 2007;97:3142–3147.

Holstege G: Descending motor pathways and the spinal motor system: limbic and non-limbic components. Prog Brain Res 1991;87:307–421.

Jones BE: Paradoxical sleep and its chemical/structural substrates in the brain. Neuroscience 1991;40:637–656.

Jones BE: From waking to sleeping: neuronal and chemical substrates. Trends Pharmacol Sci 2005;26:578–586.

Karlsson KA, Gall AJ, Mohns EJ, Seelke AM, Blumberg MS: The neural substrates of infant sleep in rats. PLoS Biol 2005;3:e143.

Meessen H, Olszewski J: A Cytoarchitectonic Atlas of the Rhombencephalon of the Rabbit. Basel, Karger, 1949.

Mesulam MM, Mufson EJ, Levey AI, Wainer BH: Atlas of cholinergic neurons in the forebrain and upper brainstem of the macaque based on monoclonal choline acetyltransferase immunohistochemistry and acetylcholinesterase histochemistry. Neuroscience 1984;12:669–686.

Nieuwenhuys R, Voogd J, van Huijzen C: The Human Central Nervous System. Berlin, Springer, 2008.

Paxinos G, Xu-Feng H, Sengul G, Watson C: Organization of the brainstem; in Mai JK, Paxinos G (eds): The Human Nervous System, ed 3. Amsterdam, Academic Press, 2012.

Simon C, Hayar A, Garcia-Rill E: Developmental changes in glutamatergic fast synaptic neurotransmission in the dorsal subcoeruleus nucleus. Sleep 2012;35:407–417.

Tracey DJ: Ascending and descending pathways in the spinal cord; in Paxinos G (ed): The Rat Nervous System, ed 2. San Diego, Academic Press, 1995.

Xi Z, Luning W: REM sleep behavior disorder in a patient with pontine stroke. Sleep Med 2009;10:143–146.

Neuromodulatory Systems – Cholinergic Nuclei (Plates 28–30, 32 and 33)

71 Laterodorsal Tegmental Nucleus (LDT)

Original name:
Not described

Location and Cytoarchitecture

The laterodorsal tegmental nucleus (LDT) lies in the ventral periaqueductal gray at the level of the inferior colliculus, its rostral pole lies at the level of the trochlear nucleus [Crosby and Woodburne, 1943]. Strands of LDT cells spread ventrally through the fascicles of the medial longitudinal fasciculus and are intermingled with the ventromedial cells of the pedunculopontine nucleus (PPT) [Koutcherov et al., 2004]. In transmitter-specific studies using choline acetyltransferase to locate cholinergic cells of the midbrain, Mesulam et al. [1989] designated the group focused around the LDT as Ch6, and the group around the PPT as Ch5. The center of the LDT lies somewhat caudal to that of the PPT. The 'Ch nomenclature' emphasizes the lack of specific cytoarchitectural boundaries of the two nuclei; in particular, Ch6 neurons were intermingled laterally with catecholaminergic neurons belonging to the locus coeruleus complex. Medial to the LDT in the pontine central gray lie the hypochromic cells of the dorsal tegmental nucleus of Gudden.

The LDT cells are medium sized and, in Nissl stains, hyperchromic (fig. 1, 2): they are slightly smaller than the PPT, but otherwise have approximately the same morphology (nucleus No. 72, fig. 1–6).

Functional Neuroanatomy

Function

The LDT, like the PPT, is part of a widespread group of cholinergic neurons with extensive ascending pathways that modulate thalamocortical activity in different behavioral states; see

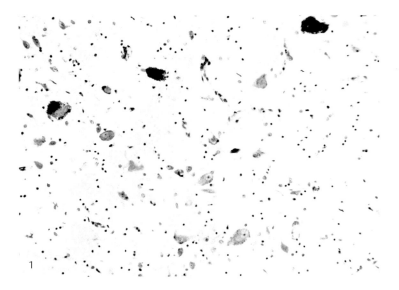

Fig. 1. Laterodorsal tegmental nucleus; note the pigmented cells of the adjacent locus coeruleus intermingled with the cells of the laterodorsal tegmental nucleus at their dorsolateral border. Magnification ×150.

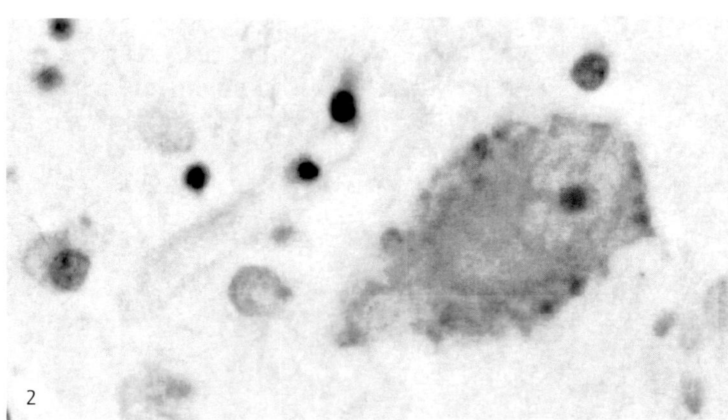

Fig. 2. Laterodorsal tegmental nucleus. Magnification ×1,000.

'Neuromodulatory Systems: Overview' [Garcia-Rill, 1991; Saper et al., 2005]. The PPT is associated more with control of motor activity, whereas the LDT is more closely associated with the modulation of limbic structures [Mena-Segovia et al., 2008; Mark et al., 2011; Dobbs and Mark, 2012].

Comment

The distribution of cholinergic neurons in the human rostral pons (Ch5–6) is similar to that of other mammalian species, but it is relatively far more expansive, and has been described as the most conspicuous and extensive cell group of the rostral brainstem [Mesulam et al., 1989].

Connections

Afferents to the LDT arise from hypothalamic and limbic basal forebrain regions, primarily from regions that have reciprocal connections to the LDT [Satoh and Fibiger, 1986].

Efferents of the LDT have been traced in the rat through two separate pathways: one has diffuse projections to the ventral tegmental area and the interpeduncular and lateral mammillary nuclei [Mena-Segovia et al., 2008], and the other long projection pathway passes through the medial forebrain bundle with branches to the tectum, pretectum, thalamus, lateral septum and medial prefrontal cortex [Satoh and Fibiger, 1986]. Other studies describe LDT efferents to the anterior thalamic nuclei and to the rostral basal forebrain regions, the cingulate and subicular cortices, but not to the parietal cortex [Mikol et al., 1984; Woolf and Butcher, 1986]. The network of structures connected with the LDT is mainly related to the limbic system.

References

Crosby EC, Woodburne RT: The nuclear pattern of the non-tectal portions of the midbrain and isthmus in primates. J Comp Neurol 1943;78:441–482.

Dobbs LK, Mark GP: Acetylcholine from the mesopontine tegmental nuclei differentially affects methamphetamine induced locomotor activity and neurotransmitter levels in the mesolimbic pathway. Behav Brain Res 2012;226: 224–234.

Garcia-Rill E: The pedunculopontine nucleus. Prog Neurobiol 1991;36:363–389.

Koutcherov Y, Huang X-F, Halliday G, Paxinos G: Organization of human brain stem nuclei; in Paxinos G, Mai JK (eds): The Human Nervous System, ed 2. Amsterdam, Elsevier Academic Press, 2004.

Mark GP, Shabani S, Dobbs LK, Hansen ST: Cholinergic modulation of mesolimbic dopamine function and reward. Physiol Behav 2011;104:76–81.

Mena-Segovia J, Winn P, Bolam JP: Cholinergic modulation of midbrain dopaminergic systems. Brain Res Rev 2008;58:265–271.

Mesulam MM, Geula C, Bothwell MA, Hersh LB: Human reticular formation: cholinergic neurons of the pedunculopontine and laterodorsal tegmental nuclei and some cytochemical comparisons to forebrain cholinergic neurons. J Comp Neurol 1989;283:611–633.

Mikol J, Menini M, Brion S, Guicharnaud L: Connections of the laterodorsal nucleus of the thalamus in the monkey. Study of efferents (in French). Rev Neurol (Paris) 1984;140:615–624.

Saper CB, Cano G, Scammell TE: Homeostatic, circadian, and emotional regulation of sleep. J Comp Neurol 2005;493:92–98.

Satoh K, Fibiger HC: Cholinergic neurons of the laterodorsal tegmental nucleus: efferent and afferent connections. J Comp Neurol 1986;253:277–302.

Woolf NJ, Butcher LL: Cholinergic systems in the rat brain. III. Projections from the pontomesencephalic tegmentum to the thalamus, tectum, basal ganglia, and basal forebrain. Brain Res Bull 1986;16:603–637.

72 Pedunculopontine Nucleus (PPT)

Original name:
Nucleus tegmenti pedunculopontinus

Alternative names:
Pedunculopontine tegmental nucleus
The pars compacta of the PPT may be referred to as nucleus of Kölliker

Location and Cytoarchitecture – Original Text

This nucleus lies in the ventrolateral part of the caudal mesencephalic tegmentum. It appears at the junction of the pons and midbrain and extends orally to the level of the caudal pole of the red nucleus. On cross-section, the nucleus occupies that portion of the tegmentum bounded medially by the superior cerebellar peduncle, laterally and ventrally by fibers of the medial lemniscus and dorsally by the cuneiform nucleus and the ventral mesencephalic reticular formation.

The pedunculopontine nucleus (PPT) is composed of medium-sized to large, oval or elongated, darkly stained cells (fig. 1–6). In many cells the nucleus is centrally placed and the darkly stained, large but rather indistinct Nissl granules are evenly distributed. In other cells the nucleus is excentric and the Nissl substance forms a dark ring about the periphery of the cell.

Two subnuclei may be distinguished within the PPT on the basis of the density of cell distribution:

the subnucleus compactus (PPTc) occupies the dorsolateral portion of the caudal half of the nucleus; its constituent cells are densely arranged and the majority of them are of the type possessing an excentric nucleus and peripherally arranged Nissl substance (fig. 1); this subnucleus is often referred to as the nucleus of Kölliker;

the subnucleus dissipatus (PPTd) constitutes the remainder of the nucleus, which is characterized by a relative paucity of cells when compared to the PPTc (fig. 2); the majority of the cells which are present possess uniformly distributed Nissl granules; in the oroventral regions of this subnucleus, a few cells may possess melanin granules within their cytoplasm.

Functional Neuroanatomy

Function
The PPT is part of a widely dispersed group of central cholinergic cells that modulate information in relation to attention, motivation and other cognitive functions. In keeping with its role as a neuromodulator, the activity of the PPT varies with the behavioral state, being active during wakefulness, less active in normal sleep, but highly active in REM sleep, which is accompanied by atonia of muscles other than those for respiration. There is evidence that the PPT is particularly associated with the modulation of motor functions; it is considered to be part of the 'mesencephalic locomotion region'. Experimentally, its stimulation produces controlled locomotion such as stepping, walking, trotting or galloping. Furthermore through its excitatory projection to the thalamic relay and the thalamic reticular nucleus, the PPT can modulate the transmission of sensory information through the thalamus to the cortex.

The locomotor role of the PPT is supported by experimental studies, which show that lesions of the PPT in monkeys produce akinesia. A significant loss of PPT cells has been reported in Parkinson's disease and supranuclear palsy, and current hypotheses propose that some of the associated locomotor and postural abnormalities could be attributed to degeneration in the PPT. For reviews on the function of the PPT, see Steckler et al. [1994], Blokland [1995], Barefoot et al. [2000], Kus et al. [2003], Hamani et al. [2007], Pahapill and Lozano [2000], Hirsch et al. [2004] and Takakusaki et al. [2004].

Comment
The PPT forms the rostral part of a column of about 20,000 cholinergic neurons [Mizukawa et al., 1986; Manaye et al., 1999; Kus et al., 2003]. Individual cell groups in the column have been given the nomenclature Ch1–6 by Mesulam and colleagues, whereby the group Ch5 is the PPT and the laterodorsal tegmental nucleus (LDT) is Ch6 [Mesulam et al., 1989; Manaye et al., 1999]. Several authors consider the two nuclei together as the PPT-LDT complex. This cholinergic system participates in the ascending activating system [Garcia-Rill, 1991; Saper et al., 2005].

Comparing the two subdivisions of the PPT, the dorsomedial PPTc contains primarily cholinergic cells (>90%), whereas in the ventrolateral PPTd 25–75% are cholinergic and in addition a significant population of the cells are glutaminergic [Lavoie and Parent, 1994]. Paxinos et al. [2012] put forward arguments for the use of the name 'retroisthmic nucleus' rather than the subnucleus PPTd.

Connections
Afferents. The connectivity of the PPT reflects its role in motor behavior through the basal ganglia, and behavioral states through thalamocortical modulation [Parent, 1996]. However, the afferents and efferents of PPTc and PPTd differ [Pahapill and Lozano, 2000]: the PPTc receives afferents from the spinal cord, but the PPTd receives its inputs from the basal ganglia, specifically the internal segment of the globus pallidus (GPi), the medial two thirds of substantia nigra pars reticulata and the subthalamic nucleus (STH) [Kim et al., 1976; Nauta and Cole, 1978; Parent and De Bellefeuille, 1982]. The anatomical association of the PPT with the superior cerebellar peduncle obtained important functional significance with the report that afferents from its fibers terminated onto PPT cells [Hazrati and Parent, 1992]. The result indicates that the deep cerebellar nuclei may provide a significant input to the PPT.
Efferents. The efferents of the PPT are typical of a neuromodulatory nucleus, with diffuse and widespread axonal branches reaching vast areas of the brain (see 'Neuromodulatory Systems: Overview'). The descending efferents of the PPTc terminate in brainstem nuclei such as the raphe nuclei, the pontine nuclei or the pontine and medullary medial reticular formation, whereas from the PPTd they descend to the spinal cord.

The ascending pathways from the PPT are more prominent. The PPTc projects to the caudate and putamen and all thalamic nuclei, but especially the intrathalamic nuclei which are known to send diffuse projections to widespread areas of the cerebral cortex and may participate in the control of the sleep-wake cycle [Saper et al., 2001]: the ascending efferents of the PPTd end strongly in the STH and substantia nigra pars compacta, and also supply the GPi [Paré et al., 1988; Heckers et al., 1992]. There are reciprocal connections between the PPT and the cerebral cortex [Hartmann-von Monakow et al., 1979].

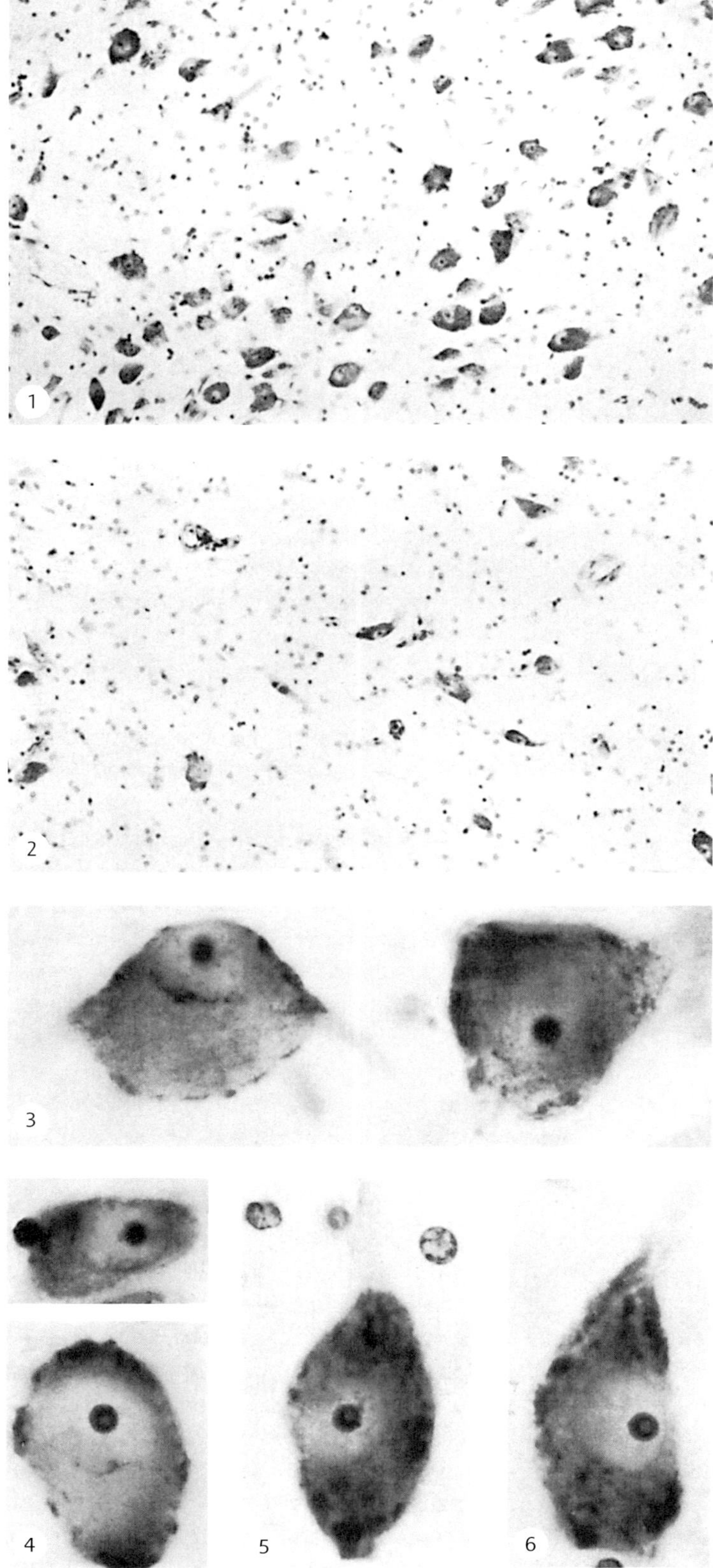

Fig. 1. Pedunculopontine nucleus, subnucleus compactus. Magnification ×150.

Fig. 2. Pedunculopontine nucleus, subnucleus dissipatus. Magnification ×150.

Fig. 3, 4. Cells from the pedunculopontine nucleus with peripherally distributed Nissl sustance. Magnification ×1,000.

Fig. 5, 6. Cells from the pedunculopontine nucleus with evenly distributed Nissl sustance. Magnification ×1,000.

References

Barefoot HC, Baker HF, Ridley RM: Synergistic effects of unilateral immunolesions of the cholinergic projections from the basal forebrain and contralateral ablations of the inferotemporal cortex and hippocampus in monkeys. Neuroscience 2000;98:243–300.

Blokland A: Acetylcholine: a neurotransmitter for learning and memory? Brain Res Rev 1995;21:285–300.

Garcia-Rill E: The pedunculopontine nucleus. Prog Neurobiol 1991;36:363–389.

Hamani C, Stone S, Laxton A, Lozano AM: The pedunculopontine nucleus and movement disorders: anatomy and the role for deep brain stimulation. Parkinsonism Relat Disord 2007;13(suppl 3):S276–S280.

Hartmann-von Monakow K, Akert K, Künzle H: Projections of precentral and premotor cortex to the red nucleus and other midbrain areas in *Macaca fascicularis.* Exp Brain Res 1979;34:91–105.

Hazrati L-N, Parent A: Projection from the deep cerebellar nuclei to the pedunculopontine nucleus in the squirrel monkey. Brain Res 1992;585:267–271.

Heckers S, Geula C, Mesulam M-M: Cholinergic innervation of the human thalamus: dual origin and differential nuclear distribution. J Comp Neurol 1992;325: 68–82.

Hirsch EC, Graybiel AM, Duyckaerts C, Javoy-Agid F: Neuronal loss in the pedunculopontine tegmental nucleus in Parkinson disease and in progressive supranuclear palsy. Proc Natl Acad Sci USA 2004;84:5976–5980.

Kim R, Nakano K, Jayaraman A, Carpenter MP: Projections of the globus pallidus and adjacent structures: an autoradiographic study in the monkey. J Comp Neurol 1976;169:263–290.

Kus L, Borys E, Chu YP, Ferguson SM, Blakely RD, Emborg ME, Kordower JH, Levey AI, Mufson EJ: Distribution of high affinity choline transporter immunoreactivity in the primate central nervous system. J Comp Neurol 2003;463:341–357.

Lavoie B, Parent A: Pedunculopontine nucleus in the squirrel monkey: distribution of cholinergic and monoaminergic neurons in the mesopontine tegmentum with evidence for the presence of glutamate in cholinergic neurons. J Comp Neurol 1994;344:190–209.

Manaye KF, Zweig R, Wu D, Hersh LB, De Lacalle S, Saper CB: Quantification of cholinergic and select non-cholinergic mesopontine neuronal populations in the human brain. Neuroscience 1999;89:759–770.

Mesulam MM, Geula C, Bothwell MA, Hersh LB: Human reticular formation: cholinergic neurons of the pedunculopontine and laterodorsal tegmental nuclei and some cytochemical comparisons to forebrain cholinergic neurons. J Comp Neurol 1989;283:611–633.

Mizukawa K, McGeer PL, Tago H, Peng JH, McGeer EG, Kimura H: The cholinergic system of the human hindbrain studied by choline acetyltransferase immunhistochemistry and acetylcolinesterase histochemistry. Brain Res 1986;379: 39–55.

Nauta HJ, Cole M: Efferent projections of the subthalamic nucleus: an autoradiographic study in monkey and cat. J Comp Neurol 1978;180:1–16.

Pahapill PA, Lozano AM: The pedunculopontine nucleus and Parkinson's disease. Brain 2000;123:1767–1783.

Paré D, Smith Y, Parent A, Steriade M: Projections of brainstem core cholinergic and non-cholinergic neurons of cat to intralaminar and reticular thalamic nuclei. Neuroscience 1988;25:69–86.

Parent A: Carpenter's Human Neuroanatomy. Baltimore, Williams & Wilkins, 1996.

Parent A, de Bellefeuille L: Organization of efferent projections from the internal segment of globus pallidus in primate as revealed by fluorescence retrograde labeling method. Brain Res 1982;245:201–213.

Paxinos G, Huang X-F, Sengul G, Watson C: Organization of the brainstem; in Mai JK, Paxinos G (eds): The Human Nervous System, ed 3. Amsterdam, Academic Press, 2012.

Saper CB, Chou TC, Scammell TE: The sleep switch: hypothalamic control of sleep and wakefulness. Trends Neurosci 2001;24:726–731.

Saper CB, Scammell TE, Lu J: Hypothalamic regulation of sleep and circadian rhythms. Nature 2005;437:1257–1263.

Steckler T, Inglis W, Winn P, Sahgal A: The pedunculopontine tegmental nucleus: a role in cognitive processes? Brain Res Rev 1994;19:298–318.

Takakusaki K, Habaguchi T, Saitoh K, Kohyama J: Changes in the excitability of hindlimb motoneurons during muscular atonia induced by stimulating the pedunculopontine tegmental nucleus in cats. Neuroscience 2004;124:467–480.

73 Substantia Nigra (SN)

Original name:
Nucleus substantiae nigrae

Location and Cytoarchitecture – Original Text

This nuclear group extends throughout the whole length of the midbrain as a broad band of cells which intervenes between the cerebral peduncles ventrolaterally and the tegmental nuclei dorsomedially. The nucleus commences just caudal to the rostral limits of the pontine nuclei and extends orally to the level of the subthalamic nucleus. On cross-section, the long axis of the nucleus is directed dorsolaterally.

It is customary to subdivide the substantia nigra (SN) into (a) subnucleus compactus (SNc), and (b) subnucleus reticulatus (SNr). Such a subdivision is based not only on morphological differences between the two subnuclei, but also on ontogenetic, histochemical and pathological evidence.

The SNc forms the bulk of the SN and lies dorsal to the SNr. It is composed of large, elongated or plump, multipolar cells which possess long dendrites, prominent nuclei and nucleoli and large, discrete, darkly stained Nissl granules (fig. 1). The majority of cells characteristically contain a large amount of melanin pigment (fig. 6–10). The amount of pigment present in a single cell may vary (fig. 2–5) from only a few scattered particles to an amount which completely obscures the internal architecture of the cell (fig. 6, 8, 10). The cells which compose the medial and lateral poles of the SNc are smaller than those in the intermediate portion.

The SNc may be further subdivided into pars α, β and γ on the basis of cell arrangement (plates 34, 36, 38 and 40):

pars α is characterized by the fact that its cells tend to congregate into relatively large, irregularly shaped cell groups which are constant in position; these cell groups comprise the ventral portion of the SNc;

pars β lies dorsal to pars α and is composed of scattered and irregularly oriented neurons which do not congregate into cell groups;

pars γ appears dorsal to pars β at the level of the greatest development of the red nucleus (RN), see 'Comment' below; it is composed of melanin-containing cells which lie with their long axes directed dorsolaterally among the most ventral fibers of the RN capsule.

The SNr lies ventral to, and begins and ends at slightly more oral levels than, the SNc. Like the SNc, the SNr is composed of both pigmented and unpigmented cells. In general the pigmented cells (fig. 7) are smaller, and possess fewer discrete Nissl granules and relatively less melanin pigment than do the cells of the SNc. The unpigmented cells of the SNr (fig. 2, 4) are morphologically similar to the cells which compose the globus pallidus, and it should be noted that these two structures are continuous through the fields of Sano [the fields of Sano are striatonigral connections that pierce through the cerebral peduncle, also called 'Edinger's Kammsystem']. [...]

[...] The melanin pigmentation of the cells of the SN is a metabolic by-product associated with dopamine (catecholamine) synthesis [Double et al., 2008]. It is not found in children under the age of about 5 years. [...] In subhuman forms, however, the degree of pigmentation is always less that that seen in the adult human. [...]

Comment

Nowadays, the SNc pars γ of Olszewski and Baxter is included in the parabrachial pigmented nucleus (PBP, nucleus No. 78) of the perirubral capsule, and the SNc is divided into only 2 tiers: a dorsal tier corresponding to pars β, and a ventral tier representing pars α [Halliday et al., 2012]. The 3 sheets of neurons in the SN, SNc dorsal tier, SNc ventral tier and SNr, become compressed and intermingled at the ends, where they form the SN medial part and SN lateral part.

The SNc and SNr stem from different neurogenic sources, have different histochemical profiles and different connections [Parent, 1996; Verney et al., 2001]. Whereas the SNr contains mainly GABAergic neurons, similar to the internal segment of the globus pallidus[1], the SNc is mainly composed of dopaminergic cells. The midbrain dopaminergic neurons form a continuum of cells, in which the neurons of the SNc are designated as A9 by Dahlström and Fuxe [1964], those in the medially adjacent ventral tegmental area (of Tsai, VTA) as A10, while the cells of the PBP form the A8 group. These midbrain dopaminergic cell groups form a massive ascending network that can modulate the activity of the striatum and frontal cortex; the fibers from the SNc target the whole striatum, those of the VTA the frontal cortex [Parent, 1996; Schultz, 1998]. Electrotonic coupling has been demonstrated between the dopamine cells of the SNc, and hence the existence of rapid communication within the SNc, but no such coupling was found in the SNr [Vandecasteele et al., 2005].

Functional Neuroanatomy

Function

The SN is considered to be an integral part of the basal ganglia, which play a fundamental role in the generation and control of voluntary motor actions. Diseases of the basal ganglia are characterized by movement disorders. Parkinson's disease, for example, is associated with a massive loss of the dopaminergic pigmented cells in the SNc [Damier et al., 1999], and lesions of the SN lead to reduction in dopamine in the striatum and motor abnormalities [Bézard et al., 2001]. However, more recently, it has been recognized that basal ganglia disorders are also associated with disturbances of cognition, emotional behavior, addiction, reward and reinforcement learning [Parent, 1996; Schultz, 1998; Middleton and Strick, 2000; Redgrave, 2007]. Furthermore aberrant dopamine neurotransmitter signaling is implicated in schizophrenia and addictive behavioral disorders. A recent compelling hypothesis explaining the wide diversity of functions associated with the basal ganglia is summarized by Redgrave [2007]. For further reviews, see Parent [1996] and Halliday [2004].

The main input to the basal ganglia from the cerebral cortex is channeled through the striatum, which is the origin of the striatonigral pathway, targeting mainly the SNr. A massive feedback to the striatum – the nigrostriatal pathway – arises,

[1] Similarities in the cytology of the SNr and globus pallidus are described in the 'Original Text'.

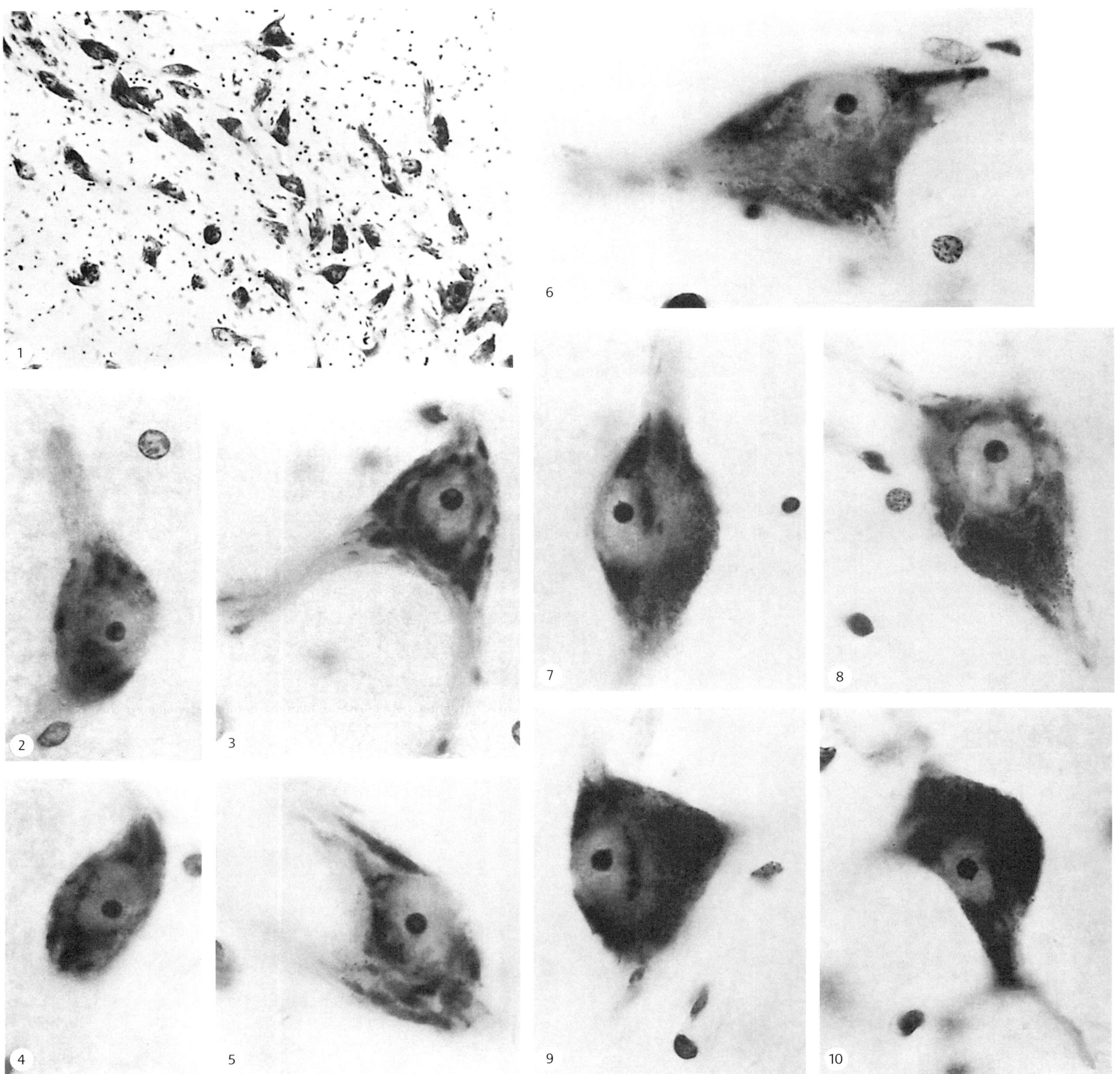

Fig. 1. Substantia nigra. Magnification ×150.
Fig. 2–5. Unpigmented cells from the substantia nigra. Magnification ×1,000.

Fig. 6–10. Cells from the substantia nigra showing various degrees of pigmentation. Magnification ×1,000.

not from the SNr, but from the dopaminergic SNc, while the SNr provides an important output from the basal ganglia network through its inhibitory projections to the thalamus, superior colliculus and other brainstem structures. The basal ganglia are interconnected with cortical and subcortical loops; for example, the 'corticostriatal-basal ganglia/SN-thalamic-motor cortex loop', which controls motor patterns; but similar loops from other functional territories such as sensory, limbic or association cortices to the basal ganglia have also been described, indicating that normally the SN is involved not only in the control of motor actions, but can also modulate sensory experience, emotional behavior and cognition [Alexander et al., 1986; Haber and Gdowski, 2004; Redgrave, 2007]. Due to the strict organization of the cortical, striatal, and nigral pathways, these functional roles are even topographically organized within the SN (fig. 21.20, 21.22 and 21.23 in Haber and Gdowski [2004]), whereby the lateral SN is associated with visuomotor functions [Harting and Updyke, 2006].

Connections

Afferents. The inputs to the SN arise from 6 main areas: the striatum, globus pallidus external segment (GPe), the subthalamic nucleus, the pedunculopontine nucleus (PPT), the dorsal raphe nucleus and the cerebral cortex. Most prominent is the striatal projection, which targets mainly the SNr, through GABAergic fibers coexpressing a combination of substance P, enkephalin and dynorphin. The projection to the SNr arises from the matrix compartment of the striatum, whereas the striosomal compartment provides afferents to the SNc [Nieuwenhuys et al., 2008]. Second, the GPe similarly projects through GABAergic fibers to the SNr. The projection from the subthalamic nucleus also terminates in the SNr and is excitatory, using glutamate as the transmitter [Rinvik and Ottersen, 1993]. Serotonergic fibers from the raphe nuclei supply mainly the SNr, whereas the PPT targets the SNc. Finally the existence of cortical afferents to the SN is uncertain in primates [Künzle, 1978].

Efferents. The dopaminergic output from the SNc to the striatum targets the striosomal compartment and is topographically organized [Haber et al., 2000]. The authors propose that this anatomical arrangement creates an ascending spiral of information flow across the whole striatonigrostriatal complex, beginning with the influence from the limbic VTA. The SNr provides an important output from the basal ganglia. It arises from the GABAergic projection neurons that have branches terminating in the superior colliculus, thalamus and PPT. The nigrotectal projection terminates in 3 different levels (tiers) in intermediate layers of the superior colliculus [Harting and Updyke, 2006]; the thalamic projection centers on the thalamic nuclei associated with the prefrontal, and in primates motor cortex [Ilinsky et al., 1985]. The inhibitory projection from the SNr to the PPT is supported by 60% of SNr projection neurons, and a robust reciprocal projection from the PPT has been described back to both the SNc and SNr [Mena-Segovia et al., 2004].

For recent and extensive reviews, see Nieuwenhuys et al. [2008, chapter 14], Haber and Gdowski [2004], Parent [1996] and Halliday et al. [2012].

References

Alexander GE, Delong MR, Strick PL: Parallel organization of functionally segregated circuits linking basal ganglia and cortex. Annu Rev Neurosci 1986;9:357–381.

Bézard E, Dovero S, Prunier C, Ravenscroft P, Chalon S, Guilloteau D, Crossman AR, Bioulac B, Brotchie JM, Gross CE: Relationship between the appearance of symptoms and the level of nigrostriatal degeneration in a progressive 1-methyl-4-phenyl-1,2,3,6-tetrahydropyridine-lesioned macaque model of Parkinson's disease. J Neurosci 2001;21:6853–6861.

Dahlström A, Fuxe K: Evidence for the existence of monoamine neurons in the central nervous system. I. Demonstration of monoamines in the cell bodies of brain stem neurons. Acta Physiol Scand 1964;62:1–55.

Damier P, Hirsch EC, Agid Y, Graybiel AM: The substantia nigra of the human brain. II. Patterns of loss of dopamine-containing neurons in Parkinson's disease. Brain 1999;122:1437–1448.

Double KL, Dedov VN, Fedorow H, Kettle E, Halliday GM, Garner B, Brunk UT: The comparative biology of neuromelanin and lipofuscin in the human brain. Cell Mol Life Sci 2008;65:1669–1682.

Haber SN, Fudge JL, McFarland NR: Striatonigrostriatal pathways in primates form an ascending spiral from the shell to the dorsolateral striatum. J Neurosci 2000;20:2369–2382.

Haber SN, Gdowski MJ: The basal ganglia; in Paxinos G, Mai JK (eds): The Human Nervous System, ed 2. Amsterdam, Elsevier Academic Press, 2004.

Halliday G: Substantia nigra and locus coeruleus; in Paxinos G, Mai JK (eds): The Human Nervous System, ed 2. Amsterdam, Elsevier Academic Press, 2004.

Halliday G, Reyes S, Double K: Substania nigra, ventral tegmental area and retrorubral fields; in Mai JK, Paxinos G (eds): The Human Nervous System, ed 3. Amsterdam, Academic Press, 2012.

Harting JK, Updyke BV: Oculomotor-related pathways of the basal ganglia. Prog Brain Res 2006;151:441–460.

Ilinsky IA, Jouandet ML, Goldman-Rakic PS: Organization of the nigrothalamocortical system in the rhesus monkey. J Comp Neurol 1985;236:315–330.

Künzle H: Cortico-cortical efferents of primary motor and somatosensory regions of the cerebral cortex in *Macaca fascicularis*. Neuroscience 1978;3:25–39.

Mena-Segovia J, Bolam JP, Magill PJ: Pedunculopontine nucleus and basal ganglia: distant relatives or part of the same family? Trends Neurosci 2004;27:585–588.

Middleton FA, Strick PL: Basal ganglia output and cognition: evidence from anatomical, behavioral, and clinical studies. Brain Cogn 2000;42:183–200.

Nieuwenhuys R, Voogd J, van Huijzen C: The Human Central Nervous System. Berlin, Springer, 2008.

Parent A: Carpenter's Human Neuroanatomy. Baltimore, Williams & Wilkins, 1996.

Redgrave P: Basal ganglia. Scholarpedia 2007;2:1825.

Rinvik E, Ottersen OP: Terminals of subthalamonigral fibres are enriched with glutamate-like immunoreactivity: an electron microscopic, immunogold analysis in the cat. J Chem Neuroanat 1993;6:19–30.

Schultz W: Predictive reward signal of dopamine neurons. J Neurophysiol 1998;80:1–27.

Vandecasteele M, Glowinski J, Venance L: Electrical synapses between dopaminergic neurons of the substantia nigra pars compacta. J Neurosci 2005;25:291–298.

Verney C, Zecevic N, Puelles L: Structure of longitudinal brain zones that provide the origin for the substantia nigra and ventral tegmental area in human embryos, as revealed by cytoarchitecture and tyrosine hydroxylase, calretinin, calbindin, and GABA immunoreactions. J Comp Neurol 2001;429:22–44.

Neuromodulatory Systems – Dopaminergic Nuclei (Plates 32, 34–36 and 38, not always labelled)

74 Ventral Tegmental Area (of Tsai) (VTA)

Original name:
Not presented in the 2nd edition as an entity, but comprising the region containing the paranigral nucleus, pigmented parabrachial nucleus and substantia perforata posterior

Nuclei of the ventral tegmental area:
Caudal linear nucleus
Rostral linear nucleus
Paranigral nucleus
Interfascicular nucleus
Parabrachial pigmented nucleus

Nomenclature and Transmitters

The triangular region of the ventral midbrain lying medial to the substantia nigra (SN), and either side of the interpeduncular fossa, is punctured with blood vessels and the rootlets of the oculomotor nerve. Tsai [1925] first called it the ventral tegmental area (VTA); Hassler preferred to classify the area as a subgroup of the SN, but Tsai disagreed with him on the grounds of cytoarchitectural differences [Saper, 1999]. Later the area was recognized as an entity by Nauta [1958] and called the VTA of Tsai [Oades and Halliday, 1987; Halliday et al., 2012]. The borders of the VTA are poorly defined, but in all mammals the area is bounded caudally by the interpeduncular nucleus and rostrally by the mammillary bodies and the posterior hypothalamus; dorsally lie the oculomotor nucleus and the red nucleus; laterally the area extends into the retrorubral fields (RRF), which are the dorsolateral extension of the lateral SN into the caudal mesencephalic reticular formation (plates 36 and 37). The VTA includes 3 unpaired nuclei at the midline, the rostral linear nucleus (RLi), central linear nucleus (CLi) and the interfascicular nucleus (IFN), and 2 paired nuclei, the paranigral nucleus (PNg) and parabrachial pigmented nucleus (PBP) [Halliday and Törk, 1986]. Each of these nuclei has different afferent and efferent connections [Oades and Halliday, 1987; Scheibner and Törk, 1987].

The VTA corresponds to the dopaminergic area A10, and also includes the serotonergic B8 raphe group in the CLi and RLi [Dahlström and Fuxe, 1964; Swanson, 1982]. The adjacent SN (A9) and the RRF (A8) contain dopaminergic neurons as well, but are not included in the definition of the VTA. About 45% of the neurons in the VTA are nondopaminergic, and nondopaminergic neurons are intermixed with dopaminergic ones throughout most of the VTA [Margolis et al., 2006]. The caudal VTA has a large population of GABAergic neurons [Kaufling et al., 2009].

Apart from the dopaminergic network associated with the VTA, this region contains other nuclei concerned with different functions, for example the ventral part of the medial terminal nucleus of the accessory optic tract and also the ventral tegmental relay zone (see nucleus No. 32). The latter is a component of the accessory optic pathways, in this case relaying vertical optokinetic information to the inferior olive and other parts of the optokinetic network [Giolli et al., 2006].

Functional Neuroanatomy

Function

The VTA neurons are the source of dopamine projections to the forebrain, and have been implicated in attention, memory, reward, drug abuse and motivation [Wise, 2004]. There is anatomical evidence supporting the association of the VTA with 'rewarding behaviors' including ejaculation [McBride et al., 1999; Holstege et al., 2003]. It is not clear what contribution the nondopaminergic fibers make to these processes [Margolis et al., 2006]. However, there are regional differences within the VTA: individual nuclei of the VTA project to different rostral targets, and pharmacological manipulations of the VTA subregions can give rise to different behaviors. In an attempt to understand the distinctions between the individual nuclei of the VTA, some of them (CLi, RLi, and PBP) are considered separately here in individual chapters [Margolis et al., 2006; Halliday et al., 2012].

A difference in calbindin function in VTA cells compared with SN cells has been used as a possible explanation for the vulnerability of the SN in Parkinson's disease [Pan and Ryan, 2012].

Connections

The VTA nuclei can be divided into two main clusters: the ventromedial tier, consisting of PNg, IFN and CLi, and the dorsolateral tier including PBP and RLi. A third nuclear cluster is formed by the RRF and the adjacent lateral SN, which are not part of the VTA [Halliday et al., 2012]. The ventromedial tier sends an ascending limbic pathway and a thalamic pathway to the forebrain. The dorsolateral tier sends a striatal and a cortical pathway to the forebrain. From the third cluster, the lateral SN projects to the superior colliculus and is important for eye movement control, whereas the RRF projects to the SN and VTA for internal regulation. For details of these pathways, see Halliday et al. [2012].

References

Dahlström A, Fuxe K: Evidence for the existence of monoamine neurons in the central nervous system. I. Demonstration of monoamines in the cell bodies of brain stem neurons. Acta Physiol Scand 1964;62:1–55.

Giolli RA, Blanks RHI, Lui F: The accessory optic system: basic organization with an update on connectivity, neurochemistry, and function. Prog Brain Res 2006; 151:407–440.

Halliday G, Reyes S, Double K: Substania nigra, ventral tegmental area and retrorubral fields; in Mai JK, Paxinos G (eds): The Human Nervous System, ed 3. Amsterdam, Academic Press, 2012.

Halliday GM, Törk I: Comparative anatomy of the ventromedial mesencephalic tegmentum in the rat, cat, monkey and human. J Comp Neurol 1986;252:423–445.

Holstege G, Georgiadis JR, Paans AMJ, Meiners LC, van der Graaf FHCE, Reinders AATS: Brain activation during human male ejaculation. J Neurosci 2003;23: 9185–9193.

Kaufling J, Veinante P, Pawlowski SA, Freund-Mercier MJ, Barrot M: Afferents to the GABAergic tail of the ventral tegmental area in the rat. J Comp Neurol 2009;513:597–621.

Margolis EB, Lock H, Hjelmstad GO, Fields HL: The ventral tegmental area revisited: is there an electrophysiological marker for dopaminergic neurons? J Physiol 2006;577:907–924.

McBride WJ, Murphy JM, Ikemoto S: Localization of brain reinforcement mechanisms: intracranial self-administration and intracranial place-conditioning studies. Behav Brain Res 1999;101:129–152.

Nauta WJ: Hippocampal projections and related neural pathways to the midbrain in the cat. Brain 1958;81:319–340.

Oades RD, Halliday GM: Ventral tegmental (A10) system: neurobiology. 1. Anatomy and connectivity. Brain Res 1987;434:117–165.

Pan PY, Ryan TA: Calbindin controls release probability in ventral tegmental area dopamine neurons. Nat Neurosci 2012;15:813–815.

Saper CB: 'Like a thief in the night': the selectivity of degeneration in Parkinson's disease. Brain 1999;122:1401–1402.

Scheibner T, Törk I: Ventromedial mesencephalic tegmental (VMT) projections to ten functionally different cortical areas in the cat: topography and quantitative analysis. J Comp Neurol 1987;259:247–265.

Swanson LW: The projections of the ventral tegmental area and adjacent regions: a combined fluorescent retrograde tracer and immunofluorescence study in the rat. Brain Res Bull 1982;9:321–353.

Tsai C: The optic tracts and centres of the opossum, *Didelphis virginiana*. J Comp Neurol 1925;39:173–215.

Wise RA: Dopamine, learning and motivation. Nat Rev Neurosci 2004;5:483–494.

75 Caudal Linear Nucleus (CLi)

Original name:
The region of the caudal linear nucleus (CLi) was originally included in the parabrachial pigmented nucleus (see p. 187 of the 2nd edition): compare with the CLi in figure 7 of nucleus No. 68, or Pearson et al. [1990]

Alternative names:
Central linear nucleus [Berman, 1968]
Nucleus linearis intermedius [Taber, 1961]
A nucleus of the ventral tegmental area

Location and Cytoarchitecture

The caudal linear nucleus (CLi) lies at the anterior border of the decussation of the superior cerebellar peduncle (SCP), and forms the rostral continuation of the median raphe nucleus (MnR). However, it is usually included in the term 'ventral tegmental area of Tsai' (VTA). The caudal CLi cells intermingle with the rostral pole of the MnR in the crossing fiber bundles of the rostral SCP, and further dorsally the caudal CLi merges with the interfascicular part of the dorsal raphe nucleus (DR). The CLi lies on the midline stretching from the medial longitudinal fasciculus dorsally down to the interpeduncular nucleus (IPN) on the ventral surface of the brain. Like the MnR there is a medial zone and a lateral zone in the CLi, each with different properties [Pearson et al., 1990; Austin et al., 1997; Hornung, 2012]: the median zone is acetylcholinesterase (AChE) negative and contains many serotonergic cells, whereas the lateral zone is AChE positive and is scattered with pigmented dopaminergic cells, typical of the VTA (A10). Rostrally around the caudal borders of the red nucleus, the lateral zone of the CLi joins with the caudal part of the rostral linear nucleus [Halliday et al., 2012; Paxinos et al., 2012].

The median CLi cells are small to medium in size, with dendrites oriented rostrocaudally (fig. 1–3). The cells of the lateral zone are larger with round or triangular somata. At the junction of the IPN with the ventral CLi the serotonergic cells develop a different character, becoming larger and multipolar [Hornung and Fritschy, 1988; Pearson et al., 1990].

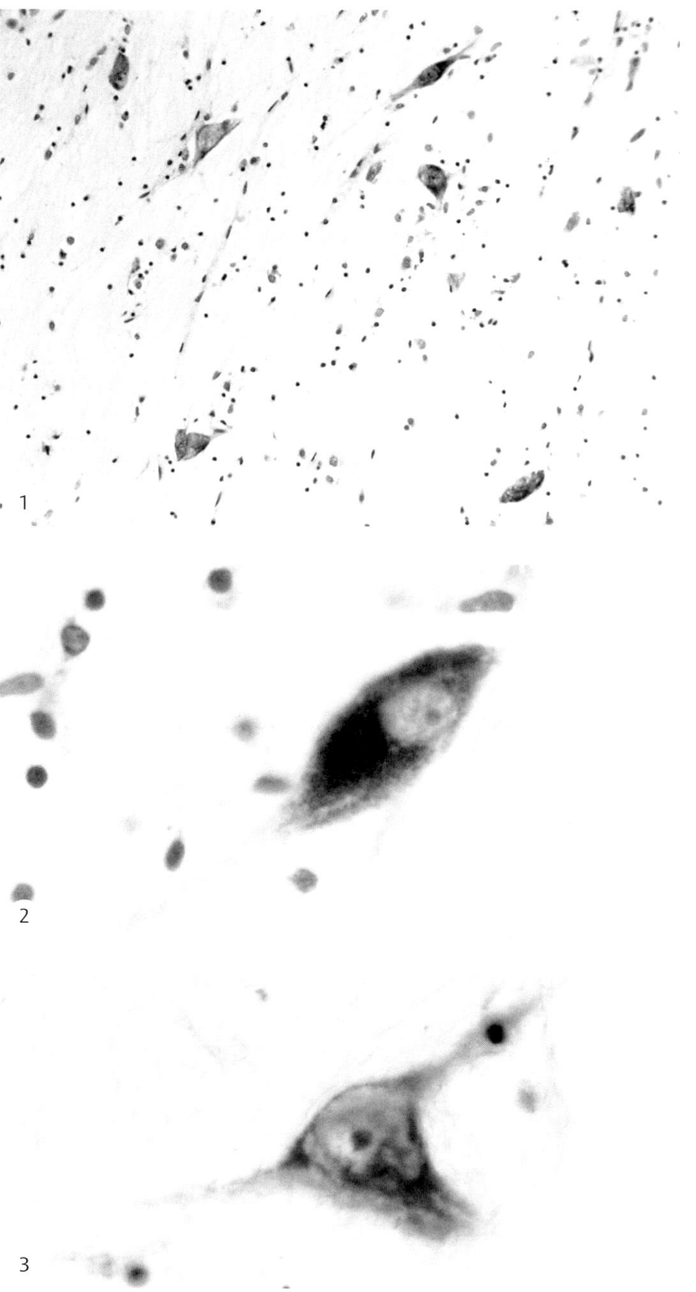

Fig. 1. Caudal linear nucleus. Note the presence of some pigmented dopaminergic cells. Magnification ×150.
Fig. 2, 3. Cells of the caudal linear nucleus. Note the cell pigment in figure 2. Magnification ×1,000.

Functional Neuroanatomy

Function
The CLi, together with the MnR and DR, belong to the rostral group of serotonergic raphe nuclei. It is the most rostral raphe nucleus, containing many serotonergic cells that were designated as group B8 [Dahlström and Fuxe, 1964]. The rostral group nuclei have a large number of afferents and efferents in common. Serotonin is reported to play a key role in depression and other mood disorders; see the other chapters on raphe nuclei and Hornung [2012].

The major difference between the CLi and the other nuclei is its population of dopaminergic cells, so characteristic of the VTA, and designated as A10 [Pearson et al., 1990; Hornung, 2012]. Through the dopaminergic projections the CLi is considered to play a specific role in the 'dopamine reward system' associated with the VTA; for reviews, see Ikemoto [2007], Pierce and Kumaresan [2006] and Schultz [1998]. Along with the adjacent VTA, the CLi projects to the limbic part of the striatum, the ventromedial striatum, and participates in the regulation of arousal, important for goal-directed learning [Ikemoto, 2007]. This medial VTA projection pathway is characterized by 'affect and drive', suggested to play a different role in goal-directed learning from the projection system from the lateral VTA.

In relation to the dopaminergic cells of the CLi, it has been shown that the administration of nicotine triggers reinforcing effects that mediate 'nicotine reward', and the dopaminergic networks play a key role in this function [Di Chiara, 2000].

Connections

Efferents. The rostral group of raphe nuclei, the CLi, MnR and DR, all have similar neural connections, associated with the limbic system [Imai et al., 1986]. Through their widespread ascending serotonergic axons they can modulate activity in most of the forebrain. However, there is some evidence that the CLi is developmentally and morphologically closer to the DR, and not to the MnR [Jacobs and Azmitia, 1992]. A distinct topographical organization of midbrain raphe efferents has been reported in rats, and it is not unreasonable to expect a similar organization in humans, since the serotonergic system does not vary widely between species [Azmitia and Gannon, 1986].

It was shown that the rostral midbrain raphe nuclei, including the CLi, project to the basal ganglia and related structures, the intermediate raphe nuclei send efferents to the amygdala, and the caudal midbrain raphe nuclei target limbic structures [Imai et al., 1986]. In accordance with this study, extensive studies demonstrated that the caudomedial VTA and CLi are the origin of a medial striatal pathway that selectively projects to the (limbic) ventromedial striatum, specifically to the medial olfactory tubercle and medial nucleus accumbens shell [Ikemoto, 2007].

An important component of the CLi is its connectivity with local nuclei in the VTA; here the caudomedial nuclei, the interfascicular and paranigral nuclei and CLi, are closely interconnected. In addition, the CLi projects to the lateral VTA, substantia nigra pars compacta and the retrorubral fields, and through these may be able to modulate the interaction of the limbic ventral striatum with the lateral somatic striatum [Ferreira et al., 2008].

Afferents. One of the main inputs to raphe nuclei is thought to be their own reciprocal serotonergic projections [Jacobs and Azmitia, 1992]. In general the afferents to the raphe nuclei are from the limbic structures, such as the medial septal area, diagonal band nuclei, the lateral habenula and several regions of the hypothalamus [Nieuwenhuys et al., 2008; Hornung, 2012]. There is strong autoradiographic labeling in the CLi after injections into the medial preoptic area of the anterior hypothalamus [Holstege, 1987]. Relatively few brainstem regions project to the CLi; however, considerably less attention is paid to the small CLi nucleus than to the DR, MnR or the caudal raphe nuclei [Arnsten and Goldman-Rakic, 1984; Behzadi et al., 1990; Jacobs and Azmitia, 1992].

References

Arnsten AF, Goldman-Rakic PS: Selective prefrontal cortical projections to the region of the locus coeruleus and raphe nuclei in the rhesus monkey. Brain Res 1984;306:9–18.

Austin MC, Rhodes JL, Lewis DA: Differential distribution of corticotropin-releasing hormone immunoreactive axons in monoaminergic nuclei of the human brainstem. Neuropsychopharmacology 1997;17:326–341.

Azmitia EC, Gannon PJ: The primate serotonergic system: a review of human and animal studies and a report on *Macaca fascicularis*. Adv Neurol 1986;43:407–468.

Behzadi G, Kalen P, Parvopassu F, Wiklund L: Afferents to the median raphe nucleus of the rat: retrograde cholera toxin and wheat germ conjugated horseradish peroxidase tracing, and selective D-[^{3}H]aspartate labelling of possible excitatory amino acid inputs. Neuroscience 1990;37:77–100.

Berman AL: The Brain Stem of the Cat: A Cytoarchitectonic Atlas with Stereotaxic Coordinates. Madison, University of Wisconsin Press, 1968.

Dahlström A, Fuxe K: Evidence for the existence of monoamine neurons in the central nervous system. I. Demonstration of monoamines in the cell bodies of brain stem neurons. Acta Physiol Scand 1964;62:1–55.

Di Chiara G: Role of dopamine in the behavioural actions of nicotine related to addiction. Eur J Pharmacol 2000;393:295–314.

Ferreira JG, Del-Fava F, Hasue RH, Shammah-Lagnado SJ: Organization of ventral tegmental area projections to the ventral tegmental area-nigral complex in the rat. Neuroscience 2008;153:196–213.

Halliday G, Reyes S, Double K: Substania nigra, ventral tegmental area and retrorubral fields; in Mai JK, Paxinos G (eds): The Human Nervous System, ed 3. Amsterdam, Academic Press, 2012.

Holstege G: Some anatomical observations on the projections from the hypothalamus to brainstem and spinal cord: an HRP and autoradiographic tracing study in the cat. J Comp Neurol 1987;260:98–126.

Hornung J-P: Raphe nuclei; in Mai JK, Paxinos G (eds): The Human Nervous System, ed 3. Amsterdam, Academic Press, 2012.

Hornung JP, Fritschy JM: Serotoninergic system in the brainstem of the marmoset: a combined immunocytochemical and three-dimensional reconstruction study. J Comp Neurol 1988;270:471–487.

Ikemoto S: Dopamine reward circuitry: two projection systems from the ventral midbrain to the nucleus accumbens-olfactory tubercle complex. Brain Res Rev 2007;56:27–78.

Imai H, Steindler DA, Kitai ST: The organization of divergent axonal projections from the midbrain raphe nuclei in the rat. J Comp Neurol 1986;243:363–380.

Jacobs BL, Azmitia EC: Structure and function of the brain serotonin system. Physiol Rev 1992;72:165–229.

Nieuwenhuys R, Voogd J, van Huijzen C: The human central nervous system. Berlin, Springer, 2008.

Paxinos G, Huang X-F, Sengul G, Watson C: Organization of the brainstem; in Mai JK, Paxinos G (eds): The Human Nervous System, ed 3. Amsterdam, Academic Press, 2012.

Pearson J, Halliday G, Sakamoto N, Michel J-P: Catecholaminergic neurons; in Paxinos G (ed): The Human Nervous System, ed 1. San Diego, Academic Press, 1990.

Pierce RC, Kumaresan V: The mesolimbic dopamine system: the final common pathway for the reinforcing effect of drug of abuse? Neurosci Biobehav Rev 2006;30:215–238.

Schultz W: Predictive reward signal of dopamine neurons. J Neurophysiol 1998;80:1–27.

Taber E: The cytoarchitecture of the brain stem of the cat. I. Brain stem nuclei of the cat. J Comp Neurol 1961;116:27–70.

76 Rostral Linear Nucleus (RLi)

Original name:
Nucleus intracapsularis

Alternative name:
A nucleus of the ventral tegmental area of Tsai

Location and Cytoarchitecture – Original Text

Immediately ventrolateral to the interstitial nucleus of Cajal in the oral mesencephalon, large pigmented cells appear among the fibers of the capsule of the red nucleus (RN). These cells, which may form small compact groups, were delineated as nucleus intracapsularis [but in the new edition are consigned to the rostral linear nucleus (RLi) on account of the pigmented character of the cells, typical of the dopaminergic A10 group associated with the ventral tegmental area (VTA). The RLi extends further rostrally from plate 38 and 39 forming a shell around the medial RN and remaining close to the midline. At caudal levels, the RLi may lie adjacent to parts of the magnocellular component of the red nucleus (RNm) in humans, but unlike the RLi, RNm cells [Onodera and Hicks, 2009] are not pigmented]. Occasionally, RLi-like cells are found among the fibers of the capsule of the RN lateral to the oculomotor complex. These cells extend ventrocaudally towards the dorsal pole of the parabrachial pigmented nucleus (PBP). Thus, the cell group may represent an orodorsal extension of the PBP [Pearson et al., 1990; Koutcherov et al., 2004].

The most conspicuous cells of the RLi are large multipolar cells which possess a centrally placed nucleus and distinct, darkly stained, evenly distributed Nissl granules. Many of these cells also contain intracytoplasmic melanin pigment (fig. 1–3). With the exception of their larger size, these cells resemble the large cells of the PBP. Lightly stained, small or medium-sized, oval or triangular neurons may be found scattered among these larger cells.

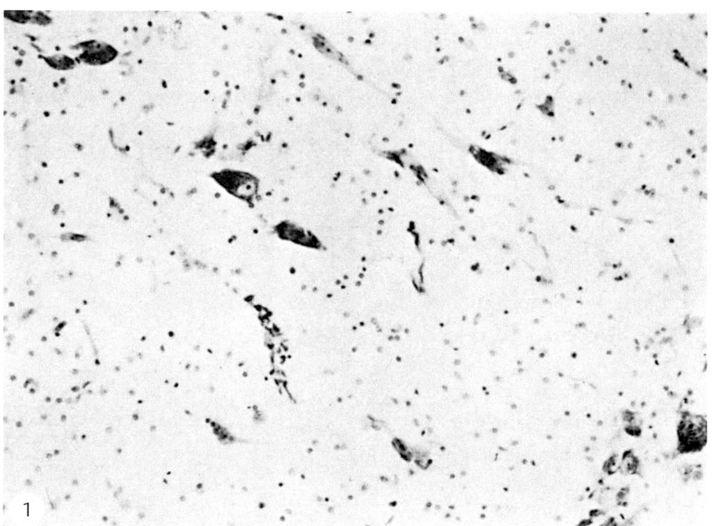

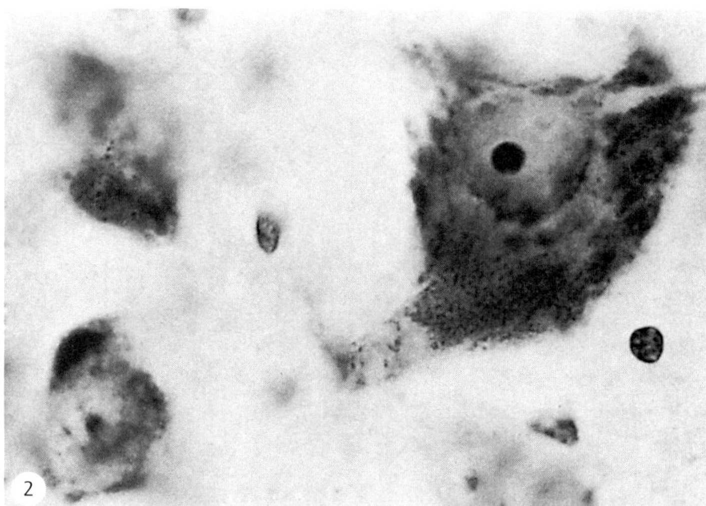

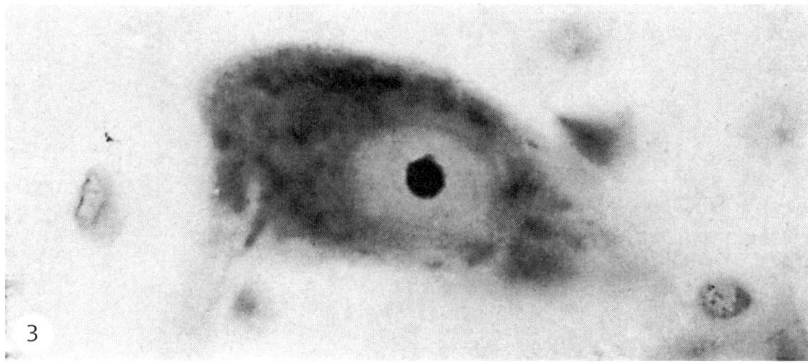

Fig. 1. Rostral linear nucleus. Magnification ×150.
Fig. 2, 3. Cells from the rostral linear nucleus. Note the fine grains of pigment in the cells. Magnification ×1,000.

Functional Neuroanatomy

Function

The RLi is a midline nucleus of the VTA and part of the 'dorsolateral tier' [Halliday et al., 2012]. These VTA nuclei form a mesocorticolimbic dopamine system, whose neuromodulatory functions are related to the reward system, motivation and cognition (see 'Neuromodulatory Systems: Overview'). The diffuse dopaminergic projections of the RLi contribute mainly to the mesocortical projection, considered important for the regulation of movement and modulation of cognitive functions [McRitchie et al., 1997; Williams and Goldman-Rakic, 1998; for a review, see Halliday et al., 2012].

Connections

Afferents. The connectivity of the dopaminergic cell groups of the VTA differs considerably between species, but usually the projections are reciprocal. In the human, there is very little evidence of direct inputs from cortical regions [Frankle et al., 2006], but a significant projection to the RLi arises from the midbrain cholinergic nuclei – the pedunculopontine nucleus and the laterodorsal tegmental nucleus [Oakman et al., 1995; Garzon et al., 1999]. Additional afferents arise from the locus coeruleus, the raphe nuclei, hypothalamus, nucleus accumbens and other limbic regions [Halliday et al., 2012].

Efferents. The efferents of the dorsolateral tier of the VTA, which include RLi and PBP, target mainly the frontal cortex in primates, specifically the prelimbic and infralimbic areas, and to a lesser extent the dorsal frontal cortex (areas 46, BB/6M and 4) [Williams and Goldman-Rakic, 1998].

The RLi is allotted to the A10 group of dopaminergic neurons according to Dahlström and Fuxe [1964], but the group forms a continuum with A8 (retrorubral fields) and A9, the substantia nigra [Dahlström and Fuxe, 1964]. Many cells in the VTA region are nondopaminergic; in rodents most of the mesocortical projection cells are GABAergic; however, recent evidence suggests that from the RLi the mesocortical pathway uses glutamate with a dopaminergic component [Margolis et al., 2006; Olson and Nestler, 2007; Gorelova et al., 2012].

References

Dahlström A, Fuxe K: Evidence for the existence of monoamine neurons in the central nervous system. I. Demonstration of monoamines in the cell bodies of brain stem neurons. Acta Physiol Scand 1964;62:1–55.

Frankle WG, Laruelle M, Haber SN: Prefrontal cortical projections to the midbrain in primates: evidence for a sparse connection. Neuropsychopharmacology 2006;31:1627–1636.

Garzon M, Vaughan RA, Uhl GR, Kuhar MJ, Pickel VM: Cholinergic axon terminals in the ventral tegmental area target a subpopulation of neurons expressing low levels of the dopamine transporter. J Comp Neurol 1999;410:197–210.

Gorelova N, Mulholland PJ, Chandler LJ, Seamans JK: The glutamatergic component of the mesocortical pathway emanating from different subregions of the ventral midbrain. Cereb Cortex 2012;22:327–336.

Halliday G, Reyes S, Double K: Substania nigra, ventral tegmental area and retrorubral fields; in Mai JK, Paxinos G (eds): The Human Nervous System, ed 3. Amsterdam, Academic Press, 2012.

Koutcherov Y, Huang X-F, Halliday G, Paxinos G: Organization of human brain stem nuclei; in Paxinos G, Mai JK (eds): The Human Nervous System, ed 2. Amsterdam, Elsevier Academic Press, 2004.

Margolis EB, Lock H, Hjelmstad GO, Fields HL: The ventral tegmental area revisited: is there an electrophysiological marker for dopaminergic neurons? J Physiol 2006;577:907–924.

McRitchie DA, Cartwright HR, Halliday GM: Specific A10 dopaminergic nuclei in the midbrain degenerate in Parkinson's disease. Exp Neurol 1997;144:202–213.

Oakman SA, Faris PL, Kerr PE, Cozzari C, Hartman BK: Distribution of pontomesencephalic cholinergic neurons projecting to substantia nigra differs significantly from those projecting to ventral tegmental area. J Neurosci 1995;15;5859–5869.

Olson VG, Nestler EJ: Topographical organization of GABAergic neurons within the ventral tegmental area of the rat. Synapse 2007;61:87–95.

Onodera S, Hicks TP: A comparative neuroanatomical study of the red nucleus of the cat, macaque and human. PLoS One 2009;4;e6623.

Pearson J, Halliday G, Sakamoto N, Michel J-P: Catacholaminergic neurons; in Paxinos G (ed): The Human Nervous System, ed 1. San Diego, Academic Press, 1990.

Williams SM, Goldman-Rakic PS: Widespread origin of the primate mesofrontal dopamine system. Cereb Cortex 1998;8;321–345.

Neuromodulatory Systems – Dopaminergic Nuclei (Plates 32, 34–36 and 38)

77 Paranigral Nucleus (PNg)

Original name:
Nucleus paranigralis

Associated name:
A nucleus of the ventral tegmental area

Location and Cytoarchitecture – Original Text

This nucleus comprises groups of pigmented cells which are associated with the medial extremity of the substantia nigra (SN). It extends from the level of the caudal pole of the SN to the level of the most oral rootlets of the third cranial nerve (NIII), a distance of approximately 5 mm.

On cross-section, the paranigral nucleus (PNg) appears as an irregularly triangular area with its base directed towards the SN and the apex directed medially. The caudal part of the nucleus is related medially to the interpeduncular nucleus (IPN), ventrally to the subpial layers of the interpeduncular fossa, laterally to the SN and dorsally to the parabrachial pigmented nucleus (PBP). A narrow bridge of cells passing dorsally to the IPN connects the PNg with one another at the level of the caudal pole of the oculomotor complex. The oral part of the PNg is related laterally to the SN and in all other directions to the posterior perforated substance. Numerous rootlets of the NIII traverse the PNg at this level.

The cells of the PNg are small to medium sized, slender and multipolar in outline and possess long dendrites. The majority contain a varying amount of melanin in their cytoplasm. The cells are irregularly oriented and in most regions are uniformly distributed (fig. 1–6). In some areas they tend to congregate into small clusters. [The PNg can be clearly differentiated from the adjacent SN pars compacta (SNc) by its smaller and densely packed cells [Halliday and Törk, 1986].]

Functional Neuroanatomy

Transmitters

The PNg is part of the ventral tegmental area (VTA). The network of dopaminergic cells in the VTA corresponds to cell group A10 in the rodent classification of catecholamine cell groups [Dahlström and Fuxe, 1964]. These cells are pigmented and are visible without immunohistochemical stains. In 'dopamine-stained' (tyrosine hydroxylase) material, the PNg stands out clearly through its intense labeling [Pearson et al., 1983, 1990]. Calbindin and GIRK2 are also prominent in the PNg [Reyes et al., 2012].

Amongst the dopaminergic neurons of the PNg and PBP nonpigmented neurons have also been identified [Bogerts et al., 1983; Gasper et al., 1983]. Approximately 70–75% of the neu-

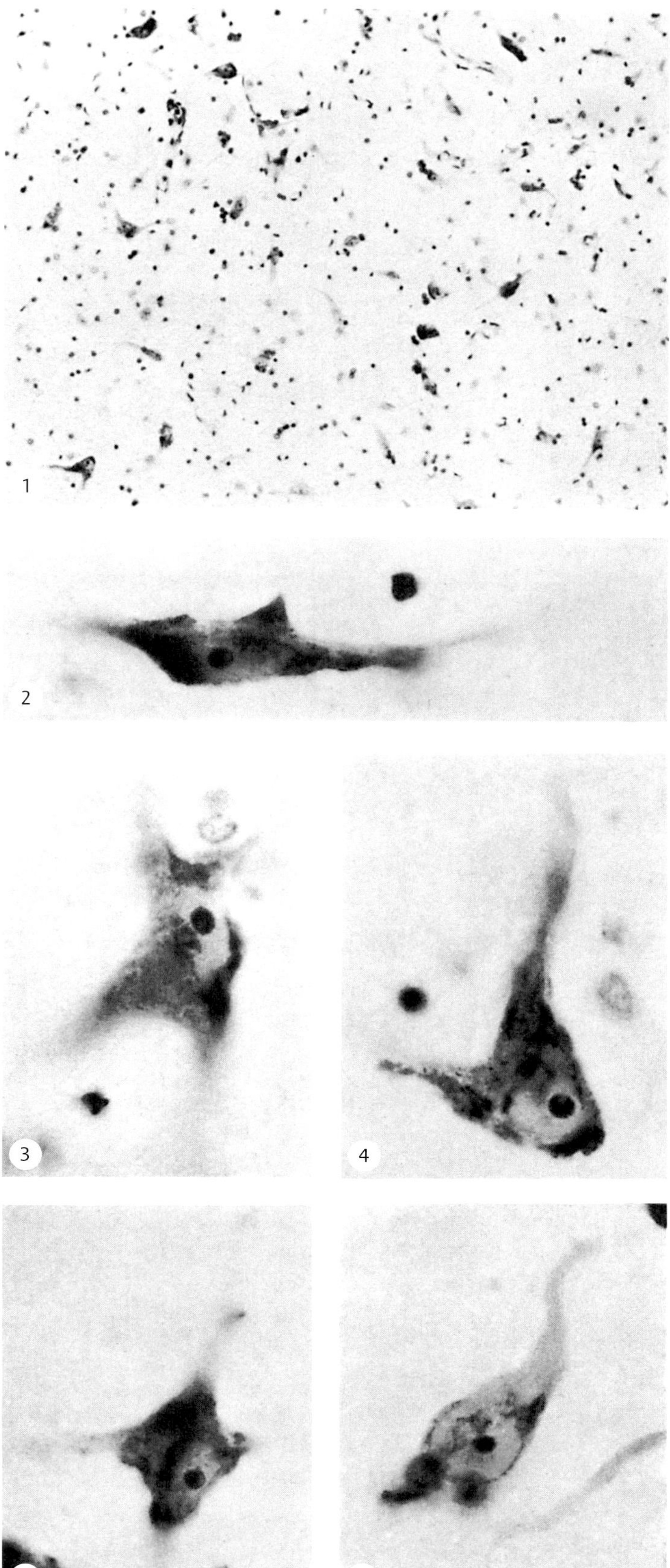

Fig. 1. Paranigral nucleus. Magnification ×150.
Fig. 2–6. Cells from the paranigral nucleus. Note that the cells in figures 2 and 5 contain pigment. Magnification ×1,000.

rons within the PNg are dopaminergic [McRitchie et al., 1997]. In contrast to other nuclei of the VTA, the expression of the peptide cholecystokinin and neurotensin, which affects the depolarization-induced release of dopamine, is rather low in the PNg [Yamada et al., 1995]. However, injections of neurotensin in the rostral PNg (rat) lead to behavioral hyperactivity [Kalivas, 1984]. The PNg contains medium-sized neurons (20–25 μm) expressing nicotinamide adenine dinucleotide phosphate (NADPH)-diaphorase reactivity as one of the few positive cell groups in the brainstem [Kowall and Mueller, 1988].

Function

The VTA has been implicated in motivation and 'reward behaviors', leading to the acquisition of 'learned appetitive behaviors', which include the addictive power of drugs such as opioids, psychostimulants, nicotine, and alcohol [Schultz, 1998; Wise, 2004]. The dopaminergic neural network plays a critical role in these behaviors but the individual contribution of the PNg is not clear.

As part of the dopaminergic system it is surprising that the PNg is only minimally affected in Parkinson's disease showing no accumulation of Lewy bodies or cell loss [McRitchie et al., 1997; Braak and Del Tredici, 2009]. The PNg can be affected in Alzheimer's disease [German et al., 1987].

Connections

Efferents. The PNg is part of the VTA ventromedial tier and provides an ascending mesolimbic pathway which is dopaminergic, and a mesothalamic one which is both dopaminergic and GABAergic [Halliday et al., 2012]. The mesolimbic projections target mainly the ventromedial striatum and ventral pallidum, as well as the olfactory tubercle, the septum, amygdala, entorhinal cortex and hippocampus [Szabo, 1980]. These ascending projections from the VTA showed some topography [Klitenick et al., 1992].

Dopaminergic fibers of the PNg, part of A10, and the adjacent A11 regions descend to innervate the limbic regions of the dorsal raphe nucleus, the ventral periaqueductal gray, the parabrachial nuclei and locus coeruleus.

Afferents. The components of the ventromedial tier of the VTA (PNg, interfascicular nucleus, and caudal linear nucleus, CLi) are interconnected with each other. In addition, especially the PNg and CLi innervate the lateral VTA, and, to a lesser degree, the SNc and retrorubral fields [Ferreira et al., 2008]. Through these projections the PNg could act as a distributer, relaying input signals to other regions of the VTA, for example the specific PNg afferents from the medial part of the lateral habenular nucleus [Goncalves et al., 2012].

References

Bogerts B, Hantsch J, Herzer M: A morphometric study of the dopamine-containing cell groups in the mesencephalon of normals, Parkinson patients, and schizophrenics. Biol Psychiatry 1983;18:951–969.

Braak H, del Tredici K: Neuroanatomy and Pathology of Sporadic Parkinson's Disease. Berlin, Springer, 2009.

Dahlström A, Fuxe K: Evidence for the existence of monoamine neurons in the central nervous system. I. Demonstration of monoamines in the cell bodies of brain stem neurons. Acta Physiol Scand 1964;62:1–55.

Ferreira JG, Del-Fava F, Hasue RH, Shammah-Lagnado SJ: Organization of ventral tegmental area projections to the ventral tegmental area-nigral complex in the rat. Neuroscience 2008;153:196–213.

Gasper P, Gay BBM, Hamon M, Cesselin F, Vigny A, Javoy-Agid F, Agid Y: Tyrosine hydroxylase and methionine-enkephalin in the human mesencephalon. J Neurol Sci 1983;58:247–267.

German DC, White CLR, Sparkman DR: Alzheimer's disease: neurofibrillary tangles in nuclei that project to the cerebral cortex. Neuroscience 1987;21:305–312.

Goncalves L, Sego C, Metzger M: Differential projections from the lateral habenula to the rostromedial tegmental nucleus and ventral tegmental area in the rat. J Comp Neurol 2012;520:1278–1300.

Halliday GM, Törk I: Comparative anatomy of the ventromedial mesencephalic tegmentum in the rat, cat, monkey and human. J Comp Neurol 1986;252:423–445.

Halliday G, Reyes S, Double K: Substania nigra, ventral tegmental area and retrorubral fields; in Mai JK, Paxinos G (eds): The Human Nervous System, ed 3. Amsterdam, Academic Press, 2012.
Kalivas PW: Neurotensin in the ventromedial mesencephalon of the rat: anatomical and functional considerations. J Comp Neurol 1984;226:495–507.
Klitenick MA, Deutch AY, Churchill L, Kalivas PW: Topography and functional role of dopaminergic projections from the ventral mesencephalic tegmentum to the ventral pallidum. Neuroscience 1992;50:371–386.
Kowall NW, Mueller MP: Morphology and distribution of nicotinamide adenine dinucleotide phosphate (reduced form) diaphorase reactive neurons in human brainstem. Neuroscience 1988;26:645–654.
McRitchie DA, Cartwright HR, Halliday GM: Specific A10 dopaminergic nuclei in the midbrain degenerate in Parkinson's disease. Exp Neurol 1997;144:202–213.
Pearson J, Goldstein M, Markey K, Brandeis L: Human brainstem catecholamine neuronal anatomy as indicated by immunocytochemistry with antibodies to tyrosine hydroxylase. Neuroscience 1983;8:3–32.
Pearson J, Halliday G, Sakamoto N, Michel J-P: Catecholaminergic neurons; in Paxinos G (ed): The Human Nervous System, ed 1. San Diego, Academic Press, 1990.
Reyes S, Fu Y, Double K, Thompson L, Kirik D, Paxinos G, Halliday GM: GIRK2 expression in dopamine neurons of the substantia nigra and ventral tegmental area. J Comp Neurol 2012;520:2591–2607.
Schultz W: Predictive reward signal of dopamine neurons. J Neurophysiol 1998;80: 1–27.
Szabo J: Organization of the ascending striatal afferents in monkeys. J Comp Neurol 1980;189:307–321.
Wise RA: Dopamine, learning and motivation. Nat Rev Neurosci 2004;5:483–494.
Yamada M, Yamada M, Richelson E: Heterogeneity of melanized neurons expressing neurotensin receptor messenger RNA in the substantia nigra and the nucleus paranigralis of control and Parkinson's disease brain. Neuroscience 1995; 64:405–417.

Neuromodulatory Systems – Dopaminergic Nuclei (Plates 34–36, 38 and 40)

78 Parabrachial Pigmented Nucleus (PBP)

Original name:
Nucleus parabrachialis pigmentosus

Associated name:
A nucleus of the ventral tegmental area

Location and Cytoarchitecture – Original Text

At a level immediately oral to the decussation of the superior cerebellar peduncle (SCP), before the first cells of the red nucleus appear, a curved band of large pigmented cells lies adjacent to the ventral half of the SCP. These cells are referred to as the parabrachial pigmented nucleus (PBP). On cross-section, the nucleus is horseshoe shaped in outline with its hilus directed dorsally. The hilus is occupied by fibers of the SCP, and the external surface of the 'horseshoe' is related laterally to the substantia nigra (SN), ventrally to the paranigral nucleus (PNg), and medially to the oroventral extension of the dorsal raphe nucleus (DR). A bridge of cells passing ventral to the DR and dorsal to the midline portion of the PNg connects the PBP with the corresponding nucleus of the opposite side.

The majority of cells of the PBP are elongated, multipolar and possess long dendrites and darkly stained, large, evenly distributed Nissl bodies. Most of these cells contain a variable amount of melanin pigment (fig. 1–4). In addition to these large cells, numerous small to medium-sized, fusiform or triangular cells which stain with moderate intensity are present. Many of the medium-sized cells also contain melanin pigment. The long axes of the majority of the elongated cells are directed parallel to the surface of the SCP.

Functional Neuroanatomy

Transmitters
In all laboratory species the PBP is the largest nucleus of the ventral tegmental area (VTA). In the PBP 25–45% of the neurons are pigmented, presumed to be dopaminergic, and correspond to the cell group A10 in rodents [Dahlström and Fuxe, 1964; Oades and Halliday, 1987; McRitchie et al., 1997; Halliday et al., 2012]. In addition there are GABAergic and glutaminergic cells in the region of the PBP, but they tend to lie more caudally and medially, respectively [Yamaguchi et al., 2011]. The latter authors' rigorous analysis of the cell populations of the PBP showed that mainly dopaminergic neurons target the nucleus accumbens (ACC), and glutamate neurons the prefrontal cortex (PFC). A high-level expression of cholecystokinin and neurotensin mRNA is present in many pigmented neurons [Jayaraman et al., 1990].

Function
The dopamine projections from the VTA to the forebrain play important roles in reward, motivation, learning, memory, and movement. This system arises from the A10 region, comprising the VTA. Given the distribution of dopaminergic terminals from the PBP to the ACC and PFC, described above, it is interesting that dopamine release in the ACC has been linked to the importance of 'unconditioned rewards', whereas dopamine release in other structures was implicated in the 'formation of memory' that attaches motivational importance to otherwise neutral stimuli [Wise, 2004]. It seems that the PBP must participate in these processes, but a more specific functional role has not been attributed to it. The ACC is a major part of the 'ventral striatum', which is considered to be the limbic part of the striatum, and present in all mammals [Parent, 1996]. A role of the PBP in the limbic system is undisputed.

Dysfunction of the working memory as in schizophrenia has been associated with dopamine disorders and altered D_1 receptor signaling within the PFC [Goldman-Rakic et al., 2004]. The PBP is affected in Parkinson's disease but not as severely as the subjacent SN pars compacta [McRitchie et al., 1997].

Connections
Efferents. The PBP belongs to the dorsolateral tier of the VTA, which includes the RLi and the region of the VTA immediately

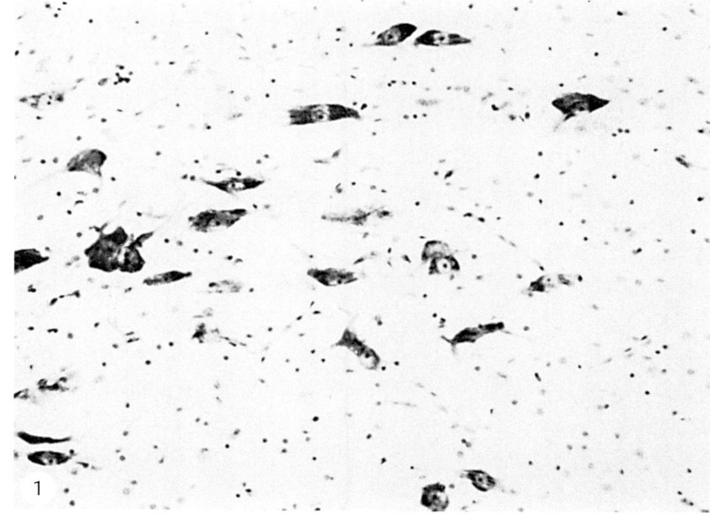

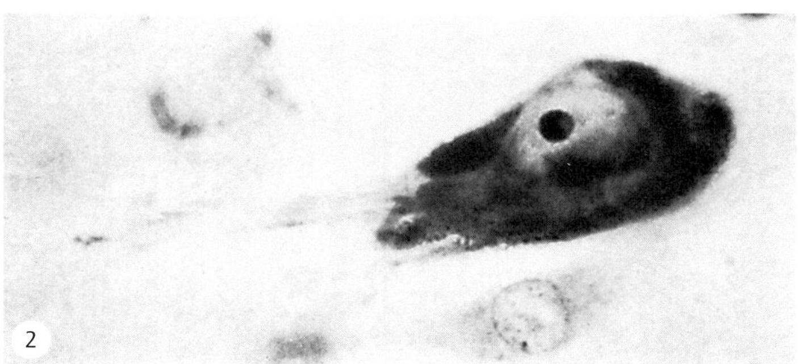

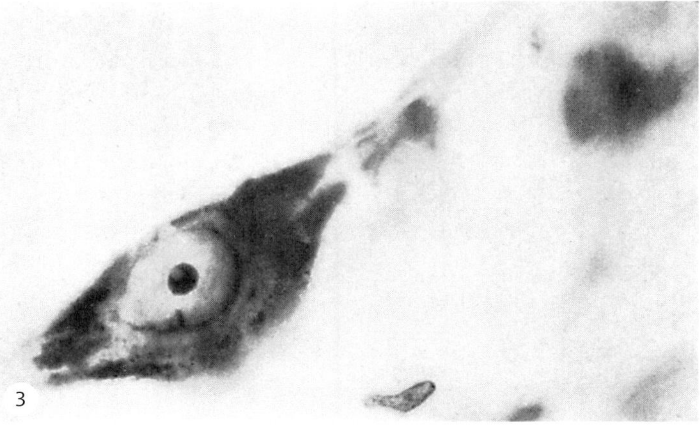

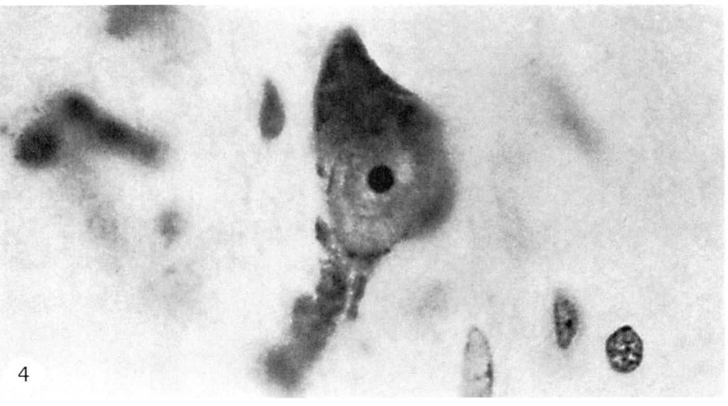

Fig. 1. Parabrachial pigmented nucleus. Magnification ×150.

Fig. 2–4. Cells from the parabrachial pigmented nucleus. Magnification ×1,000.

medial to the PBP, which is sometimes called the ventral tegmental *nucleus* [Halliday et al., 2012]. These regions are interconnected with themselves, the SN pars compacta and also the retrorubral fields, but their connections largely avoid the ventromedial VTA tier (that is the PNg, the caudal linear nucleus and medial SN) [Halliday et al., 2012].

The main projections from the dorsolateral tier are the ascending mesostriatal and mesocortical pathways. For an extensive review of VTA connections, see Oades and Halliday [1987]. The striatal inputs from the PBP were shown in rats to display a topography, where the medial PBP projects to the limbic ventromedial striatum (ACC), and mid PBP to the striatal core and anterolateral to the dorsal striatum [Ikemoto, 2007]. The PBP cortical projections show that those to the PFC are most prominent, but many other regions of the cortex also receive input from the PBP [Scheibner and Törk, 1987]. These inputs contain parallel pathways of dopaminergic, GABAergic and glutamate axons [Yamaguchi et al., 2011]. The thalamic and other limbic projections from the VTA mostly arise from the ventromedial tier, and not the PBP [Halliday et al., 2012].

Afferents. There are reciprocal connections between the PBP and several other neuromodulatory nuclei, for example the noradrenergic locus coeruleus, the serotonergic DR and median raphe nucleus, and the pedunculopontine and laterodorsal tegmental nuclei containing cholinergic and GABAergic cells. Other inputs to the VTA, but not specifically reported for the PBP, originate in the hypothalamus, the ventral (limbic) striatum (but not the somatic striatum), or the cortex (but not the hippocampal complex) [for references, see Halliday et al., 2012].

References

Dahlström A, Fuxe K: Evidence for the existence of monoamine neurons in the central nervous system. I. Demonstration of monoamines in the cell bodies of brain stem neurons. Acta Physiol Scand 1964;62:1–55.

Goldman-Rakic PS, Castner SA, Svensson TH, Siever LJ, Williams GV: Targeting the dopamine D_1 receptor in schizophrenia: insights for cognitive dysfunction. Psychopharmacology (Berl) 2004;174:3–16.

Halliday G, Reyes S, Double K: Substania nigra, ventral tegmental area and retrorubral fields; in Mai JK, Paxinos G (eds): The Human Nervous System, ed 3. Amsterdam, Academic Press, 2012.

Ikemoto S: Dopamine reward circuitry: two projection systems from the ventral midbrain to the nucleus accumbens-olfactory tubercle complex. Brain Res Rev 2007;56:27–78.

Jayaraman A, Nishimori T, Dobner P, Uhl GR: Cholecystokinin and neurotensin mRNAs are differentially expressed in subnuclei of the ventral tegmental area. J Comp Neurol 1990;296:291–302.

McRitchie DA, Cartwright HR, Halliday GM: Specific A10 dopaminergic nuclei in the midbrain degenerate in Parkinson's disease. Exp Neurol 1997;144:202–213.

Oades RD, Halliday GM: Ventral tegmental (A10) system: neurobiology. 1. Anatomy and connectivity. Brain Res 1987;434:117–165.

Parent A: Carpenter's Human Neuroanatomy. Baltimore, Williams & Wilkins, 1996.

Scheibner T, Törk I: Ventromedial mesencephalic tegmental (VMT) projections to ten functionally different cortical areas in the cat: topography and quantitative analysis. J Comp Neurol 1987;259:247–265.

Wise RA: Dopamine, learning and motivation. Nat Rev Neurosci 2004;5:483–494.

Yamaguchi T, Wang HL, Li X, Ng TH, Morales M: Mesocorticolimbic glutamatergic pathway. J Neurosci 2011;31:8476–8490.

79 Retroambiguus Nucleus (RAm)

Original name:
Nucleus retroambigualis

Alternative name:
Nucleus retroambiguus

Location and Cytoarchitecture – Original Text

This nucleus is represented by an oral prolongation of the intermediolateral column of the cervical cord into the medulla. It extends to the level of the caudal pole of the inferior olivary complex.

In cross-section through the caudal part of the medulla, the nucleus lies dorsolaterally in that portion of the ventral gray matter which is isolated by the decussating corticospinal fibers. Oral to the pyramidal decussation, the nucleus lies in the lateral portion of the central nucleus of the medulla oblongata, approximately midway between the ventral pole of the supraspinal nucleus and the ventrolateral aspect of the spinal trigeminal complex. [The nucleus lies within the intermediate reticular zone of Paxinos; see 'Reticular Formation: Overview'.] In some instances the oral extremity of the retroambiguus nucleus (RAm) appears to be directly continuous with the caudal pole of the ambiguus nucleus (AMB).

The RAm is composed of slender, multipolar and fusiform cells containing small, darkly stained Nissl granules. A few of the cells contain melanin pigment (fig. 1–4).

That portion of the nucleus lying oral to the decussation of the pyramids is composed of a group of 15–30 compactly arranged cells. In sections through the decussation the nucleus may be represented by either a fairly compact cell group or merely by a few scattered cells. [...] It appears to us that the cells of the RAm differ distinctly from the motor neurons of the AMB in that they are smaller, more slender and contain finer Nissl granules.

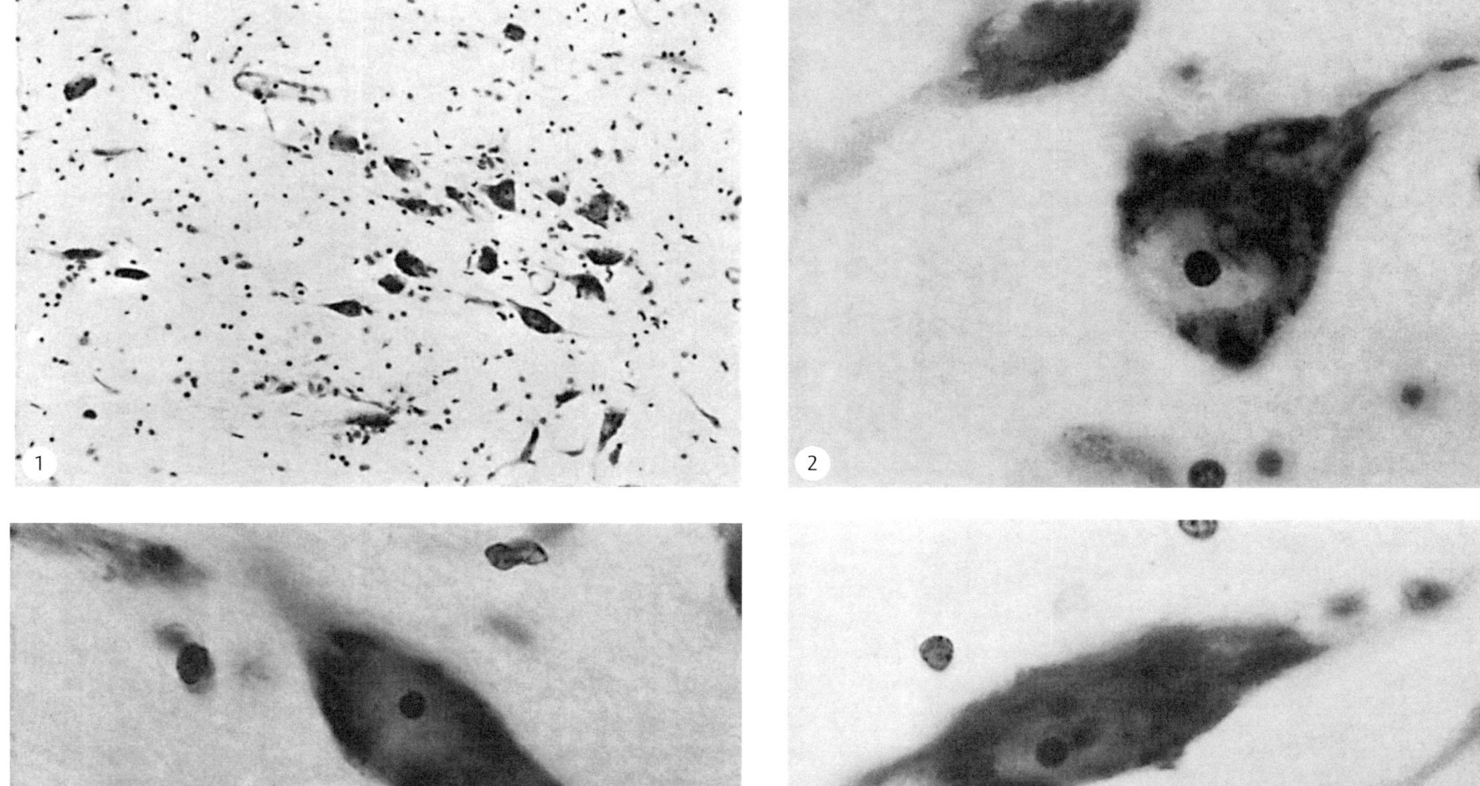

Fig. 1. Retroambiguus nucleus. Magnification ×150.

Fig. 2–4. Cells from the retroambiguus nucleus. Magnification ×1,000.

Functional Neuroanatomy

Function

The RAm is a premotor region of the lateral reticular formation, which participates in highly specific limbic or emotional motor functions: it controls motor behavior such as vocalization (the production of animal-like sounds, not speech), vomiting, mating behavior and the respiratory adjustments associated with these functions. The RAm neurons synapse on motoneurons, and can be allotted to the 'lateral component of the emotional motor system', as opposed to the 'medial component of the emotional motor system' which includes structures like the raphe nuclei with diffuse, or neuromodulatory, projections to motoneurons (see 'Neuromodulatory Systems: Overview') [Holstege et al., 2004].

Connections

Efferents. The RAm projects directly to the adjacent AMB in the lateral medulla oblongata, terminating on the laryngeal and pharyngeal motoneurons essential for vocalization [Holstege, 1989; Boers et al., 2002]. From rostral levels of the RAm, where mainly inspiration-related neurons are found, pathways descend to the contralateral spinal cord and terminate on motoneurons of the diaphragm in C_5–C_6 (the phrenic nucleus), and on motoneurons of the external intercostal (T_1–T_{13}) and abdominal muscles (T_5–L_3) [Merrill, 1970; Holstege and Kuypers, 1982; Long and Duffin, 1986]. Similarly, caudal RAm neurons are associated more with expiration; they terminate on a specific group of motoneurons in the lumbosacral cord innervating the pelvic floor, important for abdominal straining, and the upper leg and hip muscles, essential for lordosis and mating behavior [Holstege and Tan, 1987; Vanderhorst and Holstege, 1995]. Additional pathways from the region of the RAm target the motoneurons of the lower mouth in the facial nucleus and the jaw-opening muscles in the motor trigeminal nucleus. Through these pathways the RAm provides the premotor control for many different patterns of limbic motor activation, including vomiting, coughing and sneezing, vocalization, mating or giving birth, as well as coordinating respiratory muscles with behavior [Gerrits and Holstege, 1996; Subramanian and Holstege, 2009; Wild et al., 2009].

Amazingly in cats, a transient ninefold increase in the number of premotor boutons from RAm neurons onto the lumbosacral motoneurons used in mating behavior has been described [Vanderhorst and Holstege, 1995]. The increase is seen in females only during the estrous cycle, or after the administration of estrogen.

Afferents. An important input to the RAm, controlling its premotor activity, comes from 'the midbrain limbic structure', the periaqueductal gray (PAG), mainly the lateral and ventrolateral parts [Holstege, 1991]. Inputs from other higher limbic regions were markedly absent. The importance of the PAG control over the RAm is emphasized by the fact that urogenital sensory afferents target neurons in the spinal cord which project directly to these regions of the PAG, and thus influence the premotor RAm neurons [Holstege et al., 2004].

Many caudal brainstem structures project to the RAm [Gerrits and Holstege, 1996]: for example, several respiration-related nuclei such as the Bötzinger complex, the pre-Bötzinger complex and periambigual region. These cell groups all lie in the ventrolateral part of the medulla caudal to the facial nucleus (see nucleus No. 53). From the pons, projections to the RAm originate in the region enclosing the ventrolateral parabrachial nucleus, the nucleus Kölliker-Fuse and the retrotrapezoid nucleus. Afferents also arise from the solitary nucleus and two cell groups in the ventral part of the medullary medial reticular formation, one at the level of the facial nucleus and one just rostral to the hypoglossal nucleus.

References

Boers J, Klop EM, Hulshoff AC, de Weerd H, Holstege G: Direct projections from the nucleus retroambiguus to cricothyroid motoneurons in the cat. Neurosci Lett 2002;319:5–8.

Gerrits PO, Holstege G: Pontine and medullary projections to the nucleus retroambiguus: a wheat germ agglutinin-horseradish peroxidase and autoradiographic tracing study in the cat. J Comp Neurol 1996;373:173–185.

Holstege G: Anatomical study of the final common pathway for vocalization in the cat. J Comp Neurol 1989;284:242–252.

Holstege G: Descending motor pathways and the spinal motor system: limbic and non-limbic components. Prog Brain Res 1991;87:307–421.

Holstege G, Kuypers HGJM: The anatomy of brain stem pathways to the spinal cord in cat. A labeled amino acid tracing study. Brain Res 1982;57:145–175.

Holstege G, Tan J: Supraspinal control of motoneurons innervating the striated muscles of the pelvic floor including urethral and anal sphincters in the cat. Brain 1987;110:1323–1344.

Holstege GG, Mouton LJ, Gerrits NM: Emotional motor system; in Paxinos G, Mai JK (eds): The Human Nervous System, ed 2. Amsterdam, Elsevier Academic Press, 2004.

Long S, Duffin J: The neuronal determinants of respiratory rhythm. Prog Neurobiol 1986;27:101–182.

Merrill EG: The lateral respiratory neurones of the medulla: their associations with nucleus ambiguus, nucleus retroambigualis, the spinal accessory nucleus and the spinal cord. Brain Res 1970;24:11–28.

Subramanian HH, Holstege G: The nucleus retroambiguus control of respiration. J Neurosci 2009;29:3824–3832.

Vanderhorst VGJM, Holstege G: Caudal medullary pathways to lumbosacral motoneuronal cell groups in the cat: evidence for direct projections possibly representing the final common pathway for lordosis. J Comp Neurol 1995;359: 457–475.

Wild JM, Kubke MF, Mooney R: Avian nucleus retroambigualis: cell types and projections to other respiratory-vocal nuclei in the brain of the zebra finch *(Taeniopygia guttata).* J Comp Neurol 2009;512:768–783.

80 Medial Parabrachial Nucleus (PBM)

Original name:
Nucleus parabrachialis medialis

Associated nuclei:
Lateral parabrachial nucleus
Kölliker-Fuse nucleus (see nucleus No. 81)
The term parabrachial nuclei includes medial parabrachial nucleus, lateral parabrachial nucleus and Kölliker-Fuse nucleus

Location and Cytoarchitecture – Original Text

This nucleus is located in the lateral part of the oral pontine tegmentum and extends from the level of the oral pole of the superior vestibular nucleus to the level of the trochlear decussation. It appears on cross-section as a vertically directed, broad cell group which lies along the medial surface of the superior cerebellar peduncle (SCP). Caudal to the level at which the decussation of the SCP commences, the cells of the medial parabrachial nucleus (PBM) lie ventral as well as medial to the peduncle. The medial relations of the PBM include the oral pole of the motor trigeminal nucleus, the nucleus subcoeruleus, the locus coeruleus, and the mesencephalic tract and nucleus of the trigeminal complex.

The PBM is composed of densely arranged, small, oval or spindle-shaped cells which stain with moderate intensity (fig. 1–4). The long axes of the majority of the spindle-shaped cells are directed in a dorsomedial direction. An occasional larger cell which may contain melanin may also be present.

Functional Neuroanatomy

Function
The parabrachial nuclei (PB) operate as an interface between medullary regions controlling viscerosensory functions, such as taste, pain, respiration and cardiovascular function, and forebrain structures, which are involved in central autonomic systems. The PBM receives mainly gustatory signals, as opposed to the visceral afferents to the lateral parabrachial nucleus (PBL), and the PBM sends significant ascending pathways to limbic structures. These pathways are considered to participate in complex taste-guided behaviors, like 'conditioned taste aversion' and sodium appetite [Lundy and Norgren, 2004]. The PB has been well studied in rats and mice where many subdivisions have been defined by histochemical methods [Fulwiler and Saper, 1984; Herbert and Saper, 1990], and some of these subdivisions have been associated with specific behavioral patterns [Hashimoto et al., 2009], but in human the subdivisions are not so clear [Paxinos and Huang, 1995; Pritchard and Norgren, 2004; Paxinos et al., 2012].

Radical blockage of PB activity has been shown to prevent the learning of 'conditioned taste aversion' in rats [Ivanova and Bures, 1990]. For a critical review of clinical symptoms associated with lesions of PB see Pritchard [2012].

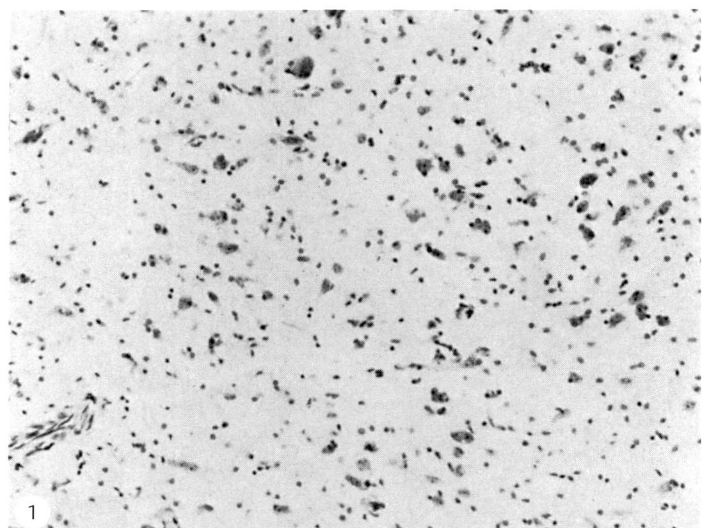

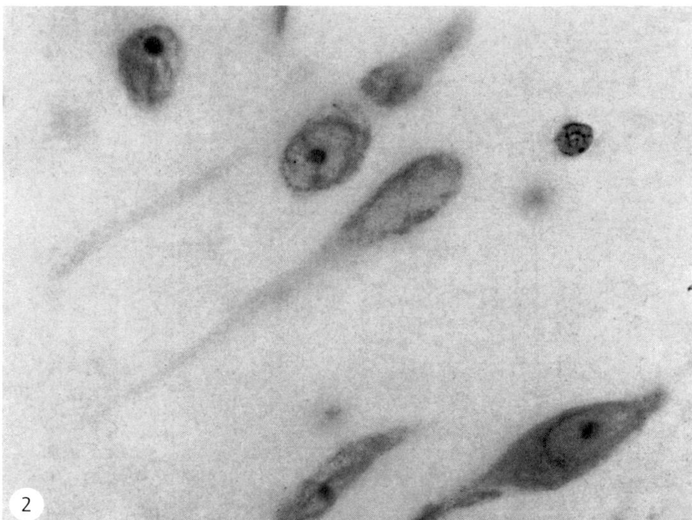

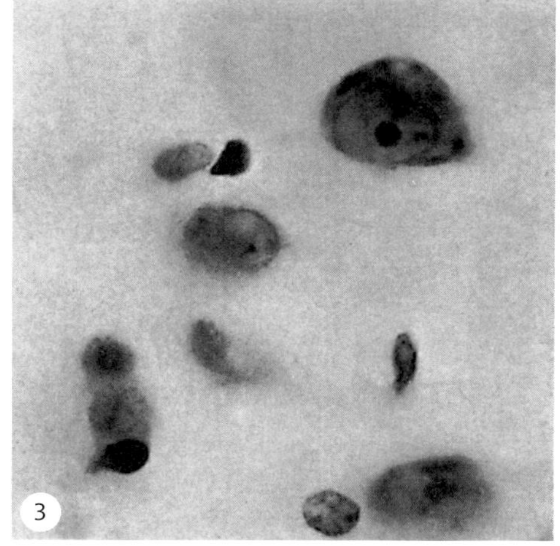

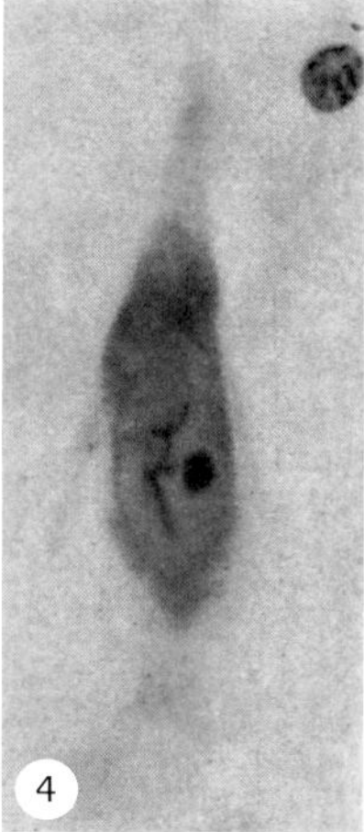

Fig. 1. Medial parabrachial nucleus. Magnification ×150.
Fig. 2–4. Cells from the medial parabrachial nucleus. Magnification ×1,000.

Connections

Afferents. Taste and visceral afferents from the seventh, ninth and tenth cranial nerves enter the medulla and terminate in the solitary nucleus (SOL). The gustatory system is unlike other sensory systems, because the signals ascend from the SOL to the thalamus without crossing the midline; they travel ipsilaterally through the brainstem in the central tegmental tract. The SOL efferents either terminate directly in the thalamus bypassing the PB, or they relay in the PB. The connections of the PBM are similar to those of the PBL (see nucleus No. 81), but with some clear exceptions. The PBM receives inputs from the rostral gustatory division of the SOL, in contrast to the PBL which receives visceral inputs from the caudal division of the SOL.

A variety of neuromodulators are found in both PBM and PBL, for example enkephalin and substance P. Other neuronal populations in PB can be defined by their afferents with stains for catecholamine, cholecystokinin, galanin and corticotrophin-releasing hormone immunoreactivity [Herbert and Saper, 1990; Paxinos et al., 2012].

Efferents. Both PBM and PBL project to the thalamus, more specifically to the medial part of the ventrobasal complex, the parvicellular part of the thalamic posteromedial ventral nucleus, and to the intralaminar and midline nuclei [Norgren, 1995]. In contrast to the PBL, the PBM also projects directly to 4 regions of the cerebral cortex: the granular insula cortex (primary taste cortex), the deep layers of the frontal cortex, the septo-olfactory area and the infralimbic cortex [Saper and Loewy, 1980; Pritchard, 2012].

The ascending pathways from both PBM and PBL also target limbic structures such as the central nucleus of the amygdala, the lateral hypothalamus and the bed nucleus of the stria terminalis. Each of these regions, and the insula, project directly back to PB, in some cases specifically targeting the PBM [Veening et al., 1984; Holstege, 1991; Whitehead et al., 2000]. The stimulation of these limbic forebrain areas modulates the response of PB neurons, and may play an important role in controlling behavior with respect to taste [Lundy and Norgren, 2004; Hashimoto et al., 2009].

The PB projects to the nucleus raphe magnus, in keeping with a possible role in nociception [Holstege, 1988]. It also provides inputs to several medullary nuclei: the ventrolateral medulla involved in cardiovascular and respiratory control, the region of the A1 and A5 catecholamine cell groups which send influences down to thoracic visceromotor preganglionic neurons, the ventrolateral reticular formation and the pontomedullary reticular formation that contains parasympathetic preganglionic neurons [Saper and Loewy, 1980]. Although the ventral PB does not contain a 'respiratory center', it does have influence over respiration, which is reflected in its direct projection to C_5–C_6, terminating on the motoneurons of the phrenic nerve controlling the diaphragm [Goldberg et al., 1990; Holstege, 1991; Dobbins and Feldman, 1994; Blessing, 2004]. The ventral PB and Kölliker-Fuse nucleus also have connections to motoneurons of the jaw closers, perioral musculature, orbicularis oculi and the tongue protruders [Holstege, 1991]. Finally there is a diffuse modulatory projection from ventral PB to the spinal cord targeting mainly the ventral horn [Holstege, 1991]. A final difference between PBL and PBM is that only the PBM projects heavily to the ambiguus nucleus which contains motoneurons controlling the pharynx, larynx and esophagus (nucleus No. 38).

For details of PB connections, see Saper and Loewy [1980], Holstege [1991], Tokita et al. [2009] and Pritchard [2012].

References

Blessing WW: Lower brain stem regulation of visceral, cardiovascular, and respiratory function; in Paxinos G, Mai JK (eds): The Human Nervous System, ed 2. Amsterdam, Elsevier Academic Press, 2004, pp 464–478.

Dobbins EG, Feldman JL: Brain stem network controlling descending drive to phrenic motoneurons in rat. J Comp Neurol 1994;347:64–86.

Fulwiler CE, Saper CB: Subnuclear organization of the efferent connections of the parabrachial nucleus in the rat. Brain Res Rev 1984;7:229–259.

Goldberg JM, Desmadryl G, Baird RA, Fernandez C: The vestibular nerve of the chinchilla. V. Relation between afferent discharge properties and peripheral innervation patterns in the utricular macula. J Neurophysiol 1990;63:791–804.

Hashimoto K, Obata K, Ogawa H: Characterization of parabrachial subnuclei in mice with regard to salt tastants: possible independence of taste relay from visceral processing. Chem Senses 2009;34:253–267.

Herbert H, Saper CB: Cholecystokinin-, galanin-, and corticotropin-releasing factor-like immunoreactive projections from the nucleus of the solitary tract to the parabrachial nucleus in the rat. J Comp Neurol 1990;293:581–598.

Holstege G: Anatomical evidence for a strong ventral parabrachial projection to nucleus raphe magnus and adjacent tegmental field. Brain Res 1988;447:154–158.

Holstege G: Descending motor pathways and the spinal motor system: limbic and non-limbic components. Prog Brain Res 1991;87:307–421.

Ivanova SF, Bures J: Acquisition of conditioned taste aversion in rats is prevented by tetrodotoxin blockade of a small midbrain region centered around the parabrachial nuclei. Physiol Behav 1990;48:543–549.

Lundy RF Jr, Norgren R: Activity in the hypothalamus, amygdala, and cortex generates bilateral and convergent modulation of pontine gustatory neurons. J Neurophysiol 2004;91:1143–1157.

Norgren R: Gustatory system; in Paxinos G (ed): The Rat Nervous System, ed 2. San Diego, Academic Press, 1995, pp 751–771.

Paxinos G, Huang X-F: Atlas of the Human Brainstem. San Diego, Academic Press, 1995.

Paxinos G, Huang X-F, Sengul G, Watson C: Organization of the brainstem; in Mai JK, Paxinos G (eds): The Human Nervous System, ed 3. Amsterdam, Academic Press, 2012, pp 260–327.

Pritchard TC: Gustatory system; in Mai JK, Paxinos G (eds): The Human Nervous System, ed 3. Amsterdam, Academic Press, 2012, pp 1187–1186.

Pritchard TC, Norgren R: Gustatory system; in Paxinos G, Mai JK (eds): The Human Nervous System, ed 2. Amsterdam, Elsevier Academic Press, 2004, pp 1171–1196.

Saper CB, Loewy AD: Efferent connections of the parabrachial nucleus in the rat. Brain Res 1980;197:291–317.

Tokita K, Inoue T, Boughter JD Jr: Afferent connections of the parabrachial nucleus in C57BL/6J mice. Neuroscience 2009;161:475–488.

Veening JG, Swanson LW, Sawchenko PE: The organization of projections from the central nucleus of the amygdala to brainstem sites involved in central autonomic regulation: a combined retrograde transport-immunohistochemical study. Brain Res 1984;303:337–357.

Whitehead MC, Bergula A, Holliday K: Forebrain projections to the rostral nucleus of the solitary tract in the hamster. J Comp Neurol 2000;422:429–447.

81 Lateral Parabrachial Nucleus (PBL)

Original name:
Nucleus parabrachialis lateralis

Associated nuclei:
Kölliker-Fuse nucleus (see below)
Medial parabrachial nucleus

Location and Cytoarchitecture – Original Text

The lateral parabrachial nucleus (PBL) is formed by a narrow band of cells which is closely applied to the lateral and ventrolateral surfaces of the superior cerebellar peduncle (SCP) in the oral part of the pons. It commences caudally approximately 4 mm rostral to the oral pole of the motor trigeminal nucleus (MoV) and extends to the level of the junction of the pons and midbrain. Throughout its extent the nucleus is related laterally to the fibers of the medial and lateral lemnisci.

The PBL increases progressively in size from its caudal pole to the level of the dorsal nucleus of the lateral lemniscus. Oral to this level, the nucleus diminishes rapidly in size and is replaced by the pedunculopontine nucleus.

The PBL is composed of densely arranged cells which are similar to those of the medial parabrachial nucleus (PBM) with the exception that they are smaller and more darkly stained (fig. 1–3).

[Many subdivisions have been defined in the parabrachial complex of the rat by histochemical methods [Fulwiler and Saper, 1984], but in the human the nuclei are less prominent, and here only 4 divisions have been recognized within the PBL: a dorsal, ventral, central and external subnucleus [Paxinos and Huang, 1995]. The PBL external division probably corresponds to the subpeduncular pigmented nucleus described by Ohm and Braak [1987]. Finally the Kölliker-Fuse nucleus (KF), described below, is usually included in the term parabrachial nuclei (PB): it lies ventral to the PBL, SCP and PBM, adjoining all three.]

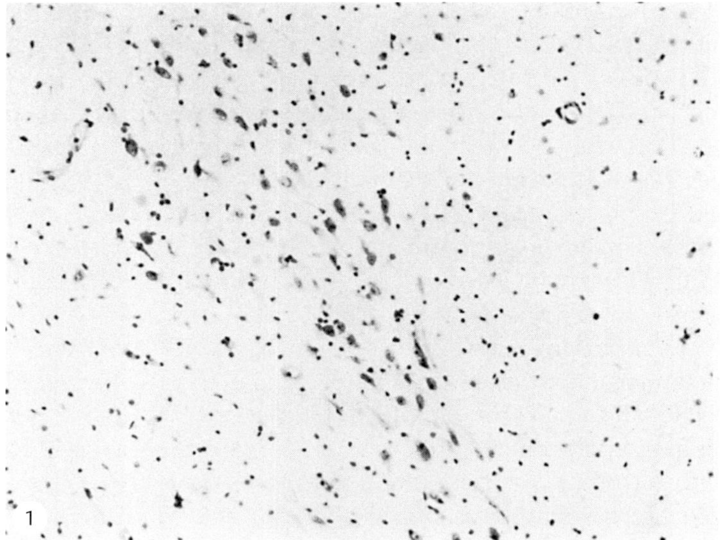

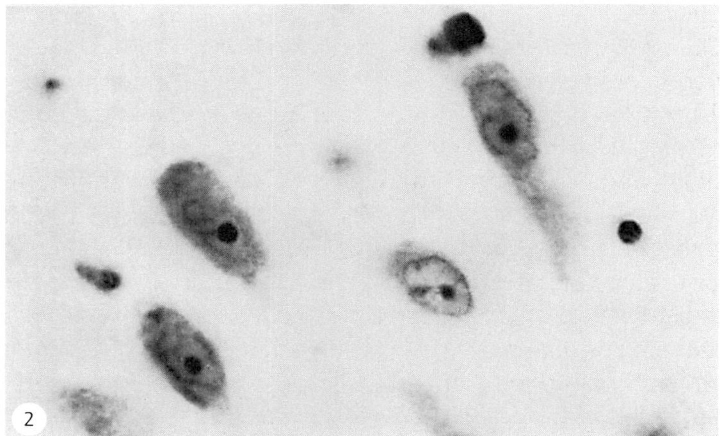

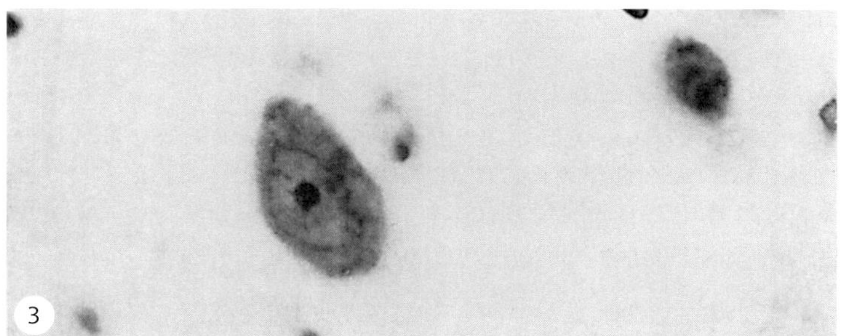

Fig. 1. Lateral parabrachial nucleus. Magnification ×150.
Fig. 2, 3. Cells from the lateral parabrachial nucleus. Magnification ×1,000.

Functional Neuroanatomy

Function

The PB operate as an interface between medullary regions controlling viscerosensory functions, such as taste, pain, respiration and cardiovascular function, and forebrain structures which are involved in central autonomic systems. The PBL receives general visceral sensory input from the caudal solitary nucleus (SOL), and is thought to play a pivotal role in integrating signals for the control of water and food intake (anorexic behavior) [Wu et al., 2012]. Other experiments show that PBL lesions shorten the duration of expiration during hypoxia [Song and Poon, 2009].

Connections

Afferents. Visceral receptors, arterio-, cardio- and pulmonary baroreceptors send excitatory and inhibitory signals to the SOL; these along with serotonergic inputs from the area postrema provide a major component of the afferents to the PB [Lanca and van der Kooy, 1985; Herbert and Saper, 1990]. The caudal SOL sends general visceral information to the PBL, and the PBM receives gustatory inputs from the more rostral SOL. Further, sensory inputs to the PBL arise from the spinal trigeminal nuclei (SpV) and from lamina I neurons of the spinal cord ascending in the spinothalamic tract: surprisingly the PBL receives a much larger projection from the spinothalamic tract than the thalamus itself [Klop et al., 2005]. The PBL also receives many forebrain descending influences: the three major sources are the cerebral cortex, the hypothalamus and the basal forebrain. Their inputs to the PBL are topographically arranged, and described in detail by Moga et al. [1990a, b]. The projections of PB are in general of a reciprocal nature.

Efferents. The PBL projects to the hypothalamus, specifically the lateral hypothalamic area and the paraventricular nucleus, and to the central nucleus of the amygdala and the preoptic area [Fulwiler and Saper, 1984]. Parabrachial neurons also provide the ventroposterior thalamic complex with both visceral and pain inputs, which are relayed to different regions of the insular cortex [Cechetto and Saper, 1987]. As part of its function in controlling viscero- and somatomotor behavior, the ventral part of the PBL and the adjacent area send premotor fibers to motoneurons in C_5–C_6 controlling the phrenic nerve, to tongue, perioral and orbicularis oculi (VII) motoneurons, and to the spinal intermediolateral nucleus of T_1–T_3 containing sympathetic preganglionic neurons [Holstege, 1991]. The PB send diffuse, noradrenergic, neuromodulatory fibers to the spinal cord. A variety of neuromodulators are found in the PBL, including enkephalin and substance P [Herbert and Saper, 1990].

Kölliker-Fuse Nucleus. The ventral parts of the PBL and PBM merge into the KF (plate 26), an area closely related functionally to the ventral parabrachial nuclei, but with different projections. It has been outlined in humans by Paxinos and Huang [1995], and well investigated experimentally in lower animals [for a review, see Dutschmann et al., 2004]. It has been found to mediate respiratory reflexes, and it may modulate respiratory behavior, it plays a role in cardiovascular control [Shafei and Nasimi, 2011], and it responds strongly to noxious stimuli. The nucleus has extensive afferent and efferent connections with autonomic and limbic systems reflecting these functions. For details of the pathways, see Holstege [1991] and Saper and Loewy [1980]. In summary afferent projections to the KF and ventral parabrachial area arise from the SOL, SpV (carrying sensory information from the upper airways), and also the nucleus retroambiguus. All but the latter are reciprocal connections. Ascending projections from the KF target limbic structures such as the lateral hypothalamic area, the lateral preoptic area and the central nucleus of the amygdala. There are dense reciprocal projections with the caudally adjacent 'ventral respiratory group'. The descending projections of the KF terminate on the phrenic motoneurons in C_5–C_6, for control of respiration [Dobbins and Feldman, 1994], and on the sympathetic preganglionic intermediolateral nucleus in T_1–T_3 of the spinal cord. In addition diffuse noradrenergic fibers from the KF to the spinal cord terminate predominantly in the ventral horn. The KF also innervates mouth-closing motoneurons in the MoV, perioral motoneurons in the VII and hypoglossal motoneurons for tongue protrusion.

References

Cechetto DF, Saper CB: Evidence for a viscerotopic sensory representation in the cortex and thalamus in the rat. J Comp Neurol 1987;262:27–45.

Dobbins EG, Feldman JL: Brain stem network controlling descending drive to phrenic motoneurons in rat. J Comp Neurol 1994;347:64–86.

Dutschmann M, Morschel M, Kron M, Herbert H: Development of adaptive behaviour of the respiratory network: implications for the pontine Kolliker-Fuse nucleus. Respir Physiol Neurobiol 2004;143:155–165.

Fulwiler CE, Saper CB: Subnuclear organization of the efferent connections of the parabrachial nucleus in the rat. Brain Res Rev 1984;7:229–259.

Herbert H, Saper CB: Cholecystokinin-, galanin-, and corticotropin-releasing factor-like immunoreactive projections from the nucleus of the solitary tract to the parabrachial nucleus in the rat. J Comp Neurol 1990;293:581–598.

Holstege G: Descending motor pathways and the spinal motor system: limbic and non-limbic components. Prog Brain Res 1991;87:307–421.

Klop EM, Mouton LJ, Hulsebosch R, Boers J, Holstege G: In cat four times as many lamina I neurons project to the parabrachial nuclei and twice as many to the periaqueductal gray as to the thalamus. Neuroscience 2005;134:189–197.

Lanca AJ, van der Kooy D: A serotonin-containing pathway from the area postrema to the parabrachial nucleus in the rat. Neuroscience 1985;14:1117–1126.

Moga MM, Herbert H, Hurley KM, Yasui Y, Gray TS, Saper CB: Organization of cortical, basal forebrain, and hypothalamic afferents to the parabrachial nucleus in the rat. J Comp Neurol 1990a;295:624–661.

Moga MM, Saper CB, Gray TS: Neuropeptide organization of the hypothalamic projection to the parabrachial nucleus in the rat. J Comp Neurol 1990b;295: 662–682.

Ohm T, Braak JW: The pigmented subpeduncular nucleus (nucleus subpeduncularis pigmentosus) in the human brain. Normal morphology and pathological changes in Alzheimer's disease. Neuroscience 1987;22:785.

Paxinos G, Huang X-F: Atlas of the Human Brainstem. San Diego, Academic Press, 1995.

Saper CB, Loewy AD: Efferent connections of the parabrachial nucleus in the rat. Brain Res 1980;197:291–317.

Shafei MN, Nasimi A: Effect of glutamate stimulation of the cuneiform nucleus on cardiovascular regulation in anesthetized rats: role of the pontine Kolliker-Fuse nucleus. Brain Res 2011;1385:135–143.

Song G, Poon CS: Lateral parabrachial nucleus mediates shortening of expiration during hypoxia. Respir Physiol Neurobiol 2009;165:1–8.

Wu Q, Clark MS, Palmiter RD: Deciphering a neuronal circuit that mediates appetite. Nature 2012;483:594–597.

82 Dorsal Tegmental Nucleus (of Gudden) (DTG)

Original name:
Nucleus compactus suprafascicularis

Alternative names:
Dorsal tegmental nucleus
Nucleus tegmentalis posterior
Posterior tegmental nucleus

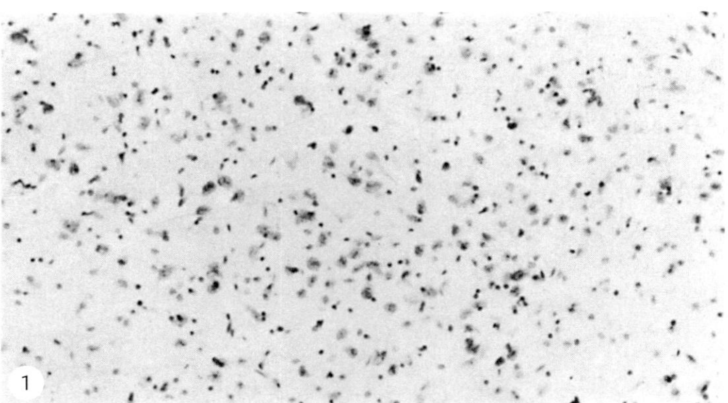

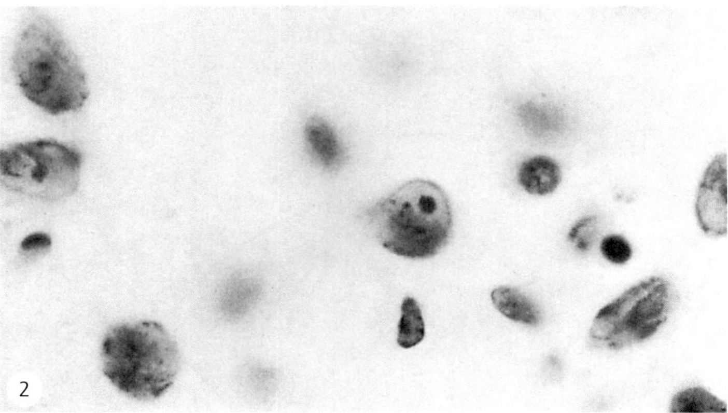

Fig. 1. Dorsal tegmental nucleus. Magnification ×150.
Fig. 2. Cells from the dorsal tegmental nucleus. Magnification ×1,000.

Location and Cytoarchitecture – Original Text

This nucleus is present at the level of the anterior medullary velum, caudal to the trochlear nucleus, and lies in the ventral part of the pontine central gray matter (PCG), directly dorsal to the medial longitudinal fasciculus. On cross-section, it is oval in outline and is related medially to the dorsal raphe nucleus, dorsally and laterally to cells of the PCG and the laterodorsal tegmental nucleus. Caudo-orally, the nucleus measures 4.5 mm.

The dorsal tegmental nucleus (DTG) is composed of compactly arranged, small, round, oval, fusiform or triangular cells which stain with moderate intensity (fig. 1, 2). The nucleus is easily delineated from the surrounding central gray matter by virtue of the compact arrangement and uniform size of its cells.

Functional Neuroanatomy

Function

The DTG and ventral tegmental nucleus of Gudden form two parallel ascending systems associated with lateral and ventral mammillary nuclei, respectively. The DTG contains head direction and angular velocity cells and has been proposed as part of an ascending 'lateral' head direction system [Sharp et al., 2001; Taube, 2007]. The principal circuit for generating these signals originates in the DTG, relays in the lateral mammillary nucleus, the anterodorsal thalamus, the postsubiculum – a parahippocampal structure – and terminates in the entorhinal cortex, which is a major gateway for memory, learning and navigation networks. Other studies have shown that the large population of GABAergic neurons in the DTG is strongly activated in active behavior and REM-like sleep [Torterolo et al., 2002].

Connections

Afferents. Central and ventromedial subdivisions have been described within the DTG of the rat, each with widespread and differential connections [Liu et al., 1984]. The central part is reciprocally connected to its counterpart on the contralateral side, and also receives projections from the lateral mammillary and interpeduncular nuclei, which are the major inputs to the DTG. The central subdivision also receives weaker inputs from the prepositus nucleus and supragenual nucleus, and various other structures which provide the DTG with multimodal information about landmarks, vestibular and self-generated movements. The ventromedial subdivision receives afferents from several limbic forebrain nuclei [Groenewegen and Van Dijk, 1984; Liu et al., 1984].

The DTG is closely associated with the poorly myelinated (and hence poorly visualized) dorsal longitudinal fascicle, or dorsal longitudinal bundle of Schütz. The latter carries visceral information from the posterior hypothalamus to the caudal medulla, and was previously, but mistakenly, thought to use the DTG as a relay station [Nieuwenhuys et al., 2008].

Efferents. The main efferents of DTG neurons target the lateral mammillary nuclei and interpeduncular nucleus, as mentioned above. Weaker projections reach the hypothalamic nuclei and the lateral habenular nucleus [Saunders et al., 2011]. Projections from the DTG to the anterior cingulate cortex – a region involved in vocalization – are found in the squirrel monkey [Jürgens, 1983].

Recent evidence on the neural mechanisms of navigation has indicated that the DTG may be functionally important in the transfer of vestibular and head direction signals to higher centers [Sharp et al., 2001; Taube, 2007]. Head direction cells are found in several areas, but a principal source of this signal originates in the neurons of the DTG.

Histochemical Properties

The DTG in primates is difficult to delineate, but the boundaries can be highlighted by the use of parvalbumin or acetylcholine esterase staining [Petrovický, 1971; Baker et al., 1990; Huang et al., 1992; Saunders et al., 2011]. The nucleus contains a large population of GABAergic cells [Wirtshafter and Stratford, 1993]. It has a dense cellular core which stains weakly for acetylcholine esterase (AChE), and a strong AChE-positive pericentral part [Huang et al., 1992]. The DTG exhibits a strong substance P immunoreactivity confined to the central part [Huang et al., 1992]. In contrast to the adjacent dorsal raphe nucleus, the DTG is completely devoid of serotonergic neurons [Hornung, 2004].

References

Baker KG, Halliday GM, Törk I: Cytoarchitecture of the human dorsal raphe nucleus. J Comp Neurol 1990;301:147–161.

Groenewegen HJ, van Dijk CA: Efferent connections of the dorsal tegmental region in the rat, studied by means of anterograde transport of the lectin *Phaseolus vulgaris* leucoagglutinin (PHA-L). Brain Res 1984;304:367–371.

Hornung JP: Raphe nuclei; in Paxinos G, Mai JK (eds): The Human Nervous System, ed 2. Amsterdam, Elsevier Academic Press, 2004.

Huang X-F, Törk I, Halliday GM, Paxinos G: The dorsal, posterodorsal, and ventral tegmental nuclei: a cyto- and chemoarchitectonic study in the human. J Comp Neurol 1992;318:117–137.

Jürgens U: Afferent fibers to the cingular vocalization region in the squirrel monkey. Exp Neurol 1983;80:395–409.

Liu R, Chang L, Wickern G: The dorsal tegmental nucleus: an axoplasmic transport study. Brain Res 1984;310:123–132.

Nieuwenhuys R, Voogd J, van Huijzen C: The Human Central Nervous System. Berlin, Springer, 2008.

Petrovický P: Structure and incidence of Gudden's tegmental nuclei in some mammals. Acta Anat (Basel) 1971;80:273–286.

Saunders RC, Vann SD, Aggleton JP: Projections from Gudden's tegmental nuclei to the mammillary body region in the cynomolgus monkey *(Macaca fascicularis)*. J Comp Neurol 2012;520:1128–1145.

Sharp PE, Tinkelman A, Cho J: Angular velocity and head direction signals recorded from the dorsal tegmental nucleus of Gudden in the rat: implications for path integration in the head direction cell circuit. Behav Neurosci 2001;115: 571–588.

Taube JS: The head direction signal: origins and sensory-motor integration. Annu Rev Neurosci 2007;30:181–207.

Torterolo P, Sampogna S, Morales FR, Chase MH: Gudden's dorsal tegmental nucleus is activated in carbachol-induced active (REM) sleep and active wakefulness. Brain Res 2002;944:184–189.

Wirtshafter D, Stratford TR: Evidence for GABAergic projections from the tegmental nuclei of Gudden to the mammillary body in the rat. Brain Res 1993;630: 188–194.

Limbic Nuclei

(Plate 27, dashed line)

83 Ventral Tegmental Nucleus (of Gudden) (VTG)

Original name:
Not described

Alternative name:
Deep tegmental nucleus of Gudden

Location and Cytoarchitecture

In Nissl-stained sections of most mammals a ventral tegmental nucleus of Gudden (VTG) can be clearly identified as a compact group of darkly staining cells lying immediately below and embedded within the lateral part of the medial longitudinal fasciculus (MLF) at the level of the rostral pontine reticular formation [Gonzalo-Ruiz et al., 1999], but the VTG is not visible in Nissl sections in primates [Petrovický, 1971; Hayakawa and Zyo, 1983]. A large-celled homolog of the VTG has been reported in humans using acetylcholinesterase staining [Huang et al., 1992], but the exact location was later questioned by the authors [Paxinos et al., 2012]. However, recently parvalbumin staining in primates was shown to be effective in selectively delineating a VTG homolog, as well as the dorsal tegmental nucleus (DTG) lying dorsal to the MLF [Saunders et al., 2011]. The VTG lies slightly further rostral than the DTG, and the authors reported 3 subdivisions within the primate VTG. On the basis of this report, an area is outlined in plate 27, just below the MLF, that may correlate with the VTG in humans.

Functional Neuroanatomy

Function

The cells of the VTG fire rhythmically, and highly coherently, in theta bursts [Kocsis et al., 2001]. The VTG neurons have dense, reciprocal connections with the medial mammillary nucleus, a structure known to be important in memory, and atrophied in amnesic patients with Korsakoff's syndrome [Kopelman, 1995]. Recent experiments imply that the VTG also plays a critical role in some aspects of memory; they showed that in the rat neurotoxic lesions of the VTG (avoiding damage to fibers of passage) impaired spatial memory [Vann, 2009].

Connections

Afferents. Projections to the VTG region in the cat originate from the medial mammillary nuclei, the lateral habenular nucleus, and to a lesser extent from the interpeduncular nucleus (IPN), lateral hypothalamus, DTG, median raphe nucleus and contralateral nucleus reticularis pontis caudalis [Leichnetz et al., 1989].

Efferents. The VTG sends a reciprocal projection back to the medial mammillary and IPN; the pathways in part use GABA and form inhibitory synapses [Petrovický, 1971; Veazey et al., 1982; Hayakawa and Zyo, 1983; Allen and Hopkins, 1990; Jones, 1995; Gonzalo-Ruiz et al., 1999]. Anterograde horseradish peroxidase tracing and autoradiographic findings demonstrate that the VTG gives rise almost exclusively to ascending projections, which largely follow the course of the mammillary peduncle and medial forebrain bundle, or the tegmentopeduncular tract. The majority of fibers terminate in the medial mammillary nuclei, IPN (especially the paramedian subnucleus), the mesencephalic ventral tegmental area, lateral hypothalamus, and the medial septum in the basal forebrain.

References

Allen GV, Hopkins DA: Topography and synaptology of mammillary body projections to the mesencephalon and pons in the rat. J Comp Neurol 1990;301:214–231.

Gonzalo-Ruiz A, Romero JC, Sanz JM, Morte L: Localization of amino acids, neuropeptides and cholinergic neurotransmitter markers in identified projections from the mesencephalic tegmentum to the mammillary nuclei of the rat. J Chem Neuroanat 1999;16:117–133.

Hayakawa T, Zyo K: Comparative cytoarchitectonic study of Gudden's tegmental nuclei in some mammals. J Comp Neurol 1983;216:233–244.

Huang X-F, Törk I, Halliday GM, Paxinos G: The dorsal, posterodorsal, and ventral tegmental nuclei: a cyto- and chemoarchitectonic study in the human. J Comp Neurol 1992;318:117–137.

Jones BE: Reticular formation: Cytoarchitecture, transmitters, and projections; in Paxinos G (ed): The Rat Nervous System, ed 2. San Diego, Academic Press, 1995.

Kocsis B, di Prisco GV, Vertes RP: Theta synchronization in the limbic system: the role of Gudden's tegmental nuclei. Eur J Neurosci 2001;13:381–388.

Kopelman MD: The Korsakoff syndrome. Br J Psychiatry 1995;166:154–173.

Leichnetz GR, Carlton SM, Katayama Y, Gonzalo-Ruiz A, Holstege G, Desalles AA, Hayes RL: Afferent and efferent connections of the cholinoceptive medial pontine reticular formation (region of the ventral tegmental nucleus) in the cat. Brain Res Bull 1989;22:665–688.

Paxinos G, Huang X-F, Sengul G, Watson C: Organization of the brainstem; in Mai JK, Paxinos G (eds): The Human Nervous System, ed 3. Amsterdam, Academic Press, 2012.

Petrovický P: Structure and incidence of Gudden's tegmental nuclei in some mammals. Acta Anat (Basel) 1971;80:273–286.

Saunders RC, Vann SD, Aggleton JP: Projections from Gudden's tegmental nuclei to the mammillary body region in the cynomolgus monkey *(Macaca fascicularis)*. J Comp Neurol, 2012;520:1128–1145.

Vann SD: Gudden's ventral tegmental nucleus is vital for memory: re-evaluating diencephalic inputs for amnesia. Brain 2009;132:2372–2384.

Veazey RB, Amaral DG, Cowan WM: The morphology and connections of the posterior hypothalamus in the cynomolgus monkey *(Macaca fascicularis)*. II. Efferent connection. J Comp Neurol 1982;207:135–156.

Limbic Nuclei

(Plates 24 and 26–29)

84 Pontine Central Gray (PCG)

Original name:
Griseum centrale pontis

Associated name:
Periaqueductal gray caudal zone

Location and Cytoarchitecture – Original Text

The pontine central gray (PCG) appears at the level of the caudal pole of the motor trigeminal nucleus and extends orally to the level of the junction of the pons and mesencephalon, where it is directly continuous with the mesencephalic central gray (PAG). It is represented, on cross-section, by a broad band of cells which lies ventral to and which follows the contours of the floor of the fourth ventricle from the median sulcus to the lateral angle. Dorsally, it is related to the subependymal layer of the fourth ventricle, ventrally to the medial longitudinal fasciculus and to the nuclei of the dorsal pontine tegmentum, like the dorsal tegmental nucleus of Gudden and the laterodorsal tegmental nucleus. At various levels, cell groups of the nucleus reticularis tegmenti pontis, the median raphe nucleus and the mesencephalic trigeminal nucleus appear intermingled with the cells of the central gray matter. The relationship of the cells of the locus coeruleus to the PCG is discussed elsewhere (nucleus No. 69).

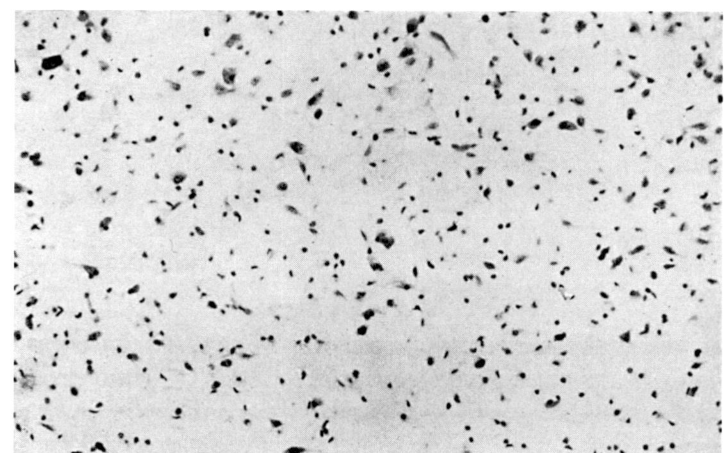

Fig. 1. Pontine central gray. Magnification ×150.

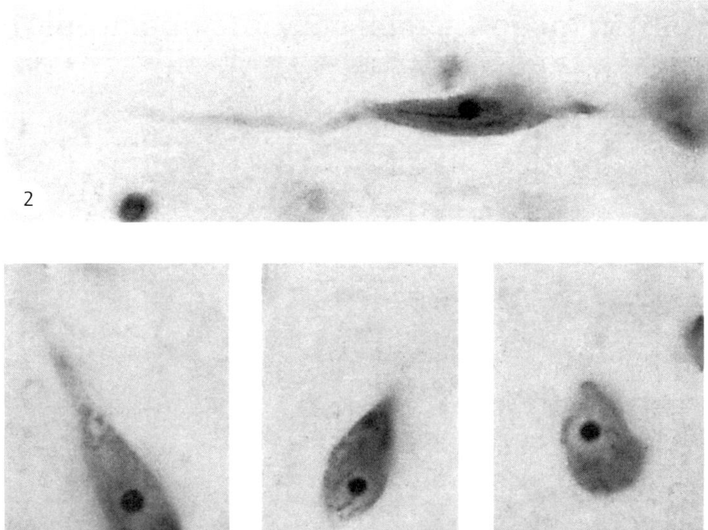

Fig. 2–5. Cells from the pontine central gray. Magnification ×1,000.

The PCG is composed of irregularly oriented, small, oval, triangular or elongated cells which stain with moderate intensity and possess indistinct Nissl granules. In general, the elongated cells stain more darkly and possess longer processes than do the other cells (fig. 1–5).

Functional Neuroanatomy

The term the 'PAG caudal zone' can be used to include the PCG as well as the PAG of the caudal mesencephalon, which are directly continuous with each other. No significant differences between their cytoarchitecture, connectivity or function have been found. For a discussion of functions associated with the 'PAG caudal zone', see the chapter on nucleus No. 85.

Limbic Nuclei

(Plates 30–34, 36, 38–39 and 41)

85 Periaqueductal Gray (PAG)

Original name:
Griseum centrale mesencephali

Alternative names:
Mesencephalic central gray
Substantia grisea centralis
Central gray
Midbrain gray matter

Associated structures:
Pontine central gray
Ependymal layer
Subcommissural circumventricular organ of the posterior commissure

Location and Cytoarchitecture – Original Text

The periaqueductal gray (PAG) is formed by a mantle of cells which envelops the dorsal and lateral aspects of the cerebral aqueduct in the caudal half of the mesencephalon and completely surrounds the aqueduct in the oral half. It is directly continuous caudally with the pontine central gray and merges orally into the subependymal gray matter of the third ventricle. Dorsally, it is related to the inferior and superior colliculi, laterally to the mesencephalic reticular formation and ventrally to the supraoculomotor area including the Edinger-Westphal complex, the interstitial nucleus of Cajal and the nucleus of Darkschewitsch, and at caudal levels the dorsal raphe nucleus.

Variation in cell type and in the density of cell distribution within the PAG is not easily visible in Nissl-stained sections, but allows the differentiation of three subnuclei – a medial, lateral and dorsal subnucleus (plates 34–39) [see also 'Additional Features' below].

The PAG medial subnucleus forms the inner half of the central gray matter and lies immediately adjacent to the subependymal layers of the cerebral aqueduct. It is formed by sparsely distributed, small, elongated, darkly stained cells which possess remarkably long processes. A very occasional darkly stained, medium-sized cell with long processes may also be found within this region (fig. 1–4).

The PAG lateral subnucleus lies dorsal and lateral to the medial zone and forms the outermost portion of the central gray matter. This zone is quite cellular, in marked contrast to the relative acellularity of the medial subnucleus. It is composed of closely congregated, small to medium-sized, plump, fusiform or triangular cells which stain with moderate intensity. A few small, darkly stained elongated cells similar to those of the medial subnucleus may also occur (fig. 8, 9).

The PAG dorsal subnucleus forms a small, oval area which lies immediately dorsal to the cerebral aqueduct in the oral half of the mesencephalon (fig. 3 in the 'Introduction'). It is related dorsally to the lateral subnucleus and on each side is related to the medial subnucleus. The dorsal subnucleus is characterized by a marked accumulation of glial nuclei, among which very occasional neurons may be found. Such neurons are small, elongated and darkly stained (fig. 5–7).

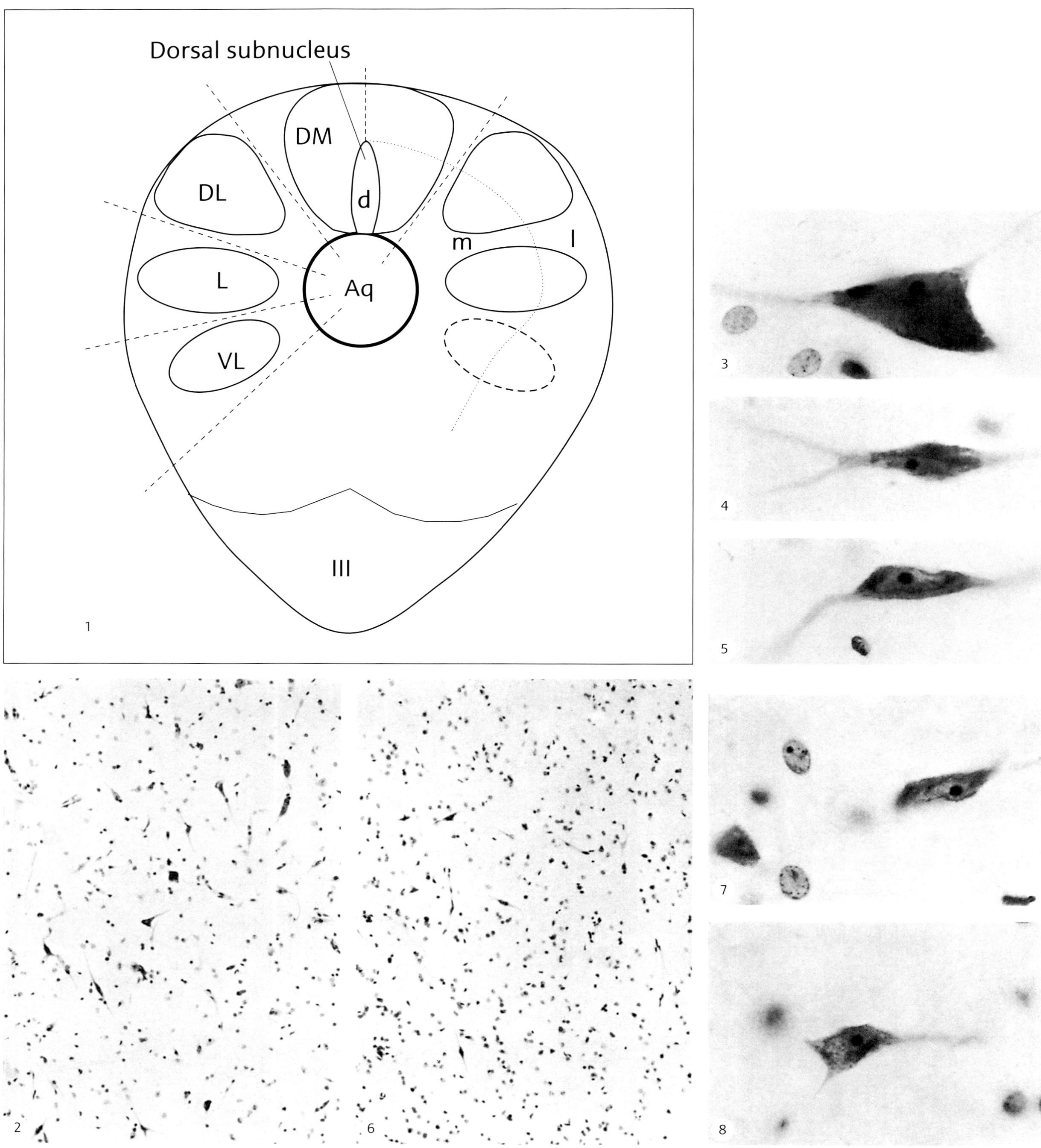

Fig. 1. Diagram showing the subdivisions of Olszewski and Baxter with dotted lines (d = dorsal nucleus; m = medial nucleus; l = lateral nucleus): these are compared with the currently used divisions of the periaqueductal gray, longitudinal columns, which are referred to as the dorsomedial column (DM), the dorsolateral column (DL), the lateral column (L), the dorsal subnucleus and the ventrolateral column (VL), which is represented within dashed lines because it is only present in the caudal PAG. Aq = Aqueduct; III = oculomotor nucleus.

Fig. 2. Periaqueductal gray, medial subnucleus. Magnification ×150.
Fig. 3–5. Cells from the periaqueductal gray, medial subnucleus. Magnification ×1,000.
Fig. 6. Periaqueductal gray, dorsal subnucleus. Magnification ×150.
Fig. 7, 8. Cells from the periaqueductal gray, dorsal subnucleus. Magnification ×1,000.

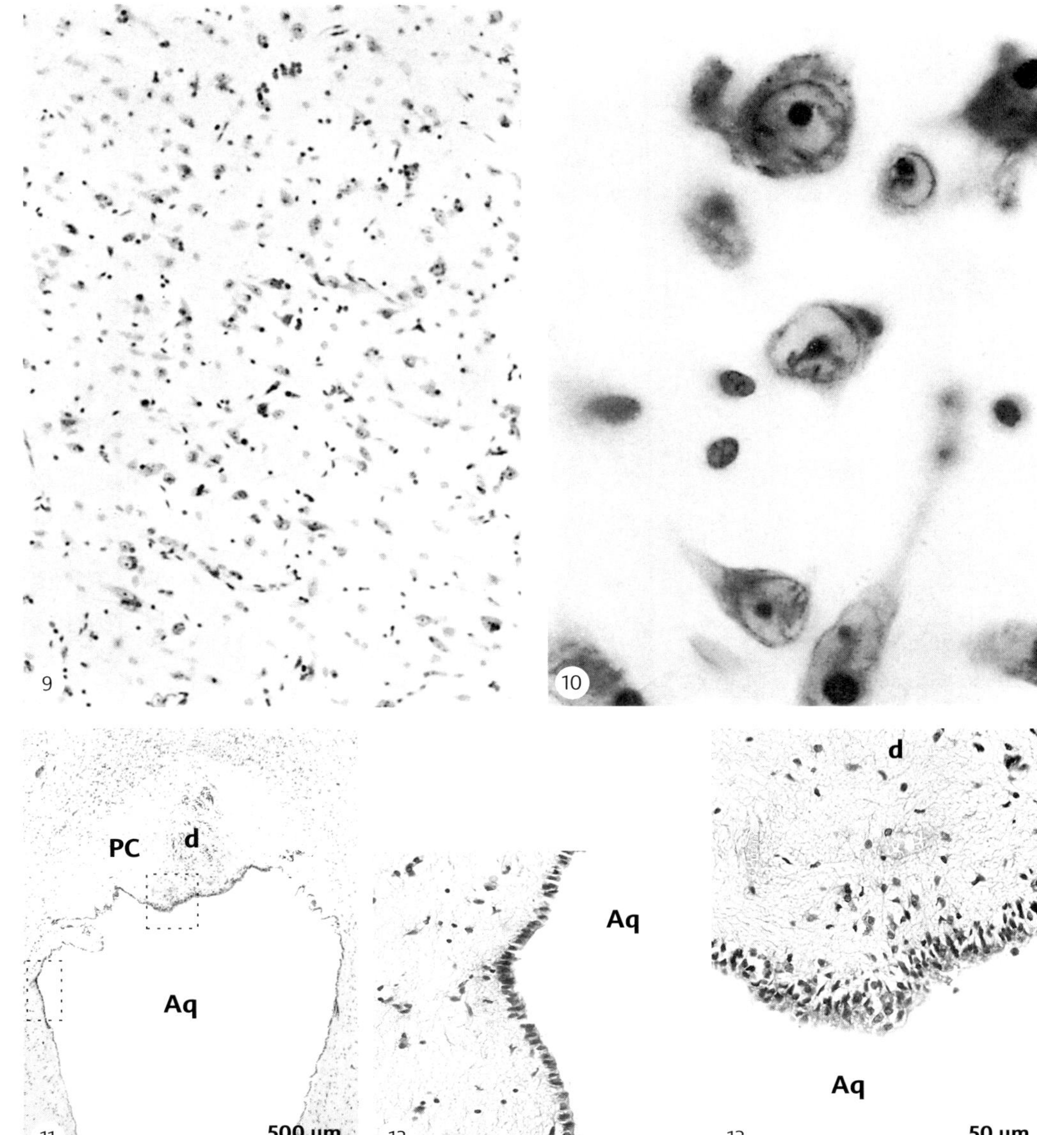

Fig. 9. Periaqueductal gray, lateral subnucleus. Magnification ×150.
Fig. 10. Cells from the periaqueductal gray, lateral subnucleus. Magnification ×1,000.
Fig. 11. Nissl section of the rostral aqueduct (Aq), at the junction with the third ventricle, showing the location of the subcommissural organ in the roof of the aqueduct and its association with the posterior commissure (PC) and the dorsal subnucleus (d). Frames outline enlargements in figures 12 and 13.
Fig. 12. The normal ependymal layer of the aqueduct (Aq) at a higher magnification.
Fig. 13. The specialization of the ependymal layer in the subcommissural organ is evident at a higher magnification. d = Dorsal subnucleus; Aq = aqueduct. Compare with figure 12.

Additional Features

The Ependymal Layer

Throughout the brain and spinal cord, the ventricular walls, including those of the cerebral aqueduct (aqueductus mesencephali), are lined by a single layer of ependymal cells (fig. 12). This epithelium is the innermost layer of the PAG. It is composed of specialized glial cells which are ciliated and joined by desmosomes, gap junctions and maculae adherentes apically, and they permit an interchange between the brain and the cerebrospinal fluid. The subependymal layer (also denoted as the subventricular layer or stratum gliosum subependymale) in the adult contains capillaries, few neurons but several undifferentiated cells that develop into glial cells. The subependymal zone is a source of neural stem cells in embryonic stages of development, but in the adult human brain this property becomes confined to the specific regions of the lateral ventricle and the subgranular layer of the hippocampus dentate gyrus [Alvarez-Buylla et al., 2002, Nieuwenhuys et al., 2008a].

Subcommissural Circumventricular Organ of the Posterior Commissure

At the most rostral tip of the PAG, in the roof of the aqueduct just below the posterior commissure, a section of the ependymal layer is modified to form the subcommissural circumventricular organ of the posterior commissure (fig. 11–13) [McKinley et al., 2004]. As in other circumventricular organs (see area postrema), the cells in this region are columnar, poorly ciliated and have projections into the ventricular lumen, which are covered with microvilli and pinocytotic vesicles on the cell surface. The morphology of circumventricular organs suggests that they are sites at which there is an active interchange of chemical and neural signals between the brain and the cerebrospinal fluid, which are poorly understood, but may amongst other things play an important role for salt and water balance. Furthermore, at these sites there is no blood-brain barrier so that the cells have access to vascular signals. However, the role of circumventricular organs is not clear.

Functional Neuroanatomy

Function

The PAG can be considered to be the most caudal part of the 'limbic brain' [Holstege et al., 2004]. It is an important relay in the emotional motor system, as well as participating in the modulation of sensory pathways, in particular those relaying pain. Many different experimental studies confirm that it is involved in the control of basic survival behaviors, including defense reactions [Carrive, 1993], reproductive behavior and mating posture [Holstege and Georgiadis, 2003], vocalization [Hannig and Jürgens, 2006], micturition [Blok et al., 1997], possibly the act of drinking, cardiovascular regulation [Bandler et al., 1991; Lovick, 1993] and pain modulation [Mason, 1999], but also fear and panic emotions.

Current studies are in agreement that the PAG can be divided up into four radial zones, or columns, on each side of the brain: lateral, ventrolateral, dorsomedial and dorsolateral columns (fig. 10) [Bandler and Shipley, 1994]. Each PAG column has different neural connections and hence different functions; however, it is beyond the scope of this article to describe these in detail. For current reviews, see Carrive and Morgan [2004] and Nieuwenhuys et al. [2008b].

Connections

Neuroanatomical studies have been able to demonstrate some of the pathways essential for the behaviors controlled by the PAG, and furthermore they show that the patterns of connections of the PAG do not vary widely across species. So it is reasonable to assume that the connectivity in humans is probably similar to that reported in higher mammals [for reviews, see Beitz, 1990; Behbehani, 1995; Holstege, 1998]. This constancy between species is supported by histochemical studies of the PAG, in particular the pattern of NADPH diaphorase which stains only the dorsolateral column across species [Carrive and Morgan, 2004].

Differences between Connectivity of Columns

As mentioned above, the connectivity of the individual columns differs. In short, it is striking that the connectivity of the dorsolateral column differs strongly from the other three columns. The dorsomedial, lateral and ventrolateral columns all project directly to the lower brainstem, whereas the dorsolateral does not [Holstege, 1991]. Spinal afferents avoid the dorsolateral column [Wiberg et al., 1987], but structures involved in oculomotor regulation, like the prepositus hypoglossi nucleus, project to this zone [Klop et al., 2005a, 2006]. The lateral and ventrolateral columns have similar though not identical connections to the lower brainstem and spinal cord [Herbert and Saper, 1992]. However, the dorsomedial zone projections to the spinal cord are most unusual [Mouton et al., 1996]. The spinal axonal branches have no synaptic contacts but end in the capillary bed of the subependymal layer around the central canal (C_4–T_8). This connection was intriguingly compared, in terms of function, to the neurosecretory hypothalamohypophysial projections of the hypophysis.

Efferents. The efferents of the dorsomedial, lateral and ventrolateral PAG columns to lower brainstem targets can be divided functionally into two categories: (1) pathways eliciting specific emotional behaviors and (2) modulator (or level setting) pathways. The first category of PAG pathways targets specific premotor and motor centers; for example, the lateral PAG targets the nucleus retroambiguus, which provide the premotor control of the pharynx, larynx and soft palate for vocalization [Holstege, 1989]. In addition the retroambiguus nucleus efferents project to spinal motoneurons which drive the postures essential for mating behavior [Vanderhorst and Holstege, 1997; Vanderhorst et al., 2000]. Projections from the PAG to the subretrofacial nucleus control cardiovascular changes via projections to the preganglionic neurons in the intermediolateral nucleus of the spinal cord. Other PAG afferents target Barrington's nucleus (the pontine micturition center or M region) and activate neurons which in turn project to the sacral column of parasympathetic preganglionic neurons, including those that stimulate the urinary bladder contraction and a separate group which controls the external urethral sphincter. These pathways have been clearly demonstrated by Holstege and colleagues [Holstege, 1990; Holstege et al., 2004].

The second category of PAG efferents projects to 'neural modulator networks', that is to nuclei such as the locus coeruleus, nucleus raphe magnus (RMg) and pallidus (RPa) (see 'Neuromodulatory Systems: Overview'). The PAG-RMg projections activate extensive serotonergic efferents from the RMg which terminate in the dorsal horn throughout the whole spinal cord, contacting pathways in lamina I. In this way, the PAG can modulate ascending sensory transmission, and specifically nociception, via the RMg [Holstege et al., 2004]. The PAG also projects to the RPa, whose neurons, in contrast to those of the RMg, target the *ventral* horn of the spinal cord and modulate motoneuron activity. It is through these descending pathways that the PAG can modulate motor activity in general.

Afferents. The hypothalamus provides the greatest descending input to the PAG, whereby the projections from many hypothalamic areas are reciprocal [Mantyh, 1982]. Other important sources of limbic afferents to the PAG are the prefrontal cortex, the insular cortex, the cingulate cortex, the central and basolateral amygdala, preoptic area and the bed nucleus of the stria terminalis [Price and Amaral, 1981; Holstege, 1990; An et al., 1998].

Inputs to the PAG associated with its sensory functions arise from the spinal cord and the caudal trigeminal nucleus [Wiberg et al., 1987]. The spino-PAG projection is unusual since it originates mainly in lamina I; it is contralateral, and it is much bigger than the well-known spinothalamic projection [Klop et al., 2005b]. Other inputs arise from the superior and inferior colliculi, and several nuclei associated with neuromodulation (see reviews for details).

References

Alvarez-Buylla A, Seri B, Doetsch F: Identification of neural stem cells in the adult vertebrate brain. Brain Res Bull 2002;57:751–758.

An X, Bandler R, Ongur D, Price JL: Prefrontal cortical projections to longitudinal columns in the midbrain periaqueductal gray in macaque monkeys. J Comp Neurol 1998;401:455–479.

Bandler R, Carrive P, Zhang SP: Integration of somatic and autonomic reactions within the midbrain periaqueductal grey: viscerotopic, somatotopic and functional organization. Prog Brain Res 1991;87:269–305.

Bandler R, Shipley MT: Columnar organization in the midbrain periaqueductal gray: modules for emotional expression? Trends Neurosci 1994;17:379–388.

Behbehani MM: Functional characteristics of the midbrain periaqueductal gray. Prog Neurobiol 1995;46:575–605.

Beitz AJ: Central gray; in Paxinos G (ed): The Human Nervous System. San Diego, Academic Press, 1990.

Blok BF, Willemsen AT, Holstege G: A PET study on brain control of micturition in humans. Brain 1997;120:111–121.

Carrive P: The periaqueductal gray and defensive behavior: functional representation and neuronal organization. Behav Brain Res 1993;58:27–47.

Carrive P, Morgan MM: Periaqueductal gray; in Paxinos G, Mai JK (eds): The Human Nervous System, ed 2. Amsterdam, Elsevier Academic Press, 2004.

Hannig S, Jürgens U: Projections of the ventrolateral pontine vocalization area in the squirrel monkey. Exp Brain Res 2006;169:92–105.

Herbert H, Saper CB: Organization of medullary adrenergic and noradrenergic projections to the periaqueductal gray matter in the rat. J Comp Neurol 1992;315:34–52.

Holstege G: Anatomical study of the final common pathway for vocalization in the cat. J Comp Neurol 1989;284:242–252.

Holstege G: Subcortical limbic system projections to caudal brainstem and spinal cord; in Paxinos G (ed): The Human Nervous System. San Diego, Academic Press, 1990.
Holstege G: Descending motor pathways and the spinal motor system: limbic and non-limbic components. Prog Brain Res 1991;87:307–421.
Holstege G: The emotional motor system in relation to the supraspinal control of micturition and mating behavior. Behav Brain Res 1998;92:103–109.
Holstege G, Georgiadis JR: Neurobiology of cat and human sexual behavior. Int Rev Neurobiol 2003;56:213–225.
Holstege GG, Mouton LJ, Gerrits NM: Emotional motor system; in Paxinos G, Mai JK (eds): The Human Nervous System, ed 2. Amsterdam, Elsevier Academic Press, 2004.
Klop EM, Mouton LJ, Ehling T, Holstege G: Two parts of the nucleus prepositus hypoglossi project to two different subdivisions of the dorsolateral periaqueductal gray in cat. J Comp Neurol 2005a;492:303–322.
Klop EM, Mouton LJ, Holstege G: Periparabigeminal and adjoining mesencephalic tegmental field projections to the dorsolateral periaqueductal grey in cat – a possible role for oculomotor input in the defensive system. Eur J Neurosci 2006; 23:2145–2157.
Klop EM, Mouton LJ, Hulsebosch R, Boers J, Holstege G: In cat four times as many lamina I neurons project to the parabrachial nuclei and twice as many to the periaqueductal gray as to the thalamus. Neuroscience 2005b;134:189–197.
Lovick TA: Integrated activity of cardiovascular and pain regulatory systems: role in adaptive behavioural responses. Prog Neurobiol 1993;40:631–644.
Mantyh PW: Forebrain projections to the periaqueductal gray in the monkey, with observations in the cat and rat. J Comp Neurol 1982;206:146–158.
Mason P: Central mechanisms of pain modulation. Curr Opin Neurobiol 1999;9: 436–441.
McKinley MJ, Clarke IJ, Oldfield BJ: Circumventricular organs; in Paxinos G, Mai JK (eds): The Human Nervous System. San Diego, Academic Press, 2004.
Mouton LJ, Kerstens L, Van der Want J, Holstege G: Dorsal border periaqueductal gray neurons project to the area directly adjacent to the central canal ependyma of the C_4–T_8 spinal cord in the cat. Exp Brain Res 1996;112:11–23.
Nieuwenhuys R, Voogd J, Van Huijzen C: Development; in Nieuwenhuys R, Voogd J, Van Huijzen C (eds): The Human Central Nervous System, ed 4. Berlin, Springer, 2008a.
Nieuwenhuys R, Voogd J, Van Huijzen C (eds): The Human Central Nervous System, ed 4. Berlin, Springer, 2008b.
Price JL, Amaral DG: An autoradiographic study of the projections of the central nucleus of the monkey amygdala. J Neurosci 1981;1:1242–1259.
Vanderhorst VG, Holstege G: Estrogen induces axonal outgrowth in the nucleus retroambiguus-lumbosacral motoneuronal pathway in the adult female cat. J Neurosci 1997;17:1122–1136.
Vanderhorst VG, Terasawa E, Ralston HJ, Holstege G: Monosynaptic projections from the lateral periaqueductal gray to the nucleus retroambiguus in the rhesus monkey implications for vocalization and reproductive behavior. J Comp Neurol 2000;424:251–268.
Wiberg M, Westman J, Blomqvist A: Somatosensory projection to the mesencephalon: an anatomical study in the monkey. J Comp Neurol 1987;264:92–117.

Limbic Nuclei

(Plates 30, 32, 34 and 35)

86 Interpeduncular Nucleus (IPN)

Original name:

Nucleus interpeduncularis

Location and Cytoarchitecture – Original Text

This is an unpaired median structure which lies immediately dorsal to the interpeduncular fossa in the caudal half of the mesencephalon. It extends from the level of the caudal pole of the substantia nigra to the caudal pole of the red nucleus, a distance of 7 mm.

On cross-section, the nucleus interpeduncularis (IPN) is roughly semicircular in outline with its base lying immediately dorsal to the subpial layers of the interpeduncular fossa. Throughout the greater part of its extent the nucleus is bordered on its lateral aspects by the paranigral nuclei (PNg). Dorsally, the caudal portion of the nucleus is related to the median raphe nucleus (MnR), whereas the oral portion is related to the median part of the PNg. Orally, the IPN is bordered by the substantia perforata posterior (see nucleus No. 68, fig. 7).

The IPN is composed of very small, elongated, lightly stained, triangular or fusiform cells (fig. 1–4). The subdivision of the nucleus into a medial and a lateral subnucleus, an obvious feature in many animals, is much less distinct in man. The medial subnucleus is present at all levels and is distinguished by the very pale staining and the irregular orientation of its cells. The lateral subnucleus is most obvious in sections through the caudal part of the nucleus and is characterized by the fact that its cells are larger and stain more intensely than those of the central portion. Further, the majority of these cells lie with their long axes directed dorsomedially. [...]

Comments

The IPN is a nonhomogeneous nucleus. Hamill and Lenn [1984] proposed a nomenclature for the subdivisions in the rat, which has been generally accepted. They subdivided the IPN into 4 paired subnuclei (dorsolateral, dorsomedial, intermediate and lateral) and 3 unpaired subnuclei (rostral, central and apical), each subnucleus having a characteristic histochemistry and connectivity [Lenn and Hamill, 1984; Hamill and Lenn, 1984; Hemmendinger and Moore, 1984; Groenewegen et al., 1986]. Although these IPN subdivisions are not so distinct in humans they have been demarcated by Paxinos et al. [2012] and Morley [1986].

Functional Neuroanatomy

Function

The IPN is an important limbic structure, receiving descending pathways from the limbic forebrain, and projecting primarily to the midbrain raphe nuclei and dorsal tegmental area. The raphe nuclei have extremely widespread serotonergic projections in the hippocampus, amygdala, striatum and cerebral cortex, which means that the IPN can modulate the activity in all these structures [Steinbusch, 1981]. The IPN has been associated with a variety of behavioral and visceral functions such as motor activity, sex, nociception, avoidance learning, stress,

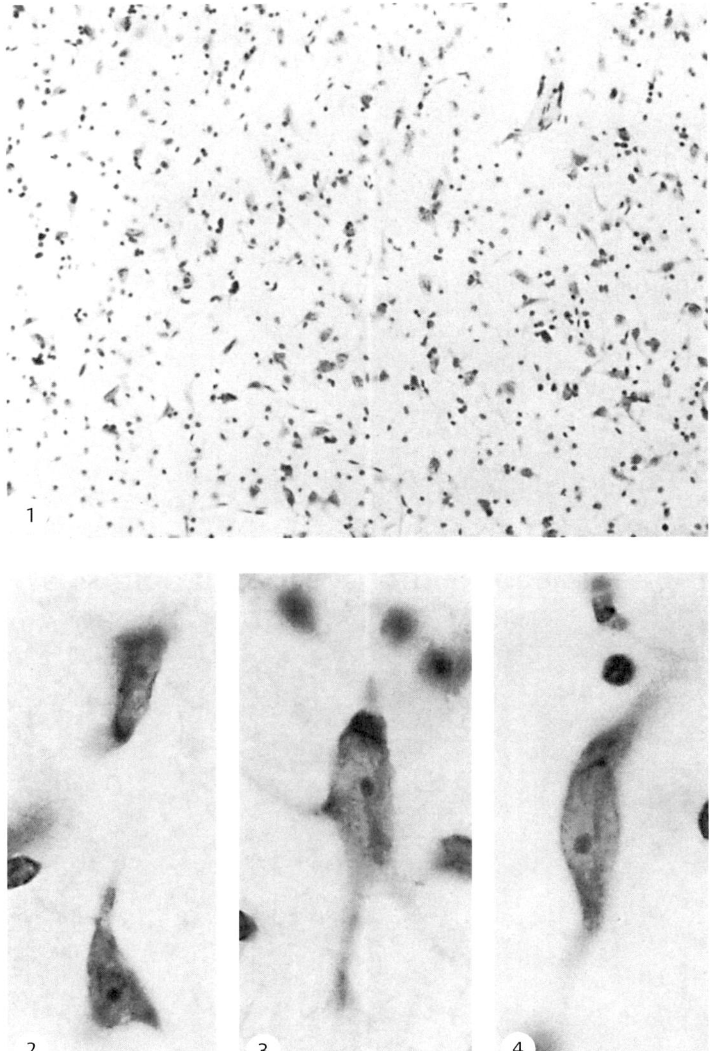

Fig. 1. Interpeduncular nucleus. Magnification ×150.
Fig. 2–4. Cells from the interpeduncular nucleus. Magnification ×1,000.

REM sleep, eating and drinking [Morley, 1986; Klemm, 2004], and recently navigation [Clark and Taube, 2009]. Current hypotheses concerning the IPN propose that it participates in the suppression of motor activity through the serotonergic, and possibly dopaminergic (see below), neuromodulatory systems, when it receives information regarding painful, stressful or aversive events [Hikosaka, 2010].

Connections

Afferents. The main input to the IPN arises from the medial habenular nucleus. The projection runs in the fasciculus or tractus retroflexus, and terminates in all IPN subdivisions except the apical subnucleus [Groenewegen et al., 1986]. The pathway is strongly cholinergic and is thought to be mainly responsible for the intense acetylcholinesterase staining of the IPN; the habenula inputs may also account for the strong substance-P-like immunoreactivity in the IPN. A second important projection to the IPN originates from the dorsal tegmental nucleus (DTG) and the adjacent central gray; all IPN subdivisions are targeted but some more intensely than others. Further afferents arise from the MnR and dorsal raphe nucleus (DR) [Vertes et al., 1999], the parabigeminal region and locus coeruleus, and weaker inputs to the IPN have been reported from forebrain limbic structures, including the mammillary bodies, the medial septum and the diagonal band of Broca [Albanese et al., 1985].

Efferents. The main output pathways from all subdivisions of the IPN travel caudally and terminate close to the midline in the MnR and DR and the dorsal tegmental region, which includes the DTG, the ventral and laterodorsal tegmental nuclei (nuclei No. 71 and 83) and the adjacent central gray matter [Groenewegen et al., 1986]. The serotonergic efferents of the raphe nuclei extend throughout the limbic forebrain and through them the IPN is able to influence a wide range of structures. A comparable and parallel pathway to the 'medial habenula – IPN – raphe nuclei' controlling the serotonergic neuromodulatory system has recently been recognized: the pathway 'lateral habenula – rostromedial tegmental nucleus (RMTg) – ventral tegmental area' similarly targets the dopaminergic (reward-related) and serotonergic systems, and may also contribute to the suppression of motor behavior [Hikosaka, 2010; Holstege, 2009]. The RMTg is a recently described area that lies above and behind the IPN [Jhou et al., 2009], and the same region has also been called the 'tail of the ventral tegmental area' [Kaufling et al., 2009]. Furthermore the reciprocal projections of the IPN with the DTG, and the direct IPN efferents to the hippocampus, provide good evidence that the IPN plays a fundamental role in path integration and navigation [Clark and Taube, 2009].

References

Albanese A, Castagna M, Altavista MC: Cholinergic and non-cholinergic forebrain projections to the interpeduncular nucleus. Brain Res 1985;329:334–339.

Clark BJ, Taube JS: Deficits in landmark navigation and path integration after lesions of the interpeduncular nucleus. Behav Neurosci 2009;123:490–503.

Groenewegen HJ, Ahlenius S, Haber SN, Kowall NW, Nauta WJ: Cytoarchitecture, fiber connections, and some histochemical aspects of the interpeduncular nucleus in the rat. J Comp Neurol 1986;249:65–102.

Hamill GS, Lenn NJ: The subnuclear organization of the rat interpeduncular nucleus: a light and electron microscopic study. J Comp Neurol 1984;222:396–408.

Hemmendinger LM, Moore RY: Interpeduncular nucleus organization in the rat: cytoarchitecture and histochemical analysis. Brain Res Bull 1984;13:163–179.

Hikosaka O: The habenula: from stress evasion to value-based decision-making. Nat Rev Neurosci 2010;11:503–513.

Holstege G: The mesopontine rostromedial tegmental nucleus and the emotional motor system: role in basic survival behavior. J Comp Neurol 2009;513:559–565.

Jhou TC, Geisler S, Marinelli M, Degarmo BA, Zahm DS: The mesopontine rostromedial tegmental nucleus: a structure targeted by the lateral habenula that projects to the ventral tegmental area of Tsai and substantia nigra compacta. J Comp Neurol 2009;513:566–596.

Kaufling J, Veinante P, Pawlowski SA, Freund-Mercier MJ, Barrot M: Afferents to the GABAergic tail of the ventral tegmental area in the rat. J Comp Neurol 2009;513:597–621.

Klemm WR: Habenular and interpeduncularis nuclei: shared components in multiple-function networks. Med Sci Monit 2004;10:RA261–FA273.

Lenn NJ, Hamill GS: Subdivisions of the interpeduncular nucleus: a proposed nomenclature. Brain Res Bull 1984;13:203–204.

Morley BJ: The interpeduncular nucleus. Int Rev Neurobiol 1986;28:157–182.

Paxinos G, Huang X-F, Sengul G, Watson C: Organization of the brainstem; in Mai JK, Paxinos G (eds): The Human Nervous System, ed 3. Amsterdam, Academic Press, 2012.

Steinbusch HWM: Distribution of serotonin-immunoreactivity in the central nervous system of the rat – cell bodies and terminals. Neuroscience 1981;6:557–618.

Vertes RP, Fortin WJ, Crane AM: Projections of the median raphe nucleus in the rat. J Comp Neurol 1999;407:555–582.

87 Peripeduncular Nucleus (PPD)

Original name:
Nucleus peripeduncularis

Alternative name:
Suprapeduncular nucleus
The peripeduncular nucleus should not be mistaken with the 'peripeduncular subdivisions of the pontine nuclei' [Schmahmann and Pandya, 1997]

Location and Cytoarchitecture – Original Text

The peripeduncular nucleus (PPD) appears in the lateral portion of the oral mesencephalic tegmentum. The caudal portion of the nucleus is related dorsally to the nucleus of the brachium of the inferior colliculus, laterally to the medial geniculate body, ventrally to the cerebral peduncles and the substantia nigra, and medially to the medial lemniscus. This nucleus extends orally beyond the extent of our serial sections and consequently its oral relations are not described [Nieuwenhuys et al., 2008].

The cells of the PPD are small to medium sized, oval or spindle shaped, and lightly stained (fig. 1–3).

Comments

The PPD lies on the border of the lateral thalamus and the midbrain as a caudal continuation of the zona incerta; it is sometimes referred to as a midbrain nucleus and at other times it is included in the 'ventral thalamus' [Nieuwenhuys et al., 1998; Jones, 1985]. It has been assumed to be part of a continuous sheet of neurons surrounding the thalamus, called the 'perithalamic nuclei', which consists of the thalamic reticular nucleus (TRN), zona incerta, the ventral subdivision of the lateral geniculate nucleus and PPD. Today we know that the perithalamic nuclei are derived from the first prosencephalic neuromere, the 'prethalamus', formerly called the ventral thalamus [Puelles and Rubenstein, 2003]. The nuclei can be divided into an outer and an inner tier, whereby the outer tier comprises a shell of neurons emerging ventrally from the subthalamic nucleus, flowing dorsally into the PPD and surrounding the thalamus as 'TRN proper' [Baldauf, 2010]. The perithalamic nuclei appear to be recessive phylogenetically, being of a smaller size in higher species [Simmons, 1976; Arcelli et al., 1997].

Functional Neuroanatomy

Function

The function of the PPD is unclear. As a perithalamic nucleus, it may have cells that are mainly GABAergic and participate in the control of transmission through the thalamus to, and from, the cerebral cortex [Guillery and Harting, 2003]. Several authors have suggested that the PPD may play a role in the modulation of sensory and motor signals, often associated with a

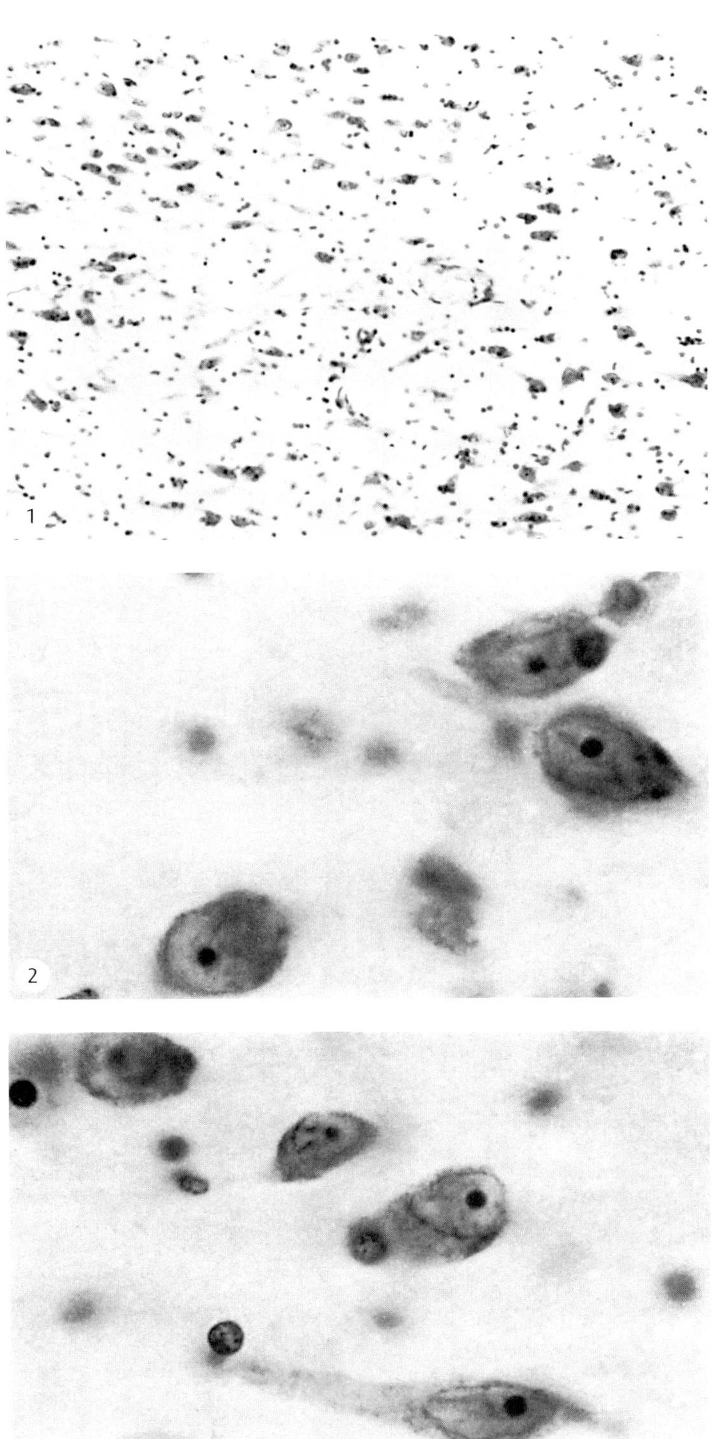

Fig. 1. Peripeduncular nucleus. Magnification ×150.
Fig. 2, 3. Cells from the peripeduncular nucleus. Magnification ×1,000.

limbic function: for example the emotional control of vocalization [Jürgens et al., 2002], visual attention [Büttner-Ennever et al., 1996], a basic feedback system associated with acoustic signals [Winer et al., 2002], sexual-related behavior and suckling stimuli [Szabo et al., 2010].

Connections

Afferents. The connections of the PPD are often bidirectional and indicate a strong association with the limbic system [Arnault and Roger, 1987]. Excitatory pathways to the PPD arise from the amygdala [Price and Amaral, 1981; Amaral and Insausti, 1992]. Inputs also emerge from the lateral and medial hypothalamus to the PPD [Jones et al., 1976; Saper et al., 1979; Kita and Oomura, 1982a, b], and the thalamic parafascicular and subparafascicular complex [Sadikot et al., 1992].

Efferents. Heavy projections from the PPD to the amygdala complex, the basal nucleus of Meynert and the lateral hypothalamus have been traced [Jones et al., 1976; Aggleton et al., 1980; Grove, 1988]. Efferents from the PPD also ascend to reach the anterior cingulate gyrus [Jürgens, 1983].

References

Aggleton JP, Burton MJ, Passingham RE: Cortical and subcortical afferents to the amygdala of the rhesus monkey *(Macaca mulatta)*. Brain Res 1980;190:347–368.

Amaral DG, Insausti R: Retrograde transport of D-[^{3}H]aspartate injected into the monkey amygdaloid complex. Exp Brain Res 1992;88:375–388.

Arcelli P, Frassoni C, Regondi MC, De Biasi S, Spreafico R: GABAergic neurons in mammalian thalamus: a marker of thalamic complexity. Brain Res Bull 1997;42:27–37.

Arnault P, Roger M: The connections of the peripeduncular area studied by retrograde and anterograde transport in the rat. J Comp Neurol 1987;258:463–476.

Baldauf ZB: Dual chemoarchitectonic lamination of the visual sector of the thalamic reticular nucleus. Neuroscience 2010;165:801–818.

Büttner-Ennever JA, Cohen B, Horn AKE, Reisine H: Efferent pathways of the nucleus of the optic tract in monkey and their role in eye movements. J Comp Neurol 1996;373:90–107.

Grove EA: Neural associations of the substantia innominata in the rat: afferent connections. J Comp Neurol 1988;277:315–346.

Guillery RW, Harting JK: Structure and connections of the thalamic reticular nucleus: advancing views over half a century. J Comp Neurol 2003;463:360–371.

Jones EG: The Thalamus. New York, Plenum Press, 1985.

Jones EG, Burton H, Saper CB, Swanson LW: Midbrain, diencephalic and cortical relationships of the basal nucleus of Meynert and associated structures in primates. J Comp Neurol 1976;15:385–419.

Jürgens U: Afferent fibers to the cingular vocalization region in the squirrel monkey. Exp Neurol 1983;80:395–409.

Jürgens U, Ehrenreich L, De Lanerolle NC: 2-Deoxyglucose uptake during vocalization in the squirrel monkey brain. Behav Brain Res 2002;136:605–610.

Kita H, Oomura Y: An HRP study of the afferent connections to rat lateral hypothalamic region. Brain Res Bull 1982a;8:63–71.

Kita H, Oomura Y: An HRP study of the afferent connections to rat medial hypothalamic region. Brain Res Bull 1982b;8:53–62.

Nieuwenhuys R, ten Donkelaar HJ, Nicholson C: The Central Nervous System of Vertebrates. Berlin, Springer, 1998.

Nieuwenhuys R, Voogd J, van Huijzen C: The Human Central Nervous System. Berlin, Springer, 2008.

Price JL, Amaral DG: An autoradiographic study of the projections of the central nucleus of the monkey amygdala. J Neurosci 1981;1:1242–1259.

Puelles L, Rubenstein JL: Forebrain gene expression domains and the evolving prosomeric model. Trends Neurosci 2003;26:469–476.

Sadikot AF, Parent A, François C: Efferent connections of the centromedian and parafascicular thalamic nuclei in the squirrel monkey: a PHA-L study of subcortical projections. J Comp Neurol 1992;315:137–159.

Saper CB, Swanson LW, Cowan WM: Some efferent connections of the rostral hypothalamus in the squirrel monkey *(Saimiri sciureus)* and cat. J Comp Neurol 1979;184:205–241.

Schmahmann JD, Pandya DN: Anatomic organization of the basilar pontine projections from prefrontal cortices in rhesus monkey. J Neurosci 1997;17:438–458.

Simmons RM: The diencephalon of the vervet monkey *(Cercopithecus aethiops)*. II. Epithalamus, subthalamus and hypothalamus. S Afr J Med Sci 1976;41:139–163.

Szabo FK, Snyder N, Usdin TB, Hoffman GE: A direct neuronal connection between the subparafascicular and ventrolateral arcuate nuclei in non-lactating female rats. Could this pathway play a role in the suckling-induced prolactin release? Endocrine 2010;37:62–70.

Winer JA, Chernock ML, Larue DT, Cheung SW: Descending projections to the inferior colliculus from the posterior thalamus and the auditory cortex in rat, cat, and monkey. Hear Res 2002;168:181–195.

88 Lateral Reticular Nucleus (LRN)

Original names:
Nucleus medullae oblongatae lateralis
Nucleus medullae oblongatae subtrigeminalis
Nucleus medullae oblongatae lateralis, subnucleus dorsalis

Subdivisions:
Lateral reticular nucleus principal (or main) part; this is equivalent to the original nucleus medullae oblongatae lateralis
Lateral reticular nucleus, subtrigeminal part; this is equivalent to the nucleus medullae oblongatae lateralis, subnucleus ventralis and subnucleus dorsalis

Alternative names:
Nucleus subtrigeminalis (for nucleus medullae oblongatae subtrigeminalis)
Nucleus reticularis lateralis
Nucleus funiculi lateralis
Nucleus of the lateral funiculus

Nomenclature

In the new edition of this atlas we have adopted the now classical nomenclature of Walberg for the lateral reticular nucleus (LRN) area [Walberg, 1952], rather than the nomenclature of Olszewski and Baxter in the previous edition. In Walberg's excellent and critical review of the comparative anatomy of the LRN 2 subgroups may be distinguished in man: (a) a principal part (LRNp), or the main part, and (b) a subtrigeminal part (LRNst). In most other mammalian species the main part of the LRN can be subdivided into a magnocellular and a parvocellular portion, but in man these cell components appear to be intermingled and form a single entity [Walberg, 1952]. The LRNp in man corresponds to Olszewski's and Baxter's *nucleus medullae oblongatae lateralis, subnucleus ventralis*, while the LRNst was described as a separate nucleus, *subtrigeminal nucleus*. Olszewski's and Baxter's *nucleus medullae oblongatae lateralis, subnucleus dorsalis,* is included here in the LRNst, although in some sections it may lie dorsally and appear separated from the LRNst (fig. 1). Further subdivisions of the LRN area are described by Koutcherov et al. [2004].

Location and Cytoarchitecture – Original Text

This nucleus is located in the lateral part of the medullary tegmentum, dorsal to the inferior olivary complex (IO) and ventrolateral to the nucleus ambiguus. The nucleus is about 9 mm in length and extends through two thirds of the IO, starting at its caudal pole.

Principal Part
The LRNp is irregularly oval in outline on a cross-section through the caudal pole. It lies close to the periphery of the medulla directly dorsal to the lateral pole of the newly formed medial accessory olive (MAO). As the principal inferior olive develops, it intervenes between LRN and MAO (plate 10). Oral to this level, the LRN becomes flattened dorsoventrally and on cross-section appears as an oblong cell group lying with its long axis directed horizontally, dorsal to the dorsal accessory olive and the amiculum of the olive. Orally, the lateral part of the LRNp disappears and is replaced by small groups and clusters of darkly stained cells of the caudal pole of the LRNst. The remaining medial portion of the LRNp may persist for a further 1–2 mm before disappearing.

The cells composing the LRNp are slender, multipolar and medium sized, with long dendrites that are confined to the nucleus, and a centrally placed nucleus. The Nissl granules of these cells are indistinct although the perikaryon is darkly stained (fig. 2–7). The cells are densely arranged throughout the nucleus except at its oral pole where they tend to become more scattered in distribution.

Subtrigeminal Part
The LRNst (in plates 10 and 12–14) occupies the ventrolateral portion of the medullary tegmentum. It extends from the level of the oral pole of the LRNp, to the level of the lateral paragigantocellular nucleus. The most caudally situated cells of the LRNst appear in small cell clusters which lie medial and lateral to the oral pole of the LRNp. Rostral to the oral pole of the LRNp, the LRNst portion increases markedly in size to form the entire ventrolateral corner of the medullary tegmentum. On cross-section, this portion of the nucleus is irregular in outline and lies lateral to the central nucleus of the medulla oblongata, ventral to the spinal trigeminal complex, and dorsal to the IO. The dorsolateral corner of the LRNst is often continuous with the cell groups of the islands of the lateral cuneate nucleus which lie between the spinal trigeminal complex and the restiform body.

The LRNst is composed of small clusters of oval, medium-sized and darkly stained cells, which characteristically possess peripherally arranged Nissl granules and nuclei which are often eccentric (fig. 8, 9). In the areas between the cell clusters one finds a few loosely scattered cells of varying form (fig. 10). The arrangement and internal architecture of the cells of the LRNst and the lateral cuneate nucleus are quite similar.

Functional Neuroanatomy

Function
The LRN is a precerebellar reticular nucleus involved in motor control. It gathers sensory and motor information from the vestibular nuclei, red nucleus and spinal cord, and conveys it exclusively to the cerebellum, via mossy fibers [Nieuwenhuys et al., 2008, fig. 21.17; Voogd and Ruigrok, 2012]. The mossy fiber input targets specifically Golgi cells and it is suggested to participate in the long-term depression of cerebellar signals [Xu and Edgley, 2010].

The LRN is surrounded by cell groups (C1 and A1) involved in cardiovascular and respiratory regulation, and by the fiber systems of the lateral funiculus, the anterolateral pathway and the rubrobulbar-spinal tract, which runs directly through the LRNst; at least some of these systems influence the LRN neurons [Robbins et al., 2005].

The LRN is severely degenerated in spinocerebellar ataxia and Machado-Joseph disease (spinocerebellar ataxia type 3) [Rüb et al., 2002].

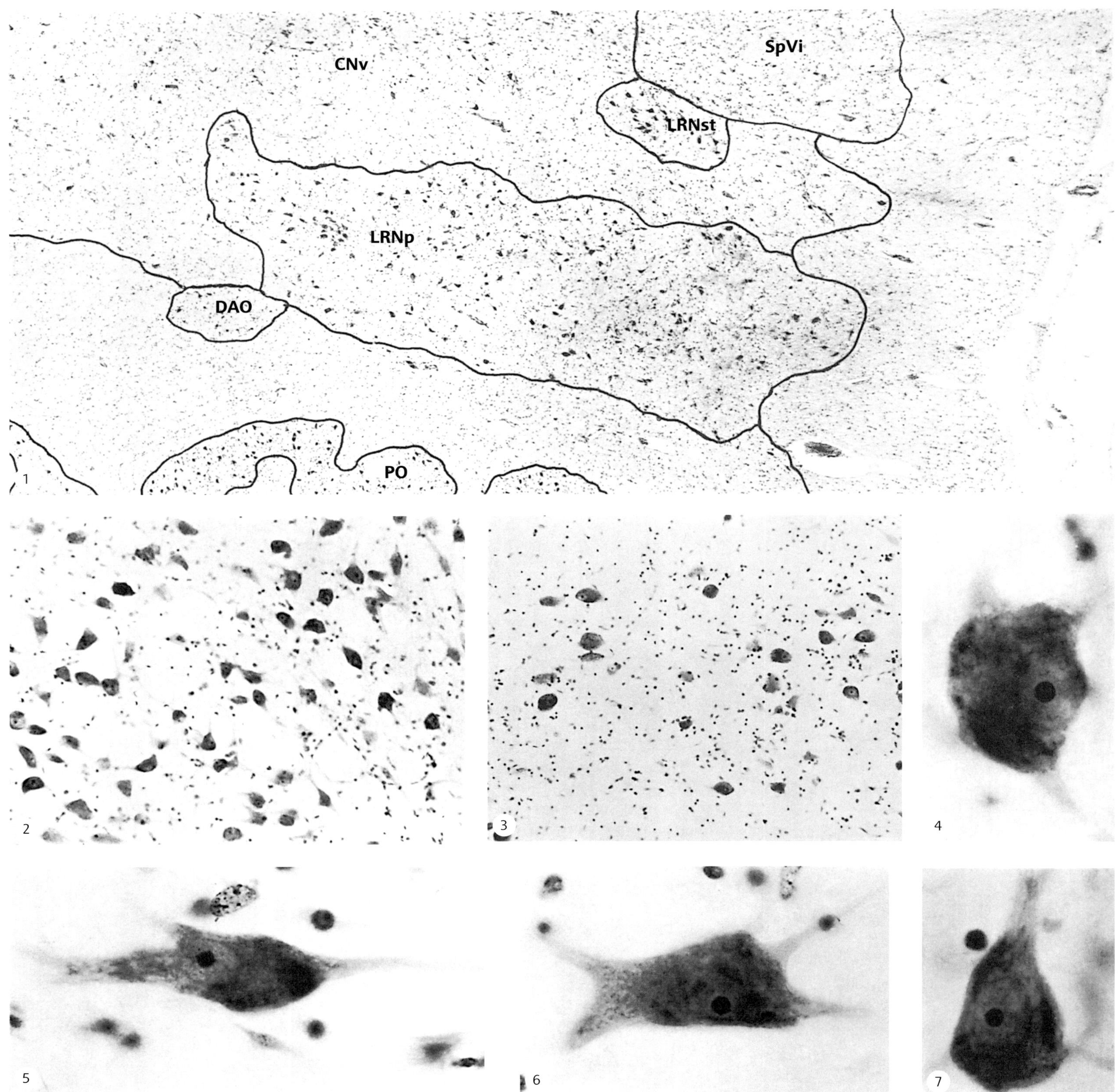

Fig. 1. Lateral reticular nucleus. CNv = Central nucleus of the medulla oblongata, ventral subnucleus; LRNp = lateral reticular nucleus, principal part; LRNst = lateral reticular nucleus, subtrigeminal part; DAO = dorsal olivary nucleus; PO = principal olivary nucleus; spVi = spinal trigeminal nucleus, interpolar part. Magnification ×40.

Fig. 2. Lateral reticular nucleus, principal part. Magnification ×150.
Fig. 3. Lateral reticular nucleus, subtrigeminal part. Magnification ×150.
Fig. 4–7. Cells from the lateral reticular nucleus, principal part. Magnification ×1,000.

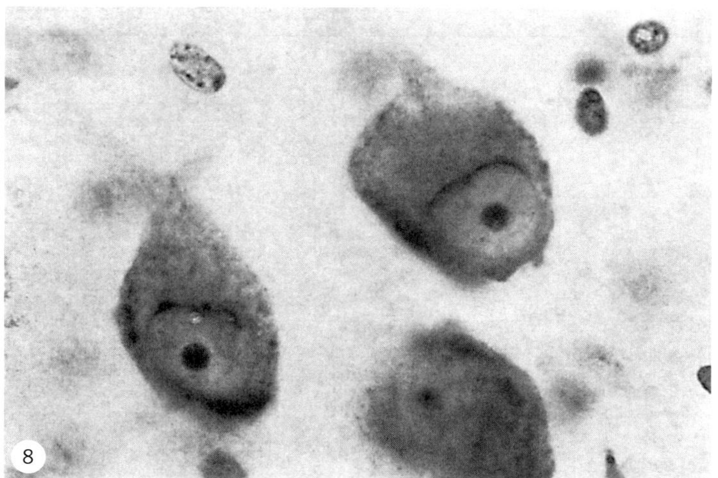

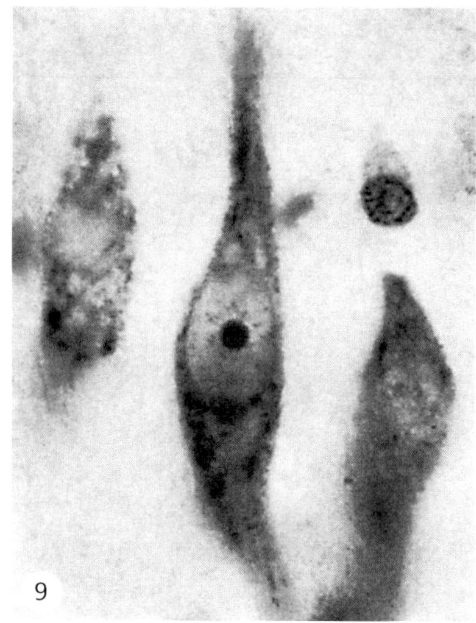

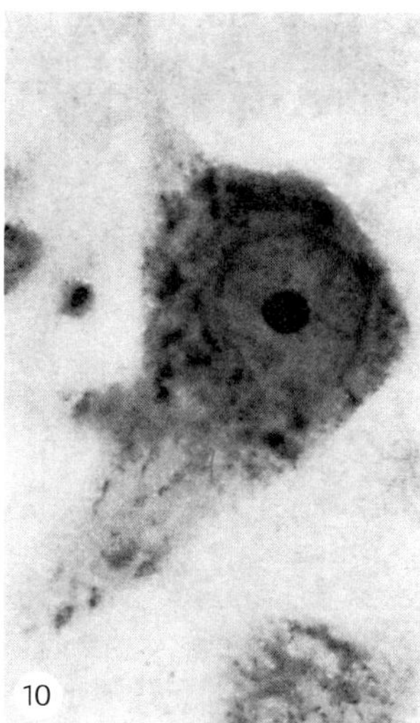

Fig. 8–10. Cells from the lateral reticular nucleus, subtrigeminal part. Magnification ×1,000.

Connections

Efferents. The cerebellar efferents from the LRN terminate in a zonal fashion, similar to the sagittal zones from the IO climbing fiber inputs (see nucleus No. 92) [Künzle, 1975; Wu et al., 1999]. The regions of the cerebellum that receive projections from the LRN are the skeletomotor and oculomotor areas [Voogd and Ruigrok, 2012]: that is the LRN projects to the anterior lobe most strongly, and also to the paramedian lobe, and it also sends terminals to the pyramis, uvula, nodulus and floccular region. The LRN also projects strongly to the vestibular and cerebellar nuclei in the rat [Ruigrok et al., 1995]. Projections from the LRN were found to target the granule cell domain of the cochlear nuclei and thus provide the dorsal cochlear nucleus with further nonauditory inputs [Zhan and Ryugo, 2007].

Afferents. The LRN receives spinal afferents from axon collaterals of propriospinal neurons of C3 and C4 ascending in spinoreticular pathways [Künzle, 1973]. The nerve cells are also responsive to head tilt [Pompeiano and Hoshino, 1977]. This otolith response may come in part from direct projections from the vestibular nuclei, originating in the lateral and superior vestibular nuclei [Carleton and Carpenter, 1983]; but the response could also come from vestibulospinal excitation of spinoreticular neurons terminating in the LRN. In addition, information from the sensorimotor cortex, the tectum and red nucleus converge on the LRN [Nieuwenhuys et al., 2008, fig. 21.17]. These signals, combined with otolith information in the LRN, are carried to the cerebellar cortex, by mossy fibers in mainly the ipsilateral restiform body (inferior cerebellar peduncle).

References

Carleton SC, Carpenter MB: Afferent and efferent connections of the medial, inferior and lateral vestibular nuclei in the cat and monkey. Brain Res 1983;278:29–51.

Koutcherov Y, Huang X-F, Halliday G, Paxinos G: Organization of human brain stem nuclei; in Paxinos G, Mai JK (eds): The Human Nervous System, ed 2. Amsterdam, Elsevier Academic Press, 2004.

Künzle H: The topographic organization of spinal afferents to the lateral reticular nucleus of the cat. J Comp Neurol 1973;149:103–115.

Künzle H: Autoradiographic tracing of the cerebellar projections from the lateral reticular nucleus in the cat. Exp Brain Res 1975;22:255–266.

Nieuwenhuys R, Voogd J, Van Huijzen C: The Human Central Nervous System, Berlin, Springer, 2008.

Pompeiano O, Hoshino K: Responses to static tilts of lateral reticular neurons mediated by contralateral labyrinthine receptors. Arch Ital Biol 1977;115:211–236.

Robbins MT, Uzzell TW, Aly S, Ness TJ: Visceral nociceptive input to the area of the medullary lateral reticular nucleus ascends in the lateral spinal cord. Neurosci Lett 2005;381:329–333.

Rüb U, De Vos RAI, Schultz C, Brunt ER, Paulson H, Braak H: Spinocerebellar ataxia type 3 (Machado-Joseph disease): severe destruction of the lateral reticular nucleus. Brain 2002;125:2115–2124.

Ruigrok TJH, Cella F, Voogd J: Connections of the lateral reticular nucleus to the lateral vestibular nucleus in the rat. An anterograde tracing study with *Phaseolus vulgaris* leucoagglutinin. Eur J Neurosc 1995;7:1410–1413.

Voogd J, Ruigrok T: Cerebellum and precerebellar nuclei; in Mai JK, Paxinos G (eds): The Human Nervous System, ed 3. Amsterdam, Academic Press, 2012.

Walberg F: The lateral reticular nucleus of the medulla oblongata in mammals. J Comp Neurol 1952;96:283–343.

Wu HS, Sugihara I, Shinoda Y: Projection patterns of single mossy fibers originating from the lateral reticular nucleus in the rat cerebellar cortex and nuclei. J Comp Neurol 1999;411:97–118.

Xu W, Edgley SA: Cerebellar Golgi cells in the rat receive convergent peripheral inputs via a lateral reticular nucleus relay. Eur J Neurosci 2010;32:591–597.

Zhan X, Ryugo DK: Projections of the lateral reticular nucleus to the cochlear nucleus in rats. J Comp Neurol 2007;504:583–598.

89 Arcuate Nucleus (ARC)

Original name:
Nucleus arcuatus

Location and Cytoarchitecture – Original Text

The arcuate nuclei (ARC) are found in cross-sections through that portion of the medulla between the caudal pole of the medial accessory olive and the caudal border of the pons. In the caudal two thirds of this extent the nuclei usually appear as small, plaque-like cell groups lying among the ventral external arcuate fibers medial and ventral to the pyramids and ventral to the inferior olivary complex (IO). Similar small groups may also be found lateral to what is usually described as ARC, but too medial to be called pontobulbar nuclei. Approaching the caudal border of the pons, ARC increase markedly in size to form conspicuous, crescent-shaped structures on the medial and ventral aspects of the pyramids. At this level, the convex surfaces of the nuclei of the two sides may fuse dorsal to the anterior median sulcus. Tongue-like processes may extend dorsolaterally from the concave surfaces of the nuclei to encircle small bundles of the adjacent pyramidal fibers. As one enters the pons proper ARC become directly continuous with the medial and ventral groups of pontine nuclei (PN).

The cells composing ARC are in all respects similar to those of PN. They are medium sized, closely packed, plump, polygonal or oval with relatively large nuclei, distinct nucleoli and long dendrites. The cytoplasm of these cells stains darkly and contains indistinct Nissl granules (fig. 1–3). Particularly at the medial and lateral extremities of the caudally situated cell groups, some cells tend to be elongated with their long axes parallel to the course of the ventral external arcuate fibers.

On the basis of both embryological and morphological evidence it appears justifiable to regard ARC nuclei as caudal extensions of PN [Essick, 1912]. Nonetheless, the functional significance as well as the afferent and efferent connections of these nuclei remain a matter of controversy. It is, however, logical to assume that they receive corticopontine fibers, either directly from the pyramidal region or via some more circuitous pathway such as the pontobulbar body. Evidence is available that the axons of the ARC pass to both the homolateral and contralateral cerebellar cortex [Swank, 1934]. [...]

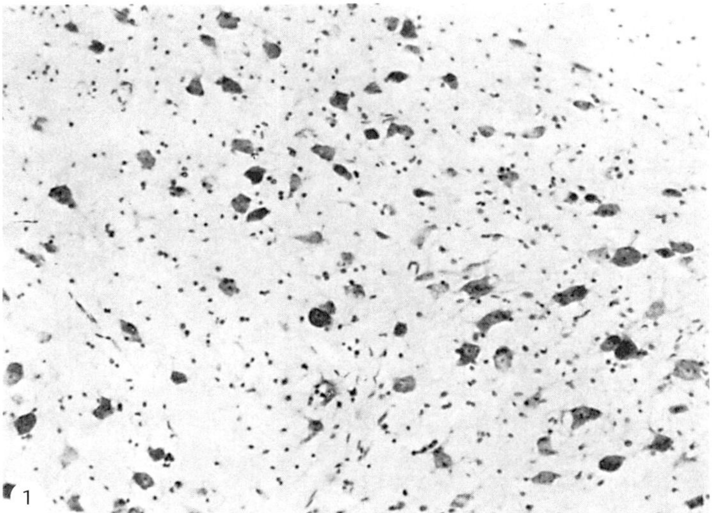

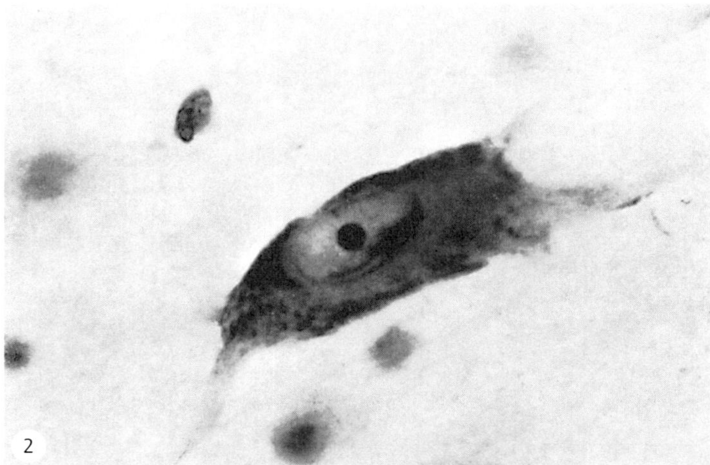

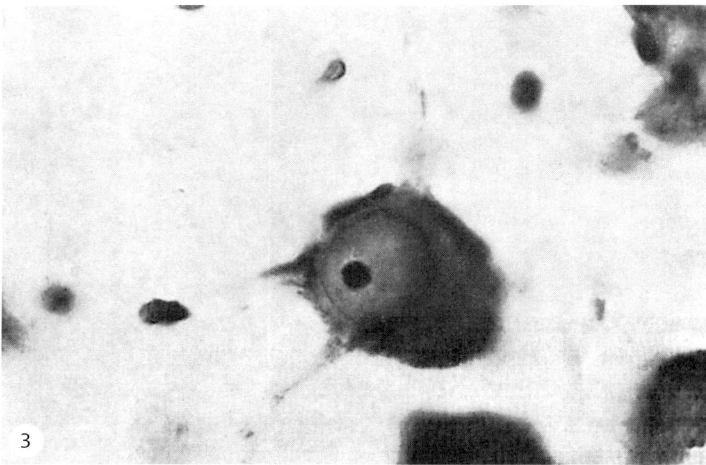

Fig. 1. Arcuate nucleus. Magnification ×150.
Fig. 2, 3. Cells from the arcuate nucleus. Magnification ×1,000.

Functional Neuroanatomy

Function

It is undisputed that the ARC project to the cerebellum, but their functional role is not known. Some studies provide evidence suggesting an involvement in central chemoregulation and respiratory control [Zec et al., 1997; for a review, see Fu and Watson, 2012]. Current hypotheses, in association with 'sudden infant death syndrome', suggest that a hypodevelopment of the brainstem, leading to cardiovascular and respiratory deficits, could explain the neuropathological changes found in ARC as well as other brainstem nuclei [Folkerth et al., 2008].

Connections

The efferent fibers from ARC neurons enter the external arcuate pathway, which travels around the base of the brain to enter the inferior cerebellar peduncle and terminate in the cerebellum. This was reported by Deiters [1865], Essick [1912] and Riley [1943] many years ago. The nucleus does not have any proven counterpart in lower animals, so its connectivity could not be investigated in animal models. Recently tract-tracing with DiI was used to label ARC afferents in the fetus. Afferents from caudal raphe nuclei were demonstrated, as were ARC efferents to the cerebellum [Zec et al., 1997]. Some studies have reported serotonergic cells within ARC, but this is a matter of controversy [Fu and Watson, 2012].

Comment

On the basis of cytoarchitectural similarities the ARC are suggested to be displaced PN, both here and by earlier authors [Deiters, 1865; Riley, 1943]. The ARC and PN (nucleus No. 100) also have similar neurohistochemical profiles: they both stain positively for calretinin and nitric oxide synthase [Baizer and Broussard, 2010]. However, on the basis of one important difference, namely that the ARC are parvalbumin positive but the PN are not, Baizer et al. [2012] argue that ARC are not displaced PN. Neither did they consider ARC to be a derivative of the IO or the nuclei pararaphales [Baizer et al., 2012; Fu and Watson, 2012].

A possible homolog of the ARC in humans has been reported in the rodent [Watson and Paxinos, 2010], but a study using histochemical and tract-tracing methods showed that in the rodent the nucleus was immunopositive for calbindin – a reliable marker for the olivocerebellar system [Servais et al., 2005] – and therefore likely to be a ventral extension of the IO rather than a displaced PN [Fu and Watson, 2012].

References

Baizer JS, Broussard DM: Expression of calcium-binding proteins and nNOS in the human vestibular and precerebellar brainstem. J Comp Neurol 2010;518:872–895.

Baizer JS, Weinstock N, Witelson SF, Sherwood CC, Hof PR: The nucleus pararaphales in the human, chimpanzee, and macaque monkey. Brain Struct Funct 2013;218:389–403.

Deiters O: Untersuchungen über Gehirn und Rückenmark. Braunschweig, Vieweg, 1865.

Essick CR: The development of the nuclei pontis and the nucleus arcuatus in man. Am J Anat 1912;13:25–54.

Folkerth RD, Zanoni S, Andiman SE, Billiards SS: Neuronal cell death in the arcuate nucleus of the medulla oblongata in stillbirth. Int J Dev Neurosci 2008;26: 133–140.

Fu YH, Watson C: The arcuate nucleus of the C57BL/6J mouse hindbrain is a displaced part of the inferior olive. Brain Behav Evol 2012;79:191–204.

Riley HA: An Atlas of the Basal Ganglia, Brain Stem and Spinal Cord. Baltimore, Williams & Wilkins, 1943, p 708.

Servais L, Bearzatto B, Schwaller B, Dumont M, de Saedeleer C, Dan B, Barski JJ, Schiffmann SN, Cheron G: Mono- and dual-frequency fast cerebellar oscillation in mice lacking parvalbumin and/or calbindin D-28k. Eur J Neurosci 2005; 22:861–870.

Swank RL: The relationship between the circumolivary pyramidal fascicles and the pontobulbar body in man. J Comp Neurol 1934;60:309–317.

Watson C, Paxinos G: Chemoarchitectonic Atlas of the Mouse Brain. San Diego, Academic Press, 2010.

Zec N, Filiano JJ, Kinney HC: Anatomic relationships of the human arcuate nucleus of the medulla: a DiI-labeling study. J Neuropathol Exp Neurol 1997;56:509–522.

Precerebellar Nuclei

(Plates 8–13)

90 Intercalated Nucleus (of Staderini) (INSt)

Original name:
Nucleus intercalatus

Alternative name:
Intercalated nucleus of the medulla (as opposed to the intercalated nucleus of the amygdala)

Location and Cytoarchitecture – Original Text

The nucleus is formed by an elongated cell group which intervenes between the oral 6–7 mm of the hypoglossal nucleus (XII) and the dorsal motor nucleus of the vagus (DMX). The caudal pole of the intercalated nucleus (INSt) is indistinct and lies just caudal to the level of the obex. Orally this nucleus is directly continuous with the nucleus prepositus (PrP).

On cross-section, the INSt is rectangular or wedge shaped in outline and lies with its long axis directed dorsomedially. It is related ventromedially to the XII, ventrolaterally to the central nucleus of the medulla oblongata, dorsolaterally to the DMX and dorsomedially to the stratum gliosum.

The caudal portion of the INSt is composed of very small, sparsely distributed, round, oval or spindle-shaped cells which contain darkly stained but indistinct Nissl granules (fig. 4). Other cells characteristic of either XII or DMX may be found scattered among these smaller cells. The oral 2–3 mm of the nucleus are characterized by the appearance of numerous medium-sized cells which lie scattered among the smaller ones (fig. 1). The medium-sized cells are plump, multipolar and possess darkly stained, indistinct Nissl granules which tend to congregate at the periphery of the cell body (fig. 2, 3). The cells composing this oral portion of the INSt are much more compactly arranged than those of the caudal portion.

Functional Neuroanatomy

Function

The INSt is the caudal continuation of the PrP. For this reason it is included with the PrP in the precerebellar section, even though it has no clear cerebellar projections. Both nuclei are considered to play a role in the control of gaze, or more specifically the maintenance of eye and head position.

There is compelling evidence from lesion studies that the PrP plays a key role in the maintaining of the position of the eye and head in gaze [Cheron et al., 1986; Cannon and Robinson, 1987; Cheron and Godaux, 1987]. It converts gaze velocity signals into tonic position signals suitable for driving motoneurons, and in this way acts as a horizontal neural integrator,

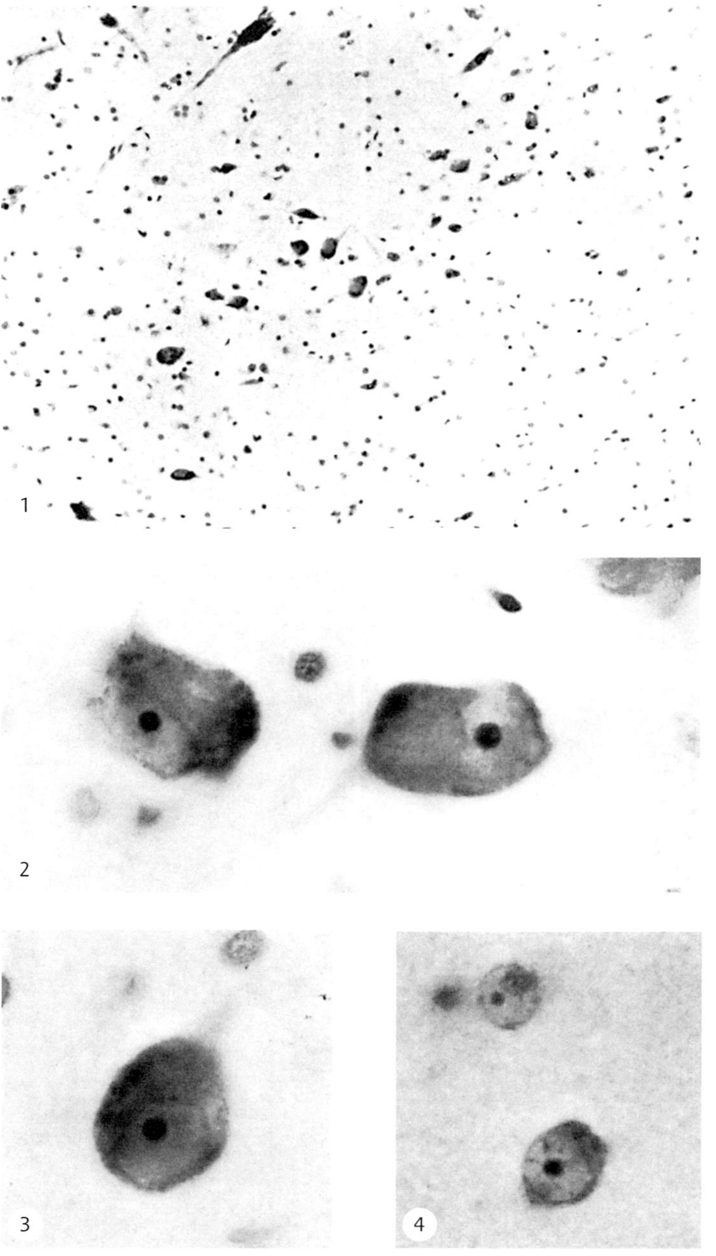

Fig. 1. Intercalated nucleus. Magnification ×150.
Fig. 2–4. Cells from the intercalated nucleus. Magnification ×1,000.

in the mathematical sense. Some results indicate that the INSt may participate in a similar role for vertical gaze. A patient with an infarct of the medial medulla oblongata involving the region of the INSt was reported to have upbeat nystagmus, and hence a possible involvement of the neural integrator for vertical eye position [Munro et al., 1999].

Connections

Afferents. The INSt is part of the perihypoglossal complex, along with the PrP, the nucleus of Roller, the interfascicular hypoglossal nucleus and (in primates) the supragenual nucleus [Brodal, 1983]. A considerable input to the INSt arrives from the ventral ipsilateral medial vestibular nucleus (MVN), and from a circumscribed region of the dorsal paragigantocellular nucleus of the medulla just below the PrP [Mergner et al., 1977]. Inputs from the vestibular nuclei onto the INSt do not arise from secondary vestibular neurons [McCrea et al., 1987a, b]. Several authors have described afferent projections to the INSt from the paramedian pontine reticular formation [Büttner-Ennever and Büttner, 1988], in particular from eye-neck reticulospinal neurons, which make direct synaptic contact with INSt neurons [Grantyn et al., 1987].

Efferents. The efferent pathways from the INSt target the contralateral MVN and PrP [Pompeiano et al., 1978], the contralateral superior colliculus [Stechison et al., 1985], and also the ipsilateral oculomotor and contralateral abducens nuclei [Graybiel and Hartwieg, 1974; Maciewicz et al., 1977; Steiger and Büttner-Ennever, 1978]. This pattern of efferent projections is similar to that of the PrP, with the exception of the major prepositus-cerebellar projection.

The INSt displays medium acetylcholine esterase reactivity and contains some cholinergic-positive cells [Koutcherov et al., 2004].

References

Brodal A: The perihypoglossal nuclei in the macaque monkey and the chimpanzee. J Comp Neurol 1983;218:257–269.

Büttner-Ennever JA, Büttner U: The reticular formation; in Büttner-Ennever JA (ed): Neuroanatomy of the Oculomotor System. Reviews of Oculomotor Research. Amsterdam, Elsevier, 1988, vol 2.

Cannon SC, Robinson DA: Loss of the neural integrator of the oculomotor system from brain stem lesions in monkey. J Neurophysiol 1987;57:1383–1409.

Cheron G, Godaux E: Disabling of the oculomotor neural integrator by kainic acid injections in the prepositus-vestibular complex of the cat. J Physiol 1987;394: 267–290.

Cheron G, Godaux E, Laune JM, Vanderkelen B: Lesions in the cat prepositus complex: effects on the vestibulo-ocular reflex and saccades. J Physiol 1986;372: 75–94.

Grantyn A, Ong Meang Jacques V, Berthoz A: Reticulo-spinal neurons participating in the control of synergic eye and head movements during orienting in the cat. II. Morphological properties as revealed by intra-axonal injections of horseradish peroxidase. Exp Brain Res 1987;66:355–377.

Graybiel AM, Hartwieg EA: Some afferent connections of the oculomotor complex in the cat: an experimental study with tracer techniques. Brain Res 1974;81: 543–551.

Koutcherov Y, Huang X-F, Halliday G, Paxinos G: Organization of human brain stem nuclei; in Paxinos G, Mai JK (eds): The Human Nervous System, ed 2. Amsterdam, Elsevier Academic Press, 2004.

Maciewicz RJ, Eagen K, Kaneko CR, Highstein SM: Vestibular and medullary brain stem afferents to the abducens nucleus in the cat. Brain Res 1977;123:229–240.

McCrea RA, Strassman A, Highstein SM: Anatomical and physiological characteristics of vestibular neurons mediating the vertical vestibulo-ocular reflex of the squirrel monkey. J Comp Neurol 1987a;264:571–594.

McCrea RA, Strassman E, May E, Highstein SM: Anatomical and physiological characteristics of vestibular neurons mediating the horizontal vestibulo-ocular reflex of the squirrel monkey. J Comp Neurol 1987b;264:547–570.

Mergner T, Pompeiano O, Corvaja N: Vestibular projections to the nucleus intercalatus of Staderini mapped by retrograde transport of horseradish peroxidase. Neurosci Lett 1977;5:309–313.

Munro N, Bronstein A, Larner A, Janssen J, Farmer S: The role of the nucleus intercalatus in vertical gaze holding. J Neurol Neurosurg Psychiatry 1999;66:552–553.

Pompeiano O, Mergner T, Corvaja N: Commissural, perihypoglossal and reticular afferent projections to the vestibular nuclei in the cat: an experimental anatomical study with horseradish peroxidase. Arch Ital Biol 1978;116:130–172.

Stechison MT, Saint-Cyr JA, Spence SJ: Projections from the nuclei prepositus hypoglossi and intercalatus to the superior colliculus in the cat: an anatomical study using WGA-HRP. Brain Res 1985;59:139–150.

Steiger HJ, Büttner-Ennever JA: Relationship between motoneurons and internuclear neurons in the abducens nucleus: a double retrograde tracer study in the cat. Brain Res 1978;148:181–188.

91 Nucleus of Roller (Ro)

Original name:
Nucleus of Roller

Alternative name:
None (but in the older literature the inferior vestibular nucleus was sometimes given the name 'Roller's nucleus', or the 'inferior vestibular nucleus of Roller')

Location and Cytoarchitecture – Original Text

The nucleus of Roller (Ro) is a small cell group which lies immediately ventral to the oral 3–4 mm of the hypoglossal nucleus (see nucleus No. 106, fig. 1, 2). On cross-section, the nucleus is round in outline and is related medially to the medial longitudinal fasciculus, ventrally to the nucleus interfascicularis hypoglossi, and laterally to the central nucleus of the medulla oblongata. The nucleus attains its greatest diameter at its midportion and tapers orally and caudally.

The Ro is composed of closely congregated, medium-sized, oval or polygonal multipolar cells which possess distinct nuclei and nucleoli, and finely granulated, darkly stained Nissl granules. In many cells the Nissl granules tend to collect about the periphery of the cell body or about the nuclear membrane (fig. 1–3).

Functional Neuroanatomy

Function

The Ro is usually considered to be part of the perihypoglossal complex, along with the prepositus nucleus (PrP) and the intercalated nucleus of Staderini [Brodal, 1983; McCrea and Horn, 2006]. These 3 nuclei receive afferents from premotor structures controlling head and eye movements, and are thought to play a role in maintaining eye and head position, i.e. the integration (in the mathematical sense) of eye and head movements. This function is so far supported by the evidence reported for the properties of the neurons in the Ro; for example, the injection of a *transsynaptic* neuroanatomical tracer into the suboccipital neck muscles retrogradely labeled neurons only in the Ro [Edney and Porter, 1986].

Connections

Efferents. The cells of the Ro and PrP project to the vermis of the cerebellum (lobules VI and VII), to the flocculus, ventral paraflocculus and uvula [Brodal and Brodal, 1983]. Some of the cells are cholinergic [Barmack et al., 1992], and a large proportion of the projection neurons contain cholecystokinin, which may function as a neuromodulator at the mossy fiber terminals in the vermis [Lee et al., 2002].

Early studies of the perihypoglossal nuclei suggested that they were connected to the hypoglossal nucleus (XII). This was superseded by more current theories described above, but recent stimulation and immunohistochemical studies have

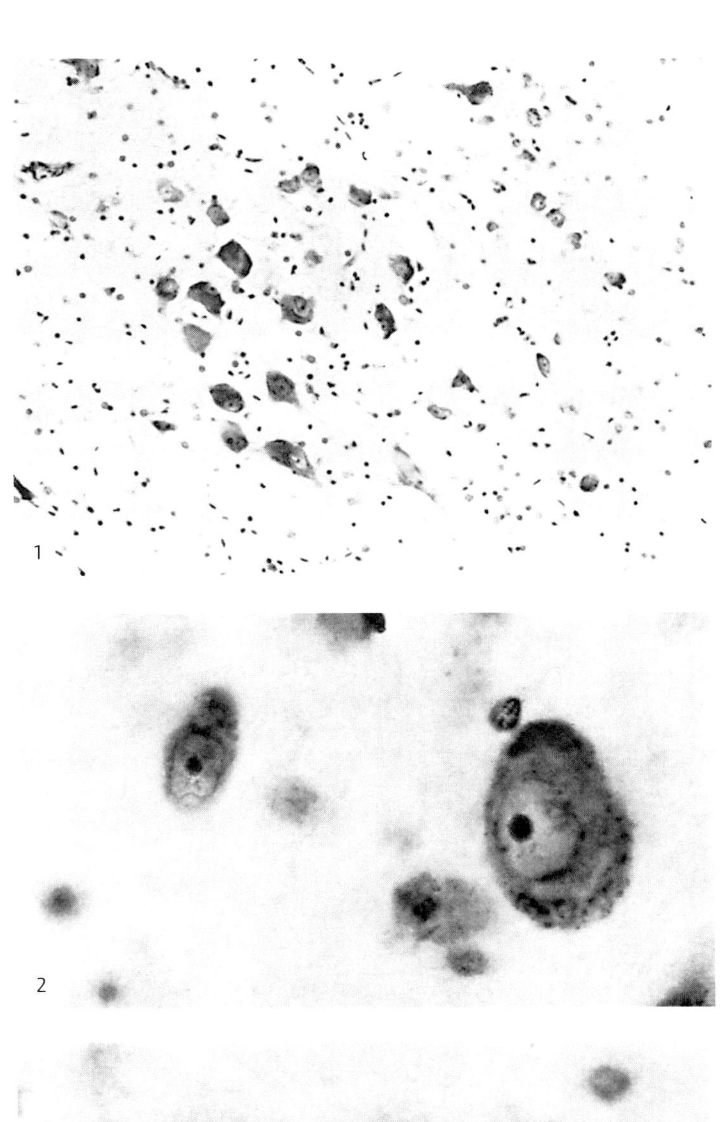

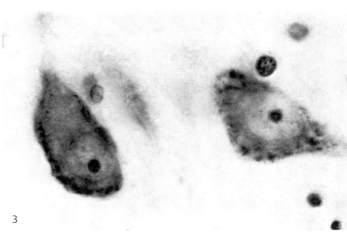

Fig. 1. Nucleus of Roller. Magnification ×150.
Fig. 2, 3. Cells from the nucleus of Roller. Magnification ×1,000.

shown that small GABAergic cells in the Ro do indeed project to the XII, and may provide a tonic inhibitory influence on tongue motoneurons [Aldes et al., 1988; van Brederode et al., 2011].

Afferents. Secondary vestibular neurons, activated from vertical and horizontal semicircular canals, have been shown to send collaterals to the Ro [McCrea et al., 1987a, b]. The interstitial nucleus of Cajal (INC), which controls vertical head and eye position, projects to the perihypoglossal complex including the Ro [Aldes and Boone, 1984; Fukushima et al., 1992], but the adjacent XII receives no such INC afferents.

References

Aldes LD, Boone TB: Does the interstitial nucleus of Cajal project to the hypoglossal nucleus in the rat? J Neurosci Res 1984;12:553–561.

Aldes LD, Chronister RB, Marco LA: Distribution of glutamic acid decarboxylase and gamma-aminobutyric acid in the hypoglossal nucleus in the rat. J Neurosci Res 1988;19:343–348.

Barmack NH, Baughman RW, Eckenstein FP, Shojaku H: Secondary vestibular cholinergic projection to the cerebellum of rabbit and rat as revealed by choline acetyltransferase immunohistochemistry, retrograde and orthograde tracers. J Comp Neurol 1992;317:250–270.

Brodal A: The perihypoglossal nuclei in the macaque monkey and the chimpanzee. J Comp Neurol 1983;218:257–269.

Brodal A, Brodal P: Observations on the projection from the perihypoglossal nuclei onto the cerebellum in the macaque monkey. Arch Ital Biol 1983;121:151–166.

Edney DP, Porter JD: Neck muscle afferent projections to the brainstem of the monkey: implications for the neural control of gaze. J Comp Neurol 1986;250: 389–398.

Fukushima K, Kaneko CRS, Fuchs AF: The neuronal substrate of integration in the oculomotor system. Prog Neurobiol 1992;39:606–639.

Lee S-H, Tseng CY, Wen C-Y, Shieh J-Y: Distribution of precerebellar cholecystokininergic neurons in the perihypoglossal nuclei of the gerbil. Zool Stud 2002; 41:162–169.

McCrea RA, Horn AKE: Nucleus prepositus. Prog Brain Res 2006;151:205–230.

McCrea RA, Strassman A, Highstein SM: Anatomical and physiological characteristics of vestibular neurons mediating the vertical vestibulo-ocular reflex of the squirrel monkey. J Comp Neurol 1987a;264:571–594.

McCrea RA, Strassman E, May E, Highstein SM: Anatomical and physiological characteristics of vestibular neurons mediating the horizontal vestibulo-ocular reflex of the squirrel monkey. J Comp Neurol 1987b;264:547–570.

Van Brederode JF, Yanagawa Y, Berger AJ: GAD67-GFP+ neurons in the nucleus of Roller: a possible source of inhibitory input to hypoglossal motoneurons. I. Morphology and firing properties. J Neurophysiol 2011;105:235–248.

Precerebellar Nuclei

(Plates 8, 10, 12–14, 16 and 18)

92 Inferior Olive (IO)

Original name:
Nucleus olivaris inferior

Major subnuclei:
Principal olive
Medial accessory olive
Dorsal accessory olive

Location and Cytoarchitecture – Original Text

This nuclear complex is composed of the principal olive (PO), the medial accessory olive (MAO), and the dorsal accessory olive (DAO). The complex is located in the oral 1.5 cm of the medulla, deep to the olivary prominence, dorsal to the pyramids, and lateral to the medial lemniscus (ML).

The major component of the inferior olive (IO) complex is the PO. This structure extends from the oral end of the pyramidal decussation to the caudal pole of the facial nucleus. The outline of the nucleus is that of a crumpled sac which is tapered caudally and orally and whose hilus is directed medially. Thus, in cross-sections through the intervening area the nucleus appears as an irregular, tortuous band of gray matter so shaped as to form a dorsal lamina, ventral lamina and a lateral lamella (fig. 1). The PO is surrounded by a dense mass of fibers which form the amiculum or capsule. The area partially enclosed by the olive is occupied predominantly by olivocerebellar fibers.

The MAO appears about 1 mm caudal to the PO and extends orally for a distance of 1 cm. The nucleus is composed of several cell groups or subnuclei, 4 of which form a rather distinct unit caudally (fig. 1). They lie end to end in such a manner as to form a band of gray matter which is curved upon itself to form an obtuse angle directed dorsolaterally towards the PO. Group β and the dorsal cap of Kooy (dc) form the perpendicular arm which lies lateral to the ML. Three millimeters oral to the caudal pole of the complex, the groups of the horizontal arm of the MAO (often referred to as 'groups a and b') merge together, and 2 mm farther orally, the perpendicular arm, composed of group β and the dc, disappears. The single MAO group then enlarges progressively and moves dorsomedially to occupy a position just medial to the hilus of the PO, where it persists to the level of the junction of the middle and upper thirds of the complex.

The ventrolateral outgrowth (vlo) is composed of a narrow, vertically placed band of cells which appears just oral to the oral pole of group β and dc, lies usually lateral to the MAO, and extends orocaudally for more than 1 mm.

Immediately lateral to the oral portion of the MAO there is often a narrow perpendicularly placed cell group lying in the hilus of the PO. [According to Nieuwenhuys et al. [2008] it may represent the medial lamina of the principal olive. This group extends orocaudally for 2 mm, and is referred to as 'group g' in the 2nd edition.]

The dorsomedial cell column (dmcc, or 'group h' in the 2nd edition) appears at the junction of the upper and middle thirds of the olivary complex and extends for 2 mm. It lies medial to the medial end of the superior lamella of the PO.

In the oral third of the IO complex, small groups of cells labeled appear in the hilus of the PO and ventral to the dmcc. Caudally, these groups are continuous with the MAO. They are inconstant in number and shape, and no attempt has been made to identify each cluster individually. Some of the more ventrally placed of these groups may easily be confused with groups of the conterminal nucleus (CT). The distinction can usually be made, however, on the basis of the facts that the cells of the CT are usually larger, and that, in older individuals particularly, the cells of the IO, but not the CT, tend to contain excessive amounts of lipofuscin.

The DAO commences caudally at the same level as the MAO and continues orally throughout the whole length of the IO complex. It is rep-

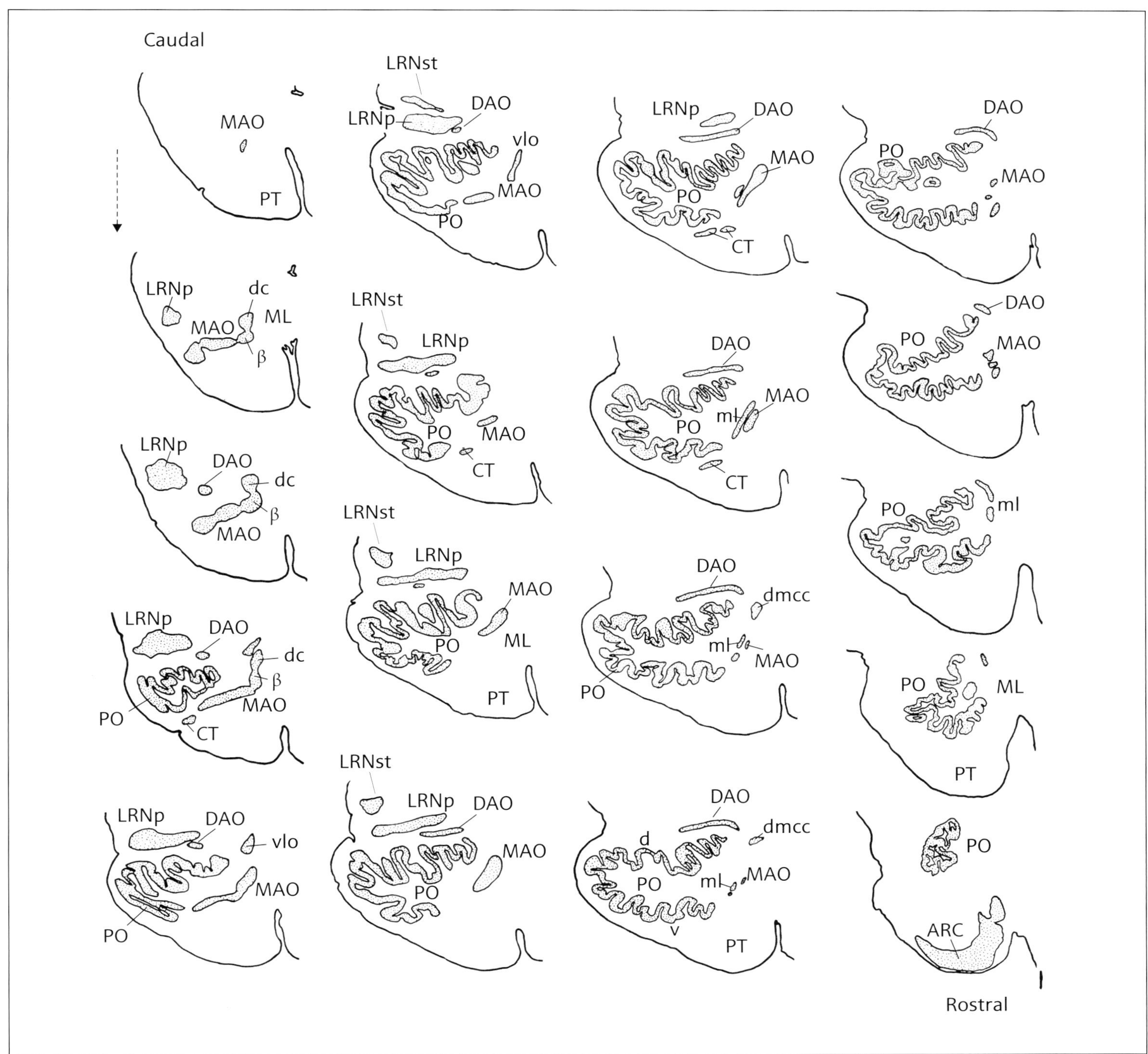

Fig. 1. Semischematic drawings of serial cross-sections through the inferior olivary complex. See text for descriptions of the individual cell groups of the complex. ARC = Arcuate nucleus; β = group β of the MAO; CT = conterminal nucleus; d = dorsal lamina of PO; DAO = dorsal accessory olive; dc = dorsal cap of Kooy; dmcc = dorsomedial cell column; LRNp = lateral reticular nucleus, principal part; LRNst = lateral reticular nucleus, subthalamic part; MAO = medial accessory olive; ml = medial lamina of PO; ML = medial lemniscus; PO = principal olive; PT = pyramidal tract; v = ventral lamina of PO; vlo = ventrolateral outgrowth.

resented by a flat, horizontally placed group of cells lying dorsal to the PO. The nucleus is best developed in sections through the oral half of the olivary complex. Caudal to this level it is represented by only a few cells in each cross-section. [However, Koutcherov et al. [2004] recognize a caudal DAO subnucleus.] Approaching the oral pole of the DAO, the nucleus diminishes gradually in size and moves medially to form a comma-shaped structure dorsomedial to the medial end of the dorsal lamina of the PO.

The cells composing all 3 units of the IO have similar characteristics. They are medium sized, plump and multipolar, and have short dendrites. The cells stain darkly and contain indistinct Nissl granules and nuclei which are often excentric (fig. 2–4). These cells tend to accumulate lipofuscin very early in life and the Nissl-stained section of 'normal' persons over middle age may show a striking paucity of cells. The cell density and arrangement vary slightly in different parts of the complex. [...]

Further reviews of the subnuclei of the IO can be found in Whitworth and Haines [1986], Paxinos et al. [2012], Nieuwenhuys et al. [1998, vol. 3] and in Voogd and Ruigrok [2012].

Functional Neuroanatomy

Function

The cerebellar cortex receives two different types of neural afferents, *mossy fibers* and *climbing fibers* (CF): the nucleus of the IO is the exclusive source of CF to the cerebellum [Szentagothai and Rajkovits, 1959; Desclin, 1974]. Every cell in the IO gives rise to an efferent CF which terminates on the dendrites of Pur-

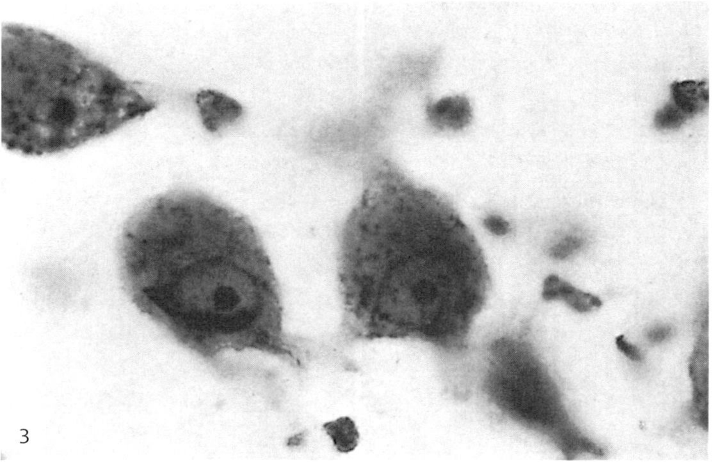

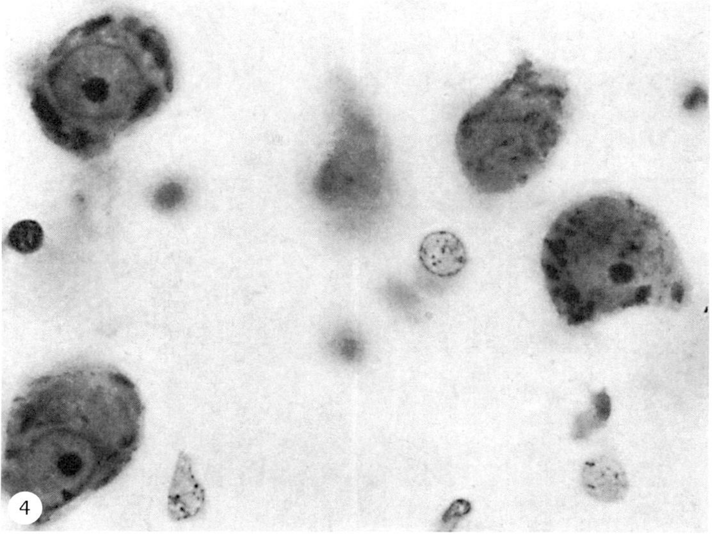

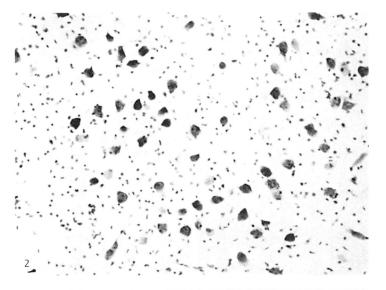

Fig. 2. Inferior olive. Magnification ×150.
Fig. 3, 4. Cells from the inferior olive. Magnification ×1,000.

kinje cells (PC) in the molecular layer of the cerebellar cortex; for reviews see Nieuwenhuys et al. [2008] and Voogd and Ruigrok [2012]. These olivocerebellar afferents are called CF, because they pass over the PC soma without synapsing, then divide and climb over the proximal dendritic tree, terminating densely all over it. Through CF, the IO can modulate the PC activity in every part of the cerebellum, and, like the cerebellum, the IO has been implicated in functions such as motor learning, coordination or refinement of movements, and motor error correction. For reviews, see Voogd [1995], Nieuwenhuys et al. [2008] and Llinás [2009].

Every time a CF discharges, it evokes multipeaked action potentials in PC, termed 'CF responses', or 'a complex spike'. It reflects the direct action of a single CF on a PC, modulating its axonal output, i.e. the output of the cerebellar cortex. The mossy fiber input onto the PC is carried by the parallel fibers, which give rise to 'single-spike activity' in the PC. The molecular basis of the cerebellar plasticity or 'learning' process is the long-term depression of the parallel fiber synapses onto PC dendrites, which results from the simultaneous firing of the CF and mossy fiber input to the same PC [Nieuwenhuys et al., 2008; Lüscher and Huber, 2010].

Some regions of the cerebellum have been studied in more detail than others, and their functions are better understood. One such area is the vestibulocerebellum (flocculus, nodulus and ventral paraflocculus) which is modulated by CF from the dc and vlo regions of the IO [Barmack, 2006, fig. 1; Voogd and Barmack, 2006, fig. 1]. These CF projections establish the highly organized optokinetic and vestibular coordinate system (essential for the orientation of the animal in space) onto the vestibulocerebellar PC, and the cerebellar nuclei. The cerebellar PC activity is structured by this coordinate system, and, via its efferent projections to the cerebellar and the vestibular nuclei, transmits finely tuned coordinate signals into other networks. In contrast to the vestibulocerebellum, the function of the cerebellar hemispheres, and their IO modulation, is less clear, and currently several alternative hypotheses for their function are being considered [Schmahmann, 1997; Glickstein, 2007; Glickstein and Doron, 2008; Stoodley and Schmahmann, 2010]. In addition to its role in motor control, reflex adaptation, and motor learning, evidence has been put forward to support the idea that the cerebellum may also be involved in modulating cognition.

The internal structure of the IO appears relatively homogeneous, but is thought to operate in groups of synchronized neuronal clusters [for a review, see Llinás, 2009]. The dendritic and axonal processes of the neurons form characteristic glomeruli, in which the neurons are electronically coupled by gap junctions between their spines [Sotelo et al., 1974; De Zeeuw et al., 1998; Llinás, 2009]. Through this coupling the electrical activity of clusters of IO neurons can be synchronized, and generate finely tuned output signals in CF [Barmack, 2006]. Recently Llinás [2009] reviewed the evidence suggesting that the internal coupling of IO clusters may provide the basis for an oscillatory function required by the cerebellar cortex for 'motor error correction'.

Connections

Efferents. The IO subnuclei described above, and in figure 1, can also be defined on the basis of their afferent and efferent projections, which show a high degree of topographical organization. An axon from each olivary cell crosses the midline, and ascends in the inferior cerebellar peduncle (or restiform body) to the cerebellar cortex [Voogd and Ruigrok, 2012, fig. 15.1]. In the cerebellar cortex each axon terminates as a CF on about 10 PC, which are distributed across several cerebellar folia in a sagittal transfolial array [Sugihara et al., 1999; Sugihara, 2001]. Each PC receives only 1 CF. In addition to terminating on PC, the CF send collaterals to the cerebellar nuclei. The PO projects to the dentate nucleus and cerebellar hemispheres; the caudal MAO projects to the fastigial nucleus and the vermis, the rostral MAO projects to the globose nucleus and the intermediate cerebellar cortex; the rostral DAO projects to the emboliform nucleus and the intermediate cerebellar cortex, and the caudal DAO projects to the lateral vestibular nucleus (LVN) [for reviews, see Voogd, 1995; Voogd and Wylie, 2004; Nieuwenhuys et al., 2008].

The mapping of the IO CF onto the cerebellar folia forms discrete and narrow, strictly parasagittal (transfolial) longitudinal zones, with an exquisite topography. This is clearly different from the more diffuse mossy fiber projection to the cerebellum [Nieuwenhuys et al., 2008, fig. 20.5, 20.7, 17.7; see also Wu et al., 1999]. The olivocerebellar zones have been carefully studied in several species, and are evident in anatomical studies, histochemical experiments (using zebrin and acetylcholinesterase staining techniques), physiological and embryological investigations [for reviews, see Voogd, 1995; Voogd and Wylie, 2004; Nieuwenhuys et al., 2008]. The PC efferents imprint this topography on the cerebellar nuclei, which are, in turn, connected through reciprocal connections to the corresponding regions of the IO [for reviews, see Voogd, 1995, 2004; Nieuwenhuys et al., 2008]. This recurrent multisynaptic olivocerebellar loop is just one of several such loops, superimposed on the olivocerebellar system [Voogd and Ruigrok, 2012]. The olivocerebellar pathways and loops have been intensely investigated in experimental animals rather than humans, but these circuits appear to be similar in humans. It has been proposed that the recurrent loops contribute to the hypertrophy of IO cells seen in both animal models and humans (accompanied by palatal myoclonus), after lesions affecting the dentate nucleus [Hong et al., 2008; Shaikh et al., 2010].

Afferents. The IO receives 3 different categories of afferents: (1) excitatory inputs from the spinal cord, from brainstem nuclei and cerebral cortex: (2) inhibitory GABAergic ones from the cerebellar nuclei and the vestibular nuclei, and lastly (3) monoaminergic afferents (dopaminergic, adrenergic) which serve as level setters modulating the cerebellar activity, whereby the serotonergic innervation mainly targets the accessory olives [Voogd et al., 1996]; see chapters in 'Neuromodulatory Systems'. The afferent systems regulate the degree of coupling between the cells within the IO clusters [Ruigrok and Voogd, 1995].

Excitatory inputs to the IO carry information from motor, somatosensory, vestibular, visual and optokinetic systems [for a review, see Ruigrok and Cella, 2004]. The parvocellular part of the red nucleus (RNp, which plays a role in motor control) projects via the massive *central* tegmental tract mainly to the PO (also to the rostral MAO and caudal DAO); the laterocaudal subdivision of the RNp innervates the dorsal lamina of the PO, and the dorsolateral RNp region innervates the ventral PO lamina [for reviews, see Voogd, 2004; Nieuwenhuys et al., 2008]. The nucleus of Darkschewitsch sends efferents to the IO, targeting the rostral MAO; the fibers descend more medially than the RNp fibers, in the *medial* tegmental tract close to the medial longitudinal fasciculus. The visuomotor pretectum projects to DAO nuclei. Somatosensory information is conveyed topographically from the spinal cord, from the dorsal column nuclei and the trigeminal complex, to the entire DAO, the adjacent dorsal lamina of the PO and caudal MAO (but not group β). Parts of the vestibular nuclei (parasolitary nucleus, PSol; superior vestibular nucleus; Y group) project to subnuclei of the MAO, specifically the group β, and dmcc [Nieuwenhuys et al., 2008, fig. 17.7]. The pathways carrying vestibular canal and otolith signals are reviewed by Voogd [1995] and Barmack [2006]. Efferents from the superior colliculus carrying visually related signals also target the dmcc. Finally the accessory optic nuclei carrying horizontal optokinetic signals, the dorsal terminal nucleus and the nucleus of the optic tract terminate in the dc of Kooy, whereas vertical accessory optic nuclei target the adjacent vlo [for a review, see Barmack, 2006].

Small GABAergic neurons in the cerebellar nuclei send a significant crossed inhibitory connection to the IO, which is thought to maintain a strong control over the olivary function [van der Want and Voogd, 1987; van der Want et al., 1989; Ruigrok and Voogd, 1990; Nieuwenhuys et al., 2008, chapter 20; Llinás, 2009]. These inhibitory cells represent almost 50% of the cerebellar nucleus neuronal population, an indication of the importance of this feedback pathway. Recent experiments show that the projection may control the synchronicity of the IO cell clusters [Leznik et al., 2002]. There is an almost complete reciprocity between the crossed olivonuclear collaterals and the crossed nucleo-olivary pathways [for reviews, see Ruigrok et al., 1990; Voogd, 2004]. The dentate nucleus projects to the PO, the emboliform nucleus to the rostral DAO, the globose nucleus to the rostral MAO, and less numerous connections from the fastigial nucleus to the caudal MAO. The caudal DAO receives afferents from the LVN of Deiters, and some GABAergic afferents to specific regions of the IO arise from the prepositus hypoglossi, the Y group and PSol [Nelson and Mugnaini, 1989; de Zeeuw et al., 1994; Barmack et al., 1998].

References

Barmack NH: Inferior olive and oculomotor system. Prog Brain Res 2006;151:269–291.

Barmack NH, Fredette BJ, Mugnaini E: Parasolitary nucleus: a source of GABAergic vestibular information to the inferior olive of rat and rabbit. J Comp Neurol 1998;392:352–372.

De Zeeuw CI, Gerrits NM, Voogd J, Leonard CS, Simpson JI: The rostral dorsal cap and ventrolateral outgrowth of the rabbit inferior olive receive a GABAergic input from dorsal group Y and the ventral dentate nucleus. J Comp Neurol 1994;341:420–432.

De Zeeuw CI, Simpson JI, Hoogenraad CC, Galjart N, Koekkoek SK, Ruigrok TJ: Microcircuitry and function of the inferior olive. Trends Neurosci 1998;21:391–400.

Desclin JC: Histological evidence supporting the inferior olive as the major source of cerebellar climbing fibers in the rat. Brain Res 1974;77:365–384.

Glickstein M: What does the cerebellum really do? Curr Biol 2007;17:R824–R827.

Glickstein M, Doron K: Cerebellum: connections and functions. Cerebellum 2008; 7:589–594.

Hong S, Leigh RJ, Zee DS, Optican LM: Inferior olive hypertrophy and cerebellar learning are both needed to explain ocular oscillations in oculopalatal tremor. Prog Brain Res 2008;171:219–226.

Koutcherov Y, Huang X-F, Halliday G, Paxinos G: Organization of human brain stem nuclei; in Paxinos G, Mai JK (eds): The Human Nervous System, ed 2. Amsterdam, Elsevier Academic Press, 2004.

Leznik E, Makarenko V, Llinás R: Electrotonically mediated oscillatory patterns in neuronal ensembles: an in vitro voltage-dependent dye-imaging study in the inferior olive. J Neurosci 2002;22:2804–2815.

Llinás RR: Inferior olive oscillation as the temporal basis for motricity and oscillatory reset as the basis for motor error correction. Neurosci 2009;162:797–804.

Lüscher C, Huber KM: Group 1 mGluR-dependent synaptic long-term depression: mechanisms and implications for circuitry and disease. Neuron 2010;65:445–459.

Nelson BJ, Mugnaini E: Origins of GABAergic inputs to the inferior olive; in Strata P (ed): The Olivocerebellar System in Motor Control. Berlin, Springer, 1989.

Nieuwenhuys R, ten Donkelaar HJ, Nicholson C: The Central Nervous System of Vertebrates. Berlin, Springer, 1998.

Nieuwenhuys R, Voogd J, van Huijzen C: The Human Central Nervous System. Berlin, Springer, 2008.

Paxinos G, Huang X-F, Sengul G, Watson C: Organization of the brainstem; in Mai JK, Paxinos G (eds): The Human Nervous System, ed 3. Amsterdam, Academic Press, 2012.

Ruigrok TJ, Cella F: Precerebellar nuclei and red nucleus; in Paxinos G (ed): The Rat Nervous System, ed 2. Amsterdam, Elsevier, 2004.

Ruigrok TJ, de Zeeuw CI, Voogd J: Hypertrophy of inferior olivary neurons: a degenerative, regenerative or plasticity phenomenon. Eur J Morphol 1990;28:224–239.

Ruigrok TJ, Voogd J: Cerebellar nucleo-olivary projections in the rat: an anterograde tracing study with *Phaseolus vulgaris* leucoagglutinin (PHA-L). J Comp Neurol 1990;298:315–333.

Ruigrok TJ, Voogd J: Cerebellar influence on olivary excitability in the cat. Eur J Neurosci 1995;7:679–693.
Schmahmann DJ: The Cerebellum and Cognition. International Review of Neurobiology. San Diego, Academic Press, 1997, vol 41.
Shaikh AG, Hong S, Liao K, Tian J, Solomon D, Zee DS, Leigh RJ, Optican LM: Oculopalatal tremor explained by a model of inferior olivary hypertrophy and cerebellar plasticity. Brain 2010;133:923–940.
Sotelo C, Llinás R, Baker R: Structural study of inferior olivary nucleus of the cat: morphological correlates of electrotonic coupling. J Neurophysiol 1974;37:541–559.
Stoodley CJ, Schmahmann JD: Evidence for topographic organization in the cerebellum of motor control versus cognitive and affective processing. Cortex 2010;46:831–844.
Sugihara I: Efferent innervation in the goldfish saccule examined by acetylcholinesterase histochemistry. Hearing Res 2001;153:91–99.
Sugihara I, Wu HS, Shinoda Y: Morphology of single olivocerebellar axons labeled with biotinylated dextran amine in the rat. J Comp Neurol 1999;414:131–148.
Szentagothai J, Rajkovits K: Über den Ursprung der Kletterfasern des Kleinhirns. Z Anat Entwicklungsgesch 1959;121:130–141.
Van der Want JJ, Voogd J: Ultrastructural identification and localization of climbing fiber terminals in the fastigial nucleus of the cat. J Comp Neurol 1987;258:81–90.
Van der Want JJ, Wiklund L, Guegan M, Ruigrok T, Voogd J: Anterograde tracing of the rat olivocerebellar system with *Phaseolus vulgaris* leucoagglutinin (PHA-L). Demonstration of climbing fiber collateral innervation of the cerebellar nuclei. J Comp Neurol 1989;288:1–18.
Voogd J: Nervous system – cerebellum; in Williams PL, Bannister LH, Berry MM, Collins P, Dyson M, Dussek JE, Ferguson MWJ (eds): Gray's Anatomy, ed 38. London, Churchill Livingstone, 1995.
Voogd J: Cerebellum and precerebellar nuclei; in Paxinos G, Mai JK (eds): The Human Nervous System, ed 2. Amsterdam, Elsevier Academic Press, 2004.
Voogd J, Barmack NH: Oculomotor cerebellum. Prog Brain Res 2006;151:231–268.
Voogd J, Jaarsma D, Marani E: The cerebellum; in Hokfelt T (ed): Chemoarchitecture and Anatomy. Handbook of Chemical Neuroanatomy. Amsterdam, Elsevier, 1996.
Voogd J, Ruigrok T: Cerebellum and precerebellar nuclei; in Mai JK, Paxinos G (eds): The Human Nervous System, ed 3. San Diego, Academic Press, 2012.
Voogd J, Wylie DRW: Functional and anatomical organization of floccular zones: a preserved feature in vertebrates. J Comp Neurol 2004;470:107–112.
Whitworth RH Jr, Haines DE: On the question of nomenclature of homologous subdivisions of the inferior olivary complex. Arch Ital Biol 1986;124:271–317.
Wu HS, Sugihara I, Shinoda Y: Projection patterns of single mossy fibers originating from the lateral reticular nucleus in the rat cerebellar cortex and nuclei. J Comp Neurol 1999;411:97–118.

Precerebellar Nuclei (Plate 12)

93 Interfascicular Hypoglossal Nucleus (IFH)

Original name:

Nucleus interfascicularis hypoglossi

Location and Cytoarchitecture – Original Text

In cross-sections through the level of the nucleus of Roller (Ro) one finds numerous cells scattered among the vertically coursing fibers of the hypoglossal nerve. These cells constitute the interfascicular hypoglossal nucleus (IFH). They lie ventral to the Ro, lateral to the medial longitudinal fasciculus (MLF) and medial to the central nucleus of the medulla oblongata. Dorsally the IFH is often directly continuous with the Ro.

Scattered among the fibers of the MLF and of the tectospinal tracts, from the level of the caudal pole of the hypoglossal nucleus to the level of the nucleus raphe magnus (RMg), one finds a variable number of small or medium-sized nerve cells. These are most numerous at the level of the IFH. We have considered these cells to be scattered elements derived from the adjacent nuclei, and thus did not delineate them as a distinct nuclear group. They lie along the lateral border of the cell clusters in and around the MLF called the paramedian reticular nuclei by Brodal [1983].

The cells which compose this nucleus are similar in their morphological characteristics to those of the Ro (fig. 1–5). This similarity of morphology suggests a probable similarity of connections.

Functional Neuroanatomy

Function

The function of the IFH is not clear. However, as suggested above, some similarity between IFH and Ro might be expected on account of their comparable morphology. This has been to some extent confirmed by the similarity of their connectivity: both nuclei project to the cerebellum, the spinal cord and perhaps the prepositus nucleus [for a review, see Büttner-Ennever and Büttner, 1988]. It is reasonable to suppose that the IFH, like the Ro, participates in motor control, possibly related to head and eye movements. In addition the IFH and Ro both appear to have some neuromodulatory neurons scattered within their borders (see below).

In keeping with these concepts the IFH is found to be degenerated, along with several other precerebellar nuclei, in cases of spinal cerebellar ataxia (types 2 and 3) [Rüb et al., 2005].

Connections

The cerebellar projections from the IFH terminate in regions which include the oculomotor vermis (lobule VII), and the flocculus [Brodal and Brodal, 1983; Yamada and Noda, 1987]. The relationship between these IFH cells and the cell groups of the paramedian tracts (nucleus No. 96) and the paramedian reticular nucleus is not clear. Each cell group projects to the cerebellum, but it seems that the cells close to the midline around the paramedian tracts control eye movements whereas those further laterally, like the IFH and the paramedian reticular nucleus, control gaze movements [Grantyn et al., 1987; Büttner-Ennever and Büttner, 1988].

Some neurons in the IFH project directly to the spinal cord, and these cells have been specifically shown to receive afferent terminals from the dorsal posterior hypothalamus [Ho-

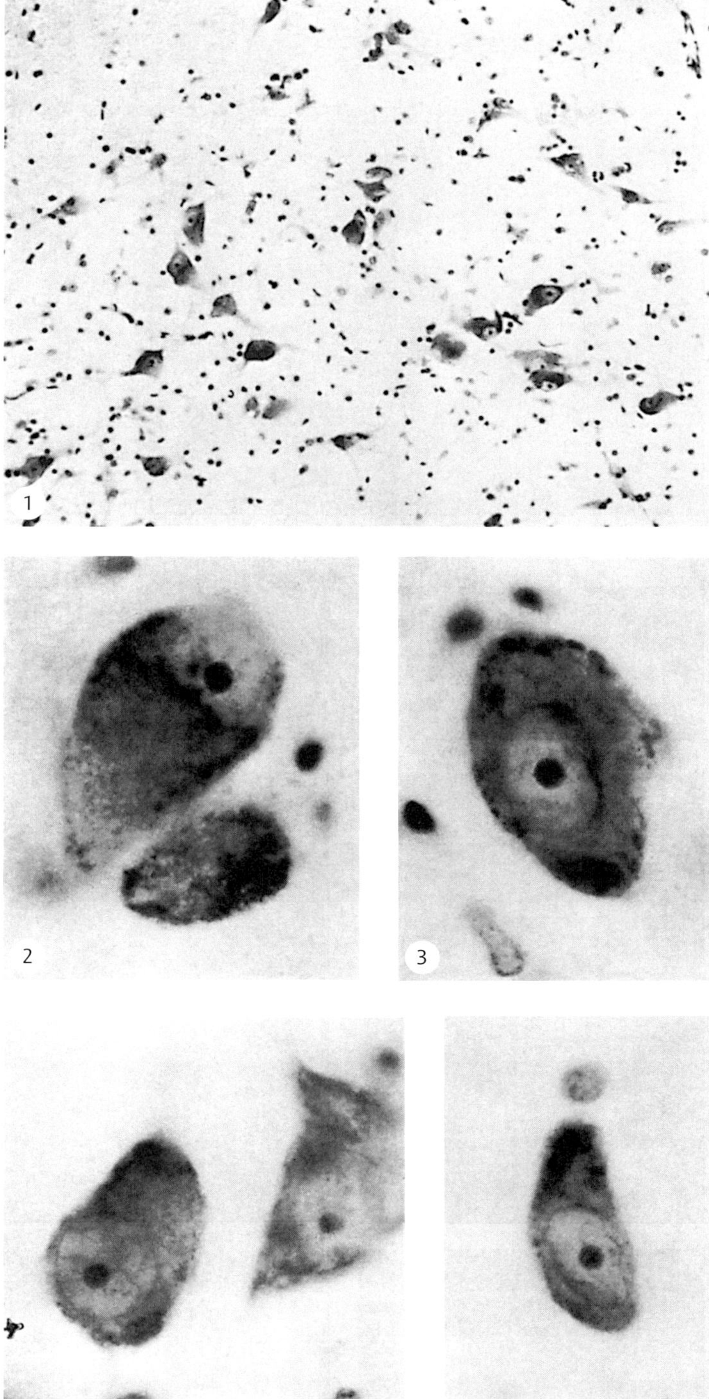

Fig. 1. Interfascicular hypoglossal nucleus. Magnification ×150.
Fig. 2–5. Cells from the interfascicular hypoglossal nucleus. Magnification ×1,000.

soya, 1985]. Other experiments have demonstrated that there are substance-P-containing neurons in the IFH which project to the spinal cord, and terminate in the intermediate and ventral horn [Helke et al., 1982; Menetrey and Basbaum, 1987]. Neurons within the IFH have also been shown to contain serotonin and met-enkephalin [Hancock, 1982]. In rats the IFH lies adjacent to the nucleus raphe pallidus, the serotonergic B1 cell group [Dahlström and Fuxe, 1964], but in humans it lies closer to the B2 group of the nucleus raphe obscurus; it seems likely that the IFH contains laterally scattered serotonergic cells belonging to the B1 and B2 groups, in a similar way to the gigantocellular nucleus, pars α, adjacent to the RMg, and suggests that the IFH may function, at least in part, as a neural modulator of the motor output like the adjacent raphe nuclei [Holstege et al., 2004].

There are relatively few reports of afferents to the IFH, apart from the limbic inputs from the hypothalamus to spinal projecting cells described above, and these are probably not associated with the cerebellar projections of the IFH, but rather the serotonergic modulatory system. However, inputs from the dorsomedial medullary reticular formation (nucleus No. 55), an area related to eye and head movements, have been reported to target the IFH [Graybiel, 1977].

References

Brodal A: The perihypoglossal nuclei in the macaque monkey and the chimpanzee. J Comp Neurol 1983;218:257–269.

Brodal A, Brodal P: Observations on the projection from the perihypoglossal nuclei onto the cerebellum in the macaque monkey. Arch Ital Biol 1983;121:151–166.

Büttner-Ennever JA, Büttner U: The reticular formation; in Büttner-Ennever JA (ed): Neuroanatomy of the Oculomotor System. Reviews of Oculomotor Research. Amsterdam, Elsevier, 1988, vol 2.

Dahlström A, Fuxe K: Evidence for the existence of monoamine neurons in the central nervous system. I. Demonstration of monoamines in the cell bodies of brain stem neurons. Acta Physiol Scand 1964;62:1–55.

Grantyn A, Ong Meang Jacques V, Berthoz A: Reticulo-spinal neurons participating in the control of synergic eye and head movements during orienting in the cat. II. Morphological properties as revealed by intra-axonal injections of horseradish peroxidase. Exp Brain Res 1987;66:355–377.

Graybiel AM: Direct and indirect preoculomotor pathways of the brainstem: an autoradiographic study of the pontine reticular formation in the cat. J Comp Neurol 1977;175:37–78.

Hancock M: Visualization of peptide-immunoreactive processes on serotonin-immunoreactive cells using two-color immunoperoxidase staining. J Histochem Cytochem 1982;32:311–314.

Helke CJ, Neil JJ, Massari VJ, Loewy AD: Substance P neurons project from the ventral medulla to the intermediolateral cell column and ventral horn in the rat. Brain Res 1982;243:147–152.

Holstege GG, Mouton LJ, Gerrits NM: Emotional motor system; in Paxinos G, Mai JK (eds): The Human Nervous System, ed 2. Amsterdam, Elsevier Academic Press, 2004.

Hosoya Y: Hypothalamic projections to the ventral medulla oblongata in the rat, with special reference to the nucleus raphe pallidus: a study using autoradiographic and HRP techniques. Brain Res 1985;344:338–350.

Menetrey D, Basbaum A: The distribution of substance P-, enkephalin- and dynorphin-immunoreactive neurons in the medulla of the rat and their contribution to bulbospinal pathways. Neuroscience 1987;23:173–187.

Rüb U, Gierga K, Brunt ER, de Vos AI, Bauer M, Schöls L, Bürk K, Auburger G, Bohl J, Schultz C: Spinocerebellar ataxias types 2 and 3: degeneration of the precerebellar nuclei isolates the three phylogenetically defined regions of the cerebellum. J Neural Transm 2005;112:1523–1545.

Yamada J, Noda H: Afferent and efferent connections of the oculomotor cerebellar vermis in the macaque monkey. J Comp Neurol 1987;265:224–241.

94 Prepositus Nucleus (PrP)

Original name:
Nucleus praepositus hypoglossi

Alternative names:
Prepositus hypoglossal nucleus
Nucleus prepositus

Location and Cytoarchitecture – Original Text

This nucleus lies beneath an elevation on the medial part of the floor of the fourth ventricle in the oral medulla and caudal pons. It extends from the oral pole of the hypoglossal nucleus caudally to the level of the abducens nucleus (VI) orally – a distance of 6–7 mm. The caudal pole of the nucleus merges with the oral pole of the intercalated nucleus (INSt).

On cross-section, the nucleus prepositus (PRP) is egg shaped in outline with the long axis directed horizontally. Its caudal pole lies medial to the dorsal motor nucleus of the vagus (DMX), dorsal to the dorsal paragigantocellular nucleus, ventral to the stratum gliosum and lateral to the extrarapheal portion of the nucleus raphe obscurus and to the dorsal paramedian nucleus. Farther orally, the DMX and the solitary nucleus migrate ventrally away from the floor of the ventricle. The PrP is then related laterally to the 'marginal zone' which is the most medial part of the medial vestibular nucleus (MVN). The oral pole of the PrP is related laterally directly to the MVN, and ventrally to the nucleus reticularis pontis caudalis (NRPC). Approaching the level of the VI, the PrP diminishes somewhat in size and terminates with the appearance of the former structure. On cross- and sagittal sections it is apparent that this oral pole of the PrP is directly continuous with the supragenual nucleus (SG).

Cells of two types are found within the PrP:

(a) medium-sized to large, plump multipolar cells which possess long dendrites and darkly stained, indistinct Nissl granules; in many of these cells the Nissl substance may be peripherally arranged (fig. 4–7);

(b) small to medium-sized, plump, oval, triangular or fusiform cells which possess indistinct Nissl granules; the intensity of staining of these Nissl granules varies from light to medium (fig. 8–13).

The distribution of these two cell types is not uniform and, accordingly, caudal, intermediate and oral portions of the nucleus can be distinguished. The caudal portion (fig. 1) is characterized by predominance of large cells (type a). These cells usually form two compact groups which, in cross-sections, lie one medially and one laterally. One such group is represented on plates 14 and 15. In the intermediate portion (fig. 2), the two cell types are equally represented. Finally, the oral portion (fig. 3) is composed exclusively of small cells (type b).

As may be noted from the foregoing descriptions, the large cells of the INSt and PrP bear a marked resemblance to the cells which compose the nucleus of Roller (Ro) and the nucleus interfascicularis hypoglossi. [...] The fact that morphologically similar cells are found in all 4 of these nuclei suggests that they possess biological properties in common. [...]

Additional Comments

The reader is referred to the studies by Brodal and McCrea for more recent reviews of the cytoarchitecture [Brodal, 1983; McCrea and Baker, 1985; McCrea, 1988; McCrea and Horn, 2006]. In these studies 3, instead of 2, different types of neurons are distinguished: (1) 'multidendritic' cells in the caudomedial PrP,

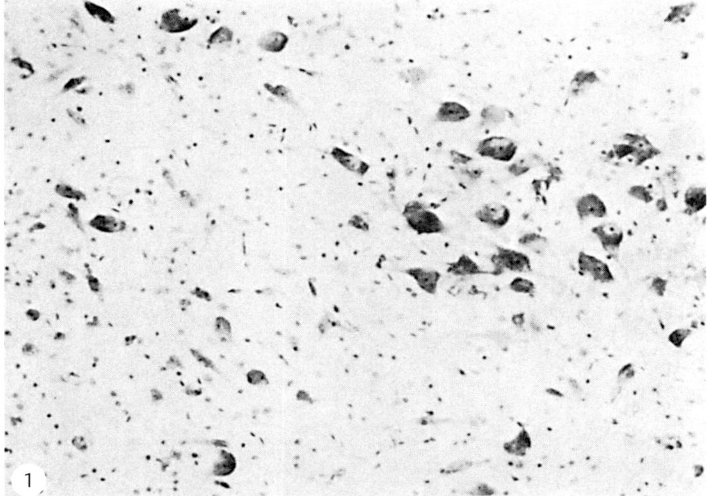

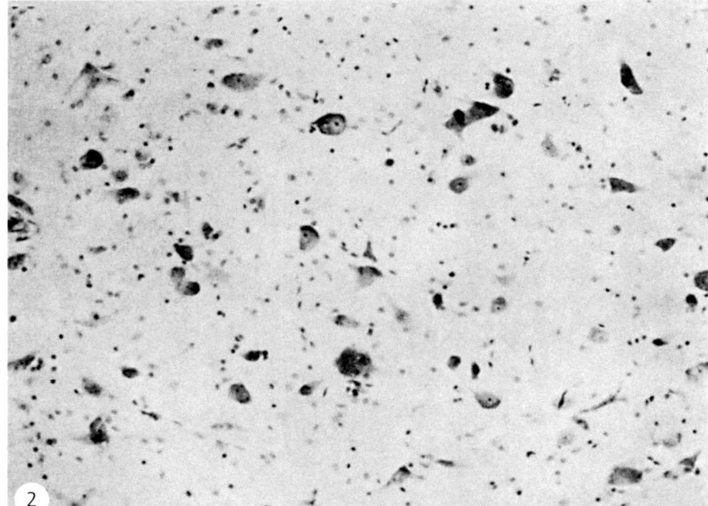

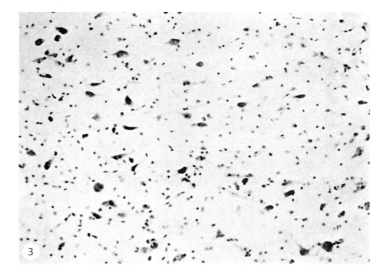

Fig. 1. Prepositus nucleus, caudal portion. Magnification ×150.
Fig. 2. Prepositus nucleus, midportion. Magnification ×150.
Fig. 3. Prepositus nucleus, oral portion. Magnification ×150.

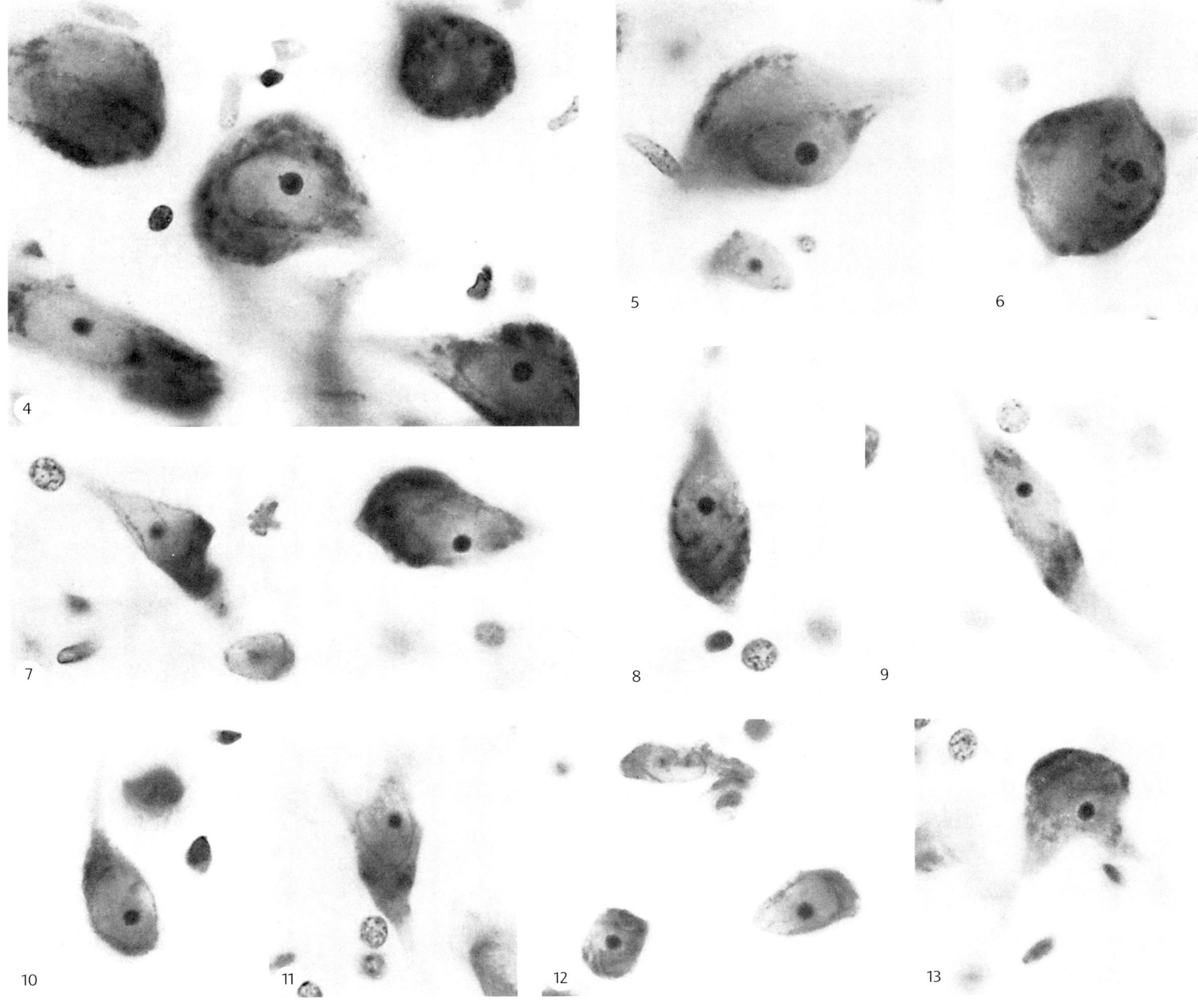

Fig. 4–13. Different cell types from the prepositus nucleus. Magnification ×1,000.

which are included in Olszewski and Baxter 'type a' neurons; (2) 'small' cells in the dorsolateral PrP, which are included in Olszewski and Baxter 'type b' neurons; (3) medium-sized 'principal' cells in the rostral, central region that include both 'type a' and 'type b'.

In plates 14 and 15 the 'multidendritic' cells of McCrea and Baker [1985] are seen medially, the 'small' cells dorsolaterally, and the 'principal' cells in the middle of the PrP. In these plates, 2 distinct regions of the caudal PrP can be seen, one consisting of larger neurons, which merge into the Ro, and the second a smaller-celled region which joins with the INSt. The small cells of the PrP are predominantly commissural or interneurons, and the larger cells give rise to projections to oculomotor-related cell groups. These major cytological features of the nucleus appear to be present in all species, including man.

In primates a subnucleus of the MVN, the 'marginal zone' forms the lateral border of the rostral half of the PrP [Langer et al., 1986]. On the medial border of the PrP lies the nucleus paramedianus dorsalis (nucleus No. 106) and (not described here) the epifascicular group, the 'EF' cell group of Koutcherov et al. [2004]. The EF cell group is a main source of afferent input to the locus coeruleus, providing a GABAergic inhibition of the neurons [Ennis and Aston-Jones, 1989].

Functional Neuroanatomy

Function

The PrP is a site of interaction between visual and vestibular systems. It is thought to contribute to the maintenance of 'eye and head position' (gaze) in the horizontal plane. It functions as a neural integrator (in the mathematical sense) converting eye and head velocity signals into position signals, and plays an important role in stabilizing horizontal components of gaze [Godaux et al., 1993; Kaneko, 1997, 2005; McCrea and Horn, 2006]. Lesions of the PrP cause a severe horizontal integrator deficit (i.e. gaze-holding deficit), resulting in gaze-evoked nystagmus; vertical gaze-holding is only partly affected and saccades remain intact [Cannon and Robinson, 1987]. For a review see Leigh and Zee [2006].

Connections

Efferents. The PrP has widespread projections to the brainstem and cerebellum, which are reviewed in detail by McCrea and Horn [2006]. The PrP sends fibers to almost all oculomotor structures: it projects to the VI, the oculomotor nucleus, the pontomedullary reticular formation and the superior colliculus. The PrP neurons which project to the ipsi- and contralateral vestibular complex seem to be located peripherally in the caudal half of the nucleus.

PrP efferent projections also supply the vestibulo- and oculomotor-cerebellar cortex, including the vermis (lobuli VII, IX and X), the flocculus, ventral paraflocculus (in humans called the tonsilla) and fastigial nuclei [Belknap and McCrea, 1988; McCrea and Horn, 2006]. Projections of the PrP convey vestibular and eye movement information also to the dorsal cap of the inferior olive, predominantly contralaterally [Barmack, 2006]. Finally this nucleus, along with the adjacent SG, sends vestibular-related information in ascending pathways to the dorsal tegmental nucleus and the lateral mammillary nucleus, both of which are part of the head direction system [Brown et al., 2005; Biazoli et al., 2006].

Afferents. The most important inputs to the PrP arise from brainstem areas involved in the control of eye and head movements [Belknap and McCrea, 1988]. More specifically afferents arise from the perihypoglossal nuclei, including the INSt, and there are particularly strong connections from the contralateral PrP, MVN and the descending vestibular nucleus, and bilaterally from a compact region of the ventrolateral vestibular complex, which is otherwise inconspicuous [McCrea et al., 1987]. Further inputs come from the paramedian medullary reticular formation, the regions contralateral to the PrP that contain inhibitory saccadic burst neurons (nucleus No. 55), the ipsilateral NRPC, the vertical premotor cell groups of the mesencephalon (the ipsilateral rostral interstitial nucleus of the medial longitudinal fasciculus and the interstitial nucleus of Cajal), the peri-extraocular motor nuclei and the surrounding regions, the vestibulocerebellum [Langer et al., 1985] and the fastigial nucleus [Noda et al., 1990]. In addition the PrP receives significant visuomotor brainstem pathways from areas such as the ventral lateral geniculate nucleus and the accessory optic nuclei [McCrea and Horn, 2006].

Histochemistry

The PrP stands out histochemically owing to its high content of GABA, and other properties such as its strong nitric oxide (NO) synthase immunoreactivity [McCrea and Horn, 2006]. NO is a diffusible gas, which was shown to act as an intercellular messenger participating in many functional roles, i.e. in ischemia, neurotoxicity and neurodegenerative processes, and modulation of sensory function [Cudeiro and Rivadulla, 1999]. It is converted from L-arginine to NO by the NADPH-dependent NO synthase. Interestingly, the 'marginal zone' of the MVN adjacent to the PrP contains numerous strongly NO-sensitive neurons, but is devoid of NO-releasing i.e. NO synthase neurons. The functional role of the NO system is not fully understood, but a pharmacological study in the alert cat demonstrated that the balanced production of NO by the PrP is necessary for the correct performance of eye movements, since unilateral injections of NO synthase inhibitors into the PrP resulted in a severe long-lasting nystagmus [Moreno-Lopez et al., 2001].

References

Barmack NH: Inferior olive and oculomotor system. Prog Brain Res 2006;151:269–291.

Belknap DB, McCrea RA: Anatomical connections of the prepositus and abducens nuclei in the squirrel monkey. J Comp Neurol 1988;268:13–28.

Biazoli CEJ, Goto M, Campos AM, Canteras NS: The supragenual nucleus: a putative relay station for ascending vestibular signs to head direction cells. Brain Res 2006;1094:138–148.

Brodal A: The perihypoglossal nuclei in the macaque monkey and the chimpanzee. J Comp Neurol 1983;218:257–269.

Brown JE, Card JP, Yates BJ: Polysynaptic pathways from the vestibular nuclei to the lateral mammillary nucleus of the rat: substrates for vestibular input to head direction cells. Exp Brain Res 2005;161:47–61.

Cannon SC, Robinson DA: Loss of the neural integrator of the oculomotor system from brain stem lesions in monkey. J Neurophysiol 1987;57:1383–1409.

Cudeiro J, Rivadulla C: Sight and insight – on the physiological role of nitric oxide in the visual system. Trends Neurosci 1999;22:109–116.

Ennis M, Aston-Jones G: GABA-mediated inhibition of locus coeruleus from the dorsomedial rostral medulla. J Neurosci 1989;9:2973–2981.

Godaux E, Mettens P, Cheron G: Differential effect of injections of kainic acid into the prepositus and the vestibular nuclei of the cat. J Physiol 1993;472:459–482.

Kaneko CR: Eye movement deficits following ibotenic acid lesions of the nucleus prepositus hypoglossi in monkey. I. Saccades and fixation. J Neurophysiol 1997;78:1753–1768.

Kaneko CR: Eye movement deficits following ibotenic acid lesions of the nucleus prepositus hypoglossi in monkeys II. Pursuit, vestibular, and optokinetic responses. J Neurophysiol 2005;81:668–681.

Koutcherov Y, Huang X-F, Halliday G, Paxinos G: Organization of human brain stem nuclei; in Paxinos G, Mai JK (eds): The Human Nervous System, ed 2. Amsterdam, Elsevier Academic Press, 2004.

Langer TP, Fuchs AF, Chubb MC, Scudder CA, Lisberger SG: Floccular efferents in the rhesus macaque as revealed by autoradiography and horseradish peroxidase. J Comp Neurol 1985;235:26–37.

Langer TP, Kaneko CR, Scudder CA, Fuchs AF: Afferents to the abducens nucleus in the monkey and cat. J Comp Neurol 1986;245:379–400.

Leigh RJ, Zee DS: The Neurology of Eye Movements. New York, Oxford University Press, 2006.

McCrea RA: The nucleus prepositus; in Büttner-Ennever JA (ed): Neuroanatomy of the Oculomotor System. Reviews of Oculomotor Research. Amsterdam, Elsevier, 1988, vol 2.

McCrea RA, Baker R: Cytology and intrinsic organization of the perihypoglossal nuclei in the cat. J Comp Neurol 1985;237:360–376.

McCrea RA, Horn AKE: Nucleus prepositus. Prog Brain Res 2006;151:205–230.

McCrea RA, Strassman E, May E, Highstein SM: Anatomical and physiological characteristics of vestibular neurons mediating the horizontal vestibulo-ocular reflex of the squirrel monkey. J Comp Neurol 1987;264:547–570.

Moreno-Lopez B, Escudero M, de Vente J, Estrada C: Morphological identification of nitric oxide sources and targets in the cat oculomotor system. J Comp Neurol 2001;435:311–324.

Noda H, Sugita S, Ikeda Y: Afferent and efferent connections of the oculomotor region of the fastigial nucleus in the macaque monkey. J Comp Neurol 1990; 302:330–348.

95 Nuclei Pararaphales (PRA)

Original name:
Nuclei pararaphales

Associated name:
Paramedian tract cell group 2 [Buresch, 2005], see nucleus No. 96

Location and Cytoarchitecture – Original Text

These nuclei are located in the oral 2–3 mm of the medulla at the level of the maximal development of the medullary striae (see nucleus No. 106, fig. 7). They are formed by elongated, vertically directed, spindle-shaped cell groups situated on either side of the median raphe (midline; fig. 1). These groups may be continuous ventrally with the arcuate nuclei (ARC) and they extend dorsally to the region of the paramedian dorsal nucleus. The size of these cell groups varies considerably from one individual to another, and in those brainstems in which the medullary striae are not well developed the nuclei pararaphales (PRA) often cannot be identified.

The cells of the PRA should probably be regarded as caudal representatives of the pontine nuclei. They are similar to, but smaller and more lightly stained than the cells of the ARC (nucleus No. 89).

Functional Neuroanatomy

Function

Langer et al. [1985] proposed that the PRA in humans is part of a series of cell clusters also found in lower species, called the paramedian tract cell groups (PMT; nucleus No. 96). The PMT cell groups carry eye movement-related signals to the cerebellar flocculus, and are considered to contribute to the stabilization of eye movement or gaze-holding [Langer et al., 1985, fig. 4, 11; Büttner-Ennever and Horn, 1996; Dean and Porrill, 2008]. Midline lesions of the PRA region in humans can result in upbeat nystagmus [Baloh and Yee, 1989; Leigh and Zee, 2006].

Contrary to the hypothesis of Langer et al. [1985], a recent study of human PRA considered the nucleus to be histochemically different from the PMT-2 cell groups in the macaque and chimpanzee, and to be more like the ARC neurons lying immediately ventral to the PRA (fig. 1) [Baizer et al., 2012]. The differing morphology of PRA and presumed PMT cell group neurons shown in figures 2–5 here supports their hypothesis. If this is the case, and the PRA is not a PMT cell group, then nothing is known of its function, and all that can be said of the PRA is that there is general agreement that it is a precerebellar nucleus, like the ARC.

Comment

Two studies have compared PRA in humans with PMT cell groups in monkeys [Buresch, 2005; Baizer et al., 2012]. The PRA in humans is usually much larger than the equivalent PMT cell group in monkeys, which is PMT-2 in the terminology of Buresch. Baizer et al. conclude that the PRA is not a PMT cell group since it was immunoreactive to calretinin and the PMT-2 in monkeys was not. However, the PRA does have some similari-

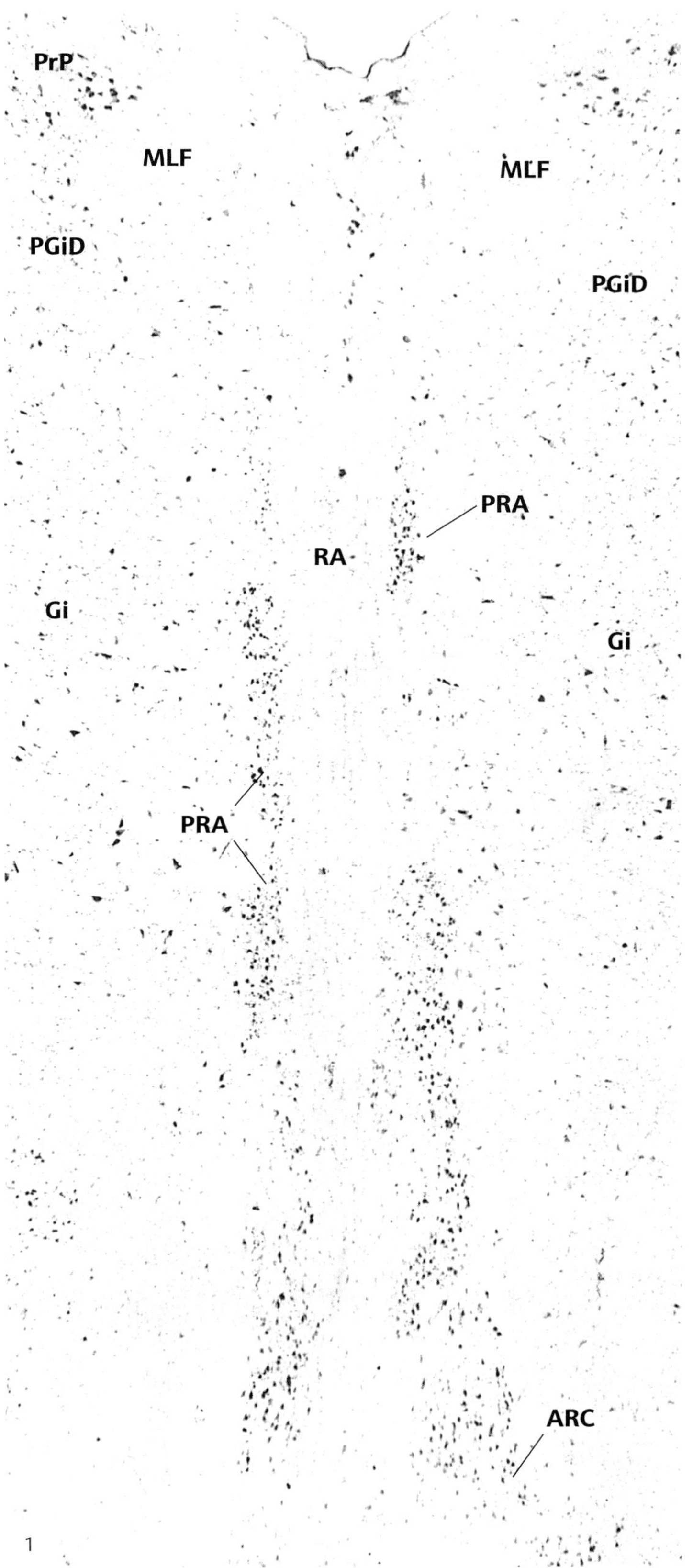

Fig. 1. Nuclei pararaphales. ARC = Arcuate nucleus; Gi = gigantocellular nucleus; MLF = medial longitudinal fasciculus; PGiD = dorsal paragigantocellular nucleus; PRA = nuclei pararaphales; PrP = prepositus nucleus; RA = median raphe (midline). Magnification ×30.

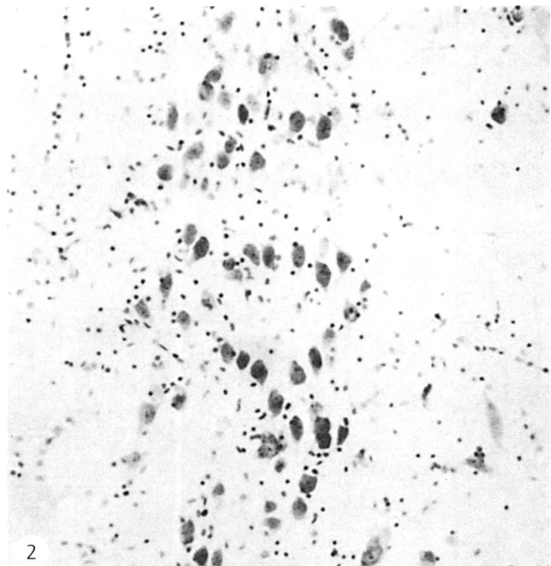

Fig. 2. Nuclei pararaphales. Magnification ×150.

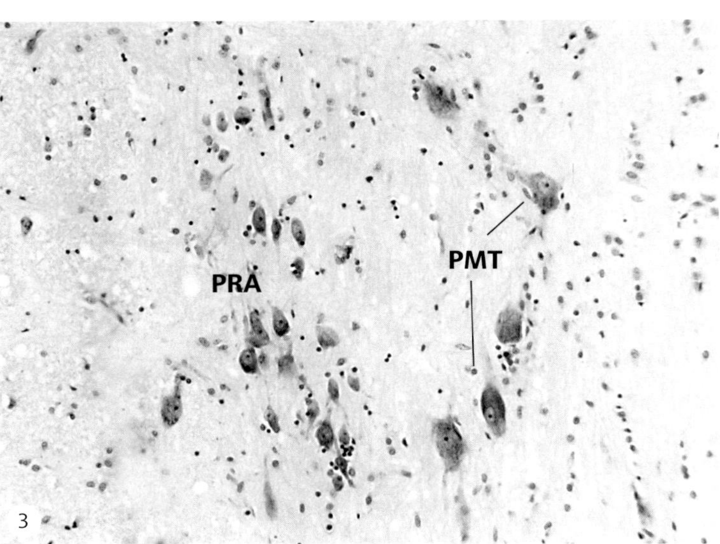

Fig. 3. Photo comparing the cells of the PRA and a presumed PMT-2 cell group (nucleus No. 96). Note the neurons of the PMT-2 cell group are larger with more prominent Nissl substance. Magnification ×150.

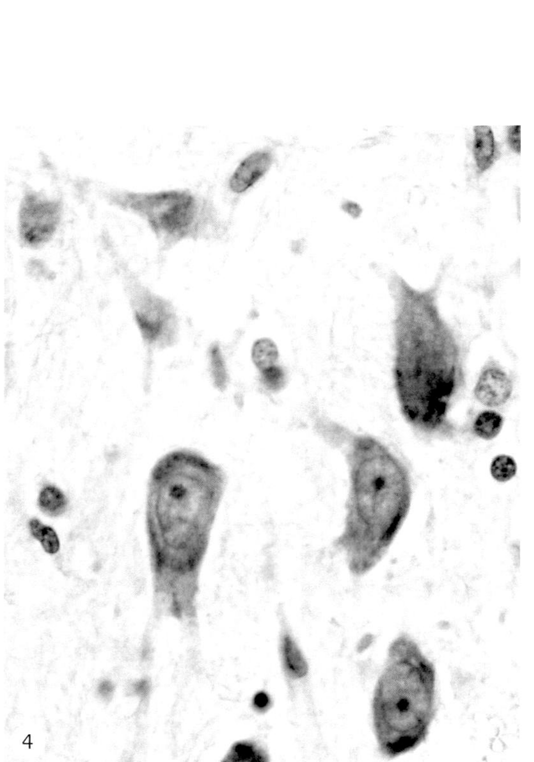

Fig. 4. Cells from the nuclei pararaphales. Magnification ×1,000.

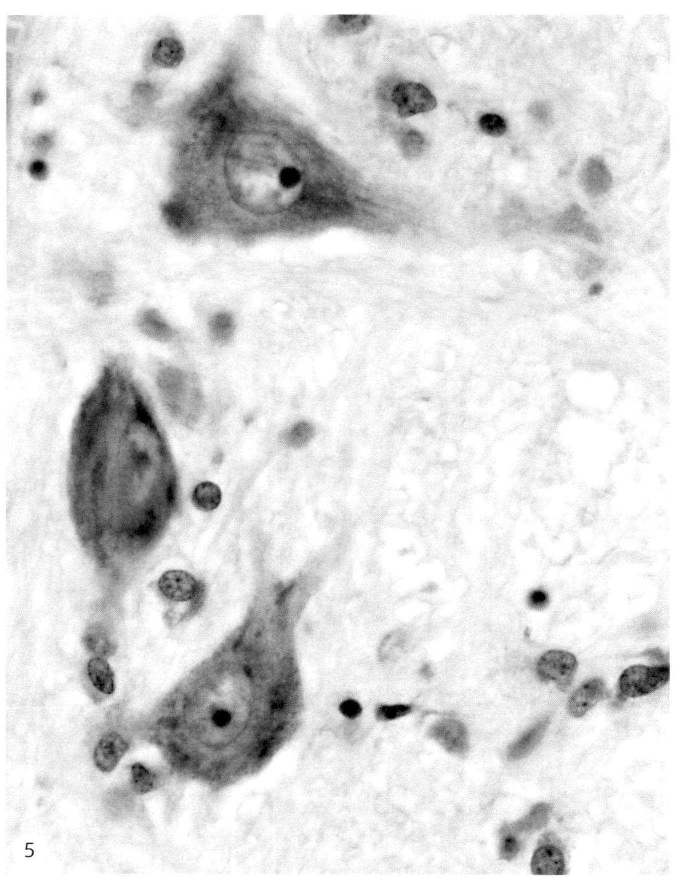

Fig. 5. Cells from the presumed PMT-2 cell group (nucleus No. 96). Magnification ×1,000.

ties with the monkey PMT-2 in terms of location, and other histochemical staining properties: for example neither contains serotonergic neurons. Both nuclei are outlined with acetylcholinesterase stains which label the neuropile strongly, both PRA and PMT-2 have cells that are cytochrome-C positive and they both stain for nonphosphorylated neurofilament protein.

Connections

There is no information on the connectivity of PRA in humans. However, if the PRA is a homolog of the PMT-2 reported in monkeys, then several studies have considered their connections [Büttner-Ennever et al., 1989; Büttner-Ennever and Horn, 1996; Buresch, 2005]. The efferent fibers of the medullary PMT cell groups travel bilaterally to the inferior cerebellar peduncle and flocculus region via the medullary striae, contributing to the internal and external arcuate pathways (see nucleus No. 106, fig. 7). In monkeys the PMT-2 receives inputs from vertical premotor areas for eye movements [Büttner-Ennever and Büttner, 1988; Buresch, 2005].

References

Baizer JS, Weinstock N, Witelson SF, Sherwood CC, Hof PR: The nucleus paraphales in the human, chimpanzee, and macaque monkey. Brain Struct Funct 2013;218:389–403.

Baloh RW, Yee RD: Spontaneous vertical nystagmus. Rev Neurol (Paris) 1989;145: 527–532.

Buresch N: Neuroanatomische Charakterisierung blickstabilisierender Neurone an der Hirnstammmittellinie der Primaten, einschliesslich des Menschen; dissertation, LMU München, 2005.

Büttner-Ennever JA, Büttner U: The reticular formation; Büttner-Ennever JA (ed): Neuroanatomy of the Oculomotor System. Reviews of Oculomotor Research. Amsterdam, Elsevier, 1988, vol 2.

Büttner-Ennever JA, Horn AKE: Pathways from cell groups of the paramedian tracts to the floccular region. Ann NY Acad Sci 1996;781:532–540.

Büttner-Ennever JA, Horn AKE, Schmidtke K: Cell groups of the medial longitudinal fasciculus and paramedian tracts. Rev Neurol (Paris) 1989;145:533–539.

Dean P, Porrill J: Oculomotor anatomy and the motor-error problem: the role of the paramedian tract nuclei. Prog Brain Res 2008;171:177–186.

Langer TP, Fuchs AF, Scudder CA, Chubb MC: Afferents to the flocculus of the cerebellum in the rhesus macaque as revealed by retrograde transport of horseradish peroxidase. J Comp Neurol 1985;235:1–25.

Leigh RJ, Zee DS: The Neurology of Eye Movements. New York, Oxford University Press, 2006.

Precerebellar Nuclei (Not on plates)

96 Cell Groups of the Paramedian Tract (PMT Cell Groups)

Individual and alternative names:

In monkeys, listed from rostral to caudal [Langer et al., 1985]

Dorsal midline pontine group:
PMT-6 [Buresch, 2005]

Intrafascicular nucleus of the preabducens area:
PMT-5a [Buresch, 2005]
nucleus incertus [Cheron et al., 1995]

Dorsal subnucleus of the nucleus raphe pontis:
PMT-5b and PMT-5c [Buresch, 2005]
dorsal nucleus of the raphe [Sato et al., 1983]

Rostral cap of the abducens nucleus:
PMT-4a [Buresch, 2005]

Nucleus supragenualis (nucleus No. 98):
PMT-3 [Buresch, 2005]

Nuclei pararaphales (nucleus No. 95):
PMT-2 [Buresch, 2005]

Medullary interfascicular nucleus:
paramedian reticular nucleus [Brodal, 1953, 1981; Sato et al., 1983]
PMT-1 [Buresch, 2005]

Location and Cytoarchitecture

The cell groups of the paramedian tracts (PMT cell groups) are a collection of 6 or more clusters of neurons, identified on the basis of their common projection to the flocculus region. They lie scattered in and around the medial paramedian fiber tracts (PMT, e.g. medial longitudinal fasciculus, MLF, reticulospinal and tectospinal tracts) throughout the whole length of the pons and medulla. They have been described in several species, and they are considered to form a continuum of flocculus-projecting neurons with a common functional role; for reviews, see Horn [2006], Büttner-Ennever and Horn [1996], Langer et al. [1985], Buresch [2005] and Goldberg et al. [2012]. In the rat, cat, monkey and man, the PMT cell groups have been given different names. The nomenclature of Langer et al. for the individual cell groups in the nonhuman primate has been adopted here since it is most easily related to humans. Buresch [2005] has shown that homologous cell groups can be identified in man. He proposes a new simplified terminology, naming the cell groups PMT-1 to PMT-6.

The most rostral group PMT-6, the 'dorsal midline pontine group', lies ventral to the nucleus reticularis tegmenti pontis, but this group remains unclear in humans. Further caudally (PMT-5a) the 'intrafascicular nucleus of the preabducens area' lies scattered between the fascicles of the MLF, and is continuous with PMT-5b, a small region ventral to the MLF and called 'nucleus L' by Paxinos et al. [2000]. The next group of PMT neurons, caudally, is PMT-4a or the 'rostral cap of the abducens nucleus', which lies within the boundaries of the classical abducens nucleus. However, unlike motoneurons, the PMT neurons are not cholinergic [Buresch, 2005]. At the same level another PMT cell group, PMT-3 or the 'supragenual nucleus', lies dorsal to the facial genu and the abducens nucleus. In the rostral medulla oblongata, ventral to the caudal prepositus nucleus (PrP), lies the 'pararaphales nucleus', which is very large in humans and is probably not equivalent to PMT-2 (see nucleus No. 95 and figs.), and the most caudal PMT group (PMT-1), ventral to the hypoglossal nucleus, is the 'medullary interfascicular nucleus'. The latter correlates with Brodal's precerebellar nucleus, the paramedian reticular nucleus, with its dorsal, ventral and accessory subdivisions in the cat [Brodal, 1953, 1981].

The neurons in PMT cell groups have a common morphology, with medium-sized chromophilic neurons possessing a large and usually clearly visible nucleus. Their cell axes are oriented regularly in a dorsoventral direction. PMT cell groups often display bridges of neuropile to other nests of neurons lying laterally around the PMT fiber tracts. The cells of PMT-2 are illustrated in figures 3 and 5 of nucleus No. 95.

The neuropile of the PMT cell groups contains high levels of cytochrome-C oxidase and acetylcholinesterase, which can be used to delineate the groups within the reticular formation. The individual PMT cell groups are illustrated with these techniques in the monkey by Horn [2006], and in the monkey and man by Buresch [2005]. The transmitter of the PMT neurons themselves is not known, but they are not serotonergic, as are the adjacent raphe neurons.

Functional Neuroanatomy

Function

The current hypothesis on the functional role of the PMT cell groups is that they send preoculomotor signals to the flocculus (a motor-like feedback signal), which together with the neural networks of the vestibular nucleus and PrP participate in the neural integration of eye velocity to eye position, and stabilize vertical or horizontal gaze [Büttner-Ennever and Horn, 1996; Dean and Porrill, 2008; Horn and Adamczyk, 2012]. Single-cell recordings from PMT cell groups confirm these concepts, as do lesion experiments which impair gaze-holding and lead to gaze-evoked nystagmus [Cheron et al., 1995; Nakamagoe et al., 2000]. A few clinical studies have discussed eye movement deficits with respect to PMT cell groups [Maas et al., 1991; Büttner et al., 1995; Leigh and Zee, 2006].

Connections

Efferents. The scattered PMT cell groups are specifically defined by their projections to the flocculus and ventral paraflocculus [Langer et al., 1985; Büttner-Ennever and Büttner, 1988]. The efferents of the PMT cell groups enter the medullary striae and follow either the internal or external arcuate pathways into the inferior cerebellar peduncle [Büttner-Ennever and Horn, 1996].

Afferents. Detailed examination of many other studies showed that in the monkey all PMT cell groups receive oculomotor premotor inputs [Büttner-Ennever and Büttner, 1988], although the inputs have often been mistakenly described in the literature as projections to the raphe nuclei, e.g. the raphe obscurus [McCrea et al., 1987a, b]. The inputs to the PMT cell groups arise from regions such as the interstitial nucleus of Cajal, the rostral interstitial nucleus of the MLF, from abducens internuclear neurons, PrP and vestibular nuclei [Büttner-Ennever and Horn, 1996; Goldberg et al., 2012]. Alone from the connectivity it seems that some of the PMT cell groups may subserve horizontal eye movements, and others vertical eye movements: this has been verified by physiological recordings [Sato et al., 1983; Cheron et al., 1995; Nakamagoe et al., 2000].

In the medulla some nests of medial PMT neurons merge with those of the more lateral ones (paramedian reticular nucleus of Brodal). These neurons are the target of head-related reticulospinal neurons, rather than oculomotor-related afferents, and indicate that the PMT cell groups may represent a continuum of precerebellar neurons coordinating the eye and head [Brodal, 1953; Grantyn et al., 1987, 1992].

References

Brodal A: Reticulo-cerebellar connections in the cat: an experimental study. J Comp Neurol 1953;98:113–153.

Brodal A: Neurological Anatomy. Oxford, Oxford University Press, 1981.

Buresch N: Neuroanatomische Charakterisierung blickstabilisierender Neurone an der Hirnstammmittellinie der Primaten, einschliesslich des Menschen; dissertation, LMU München, 2005.

Büttner-Ennever JA, Büttner U: The reticular formation; in Büttner-Ennever JA (ed): Neuroanatomy of the Oculomotor System. Reviews of Oculomotor Research. Amsterdam, Elsevier, 1988, vol 2.

Büttner-Ennever JA, Horn AKE: Pathways from cell groups of the paramedian tracts to the floccular region. Ann NY Acad Sci 1996;781:532–540.

Büttner U, Helmchen C, Büttner-Ennever JA: The localizing value of nystagmus in brainstem disorders. Neuroophthalmology 1995;15:283–290.

Cheron G, Saussez S, Gerrits NM, Godaux E: Existence in the nucleus incertus of the cat of horizontal-eye-movement-related neurons projecting to the cerebellar flocculus. J Neurophysiol 1995;74:1367–1372.

Dean P, Porrill J: Oculomotor anatomy and the motor-error problem: the role of the paramedian tract nuclei. Prog Brain Res 2008;171:177–186.

Goldberg JM, Wilson VJ, Cullen KE, Angelaki DE, Broussard DM, Büttner-Ennever JA, Fukushima K, Minor LB: The Vestibular System – A Sixth Sense. Oxford, Oxford University Press, 2012.

Grantyn A, Berthoz A, Hardy O, Gourdon A: Contribution of reticulospinal neurons to the dynamic control of head movements: presumed neck bursters; in Berthoz A, Graf W, Vidal PP (eds): The Head-Neck Sensory Motor System. New York, Oxford University Press, 1992.

Grantyn A, Ong Meang Jacques V, Berthoz A: Reticulo-spinal neurons participating in the control of synergic eye and head movements during orienting in the cat. II. Morphological properties as revealed by intra-axonal injections of horseradish peroxidase. Exp Brain Res 1987;66:355–377.

Horn AK, Adamczyk C: Reticular formation: eye movements, gaze and blinks; in Mai JK, Paxinos G (eds): The Human Nervous System, ed 3. San Diego, Academic Press, 2012.

Horn AKE: The reticular formation. Prog Brain Res 2006;151:127–155.

Langer TP, Fuchs AF, Scudder CA, Chubb MC: Afferents to the flocculus of the cerebellum in the rhesus macaque as revealed by retrograde transport of horseradish peroxidase. J Comp Neurol 1985;235:1–25.

Leigh RJ, Zee DS: The Neurology of Eye Movements. New York, Oxford University Press, 2006.

Maas EF, Ashe J, Spiegel P, Zee DS, Leigh RJ: Acquired pendular nystagmus in toluene addiction. Neurology 1991;41:282–285.

McCrea RA, Strassman A, Highstein SM: Anatomical and physiological characteristics of vestibular neurons mediating the vertical vestibulo-ocular reflex of the squirrel monkey. J Comp Neurol 1987a;264:571–594.

McCrea RA, Strassman E, May E, Highstein SM: Anatomical and physiological characteristics of vestibular neurons mediating the horizontal vestibulo-ocular reflex of the squirrel monkey. J Comp Neurol 1987b ;264:547–570.

Nakamagoe K, Iwamoto Y, Yoshida K: Evidence for brainstem structures participating in oculomotor integration. Science 2000;288:857–859.

Paxinos G, Huang X-F, Toga AW: The Rhesus Monkey Brain in Stereotaxic Coordinates. San Diego, Academic Press, 2000.

Sato Y, Kawasaki T, Ikarashi K: Afferent projections from the brainstem to the three floccular zones in cats. II. Mossy fiber projections. Brain Res 1983;272:37–48.

97 Nucleus of the Pontobulbar Body (PBu)

Original name:
Nucleus corporis pontobulbaris

Alternative name:
Nucleus 'k' [Meessen and Olszewski, 1949; Grottel et al., 1986]

Location and Cytoarchitecture – Original Text

This nucleus comprises the gray matter of the pontobulbar body, a structure which consists essentially of aberrant corticopontine fibers and displaced groups of pontine nuclei (PN). [...] The nucleus of the pontobulbar body (PBu) can be visualized as a structure having the form of a crooked staff. The straight portion of this staff lies in the ventral part of the middle cerebellar peduncle (MCP), and is separated from the main PN by pontocerebellar fibers. This straight portion extends caudally and enters the oral part of the medulla by passing medially to the ventral cochlear nucleus (VCN). At this level, the crooked portion of the 'staff' begins. It passes ventrally to, and then arches about the lateral and superior surfaces of the restiform body (inferior cerebellar peduncle, ICP) to reach the floor of the fourth ventricle.

The shape and position of the PBu vary considerably in cross-sections through different regions. The cells which compose its oral pole fuse with those of the pontine gray at the level of the motor trigeminal nucleus. In sections through the caudal part of the pons the nucleus appears as 1 or 2 irregularly shaped cell groups lying close to the ventral surface of the arch formed by the MCP, and surrounded laterally, dorsally and medially by pontocerebellar fibers. At the level of the junction of the pons and medulla the PBu is usually represented by a triangle-shaped cell group and several smaller, inconstant, irregularly shaped groups. All of these groups lie ventral to, or among the fibers of the ICP, and ventromedial to the VCN. In cross-sections between the caudal poles of the cochlear nuclei and the caudal pole of the medial vestibular nucleus (MVN), the cells of the PBu stream in a thin band dorsally, laterally and caudally about the external surface of the ICP. The most dorsal and caudal cells of this crooked portion of the nucleus may reach the floor of the fourth ventricle, dorsal to the lateral cuneate and the MVN. If due attention is not paid to the cytoarchitecture of the PBu at this level, it may easily be confused with the dorsal cochlear nucleus (fig. 1, 2).

The prominence and extent of the pontobulbar body varies considerably from one individual to another.

The cells of the PBu are irregularly arranged, plump and polygonal in outline and possess long dendrites and darkly stained, indistinct, evenly distributed Nissl granules (fig. 3, 5–7). The cells of that portion of the nucleus which arches about the ICP tend to be more elongated than the cells in the more oral portions (fig. 4). It is apparent that morphologically the PBu cells possess the same characteristic as do the cells of PN. [...] [High concentrations of somatostatin-binding sites have been found in the PBu, but they are not as strong as in PN [Carpentier et al., 1996].]

Functional Neuroanatomy

Function
The PBu is often considered as part of the dorsolateral basilar pons on the grounds of their common embryogenesis and position, but Martin et al. [1977] argue that the PBu should be considered as a separate precerebellar nucleus, on the basis of its anatomical projections: these indicate that the PBu participates in somatosensory-motor coordination.

Connections
Afferents to the PBu originate from the cerebral cortex, the red nucleus (RN), the cerebellum and the spinal cord: more specifically they arise from the sensorimotor cortex, from the RN through the crossed lateral descending pathways, and probably from the fastigial nucleus [Martin and Dom, 1970; Martin et al., 1977]. The RN projections in the cat showed some somatotopic organization [Holstege and Tan, 1988]. The pattern of terminations of the afferent pathways in the PBu shows that there are regional intrinsic differences within the nucleus, and a similar observation has been seen with respect to PBu efferents [Grottel et al., 1986].

The major target of PBu neurons is the cerebellum. The efferent projections from the PBu pass probably through the internal and external arcuate pathways (see 'Original Text') and terminate as mossy fibers in the skeletocerebellum (anterior lobe, pyramis and paramedian lobules), the visual-auditory areas of the vermis and the lobus simplex as well as crus I and II of the hemispheres [Martin et al., 1977; Grottel et al., 1986].

References

Carpentier V, Vaudry H, Mallet E, Laquerriere A, Tayot J, Leroux P: Anatomical distribution of somatostatin receptors in the brainstem of the human fetus. Neuroscience 1996;73:865–879.

Grottel K, Zimny R, Jakielska D: The nucleus 'k' of Meessen and Olszewski efferents to the cerebellar paramedian lobule: a retrograde tracing histochemical (HRP) study in the rabbit and the cat. J Hirnforsch 1986;27:305–322.

Holstege G, Tan J: Projections from the red nucleus and surrounding areas to the brainstem and spinal cord in the cat. An HRP and autoradiographical tracing study. Behav Brain Res 1988;28:33–57.

Martin GF, Dom R: Rubrobulbar projections of the opossum *(Didelphis virginiana)*. J Comp Neurol 1970;139:199–214.

Martin GF, Linauts M, Walker JM: The nucleus corporis pontobulbaris of the North American opossum. J Comp Neurol 1977;175:345–372.

Meessen H, Olszewski J: Cytoarchitektonischer Atlas des Rautenhirns des Kaninchens. Basel, Karger, 1949.

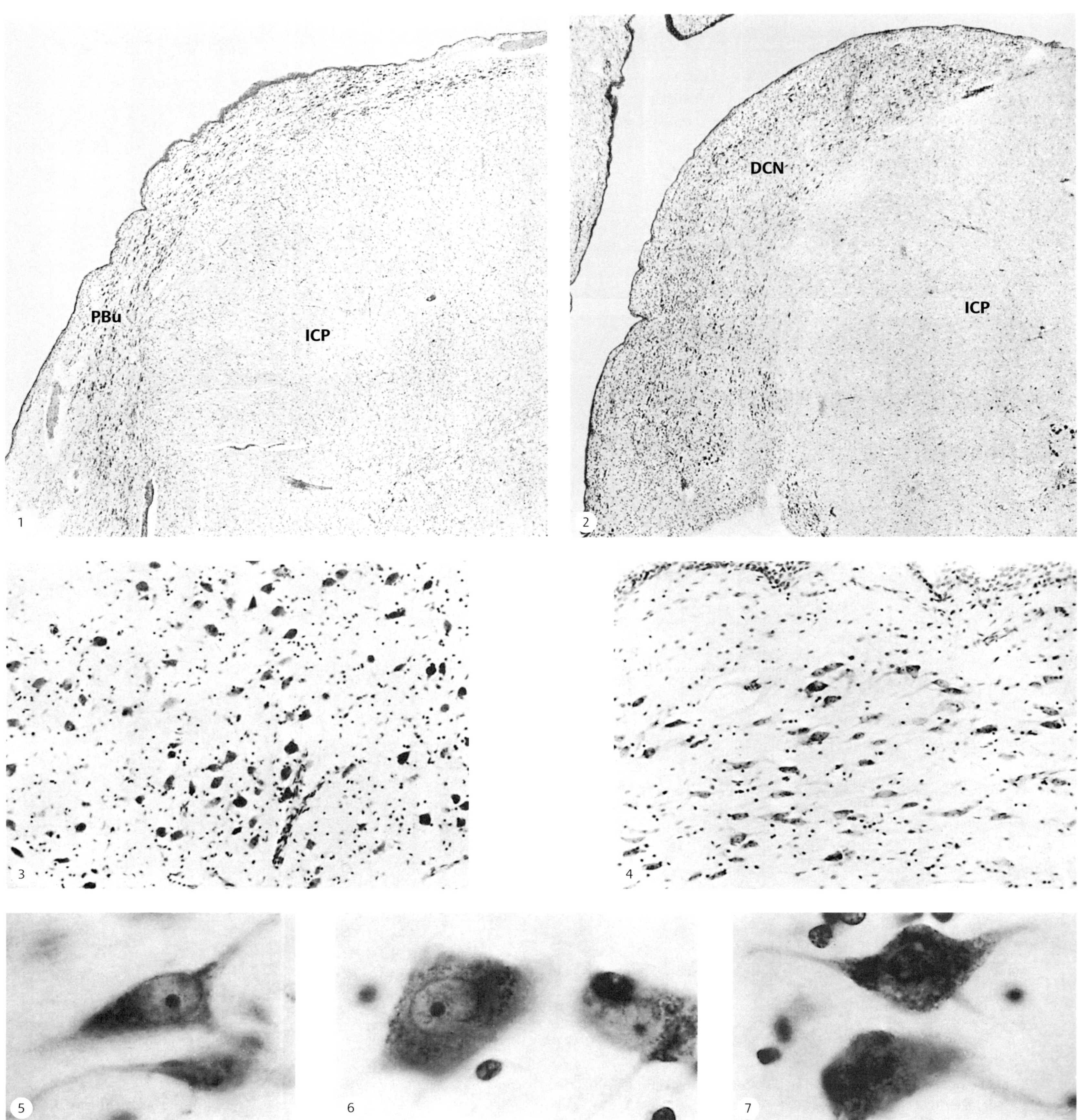

Fig. 1. Nucleus of the pontobulbar body (PBu) at the level where the structure turns caudodorsally about the inferior cerebellar peduncle (ICP). Magnification ×30.

Fig. 2. Dorsal cochlear nucleus (DCN). Note the superficial similarity to the nucleus of the pontobulbar body at this low magnification. ICP = Inferior cerebellar peduncle. Magnification ×30.

Fig. 3, 4. Nucleus of the pontobulbar body. The cells represented in figure 3 are from the oral portion of the nucleus; those in figure 4 are from that portion of the nucleus that sweeps about the inferior cerebellar peduncle. Note the elongated form of the cells in figure 4. Magnification ×150.

Fig. 5–7. Cells from the nucleus of the pontobulbar body. Magnification ×1,000.

98 Supragenual Nucleus (SG)

Original name:
Nucleus suprageniculatus

Alternative names:
Nucleus supragenualis
Paramedian tract cell group (nucleus No. 96)
Suprageniculate nucleus

Location and Cytoarchitecture – Original Text

This nucleus is formed by a small group of cells which lie dorsomedial to the abducens nucleus, between the floor of the fourth ventricle and the horizontal limb of the facial genu. On serial cross-sections it is apparent that this nucleus constitutes an oral prolongation of the prepositus nucleus (PrP). Its specific location and distinct type of nerve cell seem to justify description under a separate name. Orally, the supragenual nucleus (SG) is contiguous to the suprafascicular cells of the nucleus reticularis tegmenti pontis.

On cross-section, the nucleus is found to be composed of a very few (5–10 per 20-μm section), darkly stained, small, round or oval cells which possess a relatively large nucleus and scanty cytoplasm (fig. 1, 3). On examination of sagittal sections, it is apparent that these cells are actually spindle shaped with their long axes extending orocaudally among the fibers of the dorsal longitudinal fasciculus (fig. 2, 4).

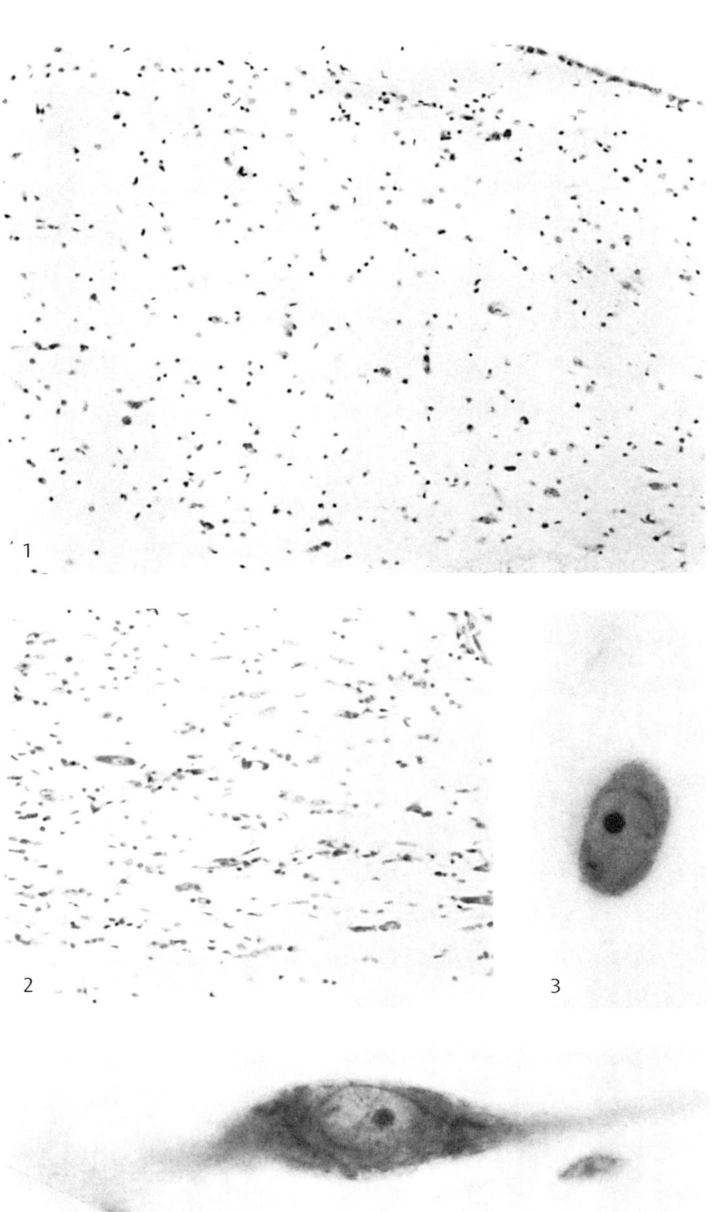

Fig. 1. Supragenual nucleus. Cross-section. Magnification ×150.
Fig. 2. Sagittal section through the supragenual nucleus. Note that the cells are elongated. Magnification ×150.
Fig. 3. Cell from the supragenual nucleus as seen on cross-section. Magnification ×1,000.
Fig. 4. Cells from the supragenual nucleus as they appear on sagittal section. Magnification ×1,000.

Functional Neuroanatomy

Function

As the rostral extension of the PrP, the SG is considered to be part of the perihypoglossal complex, and therefore thought to participate in the maintenance of eye and head position [Cheron et al., 1986; Cannon and Robinson, 1987; Cheron and Godaux, 1987]. This hypothesis is supported by the recognition of the SG as one of the clusters of neurons scattered around the paramedian tracts of the brainstem, and called PMT cell groups (nucleus No. 96) [McCrea and Horn, 2006]. The PMT cell groups play a role in the maintenance of gaze through the 'neural integrator' [Dean and Porrill, 2008]. Another function attributed to the SG is that it is the source of vestibular-related information encoding head direction, and essential for navigation centers in the brain (see below).

Connections

The SG appears to be more highly developed in the macaque monkey, than in the rabbit, cat, chimpanzee or man, on account of the presence of a population of medium-sized neurons [Brodal, 1983].

Afferents. The SG, like all PMT cell groups, receives many preoculomotor inputs [Büttner and Büttner-Ennever, 1988; Buresch, 2005; McCrea and Horn, 2006]. For example it receives afferents from the medial vestibular nucleus (MVN) [Biazoli et al., 2006], and from localized parts (dorsal and lateral regions) of the paramedian pontine reticular formation, an eye movement-related region of the reticular formation [Gerrits and Voogd, 1986]. The integration of these premotor signals will produce a motor-like activity in the SG neurons.

Efferents. Tract-tracing experiments show that SG efferents are associated with the oculomotor cerebellum, that is efferents project to the flocculus bilaterally [Langer et al., 1985; Kaufman et al., 1996], as well as the uvula and to some extent other cerebellar lobules [Brodal, 1983]. Like other regions of the perihypoglossal complex, projections from the SG to the superior colliculus have been reported [Hartwich-Young et al., 1990]. In the cat, projections to the oculomotor nucleus have been described [Graybiel, 1977; McCrea and Baker, 1985], as well as to the dorsolateral periaqueductal gray, a region possibly involved in aversive behavior [Klop et al., 2005].

Finally recent studies have demonstrated ascending projections from the SG to the dorsal tegmental nucleus of Gudden (DTG), and the lateral mammillary nucleus [Biazoli et al., 2006]. The results lend support to the hypothesis that this pathway supplies the vestibular-related information essential for navigation centers encoding head direction signals [Liu et al., 1984; Hayakawa and Zyo, 1985; Biazoli et al., 2006]. 'Head direction cells' provide signals concerning an animal's momentary directional heading, and have been demonstrated in several regions of the brain, including the postsubiculum, retrosplenial cortex, lateral dorsal thalamic nucleus, anterior dorsal thalamic nucleus, and lateral mammillary nucleus. Experiments, using transsynaptic retrograde tracers, have shown that the SG and PrP may feed vestibular information from the MVN into these circuits, via projections to the DTG, which is in turn a principal relay to the lateral mammillary nucleus [Brown et al., 2005; Biazoli et al., 2006; Bassett et al., 2007; Taube, 2007].

References

Bassett JP, Tullman ML, Taube JS: Lesions of the tegmentomammillary circuit in the head direction system disrupt the head direction signal in the anterior thalamus. Brain Res Bull 2007;25:271–284.

Biazoli CEJ, Goto M, Campos AM, Canteras NS: The supragenual nucleus: a putative relay station for ascending vestibular signs to head direction cells. Brain Res 2006;1094:138–148.

Brodal A: The perihypoglossal nuclei in the macaque monkey and the chimpanzee. J Comp Neurol 1983;218:257–269.

Brown JE, Card JP, Yates BJ: Polysynaptic pathways from the vestibular nuclei to the lateral mammillary nucleus of the rat: substrates for vestibular input to head direction cells. Exp Brain Res 2005;161:47–61.

Buresch N: Neuroanatomische Charakterisierung blickstabilisierender Neurone an der Hirnstammmittellinie der Primaten, einschliesslich des Menschen; dissertation, LMU München, 2005.

Büttner U, Büttner-Ennever JA: Present concepts of oculomotor organization; in Büttner-Ennever JA (ed): Neuroanatomy of the Oculomotor System. Reviews of Oculomotor Research. Amsterdam, Elsevier, 1988, vol 2.

Cannon SC, Robinson DA: Loss of the neural integrator of the oculomotor system from brain stem lesions in monkey. J Neurophysiol 1987;57:1383–1409.

Cheron G, Godaux E: Disabling of the oculomotor neural integrator by kainic acid injections in the prepositus-vestibular complex of the cat. J Physiol 1987;394: 267–290.

Cheron G, Godaux E, Laune JM, Vanderkelen B: Lesions in the cat prepositus complex: effects on the vestibulo-ocular reflex and saccades. J Physiol 1986;372:75–94.

Dean P, Porrill J: Oculomotor anatomy and the motor-error problem: the role of the paramedian tract nuclei. Prog Brain Res 2008;171:177–186.

Gerrits NM, Voogd J: The nucleus reticularis tegmenti pontis and the adjacent rostral paramedian reticular formation: differential projections to the cerebellum and the caudal brain stem. Exp Brain Res 1986;62:29–45.

Graybiel AM: Direct and indirect preoculomotor pathways of the brainstem: an autoradiographic study of the pontine reticular formation in the cat. J Comp Neurol 1977;175:37–78.

Hartwich-Young R, Nelson JS, Sparks DL: The perihypoglossal projection to the superior colliculus in the rhesus monkey. Vis Neurosci 1990;4:29–42.

Hayakawa T, Zyo K: Afferent connections of Gudden's tegmental nuclei in the rabbit. J Comp Neurol 1985;235:169–181.

Kaufman GD, Mustari MJ, Miselis RR, Perachio AA: Transneuronal pathways to the vestibulocerebellum. J Comp Neurol 1996;370:501–523.

Klop EM, Mouton LJ, Ehling T, Holstege G: Two parts of the nucleus prepositus hypoglossi project to two different subdivisions of the dorsolateral periaqueductal gray in cat. J Comp Neurol 2005;492:303–322.

Langer TP, Fuchs AF, Scudder CA, Chubb MC: Afferents to the flocculus of the cerebellum in the rhesus macaque as revealed by retrograde transport of horseradish peroxidase. J Comp Neurol 1985;235:1–25.

Liu R, Chang L, Wickern G: The dorsal tegmental nucleus: an axoplasmic transport study. Brain Res 1984;310:123–132.

McCrea RA, Baker R: Anatomical connections of the nucleus prepositus of the cat. J Comp Neurol 1985;237:377–407.

McCrea RA, Horn AKE: Nucleus prepositus. Prog Brain Res 2006;151:205–230.

Taube JS: The head direction signal: origins and sensory-motor integration. Annu Rev Neurosci 2007;30:181–207.

Precerebellar Nuclei

(Plates 24 and 26–28)

99 Nucleus Reticularis Tegmenti Pontis (NRTP)

Original name:
Nucleus papilloformis

Location and Cytoarchitecture – Original Text

This nucleus lies on either side of the median raphe of the oral pons. It begins at the level of the caudal pole of the motor trigeminal nucleus and extends orally for a distance of 6 mm.

When fully developed, the nucleus reticularis tegmenti pontis (NRTP) is, on cross-section, dumbbell shaped with its long axis directed dorsoventrally. The midline (raphe) forms the midline of the nucleus. The ventral enlargement lies directly dorsal to the pontine nuclei (PN) and extends laterally [over the medial lemniscus. This region was considered in the previous edition to be part of the pontine nuclei; see group c in the chapter on nucleus No. 100]. The dorsal enlargement is formed by cells which accumulate about the medial longitudinal fasciculi (MLF). [An area is outlined just below the MLF, in the NRTP, that may correlate with the ventral tegmental nucleus of Gudden in humans (see dashed line in plate

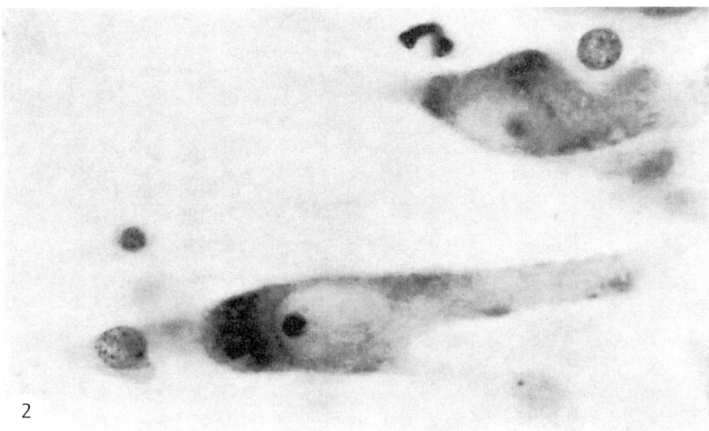

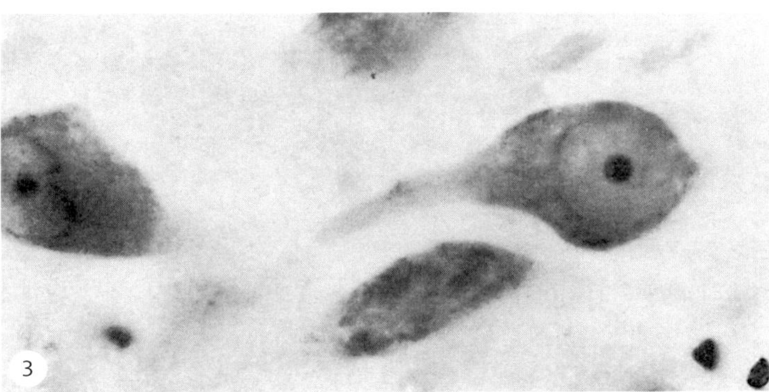

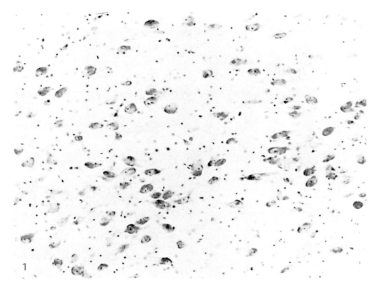

Fig. 1. Nucleus reticularis tegmenti pontis. Magnification ×150.
Fig. 2, 3. Cells from the nucleus reticularis tegmenti pontis. Magnification ×1,000.

27).] A narrow band of cells intervenes between each MLF and the floor of the fourth ventricle; a larger accumulation of cells lies ventrolateral to each fasciculus. The intermediate portion of the 'dumbbell' is formed by small groups of cells scattered on either side of the midline.

Caudally, the NRTP appears as two oval cell groups lying on either side of the midline and dorsal to the fibers of the medial lemniscus. As one proceeds orally, these cell groups fuse with the ventral portion of the fully developed nucleus. Orally, the NRTP is replaced by the median raphe nucleus, the cells of which first appear about the midline and dorsomedially to the MLF. Laterally, the NRTP is bordered, on each side, by the nucleus reticularis pontis oralis.

The NRTP is composed of densely arranged, darkly stained, plump, fusiform or polygonal cells. They possess long dendrites, a relatively large nucleus, and fine evenly distributed Nissl granules (fig. 1–3).

Functional Neuroanatomy

Function

The NRTP projects to the cerebellum, and has several features in common with the PN in terms of input and output relationships. However, the NRTP receives optokinetic and premotor afferents largely independent of the PN [Schwarz and Thier, 1996; Thier and Möck, 2006]. In lower animals many NRTP neurons give a slow response to optokinetic and vestibular stimulation; however, these cells are very scarce in primates. Here, visual and oculomotor cells are found, with saccade-, eye position- and vergence-related cells lying caudally, and pursuit-related neurons lying rostrally [Gamlin and Clarke, 1995; Suzuki et al., 2003]. Taken together these studies imply that the NRTP is 'nearer the motoneuron than PN'.

A current hypothesis suggests that the caudal NRTP is a key element in a neural circuit downstream from the superior colliculus (SC) for coordinating eye muscle activity during saccades [Thier and Möck, 2006]. No studies have been performed to estimate what proportion of nonvisual and nonoculomotor neurons is to be found in the NRTP, but given that recording oculomotor units in the NRTP is relatively difficult, the proportion may well be quite high.

Connections

Afferents. The cerebral cortex afferents to the NRTP arise from more restricted areas than those to PN: the strongest input is from the primary motor cortex, which targets the central part of the NRTP and coincides with afferents from the anterior interposed (emboliform) nucleus of the cerebellum [Brodal, 1980a, b; Hartmann-von Monakow et al., 1981]. Parietal afferents project to the lateral NRTP. Inputs from cerebral oculomotor areas, like the supplementary and frontal eye fields, terminate medially and dorsomedially, and have been well documented [Thier and Möck, 2006]: unlike their projection to PN they terminate bilaterally in the NRTP. The SC is a significant source of NRTP afferents [Harting, 1977], along with pretectal inputs from the nucleus of the optic tract [Giolli et al., 2006], and an input from the oculomotor part of the fastigial nucleus [Noda et al., 1990].

Efferents. The NRTP projects bilaterally, as mossy fibers, to widespread regions of the cerebellar cortex. In primates the visually related medial and dorsomedial NRTP project to the visual vermis (lobules VI and VII), the flocculus and ventral paraflocculus [Glickstein et al., 1994; Thier and Möck, 2006]. Collaterals of these NRTP mossy fibers terminate in the cerebellar nuclei, specifically the caudal oculomotor fastigial nucleus, the caudal posterior interposed (globose) nucleus and dentate nucleus. A summary of these connections is shown diagrammatically in figure 19.17 in Nieuwenhuys et al. [2008]. An intriguing cluster of neurons in the NRTP has been highlighted in various tract-tracing experiments; they project directly to the abducens motoneurons [Horn and Adamczyk, 2012].

References

Brodal P: The cortical projection to the nucleus reticularis tegmenti pontis in the rhesus monkey. Exp Brain Res 1980a;38:19–27.

Brodal P: The projection from the nucleus reticularis tegmenti pontis to the cerebellum in the rhesus monkey. Exp Brain Res 1980b;38:29–36.

Gamlin PDR, Clarke RJ: Single-unit activity in the primate nucleus reticularis tegmenti pontis related to vergence and ocular accommodation. J Neurophysiol 1995;73:2115–2119.

Giolli RA, Blanks RHI, Lui F: The accessory optic system: basic organization with an update on connectivity, neurochemistry, and function. Prog Brain Res 2006; 151:407–440.

Glickstein M, Gerrits NM, Kralj-Hans I, Mercier B, Stein J, Voogd J: Visual pontocerebellar projections in the macaque. J Comp Neurol 1994;349:51–72.

Harting JK: Descending pathways from the superior colliculus: an autoradiographic analysis in the rhesus monkey *(Macaca mulatta)*. J Comp Neurol 1977; 173:583–612.

Hartmann-von Monakow K, Akert K, Künzle H: Projection of precentral, premotor and prefrontal cortex to the basilar pontine grey and to nucleus reticularis tegmenti pontis in the monkey *(Macaca fascicularis)*. Schweiz Arch Neurol Neurochir Psychiatr 1981;129:189–208.

Horn AK, Adamczyk C: Reticular formation: eye movements, gaze and blinks; in Mai JK, Paxinos G (eds): The Human Nervous System, ed 3. Amsterdam, Academic Press, 2012.

Nieuwenhuys R, Voogd J, van Huijzen C: The Human Central Nervous System. Berlin, Springer, 2008.

Noda H, Sugita S, Ikeda Y: Afferent and efferent connections of the oculomotor region of the fastigial nucleus in the macaque monkey. J Comp Neurol 1990; 302:330–348.

Schwarz C, Thier P: Comparison of projection neurons in the pontine nuclei and the nucleus reticularis tegmenti pontis of the rat. J Comp Neurol 1996;376: 403–419.

Suzuki DA, Yamada T, Yee RD: Smooth-pursuit eye-movement-related neuronal activity in macaque nucleus reticularis tegmenti pontis. J Neurophysiol 2003; 89:2146–2158.

Thier P, Möck M: The oculomotor role of the pontine nuclei and the nucleus reticularis tegmenti pontis. Prog Brain Res 2006;151:293–320.

Precerebellar Nuclei

(Plates 20–22, 24–26, 28, 30 and 32)

100 Pontine Nuclei (PN)

Original name:

Griseum pontis

Location and Cytoarchitecture – Original Text

The pontine nuclei (PN) form the great mass of gray substance which occupies the basilar portion of the pons throughout its entire extent. The nuclei are broken up into cell clusters of irregular shape and size by the longitudinally coursing corticobulbar and corticospinal fibers, and by the horizontally coursing pontocerebellar fibers. With 3 exceptions no attempt has been made to classify or to describe in detail the relations of the various groups of PN. The exceptions are:

(a) the cell groups situated between the medial lemnisci (ML);

(b) the supralemniscal process of PN (plates 26–28), and

(c) the lateral tegmental process of PN (plate 24).

(a) In the caudal half of the pons (plates 20 and 22), the fibers of ML border PN dorsally, and the medial extremities of the 2 ML approach each other in the midline. In the oral half of the pons (plates 24 and 26), the ML migrate laterally and the interval between their medial extremities becomes filled with groups of PN [the intralemniscal or dorsomedial PN (DMPN)]. These groups maintain this position throughout the oral half of the pons and lie, at first, ventral to the nucleus reticularis tegmenti pontis (NRTP) and, more orally, ventral to the median raphe nucleus. In the oral quarter of the pons (plates 28 and 30), these midline groups become separated ventrally from the main mass of PN by thick bundles of transversely coursing pontocerebellar fibers.

(b) The supralemniscal process of PN is formed by a long, tongue-like extension of PN which lies dorsal to the fibers of ML in the oral half of the pons (plates 26–28). Dorsally, this process is related to the cells of the nucleus reticularis pontis oralis and medially it is continuous with the PN which lie between the ML. Small groups of pontine cells originating from both the supralemniscal process dorsally, and from the main mass of PN ventrally, infiltrate among the fibers of the ML. [The supralemniscal process is now considered to be part of the NRTP (nucleus No. 99) since it lies above the ML.]

(c) The lateral tegmental process of PN [also called the dorsolateral PN (DLPN)] (plates 22, 24 and 25) is formed by a wedge-shaped extension of PN which projects into the ventrolateral portion of the pontine tegmentum. The process appears caudally at the level of the oral pole of the facial nucleus and extends to a point 5 mm rostral to the oral pole of the motor trigeminal nucleus (MoV). Caudal to the oral pole of the MoV the process is related ventrally to the main mass of PN, medially to fibers of the lateral lemniscus (LL), dorsomedially to the nucleus reticularis pontis caudalis, dorsally to the main sensory trigeminal nucleus and MoV, and laterally to the fibers of the fifth cranial nerve and the middle cerebellar peduncle (MCP). Immediately rostral to the MoV the process is related ventromedially to fibers of the LL, dorsomedially to cells of the nucleus subcoeruleus, dorsally to cells of the parabrachial medial nucleus and laterally to fibers of the MCP. Approximately 3 mm rostral to the MoV the lateral tegmental process rapidly diminishes in size, loses its direct connection with the main mass of PN and is represented by several small cell groups arranged in a linear fashion along the lateral border of the LL and the lateral parabrachial nucleus. (These groups are represented but not labeled on plate 26.)

The PN are composed of irregularly arranged, closely packed, medium-sized, plump, polygonal or triangular cells which possess long dendrites, centrally placed nuclei and indistinct, darkly stained Nissl granules (fig. 1–4). There is considerable variability in regard to both cell size and intensity of staining. Thus one finds many cells which though still of medium size are larger than the average cell of the PN. This latter cell type is most numerous within the lateral tegmental process (DLPN), where they tend to congregate in groups. Otherwise the cells of the supralemniscal (part of the NRTP) and lateral tegmental processes (DLPN) resemble those of the remainder of the PN. [...].

Functional Neuroanatomy

Function

The general function of the PN is the sensory guidance, or modulation, of movement. The PN receive a massive input from the cerebral cortex carrying signals associated with motor control, sensory information and spatial perception, particularly related to the 'dorsal visual stream' which encodes speed and direction of stimuli [Nieuwenhuys et al., 2008; de Haan and Cowey, 2011]. Unit activity in PN reflects very closely the cell responses in the afferent cortical fields [Tziridis et al., 2012]. The PN relay the activity exclusively to the cerebellum, and the ensuing output signals from the cerebellar nuclei are channeled back only to the frontal regions of the cerebral cortex, basal ganglia and red nucleus (nucleus No. 101) to carry out the modified motor program.

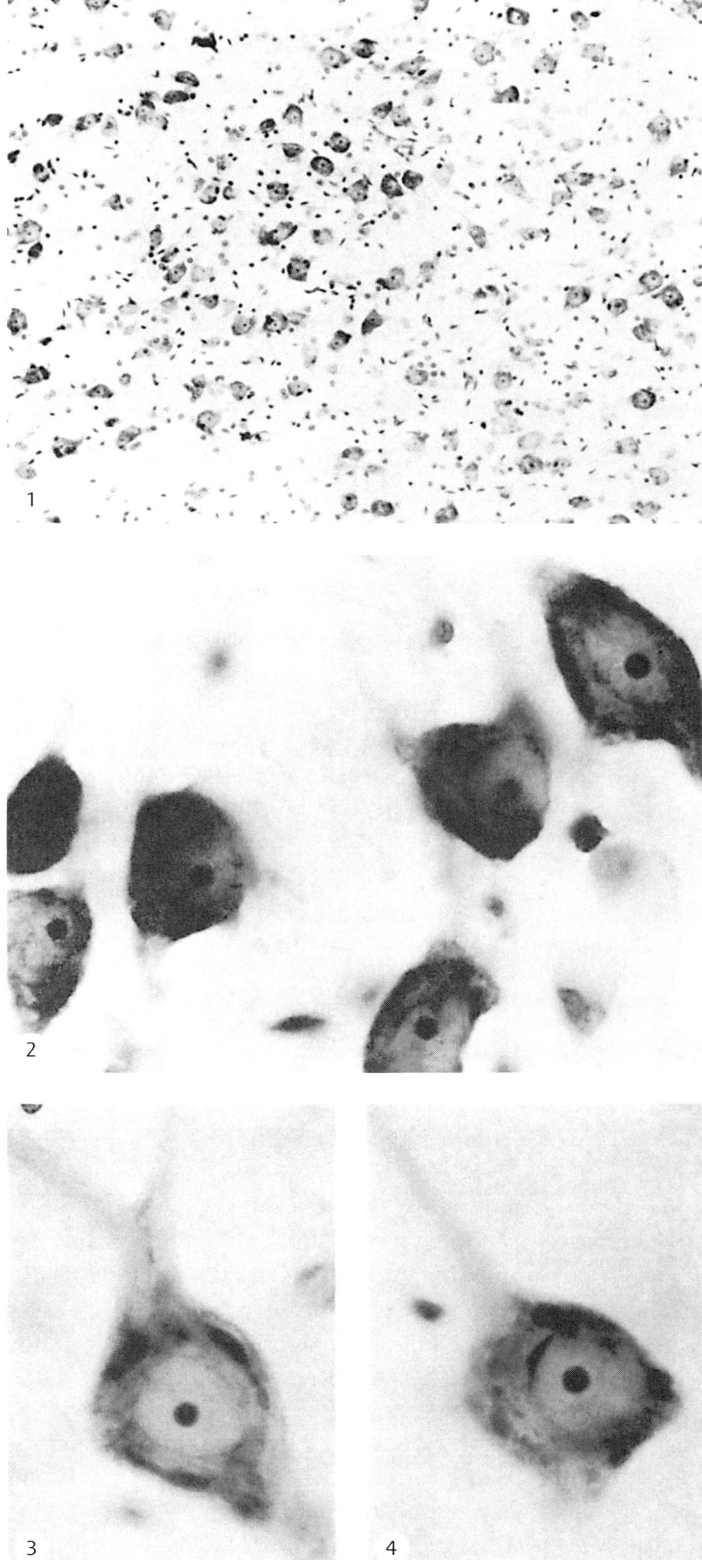

Fig. 1. Pontine nuclei. Magnification ×150.
Fig. 2–4. Cells from the pontine nuclei. Magnification ×1,000.

Currently there is an unresolved discussion about possible cognitive roles of the cerebellum, especially the lateral hemispheres. Since the PN project to the cerebellar hemispheres, these arguments are relevant here for PN function [Schmahmann, 1997; Glickstein, 2007; Voogd and Ruigrok, 2012].

Lesions confined to the PN lead to contralateral hemiparesis, which however can be attributed to damage of the corticospinal tract running through PN [Gaymard et al., 1993]. In some cases these symptoms were accompanied by dysmetria in the 'finger-to-nose test' (a cerebellar symptom), but in all cases deficits in smooth pursuit and optokinetic eye movements were found, and attributed to damage of the DLPN and lateral PN.

Connections

A useful, but loose, terminology is currently used to describe the arrangement of PN: the ventral, the median, the paramedian, the lateral, the dorsolateral, the extreme dorsolateral, the dorsal, the dorsomedial and peduncular PN. Under this terminology the subdivision described above as (a) is equivalent to the DMPN, subgroup (b) is now included in the NRTP (nucleus No. 99) and subdivision (c) is equivalent to the DLPN. It is often difficult to distinguish between ventral NRTP and dorsal PN, both physiologically and anatomically; for reviews, see Thier and Möck [2006], Voogd and Ruigrok [2012] and Nieuwenhuys et al. [2008].

Afferents. The corticopontine fibers are thought to have developed as collaterals of the corticospinal fibers and terminate on cell groups collecting around the corticospinal tracts as the core, forming concentric lamellae. Corticopontine fibers which develop later surround the already existing lamella around the core [Leergaard et al., 1995; Voogd and Ruigrok, 2012]. The motor, sensory and supplementary motor cortices develop first and occupy the central core. The prefrontal, superior temporal, posterior parahippocampal and occipital ones develop later. Inputs from the ventral prefrontal and temporal areas are scarce. The lamellae were studied in the early postnatal rat, but in the adult primate the lamellae are difficult to visualize, appearing as distributed patches surrounded by crossing fiber bundles. The topography of the corticopontine pathways has been well studied in primates [Glickstein et al., 1980; Schmahmann and Pandya, 1997]. The corticopontine projections of adjacent cortical regions show very little overlap of terminal fields within PN. The borders of these terminal patches are also respected by PN cell dendrites, which do not venture across the terminal field borders [Schwarz and Mock, 2001].

Additional afferents to the DLPN arise from the superior colliculus [Harting, 1977] and the cerebellar nuclei [Noda et al., 1990].

Efferents. The vast majority of PN efferents cross to the contralateral side and without exception all enter the MCP (brachium pontis) to provide the largest single input to the cerebellum. Note that the cerebral cortex is contralaterally interconnected with the cerebellum through the PN, therefore the cerebellum is ipsilaterally interconnected with the spinal cord. The PN efferents terminate in the granular layer of cerebellar cortex topographically, as mossy fibers. The mossy fibers form longitudinal aggregates of terminals that are not so precisely organized as those of climbing fibers. The area supplied by PN forms the pontocerebellum, as opposed to the vestibulocerebellum and spinocerebellum. The pontocerebellum covers the apical parts of the anterior lobe (lobules I–V) and pyramid (lobule VIII), and innervates the base of the lobus simplex (lobule VII). The PN do not project to the nodulus, the flocculus and the ventral paraflocculus [Voogd and Ruigrok, 2012]. More specifically, the two main oculomotor regions of the cerebellum to receive from PN are lobule VII of the vermis, whose inputs arise mostly from clusters in the dorsal PN and DLPN, and second the dorsal paraflocculus and petrosal lobule, which also receive inputs from the dorsal PN and DLPN and other regions.

Although mossy fibers enter the cerebellum as crossed fibers, they often have terminals on both sides of the cerebellar cortex, unlike climbing fibers; this is especially true in the vermis, where the input from the PN is strongly bilateral [Thier and Möck, 2006; Voogd and Ruigrok, 2012].

Collaterals of the pontocerebellar fibers also supply the cerebellar nuclei [Noda et al., 1990]. The efferent neurons use L-glutamate as their transmitter, but about 5% of the neurons in the PN use GABA; these cells are interneurons and project locally within PN [Thier and Möck, 2006].

References

de Haan EH, Cowey A: On the usefulness of 'what' and 'where' pathways in vision. Trends Cogn Sci 2011;15:460–466.

Gaymard B, Pierrot-Deseilligny C, Rivaud S, Velut S: Smooth pursuit eye movement deficits after pontine nuclei lesions in humans. J Neurol Neurosurg Psychiatry 1993;56:799–807.

Glickstein M: What does the cerebellum really do? Curr Biol 2007;17:R824–R827.

Glickstein M, Cohen JL, Dixon B, Gibson AR, Hollius M, La Bossiere E, Robinson F: Cortico-pontine visual projections in macaque monkeys. J Comp Neurol 1980; 190:209–229.

Harting JK: Descending pathways from the superior colliculus: an autoradiographic analysis in the rhesus monkey *(Macaca mulatta)*. J Comp Neurol 1977; 173:583–612.

Leergaard TB, Lakke EA, Bjaalie JG: Topographical organization in the early postnatal corticopontine projection: a carbocyanine dye and 3-D computer reconstruction study in the rat. J Comp Neurol 1995;361:77–94.

Nieuwenhuys R, Voogd J, van Huijzen C: The Human Central Nervous System. Berlin, Springer, 2008.

Noda H, Sugita S, Ikeda Y: Afferent and efferent connections of the oculomotor region of the fastigial nucleus in the macaque monkey. J Comp Neurol 1990; 302:330–348.

Schmahmann JD: The Cerebellum and Cognition. International Review of Neurobiology. San Diego, Academic Press, 1997, vol 41.

Schmahmann JD, Pandya DN: Anatomic organization of the basilar pontine projections from prefrontal cortices in rhesus monkey. J Neurosci 1997;17:438–458.

Schwarz C, Mock M: Spatial arrangement of cerebro-pontine terminals. J Comp Neurol 2001;435:418–432.

Thier P, Möck M: The oculomotor role of the pontine nuclei and the nucleus reticularis tegmenti pontis. Prog Brain Res 2006;151:293–320.

Tziridis K, Dicke PW, Thier P: Pontine reference frames for the sensory guidance of movement. Cereb Cortex 2012;22:345–362.

Voogd J, Ruigrok T: Cerebellum and precerebellar nuclei; in Mai JK, Paxinos G (eds): The Human Nervous System, ed 3. Amsterdam, Academic Press, 2012.

Precerebellar Nuclei (Plates 36 and 38–42)

101 Red Nucleus (RN)

Original name:
Nucleus ruber

Associated name:
Medial accessory nucleus of Bechterew

Location and Cytoarchitecture – Original Text

This round, ball-shaped nucleus surrounded by its capsule of longitudinally coursing fibers occupies a considerable proportion of the tegmentum of the midbrain. It extends from the level of the caudal pole of the oculomotor nucleus to the oral limits of the mammillary bodies in the caudal diencephalon. The perirubral capsule surrounding the red nucleus (RN) is continuous caudally with the brachium conjunctivum (or superior cerebellar peduncle, SCP) and orally with the fields H_1 and H_2 of Forel.

On cross-section the RN is round in outline (as it is in all planes of section) and with its capsular fibers occupies most of the tegmentum which lies between the mesencephalic central gray matter dorsomedially and the substantia nigra ventrolaterally. The caudal portion of the RN is traversed by vertically coursing fibers of the oculomotor nerve. Much farther orally, at the level of the posterior commissure (PC), the dorsomedial aspect of the nucleus is indented by the fibers of the habenulointerpeduncular tract.

It is usual to divide the RN into a magnocellular subnucleus (RNm) and a parvocellular subnucleus (RNp). The RNp may be further subdivided into a caudal part, an oral part and a dorsomedial part on the basis of cell density and the presence of lamellae.

The RNm forms the caudal 1 mm of the RN and is composed of very large, multipolar neurons which possess prominent nuclei and nucleoli and large, darkly stained, evenly distributed Nissl granules. These cells possess the characteristics of large, motor-type neurons (fig. 1–3). They are few in number (only 5–10 per 20-μm cross-section) and are gathered largely into a dorsomedial and a ventral group with an occasional single cell scattered elsewhere among the fibers of the SCP.

The remainder of the rubral complex is formed by the RNp. Caudal to the level of the PC, the RNp is composed entirely of the 'caudal part' (fig. 11, 12). The caudal part is traversed by many fiber bundles among which its constituent cells are loosely and irregularly arranged. At the level of the habenulointerpeduncular tract the cells composing the oro-dorsomedial aspect of the RN become separated from the remainder of the nucleus by a lamella of myelinated fibers. This isolated portion is the dorsomedial part (fig. 9–12). A second lamella isolates the cells which form the oral pole of the RN, i.e. the 'oral part' (fig. 9–12), from the more caudally situated 'dorsomedial' and 'caudal part'.

All 3 subdivisions of the RNp (caudal, dorsomedial and oral) are composed of medium-sized to large, triangular or irregularly multipolar neurons. The Nissl substance of these cells stains moderately darkly but does not form distinct granules; rather it is arranged into indistinct masses which are often accumulated about the periphery of the cell body. The cells of the RNp are further characterized by the fact that many glial nuclei tend to congregate in the vicinity of the nerve cells (fig. 4–8).

The cells of the oral and dorsomedial part differ from those of the caudal part only in that they are smaller, more closely congregated and more darkly stained (fig. 4, 5). [...]

Alternative Model of Red Nucleus

Instead of the above model of the RNp described as 3 subnuclei, a recent interpretation of the RN cytoarchitecture and its adjacent structures has been put forward in an excellent re-

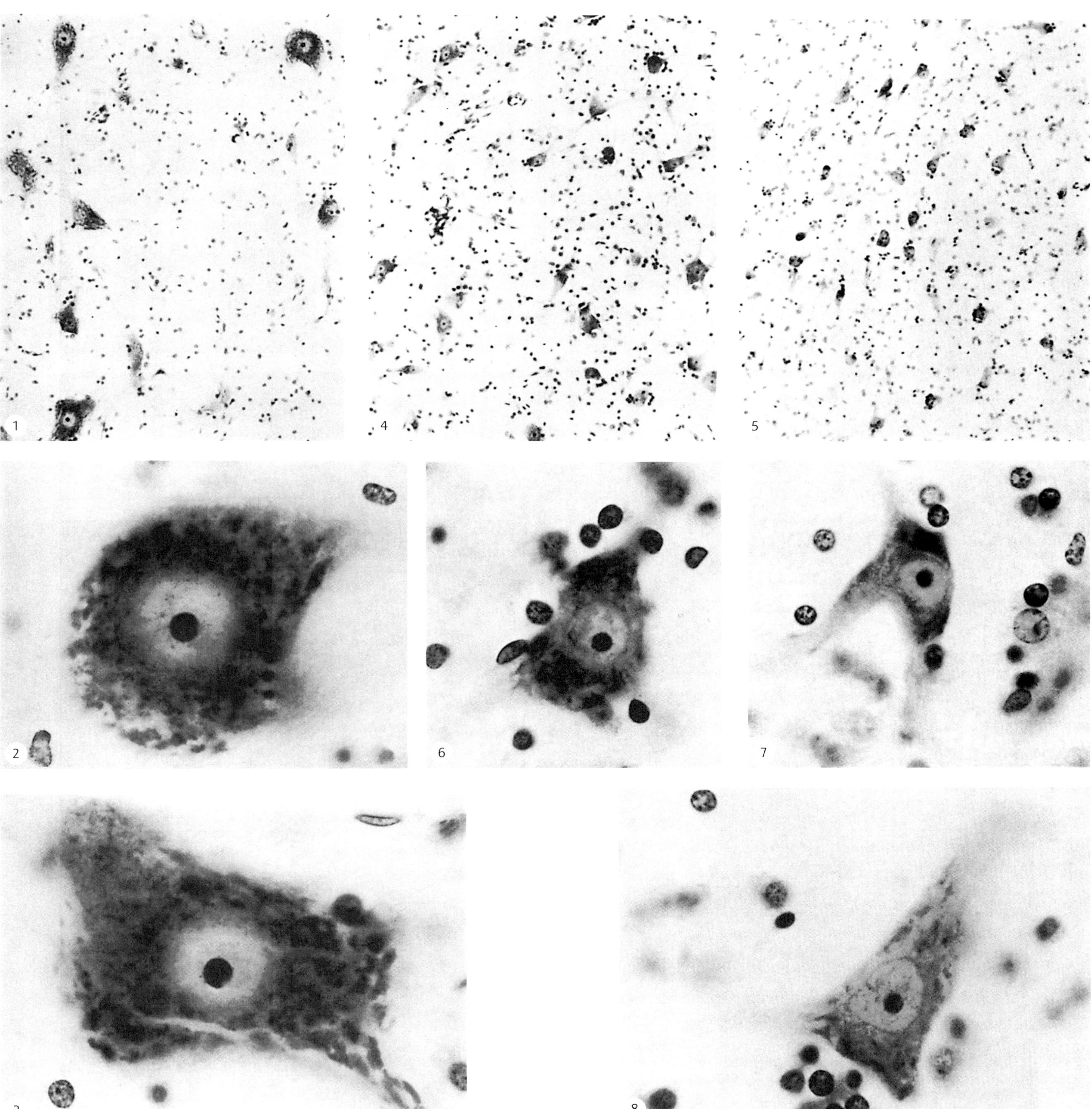

Fig. 1. Red nucleus, magnocellular subnucleus. Magnification ×150.
Fig. 2, **3.** Cells from the red nucleus, magnocellular subnucleus. Magnification ×1,000.
Fig. 4. Red nucleus, parvocellular subnucleus, caudal part. Magnification ×150.

Fig. 5. Red nucleus, parvocellular subnucleus, dorsomedial part; in the case of man, this region is equivalent to the medial accessory nucleus of Bechterew. Magnification ×150.
Fig. 6–8. Cells from the red nucleus, parvocellular subnucleus. Note the intense degree of glial satellitosis. Magnification ×1,000.

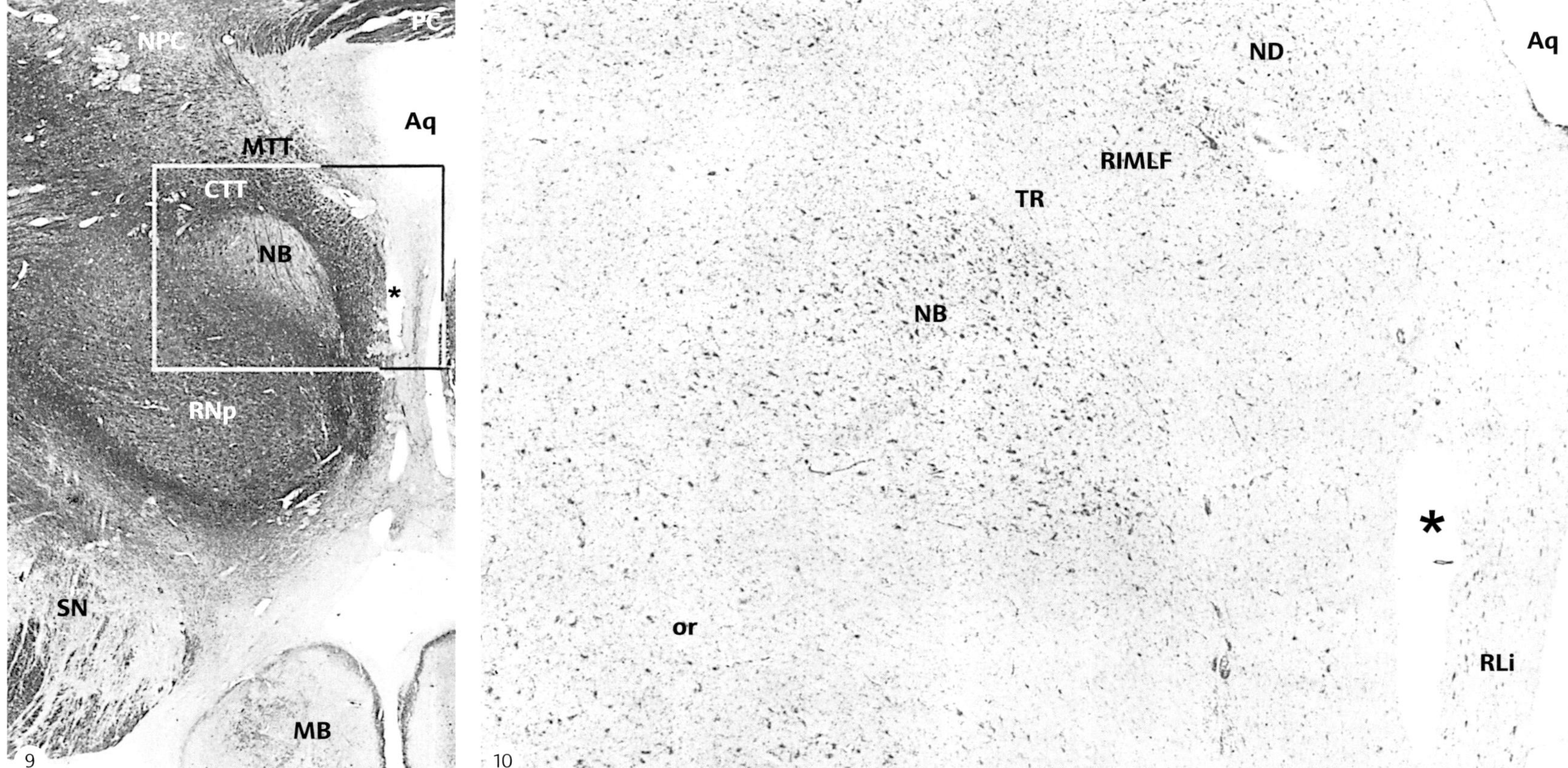

Fig. 9. Cross-section through the mesencephalon at the level of the parvocellular red nucleus to show the subdivision of the nucleus into an oral part and a dorsomedial part, which contains the medial accessory nucleus of Bechterew. The asterisk marks the posterior thalamosubthalamic paramedian artery. Aq = Sylvian aqueduct; CTT = central tegmental tract; MB = mammillary bodies; MTT = medial tegmental tract; NB = medial accessory nucleus of Bechterew; NPC = nucleus of the posterior commissure; PC = posterior commissure; RNp = red nucleus, parvocellular subnucleus, oral part; SN = substantia nigra. Heidenhain stain. Magnification ×8.

Fig. 10. Cross-section through the mesencephalon at the level of the parvocellular red nucleus to show the subdivision into an oral and a dorsomedial part, containing the medial accessory nucleus of Bechterew. The equivalent area of the adjacent myelin-stained section is delineated by a square in figure 9. The asterisk marks the posterior thalamosubthalamic paramedian artery. Aq = Sylvian aqueduct; NB = medial accessory nucleus of Bechterew; ND = nucleus of Darkschewitsch; or = red nucleus, parvocellular subnucleus, oral part; RIMLF = rostral interstitial nucleus of the medial longitudinal fasciculus; RLi = rostral linear nucleus; TR = tractus retroflexus. Magnification ×30.

view by Onodera and Hicks [2009]. On the basis of comparative studies and tract-tracing experiments they describe an RN 'rolled-sheet' model, in which the dense cellular layer of RNp cells envelops the massive superior cerebellar peduncle, as it ascends to the thalamus, giving axon collaterals off to the RNp. The caudal pole has a large ventrolateral part of RNp and a thinner dorsomedial part as the neuronal 'sheet' curves dorsomedially. Further rostrally the sheet merges with the caudal extension of the medial accessory nucleus of Bechterew (NB). The NB is not considered as a separate chapter here; for further details, see Onodera and Hicks [2009]. The region of the NB is shown in figures 9 and 10, the area is usually described as the dorsomedial subdivision of the RNp, but in the sheet model it has been convincingly shown to contain the NB and RNp. In addition the habenulointerpeduncular tract (also called tractus retroflexus) passes through this dorsomedial region at the rostral pole of the NB.

The NB in the cat and monkey lies more medial in the periaqueductal gray, and merges further dorsally into the nucleus of Darkschewitsch (ND) but not in humans. This model emphasizes that the RNp, NB and ND form a functional continuum, specifically with respect to projections to the inferior olive (IO, see below). The rudimentary nature of the RNm in humans is confirmed in this study.

Functional Neuroanatomy

Function

The RN participates in motor coordination. The RNm is the origin of the rubrospinal tract, and like the corticospinal tract, it controls the distal limb muscles rather than the proximal axial musculature. The rubrospinal tract mainly participates in learned, automatically performed actions such as arm swinging, the conditioned-blink response or crawling of infants, whereas the corticospinal system is most active during the learning of new movements [Holstege, 1991; Horn and Adamczyk, 2012]. The RNp is the major origin of the rubro-olivary projections, which form part of a feedback pathway involving the olivocerebellar-RN-motor cortex neural loops. These are also considered to be important for motor learning. Through these circuits the RN can indirectly influence corticospinal pathways. It has been proposed that the rubro-olivary pathways could serve as a switch between the use of 'conscious corticospinal' and 'automated rubrospinal' programs [Kennedy, 1990]. The cerebellar-rubral-olivary loop has been referred to as the Guillain-Mollaret triangle in clinical circles, and it is thought to play a role in oculopalatal and palatal tremor [Shaikh et al., 2010].

Rubrospinal pathways increase the muscle tone in proximal limb flexors. Limb movement deficits resulting from lesions of

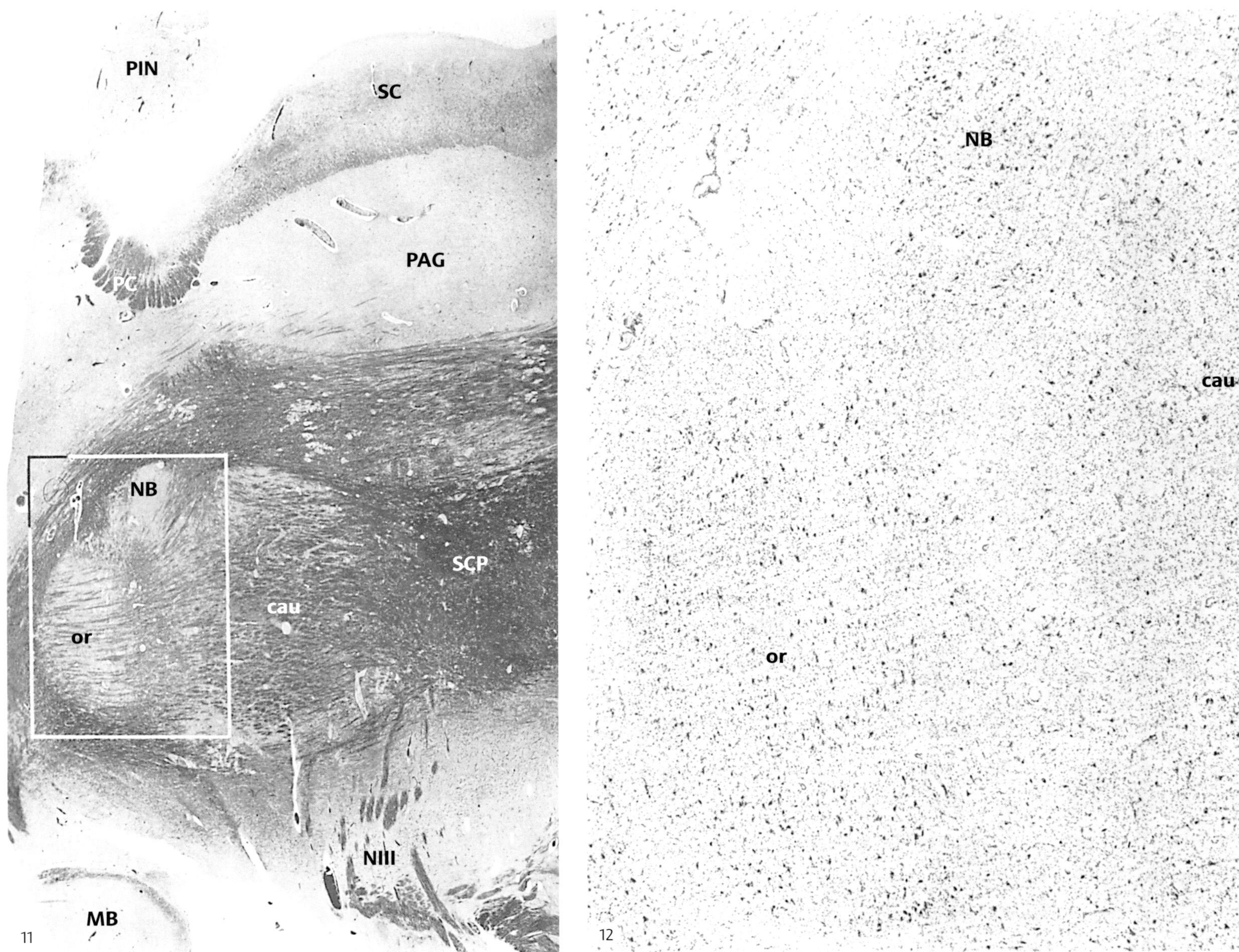

Fig. 11. Sagittal section through the mesencephalon at the level of the red nucleus to show the subdivision of the parvocellular subnucleus into an oral, dorsomedial and caudal part. cau = Red nucleus, parvocellular subnucleus, caudal part; MB = mammillary body; NB = medial accessory nucleus of Bechterew; NIII = rootlets of the oculomotor nerve; or = red nucleus, parvocellular subnucleus, oral part; PAG = periaqueductal gray; PC = posterior commissure; PIN = pineal body; SC = superior colliculus; SCP = superior cerebellar peduncle. Heidenhain stain. Magnification ×8.

Fig. 12. Sagittal section through the mesencephalon at the level of the red nucleus to show the parvocellular subdivision into the oral and caudal parts, as well as the dorsomedial part containing the medial accessory nucleus of Bechterew. The corresponding area of the adjacent myelin-stained section is delineated by a square in figure 11. cau = Red nucleus, parvocellular subnucleus, caudal part; NB = medial accessory nucleus of Bechterew; or = red nucleus, parvocellular subnucleus, oral part. Magnification ×30.

the corticospinal pathways can be compensated for to some extent by RN pathways. However, this is not the case in humans where the corticospinal tract is dominant and the rubrospinal tract is poorly developed [Holstege, 1991]. Lesions of the RN in humans lead to contralateral motor deficits such as intention tremor, ataxia and choreiform movements, but some of the symptoms are due to damage of fibers of passage [Levesque and Fabre-Thorpe, 1990]. Recent MRI studies in humans suggest that the RN might also play a role in sensorimotor processing [Liu et al., 2000].

Connections

Strictly speaking RN is a 'cerebellar associated nucleus' and not a direct precerebellar nucleus. However, it is included under the 'Precerebellar Nuclei' because of its strong influence on the cerebellum through projections to the IO. Thus, it is an indirect precerebellar nucleus.

Magnocellular RN. The RNm, albeit rudimentary in humans, receives 2 main inputs: one from collaterals of the pyramidal tract, arising from the ipsilateral motor cortex, and the second from collaterals of the SCP arising from the contralateral emboliform nucleus (in animals, the anterior interposed nucleus). This cerebellar nucleus receives highly organized zonal inputs from the cerebellar cortex of the anterior lobe and paramedian lobule [Nieuwenhuys et al., 2008, chapter 20]. The cells in the RNm are the origin of the rubrospinal tract. This topographically organized fiber pathway exits the nucleus medially, crosses in the ventral part of the tegmental decussation of Forel, descends in the ventrolateral pons and medulla giving collaterals to the facial and lateral reticular nuclei, then descends in the dorsolateral funiculus of the spinal cord. It terminates in lateral parts of layers V–VII, and layer XI in primates, targeting mainly distal limb muscles; for reviews, see Holstege and Tan [1988], Shinoda et al. [2006] and Burman et al. [2000].

Parvocellular RN. The RNp is topographically organized (face dorsal, upper extremities medial, lower extremities ventrolateral). Major inputs to the RNp arise from wider areas of the cerebral cortex than those to the RNm; for a review, see Nieuwenhuys et al. [2008], chapter 20. The projections from the motor and supplementary motor cortex to the rostral and lateral RNp are also somatotopically organized. The dorsomedial RNp (including the NB, see above) receives afferents from the rostral premotor area and the supplementary and frontal eye fields. The second major input to the RNp is from the contralateral cerebellar nuclei; the rostral dentate nucleus projects to the lateral RNp and the caudal dentate to the dorsomedial subdivision (including the NB, see above). This connection is an important part of the feedback loop between the 'RN – IO – cerebellum – RN'.

A key efferent projection from the RNp is to the IO, and it is also part of the 'rubro-olivary-cerebellar' loop. These pathways have a similar topography in the rat, cat and monkey, whereby the ventrolateral RNp projects to the dorsal lamella of the principal olive (PO), the dorsomedial RNp projects further rostrally to the same lamella, the NB projects to the lateral bend and the ventral lamella of the PO, and the ND projects to the rostral medial accessory olive [Onodera and Hicks, 2009]. The RN and NB pathways travel in the central tegmental tract, but those of the ND travel medially in the medial tegmental tract (fig. 9). The IO projects via climbing fibers to the contralateral cerebellar cortex, which in turn innervates the cerebellar nuclei whose efferents supply the RNp of the opposite side, as described above.

Additional efferents of the RNp target the pontine and medullary 'blink premotor centers' [Holstege, 1991]. Furthermore the dorsolateral RNp receives afferents from the trigeminal complex, and the cerebellar interpositus nucleus (important for motor learning), thereby providing a neural network that may support the conditioned-blink response [Morcuende et al., 2002].

References

Burman K, Darian-Smith C, Darian-Smith I: Macaque red nucleus: origins of spinal and olivary projections and terminations of cortical inputs. J Comp Neurol 2000;423:179–196.

Holstege G: Descending motor pathways and the spinal motor system: limbic and non-limbic components. Prog Brain Res 1991;87:307–421.

Holstege G, Tan J: Projections from the red nucleus and surrounding areas to the brainstem and spinal cord in the cat. An HRP and autoradiographical tracing study. Behav Brain Res 1988;28:33–57.

Horn AK, Adamczyk C: Reticular formation: eye movements, gaze and blinks; in Mai JK, Paxinos G (eds): The Human Nervous System, ed 3. Amsterdam, Academic Press, 2012.

Kennedy PR: Corticospinal, rubrospinal and rubro-olivary projections: a unifying hypothesis. Trends Neurosci 1990;13:474–479.

Levesque F, Fabre-Thorpe M: Motor deficit induced by red nucleus lesion: re-appraisal using kainic acid destructions. Exp Brain Res 1990;81:191–198.

Liu Y, Pu Y, Gao JH, Parsons LM, Xiong J, Liotti M, Bower JM, Fo PT: The human red nucleus and lateral cerebellum in supporting roles for sensory information processing. Hum Brain Mapp 2000;10:147–159.

Morcuende S, Delgado-García JM, Ugolini G: Neuronal premotor networks involved in eyelid responses: retrograde transneuronal tracing with rabies virus from the orbicularis oculi muscle in the rat. J Neurosci 2002;22:8808–8818.

Nieuwenhuys R, Voogd J, van Huijzen C: The Human Central Nervous System. Berlin, Springer, 2008.

Onodera S, Hicks TP: A comparative neuroanatomical study of the red nucleus of the cat, macaque and human. PLoS One 2009;4:e6623.

Shaikh AG, Hong S, Liao K, Tian J, Solomon D, Zee DS, Leigh RJ, Optican LM: Oculopalatal tremor explained by a model of inferior olivary hypertrophy and cerebellar plasticity. Brain 2010;133:923–940.

Shinoda Y, Sugiuchi Y, Izawa Y, Hata Y: Long descending motor tract axons and their control of neck and axial muscles. Prog Brain Res 2006;151:527–563.

Precerebellar Nuclei

(Plate 41)

102 Nucleus of Darkschewitsch (ND)

Original name:
Nucleus Darkschewitsch

Alternative name:
Nucleus ellipticus (Cetacea and Proboscidea)

Location and Cytoarchitecture – Original Text

The nucleus of Darkschewitsch (ND) is formed by a rather indistinct cell group found just inside the ventrolateral border of the periaqueductal gray matter (PAG) in the rostral mesencephalon. Orocaudally, the nucleus measures approximately 2 mm. On cross-section, it is elongated in outline and lies with its long axis directed dorsolaterally. It is related ventrolaterally to the interstitial nucleus of Cajal (INC) (fig. 7), the rostral interstitial nucleus of the medial longitudinal fasciculus (MLF; see nucleus No. 101, fig. 10) and the MLF, and in all other directions to cells of the PAG.

On myelin-stained sections, the background of the ND stains less intensely than that of the adjacent INC, and finely myelinated fibers may be seen coursing ventromedially from the ND in the direction of the MLF. A variable number of myelinated fibers intervenes between the ND and INC.

The cells of the ND are loosely arranged, medium-sized, elongated, darkly stained and possess indistinct Nissl granules (fig. 1–6). Their long axes are directed dorsolaterally, parallel to the border of the PAG [resembling a swarm of tiny fish]. Smaller cells, similar to those of the PAG, are found scattered among the medium-sized elements. Due to the paucity of characteristic cells and to the admixture of small cells, the delineation of the ND from the PAG may be difficult.

In some of our preparations, one could distinguish within the boundaries of the oral part of the ND a distinct, elongated cell group oriented parallel to the border of the PAG and separated from the INC by a distinct fiber bundle. [...]

The problem of delineation of the ND and INC was critically reviewed by Ingram and Ranson [1935]. According to these authors, Darkschewitsch, in his original articles, described only one cell group which included both of the above-mentioned nuclei. Twenty-six years later, Ramón y Cajal [1909] designated the ventrolateral part of the complex the 'interstitial nucleus' and retained the name 'nucleus of Darkschewitsch' for the dorsomedial cell group. [...]

Functional Neuroanatomy

Function

The ND participates in visuomotor and skeletomotor control. The ND receives inputs from the motor cortex arm and leg areas, the cortical eye fields, as well as visual inputs from the accessory optic system, and it projects to the inferior olive (IO), as do the adjacent nucleus of Bechterew (NB) and the red nucleus (RN). The ND, NB and RN form a topographically organized group of nuclei at the mesodiencephalic junction which are part of a series of recurrent olivocerebellar-cerebellar nuclei loops. The loops are controlled by superimposed cortical loops, and are thought to contribute to timing and learning in motor operations; for reviews, see Nieuwenhuys et al. [2008], Onodera and Hicks [2009], Voogd and Ruigrok [2012] and De Zeeuw et al. [1998].

The mesodiencephalic-olivary pathways show striking changes in the evolutionary progression from lower mammals to primates, and humans [Onodera and Hicks, 2009; Hicks and Onodera, 2012]. These authors put forward an attractive hypothesis suggesting that the ND olivary loops may play a role in the fine manipulative skills of bipeds, whereas the inferior frontal cortex (Broca's area 44)-NB-olive-lateral cerebellar nuclei loop in humans may expand in association with their development of language skills, and in elephants the large ND (nucleus ellipticus) and well-developed olivary target regions could reflect the prehensile skills of the elephant's trunk.

Comment

In some studies the term 'accessory oculomotor nuclei' has been used to describe nuclei lying close to the oculomotor nucleus and thought to be associated with its function. The term referred to the ND, Edinger-Westphal nucleus, nucleus of the posterior commissure and INC, collectively. Since the functional role of the ND is precerebellar and not premotor to the oculomotor nuclei, this term is considered to be no longer useful.

Comprehensive comparative studies of Onodera and Hicks [2009] have related the ND to a crescent of neurons, of which the ND is the dorsal tip followed by the NB and the parvocellular RN. This crescent of neurons provides a major input to the ipsilateral IO in all species, but in evolutionary development the ND of humans becomes separated from the NB, which sinks ventrolaterally out of the PAG into the 'reticular formation' and forms the dorsomedial part of the RN (see nucleus No. 101). Thus, in humans the ND is separated from the NB by the INC, MLF, and the medial tegmental tract (MTT). The authors call the hypothesis 'the rolled-sheet model of the red nucleus' and it is useful for the understanding of the organization of the mesodiencephalic nuclei and evolutionary changes seen in this region [Onodera and Hicks, 2009].

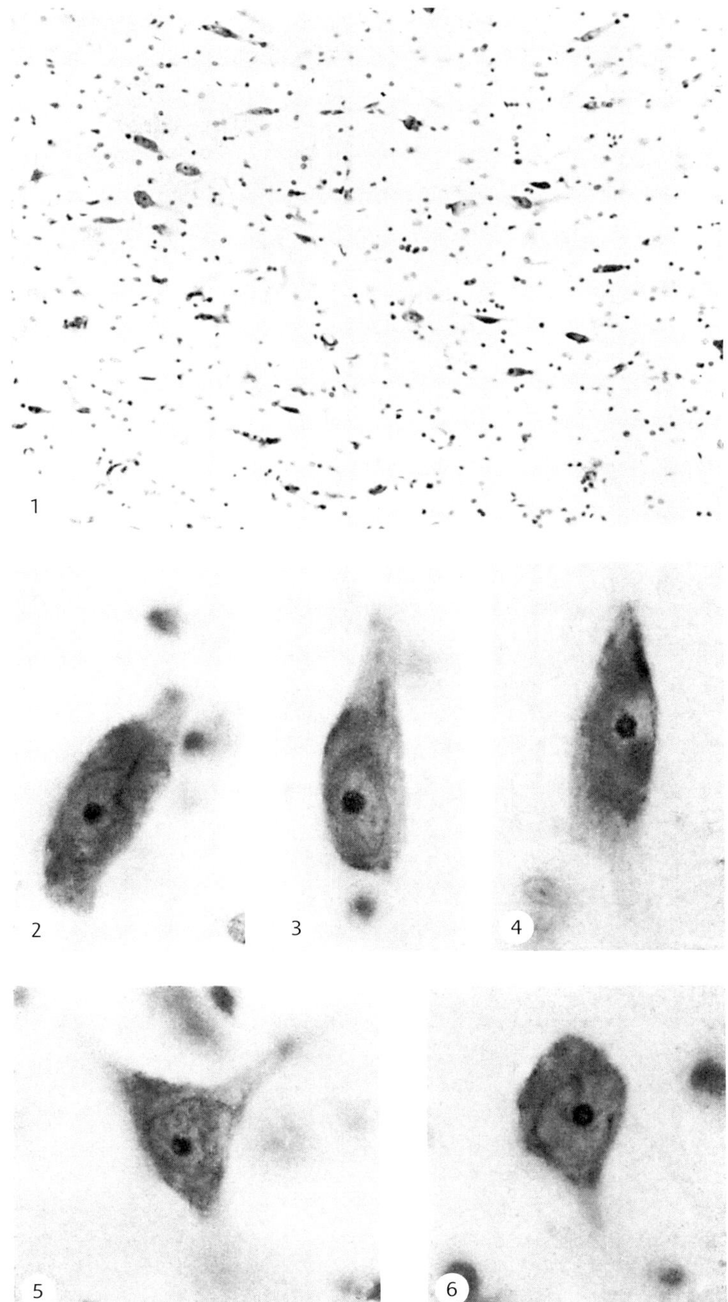

Fig. 1. Nucleus of Darkschewitsch. Magnification ×150.
Fig. 2–6. Cells from the nucleus of Darkschewitsch. Magnification ×1,000.

Connections

Efferents. Experimental evidence shows that the main efferent pathway from the ND is a topographically organized projection through the MTT to the rostral medial accessory olive (MAOr) [Onodera, 1984; Voogd and Ruigrok, 2012]. The MAOr innervates the C_2 zone of the cerebellar cortex in the medial parts of the flocculus and ventral paraflocculus. The C_2 zone, in turn, projects to the posterior interposed (globose) cerebellar nucleus, which sends typically excitatory signals back to the ND and other brainstem nuclei, but sends GABAergic inhibitory projections to the IO, specifically to the MAOr, thereby forming 2 of many recurrent cerebello-olivary loops [De Zeeuw and Ruigrok, 1994; De Zeeuw et al., 1998; Voogd and Ruigrok, 2012].

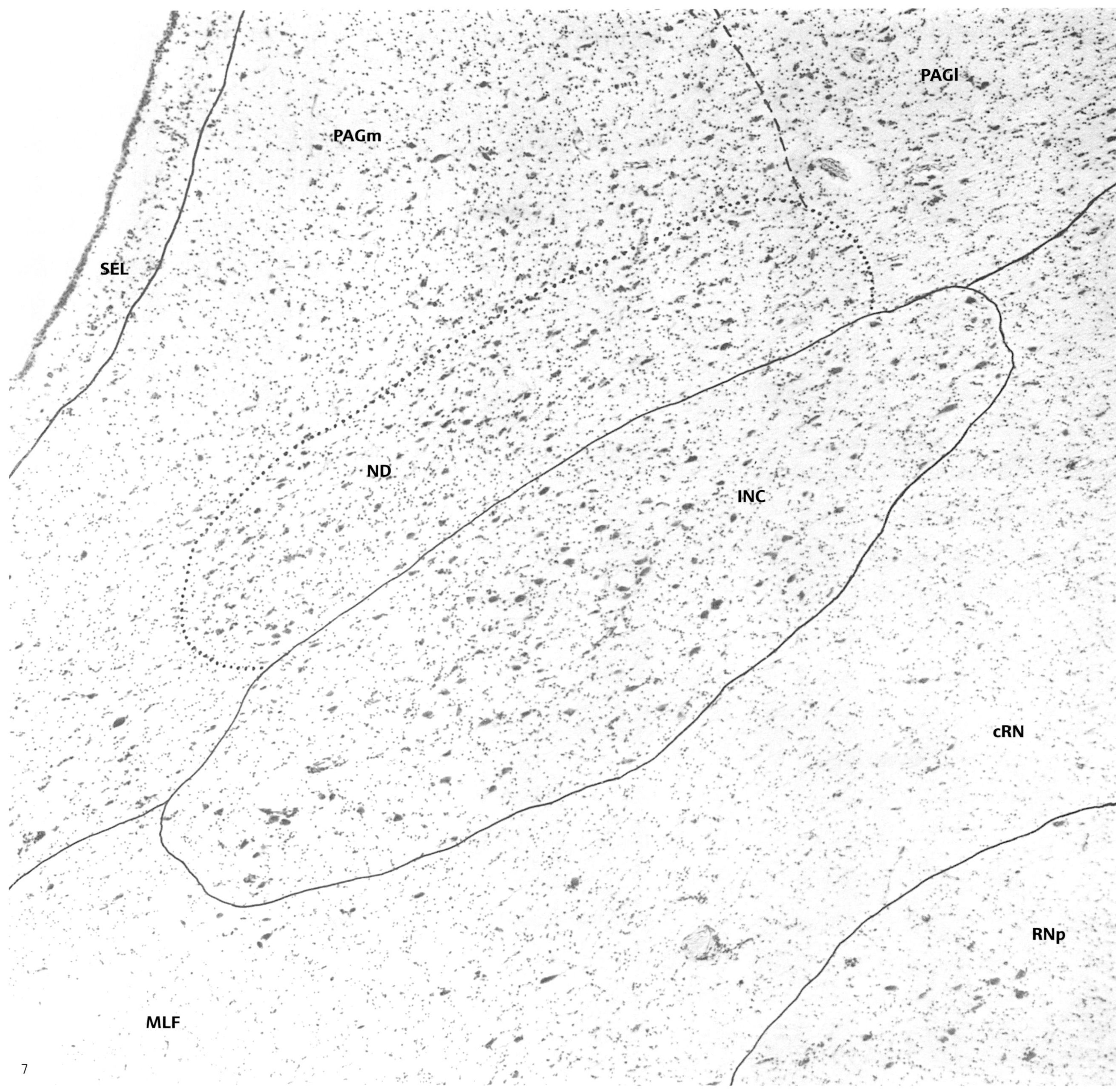

Fig. 7. Cross-section through the oral mesencephalic tegmentum. cRN = Perirubral capsule; ND = nucleus of Darkschewitsch; MLF = medial longitudinal fasciculus; PAGl = periaqueductal gray, lateral part; PAGm = periaqueductal gray, medial part; INC = interstitial nucleus of Cajal; RNp = red nucleus, parvocellular part; SEL = subependymal layer. Magnification ×80.

Afferents. The ND receives projections from several regions of the cerebral cortex: the arm and leg areas of the motor cortex, weak projections from the premotor area and stronger input from the frontal eye field [Leichnetz et al., 1984; Huerta et al., 1986; Stanton et al., 1988] and the posterior area 7 [Faugier-Grimaud and Ventre, 1989].

Afferents from the spinal cord have also been reported to terminate in the ND [Bjorkeland and Boivie, 1984]. Several studies have reported efferents to the ND, and its adjacent cell groups, from the accessory optic system [Blanks et al., 1995], but there are conflicting results concerning inputs from the nucleus of the optic tract [Benevento et al., 1977; Weber and Harting, 1980; Büttner-Ennever et al., 1996].

References

Benevento LA, Rezak M, Santos-Anderson R: An autoradiographic study of the projections of the pretectum in the rhesus monkey *(Macaca mulatta)*: evidence for sensorimotor links to the thalamus and oculomotor nuclei. Brain Res 1977; 127:197–218.

Bjorkeland M, Boivie J: The termination of spinomesencephalic fibers in cat. An experimental anatomical study. Anat Embryol 1984;170:265–277.

Blanks RHI, Clarke RJ, Lui F, Giolli RA, Vanpham S, Torigoe Y: Projections of the lateral terminal accessory optic nucleus of the common marmoset *(Callithrix jacchus)*. J Comp Neurol 1995;354:511–532.

Büttner-Ennever JA, Cohen B, Horn AK, Reisine H: Efferent pathways of the nucleus of the optic tract in monkey and their role in eye movements. J Comp Neurol 1996;373:90–107.

De Zeeuw CI, Ruigrok TJ: Olivary projecting neurons in the nucleus of Darkschewitsch in the cat receive excitatory monosynaptic input from the cerebellar nuclei. Brain Res 1994;653:345–350.

De Zeeuw CI, Simpson JI, Hoogenraad CC, Galjart N, Koekkoek SK, Ruigrok TJ: Microcircuitry and function of the inferior olive. Trends Neurosci 1998;21:391–400.

Faugier-Grimaud S, Ventre J: Anatomic connections of inferior parietal cortex (area 7) with subcortical structures related to vestibulo-ocular function in a monkey *(Macaca fascicularis)*. J Comp Neurol 1989;280:1–14.

Hicks TP, Onodera S: The mammalian red nucleus and its role in motor systems, including the emergence of bipedalism and language. Prog Neurobiol 2012;96: 165–175.

Huerta MF, Krubitzer LA, Kaas JH: Frontal eye field as defined by intracortical microstimulation in squirrel monkeys, owl monkeys, and macaque monkeys. I. Subcortical connections. J Comp Neurol 1986;253:415–439.

Ingram WR, Ranson SW: The nucleus of Darkschewitsch and nucleus interstitialis in the brain of man. J Nerv Mental Dis 1935;81:125–137.

Leichnetz GR, Spencer RF, Smith DJ: Cortical projections to nuclei adjacent to the oculomotor complex in the medial dien-mesencephalic tegmentum in the monkey. J Comp Neurol 1984;228:359–387.

Nieuwenhuys R, Voogd J, van Huijzen C: The Human Central Nervous System. Berlin, Springer, 2008.

Onodera S: Olivary projections from the mesodiencephalic structures in the cat studied by means of axonal transport of horseradish peroxidase and tritiated amino acids. J Comp Neurol 1984;227:37–49.

Onodera S, Hicks TP: A comparative neuroanatomical study of the red nucleus of the cat, macaque and human. PLoS One 2009;4:e6623.

Ramón y Cajal S: Histologie du système nerveux de l'homme et des vertébrés. Paris, Maloine, 1909.

Stanton GB, Goldberg ME, Bruce CJ: Frontal eye field efferents in the macaque monkey: II. Topography of terminal fields in midbrain and pons. J Comp Neurol 1988;271:493–506.

Voogd J, Ruigrok T: Cerebellum and precerebellar nuclei; in Mai JK, Paxinos G (eds): The Human Nervous System, ed 3. Amsterdam, Academic Press, 2012.

Weber JT, Harting JK: The efferent projections of the pretectal complex: an autoradiographic and horseradish peroxidase analysis. Brain Res 1980;194:1–28.

103 Conterminal Nucleus (CT)

Original name:
Nucleus conterminalis

Alternative name:
Nucleus gliosis [Kooy, 1916]

Location and Cytoarchitecture – Original Text

This nucleus is composed of numerous plate-like cell groups which lie interposed between the pyramid ventrally and the olivary amiculum dorsally [the fiber capsule surrounding the inferior olive], at levels between the caudal and oral extremities of the inferior olivary complex (IO). The caudal extent of a single cell group is usually very short.

On cross-section these cell groups are elongated and lie with their long axes parallel to the dorsal surface of the pyramidal tract (PT) (fig. 1–4). They may be situated at any point between the medial and lateral extremities of the dorsal surface of the PT. In general the cell groups of the conterminal nucleus (CT) are most numerous in sections through the oral two thirds of the IO. However, due to the discontinuity and inconstancy of position of these cell groups, many cross-sections through the IO will contain no representatives of the CT.

The CT is composed of irregularly arranged, medium-sized, plump or polygonal cells which possess long dendrites, large nuclei and darkly stained Nissl granules. A few cells tend to be elongated but are otherwise similar to the foregoing.

It is thus apparent that the cells of the CT resemble those of the arcuate nuclei (ARC) which may appear on the same cross-sections in the ventral aspect of the PT. The distinction between the cells of the CT and those of the IO may be difficult in some instances. The differentiation can usually be made, however, on the basis of the observations that the cells of the CT are larger, and stain more intensely than those of the IO. Further, in persons over middle age, the cells of the CT will contain much less lipofuscin than the cells of the IO.

Functional Neuroanatomy

Function
The CT cell groups appear to be well developed only in humans, and their function is not known. It has been suggested that the CT and the adjacent ARC may be the homologs of respiratory chemosensitive fields investigated in the cat and human [Filiano et al., 1990].

Additional Features
The cells of the CT characteristically lie at the hilus of the IO. In Nissl sections they resemble ARC cell groups more closely than (displaced) pontine nuclei; but in the aldehyde fuchsin stains for lipofuscin used by Braak [1970], which leave glial cells in the CT unstained, and hence the CT neurons more clearly visible, the CT neurons were described as resembling those of the lateral reticular nucleus.

There is no evidence available on the projections of the CT, even though it is sometimes assumed to be precerebellar, presumably because of its proximity to the IO. However, along with other precerebellar nuclei, the cells of the CT are degener-

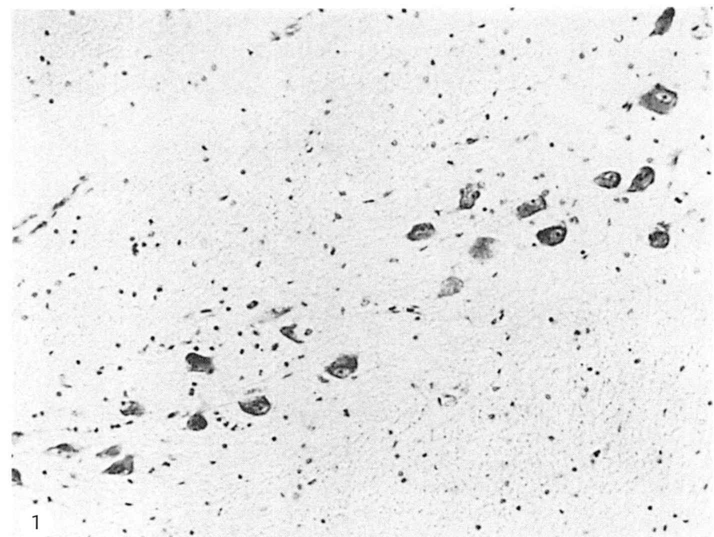

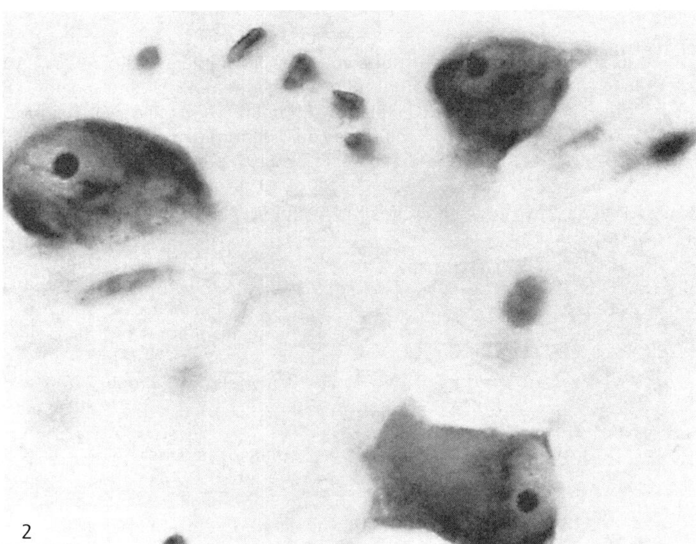

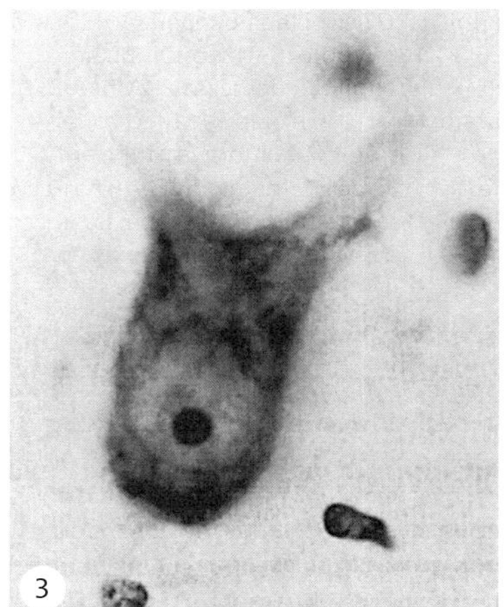

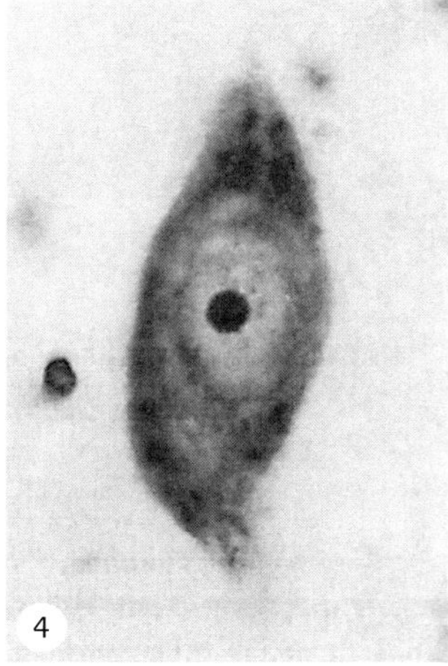

Fig. 1. Conterminal nucleus. Magnification ×150.
Fig. 2–4. Cells from the conterminal nucleus. Magnification ×1,000.

ated in spinocerebellar ataxia (types 2 and 3) and in multisystem atrophy [Braak et al., 2003; Rüb et al., 2005]. The CT is intensely acetylcholinesterase positive, a property that assists its identification along with the distinctive cytoarchitectural features described above [Koutcherov et al., 2004].

References

Braak H: Nuclei of the human brain stem. I. Inferior olive, conterminal nucleus and vermiform nucleus of the restiform body (in German). Z Zellforsch Mikrosk Anat 1970;105:442–456.

Braak H, Rüb U, del Tredici K: Involvement of precerebellar nuclei in multiple system atrophy. Neuropathol Appl Neurobiol 2003;29:60–76.

Filiano JJ, Choi JC, Kinney HC: Candidate cell populations for respiratory chemosensitive fields in the human infant medulla. J Comp Neurol 1990;293:448–465.

Kooy FH: The inferior olive in vertebrates. Folia Neurobiol 1916;10:205–369.

Koutcherov Y, Huang X-F, Halliday G, Paxinos G: Organization of human brain stem nuclei; in Paxinos G, Mai JK (eds): The Human Nervous System, ed 2. Amsterdam, Elsevier Academic Press, 2004.

Rüb U, Gierga K, Brunt ER, de Vos AI, Bauer M, Schöls L, Bürk K, Auburger G, Bohl J, Schultz C: Spinocerebellar ataxias types 2 and 3: degeneration of the precerebellar nuclei isolates the three phylogenetically defined regions of the cerebellum. J Neural Transm 2005;112:1523–1545.

Nuclei of Unknown Function

(Plates 16 and 17)

104 Supravestibular Nucleus (SPV)

Original name:
Nucleus supravestibularis

Location and Cytoarchitecture – Original Text

This nucleus lies dorsal to the medial and descending vestibular nuclei and beneath the floor of the lateral angle of the fourth ventricle. It appears at the level of the caudal pole of the dorsal cochlear nucleus (DCN) and extends orally for only 1 mm.

On cross-section the supravestibular nucleus (SPV) is spindle shaped in outline and lies with its long axis directed horizontally. It is related ventrally to the vestibular complex and dorsally to the ependyma of the floor of the ventricle. The lateral extremity of the nucleus may extend dorsally to the inferior cerebellar peduncle where it is contiguous to the dorsomedial extremity of the DCN.

The cells which compose the SPV are medium sized, elongated and multipolar in outline, and possess darkly stained, indistinct, evenly distributed Nissl granules (fig. 1–5). These cells are more lightly stained than those of the neighboring, vestibular complex and are larger than the cells of the DCN. Scattered among these medium-sized neurons are small, lightly stained cells of varying form.

Functional Neuroanatomy

The SPV lies lateral to the subventricular nuclei (SBV) and medial to the cells of the Y group, which cover the dorsal border of the restiform body. The SPV like the SBV extend as islands enclosed by the bundles of the medullary striae (see nucleus No. 106, fig. 7), and show different degrees of development in different individuals [Suarez et al., 1997]. The function of the SPV is not known, there is no evidence that it receives vestibular nerve afferents; but tract-tracing studies in the cat indicate that it may send efferent fibers to the cerebellum [Gould, 1980], the facial nucleus [Shaw and Baker, 1983], and to the thalamus [Kotchabhakdi et al., 1980; Shaw and Baker, 1983].

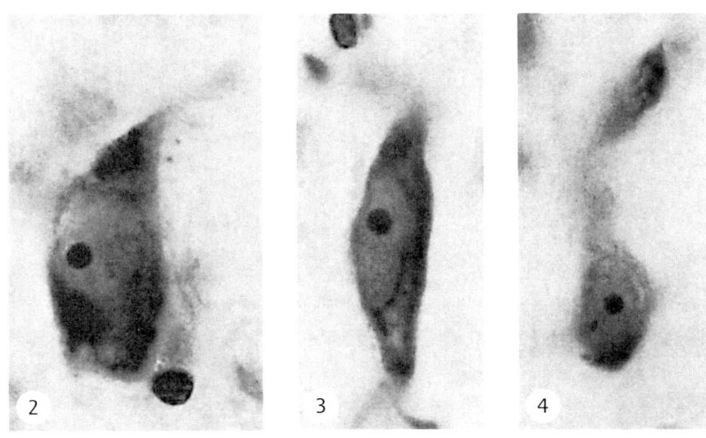

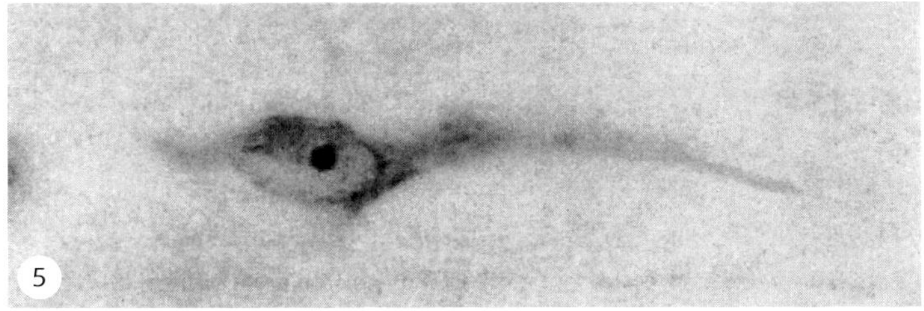

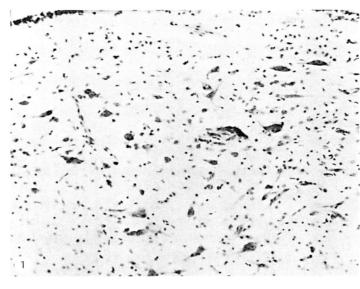

Fig. 1. Supravestibular nucleus. Magnification ×150.
Fig. 2–5. Cells from the supravestibular nucleus. Magnification ×1,000.

References

Gould BB: Organization of afferents from the brain stem nuclei to the cerebellar cortex in the cat. Adv Anat Embryol Cell Biol 1980;62:1–90.

Kotchabhakdi N, Rinvik E, Walberg F, Yingcharoen K: The vestibulothalamic projections in the cat studied by retrograde axonal transport of horseradish peroxidase. Exp Brain Res 1980;40, 405–418.

Shaw MD, Baker R: Direct projections from vestibular nuclei to facial nucleus in cats. J Neurophysiol 1983;50:1265–1280.

Suarez C, Diaz C, Tolivia J, Alvarez JC, Gonzalez del Rey C, Navarro A: Morphometric analysis of the human vestibular nuclei. Anat Rec 1997;247:271–288.

Nuclei of Unknown Function

(Not on plates)

105 Subventricular Nuclei (SBV)

Original name:
Nuclei subventriculares

Associated name:
Possibly laterally dispersed cell groups of the nucleus paramedianus dorsalis

Location and Cytoarchitecture – Original Text

The subventricular nuclei (SBV) are composed of small cell groups which lie among the fibers of the medullary striae in the floor of the fourth ventricle, at the level of the oral part of the nucleus paramedianus dorsalis (PMD; see nucleus No. 106, fig. 4, 5, 7). Fibers of the medullary striae arise from the arcuate nuclei on the ventral surface of the medullary pyramid, traverse the floor of the fourth ventricle, and enter the inferior cerebellar peduncle (ICP). On cross-section, these small cell groups are oval in outline and may be situated at any point in the floor of the ventricle between the lateral border of the PMD and a point dorsal to the fibers of the ICP (fig. 1, 2). Several of these nuclei may appear on each cross-section.

The cells of the SBV resemble those of the PMD in all particulars. It would thus appear that these nuclei represent laterally displaced cell groups of the PMD, which, like the latter, probably receive aberrant afferent fibers and give rise to aberrant pontocerebellar fibers. The degree of development of SBV is quite variable from one individual to another. In general, they are most prominent in those specimens in which the medullary striae and PMD are well developed.

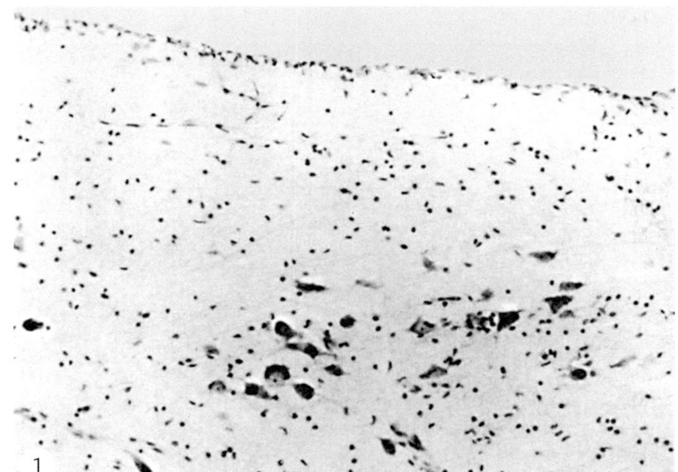

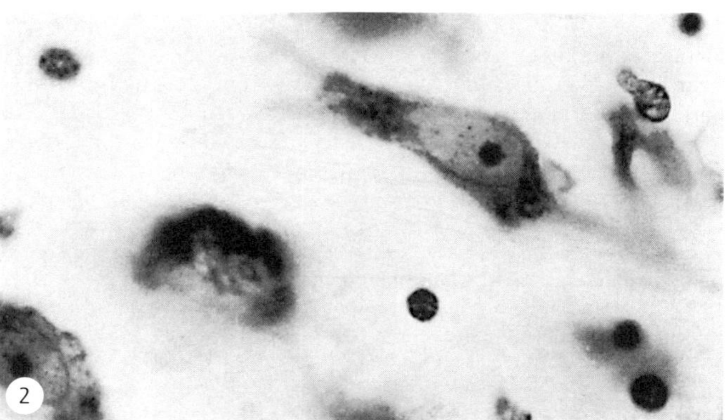

Fig. 1. Subventricular nuclei. Magnification ×150.
Fig. 2. Cells from the subventricular nuclei. Magnification ×1,000.

Functional Neuroanatomy

Function

The function of the cell groups is not known, but in patients with multiple system atrophy the SBV, along with other precerebellar cell groups, show a significant number of inclusion bodies [Braak et al., 2003].

Reference

Braak H, Rüb U, del Tredici K: Involvement of precerebellar nuclei in multiple system atrophy. Neuropathol Appl Neurobiol 2003;29:60–76.

106 Nucleus Paramedianus Dorsalis (PMD)

Original name:
Nucleus paramedianus dorsalis, subnuclei oralis and caudalis

Alternative names:
Nucleus paramedianus posterior
Posterior paramedian nucleus
Dorsal paramedian nucleus

Associated name:
Subventricular nucleus (nucleus No. 105)

Location and Cytoarchitecture – Original Text

This nuclear complex lies immediately beneath the floor of the fourth ventricle and adjacent to the midline, or median raphe, in the oral half of the medulla and caudal few millimeters of the pons. Two subdivisions of this complex may be distinguished: (a) the subnucleus paramedianus dorsalis caudalis (PMDc), and (b) the subnucleus paramedianus dorsalis oralis (PMDo). The degree of development of these subnuclei varies remarkably from one individual to another.

When well developed, the PMDc extends from the level of the obex to the rostral pole of the hypoglossal nucleus (XII). The caudal portion of the PMDc is represented by an oval, elongated cell group which intervenes between the medial border of the XII and the dorsal part of the median raphe. Farther orally, approaching the oral pole of the XII the PMDc migrates dorsolaterally to occupy a position between the dorsomedial border of the XII and the ependyma of the fourth ventricle (fig. 1, 2). It is at this level that the PMDc exhibits its maximal development. Oral to the level at which XII neurons disappear, the PMDc rapidly diminishes in size and is represented by only a few cells lying beneath the ependyma of the fourth ventricle, dorsal to the extrarapheal cells of the nucleus raphe obscurus (RObe) and medial to the caudal pole of the prepositus nucleus (PrP).

In contrast to the above, the PMDc is represented in many human brainstems by only a few scattered cells which lie dorsomedial to the oral pole of the XII. The degree of development of the PMDc appears to parallel that of the PMDo.

The PMDo appears approximately 2.5 mm rostral to the oral pole of the XII and, if well developed, may extend to the level of the oral pole of the PrP. Caudally the PMDo is in direct continuity with the PMDc. On cross-section, the PMDo is oval or egg shaped in outline. It occupies an area beneath the medial part of the floor of the fourth ventricle, medial to the PrP and dorsal to the RObe and to the medial longitudinal fasciculus (fig. 4, 5; see also nucleus No. 65, fig. 1).

The PMD is composed of closely arranged, medium-sized, polygonal or triangular, multipolar cells which possess long dendrites, a relatively large nucleus and darkly stained, indistinct Nissl granules. The cells of the two subnuclei are similar with the exception of the fact that the cells composing the PMDc are slightly smaller than those of the PMDo (fig. 3, 6, 8–12).

[...] It is apparent from examination of figures 5 and 7 that the PMD is closely associated with the striae medullares and it is our impression that the degree of development of the PMD and the striae parallel each other. Further, the cells of the PMD are morphologially similar to those of the pontine nuclei (PN). On the basis of these observations one might hypothesize that the PMD represents a group of caudally and dorsally displaced PN which send their axons to the contralateral cerebellum either directly via the medullary striae or via the more circuitous pathway of the ventral external arcuate fibers (fig. 7).

Functional Neuroanatomy

Function
The function of the PMD remains unknown.

Connections and Comments
The PMD is usually associated topographically with the PrP [Brodal, 1983; Baizer et al., 2007]. These authors confirmed that the PMD is variable in size in humans; in the chimpanzee and monkey it is less developed, and in the cat it is barely discernible [Brodal, 1983]. The suggestion in the 'Original Text' that the PMD is a displaced PN has been partially supported by retrograde tracer injections into the flocculus in monkeys which consistently labeled a small percentage of PMD neurons [Langer et al., 1985].

Histochemistry
In humans the PMD was distinctly outlined by immunochemical stains for SMI-32, which labels nonphosphorylated neurofilaments, and it is also stained by the synthetic enzyme for nitric oxide and calretinin [Baizer et al., 2007]. In addition the PMD is reported to be acetylcholinesterase positive [Koutcherov et al., 2004]. It is possible that this nucleus is part of the complex called the paramedian tract cell groups (nucleus No. 96).

References

Baizer JS, Baker JF, Haas K, Lima R: Neurochemical organization of the nucleus paramedianus dorsalis in the human. Brain Res 2007;1176:45–52.
Brodal A: The perihypoglossal nuclei in the macaque monkey and the chimpanzee. J Comp Neurol 1983;218:257–269.
Koutcherov Y, Huang X-F, Halliday G, Paxinos G: Organization of human brain stem nuclei; in Paxinos G, Mai JK (eds): The Human Nervous System, ed 2. Amsterdam, Elsevier Academic Press, 2004.
Langer TP, Fuchs AF, Scudder CA, Chubb MC: Afferents to the flocculus of the cerebellum in the rhesus macaque as revealed by retrograde transport of horseradish peroxidase. J Comp Neurol 1985;235:1–25.

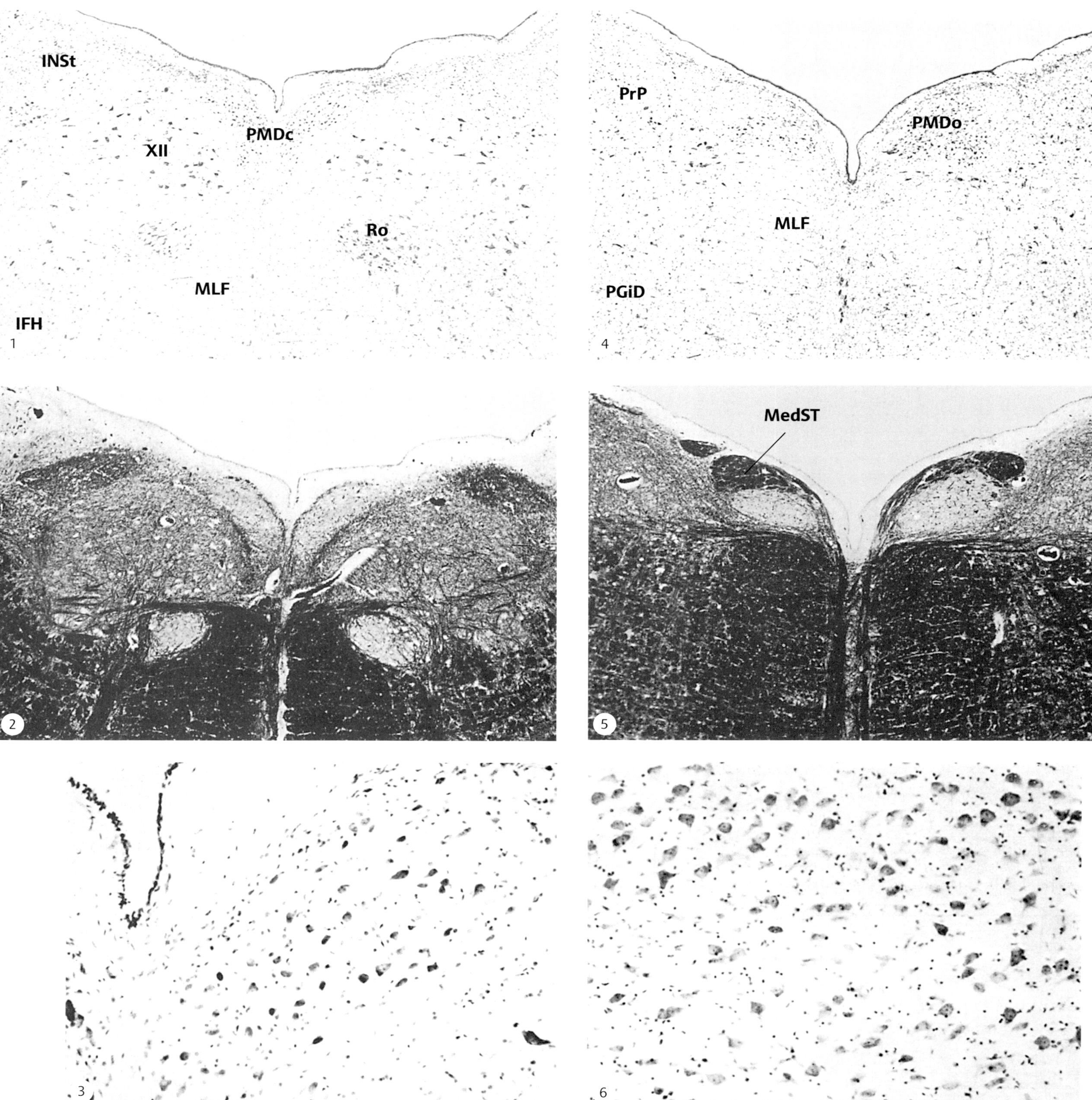

Fig. 1. Cross-section through the dorsal part of the tegmentum of the medulla at the level of the nucleus paramedianus dorsalis, subnucleus caudalis. IFH = Interfascicular hypoglossal nucleus; INSt = intercalated nucleus (of Staderini); MLF = medial longitudinal fasciculus; PMDc = nucleus paramedianus dorsalis, caudal subnucleus; Ro = nucleus of Roller; XII = hypoglossal nucleus. Magnification ×30.

Fig. 2. Cross-section through the dorsal part of the tegmentum of the medulla at the level of the nucleus paramedianus dorsalis, caudal subnucleus. This section is adjacent to that shown in figure 1. Heidenhain stain. Magnification ×30.

Fig. 3. Nucleus paramedianus dorsalis, caudal subnucleus. Magnification ×150.

Fig. 4. Cross-section through the dorsal part of the tegmentum of the medulla at the level of the nucleus paramedianus dorsalis, oral subnucleus. MLF = Medial longitudinal fasciculus; PGiD = dorsal paragigantocellular nucleus; PMDo = nucleus paramedianus dorsalis, oral subnucleus; PrP = prepositus nucleus. Magnification ×30.

Fig. 5. Cross-section through the dorsal part of the tegmentum of the medulla at the level of the nucleus paramedianus dorsalis, subnucleus oralis. Note the medullary striae on the floor of the fourth ventricle. MedST = Medullary striae. This section is adjacent to that shown in figure 4. Heidenhain stain. Magnification ×30.

Fig. 6. Nucleus paramedianus dorsalis, oral subnucleus. Magnification ×150.

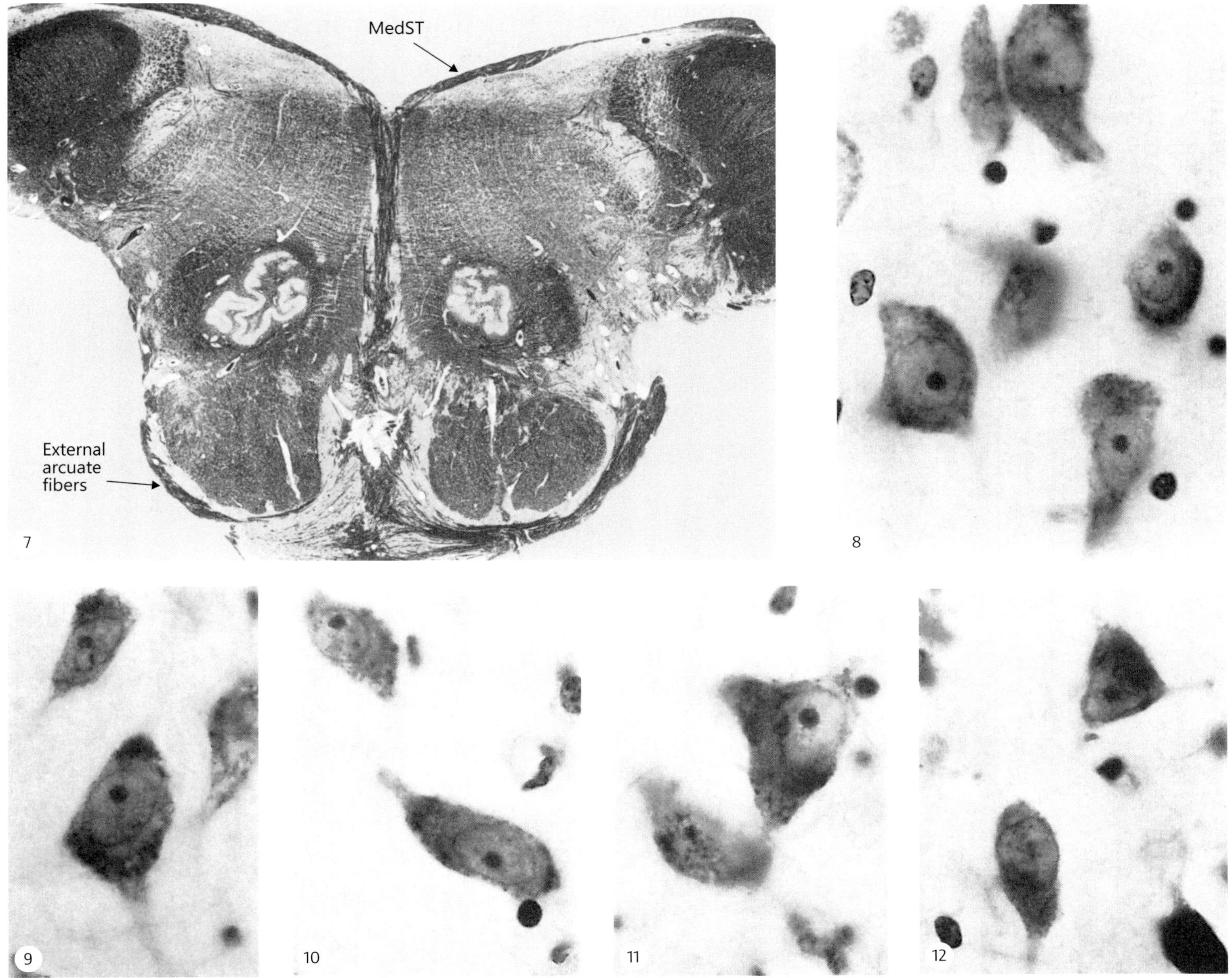

Fig. 7. Cross-section through the oral part of the medulla oblongata showing well-developed medullary striae (MedST) on the midline and on the floor of the fourth ventricle. Heidenhain stain. Magnification ×6.

Fig. 8–12. Cells from the nucleus paramedianus dorsalis. Magnification ×1,000.

107 Compact Interfascicular Nucleus (CIF)

Original name:
Nucleus compactus interfascicularis

Location and Cytoarchitecture – Original Text

This is a small, medially situated nucleus which lies interposed between the medial longitudinal fasciculi at the level of the trochlear nuclei. On cross-section the nucleus is oblong in outline and is surrounded by cells composing the caudal pole of the dorsal raphe nucleus. Caudo-orally the nucleus measures 1.5–2 mm.

The compact interfascicular nucleus (CIF) is composed of closely congregated, small, lightly stained, oval or fusiform cells (fig. 1, 2).

Functional Neuroanatomy

Function

The function of this small cell group is unknown. However, because of its name, the CIF should not be confused with the *interfascicular* nucleus of the mesencephalic ventral tegmental area, or with the serotonergic *interfascicular* subgroup of the dorsal raphe nucleus, immediately rostral to the CIF, or the *interfascicular* hypoglossal nucleus clustered around the hypoglossal nerve.

Connections

No literature was found on CIF projections, or on its histochemistry.

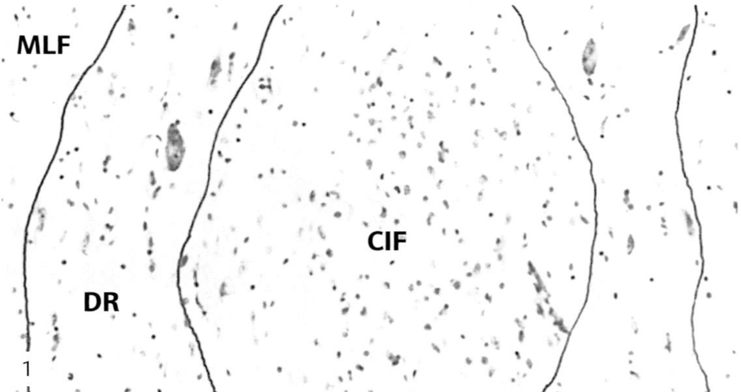

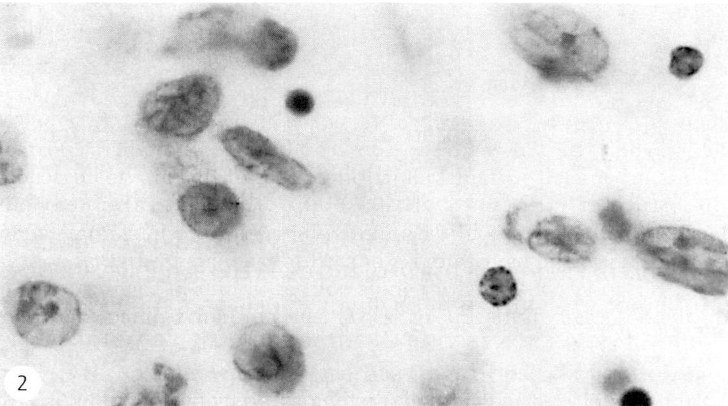

Fig. 1. Compact interfascicular nucleus (CIF). DR = Dorsal raphe nucleus; MLF = medial longitudinal fasciculus. Magnification ×150.
Fig. 2. Cells from the compact interfascicular nucleus. Magnification ×1,000.

Subject Index

Page numbers in bold indicate chapters